ELECTROMAGNETIC WAVES

Edited by **Vitaliy Zhurbenko**

Electromagnetic Waves
Edited by Vitaliy Zhurbenko

Published by InTech
Janeza Trdine 9, 51000 Rijeka, Croatia

Publishing Process Manager Iva Lipovic

Technical Editor Teodora Smiljanic

Cover Designer InTech Design Team

Additional hard copies can be obtained from orders@intechopen.com

Electromagnetic Waves, Edited by Vitaliy Zhurbenko
p. cm.
ISBN 978-953-307-304-0

For EU product safety concerns:
IN TECH d.o.o., Prolaz Marije Krucifikse Kozulić 3, 51000 Rijeka, Croatia,
info@intechopen.com or visit our website at intechopen.com.

Meet the editor

Dr. Vitaliy Zhurbenko obtained the B.Sc. and M.Sc. degrees from the Kharkiv National University of Radio Electronics, Kharkiv, Ukraine, in 2000 and 2001, respectively, and the Ph.D. degree from the Technical University of Denmark, Copenhagen, Denmark, in 2008, all in electrical engineering. From November 2000 to June 2005 he was a Metrology Engineer with the Laboratory of Metrology, Kharkiv, Ukraine. In 2004 he became a Junior Member of the Teaching Staff with the Kharkiv National University of Radio Electronics. In 2005 he joined the Technical University of Denmark, where he is currently an Assistant Professor. His current teaching and research interests include wireless communications, microwave and millimeter-wave sensing for biomedical and security applications, microwave and millimeter-wave devices and integrated circuits for instrumentation applications, antenna and passive circuit design and characterization, terahertz technologies

Contents

Preface XI

Part 1 The Physics of Electromagnetic Fields 1

Chapter 1 **The Fundamental Physics of Electromagnetic Waves 3**
Juliana H. J. Mortenson

Chapter 2 **Modern Classical Electrodynamics and Electromagnetic Radiation – Vacuum Field Theory Aspects 27**
Nikolai N. Bogolubov (Jr.), Anatoliy K. Prykarpatsky

Chapter 3 **Electromagnetic-wave Contribution to the Quantum Structure of Matter 57**
Burke Ritchie

Chapter 4 **Gouy Phase and Matter Waves 71**
Irismar G. da Paz, Maria C. Nemes and José G. P. de Faria

Part 2 Methods of Computational Analysis 97

Chapter 5 **Simulation and Analysis of Transient Processes in Open Axially-symmetrical Structures: Method of Exact Absorbing Boundary Conditions 99**
Olena Shafalyuk, Yuriy Sirenko and Paul Smith

Chapter 6 **Fractional Operators Approach and Fractional Boundary Conditions 117**
Eldar Veliev, Turab Ahmedov, Maksym Ivakhnychenko

Part 3 Electromagnetic Wave Propagation and Scattering 137

Chapter 7 **Atmospheric Refraction and Propagation in Lower Troposphere 139**
Martin Grabner and Vaclav Kvicera

Chapter 8 **Atmospheric Attenuation due to Humidity 157**
Milda Tamošiūnaitė, Mindaugas Žilinskas,
Milda Tamošiūnienė and Stasys Tamošiūnas

Chapter 9 **Effects of Interaction of Electromagnetic Waves in Complex Particles 173**
Ludmilla Kolokolova, Elena Petrova and Hiroshi Kimura

Chapter 10 **Models for Scattering from Rough Surfaces 203**
F. Ticconi, L. Pulvirenti and N. Pierdicca

Chapter 11 **Electromagnetic Wave Propagation in Circular Tunnels 227**
Osama M. Abo-Seida

Part 4 Analysis and Applications of Periodic Structures and Waveguide Components 233

Chapter 12 **Propagation of Electromagnetic Waves in Thin Dielectric and Metallic Films 235**
Luc Lévesque

Chapter 13 **Quasi-optical Systems Based on Periodic Structures 257**
Gennadij Vorobjov, Yulya Shulga and Vitaliy Zhurbenko

Chapter 14 **Waveguide Mode Converters 283**
Yoshihiro Kokubo

Part 5 Electromagnetic Material Analysis and Characterization 297

Chapter 15 **Resonance Properties of Scattering and Generation of Waves on Cubically Polarisable Dielectric Layers 299**
Lutz Angermann and Vasyl V. Yatsyk

Chapter 16 **Cholesteric Elastomers with Mechanical Control of Optical Spectra 341**
J. Adrián Reyes, Laura O. Palomares and Carlos G. Avendaño

Chapter 17 **Time Domain Reflectometry: Temperature-dependent Measurements of Soil Dielectric Permittivity 369**
Wojciech Skierucha

Chapter 18 **The Temperature Behavior of Resonant and Non-resonant Microwave Absorption in Ni-Zn Ferrites 387**
Raúl Valenzuela

Chapter 19 **Complex Permittivity Measurement of High Loss Liquids and its Application to Wine Analysis 403**
Z.E. Eremenko, V.N. Skresanov, A.I. Shubnyi, N.S. Anikina, V.G. Gerzhikova and T.A. Zhilyakova

Part 6 Applications of Plasma 423

Chapter 20 **EMI Shielding using Composite Materials with Plasma Layers 425**
Ziaja Jan and Jaroszewski Maciej

Chapter 21 **Reduction of Reflection from Conducting Surfaces using Plasma Shielding 449**
Çiğdem Seçkin Gürel and Emrah Öncü

Part 7 Biological Effects and Medical Imaging 471

Chapter 22 **Electromagnetic Waves and Human Health 473**
Feyyaz Özdemir and Aysegül Kargi

Chapter 23 **Image Resolution and Sensitivity Improvements of a Molecular Imaging Technique Based on Magnetic Nanoparticles 493**
Yasutoshi Ishihara, Tsuyoshi Kuwabara and Naoki Wadamori

Preface

This book is dedicated to various aspects of electromagnetic wave theory and its applications in science and technology. The covered topics include the fundamental physics of electromagnetic waves, theory of electromagnetic wave propagation and scattering, methods of computational analysis, material characterization, electromagnetic properties of plasma, analysis and applications of periodic structures and waveguide components, and finally, the biological effects and medical applications of electromagnetic fields. Even though the classical electromagnetic theory is well-established and experimentally verified, it is far from being a closed subject. In spite of the fact that the theory is capable of providing explanations for all (classical) electromagnetic effects, there are several fundamental problems that remain open. These problems mainly concern the electromagnetic waves behaving like quantum particles. In order to complete the theory of electromagnetic waves, a new fundamental physics emerged suggesting novel concepts to explain observed physical phenomena. The first part of this book is dedicated to the research in this field including various aspects of vacuum field theory, electromagnetic wave contribution to the quantum structure of matter, and matter waves.

Modelling and computations in electromagnetics is a fast-growing research area. The general interest in this field is driven by the increased demand for analysis and design of non-canonical electromagnetic structures and rapid increase in computational power for calculation of complex electromagnetic problems. The second part of this book is devoted to the advances in the analysis techniques such as the method of exact absorbing boundary conditions, fractional operator approach, and fractional boundary conditions. The problems of diffraction on infinitely thin surfaces are considered, and the difficulties in the analysis of axially-symmetrical open resonators are addressed.

The third part of the book deals with electromagnetic wave propagation and scattering effects. The main focus is made on atmospheric refraction and propagation in the lower troposphere, atmospheric attenuation due to the humidity, interaction of electromagnetic waves with inhomogeneous media composed of complex particles, modelling of scattering from random rough surfaces, and the problems of propagation in waveguides with imperfectly reflecting boundaries.

Waveguides are essential parts of millimetre and submillimetre-wave devices and systems. They are used for guiding electromagnetic energy between the components of the system. In the mentioned frequency band, periodic structures are also often used for wave guiding as well as for realization of delay lines, filter elements, and interaction structures in vacuum electron devices. The fourth part of the book starts with the description of the method of matrix formalism and its application to the analysis of planar waveguides and periodic structures. Then, the open resonators and open waveguides employing periodic structures and their implementation in vacuum electron devices are considered. The fourth part concludes with a chapter on waveguide mode converters.

The fifth part of the book is dedicated to interaction of electromagnetic waves with materials and implementation of electromagnetic methods for material analysis and characterisation. This includes scattering and generation of waves on cubically polarisable dielectrics, electromagnetic properties of elastomers, temperature behaviour of microwave absorption in ferrites and permittivity of soil. Time and frequency domain measurement techniques are also considered here.

Plasma technology is becoming increasingly attractive for radio communications, radio astronomy and military (stealth) applications due to electromagnetic properties of plasma medium. The shielding properties of plasma are investigated in the sixth part of this book. The final (seventh) part of this book deals with biological effects of electromagnetic radiation and its implementation to medical imaging, particularly, sensitivity and resolution improvement of molecular imaging using magnetic nanoparticles.

The presented material in this book is based on recent research work conducted by the authors working within the covered topics, who deserve all the credits for the presented scientific results.

Vitaliy Zhurbenko
Technical University of Denmark,
Denmark

Part 1

The Physics of Electromagnetic Fields

1

The Fundamental Physics of Electromagnetic Waves

Juliana H. J. Mortenson
General Resonance, LLC
USA

1. Introduction

A new foundational physics is emerging which radically changes our concepts of electromagnetic waves. The original quantum ideas of Max Planck and Albert Einstein from the turn of the twentieth century, are undergoing an impressive renaissance now at the turn of the twenty-first century. The result is a fundamental physics of electromagnetic waves that is both new and classical. Einstein's insistence that quantum mechanics was incomplete - that "hidden variables" were yet to be discovered - was correct. The recent discovery of those variables is the driving force behind this rebirth of the foundations of quantum mechanics and the fundamental physics of electromagnetic ("EM") waves.

The new quantum variables have led to the discovery of new universal constants for EM waves. The new constants have revealed an elegant simplicity in quantum concepts, that requires no paradoxical explanations and imposes no uncertainties or limits. Instead, the new physics provides a more realistic understanding of physical concepts related to EM waves. The old paradigm is disappearing, and yielding to a new paradigm which is both more understandable and more powerful.

2. Background

It is often said that to successfully navigate the future one must understand the past. The fundamental physics of electromagnetic waves are no exception to this wisdom. In fact, an understanding of the origins of 20th century physics regarding electromagnetic waves is of vital importance to understanding the scientific revolution that is currently taking place.

2.1 Physics in the ages of reason and enlightenment

Galileo Galilei (1564 - 1642) was one of the most influential scientists of the millennium, however he lived during a time when the protestant reformation was gaining momentum and Europe was in turmoil. The Catholic Church was losing its hold on much of northern Europe and the Thirty Years' War raged. Galileo resided on the Italian peninsula, where the Church maintained a strong hold, and he could not rely on the protection of reformers in other parts of Europe. None-the-less, even though "pagan" beliefs associated with frequency and resonance-related phenomena had been banned by the Church for centuries, Galileo performed research on natural resonant frequencies in a pendulum system. (Mortenson, 2010b).

In 1632, Galileo published his "Dialogue" and in a daring move described the mechanics of natural resonant frequencies writing, "*the Pendulum makes its vibrations with one and the same frequency*" and "*every Pendulum hath the Time of its Vibrations...pre-fixed...[and] it is impossible to make it move under any other Period, than that ...which is natural unto it.*" (Galilei, 1632) He described the resonant accelerating forces produced by precisely time puffs of his breath stating, "*by blowing upon [the Pendulum one may] confer a Motion, and a Motion considerably great by reiterating the blasts, but only under the Time properly belonging to its Vibrations*". Galileo thus provided one of the first documented descriptions of resonance, namely the increase in amplitude and energy of a system's vibrations when an applied vibration, motion or energy matches the natural frequency of the system. Unfortunately, the Church was less accommodating than Galileo had anticipated. He was convicted of heresy and placed under house arrest for the rest of his life.

Pierre de Fermat (1601 - 1665) was a French attorney who was in his mid-thirties when Galileo was accused of heresy. Although Fermat's personal passion was mathematics, he was well aware that pursuit of certain mathematical subjects could be very dangerous. Thus Fermat engaged in his passion in secret, scribbling notes in the margins of books in his private library. One set of notes was a resonance equation, demonstrating that as the rate of a mechanical vibration (e.g., a puff of breath) neared the natural vibratory rate of a body (e.g., the swing of a pendulum), the amplitude of vibrations in the body increased (also see Figure 1., below):

$$y = 1 / \left(1+x^2\right) \quad (1)$$

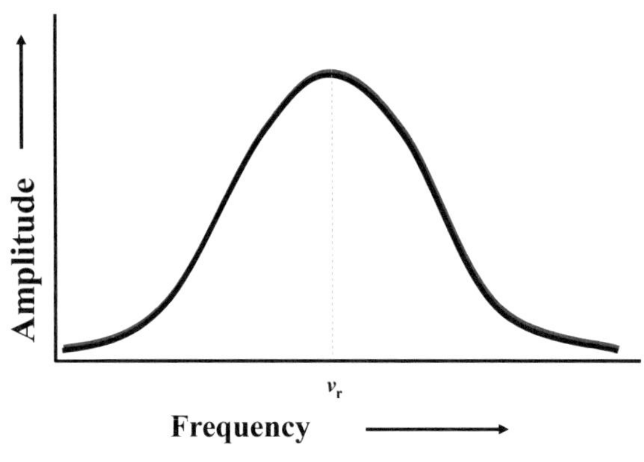

Fig. 1. Fermat's resonance curve showing an increase in vibration amplitude when forces are applied at natural resonant frequencies ("v_r").

The brilliant young Isaac Newton (1643 - 1727) wrote his famous *Principia,* describing his three (3) laws of motion around the time of Fermat's death. (Newton, 1898) The religious climate in England was quite chaotic at the time, and Newton waited another twenty (20) years to actually publish his *Principia.* His second law (force equals mass times acceleration) provided the basis for yet another resonance equation:

$$A = \frac{a}{\left|v_r^{\,2} - v_o^{\,2}\right|} \quad (2)$$

where "A" is the amplitude of the system's oscillations, "a" is the acceleration in the system's oscillation (caused in Galileo's case by the force of his small puffs of breath), "ν_r" is the resonant or natural frequency of the system, and "ν_o" is the frequency of the outside force applied to the system. As this second resonance equation shows, an outside force applied at a frequency which is either much higher or much lower than the natural resonant frequency of the system, produces a large denominator and hence a small amplitude. Conversely, the closer the frequency of the outside force is to the resonant natural frequency, the smaller the denominator becomes. Very large amplitudes are produced. When the outside frequency exactly *matches* the resonant frequency of the system the amplitude is theoretically *infinite* (Figure 2.).

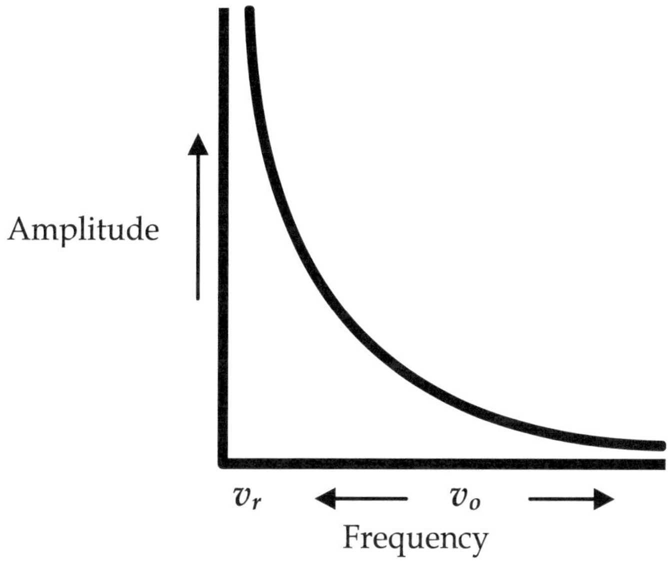

Fig. 2. Graphical representation of resonant amplitude equation (Eq. 2). The resonant frequency "v_r" is at the origin, and input frequency of the outside force "v_o" varies. As the input frequency approaches the resonant frequency, amplitude approaches infinity.

Newton distinguished the force exerted by an accelerating body, from the energy of a body simply in motion (which he referred to as vis viva) the product of mass and velocity:

$$\textit{vis viva} = \text{m v} \tag{3}$$

where "m" is mass and "v" is velocity. This led to the great *vis viva* controversy several decades later (see below). By 1704 Newton had published his treatise "*Opticks*" in which he proposed the corpuscular theory of light, namely that light is composed of tiny particles that travel in straight lines. In a foreshadowing of Einstein's later work, Newton stated, "*Are not gross Bodies and Light convertible into one another, ...and may not Bodies receive much of their Activity from the Particles of Light which enter their Composition?*"

A few decades later the great *vis viva* controversy erupted with Giovanni Poleni's (1683-1761) proposal that *vis viva* energy was proportional to the product of mass and velocity squared, putting him at odds with Newton. The debate was soon joined by Leibnitz, Huygens, and others. Dutch physicist Willem Gravesande (1688 -1742) performed meticulous experiments and concluded that energy of motion, "*follow[s] the Ratio compounded of the Masses, and the* **Squares** *of the Velocities"* (underline added). (Gravesande, 1747) The noted French Newtonian scholar, Emilie du Châtelet (1706 - 1749) in her 1740 book, *"Institutions Physiques"* asserted that *vis viva* energy is proportional to the product of mass and velocity squared, based on Gravesand's painstaking experiments.

While the *vis viva* debate raged, the Italian mathematical prodigy Maria Gaetana Agnesi (1718–1799), published her 1748 book on calculus and differential equations, organizing the work of Fermat, Newton, Leibnitz and others. (Agnesi, 1748) She expanded on Fermat's resonance curve, providing a detailed geometric proof and a third resonance equation:

$$y = ha^2 / a^2 + x^2 \tag{4}$$

where "h" is the height of the curve and "a" the half-width at half-maximum. Her book was an immediate sensation throughout Europe, and resonance began to become a well known scientific principle, in spite of the English translation error that resulted in the resonance curve being known as the "Witch of Agnesi". (Spencer, 1940)

2.2 Nineteenth century physics

By the nineteenth century, the brilliant Joseph Louis Lagrange (1736 - 1813) had organized the works of nearly every known scientist on matters of velocity, inertia, force, energy, and dynamics into his *"Méchanique Analytique"*. (Lagrange, 1811) Lagrange declared that for a body at constant velocity, its energy (*vis viva*) was equal to "mv^2", resulting *"solely from the inertia forces of the bodies"*. Conversely, the energy required to accelerate a body was a function of the distance over which a force acted "$F\ \delta s$". Lagrange explained that all systems exhibited a dynamic equilibrium between the *vis viva* of constant velocity and the forces of acceleration, *"The sum of these two quantities, when equated to zero, constitutes the general formula of dynamics… when the equilibrium does not hold, the bodies must necessarily move due to all or some of the forces which act on them."* For purposes of systematically explaining analytic mechanics Lagrange stated that he had assumed that an acceleration always occurs in a time period at least as long as the unit time for velocity. His assumption effectively fixed the acceleration time interval at "one second" and excluded accelerations taking place in less than one second.

Lagrange also addressed resonance dynamics using a mathematical function: *"in the case where the same function is a maximum, the equilibrium will not be stable and once disturbed the system will begin by performing fairly small oscillations but the amplitude of the [resonant] oscillation will continually grow larger."* He included additional sections on *"harmonics [at the] nodes of vibration"*, *"the resonance of a sonorous body"*, and the resonance dynamics of pendulum oscillations.

Forty years later, Gaspard-Gustave de Coriolis (1792–1843) borrowed heavily from Lagrange's work in his popular engineering textbook. (Coriolis, 1829) Coriolis adopted Lagrange's assumption regarding the acceleration time interval for simplicity's sake, and explicitly explained that this assumption excluded consideration of *"instantaneous"* effects. Without the assumption, separate time variables for velocity and acceleration would have been required. Coriolis also introduced the concept of kinetic energy as a convenience in engineering applications involving gravitational effects: *"the mass times one-half the square of the speed [$\frac{1}{2}mv^2$]…will introduce more simplicity…since the factor '$\frac{1}{2}(v^2/g)$' is nothing more than the height from which a heavy body…must fall so that it may acquire the speed 'v'"*. Acutely aware that his kinetic energy formula did not apply to objects moving at constant velocity, Coriolis wrote that when *"the speeds have become the same... [the kinetic energy] becom[es] zero"*. Coriolis' caveats were soon forgotten, however. By the time James Clerk Maxwell (1831-1879) later wrote his basic physics textbook, he errantly summarized, *"The kinetic energy of a body is the energy it has in virtue of being in motion..."*

Meanwhile, the interdisciplinary scientist Thomas Young, M.D., (1773 - 1829) began publishing physics articles anonymously (to protect the reputation of his medical practice). He eventually went public, and according to Young his greatest scientific achievement was establishment of the wave theory of light, based on his double slit experiment. Published exactly 100 years after Newton's *Opticks*, Young's reports on the wave-like interference of light eventually resulted in abandonment of Newton's light corpuscle theory. This led to development of the belief that matter was composed of small particles, and light composed of continuous waves.

Another interdisciplinary scientist - Hermann von Helmholtz, M.D., (1821 - 1894) - was an army surgeon who set up energy-related experiments on frogs in his army barracks. Those same biomechanical experiments led to his great treatise on the transformation and conservation of energy. (Helmholtz, 1889) Helmholtz's work on conservation of energy became the first law of thermodynamics, namely that energy is neither created nor destroyed, but is instead conserved and transformed from one form to another, *"...heat, electricity, magnetism, light, and chemical affinity ... from each of these different manifestations of [energy] we can set every other [manifestation] in motion"*. Helmholtz carefully differentiated between orderly work energy and disorderly thermal energy, and taught that the total energy of a system was their sum:

$$U = A + TS \tag{5}$$

where "U" is the internal energy of a system, "A" the work (Helmholtz) energy, "T" temperature, "S" entropy, and the product "TS" thermal energy.

Helmholtz also wrote extensively about resonance which is, *"always found in those bodies which when once set in motion by any impulse, continue to perform a long series of vibrations before they come to rest ... provided the periodic time of the gentle blows is precisely the same as the periodic time of the body's own vibrations, very large and powerful oscillations may result. But if the periodic time of the regular blows is different from the periodic time of the oscillations, the resulting motion will be weak or quite insensible."* (Helmholtz, 1862) He also described resonant coupling as *"sympathetic resonance"*. Helmholtz eventually rose to the highest physics position in Germany at the University of Berlin, where he influenced many young students including Max Planck (1858 - 1947) and Heinrich Hertz (1857 - 1894). (Helmholtz, 1896 and 1904)

After Helmholtz challenged Hertz to prove the existence of Maxwell's theoretical EM waves, Hertz succeeded brilliantly. The new EM waves were called *"resonant Hertzian waves"*, based on the resonant electrical processes Hertz used to transmit and receive them.

2.3 The quantum revolution

By the late 1800's, the young Max Planck was himself a professor at the University of Berlin and was doing theoretical work on Hertz's electromagnetic waves. (Planck 1896 and 1897) Planck modeled the EM waves on the one hand as resonant waves capable of producing orderly work energy "A", and on the other hand as EM waves produced by random chaotic motions based solely on temperature "TS" (blackbody radiation). (Planck, 1900) Late in 1900 Planck met with success regarding the random thermal EM waves when he empirically determined the correct formula for blackbody radiation. A proper derivation of that empirical equation, however, was another matter altogether and according to Planck was the hardest work of his life. (Planck, 1901 and 1920)

Planck started with the Helmholtz equation (U = A + TS) and then introduced his non-controversial resonance hypothesis: EM "resonant Hertzian waves" are orderly and are thus completely free to be converted into work, and thereby constitute work energy, "A". Planck next explained that, because the blackbody apparatus used in the laboratory had been specifically designed to exclude all resonant EM waves, he could assume there was no work energy in the blackbody device. According to Planck, "A" equaled zero, and thus *"the entire problem is reduced to determining S as a function of U"*. He borrowed Wilhelm Wien's method of solving for energy density (which eliminated a time variable) and also eventually resorted to the statistical methods of his arch nemesis, Ludwig Boltzmann (1844 - 1906).

Boltzmann's kinetic mechanics were based on the limiting assumption that all the elements (e.g., molecules or atoms) in a system were moving randomly, in a completely disordered manner. Boltzmann's mechanics were restricted to the thermal energy portion, "TS", of Helmholtz's energy equation and could not be applied to orderly work energy, "A". The blackbody device and experiments were deliberately designed to exclude work energy and measure *only* disorderly, chaotic thermal energy, however. This fact allowed Planck to use Boltzmann's statistical methods in his blackbody derivation, and *"determin[e] S [solely] as a function of U"*. It also required however, that Planck introduce his quantum hypothesis - namely, that energy is quantized in small uniform amounts. Significantly, Planck assumed that those small uniform amounts of energy were *different* for each frequency, creating an *a priori* limitation which excluded consideration of a unit of *energy* for EM waves, analogous to the unit of *charge* for electrons. Mathematically Planck's quantum hypothesis took the form of the quantum formula which Planck *assumed as a given:*

$$E = h\upsilon \tag{6}$$

where "h" is Planck's action constant, 6.626 X 10^{-34} Joule seconds.

Planck also calculated a thermodynamic constant, now called the Boltzmann constant:

$$E = k_B T \tag{7}$$

where "E" is the energy of a single element (e.g., a single atom or molecule) based solely on its temperature "T", and "k_B" is the Boltzmann constant, 1.38 ×10^{-23} Joules per degree K per element. Just as Helmholtz's equation provided the energy of a macroscale system based on its temperature ("TS"), Planck's thermodynamic equation provided the energy of an individual microscale element based on its temperature ("$k_B T$"). Thus, the Boltzmann constant is the microscale equivalent of entropy. Planck never introduced a microscale equivalent of the work energy "A", however: the blackbody experiments excluded work energy. This resulted in a microscale energy formula which was necessarily incomplete for any system in which work energy was present.

While some scientists used Planck's blackbody equation for practical applications, his revolutionary quantum hypothesis received little attention - until, that is, Albert Einstein's (1879 - 1955) own revolutionary papers were published in 1905. (Einstein, 1905) Einstein had seized on Planck's quantum hypothesis and used it to provide explanations for a number of unexplained phenomena such as the photoelectric effect and ionization of gases. The interactions of EM waves and matter, he proposed, *"appear more readily understood if one assumes that the energy of light is discontinuously distributed in space"*, e.g., in small particles or packets along the lines of Newton's *"light corpuscles"*, and is absorbed in *"complete units"* or *"quanta"*.

Although highly controversial, Einstein's papers brought attention to Planck's quantum hypothesis and formula. A few years later, Niels Bohr (1885 - 1962) adopted Planck's quantum formula in his theory of the hydrogen atom. (Bohr, 1913) Controversy still raged however, and Robert Millikan (1858 - 1963) undertook a series of meticulous experiments testing the validity of Planck's constant and what he described as Einstein's *"reckless"* theories regarding energy quanta and photoelectric phenomena. (Millikan, 1916)
Millikan, well familiar with Planck's accepted resonance hypothesis distinguished the photoelectric effect as an ordered work function and not a thermal effect: *"photoelectrons do not share in the energies of thermal agitation…absorption [of EM waves] is due to resonance (and we know of no other way in which to conceive it…)"*. Echoing Galileo, Millikan stated, *"the phenomena of absorption and of emission show that…oscillators possess natural frequencies…and the characteristic waves which they emit are of these frequencies…if any particular frequency is incident upon such a substance the oscillators in it which are in tune with the impressed waves may be assumed to absorb the incident waves"*. Regarding the resonant work nature of the photoelectric effect he stated, *"emission of [electrons] from the atom…takes place especially copiously when the impressed frequency coincides with a 'natural frequency'… [It] furnishes a proof which is quite independent of the facts of black-body [thermal] radiation, of the correctness of the fundamental assumption of the quantum theory, namely, the assumption of a discontinuous…energy absorbed by the electronic constituents of atoms from [EM] waves"*. (Underline added)
The quantum revolution begun by Planck and Einstein was taking hold.

2.4 The quantum paradox

As the quantum revolution began to gain momentum, paradoxes and puzzles began cropping up. The simple model of light waves and matter particles had been disrupted. Louis de Broglie (1892 - 1987) added to the confusion in the early 1920's when he proposed that if *light* could be both a wave and a particle, then so could *matter*. (de Broglie, 1924) Pursuing that line of reasoning, de Broglie found the lack of a unit of energy for EM waves, i.e., *"an isolated quantity of energy"* particularly troublesome. Without an energy constant for light (i.e., an isolated quantity of energy), de Broglie was unable to determine the fundamental mass of light using Einstein's energy-mass equivalence equation, "$E = mc^2$". Instead, the energy of light paradoxically depended on its frequency. De Broglie made the best of a conceptually difficult situation, and instead set Einstein's mass equivalence equation equal to Planck's quantum formula and solved for the rest mass of light at a particular frequency:

$$m_0 c^2 = h\upsilon_0 \text{, } \textit{therefore } m_0 = h\upsilon_0 / c^2 \tag{8}$$

where "m_0" is the rest mass of light, and "c" the speed of light *in vacuo*. Since the number of different frequencies of EM waves are theoretically infinite, this approach produced a paradoxically infinite number of values for the rest mass of light. Unlike other particles such as the electron or proton, de Broglie could find no constant rest mass associated with light particles.
The lack of any energy or mass constants for light was quite puzzling indeed. Unbeknownst to de Broglie, Planck's limiting assumption about different quanta for each frequency excluded the very unit quantity of energy de Broglie sought. De Broglie could at least conclude however, that the rest mass of light in the visible region was quite small and in his Nobel prize speech explained, *"The general formulae…may be applied to corpuscles of [visible] light on the assumption that here the rest mass m_0 is infinitely small… the upper limit of m_0 … is approximately 10^{-24} gram.* (de Broglie, 1929)

De Broglie also used Planck's quantum formula to derive the momentum for light:

$$\rho = m_0c = hv/c = h/\lambda \quad (9)$$

finding that the momentum of light appeared to be directly proportional to its frequency, and thus inversely proportional to its wavelength "λ". Once again, De Broglie obtained a zoo of values - this time for momentum since the range of frequencies and wavelengths in the EM spectrum is infinite.

In the meantime, Neils Bohr undertook his ambitious project modeling the hydrogen atom based on Planck's quantum formula and constant. Bohr found that he could not calculate time intervals in regard to the interactions between EM waves and electrons. He was forced to model instantaneously *"jumping electrons"* instead. (Bohr, 1913 and 1920) Few (including Bohr) were satisfied with the jumping electrons however, and in the mid-1920's two new approaches to quantum mechanics were introduced. In 1925, Werner Heisenberg, introduced matrix mechanics. (Heisenberg, 1925) A year later Erwin Schrödinger began publishing a series of papers on wave equations, intended to represent the real electron waves suggested by de Broglie. (Schrödinger, 1982)

Even with these two new approaches quantum mechanics still did not make sense to many early quantum pioneers. It lacked the certainty and definiteness of classical mechanics. Efforts to compensate for the many paradoxes included additional principles such as Heisenberg's uncertainty principle, and Bohr's complementarity principle. (Heisenberg, 1920 and Bohr, 1928) Additional variables and constants of inexplicable origin were discovered, such as the dimensionless fine structure constant. Discussion and debates continued. The Bohr-Heisenberg school of probabilities and uncertainty battled the Einstein - Schrödinger school of realism and certainty. Without answers for such simple matters as an energy constant or rest mass for light, the Bohr-Heisenberg school eventually prevailed. Scientists concluded (over Schrödinger's strenuous objections) that his wave equations represented only probabilities, and not real physical waves. The consensus that finally emerged was that the classical mechanics of our macroscale world simply could not be applied to the kaleidoscopic microscale world of the quantum. According to Bohr, a classical limit existed at the very highest electron energy levels in atoms, and below that limit classical mechanics simply could not be applied.

The iconoclastic brilliance which initially led Einstein to make his *"reckless"* quantum proposals, would not allow him to join the quantum crowd and he insisted that something had been missed. He simply could not believe that God and the universe were so perversely paradoxical. In 1935, Einstein published his "EPR" paper loudly proclaiming that quantum mechanics was incomplete due to the existence of "hidden" quantum variables. (Einstein, 1935) Einstein and others such as Bohm and Bell tried to describe the hidden variables, but such a task was difficult, if not impossible. (Bohm, 1952) How does one describe a quantum variable mathematically, when the very nature of the variable is unknown? Small groups of scientists have attempted to keep Einstein's quest alive, but the scientific community as a whole abandoned efforts to find any "hidden variables". Instead, it was generally agreed that the paradoxical nature of quantum mechanics was an undeniable reality of life. Incredible efforts then went into developing more quantum models incorporating the paradoxes, such as theories of strings, super-symmetry, membranes, and the like.

Einstein's stubborn insistence that something had been missed was correct, however. The first of his "hidden variables" was discovered nearly a century later, the result of a small mathematical thread. (Brooks, 2009,a) Tracing that thread through the historical record, it led to the discovery that a minor mathematical inadvertence in Planck's brilliant blackbody work had induced him to assume an *incomplete* and *abbreviated* version of the full quantum formula. All of quantum physics was based on Planck's simple quantum formula, and that *assumed* formula was incomplete: it was missing a time variable. After restoring the time variable, Planck's constant took on new fundamental meaning. The rich quantum tapestry that emerged, revealed beautifully symmetric quantum principles grounded in reality and certainty, using the complete quantum formula and a more inclusive or complete "thermo"dynamic formula. (Brooks, 2009,b)

3. The complete quantum formula

The complete quantum formula is:

$$E=\tilde{h}\,t_m v \qquad (10)$$

where "$\tilde{h}$" is the *energy* constant for light (6.626 X 10^{-34} Joules/oscillation) and "t_m" is the measurement time variable.

3.1 The time variable

The complete quantum formula is quite similar to an energy relationship found in Planck's early theoretical electromagnetic work from the late 1890's. He converted time-based power measurements, "E/t", to total energy values by multiplying by the measurement time, "t_m". Planck's EM theory used that simple conversion in a generic relationship in which the oscillation energy of a system was proportional to the product of a generic constant "a", the measurement time variable, and frequency:

$$\delta U \approx a\,\delta t_m\, v \qquad (11)$$

A few years later, the time variable was lost in Planck's complicated blackbody derivation. Instead of multiplying time-based energy measurements by the measurement time, Planck adopted Wien's mathematical methods which converted the power measurements into energy density values by dividing by the speed of light. This caused the measurement time variable "δt_m" to be simultaneously fixed at a value of "one second", and then "hidden". Proof of these facts are found in Planck's 1901 blackbody paper, in which he described the experimental data and mathematical methods he used:

"§11. The values of both universal constants h and k may be calculated rather precisely with the aid of available measurement. F. Kurlbaum, designating the total energy radiating into air from 1 sq cm of a black body at temperature t° C in **1 sec**, *by S_t found that:*

$$S_{100} - S_0 = 0.0731\ watt / cm^2 = 7.31\ x\ 105 erg / cm^2 \underline{sec}"$$

Instead of multiplying Kurlbaum's time-based power measurement by the measurement time to obtain total energy (as Planck had done in his earlier work), he converted the power measurement to energy density by dividing by the speed of light "c" (3 X 10^{10} cm/sec), according to Wien's method:

"From this one can obtain the energy density of the total radiation energy in air at the absolute temperature

$$\frac{4 \bullet 7.31x10^5}{3x10^{10}(373^4 - 273^4)"} = 7.031x10^{-15} \; erg \, / \, cm^3 \, deg^4$$

The time variables in the numerator and denominator cancelled out and Planck was seemingly able to address energy independent of time. Dividing by the constant speed of light however, is the same as multiplying by time:

$$\frac{E / ts^2}{c} = \frac{E}{ts^2} x \frac{t}{s} = \frac{E}{s^3} \tag{12}$$

where "s" is distance. In this case the time value by which the power measurement was multiplied was the constant "one second" unit time of the constant speed of light. Planck seems to have been unaware that by using Wien's energy density calculation he was actually causing the infinitely variable measurement time to be fixed at a constant value of one second. He also seems to have been unaware that the fixed time variable was subsequently hidden in the final calculations of his action constant "h":

$$\mathrm{h} = 6.626 \text{ X } 10^{-34} \text{ Joule seconds} \tag{13}$$

His action constant is actually the *product* of a true universal constant - " $\tilde{\mathrm{h}}$ " - and the fixed, hidden measurement time variable, "t_m".

$$\mathrm{h} = \tilde{\mathrm{h}} \mathrm{t_m} \text{ where } t_m = 1 \text{ second} \tag{14}$$

3.2 The energy constant

When the missing time variable is restored to the quantum formula, the identity of Planck's *real* universal constant becomes apparent. The hidden constant is, in fact, a universal *energy* constant, namely the energy of a single oscillation or EM wave. This universal energy constant for light is that same *"isolated quantity of energy"* de Broglie searched for, i.e., the fundamental small quantum of light's energy:

$$\tilde{\mathrm{h}} = 6.626 \text{ X } 10^{-34} \text{ Joules / oscillation} \tag{15}$$

This fact is easily verified by solving Planck's incomplete formula for the energy of a single oscillation of light (see Brooks, 2009a for derivations). The numerical value Planck calculated for his action constant "h" is actually the numerical value of the mean oscillation energy of individual EM waves. The *"isolated quantity of energy"* hoped for by de Broglie, has been found.

The universal nature of this constant is made clear by consideration of the energy constant over a wide range of wavelengths, time periods and frequencies. The mean energy of a single EM wave remains constant regardless of whether it is a radio wave, microwave, infrared, visible or ultraviolet wave. For low frequency and long wavelength EM waves such as radio waves, the constant mean oscillation energy is spread out diffusely over a large volume of space. At higher frequencies and shorter wavelengths, the energy becomes more concentrated in a smaller volume of space. In the ultraviolet region, the energy of an oscillation becomes extremely dense, being confined to a very small region of space, around 100 nanometers or so in dimension. The amount of energy in a single oscillation is the same, however, regardless of the volume or time period it occupies.

The constancy of the energy of a single EM wave over a variety of wavelengths and time periods means that the elementary quantum of light is constant over a shift in time or space. When a property is constant over a shift in time or space, that property is conserved and represents a universal property. The fundamental relationships are now clear. Just as the electron has a fundamental unit of charge which is conserved and represents a universal constant for electrons, light has a fundamental unit of *energy,* "$\tilde{h}$", which is conserved and represents a universal constant for EM waves.

3.3 The frequency variable

Planck's quantum formula was incomplete, and as a result did not contain the oscillation energy constant. This in turn resulted in a quantum formula in which the units did not balance:

$$\text{E (Joules)} = \text{h (Joule seconds)}\, v\text{(oscillations per second)}, \text{ but Joules} \neq \text{Joules oscillations} \tag{16}$$

Scientists found they were unable to balance the quantum equations and use complete mathematical notation for frequency, namely cycles, waves or oscillations per second. As a result, mathematically incomplete notation, which omitted descriptive units for frequency's numerator, was adopted instead. Frequency is currently described in the International System of Units ("SI") as "1/sec" or "sec^{-1}". This incomplete SI notation for frequency removes an essential mathematical element of reality in quantum mechanics.

Incomplete mathematical notation for frequency is no longer required to compensate for the deficiencies of the incomplete quantum formula. With the recognition of the energy constant - energy per *oscillation* - frequency can once again be correctly and completely notated as *oscillations* per unit time. The use of complete mathematical notation in quantum mechanics restores a vital aspect of mathematical reality. Recognition of "oscillations" in the numerator of frequency measurements provides a theoretical element corresponding to each element of reality in the complete quantum formula. As Einstein argued, such a correspondence is a critical requirement of a complete quantum mechanics.

3.4 The photon

In 1926, Gilbert Lewis coined the term "photon" for Einstein's light quantum. The energy of the photon was calculated with Planck's (incomplete) quantum formula, "$E = hv$". Questions have been raised from time to time since then, as to whether the "photon" is truly an indivisible particle of light. The answer to that question is now clearly, "No". The photon as previously defined is *not* an indivisible elementary particle.

The fixed time variable and energy constant had been hidden in Planck's "action" constant, and so it was not apparent to Lewis or others that what they were calling the '"photon" was actually a *time-based* quantity of light energy, which relied on a fixed and arbitrary one second measurement time interval. A time-based amount of energy which relies on an arbitrarily defined time interval cannot be a fundamental or elementary particle of light. The photon is not an elementary particle of light.

What is the elementary particle of light, then? As identified by the universal energy constant, the elementary particle of light is the single oscillation of EM energy, i.e., a single cycle or wave of light. The elementary particle of light possesses the constant energy of 6.626×10^{-34} Joules. It is the smallest known quantum of energy in the universe. What was labeled a "photon" by 20th century physics is actually a *collection* or *ensemble* of these small elementary light particles. Each individual oscillation is a *"complete unit"* of light and can be emitted or absorbed as a complete and discreet unit.

The "photon" is *not* an indivisible particle of light, and is in fact a collection or ensemble of light oscillations, which can act separately and individually as complete energy units. Upon absorption by a detector or object, the energy of a collection of discreet oscillations can spread over several atoms or molecules, resulting in a multi-atom, energy distribution state known as "entanglement". (Brooks, 2009, c) This entanglement of EM energy can take place in different patterns or distributions, depending on the nature of the absorbing or detecting material. Similarly, emission of light energy can occur from an "entangled" energy state shared by multiple atoms or molecules in the emitter. An ensemble of EM waves with fewer than "N" oscillations (where "v = N/sec") results in a "sub-photonic" collection of EM waves. The ultimate sub-photonic particle is the elementary particle of light, the single EM oscillation.

3.5 The mass of light

De Broglie bemoaned the absence of *"an isolated quantity of energy"* with which he could calculate the constant rest mass of light. Using the energy constant for light, it is now possible to complete de Broglie's calculations and determine the rest mass of a single quantum of light. Under de Broglie's original formulation using Einstein's energy-mass equivalence equation of "E = mc^2", the rest mass of light is readily determined:

$$m_0 = 7.372 \text{ X } 10^{-51} \text{ kg / oscillation} \tag{17}$$

This value is within the same order of magnitude as the most recent and reliable estimates for the upper limits of the rest mass of light. Since the energy of a single oscillation of light is constant, regardless of its wavelength, time period or frequency, its mass is also constant regardless of its wavelength, time period or frequency. Hence, the mass of light is constant over a shift in time or space. The mass of light is thus conserved and represents another universal constant for light (Mortenson, 2011).

Just as the density of light's constant wave energy varies with the length and volume the wave occupies, the density of its mass varies as well. The mass of long EM radio waves, spread over a distance and volume of hundreds of meters, is low in density. The identical mass, when confined to the small wavelength and volume of an X-ray oscillation (on the order of 10^{-8} to 10^{-11} meters) is trillions of times more dense. High density X-ray oscillations, with their intensely concentrated mass and energy, can create interactions not typically seen with low density radio waves, and give rise to effects such as X-ray scattering and particle-like properties.

3.6 The momentum of light

Momentum is classically calculated as the product of an object's mass and its speed. Using the constant mass of an EM oscillation as calculated above, and the constant speed of light (2.99 X 10^8 m/sec), De Broglie's calculation for the momentum of light can be completed:

$$\rho = m_0 c = 2.21 \text{ X } 10^{-42} \text{ kg m / sec per osc} \tag{18}$$

As with mass, the momentum of a single oscillation of light is constant, rather than being infinitely variable. The momentum of an EM wave is constant regardless of its wavelength, time period or frequency. Thus the momentum of light is constant over a shift in time or space, and is a conserved property.

In terms of de Broglie's earlier calculations for the masses and momenta of photons, the mass and momentum constants for EM waves are not contradictory or confounding. It should be remembered that the photon of 20th century concepts was actually a *collection* of elementary light particles, i.e., EM oscillations. Collections of masses and momenta can be additive. Summation of the constant mass and momentum of single oscillations (based on the number of oscillations "N" in a one second "photon") yields the same collective mass and momentum that de Broglie obtained with his photon-based calculations. Although de Broglie's mass and momentum calculations provided infinitely variable results, it is now recognized that his variable results were an artifact of the missing energy, mass and momentum constants. A previously unrecognized symmetry for conservation becomes apparent. Energy, mass and momentum are all conserved for both light and matter, completing the triad of conservation relationships outlined earlier by Helmholtz, Einstein and de Broglie.

3.7 The force of light

Energy, mass and momentum are all constant and conserved for light. Using classical mechanics, however, it is easily discerned that the *force* exerted by light is not constant. According to Lagrange, force is the product of mass and the change in velocity *"during the instant dt"* when the velocity changes:

$$F = m\,v \;/\; dt \tag{19}$$

For changes in velocity occurring in an interval of time equal to or greater then the velocity unit time, the same time variable for both velocity and acceleration can be used. If, on the other hand, the acceleration (or deceleration) occurs in a time interval much smaller than the velocity unit time (i.e., an "instantaneous" event), a *second* time variable, "t_a", must be used for the acceleration time interval. When an EM oscillation is emitted by an object, a small bit of mass of 7.372×10^{-51} kg is instantaneously accelerated to the speed of light, "c". Likewise, when a light wave is absorbed by an object, its mass is instantaneously decelerated. The acceleration or deceleration occurs *"during the instant dt"* which is the time period "τ" of the EM wave. The force that accelerates an EM oscillation at its emission (or that is exerted by an oscillation when it is absorbed) is thus:

$$F = m\,c \;/\; t_a \text{ where } t_a = \tau \tag{20}$$

The time periods of EM waves are infinitely variable, as are their frequencies ($\tau = 1/\upsilon$). Thus, although the mass and velocity of EM waves are constant, the forces which they exert are not. The forces associated with light oscillations vary inversely with their time periods, and directly with their frequencies ("$F = m\,c\,\upsilon$").

The energy and mass of a radio wave, distributed over a comparatively long period of time, exert relatively little force on an absorbing detector. The energy and mass of an X-ray or gamma ray oscillation, on the other hand, are concentrated in a minute period of time and exert tremendously large forces on an absorbing object.

These EM light forces are additive, and given sufficient accumulation the forces can be quite large and result in the physical acceleration of absorbing matter. (Liu et al, 2010) The force of light is the operative mechanism behind "space sails" which are now being employed on space craft. The sails of the ancient mariners were pushed by the forces of the wind which filled them. The sails of modern space explorers are now filled by the forces of light which impinge on them. Likewise, an object emitting light experiences a recoil force proportional to the emission force of the EM waves. (She et al, 2008)

3.8 Classical limit

The quantum pioneers anticipated that classical mechanics would be used to provide a description of physical processes at very small length and energy scales. Numerous roadblocks were encountered, however, due to the hidden quantum variables and constants. The quantum mechanics developed by Heisenberg and Schrödinger provided a mathematical framework for low energy kinetics, however they were unable to obtain the certainty and definitiveness provided by classical mechanics. Without the mass constant for EM waves, it was impossible to use classical properties of position, time, and mass in any meaningful way. Heisenberg and Bohr found that they were limited to finding just probabilities, and that they could apply classical mechanics only at very high electron energy levels. The region where the classical and quantum mechanics formed a boundary zone, was deemed the "classical limit" by Bohr. (Bohr, 1920) Above the limit, classical mechanics could be applied with reality and certainty, while below the limit all was uncertain and only quantum mechanics could be applied.

Using the new quantum variables and constants, the classical limit/boundary zone between quantum and classical mechanics is disappearing. (Mortenson, 2010,a) It is now possible to use classical mechanics at the smallest possible energy levels for light, equivalent to fractions of a percentage of the lowest known electron energy levels. The kinetics of energy absorption for a single EM oscillation, namely 6.626×10^{-34} Joules, are now fully describable using classical mechanics. In this regard, the classical limit previously theorized by Bohr, is being recognized as an artifact of the missing quantum variables and constants.

The application of classical physics at the smallest known energy levels, is made possible with the use of the *second* hidden time variable, Lagrange's acceleration time variable, "t_a". The absorption or emission of an EM oscillation in the visible light region takes place in 10^{-10} seconds. This results in a near instantaneous deceleration or acceleration of light's mass. The energy required to accelerate a body is a function of the distance over which the force acts, "$F\ \delta s$". In the case of an individual EM oscillation, the distance over which the force acts is the wavelength, "λ" of the oscillation. Multiplying the variable force for light by its wavelength, i.e., "$F\ \delta s = (m\ c\ v)\ \lambda$", results in constant energy of "mc^2", or in other words 6.626×10^{-34} Joules/osc. The energy constant for light is thus quickly derived from first principles of position, time and mass.

Lack of appreciation, for the caveats of Lagrange and Coriolis regarding acceleration time intervals and instantaneous events, contributed to the perception that a barrier or limit existed between classical and quantum mechanics. The new fundamental physics of EM waves reveals that particle mechanics can be described at both the macroscale and microscale levels using the certainty, realism and determinism of classical mechanics.

3.9 The uncertainty principle

Heisenberg suggested the uncertainty principle as a response to the inability of early quantum pioneers to determine quantum properties related to time or energy with any certainty. He proposed that changes in energy and time are uncertain to the extent that their product must always be greater than or equal to Planck's constant ($\Delta E\ \Delta t\ \geq\ h$). That principle included, of course, the incomplete quantum constant "h", which hid an energy constant and a fixed time variable. Heisenberg's uncertainty principle cured a multitude of quantum paradoxes, and as David Bohm wrote a generation later, *"the physical interpretation of the quantum theory centers around the uncertainty principle"*. When "h" is properly replaced with the energy constant and measurement time however, the physical interpretation of quantum theory is changed dramatically and centers around *certainty* and *constancy,* where the change in energy is the energy of a single EM wave, and the change in time "Δt" and measurement time "t_m" are equal to the time period "τ" for the oscillation.:

$$\Delta E\,\Delta t \geq \tilde{h}\,t_m \text{ and } \Delta E \geq \tilde{h} \tag{21}$$

The smallest possible change in energy is the energy of a single wave of light.
This concept was obscured in the past due to the absence of a separate energy constant and time variable in Planck's quantum formula. Under the circumstances, it was inevitable that calculations of quantities involving time and energy, would yield uncertain results. The uncertainty is now gone, replaced by a quantum mechanics that accommodates a more certain and realistic physical interpretation.

3.10 The fine structure constant

The fine-structure constant *"has been a mystery every since it was discovered more than fifty years ago, and all good theoretical physicists put this number up on their wall and worry about it....It's one of the greatest damn mysteries of physics: a magic number that comes to us with no understanding by man..."* (Feinman, 1988)
Using the newly discovered quantum constants and variables, the fine structure constant "α" is far less of a mystery. Examination of the fine structure constant in relation to light's action "S" and Planck's constant "h" (i.e., "α h = S"), and substitution of "h" with the energy constant "$\tilde{h}$" one finds:

$$\alpha = S\frac{1}{\tilde{h}}\text{, or in other words } \alpha = (Et)\frac{osc}{E} \text{ and } \alpha = osc\,t \tag{22}$$

The fine structure constant is not dimensionless. It represents a scaling constant between time and a single oscillation of EM energy, i.e., "osc t". As such, a theoretical element corresponding to an element of reality is now provided for the fine structure constant. This is a critical requirement for a complete quantum mechanics.

3.11 Wave – Particle duality

Two opposing models of light - particles and waves - have been debated for centuries. Some investigations suggest light is composed of waves, while others suggest particles. This conundrum led Einstein to object, *""But what is light really? Is it a wave or a shower of photons? There seems no likelihood for forming a consistent description of the phenomena of light...we must use sometimes the one theory and sometimes the other..."*. (Einstein, 1938) Bohr responded to these two contradictory pictures of reality with his complementarity principle, asserting that certain aspects of light could be viewed one way or another, but never both at the same time.
We are now presented with a picture of reality which *demands* that we view light simultaneously as a wave and a particle. The elementary *"particle"* of light is the single EM

oscillation or *"wave"*. Although the time and space the wave occupies may vary, that variance is according to the constant ratio "c", and the elementary particle's energy, mass and momentum remain constant as well.

The divergent pictures of the past resulted from relative size differences between the EM waves and the matter with which they interacted. For example, scattering studies were first performed using soft X-rays with wavelengths larger than atoms, and no clear-cut particle properties were detected. When Arthur Compton used ultra-short hard X-rays and gamma rays, however (up to two orders of magnitude *smaller* than an atom) he observed particle-like properties. (Compton, 1923) The concentrated energy and mass of the X-ray and gamma ray waves appeared as small points *relative* to the size of the atoms in the irradiated materials.

On the other hand, one and two-slit experiments demonstrate wave-like properties for light via interference bands. These wave-like properties are *also* relative, however, to the sizes of the light oscillations and the matter with which they interact. For a slit whose width is *equal* to the wavelength of the light, *no* interference bands are observed and particle-like behavior is seen. It is only when the width of the slit is increased relative to the wavelength of the light that interference bands and wave-like properties begin to appear.

Recent experiments with light slits and "single photons" reveal as much about the detecting material as they do about the light itself. A "photon" is merely a *collection* of individual EM quanta. When visible light waves (which are much larger in size (400 - 800 nm) relative to the individual atoms in the detector (0.1 - 0.5 nm)), strike a detector the energy of the light wave ensemble impinges on multiple detector atoms simultaneously. This produces an energy entanglement state in several of the detector atoms. (Brooks, 2009, c) Distribution of the light energy over several atoms excites a small point-like portion of the detector material resulting in a photonic reaction, and produces a particle-like pattern in the detector. (Roychouhuri, 2009) Although the resulting detector imaging appears to show the buildup over time of "photon" collisions, they actually show the buildup of energy entanglement states in the detector itself, which are subject to positive and negative interference within and between groups of entangled atoms.

4. Energy dynamics

The experimental data Planck used to derive the blackbody equation and thermodynamic formula did not include any measurements arising from orderly work energy. Hence, Planck did not include work energy in his thermodynamic formula, "$E = k_B T$". Instead, Planck's formulation was limited exclusively to the energy of a small system element (e.g., an atom, molecule or ion) based only on its temperature and random chaotic motion.

When orderly work energy is present in a system, more inclusive formulae must be used to represent the total energy of a system or its elements. (Brooks, 2009b and Mortenson, 2010b) Helmholtz's energy equation, "$E = A + TS$", embodies once such inclusive formula on the macroscale, and represents the total energy of a system as the sum of its work and thermal energies. This more complete formula encompasses significantly more than simple thermodynamics, and is more appropriately referred to as an *energy dynamics* formula.

4.1 Energy dynamics formula

A complete energy dynamic formula for an entire system is given by Helmholtz's energy equation, "$E = A + TS$", (Equation 4., above). While calculation of the thermal energy of a system is relatively simple and straightforward, determination of the total work energy can be considerably more involved. Work energies come in many forms, including mechanical, chemical, gravitational and resonant energies. Resonance work energy is a broad category encompassing time-varying forces and fields such as sound waves, electric or magnetic fields, and light waves. These resonant energies couple to matter via *"sympathetic resonance"* and are denoted in the fundamental energy dynamics formula as, "A_r":

$$E = A_r + TS \tag{23}$$

The fundamental principle described by Galileo in his pendulum studies holds true for resonant work energies, i.e., *"by [providing a time-varying energy one may] confer a Motion, and a Motion considerably great by reiterating…but only under the Time properly belonging to its Vibrations"*. Anyone who has pushed a child on a swing has applied a resonant mechanical energy to the child/swing system. Pushing the child at just the right time (i.e., the resonant frequency for the child/swing ensemble) increases the speed, height and excitement of the child's ride. Pushing at the wrong time, when the child is a few meters away, produces no effect on the system and may detract from the excitement of the child's ride.

In the same way, electromagnetic waves impinging on a material transfer resonant EM energy to the absorbing matter via their momentum, force, speed and mass. An acceleration of the oscillating element within the system results from the applied EM force, and an increase in the oscillation amplitude of that element results (see Fig. 1, above). Thus, *"pushing"* the system elements with EM waves at just the right time increases the amplitude (height) of the system's oscillations and excites them to higher energy levels. The amount the system's oscillation amplitude increases is a function of how close the resonant EM wave frequency is to the oscillation frequency inside the system (Eq. 2., and Fig. 2, above).

The increased oscillation amplitudes and energy levels in the system can perform work in a variety of ways, depending on which element or oscillation amplitude is increased. For example, changes in motion, chemical, material, organizational, or behavioral states may all result from a resonant energy excitement in the system.

Expressed at the microscale level, a complete energy dynamics formula for the total energy of an individual element in a system is formulated parallel to Helmholtz's system formula:

$$E = W_e + k_B T \tag{24}$$

where "W_e", is the total microscale work variable representing the total work performed on an individual element. In the case of resonance work energy, a resonance work variable, "r_A" can be used. This microscale resonance work variable represents the energy gained by an individual element *in* a system, as a result of resonance work energy, "A_r", applied *to* the system as a whole:

$$E = r_A + k_B T \tag{25}$$

4.2 Determination of system resonance work energy "Ar"

The resonance work energy (system/macroscale) and variable (element/microscale) may be determined experimentally. An aqueous solvent system under resonant conditions was compared to an identical system under thermal conditions (see Table 1., below): [1]

---

[1] *Experimental Procedure – Distilled water (500 ml at 20° C) was placed in each of two 1,000 ml beakers. One beaker was irradiated with resonant vibrational electromagnetic frequencies of water for three (3) hours, by a light*

	Resonant system	Thermal system
Weight Dissolved (g/100ml NaCl)	26.0	23.8
Moles Dissolved (NaCl)	4.65	4.25
Heat of solution (kJ)	17.4	16.0

Table 1. Resonant vs. Thermal aqueous solvent system

The heat of solution is a measurement of the work performed by the solvent on the dissolving solute. The work performed by the resonant system was 17.4 kJ, while the thermal system performed only 16.0 kJ of work on the NaCl solute. The energy dynamics formulae for both systems are:

$$\text{Thermal system } E_T = TS \text{, and } 16.0 \text{ kJ} = (274° \text{ K}) S \tag{26}$$

$$\text{Resonant system } E_R = A_r + TS \text{ and } 17.4 \text{ kJ} = A_r + (274° \text{ K}) S \tag{27}$$

Subtracting, one finds that the resonance work energy, "A_r", in the resonant system is 1.4 kJ of energy:

$$A_r = 1.4 \text{ kJ} \tag{28}$$

4.3 The resonance factor

The ratio of the total energy in the resonant system to the total energy in the thermal system:

$$E_R \;/\; E_T = r_f \tag{29}$$

is the resonance factor, "r_f". In the aqueous solvent system described above, the resonance factor is 1.09. There was 9% more energy available in the resonant system to perform work on the solute and to dissolve it. This resonance work energy was in addition to the thermal energy already inherent in the system as a result of its temperature.

4.4 Determination of element resonance work energy "r_A"

The amount of resonance work energy at the microscale is the resonance work variable, "r_A". In the solvent system example, individual elements in the system irradiated with resonant EM waves possessed greater energy than the elements in the thermal system. The value of the microscale resonance work energy can be calculated using Equation 25., above:

source using 2.1 kJ total energy. The other beaker was placed in an opaque incubator for three (3) hours. The water in both beakers at the end of the three (3) hours was 23° C. Sodium chloride (250 g) was added to each beaker, and the beakers were stored identically in a darkened cabinet. Twenty (20) hours later temperatures of the solutions were identical (21° C/274° K). The solutions were decanted and the dissolved weight, salinity and concentration measurements of the resonant and thermal saline solutions were made using standard methods. (Brooks et al, 2005)

$$\text{Thermal element } E_t = k_B T \text{ and } E_T = 405 \text{ X } 10^{-23}\text{ J per molecule} \quad (30)$$

$$\text{Resonant element } E_r = r_A + k_B T \text{ and } E_r = 440 \text{ X } 10^{-23}\text{ J per molecule} \quad (31)$$

Subtraction shows that each water molecule in the resonant system performed an additional 35 X 10^{-23} J of work on the solute, as a result of absorption of the resonant EM waves.

4.5 Virtual thermal effects of resonant EM waves

When resonant EM waves perform work on a system and increase one or more oscillation amplitudes within the system, that increased oscillation energy is free to be transformed into work within the system. In the case of the aqueous solvent system described in the experimental example above, the vibrational oscillations of the solvent water molecules were excited. This in turn led to a change in the *behavior* of the water as a solvent. The resonant water dissolved 26.0 g/100 ml of NaCl, while the thermal water dissolved only 23.8 g/100 ml. To what temperature would the thermal system need to be raised, in order to dissolve the same amount of NaCl that the resonant system dissolved, all else being equal? This is readily calculated by setting the total element energy in the resonant system equal to the thermal energy, and solving for temperature "T":

$$E_r = k_B T \text{ , therefore } T = E_r / k_B \quad (32)$$

$$T = 319^{\circ}K \ (46^{\circ}C)$$

In order to dissolve the same amount of NaCl in the thermal system, that the resonant system had dissolved, the thermal system would have had to be heated to 46 ° C. The water in the resonant system *behaved* as though it had been heated to 46 ° C, even though it had not. The EM waves provided a virtual thermal effect in the resonant solvent system.

4.6 Energy efficiency of resonant EM waves

As Helmholtz described many decades ago, energy can be transformed and converted from one form to another, "*...heat, electricity, magnetism, light, and chemical affinity* ". The efficiencies with which these transformations take place is not uniform across all conversions of energy. Depending on the process and desired end-result or product, the energy transformation efficiency can vary widely. For example, in the water solvent example given above, one would need to heat the water to 46 ° C, in order to dissolve the same amount of salt that the resonant water system had dissolved. Heating 500 ml of water to that higher temperature would require at least 52 kJ of energy.

On the other hand, the light source which provided the resonant vibrational EM waves to the resonant system consumed only 2.1 kJ of energy. (Mortenson & George, 2011). The total additional energy required to achieve the desired end-result or product (i.e., dissolve more salt), is far less with the resonant EM waves: ninety-six percent (96%) less, in fact.

The total extra heat of dissolution work performed by the resonant water on the solute was 1.4 kJ. The energy transformation efficiency for the resonant system was 67%. If one were to heat the thermal water to increase its temperature by 25°, using the 52 kJ of energy, the efficiency of the thermal energy conversion into heat of dissolution would be only 3%. There is more than an order of magnitude difference between the energy conversion efficiency of the resonant EM system and the thermal system.

4.7 Boltzmann weighting

Under Boltzmann mechanics for purely random and chaotic thermal systems, the elements (e.g., molecules) in a system adhere to a thermal distribution curve (Figure 3.a., below). In general terms, at low temperatures, most of the elements or molecules are at the lowest possible energy level or ground state. As the temperature in the system increases, the elements begin to leave the lower energy levels and populate the upper energy levels. At very high temperatures, several of the upper energy levels may be populated, leaving few molecules in the lowest ground state. The distribution of elements in the energy levels is determined with the Boltzmann factor:

$$e^{-E_n/k_B T}$$

This is a weighting factor that determines the probability that an element will be in the "n^{th}" energy state when the system is in thermodynamic equilibrium. The Boltzmann factor *excludes* consideration of resonant work energies (which are orderly), and assumes completely random motions in the system. Helmholtz energies are thus typically assumed to be at a minimum when Boltzmann mechanics are applied.

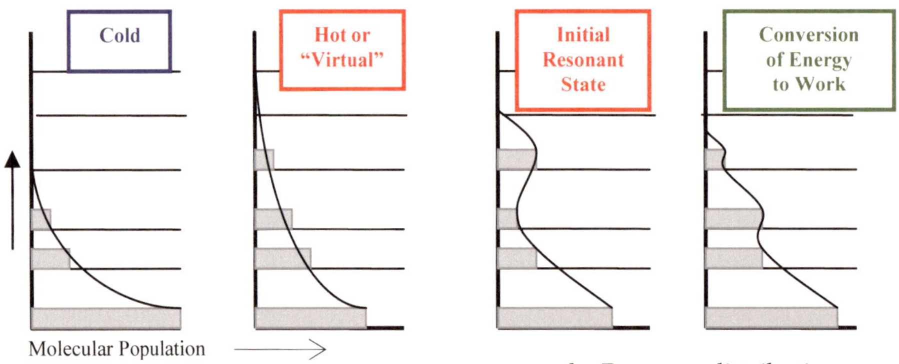

a. Boltzmann thermal distribution

b. Resonant distribution (Non-Boltzmann)

Fig. 3. Comparison of energy level population states under thermal conditions and resonant EM conditions. Upper energy level populations are increased as temperature increases. Absorption of resonant EM waves produces an irregular resonant energy distribution curve. This can result in system behavior equivalent to a "virtual" thermal distribution curve.

When a system is exposed to resonant EM waves, a "virtual" thermal effect can be produced, as in the aqueous solvent example above. In such a case, the "virtual" thermal distribution may be determined using a modification to the Boltzmann weighting factor:

$$e^{-E_n/r_f k_B T}$$

in which the resonance factor, "r_f" is included. The resulting thermal distribution curve reveals the energy state distribution curve of the thermodynamic system that would result in the same desired product or behavior that is produced by the system absorbing resonant energy.

The assumptions of randomness in the Boltzmann mechanics do not apply to resonant EM systems, with their uniform work energies, and systems exposed to resonant EM waves do not follow the smooth thermal distributions curves, however. Individual energy levels in the system may be selectively populated, changing the shape of the traditional smooth curve, to a bulging or "lumpy" energy distribution curve. (Figure 3b., above). Absorption of resonant EM waves initially results in increased population of an upper energy level. As the energy is converted to work in the system, the energy state devolves and relaxes. When all of the work energy has been spent, a thermal distribution is once again exhibited. Depending on which energy level(s) are selectively populated, the work performed will vary and can include speeding the rate of a reaction in a catalytic manner, e.g., virtual thermal effects can replace chemical activation energies. (Fukushima J. et al, 2010)

4.8 Equilibrium constant

In chemical and materials systems the work performed by the resonant EM waves can also shift the equilibrium of the system and produce dramatic changes in its chemical and material dynamics. In the dynamic equilibrium of chemical and material systems, in which reactants are transformed into products at the same rate products are transformed back into reactants, the equilibrium constant "K" indicates the point of dynamic equilibrium (product-to-reactant concentration ratio). Systems with a large equilibrium constant contain mostly product, while a low "K" indicates mainly reactants.

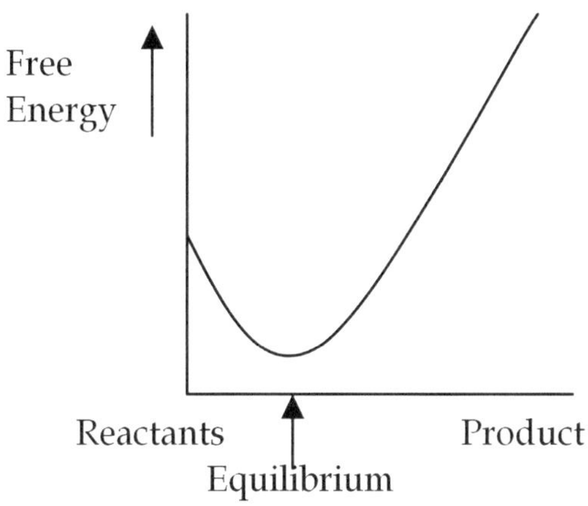

Fig. 4. a. Thermal system

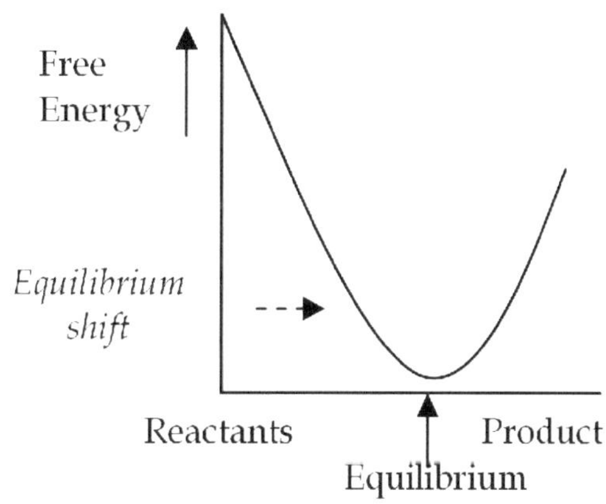

Fig. 4. b. Resonant system

In statistical thermodynamics the equilibrium constant is proportional to another natural log, K $\approx e^{-\Delta E/RT}$" (where ΔE is chemical free energy). When resonant EM waves are present in a system, the resonance factor properly appears in the denominator of the power notation, i.e., "$r_f RT$" to reflect the total increase in system energy. If the resonant EM waves increase chemical free energy, "ΔE", the ratio of chemical free energy to system energy remains the same. The equilibrium constant does not change, however the EM waves act as a catalyst and increase the *rate* of the reaction. (If the chemical free energy decreases, i.e., the resonance factor is less than "one", the EM waves will act as a negative catalyst and *slow* the rate of reaction.)

When resonant EM waves are absorbed and transformed to something *other* than chemical free energy, then $\Delta E/r_f RT < \Delta E/RT$, and the equilibrium constant will increase. Resonant EM waves that perform useful work on a system can thus increase the equilibrium constant, "K", and increase the actual concentration of desired products. Resonant EM waves can cause a shift of the equilibrium curve. (Figure 4.b., above) In other words, resonant EM waves can achieve results *not obtainable* with classical catalysis or thermodynamics. (Brooks and Abel, 2007, Blum et al, 2003, Fukushima H., 2010, and Roy et al, 2002)

5. Conclusion

The two fundamental formulae which formed the foundations of quantum mechanics and 20th century physics were both incomplete in regard to electromagnetic waves. The quantum formula was missing a time variable and energy constant, due to a minor inadvertence in Max Planck's derivation of the blackbody equation. Unforeseen consequences occurred as a result. The identity of the true elementary particle of light - the single EM oscillation - was obscured. Mathematical nomenclature for frequency became incomplete. Calculations of the mass, momentum, and force of EM waves were made impossible. Paradoxical principles including the classical limit, the uncertainty principle, and the complementarity principle were made necessary. Dimensionality of the fine structure constant was hidden, and great confusion arose over the wave vs. particle nature of light. Use of the *complete* quantum formula remedies these difficulties and provides a sound foundation for a certain and realistic quantum mechanics.

Likewise, the thermodynamic formula derived in Planck's blackbody work was not an inclusive or complete formula for energy dynamics. The complete energy dynamics formulae allow the resonant EM work energies in systems to be accounted for mathematically, both at the macroscale and microscale. Traditional Boltzmann mechanics cannot be strictly applied to EM waves, because Boltzmann mechanics assume completely random motions. Absorption of uniform EM waves requires modifications of Boltzmann weighting. Resonant EM waves can provide virtual thermal effects, decreased energy requirements, and increased energy efficiencies. Depending on how their energy is converted to work, resonant EM waves can act as catalysts - changing chemical or materials reaction rates - or they can shift reaction equilibria altogether, producing effects and products not seen or obtainable under typical thermodynamic conditions.

A new and powerful scientific paradigm is being revealed in the fundamental physics of electromagnetic waves.

6. References

Agnesi, M. G. (1748). *Instituzioni Analitiche,* Nella Regia-Ducal Corte, Milan, Italy

Blum, B.; Brooks (now Mortenson), J.; & Mortenson, M. (2003). *Methods for controlling crystal growth, crystallization, structures and phases in materials and systems,* U.S. Patent Application No. 10/508,462, 2003.

Bohm, D., (1952). A Suggested Interpretation of the Quantum Theory in Terms of "Hidden" Variables, *Physical Review,* Vol.85, No.2, pp. 166-179

Bohr, N., (1913). On the Constitution of Atoms and Molecules, Part I, *Philosophical Magazine* Vol.26, pp. 1-24

Bohr, N. , (1920). Uber die Serienspektra der Element, *Zeitschrift für Physik,* Vol.2, No.5, pp. 423- 478, English translation in *Niels Bohr Collected Works, Vol. 3: The Correspondence Principle (1918-1923),* (1976). Nielsen, (Ed.), Amsterdam: North-Holland Publ., pp. 241-282

Bohr, N., (1928). The Quantum Postulate and Atomic Theory, *Nature,* Vol.121, No.3050, pp. 580-590

Compton, A., (1923). Quantum Theory of the Scattering of X-rays by Light Elements, *The Physical Review,* Vol.21, No.5, pp. 483-502

Brooks (now Mortenson), J.; Mortenson M.; & Blum, B. (2005). *Controlling chemical reactions by spectral chemistry and spectral conditioning.* U.S. Patent Appl. No. 10/507,660, 2005.

Brooks (now Mortenson) J. & A. Abel (2007). *Methods for using resonant acoustic and/or resonant acousto-EM energy,* U.S. Patent No. 7,165,451

Brooks (now Mortenson), J. (2009). Hidden Variables: The Elementary Quantum of Light', Proc. of SPIE Optics and Photonics: *The Nature of Light: What are Photons? III,* Vol.7421, pp. 74210T-3, San Diego, California, USA, August 2009

Brooks (now Mortenson), J. (2009). Hidden Variables: The Resonance Factor, Proc. of SPIE Optics and Photonics:, *The Nature of Light: What are Photons? III,* Vol.7421, pp. 74210C, San Diego

Brooks (now Mortenson), J. (2009). Is indivisible single photon essential for quantum communications…? No. 4, Proc. of SPIE, *The Nature of Light III,* 2009, Vol. 7421, pp. 74210Y, San Diego, California, USA, August 2009

Coriolis, G. (1829). *Calculation of the Effect of Machines,* Carilian-Goeury, Paris, France

De Broglie, L. (1924). *Researches on the quantum theory*, Thesis, Paris, France

De Broglie, L. (1929). *The Wave Nature of the Electron*, Nobel Lecture, Stockholm, Sweden, 1929

Einstein, A. (1905). On a Heuristic Point of View Concerning the Production and Transformation of Light, *Annalen der Physik*, Vol.17, pp. 132-148

Einstein,A.; Podolsky, B.; & Rosen, N. (1935). Can a Quantum-Mechanical Description of Physical Reality be Considered Complete?, *Physical Review*, Vol.47, pp. 777-780

Einstein, A. & Infeld, L. (1938). *The Evolution of Physics*, Simon and Schuster Inc., New York , New York, USA pp. 262-263

Feinman, R. (1988). *QED. The Strange Theory of Light and Matter.* Princeton Univ. Press, Princeton, New Jersey, USA, pp. 129

Fukushima, H. (2010). New saccharification process of cellulosic biomass by microwave irradiation, *Proc. of Mat. Sci. & Tech. 2010,* Pittsburg, PA, USA, Oct. 2010, pp. 2859 - 2863

Fukushima, J. , Takayama, S., Matsubara, A., Kashimura, K. and Sato, M. (2010). Reduction Behavior of TiO_2 During Microwave Heating, *Proc. of MS&T 2010,* Pittsburg PA, pp. 2831 - 2836

Galilei, G. (1632). *Dialogue Concerning the Two Chief World Systems,* English translation available at: http://archimedes.mpiwg-berlin.mpg.de/cgi-bin/toc/toc.cgi? step=thumb&dir =galil _syste_065_en_1661

Gravesande, W. J. (1747). *Mathematical Elements of Natural Philosophy Confirm'd by Experiments,* Third Edition, Book II, Chapter 3, Innys, Longman and Shewell, London, pp. 193-205

Heisenberg, W. (1925). Über quantentheoretische Umdeutung kinematischer und mechanischer Beziehungen, *Zeitschrift für Physik*, Vol.33, pp. 879-893

Heisenberg, W. (1927). Über den anschaulichen Inhalt der quantentheoretischen Kinematik und Mechanik. *Zeitschrift für Physik*, Vol.43, pp. 172-198

Helmholtz, H. (1862). *On the Sensations of Tone*, Longmans, Green, and Co., New York, NY

Helmholtz, H. (1889). *Ostwald's Kalssiker der Exacten wissenschaften,* Über die Erhaltung der Kraft. Verlag von Wilhem Engelmann, Leipzig, Germany

Helmholtz, H. (1986). *The Correlation and Conservation of Forces: A Series of Expositions,* D. Appleton and Co., New York, NY, USA, pp. 211-250

Helmholtz, H. (1904). *Popular Lectures on Scientific Subjects,* Lecture VI. On the Conservation of Force, Trans. Atkinson E., Longmans, Green and Co., New York, NY, USA, pp. 306

Lagrange, J. L. (1811). *Analytical Mechanics,* Trans. Boissonnade A. & Vagliente, VN, 1997, Kluwer Academic Publishers,Norwell, MA, USA

Liu, M. , Zentgraf, T. , Liu Y. and Bartal G., Light-driven nanoscale plasmonic motors, *Nature Nanotechnology,* Published online: July 4, 2010 | DOI: 10.1038/NNANO.2010.128

Millikan, R. A. (1916). A Direct Photoelectric Determination of Planck's "h", *Phys Rev,* Vol.7, No.3

Mortenson (formerly Brooks), J (2010a). Double Your Kin. Energy: Disappearance of the Classical Limit in Quantized Classical Energy Mechanics. *Proc. MS&T 2010,* pp. 2907-2918, 2010a.

Mortenson (formerly Brooks) , J (2010b). The Fall and Rise of Resonance Science. *Proceedings of MS&T 2010,* pp. 2865 - 2875, Houston, TX, USA, October 2010

Mortenson (formerly Brooks), J and George, R. (2011). The New Physics and Global Sustainability.
IEEExplore, 978-1-4244-9216-9/11/IEEE FSNC, Vol.2, Earth Sci. & Eng., pp. 332-335

Mortenson (formerly Brooks), J (2011). The Conservation of Light's Energy, Mass and Momentum, Proc. Of SPIE Optics and Photonics 2011, In press

Newton, I. (1887). *Philosophiae Naturalis Principia Mathematica,* Royal Society, London

Planck, M. (1896). Absorption und Emission electrischer Wellen durch Resonanz, *Ann. der Phys. und Chem.,* Vol.293, No.1, pp. 1-14

Planck, M. (1897). Ueber electrische Schwingungen welche durch Resonanz erregt und durch Strahlung gedämpft warden", *Ann. der Phys. und Chem.,* Vol.296, No.4, pp. 577-599

Planck, M. (1900). Ueber irreversible Strahlungsvorgänge, *Ann.Phys.,* Vol.306, No.1, p 69-122

Planck, M. (1901). On the Law of Distribution of Energy in the Normal Spectrum", *Annalen der Physik,* Vol.309, No.3 pp. 553–63

Planck, M. (1920). The Genesis and Present State of Development of the Quantum Theory, Nobel Lectures, Physics 1901-1921, Elsevier Publishing Company, Amsterdam

Roy, R., et al. (2002). Definitive experimental evidence for Microwave Effects", *Materials Research Innovation,* Vol.6, pp. 128-140

Roychouhduri, C. (2009), Exploring divisibility and summability of 'photon' wave packets in non-linear optical phenomena, Proc. of SPIE, *The Nature of Light III,* 2009, Vol. 7421, p 74210A.

Schrödinger, E. (1982). *Collected Papers on Wave Mechanics,* Quantization as a Problem of Proper Values (Part I), Chelsea Publ. Co., New York, NY, USA, pp. 1-12.

She, W.; Yu, J.; & Feng, R.(2008). Observation of a Push Force on the End Face of a Nanometer Silica Filament Exerted by Outgoing Light, PhysReview Letters, Vol.101, No. 24

Spencer, R. C. (1940). Properties of the Witch of Agnesi - Application to Fitting the Shapes of Spectral Lines, *Journal of the Optical Society of America,* Vol.30, pp. 415 - 419

2

Modern Classical Electrodynamics and Electromagnetic Radiation - Vacuum Field Theory Aspects

Nikolai N. Bogolubov (Jr.)[1], Anatoliy K. Prykarpatsky[2]
[1]*V.A. Steklov Mathematical Institute of RAS, Moscow*
[2] *The AGH University of Science and Technology, Krakow*
[2]*Ivan Franko State Pedagogical University, Drohobych, Lviv region*
[1]*Russian Federation*
[2]*Poland*
[2]*Ukraine*

1. Introduction

"A physicist needs his equations should be mathematically sound and that in working with his equations he should not neglect quantities unless they are small"
P.A. M. Dirac

Classical electrodynamics is nowadays considered [29; 57; 80] the most fundamental physical theory, largely owing to the depth of its theoretical foundations and wealth of experimental verifications. Electrodynamics is essentially characterized by its Lorentz invariance from a theoretical perspective, and this very important property has had a revolutionary influence [29; 57; 80; 102; 111] on the whole development of physics. In spite of the breadth and depth of theoretical understanding of electromagnetism, there remain several fundamental open problems and gaps in comprehension related to the true physical nature of Maxwell's theory when it comes to describing electromagnetic waves as quantum photons in a vacuum: These start with the difficulties in constructing a successful Lagrangian approach to classical electrodynamics that is free of the Dirac-Fock-Podolsky inconsistency [53; 111; 112], and end with the problem of devising its true quantization theory without such artificial constructions as a Fock space with "indefinite" metrics, the Lorentz condition on "average", and regularized "infinities" [102] of S-matrices. Moreover, there are the related problems of obtaining a complete description of the structure of a vacuum medium carrying the electromagnetic waves and deriving a theoretically and physically valid Lorentz force expression for a moving charged point particle interacting with and external electromagnetic field. To describe the essence of these problems, let us begin with the classical Lorentz force expression

$$F := qE + qu \times B, \tag{2.1}$$

where $q \in \mathbb{R}$ is a particle electric charge, $u \in \mathbb{E}^3$ is its velocity vector, expressed here in the light speed c units,

$$E := -\partial A/\partial t - \nabla\varphi \tag{2.2}$$

is the corresponding external electric field and

$$B := \nabla \times A \tag{2.3}$$

is the corresponding external magnetic field, acting on the charged particle, expressed in terms of suitable vector $A : M^4 \to \mathbb{E}^3$ and scalar $\varphi : M^4 \to \mathbb{R}$ potentials. Here "∇" is the standard gradient operator with respect to the spatial variable $r \in \mathbb{E}^3$, "$\times$" is the usual vector product in three-dimensional Euclidean vector space $\mathbb{E}^3$, which is naturally endowed with the classical scalar product $< \cdot, \cdot >$. These potentials are defined on the Minkowski space $M^4 := \mathbb{R} \times \mathbb{E}^3$, which models a chosen laboratory reference system $\mathcal{K}$. Now, it is a well-known fact [56; 57; 70; 80] that the force expression (2.1) does not take into account the dual influence of the charged particle on the electromagnetic field and should be considered valid only if the particle charge $q \to 0$. This also means that expression (2.1) cannot be used for studying the interaction between two different moving charged point particles, as was pedagogically demonstrated in [57].

Other questionable inferences, which strongly motivated the analysis in this work, are related both to an alternative interpretation of the well-known *Lorentz condition*, imposed on the four-vector of electromagnetic potentials $(\varphi, A) : M^4 \to \mathbb{R} \times \mathbb{E}^3$ and the classical Lagrangian formulation [57] of charged particle dynamics under an external electromagnetic field. The Lagrangian approach is strongly dependent on the important Einsteinian notion of the rest reference system $\mathcal{K}_r$ and the related least action principle, so before explaining it in more detail, we first analyze the classical Maxwell electromagnetic theory from a strictly dynamical point of view.

2. Relativistic electrodynamics models revisited: Lagrangian and Hamiltonian analysis

2.1 The Maxwell equations revisiting

Let us consider the additional Lorentz condition

$$\partial\varphi/\partial t + < \nabla, A > = 0, \tag{2.4}$$

imposed *a priori* on the four-vector of potentials $(\varphi, A) : M^4 \to \mathbb{R} \times \mathbb{E}^3$, which satisfy the Lorentz invariant wave field equations

$$\partial^2\varphi/\partial t^2 - \nabla^2\varphi = \rho, \quad \partial^2 A/\partial t^2 - \nabla^2 A = J, \tag{2.5}$$

where $\rho : M^4 \to \mathbb{R}$ and $J : M^4 \to \mathbb{E}^3$ are, respectively, the charge and current densities of the ambient matter, which satisfy the charge continuity equation

$$\partial\rho/\partial t + < \nabla, J > = 0. \tag{2.6}$$

Then the classical electromagnetic Maxwell field equations [56; 57; 70; 80]

$$\begin{aligned} \nabla \times E + \partial B/\partial t = 0, \quad & < \nabla, E > = \rho, \\ \nabla \times B - \partial E/\partial t = J, \quad & < \nabla, B > = 0, \end{aligned} \tag{2.7}$$

hold for all $(t,r) \in M^4$ with respect to the chosen reference system $\mathcal{K}$.
Notice here that Maxwell's equations (2.7) do not directly reduce, via definitions (2.2) and (2.3), to the wave field equations (2.5) without the Lorentz condition (2.4). This fact is very important, and suggests that when it comes to a choice of governing equations, it may be reasonable to replace Maxwell's equations (2.7) and (2.6) with the Lorentz condition (2.4), (2.5) and the continuity equation (2.6). From the assumptions formulated above, one infers the following result.

Proposition 2.1. *The Lorentz invariant wave equations (2.5) for the potentials* $(\varphi, A) : M^4 \to \mathbb{R} \times \mathbb{E}^3$, *together with the Lorentz condition (2.4) and the charge continuity relationship (2.5), are completely equivalent to the Maxwell field equations (2.7).*

Proof. Substituting (2.4), into (2.5), one easily obtains

$$\partial^2\varphi/\partial t^2 = - < \nabla, \partial A/\partial t > = < \nabla, \nabla\varphi > + \rho, \tag{2.8}$$

which implies the gradient expression

$$< \nabla, -\partial A/\partial t - \nabla\varphi > = \rho. \tag{2.9}$$

Taking into account the electric field definition (2.2), expression (2.9) reduces to

$$< \nabla, E > = \rho, \tag{2.10}$$

which is the second of the first pair of Maxwell's equations (2.7).
Now upon applying $\nabla\times$ to definition (2.2), we find, owing to definition (2.3), that

$$\nabla \times E + \partial B/\partial t = 0, \tag{2.11}$$

which is the first of the first pair of the Maxwell equations (2.7).
Applying $\nabla\times$ to the definition (2.3), one obtains

$$\begin{aligned} \nabla \times B &= \nabla \times (\nabla \times A) = \nabla < \nabla, A > -\nabla^2 A = \\ &= -\nabla(\partial\varphi/\partial t) - \partial^2 A/\partial t^2 + (\partial^2 A/\partial t^2 - \nabla^2 A) = \\ &= \frac{\partial}{\partial t}(-\nabla\varphi - \partial A/\partial t) + J = \partial E/\partial t + J, \end{aligned} \tag{2.12}$$

leading to

$$\nabla \times B = \partial E/\partial t + J,$$

which is the first of the second pair of the Maxwell equations (2.7). The final *"no magnetic charge"* equation

$$< \nabla, B > = < \nabla, \nabla \times A > = 0,$$

in (2.7) follows directly from the elementary identity $< \nabla, \nabla\times > = 0$, thereby completing the proof. □

This proposition allows us to consider the potential functions $(\varphi, A) : M^4 \to \mathbb{R} \times \mathbb{E}^3$ as fundamental ingredients of the ambient *vacuum field medium*, by means of which we can try to describe the related physical behavior of charged point particles imbedded in space-time M^4. The following observation provides strong support for this approach:

Observation. *The Lorentz condition (2.4) actually means that the scalar potential field* $\varphi : M^4 \rightarrow \mathbb{R}$ *continuity relationship, whose origin lies in some new field conservation law, characterizes the deep intrinsic structure of the vacuum field medium.*

To make this observation more transparent and precise, let us recall the definition [56; 57; 70; 80] of the electric current $J : M^4 \rightarrow \mathbb{E}^3$ in the dynamical form

$$J := \rho v, \tag{2.13}$$

where the vector $v : M^4 \rightarrow \mathbb{E}^3$ is the corresponding charge velocity. Thus, the following continuity relationship

$$\partial\rho/\partial t + <\nabla, \rho v> = 0 \tag{2.14}$$

holds, which can easily be recast [122] as the integral conservation law

$$\frac{d}{dt}\int_{\Omega_t} \rho d^3 r = 0 \tag{2.15}$$

for the charge inside of any bounded domain $\Omega_t \subset \mathbb{E}^3$ moving in the space-time M^4 with respect to the natural evolution equation

$$dr/dt := v. \tag{2.16}$$

Following the above reasoning, we are led to the following result.

Proposition 2.2. *The Lorentz condition (2.4) is equivalent to the integral conservation law*

$$\frac{d}{dt}\int_{\Omega_t} \varphi d^3 r = 0, \tag{2.17}$$

where $\Omega_t \subset \mathbb{E}^3$ *is any bounded domain moving with respect to the evolution equation*

$$dr/dt := v, \tag{2.18}$$

which represents the velocity vector of local potential field changes propagating in the Minkowski space-time M^4.

Proof. Consider first the corresponding solutions to the potential field equations (2.5), taking into account condition (2.13). Owing to the results from [57; 70], one finds that

$$A = \varphi v, \tag{2.19}$$

which gives rise to the following form of the Lorentz condition (2.4):

$$\partial\varphi/\partial t + <\nabla, \varphi v> = 0. \tag{2.20}$$

This obviously can be rewritten [122] as the integral conservation law (2.17), so the proof is complete. □

The above proposition suggests a physically motivated interpretation of electrodynamic phenomena in terms of what should naturally be called *the vacuum potential field,* which determines the observable interactions between charged point particles. More precisely, we can *a priori* endow the ambient vacuum medium with a scalar potential field function $W := q\varphi : M^4 \to \mathbb{R}$, satisfying the governing *vacuum field equations*

$$\partial^2 W/\partial t^2 - \nabla^2 W = 0, \quad \partial W/\partial t + < \nabla, Wv >= 0, \tag{2.21}$$

taking into account that there are no external sources besides material particles possessing only a virtual capability for disturbing the vacuum field medium. Moreover, this vacuum potential field function $W : M^4 \to \mathbb{R}$ allows the natural potential energy interpretation, whose origin should be assigned not only to the charged interacting medium, but also to any other medium possessing interaction capabilities, including for instance, material particles interacting through the gravity.

This leads naturally to the next important step, which consists in deriving the equation governing the corresponding potential field $\bar{W} : M^4 \to \mathbb{R}$, assigned to the vacuum field medium in a neighborhood of any spatial point moving with velocity $u \in \mathbb{E}^3$ and located at $R(t) \in \mathbb{E}^3$ at time $t \in \mathbb{R}$. As can be readily shown [53; 54], the corresponding evolution equation governing the related potential field function $\bar{W} : M^4 \to \mathbb{R}$ has the form

$$\frac{d}{dt}(-\bar{W}u) = -\nabla \bar{W}, \tag{2.22}$$

where $\bar{W} := W(r,t)|_{r \to R(t)}$, $u := dR(t)/dt$ at point particle location $(R(t), t) \in M^4$.

Similarly, if there are two interacting point particles, located at points $R(t)$ and $R_f(t) \in \mathbb{E}^3$ at time $t \in \mathbb{R}$ and moving, respectively, with velocities $u := dR(t)/dt$ and $u_f := dR_f(t)/dt$, the corresponding potential field function $\bar{W} : M^4 \to \mathbb{R}$ for the particle located at point $R(t) \in \mathbb{E}^3$ should satisfy

$$\frac{d}{dt}[-\bar{W}(u - u_f)] = -\nabla \bar{W}. \tag{2.23}$$

The dynamical potential field equations (2.22) and (2.23) appear to have important properties and can be used as a means for representing classical electrodynamics. Consequently, we shall proceed to investigate their physical properties in more detail and compare them with classical results for Lorentz type forces arising in the electrodynamics of moving charged point particles in an external electromagnetic field.

In this investigation, we were strongly inspired by the works [81; 82; 89; 91; 93]; especially by the interesting studies [87; 88] devoted to solving the classical problem of reconciling gravitational and electrodynamical charges within the Mach-Einstein ether paradigm. First, we revisit the classical Mach-Einstein relativistic electrodynamics of a moving charged point particle, and second, we study the resulting electrodynamic theories associated with our vacuum potential field dynamical equations (2.22) and (2.23), making use of the fundamental Lagrangian and Hamiltonian formalisms which were specially devised for this in [52; 55]. The results obtained are used to apply the canonical Dirac quantization procedure to the corresponding energy conservation laws associated to the electrodynamic models considered.

2.2 Classical relativistic electrodynamics revisited

The classical relativistic electrodynamics of a freely moving charged point particle in the Minkowski space-time $M^4 := \mathbb{R} \times \mathbb{E}^3$ is based on the Lagrangian approach [56; 57; 70; 80] with Lagrangian function

$$\mathcal{L} := -m_0(1-u^2)^{1/2}, \tag{2.24}$$

where $m_0 \in \mathbb{R}_+$ is the so-called particle rest mass and $u \in \mathbb{E}^3$ is its spatial velocity in the Euclidean space $\mathbb{E}^3$, expressed here and in the sequel in light speed units (with light speed c). The least action principle in the form

$$\delta S = 0, \quad S := -\int_{t_1}^{t_2} m_0(1-u^2)^{1/2} dt \tag{2.25}$$

for any fixed temporal interval $[t_1, t_2] \subset \mathbb{R}$ gives rise to the well-known relativistic relationships for the mass of the particle

$$m = m_0(1-u^2)^{-1/2}, \tag{2.26}$$

the momentum of the particle

$$p := mu = m_0 u(1-u^2)^{-1/2} \tag{2.27}$$

and the energy of the particle

$$\mathcal{E}_0 = m = m_0(1-u^2)^{-1/2}. \tag{2.28}$$

It follows from [57; 80], that the origin of the Lagrangian (2.24) can be extracted from the action

$$S := -\int_{t_1}^{t_2} m_0(1-u^2)^{1/2} dt = -\int_{\tau_1}^{\tau_2} m_0 d\tau, \tag{2.29}$$

on the suitable temporal interval $[\tau_1, \tau_2] \subset \mathbb{R}$, where, by definition,

$$d\tau := dt(1-u^2)^{1/2} \tag{2.30}$$

and $\tau \in \mathbb{R}$ is the so-called proper temporal parameter assigned to a freely moving particle with respect to the rest reference system $\mathcal{K}_r$. The action (2.29) is rather questionable from the dynamical point of view, since it is physically defined with respect to the rest reference system $\mathcal{K}_r$, giving rise to the constant action $S = -m_0(\tau_2 - \tau_1)$, as the limits of integrations $\tau_1 < \tau_2 \in \mathbb{R}$ were taken to be fixed from the very beginning. Moreover, considering this particle to have charge $q \in \mathbb{R}$ and be moving in the Minkowski space-time M^4 under action of an electromagnetic field $(\varphi, A) \in \mathbb{R} \times \mathbb{E}^3$, the corresponding classical (relativistic) action functional is chosen (see [52; 55–57; 70; 80]) as follows:

$$S := \int_{\tau_1}^{\tau_2} [-m_0 d\tau + q < A, \dot{r} > d\tau - q\varphi(1-u^2)^{-1/2} d\tau], \tag{2.31}$$

with respect to the *rest reference system*, parameterized by the Euclidean space-time variables $(\tau, r) \in \mathbb{E}^4$, where we have denoted $\dot{r} := dr/d\tau$ in contrast to the definition $u := dr/dt$. The action (2.31) can be rewritten with respect to the laboratory reference system $\mathcal{K}$ moving with velocity vector $u \in \mathbb{E}^3$ as

$$S = \int_{t_1}^{t_2} \mathcal{L} dt, \ \mathcal{L} := -m_0(1-u^2)^{1/2} + q < A, u > -q\varphi, \tag{2.32}$$

on the temporal interval $[t_1, t_2] \subset \mathbb{R}$, which gives rise to the following [56; 57; 70; 80] dynamical expressions

$$P = p + qA, \quad p = mu, \quad m = m_0(1-u^2)^{-1/2}, \tag{2.33}$$

for the particle momentum and

$$\mathcal{E}_0 = [m_0^2 + (P - qA)^2]^{1/2} + q\varphi \tag{2.34}$$

for the particle energy, where, by definition, $P \in \mathbb{E}^3$ is the common momentum of the particle and the ambient electromagnetic field at a space-time point $(t, r) \in M^4$.
The expression (2.34) for the particle energy $\mathcal{E}_0$ also appears open to question, since the potential energy $q\varphi$, entering additively, has no affect on the particle mass $m = m_0(1-u^2)^{-1/2}$. This was noticed by L. Brillouin [59], who remarked that since the potential energy has no affect on the particle mass, this tells us that "... any possibility of existence of a particle mass related with an external potential energy, is completely excluded". Moreover, it is necessary to stress here that the least action principle (2.32), formulated with respect to the laboratory reference system $\mathcal{K}$ time parameter $t \in \mathbb{R}$, appears logically inadequate, for there is a strong physical inconsistency with other time parameters of the Lorentz equivalent reference systems. This was first mentioned by R. Feynman in [29], in his efforts to rewrite the Lorentz force expression with respect to the rest reference system $\mathcal{K}_r$. This and other special relativity theory and electrodynamics problems induced many prominent physicists of the past [29; 59; 61; 64; 80] and present [4; 5; 60; 65; 66; 68; 69; 81; 82; 87; 89; 90; 93] to try to develop alternative relativity theories based on completely different space-time and matter structure principles.
There also is another controversial inference from the action expression (2.32). As one can easily show [56; 57; 70; 80], the corresponding dynamical equation for the Lorentz force is given as

$$dp/dt = F := qE + qu \times B. \tag{2.35}$$

We have defined here, as before,

$$E := -\partial A/\partial t - \nabla\varphi \tag{2.36}$$

for the corresponding electric field and

$$B := \nabla \times A \tag{2.37}$$

for the related magnetic field, acting on the charged point particle q. The expression (2.35) means, in particular, that the Lorentz force F depends linearly on the particle velocity vector $u \in \mathbb{E}^3$, and so there is a strong dependence on the reference system with respect to which the charged particle q moves. Attempts to reconcile this and some related controversies [29; 59; 60; 63] forced Einstein to devise his special relativity theory and proceed further to creating his general relativity theory trying to explain gravity by means of geometrization of space-time and matter in the Universe. Here we must mention that the classical Lagrangian function $\mathcal{L}$ in (2.32) is written in terms of a combination of terms expressed by means of both the Euclidean rest reference system variables $(\tau, r) \in \mathbb{E}^4$ and arbitrarily chosen Minkowski reference system variables $(t, r) \in M^4$.

These problems were recently analyzed using a completely different "no-geometry" approach [6; 53; 54; 60], where new dynamical equations were derived, which were free of the controversial elements mentioned above. Moreover, this approach avoided the introduction of the well-known Lorentz transformations of the space-time reference systems with respect to which the action functional (2.32) is invariant. From this point of view, there are interesting conclusions in [83] in which Galilean invariant Lagrangians possessing intrinsic Poincaré-Lorentz symmetry are reanalyzed. Next, we revisit the results obtained in [53; 54] from the classical Lagrangian and Hamiltonian formalisms [52] in order to shed new light on the physical underpinnings of the vacuum field theory approach to the investigation of combined electromagnetic and gravitational effects.

2.3 The vacuum field theory electrodynamics equations: Lagrangian analysis

2.3.1 A point particle moving in a vacuum - an alternative electrodynamic model

In the vacuum field theory approach to combining electromagnetism and the gravity devised in [53; 54], the main vacuum potential field function $\bar{W}: M^4 \to \mathbb{R}$ related to a charged point particle q satisfies the dynamical equation (2.21), namely

$$\frac{d}{dt}(-\bar{W}u) = -\nabla\bar{W} \tag{2.38}$$

in the case when the external charged particles are at rest, where, as above, $u := dr/dt$ is the particle velocity with respect to some reference system.

To analyze the dynamical equation (2.38) from the Lagrangian point of view, we write the corresponding action functional as

$$S := -\int_{t_1}^{t_2} \bar{W}dt = -\int_{\tau_1}^{\tau_2} \bar{W}(1+\dot{r}^2)^{1/2}\, d\tau, \tag{2.39}$$

expressed with respect to the rest reference system $\mathcal{K}_r$. Fixing the proper temporal parameters $\tau_1 < \tau_2 \in \mathbb{R}$, one finds from the least action principle ($\delta S = 0$) that

$$\begin{aligned} p &:= \partial\mathcal{L}/\partial\dot{r} = -\bar{W}\dot{r}(1+\dot{r}^2)^{-1/2} = -\bar{W}u, \\ \dot{p} &:= dp/d\tau = \partial\mathcal{L}/\partial r = -\nabla\bar{W}(1+\dot{r}^2)^{1/2}, \end{aligned} \tag{2.40}$$

where, owing to (2.39), the corresponding Lagrangian function is

$$\mathcal{L} := -\bar{W}(1+\dot{r}^2)^{1/2}. \tag{2.41}$$

Recalling now the definition of the particle mass

$$m := -\bar{W} \tag{2.42}$$

and the relationships

$$d\tau = dt(1-u^2)^{1/2}, \ \dot{r}d\tau = udt, \tag{2.43}$$

from (2.40) we easily obtain the dynamical equation (2.38). Moreover, one now readily finds that the dynamical mass, defined by means of expression (2.42), is given as

$$m = m_0(1-u^2)^{-1/2},$$

which coincides with the equation (2.26) of the preceding section. Now one can formulate the following proposition using the above results

Proposition 2.3. *The alternative freely moving point particle electrodynamic model (2.38) allows the least action formulation (2.39) with respect to the "rest" reference system variables, where the Lagrangian function is given by expression (2.41). Its electrodynamics is completely equivalent to that of a classical relativistic freely moving point particle, described in Section 2.*

2.3.2 An interacting two charge system moving in a vacuum - an alternative electrodynamic model

We proceed now to the case when our charged point particle q moves in the space-time with velocity vector $u \in \mathbb{E}^3$ and interacts with another external charged point particle, moving with velocity vector $u_f \in \mathbb{E}^3$ in a common reference system $\mathcal{K}$. As shown in [53; 54], the corresponding dynamical equation for the vacuum potential field function $\bar{W} : M^4 \rightarrow \mathbb{R}$ is given as

$$\frac{d}{dt}[-\bar{W}(u-u_f)] = -\nabla\bar{W}. \tag{2.44}$$

As the external charged particle moves in the space-time, it generates the related magnetic field $B := \nabla \times A$, whose magnetic vector potential $A : M^4 \rightarrow \mathbb{E}^3$ is defined, owing to the results of [53; 54; 60], as

$$qA := \bar{W}u_f. \tag{2.45}$$

Whence, it follows from (2.40) that the particle momentum $p = -\bar{W}u$ equation (2.44) is equivalent to

$$\frac{d}{dt}(p+qA) = -\nabla\bar{W}. \tag{2.46}$$

To represent the dynamical equation (2.46) in the classical Lagrangian formalism, we start from the following action functional, which naturally generalizes the functional (2.39):

$$S := -\int_{\tau_1}^{\tau_2} \bar{W}(1+|\dot{r}-\dot{\xi}|^2)^{1/2}\, d\tau, \tag{2.47}$$

where $\dot{\xi} = u_f dt/d\tau$, $d\tau = dt(1-(u-u_f)^2)^{1/2}$, which takes into account the relative velocity of the charged point particle q with respect to the reference system $\mathcal{K}'$, moving with velocity $u_f \in \mathbb{E}^3$ in the reference system $\mathcal{K}$. It is clear in this case that the charged point particle q moves with velocity $u - u_f \in \mathbb{E}^3$ with respect to the reference system $\mathcal{K}'$ in which the external charged particle is at rest.

Now we compute the least action variational condition $\delta S = 0$ taking into account that, owing to (2.47), the corresponding Lagrangian function is given as

$$\mathcal{L} := -\bar{W}(1 + (\dot{r} - \dot{\xi})^2)^{1/2}. \tag{2.48}$$

Hence, the common momentum of the particles is

$$\begin{aligned} P &:= \partial\mathcal{L}/\partial\dot{r} = -\bar{W}(\dot{r} - \dot{\xi})(1 + (\dot{r} - \dot{\xi})^2)^{-1/2} = \\ &= -\bar{W}\dot{r}(1 + (\dot{r} - \dot{\xi})^2)^{-1/2} + \bar{W}\dot{\xi}(1 + (\dot{r} - \dot{\xi})^2)^{-1/2} = \\ &= mu + qA := p + qA, \end{aligned} \tag{2.49}$$

and the dynamical equation is given as

$$\frac{d}{d\tau}(p + qA) = -\nabla\bar{W}(1 + |\dot{r} - \dot{\xi}|^2)^{1/2}. \tag{2.50}$$

As $d\tau = dt(1 - (u - u_f)^2)^{1/2}$ and $(1 + (\dot{r} - \dot{\xi})^2)^{1/2} = (1 - (u - u_f)^2)^{-1/2}$, we obtain finally from (2.50) the dynamical equation (2.46), which leads to the next proposition.

Proposition 2.4. *The alternative classical relativistic electrodynamic model (2.44) allows the least action formulation (2.47) with respect to the "rest" reference system variables, where the Lagrangian function is given by expression (2.48).*

2.3.3 A moving charged point particle formulation dual to the classical alternative electrodynamic model

It is easy to see that the action functional (2.47) is written utilizing the classical Galilean transformations of reference systems. If we now consider the action functional (2.39) for a charged point particle moving with respect the reference system $\mathcal{K}_r$, and take into account its interaction with an external magnetic field generated by the vector potential $A : M^4 \to \mathbb{E}^3$, it can be naturally generalized as

$$S := \int_{t_1}^{t_2} (-\bar{W}dt + q < A, dr >) = \int_{\tau_1}^{\tau_2} [-\bar{W}(1 + \dot{r}^2)^{1/2} + q < A, \dot{r} >] d\tau, \tag{2.51}$$

where $d\tau = dt(1 - u^2)^{1/2}$.
Thus, the corresponding common particle-field momentum takes the form

$$\begin{aligned} P &:= \partial\mathcal{L}/\partial\dot{r} = -\bar{W}\dot{r}(1 + \dot{r}^2)^{-1/2} + qA = \\ &= mu + qA := p + qA, \end{aligned} \tag{2.52}$$

and satisfies

$$\begin{aligned} \dot{P} &:= dP/d\tau = \partial\mathcal{L}/\partial r = -\nabla\bar{W}(1 + \dot{r}^2)^{1/2} + q\nabla < A, \dot{r} >= \\ &= -\nabla\bar{W}(1 - u^2)^{-1/2} + q\nabla < A, u > (1 - u^2)^{-1/2}, \end{aligned} \tag{2.53}$$

where

$$\mathcal{L} := -\bar{W}(1 + \dot{r}^2)^{1/2} + q < A, \dot{r} > \tag{2.54}$$

is the corresponding Lagrangian function. Since $d\tau = dt(1-u^2)^{1/2}$, one easily finds from (2.53) that

$$dP/dt = -\nabla \bar{W} + q\nabla < A, u > . \tag{2.55}$$

Upon substituting (2.52) into (2.55) and making use of the well-known [57] identity

$$\nabla < a, b > = < a, \nabla > b + < b, \nabla > a + b \times (\nabla \times a) + a \times (\nabla \times b), \tag{2.56}$$

where $a, b \in \mathbb{E}^3$ are arbitrary vector functions, we obtain the classical expression for the Lorentz force F acting on the moving charged point particle q :

$$dp/dt := F = qE + qu \times B, \tag{2.57}$$

where, by definition,

$$E := -\nabla \bar{W} q^{-1} - \partial A/\partial t \tag{2.58}$$

is its associated electric field and

$$B := \nabla \times A \tag{2.59}$$

is the corresponding magnetic field. This result can be summarized as follows:

Proposition 2.5. *The classical relativistic Lorentz force (2.57) allows the least action formulation (2.51) with respect to the rest reference system variables, where the Lagrangian function is given by formula (2.54). Its electrodynamics described by the Lorentz force (2.57) is completely equivalent to the classical relativistic moving point particle electrodynamics characterized by the Lorentz force (2.35) in Section 2.*

As for the dynamical equation (2.50), it is easy to see that it is equivalent to

$$dp/dt = (-\nabla \bar{W} - qdA/dt + q\nabla < A, u >) - q\nabla < A, u >, \tag{2.60}$$

which, owing to (2.55) and (2.57), takes the following Lorentz type force form

$$dp/dt = qE + qu \times B - q\nabla < A, u >, \tag{2.61}$$

that can be found in [53; 54; 60].
Expressions (2.57) and (2.61) are equal to up to the gradient term $F_c := q\nabla < A, u >$, which reconciles the Lorentz forces acting on a charged moving particle q with respect to different reference systems. This fact is important for our vacuum field theory approach since it uses no special geometry and makes it possible to analyze both electromagnetic and gravitational fields simultaneously by employing the new definition of the dynamical mass by means of expression (2.42).

2.4 The vacuum field theory electrodynamics equations: Hamiltonian analysis

Any Lagrangian theory has an equivalent canonical Hamiltonian representation via the classical Legendre transformation[1; 2; 46; 56; 104]. As we have already formulated our vacuum field theory of a moving charged particle q in Lagrangian form, we proceed now to its Hamiltonian analysis making use of the action functionals (2.39), (2.48) and (2.51).

Take, first, the Lagrangian function (2.41) and the momentum expression (2.40) for defining the corresponding Hamiltonian function

$$\begin{aligned} H &:=< p, \dot{r} > -\mathcal{L} = \\ &= -< p, p > \bar{W}^{-1}(1 - p^2/\bar{W}^2)^{-1/2} + \bar{W}(1 - p^2/\bar{W}^2)^{-1/2} = \\ &= -p^2\bar{W}^{-1}(1 - p^2/\bar{W}^2)^{-1/2} + \bar{W}^2\bar{W}^{-1}(1 - p^2/\bar{W}^2)^{-1/2} = \\ &= -(\bar{W}^2 - p^2)(\bar{W}^2 - p^2)^{-1/2} = -(\bar{W}^2 - p^2)^{1/2}. \end{aligned} \tag{2.62}$$

Consequently, it is easy to show [1; 2; 56; 104] that the Hamiltonian function (2.62) is a conservation law of the dynamical field equation (2.38); that is, for all $\tau, t \in \mathbb{R}$

$$dH/dt = 0 = dH/d\tau, \tag{2.63}$$

which naturally leads to an energy interpretation of H. Thus, we can represent the particle energy as

$$\mathcal{E} = (\bar{W}^2 - p^2)^{1/2}. \tag{2.64}$$

Accordingly the Hamiltonian equivalent to the vacuum field equation (2.38) can be written as

$$\begin{aligned} \dot{r} &:= dr/d\tau = \partial H/\partial p = p(\bar{W}^2 - p^2)^{-1/2} \\ \dot{p} &:= dp/d\tau = -\partial H/\partial r = \bar{W}\nabla\bar{W}(\bar{W}^2 - p^2)^{-1/2}, \end{aligned} \tag{2.65}$$

and we have the following result.

Proposition 2.6. *The alternative freely moving point particle electrodynamic model (2.38) allows the canonical Hamiltonian formulation (2.65) with respect to the "rest" reference system variables, where the Hamiltonian function is given by expression (2.62). Its electrodynamics is completely equivalent to the classical relativistic freely moving point particle electrodynamics described in Section 2.*

In an analogous manner, one can now use the Lagrangian (2.48) to construct the Hamiltonian function for the dynamical field equation (2.46) describing the motion of charged particle q in an external electromagnetic field in the canonical Hamiltonian form:

$$\dot{r} := dr/d\tau = \partial H/\partial P, \qquad \dot{P} := dP/d\tau = -\partial H/\partial r, \tag{2.66}$$

where

$$\begin{aligned} H &:=< P, \dot{r} > -\mathcal{L} = \\ &=< P, \dot{\xi} - P\bar{W}^{-1}(1 - P^2/\bar{W}^2)^{-1/2} > + \bar{W}[\bar{W}^2(\bar{W}^2 - P^2)^{-1}]^{1/2} = \\ &=< P, \dot{\xi} > + P^2(\bar{W}^2 - P^2)^{-1/2} - \bar{W}^2(\bar{W}^2 - P^2)^{-1/2} = \\ &= -(\bar{W}^2 - P^2)(\bar{W}^2 - P^2)^{-1/2} + < P, \dot{\xi} >= \\ &= -(\bar{W}^2 - P^2)^{1/2} - q < A, P > (\bar{W}^2 - P^2)^{-1/2}. \end{aligned} \tag{2.67}$$

Here we took into account that, owing to definitions (2.45) and (2.49),

$$\begin{aligned} qA &:= \bar{W}u_f = \bar{W}d\xi/dt = \\ &= \bar{W}\frac{d\xi}{d\tau}\cdot\frac{d\tau}{dt} = \bar{W}\dot{\xi}(1-(u-v))^{1/2} = \\ &= \bar{W}\dot{\xi}(1+(\dot{r}-\dot{\xi})^2)^{-1/2} = \\ &= -\bar{W}\dot{\xi}(\bar{W}^2-P^2)^{1/2}\bar{W}^{-1} = -\dot{\xi}(\bar{W}^2-P^2)^{1/2}, \end{aligned} \tag{2.68}$$

or

$$\dot{\xi} = -qA(\bar{W}^2-P^2)^{-1/2}, \tag{2.69}$$

where $A: M^4 \to \mathbb{R}^3$ is the related magnetic vector potential generated by the moving external charged particle. Equations (2.67) can be rewritten with respect to the laboratory reference system $\mathcal{K}$ in the form

$$dr/dt = u, \quad dp/dt = qE + qu\times B - q\nabla < A, u >, \tag{2.70}$$

which coincides with the result (2.61).
Whence, we see that the Hamiltonian function (2.67) satisfies the energy conservation conditions

$$dH/dt = 0 = dH/d\tau, \tag{2.71}$$

for all $\tau, t \in \mathbb{R}$, and that the suitable energy expression is

$$\mathcal{E} = (\bar{W}^2-P^2)^{1/2} + q < A, P > (\bar{W}^2-P^2)^{-1/2}, \tag{2.72}$$

where the generalized momentum $P = p + qA$. The result (2.72) differs in an essential way from that obtained in [57], which makes use of the Einsteinian Lagrangian for a moving charged point particle q in an external electromagnetic field. Thus, we obtain the following result:

Proposition 2.7. *The alternative classical relativistic electrodynamic model (2.70), which is intrinsically compatible with the classical Maxwell equations (2.7), allows the Hamiltonian formulation (2.66) with respect to the rest reference system variables, where the Hamiltonian function is given by expression (2.67).*

The inference above is a natural candidate for experimental validation of our theory. It is strongly motivated by the following remark.

Remark 2.8. *It is necessary to mention here that the Lorentz force expression (2.70) uses the particle momentum $p = mu$, where the dynamical "mass" $m := -\bar{W}$ satisfies condition (2.72). The latter gives rise to the following crucial relationship between the particle energy $\mathcal{E}_0$ and its rest mass m_0 (at the velocity $u := 0$ at the initial time moment $t = 0 \in \mathbb{R}$) :*

$$\mathcal{E}_0 = m_0(1-\frac{q^2}{m_0^2}A_0^2)^{-1/2}, \tag{2.73}$$

or, equivalently,

$$m_0 = \mathcal{E}_0(\frac{1}{2} \pm \frac{1}{2}\sqrt{1-4q^2A_0^2}), \tag{2.74}$$

where $A_0 := A|_{t=0} \in \mathbb{E}^3$, *which strongly differs from the classical formulation (2.34).*

To make this difference more clear, we now analyze the Lorentz force (2.57) from the Hamiltonian point of view based on the Lagrangian function (2.54). Thus, we obtain that the corresponding Hamiltonian function

$$\begin{aligned} H &:=< P,\dot{r} > -\mathcal{L} =< P,\dot{r} > +\bar{W}(1+\dot{r}^2)^{1/2} - q < A,\dot{r} >= \\ &=< P - qA,\dot{r} > +\bar{W}(1+\dot{r}^2)^{1/2} = \\ &= - < p,p > \bar{W}^{-1}(1-p^2/\bar{W}^2)^{-1/2} + \bar{W}(1-p^2/\bar{W}^2)^{-1/2} = \\ &= -(\bar{W}^2 - p^2)(\bar{W}^2 - p^2)^{-1/2} = -(\bar{W}^2 - p^2)^{1/2}. \end{aligned} \tag{2.75}$$

Since $p = P - qA$, expression (2.75) assumes the final *"no interaction"* [12; 57; 67; 80] form

$$H = -[\bar{W}^2 - (P - qA)^2]^{1/2}, \tag{2.76}$$

which is conserved with respect to the evolution equations (2.52) and (2.53), that is

$$dH/dt = 0 = dH/d\tau \tag{2.77}$$

for all $\tau, t \in \mathbb{R}$. These equations latter are equivalent to the following Hamiltonian system

$$\begin{aligned} \dot{r} &= \partial H/\partial P = (P - qA)[\bar{W}^2 - (P - qA)^2]^{-1/2}, \\ \dot{P} &= -\partial H/\partial r = (\bar{W}\nabla\bar{W} - \nabla < qA,(P - qA) >)[\bar{W}^2 - (P - qA)^2]^{-1/2}, \end{aligned} \tag{2.78}$$

as one can readily check by direct calculations. Actually, the first equation

$$\begin{aligned} \dot{r} &= (P - qA)[\bar{W}^2 - (P - qA)^2]^{-1/2} = p(\bar{W}^2 - p^2)^{-1/2} = \\ &= mu(\bar{W}^2 - p^2)^{-1/2} = -\bar{W}u(\bar{W}^2 - p^2)^{-1/2} = u(1-u^2)^{-1/2}, \end{aligned} \tag{2.79}$$

holds, owing to the condition $d\tau = dt(1-u^2)^{1/2}$ and definitions $p := mu$, $m = -\bar{W}$, postulated from the very beginning. Similarly we obtain that

$$\begin{aligned} \dot{P} &= -\nabla\bar{W}(1-p^2/\bar{W}^2)^{-1/2} + \nabla < qA,u > (1-p^2/\bar{W}^2)^{-1/2} = \\ &= -\nabla\bar{W}(1-u^2)^{-1/2} + \nabla < qA,u > (1-u^2)^{-1/2}, \end{aligned} \tag{2.80}$$

coincides with equation (2.55) in the evolution parameter $t \in \mathbb{R}$. This can be formulated as the next result.

Proposition 2.9. *The dual to the classical relativistic electrodynamic model (2.57) allows the canonical Hamiltonian formulation (2.78) with respect to the rest reference system variables, where the Hamiltonian function is given by expression (2.76). Moreover, this formulation circumvents the "mass-potential energy" controversy associated with the classical electrodynamical model (2.32).*

The modified Lorentz force expression (2.57) and the related rest energy relationship are characterized by the following remark.

Remark 2.10. *If we make use of the modified relativistic Lorentz force expression (2.57) as an alternative to the classical one of (2.35), the corresponding particle energy expression (2.76) also gives* *rise to a different energy expression (at the velocity $u := 0 \in \mathbb{E}^3$ at the initial time $t = 0$) corresponding to the classical case (2.34); namely, $\mathcal{E}_0 = m_0$ instead of $\mathcal{E}_0 = m_0 + q\varphi_0$, where $\varphi_0 := \varphi|_{t=0}$.*

2.5 Concluding remarks

All of dynamical field equations discussed above are canonical Hamiltonian systems with respect to the corresponding proper rest reference systems $\mathcal{K}_r$, parameterized by suitable time parameters $\tau \in \mathbb{R}$. Upon passing to the basic laboratory reference system $\mathcal{K}$ with the time parameter $t \in \mathbb{R}$,naturally the related Hamiltonian structure is lost, giving rise to a new interpretation of the real particle motion. Namely, one that has an absolute sense only with respect to the proper rest reference system, and otherwise completely relative with respect to all other reference systems. As for the Hamiltonian expressions (2.62), (2.67) and (2.76), one observes that they all depend strongly on the vacuum potential field function $\bar{W} : M^4 \rightarrow \mathbb{R}$, thereby avoiding the mass problem of the classical energy expression pointed out by L. Brillouin [59]. It should be noted that the canonical Dirac quantization procedure can be applied only to the corresponding dynamical field systems considered with respect to their proper rest reference systems.

Remark 2.11. *Some comments are in order concerning the classical relativity principle. We have obtained our results without using the Lorentz transformations of reference systems - relying only on the natural notion of the rest reference system and its suitable parametrization with respect to any other moving reference systems. It seems reasonable then that the true state changes of a moving charged particle q are exactly realized only with respect to its proper rest reference system. Then the only remaining question would be about the physical justification of the corresponding relationship between time parameters of moving and rest reference systems.*

The relationship between reference frames that we have used through is expressed as

$$d\tau = dt(1 - u^2)^{1/2}, \tag{2.81}$$

where $u := dr/dt \in \mathbb{E}^3$ is the velocity with which the rest reference system $\mathcal{K}_r$ moves with respect to another arbitrarily chosen reference system $\mathcal{K}$. Expression (2.81) implies, in particular, that

$$dt^2 - dr^2 = d\tau^2, \tag{2.82}$$

which is identical to the classical infinitesimal Lorentz invariant. This is not a coincidence, since all our dynamical vacuum field equations were derived in turn [53; 54] from the governing equations of the vacuum potential field function $W : M^4 \rightarrow \mathbb{R}$ in the form

$$\partial^2 W/\partial t^2 - \nabla^2 W = \rho, \; \partial W/\partial t + \nabla(vW) = 0, \; \partial\rho/\partial t + \nabla(v\rho) = 0, \tag{2.83}$$

which is *a priori* Lorentz invariant. Here $\rho \in \mathbb{R}$ is the charge density and $v := dr/dt$ the associated local velocity of the vacuum field potential evolution. Consequently, the dynamical infinitesimal Lorentz invariant (2.82) reflects this intrinsic structure of equations (2.83). If it is rewritten in the nonstandard Euclidean form:

$$dt^2 = d\tau^2 + dr^2 \tag{2.84}$$

it gives rise to a completely different relationship between the reference systems $\mathcal{K}$ and $\mathcal{K}_r$, namely

$$dt = d\tau(1 + \dot{r}^2)^{1/2}, \tag{2.85}$$

where $\dot{r} := dr/d\tau$ is the related particle velocity with respect to the rest reference system. Thus, we observe that all our Lagrangian analysis in Section 2 is based on the corresponding functional expressions written in these "Euclidean" space-time coordinates and with respect to which the least action principle was applied. So we see that there are two alternatives - the first is to apply the least action principle to the corresponding Lagrangian functions expressed in the Minkowski space-time variables with respect to an arbitrarily chosen reference system $\mathcal{K}$, and the second is to apply the least action principle to the corresponding Lagrangian functions expressed in Euclidean space-time variables with respect to the rest reference system $\mathcal{K}_r$.

This leads us to a slightly amusing but thought-provoking observation: It follows from our analysis that all of the results of classical special relativity related to the electrodynamics of charged point particles can be obtained (in a one-to-one correspondence) using our new definitions of the dynamical particle mass and the least action principle with respect to the associated Euclidean space-time variables in the rest reference system.

An additional remark concerning the quantization procedure of the proposed electrodynamics models is in order: If the dynamical vacuum field equations are expressed in canonical Hamiltonian form, as we have done here, only straightforward technical details are required to quantize the equations and obtain the corresponding Schrödinger evolution equations in suitable Hilbert spaces of quantum states. There is another striking implication from our approach: the Einsteinian equivalence principle [29; 57; 63; 70; 80] is rendered superfluous for our vacuum field theory of electromagnetism and gravity.

Using the canonical Hamiltonian formalism devised here for the alternative charged point particle electrodynamics models, we found it rather easy to treat the Dirac quantization. The results obtained compared favorably with classical quantization, but it must be admitted that we still have not given a compelling physical motivation for our new models. This is something that we plan to revisit in future investigations. Another important aspect of our vacuum field theory no-geometry (geometry-free) approach to combining the electrodynamics with the gravity, is the manner in which it singles out the decisive role of the rest reference system $\mathcal{K}_r$. More precisely, all of our electrodynamics models allow both the Lagrangian and Hamiltonian formulations with respect to the rest reference system evolution parameter $\tau \in \mathbb{R}$, which are well suited the to canonical quantization. The physical nature of this fact still remains somewhat unclear. In fact, as far as we know [4; 5; 57; 63; 80], there is no physically reasonable explanation of this decisive role of the rest reference system, except for that given by R. Feynman who argued in [70] that the relativistic expression for the classical Lorentz force (2.35) has physical sense only with respect to the rest reference system variables $(\tau, r) \in \mathbb{E}^4$. In future research we plan to analyze the quantization scheme in more detail and begin work on formulating a vacuum quantum field theory of infinitely many particle systems.

3. The modified Lorentz force and the radiation theory

3.1 Introductory setting

Maxwell's equations may be represented by means of the electric and magnetic fields or by the electric and magnetic potentials. The latter were once considered as a purely mathematically motivated representation, having no physical significance.

The situation is actually not so simple now that evidence of the physical properties of the magnetic potential was demonstrated by Y. Aharonov and D. Bohm [92] in the formulation their "paradox" concerning the measurement of a magnetic field outside a separated region where it is vanishes. Later, similar effects were also revealed in the superconductivity theory of Josephson media. As the existence of any electromagnetic field in an ambient space can be tested only by its interaction with electric charges, the dynamics of the charged particles is very important. Charged particle dynamics was studied in detail by M. Faraday, A. Ampere and H. Lorentz using Newton's second law. These investigations led to the following representation for the Lorentz force

$$dp/dt = qE + q\frac{u}{c} \times B, \tag{2.86}$$

where $\mathbb{E}$ and $B \in \mathbb{E}^3$ are, respectively, electric and magnetic fields, acting on a point charged particle $q \in \mathbb{R}$ having momentum $p = mu$. Here $m \in \mathbb{R}_+$ is the particle mass and $u \in T(\mathbb{R}^3)$ is its velocity, measured with respect to a suitably chosen laboratory reference frame.
That the Lorentz force (2.86) is not completely correct was known to Lorentz. The defect can be seen from the nonuniform Maxwell equations for electromagnetic fields radiated by any accelerated charged particle, as easily seen from the well-known expressions for the Lienard-Wiechert potentials.
This fact inspired many physicists to "improve" the classical Lorentz force expression (2.86), and its modification was soon suggested by M. Abraham and P.A.M. Dirac, who found the so-called "radiation reaction" force induced by the self-interaction of a point charged particle:

$$\frac{dp}{dt} = qE + q\frac{u}{c} \times B - \frac{2q^2}{3e^3}\frac{d^2u}{dt^2}. \tag{2.87}$$

The additional force expression

$$F_s := -\frac{2q^2}{3c^3}\frac{d^2u}{dt^2}, \tag{2.88}$$

depending on the particle acceleration, immediately raised many questions concerning its physical meaning. For instance, a uniformly accelerated charged particle, owing to the expression (2.88) , experiences no radiation reaction, contradicting the fact that any accelerated charged particle always radiates electromagnetic waves. This "paradox" was a challenging problem during the 20th century [96–98; 100; 102] and still has not been completely explained [101]. As there exist different approaches to explanation this reaction radiation phenomenon, we mention here only some of the more popular ones such as the Wheeler-Feynman [99] "absorber radiation" theory, based on a very sophisticated elaboration of the retarded and advanced solutions to the nonuniform Maxwell equations, and Teitelbom's [95] approach which exploits the intrinsic structure of the electromagnetic energy tensor subject to the advanced and retarded solutions to the nonuniform Maxwell equations. It is also worth mentioning the very nontrivial development of Teitelbom's theory devised recently by [94] and applied to the non-abelian Yang-Mills equations, which naturally generalize the classical Maxwell equations.

3.2 Radiation reaction force: the vacuum-field theory approach

In the Section, we shall develop our vacuum field theory approach [6; 52–55] to the electromagnetic Maxwell and Lorentz theories in more detail and show that it is in complete agreement with the classical results. Moreover, it allows some nontrivial generalizations, which may have physical applications.

For the radiation reaction force in the vacuum field theory approach, the modified Lorentz force, which was derived in Section 1, acting on a charged point particle q, is

$$dp/dt = -q(\frac{1}{c}\frac{\partial A}{\partial t} + \nabla \varphi) + q\frac{u}{c} \times (\nabla \times A) - q\nabla < \frac{u}{c}, A > \tag{2.89}$$

where $(\varphi, A) \in \mathbb{R} \times \mathbb{E}^3$ is the extended electromagnetic 4-vector potential. To take into account the self-interaction of this particle, we make use of the distributed charge density $\rho : M^4 \to \mathbb{R}$ satisfying the condition

$$q = \int\limits_{\mathbb{R}^3} \rho(t,r)d^3r \tag{2.90}$$

for all $t \in \mathbb{R}$ in a laboratory reference frame $\mathcal{K}$ with coordinates $(t,r) \in M^4$. Then, owing to 2.89 and results in [96], the self-interaction force can be expressed as

$$\begin{aligned} F_s &= q\nabla \varphi_s + \frac{q}{c}\partial A_s/\partial t + < \frac{u}{c}, \nabla > A_s = \\ &= q\nabla \varphi_s + dA_s(t,r)/dt, \end{aligned} \tag{2.91}$$

where

$$\begin{aligned} \varphi_s(t,r) &= \int\limits_{\mathbb{R}^3} \frac{\rho(t',r')d^3r'}{|r-r'|}, \; A_s(t,r) = \\ &= \frac{1}{c}\int\limits_{\mathbb{R}^3} \frac{\rho(t',r')u(t')d^3r'}{|r-r'|}, \end{aligned} \tag{2.92}$$

are the well-known Lienard-Wiechert potentials, which are calculated at the retarded time parameter $t' := t - |r-r'|/c \in \mathbb{R}$. Then, taking into account the continuity equation

$$\partial \rho/\partial t + < \nabla, \rho u >= 0, \tag{2.93}$$

for the charge q, from (2.91) one finds using calculations similar to those in [96] that

$$\begin{aligned} F_s \simeq& \frac{2}{3c^2}\frac{d}{dt}[\int\limits_{\mathbb{R}^3} d^3r' \int\limits_{\mathbb{R}^3} u(t)d^3r\rho(t,r)\rho(t,r')/|r-r'|] - \\ &- \frac{2}{3c^3}\frac{d^2u}{dt^2}\int\limits_{\mathbb{R}^3} d^3r' \int\limits_{\mathbb{R}^3} d^3r\rho(t,r)\rho(t,r') + \\ &+ u < F_{es}, \frac{u}{c^2} > - \frac{1}{2c^3}\frac{du}{dt}\int\limits_{\mathbb{R}^3} d^3r' \int\limits_{\mathbb{R}^3} d^3r\rho(t,r)\rho(t,r') < \frac{(r-r')}{|r-r'|^2}, u >= \\ &= \frac{4}{3}\frac{d}{dt}[\frac{W_s}{c^2}u(t)] - \frac{2q^2}{3c^3}\frac{d^2u}{dt^2} + \Delta F_s, \end{aligned} \tag{2.94}$$

where we defined, respectively, the positive electrostatic self-interaction repulsive energy and force as

$$\bar{W}_s := \frac{1}{2}\int_{\mathbb{R}^3} d^3r \int_{\mathbb{R}^3} d^3r' \frac{\rho(t,r)\rho(t,r')}{|r-r'|}, \tag{2.95}$$

$$F_{es} := \int_{\mathbb{R}^3} d^3r \int_{\mathbb{R}^3} d^3r' \rho(t,r)\rho(t,r') \frac{(r-r')}{|r-r'|^3}, \tag{2.96}$$

and the force component corresponding to the term $< \frac{u}{c}, \nabla > A_s$ in (2.91) by ΔF_s. Assuming now that the external electromagnetic field vanishes, from (2.89) one obtains that

$$\frac{d}{dt}(mu) = -\frac{2q^2}{3c^3}\frac{d^2u}{dt^2} + \frac{4}{3}\frac{d}{dt}(m_s u) + \Delta F_s, \tag{2.97}$$

where we have made use of the inertial mass definitions

$$m := -\bar{W}/c^2, \quad m_s := W_s/c^2, \tag{2.98}$$

following from the vacuum field theory approach. From (2.97) one computes that the additional force term is

$$\Delta F_s = \frac{d}{dt}[(m - \frac{4}{3}m_s)u] + \frac{2q^2}{3c^3}\frac{d^2u}{dt^2}. \tag{2.99}$$

Then we readily infer from (2.97) that the observed charged particle mass satisfies at rest the inequality

$$m \neq m_s. \tag{2.100}$$

This expression means that the real physically observed mass strongly depends both on the intrinsic geometric structure of the particle charge distribution and on the external physical interaction with the ambient vacuum medium.

3.3 Conclusion

The charged particle radiation problem, revisited in this section, allows the explanation of the point charged particle mass as that of a compact and stable object, which should have a negative vacuum interaction potential $\bar{W} \in \mathbb{R}^3$ owing to (2.98). This negativity can be satisfied if and only if the quantity (2.99) holds, thereby imposing certain nontrivial geometric constraints on the intrinsic charged particle structure [103]. Moreover, as follows from the physically observed particle mass expressions (2.98), the electrostatic potential energy comprises the main portion of the full mass.

There exist different relativistic generalizations of the force expression (2.97), all of which suffer the same common physical inconsistency related to the no radiation effect of a charged point particle in uniform motion.

Another problem closely related to the radiation reaction force analyzed above is the search for an explanation to the Wheeler and Feynman reaction radiation mechanism, which is called the absorption radiation theory. This mechanism is strongly dependent upon the Mach type interaction of a charged point particle in an ambient vacuum electromagnetic medium. It is also interesting to observe some of the relationships between this problem and the one devised above in the context of the vacuum field theory approach, but more detailed and extended analyzes will be required to explain the connections.

4. Maxwell's equations and the Lorentz force derivation - the legacy of Feynman's approach

4.1 Poissonian analysis preliminaries

In 1948 R. Feynman presented but did not published [127; 128] a very interesting, in some respects "heretical", quantum-mechanical derivation of the classical Lorentz force acting on a charged particle under the influence of an external electromagnetic field. His result was analyzed by many authors [129–137] from different points of view, including its relativistic generalization [138]. As this problem is completely classical, we reanalyze the Feynman's derivation from the classical Hamiltonian dynamics point of view on the coadjoint space $T^*(N), N \subset \mathbb{R}^3$, and construct its nontrivial generalization compatible with results [6; 52; 53] of Section 1, based on a recently devised vacuum field theory approach [52; 55]. Upon obtaining the classical Maxwell electromagnetic equations, we supply the complete legacy of Feynman's approach to the Lorentz force and demonstrate its compatibility with the relativistic generalization presented in [52–55; 72].

Consider the motion of a charged point particle $\xi \in \mathbb{R}$ under the influence of an external electromagnetic field. For its description, following [114; 123; 124], it is convenient to introduce a trivial fiber bundle structure π: $M \to N$, $M = N \times G$, $N \subset \mathbb{R}^3$, with the abelian structure group $G := \mathbb{R}\backslash\{0\}$ equivariantly acting [1] on the canonically symplectic coadjoint space $T^*(M)$. Then we endow the bundle with *a connection one-form* $\mathcal{A}$: $M \to \Lambda^1(M) \times \mathcal{G}$ defined as

$$\mathcal{A}(q;g) := <\vartheta(q), \xi >_{\mathcal{G}} + g^{-1}dg \tag{2.101}$$

on the phase space M, where $q \in N$ and $g \in G$ and $\alpha : N \to \Lambda^1(N)$ is a differential form, constructed from the magnetic potential $A : N \to \mathbb{E}^3$ as $\vartheta(q) := <A(q), dq>_{\mathbb{E}^3} \in T^*_q(N)$. If $l : T^*(M) \to \mathcal{G}^*$ is the related momentum mapping, one can construct the reduced phase space $\bar{\mathcal{M}}_\xi := l^{-1}(\xi)/G \simeq T^*(N)$, where $\xi \in \mathcal{G} \simeq \mathbb{R}$ is taken to be fixed. This reduced space has the symplectic structure

$$\omega^{(2)}_\xi(q,p) = < dp, \wedge dq > + \xi d < A(q), dq >, \tag{2.102}$$

where we taken in to account that $\vartheta(q) = <A(q), dq>_{\mathbb{E}^3} \in T^*_q(N)$. From (2.102), one readily computes the respective reduced Poisson brackets on $T^*(N)$:

$$\{q^i, q^j\}_{\omega^{(2)}_\xi} = 0, \quad \{p_j, q^i\}_{\omega^{(2)}_\xi} = \delta^i_j, \quad \{p_i, p_j\}_{\omega^{(2)}_\xi} = \xi F_{ji}(q) \tag{2.103}$$

for $i,j = \overline{1,3}$ with respect to the reference frame $\mathcal{K}(t,q)$, characterized by the phase space coordinates $(q,p) \in T^*(N)$. If one introduces a new momentum variable $\tilde{p} := p + \xi A(q)$ on $T^*(N) \ni (q,p)$, it is easy to verify that $\omega^{(2)}_\xi \to \tilde{\omega}^{(2)}_\xi := < d\tilde{p}, \wedge dq >$, giving rise to the following "minimal coupling" canonical Poisson brackets [12; 123; 124]:

$$\{q^i, q^j\}_{\tilde{\omega}^{(2)}_\xi} = 0, \quad \{\tilde{p}_j, q^i\}_{\tilde{\omega}^{(2)}_\xi} = \delta^i_j, \quad \{\tilde{p}_i, \tilde{p}_j\}_{\tilde{\omega}^{(2)}_\xi} = 0 \tag{2.104}$$

for $i,j=\overline{1,3}$ with respect to the reference frame $\mathcal{K}_f(t,q-q_f)$, characterized by the phase space coordinates $(q,\tilde{p})\in T^*(N)$, if and only if the Maxwell field equations

$$\partial F_{ij}/\partial q_k + \partial F_{jk}/\partial q_i + \partial F_{ki}/\partial q_j = 0 \tag{2.105}$$

are satisfied on N for all $i,j,k=\overline{1,3}$ for the curvature tensor $F_{ij}(q) := \partial A_j/\partial q^i - \partial A_i/\partial q^j$, $i,j=\overline{1,3}$, $q\in N$.

4.2 The Lorentz force and Maxwell electromagnetic field equations - Lagrangian analysis

The Poisson structure (2.104) makes it possible to describe a charged particle $\xi\in\mathbb{R}$, located at point $q\in N\subset\mathbb{R}^3$, moving with a velocity $q' := u\in T_q(N)$ with respect to the reference frame $\mathcal{K}(t,q)$.The particle is under the electromagnetic influence of an external charged particle $\xi_f\in\mathbb{R}$ located at point $q_f\in N\subset\mathbb{R}^3$ and moving with respect to the same reference frame $\mathcal{K}(t,q)$ with a velocity $q'_f := u_f\in T_{q_f}(N)$, where $\frac{d}{dt}(...) := (...)'$ is the temporal derivative with respect to the temporal parameter $t\in\mathbb{R}$. More precisely, consider a new reference frame $\mathcal{K}_f(t,q-q_f)$ moving with respect to the reference frame $\mathcal{K}(t,q)$ with velocity u_f. With respect to the reference frame $\mathcal{K}_f(t,q-q_f)$, the charged particle ξ moves with the velocity $u-u_f\in T_{q-q_f}(N)$ and, respectively, the charged particle ξ_f stays in rest. Then one can write the standard *classical Lagrangian function* of the charged particle ξ with a constant mass $m\in\mathbb{R}_+$ subject to the reference frame $\mathcal{K}_f(t,q-q_f)$ as

$$\mathcal{L}_f(q,q') = \frac{m}{2}|q'-q'_f|^2 - \xi\varphi, \tag{2.106}$$

and the scalar potential $\varphi\in C^2(N;\mathbb{R})$ is the corresponding potential energy. On the other hand, owing to (2.106) and the Poisson brackets (2.104), the following equation for the charged particle ξ *canonical momentum* with respect to the reference frame $\mathcal{K}_f(t,q-q_f)$ holds:

$$\tilde{p} := p + \xi A(q) = \delta\mathcal{L}_f(q,q')/\delta q', \tag{2.107}$$

or, equivalently,

$$p + \xi A(q) = m(q'-q'_f), \tag{2.108}$$

expressed in the units when the light speed $c = 1$. Taking into account that the charged particle ξ momentum with respect to the reference frame $\mathcal{K}(t,q)$ equals $p := mu\in T^*_q(N)$, one computes from (2.108) that

$$\xi A(q) = -mu_f \tag{2.109}$$

for the magnetic vector potential $A\in C^2(N;\mathbb{R}^3)$, which was obtained in [54; 55; 126] using a vacuum field theory approach. Now, it follows from (2.106) and (2.109) one has the Lagrangian equations,

$$\frac{d}{dt}[p+\xi A(q)] = \partial\mathcal{L}_f(q,q')/\partial q = -\xi\nabla\varphi, \tag{2.110}$$

which induce the charged particle ξ dynamics

$$\begin{aligned} dp/dt &= -\xi\partial A/\partial t - \xi\nabla\varphi - \xi < u,\nabla > A = \\ &= -\xi\partial A/\partial t - \xi\nabla\varphi - \xi < u,\nabla > A + \xi\nabla < u,A > - \xi\nabla < u,A > = \\ &= -\xi(\partial A/\partial t + \nabla\varphi) + \xi u\times(\nabla\times A) - \xi\nabla < u,A > . \end{aligned} \tag{2.111}$$

As a result of (2.111), we obtain the modified Lorentz type force

$$dp/dt = \xi E + \xi u \times B - \xi \nabla < u, A >, \tag{2.112}$$

obtained in [54; 55], where

$$E := -\partial A/\partial t - \nabla \varphi, \quad B := \nabla \times A. \tag{2.113}$$

This differs from the classical Lorentz force expression

$$dp/dt = \xi E + \xi u \times B \tag{2.114}$$

by the gradient component

$$F_c := -\nabla < u, A > . \tag{2.115}$$

Remark now that the Lorentz type force expression (2.112) can be naturally generalized to the relativistic case if to take into account that the Lorentz condition

$$\partial \varphi / \partial t + < \nabla, A >= 0 \tag{2.116}$$

imposed on the electromagnetic potential $(\varphi, A) \in C^2(N; \mathbb{R} \times \mathbb{R}^3)$.
Indeed, from (2.113) one obtains the Lorentz invariant field equation

$$\partial^2 \varphi / \partial t^2 - \Delta \varphi = \rho_f, \tag{2.117}$$

where $\Delta :=< \nabla, \nabla >$ and $\rho_f : N \to \mathcal{D}'(N)$ is the generalized density function of the external charge distribution ξ_f. Employing calculations from [54; 55], derive readily from (2.117) and the charge conservation law

$$\partial \rho_f / \partial t + < \nabla, J_f >= 0 \tag{2.118}$$

the Lorentz invariant equation on the magnetic vector potential $A \in C^2(N; \mathbb{R}^3)$:

$$\partial^2 A / \partial t^2 - \Delta A = J_f. \tag{2.119}$$

Moreover, relationships (2.113), (2.117) and (2.119) imply the true classical Maxwell equations

$$\begin{gathered} \nabla \times E = -\partial B/\partial t, \ \nabla \times B = \partial E/\partial t + J_f, \\ < \nabla, E >= \rho_f, \quad < \nabla, B >= 0 \end{gathered} \tag{2.120}$$

on the electromagnetic field $(E, B) \in C^2(N; \mathbb{R}^3 \times \mathbb{R}^3)$.
Consider now the Lorentz condition (2.116) and observe that it is equivalent to the following local conservation law:

$$\frac{d}{dt} \int_{\Omega_t} \varphi d^3 q = 0. \tag{2.121}$$

This gives rise to the important relationship for the magnetic potential $A \in C^2(N; \mathbb{R}^3)$

$$A = q'_f \varphi \tag{2.122}$$

with respect to the reference frame $\mathcal{K}(t, q)$, where $\Omega_t \subset N$ is any open domain with a smooth boundary $\partial \Omega_t$, moving together with the charge distribution ξ_f in the region $N \subset \mathbb{R}^3$ with velocity q'_f. Taking into account relationship (2.109), one obtains the expression for the charged particle ξ *'inertial'* mass as

$$m = -\bar{W}, \quad \bar{W} := \lim_{q_f \to q} \xi\varphi, \tag{2.123}$$

coinciding with that obtained in [54; 55; 126]. Her we denoted the corresponding potential energy of the charged particle ξ by $\bar{W} \in C^2(N;\mathbb{R})$.

4.3 The modified least action principle and its Hamiltonian analysis

Using the representations (2.122) and (2.123), one can rewrite the determining Lagrangian equation (2.110) as

$$\frac{d}{dt}[-\bar{W}(u - u_f)] = -\nabla\bar{W}, \tag{2.124}$$

which is completely equivalent to the Lorentz type force expression (2.112) calculated with respect to the reference frame $\mathcal{K}(t,q)$.

Remark 4.1. *It is interesting to remark here that equation (2.124) does not allow the Lagrangian representation with respect to the reference frame $\mathcal{K}(t,q)$ in contrast to that of equation (2.110).*

The remark above is a challenging source of our further analysis concerning the relativistic generalization of the Lorentz type force (2.112). Namely, the following proposition holds.

Proposition 4.2. *The Lorentz type force (2.112), in the case when the charged particle ξ momentum is defined as $p = -\bar{W}u$, according to (2.123), is the exact relativistic expression allowing the Lagrangian representation with respect to the charged particle ξ rest reference frame $\mathcal{K}_r(\tau, q - q_f)$, connected with the reference frame $\mathcal{K}(t,q)$ by means of the classical relativistic proper time relationship:*

$$dt = d\tau(1 + |\dot{q} - \dot{q}_f|^2)^{1/2}. \tag{2.125}$$

Here $\tau \in \mathbb{R}$ is the proper time parameter in the rest reference frame $\mathcal{K}_r(\tau, q - q_f)$ and, by definition, the derivative $d/d\tau(...) := (\dot{...})$.

Proof. Take the following action functional with respect to the charged particle ξ rest reference frame $\mathcal{K}_r(\tau, q - q_f)$:

$$S^{(\tau)} := -\int_{t_1(\tau_1)}^{t_2(\tau_2)} \bar{W} dt = \int_{\tau_1}^{\tau_2} \bar{W}(1 + |\dot{q} - \dot{q}_f|^2)^{1/2} d\tau, \tag{2.126}$$

where the proper temporal values $\tau_1, \tau_2 \in \mathbb{R}$ are considered to be fixed. In contrast, the temporal parameters $t_2(\tau_2), t_2(\tau_2) \in \mathbb{R}$ depend, owing to (2.125), on the charged particle ξ trajectory in the phase space. The least action condition

$$\delta S^{(\tau)} = 0, \delta q(\tau_1) = 0 = \delta q(\tau_2), \tag{2.127}$$

applied to (2.126) yields the dynamical equation (2.124), which is also equivalent to the relativistic Lorentz type force expression (2.112). This completes the proof. □

Making use of the relationships between the reference frames $\mathcal{K}(t,q)$ and $\mathcal{K}_r(\tau, q - q_f)$ in the case when the external charge particle velocity $u_f = 0$, we can easily deduce the following result.

Corollary 4.3. *Let the external charge distribution ξ_f be at rest, that is the velocity $u_f = 0$. Then equation (2.124) reduces to*

$$\frac{d}{dt}(-\bar{W}u)] = -\nabla\bar{W}, \tag{2.128}$$

which implies the following conservation law:

$$H_0 = \bar{W}(1-u^2)^{1/2} = -(\bar{W}^2 - p^2)^{1/2}. \tag{2.129}$$

Moreover, equation (2.128) is Hamiltonian with respect to the canonical Poisson structure (2.104) with Hamiltonian function (2.129) and the rest reference frame $\mathcal{K}_r(\tau, q)$:

$$\left.\begin{array}{r} dq/d\tau := \partial H_0/\partial p = p(\bar{W}^2 - p^2)^{-1/2} \\ dp/d\tau := -\partial H_0/\partial q = -\bar{W}(\bar{W}^2 - p^2)^{-1/2}\nabla\bar{W} \end{array}\right\} \Rightarrow \left.\begin{array}{r} dq/dt = -p\bar{W}^{-1}, \\ dp/dt = -\nabla\bar{W} \end{array}\right\}. \tag{2.130}$$

In addition, if the rest particle mass is defined as $m_0 := -H_0|_{u=0}$, the "inertial" particle mass quantity $m \in \mathbb{R}$ has the well-known classical relativistic form

$$m = -W = m_0(1-u^2)^{-1/2}, \tag{2.131}$$

which depends on the particle velocity $u \in \mathbb{R}^3$.

As for the general case of equation (2.124), analogous results to those above hold as described in detail in [52–55]. We need only mention that the Hamiltonian structure of the general equation (2.124) results naturally from its least action representation (2.126) and (2.127) with respect to the rest reference frame $\mathcal{K}_r(\tau, q)$.

4.4 Conclusion

We have demonstrated the complete legacy of the Feynman's approach to the Lorentz force based derivation of Maxwell's electromagnetic field equations. Moreover, we have succeeded in finding the exact relationship between Feynman's approach and the vacuum field approach devised in [54; 55]. Thus, the results obtained provide deep physical backgrounds lying in the vacuum field theory approach. Consequently, one can simultaneously describe the origins of the physical phenomena of electromagnetic forces and gravity. Gravity is physically based on the particle *"inertial"* mass expression (2.123), which follows naturally from both the Feynman approach to the Lorentz type force derivation and the vacuum field approach.

5. References

[1] Abraham R. and Marsden J. Foundations of Mechanics, Second Edition, Benjamin Cummings, NY, 1978

[2] Arnold V.I. Mathematical Methods of Classical Mechanics., Springer, NY, 1978

[3] Bogolubov N.N., Logunov A.A., Oksak A.I. and Todorov I.T. *Introduction to Axiomatic Field Theory*, Massachussetts: W.A.Benjamin, Inc. Advanced Book Progarm, 1975

[4] Logunov A. A. The Theory of Gravity. Moscow. Nauka Publisher, 2000

[5] Logunov A.A. and Mestvirishvili M.A. Relativistic theory of gravitation. Moscow, Nauka, 1989 (In Russian)

[6] Blackmore D., Prykarpatsky A.K. and Samoylenko V. Nonlinear dynamical systems of mathematical physics: spectral and differential-geometric integrability analysis. NY, World Scientific, 2010

[7] Prykarpatsky A., Bogoluboiv N. (jr.), Golenia J., Taneri U. Introductive backgrounds of modern quantum mathematics with application to nonlinear dynamical systems. Available at: http://publications.ict it Preprint ICTP-IC/2007/108
[8] Prykarpatsky A., Bogolubov N. (jr.), Golenia J., Taneri U. Introductive Backgrounds to Modern Quantum Mathematics with Application to Nonlinear Dynamical Systems. Int J Theor Phys., 47, 2882–2897, 2008
[9] Dirac P.A.M. Generalized Hamiltonian dynamics. *Canad. J. of Math.*, 2, N2, 129-148, 1950
[10] Dubrovin B.A., Novikov S.P. and Fomenko A.T. *Modern geometry* Nauka, Moscow, 1986 (in Russian)
[11] Drinfeld V.G. Quantum Groups. Proceedings of the International Congress of Mathematicians, MRSI Berkeley, p. 798, 1986
[12] Kupershmidt B.A. Infinite-dimensional analogs of the minimal coupling principle and of the Poincare lemma for differential two-forms. Diff. Geom. & Appl. 1992, 2,p. 275-293.
[13] Donaldson S.K. An application of gauge theory to four dimansional topology. J. Diff. Geom., 17, 279-315, 1982
[14] Kowalski K. and Steeb W.-H. Symmetries and first integrals for nonlinear dynamical systems: Hilbert space approach. I and II. Progress of Theoretical Physics, 85, N4, 713-722, 1991, and 85, N4, 975-983, 1991
[15] Kondratiev Y.G., Streit L., Westerkamp W. and Yan J.-A. Generalized functions in infinite dimensional analysis. II AS preprint, 1995
[16] Berezansky Yu. M. Eigenfunctions expansions related with selfadjoint operators, Kiev, Nauk. Dumka Publ., 1965 (in Russian)
[17] Neumann J. von. Mathematische Grundlagen der Quanten Mechanik. J. Springer, Berlin, 1932
[18] Prykarpatsky A.K., Taneri U. and Bogolubov N.N. (jr.) Quantum field theory and application to quantum nonlinear optics. World Scientific, NY, 2002
[19] Weyl H. The Theory of Groups and Quantum Mechanics. Dover, New York, 1931
[20] Witten E. Nonabelian bozonization in two dimensions. Commun. Mathem. Physics, 92, 455-472, 1984
[21] Prytula M., Prykarpatsky A., Mykytyuk I. Fundamentals of the Theory of Differential-Geometric Structures and Dynamical Systems. Kiev, the Ministry of Educ. Publ., 1988 (in Ukrainian)
[22] Albeverio S., Kondratiev Y.G. and Streit L. How to generalize white noice analysis to non-gaussian measures. Preprint Bi-Bo-S, Bielefeld, 1992.
[23] Berezansky Y.M. A generalization of white noice analysis by means of theory of hypergroups. Reports on Math. Phys., 38, N3, 289-300, 1996
[24] Bogolubov N.N. and Bogolubov N.N. (jr.) Introduction into quantum statistical mechanics. World Scientific, NJ, 1986
[25] Bogolubov N.N. (Jr.), Prykarpatsky A.K. Quantum method of generating Bogolubov functionals in statistical physics: current Lie algebras, their representations and functional equations. Physics of Elementary Particles and Atomique Nucleus, v.17, N4, 791-827, 1986 (in Russian)
[26] Prykarpatsky A.K. and Kalenyuk P.I. Gibbs representation of current Lie algebra and complete system of quantum Bogolubov functional equations. Soviet Mathematical Doklady. 300(2), (1988), p. 346 - 349 (1988). (In Russian)
[27] Berezin F.A. The second quantization method. Nauka Puplisher, Moscow, 1986 (in Russian)

[28] Faddeev L.D., Sklyanin E.K. Quantum mechanical approach to completely integrable field theories. Proceed. of the USSR Academy of Sciences (DAN), 243, 1430-1433, 1978 (in Russian).
[29] Feynman R. Quantum mechanical computers. Found. Physics, 16, 507-531, 1986
[30] Fock V.A. Konfigurationsraum und zweite Quantelung. Zeischrift Phys. Bd. 75, 622-647, 1932
[31] Gelfand I., Vilenkin N. Generalized functions, 4, Academic Press, New York, 1964
[32] Goldin G.A. Nonrelativistic current algebras as unitary representations of groups. Journal of Mathem. Physics, 12(3), 462-487, 1971
[33] Goldin G.A. Grodnik J. Powers R.T. and Sharp D. Nonrelativistic current algebra in the N/V limit. J. Math. Phys. 15, 88-100, 1974
[34] Goldin G.A., Menikoff R., Sharp F.H. Diffeomorphism groups, gauge groups, and quantum theory. Phys. Rev. Lett. 51, 1983, p. 2246-2249
[35] Goldin G.A., Menikoff R., Sharp F.H. Representations of a local current algebra in nonsimply connected space and the Aharonov-Bohm effect. J. Math. Phys., 22(8), 1981, p. 1164-1668
[36] Grover L.K. Quantum mechanics helps in searching for a needle in a haystack. Phys. Rev. Lett. 79, 325-328, 1997
[37] Jaffe A. and Quinn F. Theoretical mathematics: toward a cultural synthesis of mathematics and theoretical physics. Bull. Amer. Math. Soc. 29, 1-13, 1993
[38] Kowalski K. Methods of Hilbert spaces in the theory of nonlinear dynamical systems. World Scientific, 1994
[39] Kowalski K. and Steeb W.-H. Non linear dynamical systems and Carleman linearization. World Scientific, 1991
[40] Lytvynov E.W. Rebenko A.L. and Shchepaniuk G.V. Wick calculus on spaces of generalized functions compound Poisson white noise. Reports on Math. Phys. 39, N2, 219-247, 1997
[41] Berezin F.A. and Shubin M.A. Schrodinger equation, Moscow, the Moscow University Publisher, 1983 (in Russian).
[42] Samoilenko A., Prykarpatsky Ya., Taneri U., Prykarpatsky A., Blackmore D. A geometrical approach to quantum holonomic computing algorithms. Mathematics and Computers in Simulation, 66, 1–20, 2004
[43] Prykarpatsky A., Bogolubov N (jr.), Golenia J. A symplectic generalization of the Peradzynski Helicity Theorem and Some Applications. Int J Theor Phys 47, 1919–1928, 2008
[44] Hentosh O.Ye., Prytula M.M. and Prykarpatsky A.K Differential-geometric integrability fundamentals of nonlinear dynamical systems on functional menifolds. (The second revised edition), Lviv University Publisher, Lviv, Ukraine, 2006
[45] Faddeev L.D., Quantum inverse scattering problem II, in Modern problems of mathematics, M: VINITI Publ.,3, 93-180, 1974 (in Russian)
[46] Holm D. and Kupershmidt B. Poisson structures of superfluids. Phys. Lett., 91A, 425-430, 1982
[47] Holm D., Marsden J., Ratiu T. and Weinstein A. Nonlinear stability of fluid and plasma equilibria. Physics Reports, 123/(1 and 2), 1-116, 1985
[48] Moffat H.K. The degree of knottedness of tangled vortex lines. Journal of Fluid Mechanics, 35/1, 117-129., 1969

[49] Owczarek R. Topological defects in superfluid Helium. Int. J. Theor. Phys., 30/12, 1605-1612, 1991
[50] Peradzyński Z. Helicity theorem and vertex lines in superfluid ^{4}He. Int. J. Theor. Phys., 29/11, 1277-1284, 1990
[51] Putterman S.J. Superfluid Hydrodynamics, North Holland, Amsterdam, 1974.
[52] Bogolubov N.N. (jr.) and Prykarpatsky A.K. The Analysis of Lagrangian and Hamiltonian Properties of the Classical Relativistic Electrodynamics Models and Their Quantization. Found Phys 40, (2010), p. 469–493
[53] Prykarpatsky A.K., Bogolubov N.N. (Jr.) and Taneri U. The vacuum structure, special relativity and quantum mechanics revisited: a field theory no-geometry approach. Theor. Math. Phys. 160(2), (2009), p. 1079–1095; (arXiv lanl: 0807.3691v.8 [gr-gc] 24.08.2008)
[54] Prykarpatsky A.K., Bogolubov N.N. (Jr.) and Taneri U. The field structure of vacuum, Maxwell equations and relativity theory aspects. Preprint ICTP, Trieste, IC/2008/051 (http://publications.ictp.it)
[55] Prykarpatsky A.K., Bogolubov N.N. (Jr.) and Taneri U. The Relativistic Electrodynamics Least Action Principles Revisited: New Charged Point Particle and Hadronic String Models Analysis. Int. J. Theor. Phys. (2010) 49, p. 798–820
[56] Thirring W. Classical Mathematical Physics. Springer, Third Edition, 1992
[57] Landau L.D. and Lifshitz E.M. Field theory. v. 2, "*Nauka*" Publisher, Moscow, 1973
[58] Landau L.D. and Lifshitz E.M. Quantum mechanics. v. 6, "Nauka" Publisher, Moscow, 1974
[59] Brillouin L. Relativity reexamined. Academic Press Publ., New York and London, 1970
[60] Repchenko O. Field physics. Moscow, "Galeria" Publ., 2005
[61] Brans C.H. and Dicke R.H. Mach's principle and a relativistic theory of gravitation. *Phys. Rev.*, 124, 1961, 925
[62] Kleinert H. Path integrals. World Scientific. Second Edition, 1995, 685
[63] Klymyshyn I.A. Relativistic astronomy. (in Ukrainian).Kyiv, "Naukova Dumka" Publisher, 1980
[64] Weinstock R. New approach to special relativity. Am. J. Phys., 33, 1965, 640-645
[65] Mermin N.D. Relativity without light. *Am. J. Phys.*, 52, 1984, 119-124.
[66] Mermin N.D. It's About Time: *Understanding Einstein's Relativity*, Princeton, NJ., Princeton University Press, 2005
[67] Prykarpatsky Y.A., Samoilenko A.M. and Prykarpatsky A.K. The geometric properties of canonically reduced symplectic spaces with symmetry and their relationship with structures on associated principal fiber bundles and some applications. *Opuscula Mathematica*, 2005, 25, No. 2, 287-298
[68] Green B. The fabric of the Cosmos. Vintage Books Inc., New York, 2004
[69] Newman R.P. The Global Structure of Simple Space-Times. Comm. Mathem. Physics, 123, 1989, 17-52
[70] Feynman R., Leighton R. and Sands M. The Feynman Lectures on Physics. Electrodynamics, v. 2, Addison-Wesley, Publ. Co., Massachusetts, 1964
[71] Feynman R., Leighton R. and Sands M. The Feynman lectures on physics. The modern science on the Nature. Mechanics. Space, time, motion. v. 1, *Addison-Wesley*, Publ. Co., Massachusetts, 1963

[72] Bogolubov N.N., Prykarpatsky A.K. The Lagrangian and Hamiltonian formalisms for the classical relativistic electrodynamical models revisited. arXiv:0810.4254v1 [gr-qc] 23 Oct 2008

[73] Bogolubov N.N. and Shirkov D.V. *Introduction to quantized field theory* . Moscow, Nauka, 1984 (in Russian)

[74] Bogolubov N.N., Shirkov D.V. Quantum Fields. "*Nauka*", Moscow, 1984

[75] Faddeev L.D. Energy problem in the Einstein gravity theory. Russian Physical Surveys, 136, N 3, 1982, 435-457 (in Russian)

[76] Barbashov B. M., Nesterenko V. V. Introduction to the relativistic string theory. *World Scientific*, Singapore, 1990

[77] Barbashov B.M., Chernikov N.A. Solution and quantization of a nonlinear Born-Infeld type model. Journal of Theoret. Mathem. Physics, 60, N 5, 1966, 1296-1308 (in Russian)

[78] Nambu Y. Strings, monopoles, and guage fields. Phys. Rev. D., 10, N 12, 1974, 4262-4268

[79] Goto T. Relativistic quantum mechanics of one-dimensional mechanical continuum and subsidiary condition of dual resonance model. Prog. Theor. Phys. 46, N 5, 1971, p.1560-1569

[80] Pauli W. Theory of Relativity. Oxford Publ., 1958

[81] Deser S., Jackiw R. Time travel? arXiv:hep-th/9206094, 1992

[82] Dunner G., Jackiw R. "Peierles substitution" and Chern-Simons quantum mechanics. arXiv:hep-th/92004057, 1992

[83] Jackiw R., Polychronakos A.P. Dynamical Poincare symmetry realized by field-dependent diffeomorhisms. arXiv:hep-th/9809123, 1998

[84] Barbashov B.M., Pervushin V.N., Zakharov A.F. and Zinchuk V.A. The Hamiltonian approach to general relativity and CNB-primordal spectrum. arXiv:hep-th/0606054, 2006

[85] Barbashov B.M. On the canonical treatment of Lagrangian constraints. arXiv:hepth/0111164, 2001

[86] Jackiw R. Lorentz violation in a diffeomorphism-invariant theory. arXiv:hep-th/0709.2348, 2007

[87] Bulyzhenkov-Widicker I.E. Einstein's gravitation for Machian relativism of nonlocal energy charges. Int. Journal of Theor. Physics, 47, 2008, 1261-1269

[88] Bulyzhenkov I.E. Einstein's curvature for nonlocal gravitation of Gesamt energy carriers. arXiv:math-ph/0603039, 2008

[89] Wilczek F. QCD and natural phylosophy. Ann. Henry Poincare, 4, 2003, 211-228

[90] Bialynicky-Birula I. Phys Rev., 155, 1967, 1414; 166,1968, 1505

[91] Glauber R.J. Quantum Theory of Optical Coherence. Selected Papers and Lectures, Wiley-VCH, Weinheim 2007

[92] Aharonov Y. and Bohm D. Phys. Rev. 115, 1959, p. 485

[93] Kulish P.I., Sklyanin E.K. Lecture Notes in Physics, 151, 1982a, 61-119, Berlin: Springer.

[94] Kosyakov B.P. Radiation in electrodynamics and in Yang-Mills theory. Sov. Phys. Usp. 35 (2) 135–142 (1992)

[95] Teitelboim C. Phys. Rev. 1970., V. D1, p. 1512

[96] Jackson J.D. Classical electrodynamics. 3-rd edition., NY, Wiley, 1999

[97] Jefimenko O.D. Causality Electromagnetic Induction and Gravitation. Electret Scientific Company, Star City, 2000

[98] Dirac P.A.M. Proc. Roy. Soc. Ser. A, 167, 1938, p. 148-169

[99] Wheeler J.B. and Feynman R.P. Interaction with the Absorber as the Mechanism of Radiation. Rev. Modern Phys., 1945, 17, N 2-3, p. 157-181
[100] Pegg D.T. Rep. Prog. Phys. 1975. V. 38. P. 1339
[101] Rohrlich F. Classical Charged Particles. Reading, Mass.: Addison-Wesley, 1965 -
[102] Bolotovskii B. M. and Stolyarov S. N. Radiation from and energy loss by charged particles in moving media. Sov. Phys. Usp. 35, N 2, 1992, p. 143–150
[103] Radiation reaction of a classical quasi-rigid extended particle. J. Phys. A: Math. Gen. ,2006, p. 3801-3816
[104] Prykarpatsky A. and Mykytyuk I. Algebraic Integrability of nonlinear dynamical systems on manifolds: classical and quantum aspects. Kluwer Academic Publishers, the Netherlands, 1998, 553p.
[105] Bogolubov N.N. (Jr.), Mykytyuk I.V. and Prykarpatsky A.K. Verma moduli over the quantum Lie current algebra on the circle. Soviet Mathematical Doklady. 314(2), (1990), p. 268-272 - 349 (In Russian)
[106] Mitropolsky Yu., Bogolubov N. (jr.), Prykarpatsky A. and Samoylenko V. Integrable dynamical system: spectral and differential-geometric aspects. Kiev, "Naukova Dunka", 1987. (in Russian)
[107] Hentosh O., Prytula M. and Prykarpatsky A. Differential-geometric and Lie-algebraic foundations of investigating nonlinear dynamical systems on functional manifolds. The Second edition. Lviv University Publ., 2006 (in Ukrainian)
[108] Godbillon C. Geometrie Differentielle et Mecanique Analytique. Hermann Publ., Paris, 1969
[109] Gillemin V., Sternberg S. On the equations of motion of a classical particle in a Yang-Mills field and the principle of general covariance. Hadronic Journal, 1978, 1, p.1-32
[110] Bjorken J.D. and Drell S.D. Relativistic Quantum Fields. Mc Graw-Hill Book Co., NY, 1965
[111] Dirac P.A.M. The Principles of Quantum Mechanics. Second edition. Oxford, Clarendon Press, 1935
[112] Dirac P.A.M, Fock W.A. and Podolsky B. Phys. Zs. Soviet. 1932, 2, p. 468
[113] Bogolubov N.N. and Prykarpatsky A.K. The Lagrangian and Hamiltonian formalisms for the classical relativistic electrodynamical models revisited. arXiv:0810.4254v1 [gr-qc] 23 Oct 2008
[114] Bogolubov (Jr.), Prykarpatsky A.K., Taneri U., and Prykarpatsky Y.A. The electromagnetic Lorentz condition problem and symplectic properties of Maxwell- and Yang–Mills-type dynamical systems. J. Phys. A: Math. Theor. 42 (2009) 165401 (16pp)
[115] Berezansky Y.M. and Kondratiev Y.G. Spectral methods in infinite dimensional analysis, v.1 and 2, Kluwer, 1995.
[116] Kummer M. On the construction of the reduced phase space of a Hamiltonian system with symmetry. Indiana University Mathem. Journal, 1981, 30,N2, p.281-281.
[117] Ratiu T., Euler-Poisson equations on Lie algebras and the N-dimensional heavy rigid body. Proc. NAS of USA, 1981, 78, N3, p. 1327-1328.
[118] Holm D., and Kupershmidt B. Superfluid plasmas: multivelocity nonlinear hydrodynamics of superfluid solutions with charged condensates coupled electromagnetically. Phys. Rev., 1987, 36A, N8, p. 3947-3956
[119] Moor J.D. Lectures on Seiberg-Witten invariants. Lect. Notes in Math., N1629, Springer, 1996.

[120] Kupershmidt B.A. Infinite-dimensional analogs of the minimal coupling principle and of the Poincare lemma for differential two-forms. Diff. Geom. & Appl. 1992, 2,p. 275-293.
[121] Marsden J. and Weinstein A. The Hamiltonian structure of the Maxwell-Vlasov equations. Physica D, 1982, 4, p. 394-406
[122] Marsden J. and Chorin A. Mathematical foundations of fluid mechanics. Springer, New York, 1993
[123] Prykarpatsky Ya.A., Samoylenko A.M. and Prykarpatsky A.K. The geometric properties of reduced symplectic spaces with symmetry, their relationship with structures on associated principle fiber bundles and some applications. Part 1. Opuscula Mathematica, Vol. 25, No. 2, 2005, p. 287-298
[124] Prykarpatsky Ya.A. Canonical reduction on cotangent symplectic manifolds with group action and on associated principal bundles with connections. Journal of Nonlinear Oscillations, Vol. 9, No. 1, 2006, p. 96-106
[125] Prykarpatsky A. and Zagrodzinski J. Dynamical aspects of Josephson type media. Ann. of Inst. H. Poincaré, Physique Theorique, v. 70, N5, p. 497-524
[126] Repchenko O. Field physics. Moscow, Galeria Publ., 2005 (in Russian)
[127] Dyson F.J. Feynman's proof of the Maxwell equations, Am. J. Phys. 58, 1990, p. 209-211
[128] Dyson F. J. Feynman at Cornell. Phys. Today 42 (2), 32-38 (1989).
[129] [3] Lee C. R. The Feynman-Dyson Proof Of The Gauge Field Equations. Phys. Lett. A 148, 1990, p. 146-148
[130] Silagadze Z.K. Feynman's derivation of Maxwell equations and extra dimensions. arXiv:hep-ph/0106235, 2002
[131] Brehme R.W. Comment on Feynman's proof of the Maxwell equations by F.J. Dyson [Am.J.Phys.58, 209-211(1990)]. Am. J. Phys. 59, 1991, p. 85-86 ;
[132] Dombey N. Comment on Feynman's proof of the Maxwell equations. by F.J. Dyson [Am.J.Phys. 58,209-211(1990)]." Am. J. Phys. 59, 1991, p. 85
[133] Anderson J. L. Comment on Feynman's proof of the Maxwell equations by F.J. Dyson [Am.J.Phys.58,209-211(1990)]. Am. J. Phys. 59, 1991, p. 86
[134] Farquar I. E. Comment on Feynman's proof of the Maxwell equations by F.J. Dyson [Am.J.Phys.58,209-211(1990)],. Am. J. Phys. 59, 1991, p. 87
[135] Vaidya A. and Farina C. Can Galilean mechanics and full Maxwell equations coexist peacefully? Phys. Lett. A 153, 1991, p. 265-267
[136] Hughes R. J. On Feynman's proof of the Maxwell equations. Am. J. Phys. 60, 1992, p.301-306
[137] Hojman S. and Shepley L. C. No Lagrangian? No quantization! J. Math. Phys. 32, 142-146 (1991)
[138] Tanimura S. Relativistic generalization and extension to the nonabelian gauge theory of Feynman's proof of the Maxwell equations. Annals Phys. 220, 1992, p. 229-247

3

Electromagnetic-wave Contribution to the Quantum Structure of Matter

Burke Ritchie
Livermore Software Technology Corporation
USA

1. Introduction

The quantum theory of matter does not describe real matter until electromagnetic theory is used to account for such diverse radiative phenomena as spontaneous emission and the shift of quantum energy levels. Classical electrodynamics fails to account quantitatively for these radiative effects in the structure of matter. Quantum electrodynamics (QED) does successfully account for radiative effects in the structure of matter once an infinite contribution to the energy, which diverges linearly with electromagnetic-wave frequency, is subtracted from the theory based on physical argument that such contribution is already included, to zeroth order in perturbation theory, in the description of a radiative as opposed to a nonradiative or bare electron. This mathematical procedure is known as mass renormalization and introduces the concept that total mass comprises both material and electromagnetic contributions, neither of which is observable by itself.

In Section II of this paper a theory is presented which describes both the material and radiative properties of matter in a single, inseparable form. We show that the time-domain relativistic-wave equation of Paul Dirac can be inferred from the Lorentz invariant obtained from the scalar product of the electron's four-momentum and an electromagnetic four-potential, once an electromagnetic carrier-wave frequency is formally identified with the rest-mass energy of the electron divided by $\hbar$, namely $\omega = \frac{mc^2}{\hbar}$. (The scalar product of two four-vectors always gives a Lorentz invariant such that the present derivation proves the Lorentz invariance of Dirac's equation in a single step. In the standard treatment [1], in which the Dirac Hamiltonian is the scalar product of two operator four-vectors, a second step is required to prove the Lorentz invariance of the wave equation itself.) Our derivation elucidates a long-studied problem in the literature of the identity of Dirac's equation with the spinorial form of Maxwell's equation [2-5]. The value of ω given above is just the cut off of the electromagnetic frequency used in QED to insure the finite value of the logarithmically-divergent contribution to the energy, which is the only divergent term remaining after the term linearly divergent in the frequency has been removed by mass renormalization. In summary Dirac's time-domain relativistic wave equation is reinterpreted to be an equation which accounts for both the material and radiative properties of matter.

In Section III we provide an analytic proof that Dirac's temporally harmonic solution of his equationis equivalent to solving temporally coupled equations by adiabatic elimination, which is a widely-used approximation method to solve temporally coupled equations in the optical-physics literature. In a wordDirac's temporally harmonic solution is approximate, but his solution ofthe resulting time-independent or energy-domain equation, which is astaple of the relativistic quantum mechanics literature, is exact.The current interpretation of Dirac's theory as describing only the material properties of matter derives from Dirac's solution of his time-domain equation using the harmonic substitution, $\Psi_D(\vec{r},t) = e^{-i\frac{E}{\hbar}t}\Psi_{D,E}(\vec{r})$, albeit this form, since it constrains all four components of his energy-domain vector wave function $\Psi_{D,E}(\vec{r})$ to oscillate in time at a single frequency $\omega = \frac{E}{\hbar}$, is obviously not the general solution. The same harmonic form however exactly solves Schroedinger's time-domain equation and thus gives a result which is compatible with the way quantum theory evolved as a matter-only theory without radiative effects until augmented by QED.

In Section IV numerical results are presented for the general solution of Dirac's time-dependent equation. Fourier analysis of the generaltime-dependent solution shows that the spectrum of quantum states for the Coulomb problem comprises coupled positive- and negative-energy states. The wave function is a mixture of bound and continuum states, with an unbound component propagating away from the atom in a manner which satisfies the Lorentz-invariant relationship or causality between position and time, r**2 - (ct)**2 = 0. The unbound behavior has long been known as Zitterbewegung for a free electron, and here we show its counterpart for the Coulomb problem. In view of the Dirac-Maxwell relationship elucidated in Section II we postulate that the physical interpretation of Zitterbewegung is the emission of a photon with energy of order 2mc**2 due to the presence of empty negative-energy states in the general time-dependent solution. Dirac's artifice of filling up the negative-energy levels with electrons to stabilize the atom is not available in the general time-dependent solution.

In Section V equations of motion for the photon are given. In Section VI subatomic bound solutions are discovered which are expected due to the temporally second-order nature of the time-domain Dirac equation. Subatomic bound solutions do not exist for Dirac's time-independent equation (hereafter called standard Dirac theory) due to his use of the single-frequency temporally harmonic form discussed in Section III. Thus the existence of a complex neutron cannot be ruled out in the case of the general time-dependent solution as it was earlier in the case of standard Dirac theory [6]. These solutions resemble known Dirac energy-domain functions for Z > 137. The spectral content of these solutions comprises a spectral peak at -mc**2 for an electron and a spectral peak at +mc**2 for a positron - yes this state exists for a positive Coulomb potential. An electron can thus make an upward transition into the positive-energy continuum with transition energy 2mc**2, as in standard Dirac theory, while a positron can make a downward transition into the negative-energy continuum with transition energy -2mc**2. The upward transition is considered to be a matter transition, while the downward transition is considered to be an anti-matter transition.

2. Maxwell-Dirac equivalency

There exists a physical equivalency between Dirac and Maxwell theories which can be stated as follows. It is well known that Lorentz' equation is the Lorentz invariant formed by taking the scalar product of the four-gradient and the electromagnetic four-potential,

$$(\frac{1}{c}\frac{\partial}{\partial t},\vec{\nabla})\cdot(\Phi,\vec{A})=\frac{1}{c}\frac{\partial\Phi}{\partial t}+\vec{\nabla}\cdot\vec{A}=0 \tag{1}$$

Recall that the scalar product of four-vectors is always a Lorentz Invariant. One may postulate that a four-potential exists for the electron, such that an electron equation of motion can be written as the Lorentz invariant formed by taking the scalar product of the electron's four-momentum and the electron's four-potential,

$$(\frac{i\hbar}{c}\frac{\partial}{\partial t}-\frac{e}{c}\Phi,i\hbar\vec{\nabla}+e\vec{A})\cdot(\Phi_e,\vec{A}_e)=(\frac{i\hbar}{c}\frac{\partial}{\partial t}-\frac{e}{c}\Phi)\Phi_e+(i\hbar\vec{\nabla}+\frac{e}{c}\vec{A})\cdot\vec{A}_e=0 \tag{2}$$

The electron scalar and vector potentials can be written in the form of carrier-wave expansions,

$$\Phi_e=\Phi_{e-}e^{-i\omega_e t}+\Phi_{e+}e^{i\omega_e t}, \tag{3a}$$

$$\vec{A}_e=\vec{A}_{e-}e^{-i\omega_e t}+\vec{A}_{e+}e^{i\omega_e t}, \tag{3b}$$

from which on substituting Eqs. (3) into Eq. (2) and separately setting the coefficients of the exponential factors equal to zero, we obtain,

$$(i\hbar\frac{\partial}{\partial t}-e\Phi-\hbar\omega_e)\Phi_{e+}+(i\hbar c\vec{\nabla}+e\vec{A})\cdot\vec{A}_{e+}=0 \tag{4a}$$

$$(i\hbar\frac{\partial}{\partial t}-e\Phi+\hbar\omega_e)\Phi_{e-}+(i\hbar c\vec{\nabla}+e\vec{A})\cdot\vec{A}_{e-}=0\,. \tag{4b}$$

On setting $\Phi_{e+}=\psi$, $\vec{A}_{e+}=\vec{\sigma}\chi$, $\Phi_{e-}=\chi$, $\vec{A}_{e-}=\vec{\sigma}\psi$ we obtain Dirac's equation
Identically if the carrier-wave energy is equal to the rest-mass energy $\hbar\omega_e=m_e c^2$,

$$(i\hbar\frac{\partial}{\partial t}-e\Phi-\hbar\omega_e)\psi+\vec{\sigma}\cdot(i\hbar c\vec{\nabla}+e\vec{A})\chi=0 \tag{5a}$$

$$(i\hbar\frac{\partial}{\partial t}-e\Phi+\hbar\omega_e)\chi+\vec{\sigma}\cdot(i\hbar c\vec{\nabla}+e\vec{A})\psi=0\,, \tag{5b}$$

where $\vec{\sigma}$ is Pauli's vector. Unlike the classical electromagnetic potentials, which are real, the electron's potentials are complex. This is obvious when we notice that the + and - envelopes are not complex or Hermitian conjugates of one another.

An electromagnetic contribution to the mass of the electron due to the quantum radiation field associated with its motion is a well known concept in QED. Indeed the carrier-wave frequency of the electron's four-potential [Eqs. (3)] is equal to mc**2/hbar, which is the high-frequency cut off for the quantum radiation field assumed in QED atomic structure calculations. The present derivation of Dirac's equation suggests thatthe total mass of the electron is electromagnetic in nature. This result is consistent with a previous result in which the charge of the electron was derived from Maxwell's equations [7].

3. Adiabatic nature of Dirac's solution of his equation

Although the time-dependent Dirac equation can be written in the Schroedinger form,

$$i\hbar\frac{\partial\psi_D}{\partial t} = H_D\psi_D \ , \tag{6}$$

where HD is the Dirac Hamiltonian and $\psi_D(\vec{r},t)$ is Dirac's four-component vector wave function, it does not follow that the energy-domain equation can be written in the Schroedinger form,

$$E\psi_{D,E} = H_D\psi_{D,E} \ , \tag{7}$$

unless one requires that all components of $\psi_D(\vec{r},t)$ oscillate in time at a single frequency $\omega = \frac{E}{\hbar}$, such that $\psi_D(\vec{r},t)$ has the harmonic form $\psi_D(\vec{r},t) = e^{-i\frac{E}{\hbar}t}\psi_{D,E}(\vec{r})$. The requirement does not hold in the general time-dependent solution of a vector wave function whose components are temporally coupled.

Eqs. (5) are rewritten in the standard Dirac form,

$$(i\hbar\frac{\partial}{\partial t} - e\Phi - mc^2)\psi + \vec{\sigma}\cdot(i\hbar c\vec{\nabla} + e\vec{A})\chi = 0 \tag{8a}$$

$$(i\hbar\frac{\partial}{\partial t} - e\Phi + mc^2)\chi + \vec{\sigma}\cdot(i\hbar c\vec{\nabla} + e\vec{A})\psi = 0 \ , \tag{8b}$$

where $\vec{\sigma}$ is Pauli's vector and ψ, χ are the large, small components of Dirac's four-component wave function ψ_D. Eq. (8b) can be eliminated exactly in favor of Eq. (8a) as follows,

$$(i\hbar\frac{\partial}{\partial t} - e\Phi - mc^2)\psi(\vec{r},t) = i\hbar c^2\vec{\sigma}\cdot\nabla\int_0^t dt' e^{\frac{i}{\hbar}(mc^2 - e\Phi)(t-t')}\vec{\sigma}\cdot\nabla\psi(\vec{r},t') \ , \tag{9}$$

where we have specialized to an electromagnetic field free problem by setting $\vec{A} = 0$. Dirac's energy-domain solution is obtained by substituting $\psi(\vec{r},t) = e^{-i\frac{E}{\hbar}t}\psi_E(\vec{r},t)$ and assuming that $\psi_E(\vec{r},t)$ is slowly varying in the time compared to the exponential factor such that the integral is evaluated approximately by holding $\psi_E(\vec{r},t')$ constant at t'=t. Then the integration is performed, and the rapidly oscillating lower-limit contribution is dropped as small compared to the stationary upper-limit contribution. Such approximations to solve coupled time-dependent equations are known in the optical-physics literature as adiabatic elimination. Dirac's second-order equation for the large component follows immediately,

$$[(E - e\Phi)^2 - (mc^2)^2]\psi_E(\vec{r}) = -(\hbar c)^2[\nabla^2 + \frac{1}{E - e\Phi + mc^2}[(\vec{\nabla}e\Phi)\cdot\vec{\nabla} + i\vec{\sigma}\cdot(\vec{\nabla}e\Phi\times\vec{\nabla})]\psi_E(\vec{r}) \tag{10}$$

where we have used the identity, $(\vec{\sigma}\cdot\vec{A})(\vec{\sigma}\cdot\vec{B})=\vec{A}\cdot\vec{B}+i\vec{\sigma}\cdot(\vec{A}\times\vec{B})$ and the time has been dropped from the argument list since the approximations to the t′ integral render the wave function stationary. Clearly Dirac's use of the Schroedinger forms $\psi(\vec{r},t)=e^{-i\frac{E}{\hbar}t}\psi_E(\vec{r})$ and $\chi(\vec{r},t)=e^{-i\frac{E}{\hbar}t}\chi_E(\vec{r})$ to write the energy-domain form of his coupled time-dependent equations [Eqs. (8)] rests on an implicit assumption that adiabatic elimination of one of these equations in favor of the other is an accurate approximation. On other words the Schroedinger form does not hold exactly in the case of a vector wave function whose components are temporally coupled.

Dirac's harmonic ansatz for his time-dependent equation gives him aenergy-domain equation which is exactly solvable for the free-electron and Coulomb problems. The Schroedinger form of the temporal solution, which is exact for Schroedinger's scalar wave equation but not for Dirac's vector wave equation, is in effect a form of calibration of Dirac theory to Schroedinger theory and has cast Dirac theory in the limited role of "correcting" Schroedinger theory primarily for relativistic effects in atomic structure. Probably as a result of its restricted use in electron physics, time-domain Dirac theory until recently had not been used to discover the a priori physical basis for Fermi-Dirac statistics [8], which is a spin-dependent phenomenon. The history of quantum mechanics instead followed a path of ensuring that Schroedinger wave functions satisfy Fermi-Dirac statistics on the basis of experimental observation and not a priori theory by using the Slater determinantal wave function to solve Schroedinger's wave equation for many electrons, even though Schroedinger theory, in which particle spin is absent, contains no physical basis for Fermi-Dirac statistics. One must instead turn to time-domain Dirac theory and the Dirac current to discover the physical basis for Fermi-Dirac statistics, which is elucidated using spin-dependent quantum trajectories [8]. Richard Feynman [9] once asked if spin is a relativistic requirement and then answered in the negative because the Klein-Gordon equation is a valid relativistic equation for a spin-0 particle. The correct answer is thatspin is a relativistic requirement to insure Lorentz invariance in a vector-wave theory such as the Dirac or Maxwell theories. In the sense that Fermi-Dirac statistics depends critically on spin and yet is a phenomenonof order $(Zc)^0$, where c is the speed of light and Z is the atomic number, it would appear that authors [10] are misguided who present the quantum theory of matter as fundamentally based on Schroedinger theory as augmented by Dirac theory for "relativistic corrections" of order Z^4c^{-2} due to the acceleration of an electron moving near a nucleus with atomic number Z.

4. Genera solution of Dirac's time-domain wave equation

In this section the general time-dependent solution is presented free of any harmonic bias. Solving the Coulomb problem ($e\Phi=V=-\frac{Ze^2}{r}$) the radial form of Eq. (9),

$$(i\hbar\frac{\partial}{\partial t}-V-mc^2)G_\kappa(r,t)=i\hbar c^2(\frac{\partial}{\partial r}-\frac{(\kappa-1)}{r})\int_0^t dt' e^{\frac{i}{\hbar}(mc^2-V)(t-t')}(\frac{\partial}{\partial r}+\frac{(\kappa+1)}{r})G_\kappa(r,t') \quad (11)$$

follows from the well-known substitutions,

$$\psi(\vec{r},t) = G_{\kappa}(r,t)\chi_{\kappa\mu}(\theta,\phi) \tag{12a}$$

$$\vec{\sigma}\cdot\vec{\nabla} = \vec{\sigma}\cdot\hat{r}(\frac{\partial}{\partial r} - \frac{1}{r}\vec{\sigma}\cdot\vec{L}) \tag{12b}$$

$$\vec{\sigma}\cdot\hat{r}\chi_{\kappa\mu} = -\chi_{-\kappa\mu} \tag{12c}$$

$$\vec{\sigma}\cdot\vec{L}\chi_{\kappa\mu} = -(\kappa+1)\chi_{\kappa\mu}, \tag{12d}$$

where the angular functions are Dirac's two-component spinors.Eq. (11) is solved numerically in the variables r and ct for the hydrogen-like ground state ($\kappa = -1$) with Z=70, starting for mathematical convenience with a Schroedinger wave function at initial time and using the trapezoid rule to evaluate the integral. It is found that the evolved wave function is insensitive to the starting function at initial time.

At the point t=t′ the Crank-Nicolson implicit integration procedure is used in order to insure that the time integration of the equation itself is unconditionally stable. Fig. 1 shows the spectrum of states calculated from the inverse temporal Fourier transform of the wave function [11-12]. The spectrum has a strong peak in the positive-energy regime and a weak peak in the negative-energy regime, which lies in the negative-energy continuum and thus accounts for the unbound tail (Fig. 2). This temporally expanding tail appears to be the Coulomb counterpart of the Zitterbewegung solution calculated by Schroedinger [13] using the time-dependent Dirac equation for a free electron.

Fig. 2 shows the real part of radial wave function times r. Notice that the wave function is unusual in that it behaves like a bound state close to the nucleus but yet is unbound with a small-amplitude tail along the r axis whose length is equal to ct. In other words the tail propagates away from the nucleus at the speed of light. Nevertheless I have normalized the wave function for unit probability of finding the electron within a sphere of radius r_{max}. The amplitude of the interior portion flows with time between the real part (Fig. 2) and imaginary part of the wave function such that the probability density is steady within the radius of the atom. $(ct)_{max}$ is chosen to be three-fourths of r_{max} in order that the propagating piece of the wave function stays well away from the grid boundary at r_{max}. Calculations show that the results are Insensitive to r_{max} and therefore to $(ct)_{max}$ as long as r_{max} is well outside the region represented by the bound piece of the wave function, that is well outside of the radius of the atom as represented by standard Dirac theory. Notice that if the dynamical calculation were extended to very large times, then the wave function would fill a verylarge volume. In principle after a sufficient time the wave function could fill a volume the size of the universe although its interior part would remain the size of an atom.

What is the physical interpretation of Zitterbewegung? In view of theMaxwell-Dirac equivalency elucidated in Section II, we postulate here thatit is a photonic energy of order 2mc**2, which is the energy gap betweenthe positive- and negative-energy electron continua and which was identified in Section II as an electromagnetic carrier-wave energy equal to $2\hbar\omega$. This amount of energy must be carried away from the atom in a continuous sense since there is no net loss of interior probability densityover time. The energy originates from the electron's simulta-neous double occupancy of both positive- and negative-energy states(Fig. 4) whose energy difference is of order 2mc**2. In standard Diractheory the positive- and negative-energy levels are dynamically uncoupledsuch that Dirac assumed that electrons exclusively occupy the positive-energy levels and that the atom was stabilized by a set of negative-energy levels - the negative-energy sea - which are totally filledwith electrons such that the Pauli Exclusion Principle forbad thedownward fall of an electron from positive- to negative-energy levelsaccompanied by the emission of a photon with energy of order 2mc**2.

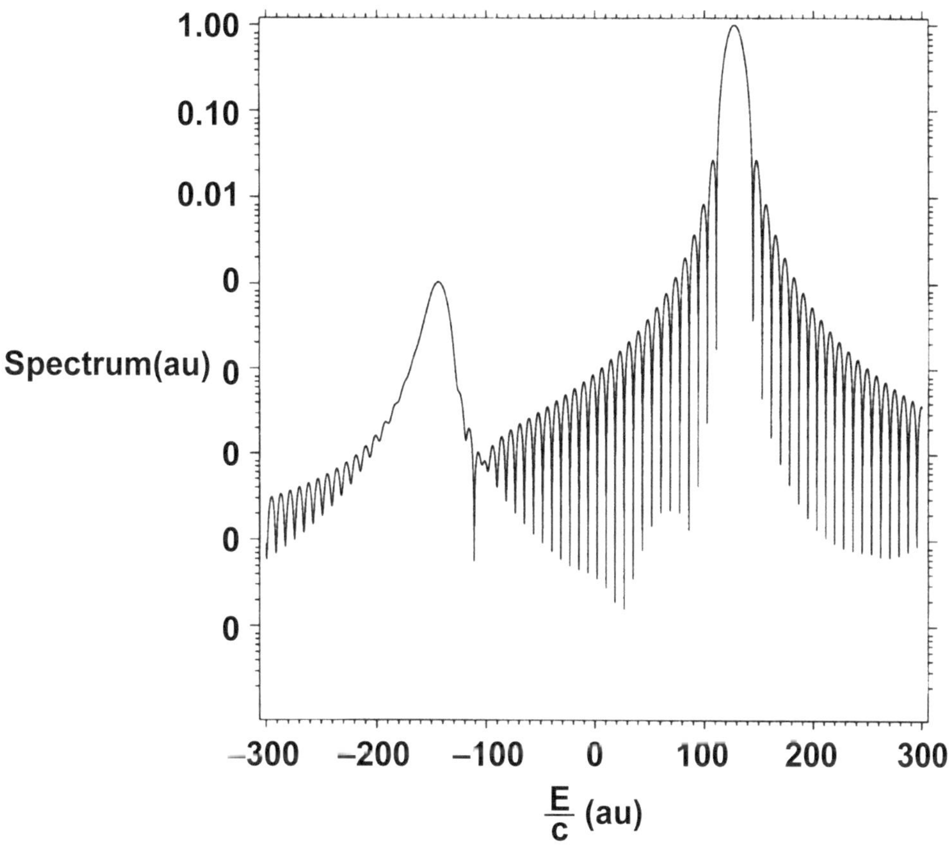

Fig. 1. Spectrum showing weak coupling of the positive- and negative-energy regions. The continuum edges are at $E / c = \pm mc$ au. The energy is obtained by multiplying the graphical numbers by c. A blow up of the positive energy peek shows good agreement with the eigenvalue at 17474.349, although the spectral calculation, because of the nature of the spectral determination of the eigenenergy, is not good to the number of significant figures shown.

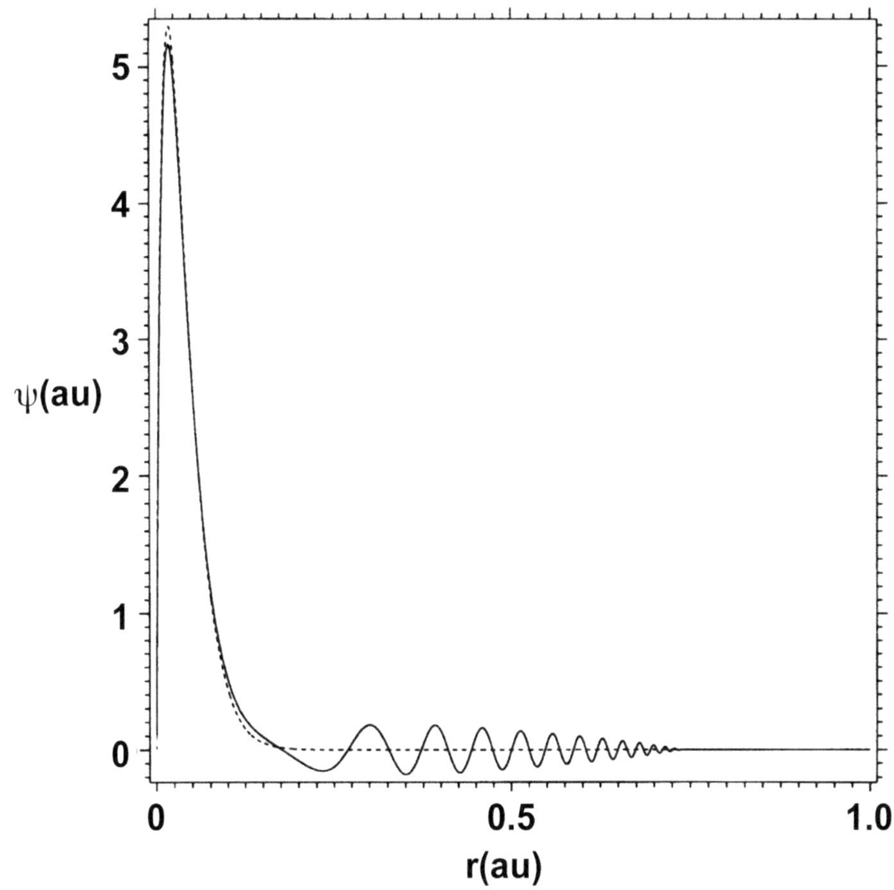

Fig. 2. Solid: imaginary part of the solution of Eq. (6) times r for Z=50.$(ct)_{max}$=0.75r_{max}=0.75 au. The number of ct, r grid points is 20K, 20K. Dotted: radial solution of Eq. (5) times r. The eigenvalue is found from the zero wronskian of forward and backward integrations and is equal to 17474.349 au to the number of significant figures shown in agreement with the analytic Dirac energy $\varepsilon = \dfrac{mc^2}{\sqrt{1+[\dfrac{\alpha Z}{\sqrt{1-(\alpha Z)^2}}]^2}}$ where $\alpha = \dfrac{e^2}{\hbar c}$ is the fine structure constant.

Dotted: wave function calculated from the radial equation inferred from Eq. (5).

In the general time-domain solution presented here it appears that theatom is self-stabilizing due to the mixed material-electromagnetic natureof the electron. Recall that in Section II we postulated that the electron'sequation of motion should be the scalar product of its material four-momentum and its electromagnetic four potential. The solution of the equation of motion shows that the electron can share two ground-state material energy levels with energy conservation and without temporal decay of its quantum state as long as the energy difference between the two ground-state levels is converted to the energy of a continuously-emitted photon.

5. Photon equations of motion

In this section equations of motion for the photon are given and used to calculate a divergence-free Lamb shift [14-15]. As in the case of the electron in Section II we assume that a complex four-potential exists for the photon such that a photon EOM can be written as the Lorentz invariant formed by taking the scalar product of the photon's four-momentum and the photon's four-potential,

$$(\frac{\hbar}{c}\frac{\partial}{\partial t}, \hbar\nabla - \frac{e\hbar}{mc^2}\vec{E},\vec{H}) \cdot (\Phi_\nu, \vec{A}_\nu) = \frac{\hbar}{c}\frac{\partial}{\partial t}\Phi_e + (\hbar\nabla - \frac{e\hbar}{mc^2}\vec{E},\vec{H}) \cdot \vec{A}_\nu = 0\,, \tag{13}$$

for either electric or magnetic fields $\vec{E},\vec{H}$. The photon four-momentum was found in [14] from $\hbar$ times a form of the four-gradient whose scalar product with the four-electromagnetic-energy density gives the electromagnetic continuity equation. This is simply the electromagnetic analog of writing the material continuity equation as the scalar product of the four-gradient and the material four-density.

The electron scalar and vector potentials can be written in the form of carrier-wave expansions,

$$\Phi_\nu = \Phi_{\nu-}e^{-i\omega_\nu t} + \Phi_{\nu+}e^{i\omega_\nu t} \tag{14a}$$

$$\vec{A}_\nu = \vec{A}_{\nu-}e^{-i\omega_\nu t} + \vec{A}_{\nu+}e^{i\omega_\nu t}\,, \tag{14b}$$

from which on substituting Eqs. (14) into Eq. (13) and separately setting the coefficients of the exponential factors equal to zero, we obtain,

$$(\frac{1}{c}\frac{\partial}{\partial t} + i\frac{\omega_\nu}{c})\Phi_{\nu+} + (\nabla - \frac{e}{mc^2}\vec{E},\vec{H}) \cdot \vec{A}_{\nu+} = 0 \tag{15a}$$

$$(\frac{1}{c}\frac{\partial}{\partial t} - i\frac{\omega_\nu}{c})\Phi_{\nu-} + (\nabla - \frac{e}{mc^2}\vec{E},\vec{H}) \cdot \vec{A}_{\nu-} = 0\,. \tag{15b}$$

On setting $\Phi_{\nu+} = \xi_{E,H}$, $\vec{A}_{\nu+} = \vec{\sigma}\zeta_{E,H}$, $\Phi_{\nu-} = \zeta_{E,H}$, $\vec{A}_{\nu-} = \vec{\sigma}\xi_{E,H}$ we obtain the Dirac form for the photon EOM presented previously assuming zero photon mass ($\hbar\omega_\nu - 0$),

$$\frac{\partial\xi_{E,H}}{c\partial t} + i\frac{\omega_\nu}{c}\xi_{E,H} + \vec{\sigma}\cdot(\nabla - \frac{e}{mc^2}\vec{E},\vec{H})\zeta_{E,H} = 0 \tag{16a}$$

$$\frac{\partial\zeta_{E,H}}{c\partial t} - i\frac{\omega_\nu}{c}\zeta_{E,H} + \vec{\sigma}\cdot(\nabla - \frac{e}{mc^2}\vec{E},\vec{H})\xi_{E,H} = 0 \tag{16b}$$

Writing $\xi_{E,H} = e^{-i\omega t}\psi_{E,H}$ and $\zeta_{E,H} = e^{-i\omega t}\chi_{E,H}$ in Eqs. (16) we derive stationary equations for $\psi_{E,H}$ and $\chi_{E,H}$; then we eliminate the equation for $\chi_{E,H}$ in favor of a second-order equation for $\psi_{E,H}$,obtaining equations for the electric and magnetic photon wave functions which have the Helmholtz form,

$$\{\nabla^2 + \frac{\omega^2 - \omega_\nu^2}{c^2} - \frac{e}{mc^2}[\nabla\cdot\vec{E} + 2\vec{E}\cdot\nabla + i\sigma\cdot(\nabla\times\vec{E}) - \frac{e}{mc^2}E^2]\}\psi_E = 0 \tag{17a}$$

$$\{\nabla^2 + \frac{\omega^2 - \omega_v^2}{c^2} - \frac{e}{mc^2}[\vec{\nabla}\cdot\vec{H} + 2\vec{H}\cdot\vec{\nabla} + i\sigma\cdot(\vec{\nabla}\times\vec{H}) - \frac{e}{mc^2}H^2]\}\psi_H = 0, \tag{17b}$$

where we have used the identity, $(\vec{\sigma}\cdot\vec{A})(\vec{\sigma}\cdot\vec{B}) = \vec{A}\cdot\vec{B} + i\vec{\sigma}\cdot(\vec{A}\times\vec{B})$ .Eq. (17b) for $\hbar\omega_v = 0$ was used in previous work to calculate the Lamb shift [14] and anomalous magnetic moment [16].

6. Subatomic bound states

Dirac's time-domain equation can be cast in the form of an equationsecond order in space and time; thus we should expect a second spatial-temporal solution to exist which is independent of the first spatial-temporal solution which we have elucidated in Section IV. I show that a regime exists in which an adiabatic solution to the time-dependent Dirac equation is not justified even in an approximate sense. The existence of the regime is easily recognized by writing Dirac equations in the form given by Eq. (11) for the large component with a reversal of charge and for the small component with no reversal of charge and then seeking solutions for which the phase in the exponential factor vanishes for all times. These equations are,

$$(i\hbar\frac{\partial}{\partial t} \mp |V| \mp mc^2)H_\kappa^\pm(r,t) = i\hbar c^2(\frac{\partial}{\partial r} \mp \frac{(\kappa\mp 1)}{r})\int_0^t dt' e^{\pm\frac{i}{\hbar}(mc^2-|V|)(t-t')}(\frac{\partial}{\partial r} \pm \frac{(\kappa\pm 1)}{r})H_\kappa^\pm(r,t') \,. \tag{21}$$

Eqs. (21) are solved numerically for Z=1 and $\kappa = \pm 1$ using the same techniques used to solve Eq. (11). The two equations for positronic or electronic binding are solved for a wave function or its complex conjugate respectively. The spectrum is found to be given simply by $E_\pm = \pm mc^2$ (Fig. 3). The real part of the wave function is shown in Fig. 4. Notice that if a bound state exists for one charge, then a bound state must also exist for the other charge by the charge-conjugation symmetry of Dirac's equation. Charge-conjugation symmetry is well known in standard time-independent Dirac theory, whose adiabatic regime does not support positronic-electronic bound states, and arises in Dirac's interpretation of the negative-energy states in which a hole or absence of an electron registers the existence of a positron or conversely in a positron world the absence of a positron would signal the existence of an electron.

Although the wave function is pulled inward toward the origin, its extent is still large compared to the radius of the proton r_p=1.3x10^{-13} cm = 2.46x10^{-5} au.

The spectral energies are those which cancel the terms $\mp mc^2$ on the left side of Eqs. (21) and for which the stationary phases on the right side occur at 2mc^2-|V| = 0. For the unit-strength Coulomb potential the radius at which the stationary-phases occur is given by $r_{sp} = \frac{e^2}{2mc^2}$, which is roughly the radius of the proton.

The bound behavior of the positronic-electronic wave function shown in Fig. 4 can be understood as follows. Recognizing that the first and third terms on the left side of Eq. (21) cancel from the spectral values $E_\pm = \pm mc^2$ (Fig. 3), one may write an equation in which zero phase of the integration factor is assumed and which is the time derivative of both sides of Eq. (21) with zero phase,

$$\mp |V| \frac{\partial f}{\partial t} = i\hbar c^2 \left(\frac{\partial^2 f}{\partial r^2} + \frac{2}{r}\frac{\partial f}{\partial r} - \frac{2}{r^2} f \right), \tag{22}$$

A solution to Eq. (22) is sought in the form $f(r,t) = e^{i\omega t} g(r)$ for the complex separation constant $\omega_r + i\omega_i$, giving the equation for g,

$$\frac{\partial^2 g}{\partial r^2} + \frac{2}{r}\frac{\partial g}{\partial r} - \frac{2}{r^2} g \pm \frac{|V|(\omega_r + i\omega_i)}{\hbar c^2} g = 0 . \tag{23}$$

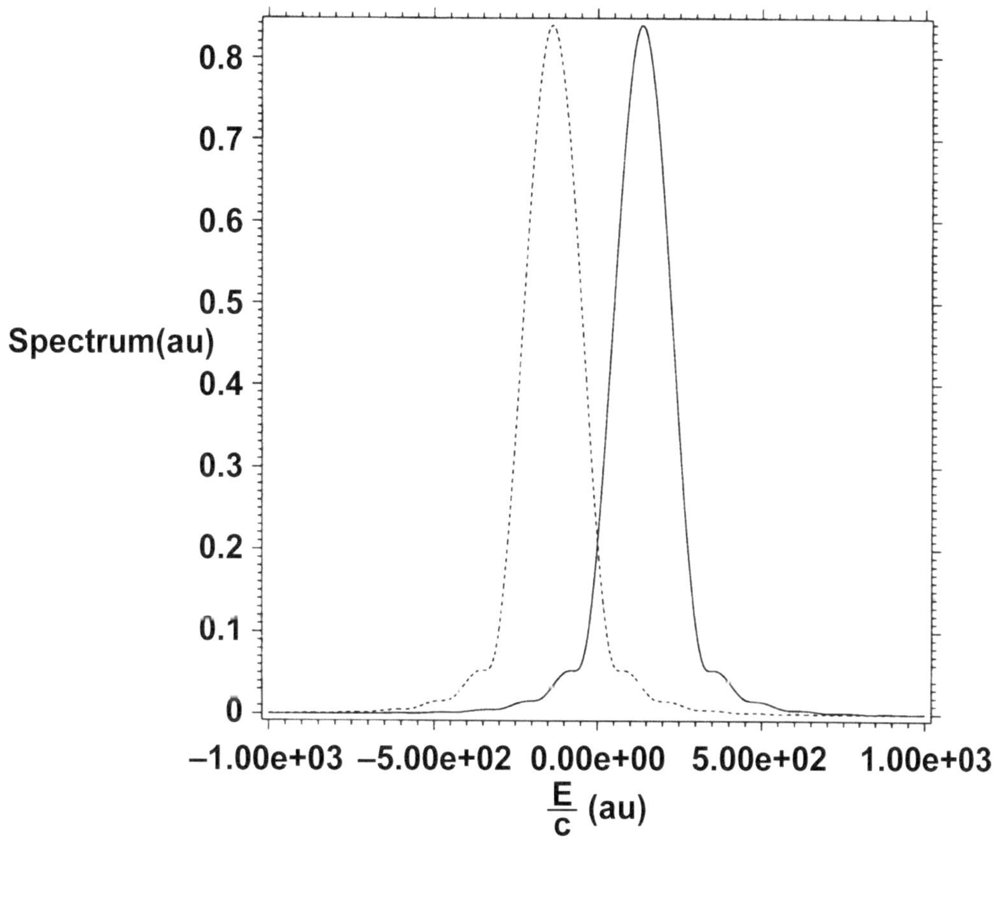

Fig. 3. Spectra from the solution of Eq. (21) using r_{max}=0.1 au. Solid: positive charge. Dashed: negative charge. The continuum edges are at $E/c = \pm mc$ au. The energy is obtained by multiplying the graphical numbers by c.

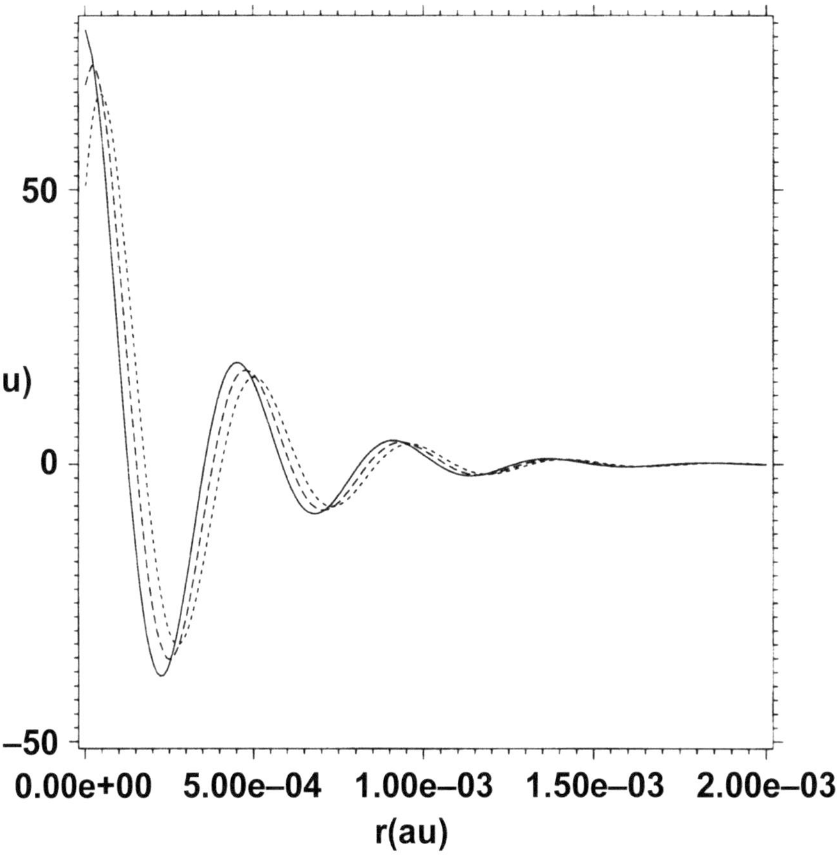

Fig. 4. Real part of the positronic or electronic solution of Eq. (21) times r at ct=0.0375 (solid), 0.0750 (dashed), and 0.1125 (dotted) au showing the convergence to a stationary solution. The initial wave function, which s hydrogenic and spread out in the domain $0.25x10^{-5}<r<0.2$ au, is pulled into the origin as shown in the figure.

Figs. 5-6 show plots of the real part of f and of the real and imaginary parts of g respectively for r=ct and $\omega_r = -92\frac{mc^2}{\hbar}$, and $\omega_i = 35\frac{mc^2}{\hbar}$. Except for the behavior near the origin the unnormalized solution of Eq. (23) is a good mimic of the solution of Eq. (21) shown in Fig. 3. Remarkably the bound positronic-electronic states in the nonadiabatic regime (Fig. 4) exhibit an altogether different form of binding than that of Schroedinger or time-independent Dirac theory. This is obvious from the spectrum (Fig. 3), in which the energies lie at the edges of the positive-and negative-energy continua. One may understand this form of binding as binding which satisfies the four-space Lorentz-invariant relationship $r^2 - (ct)^2 = 0$ between position and time . In other words the binding can occur as a temporal exponential decay in which ct = r rather than as a spatial exponential decay requiring eigenvalues which fall somewhere in the gap between the two continua. This point is clear fromFigs. 5-6 in which binding occurs in the temporal part of the function f(r,t) (Fig. 5) while the radial function g(r) is unbound (Fig. 6).

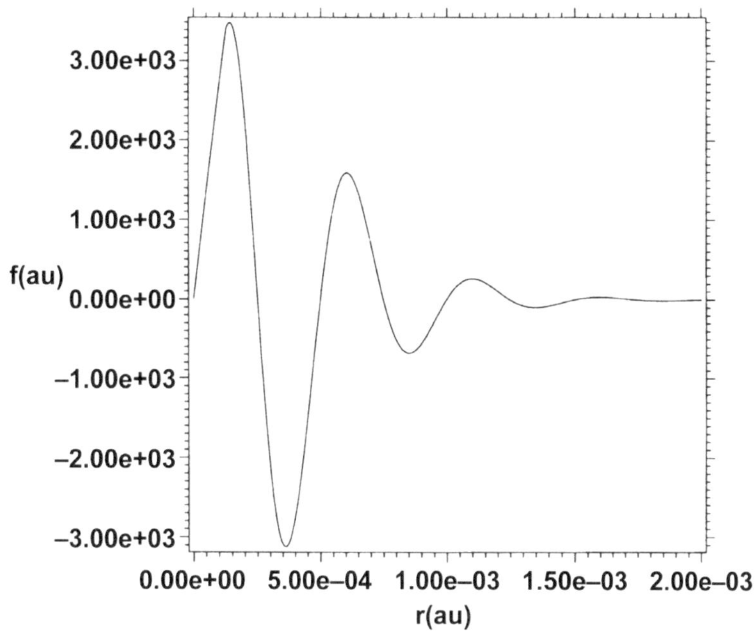

Fig. 5. Simulation using Eqs. (22)-(23) of the wave function shown in Fig. 4. The simulated wave function is unnormalized.

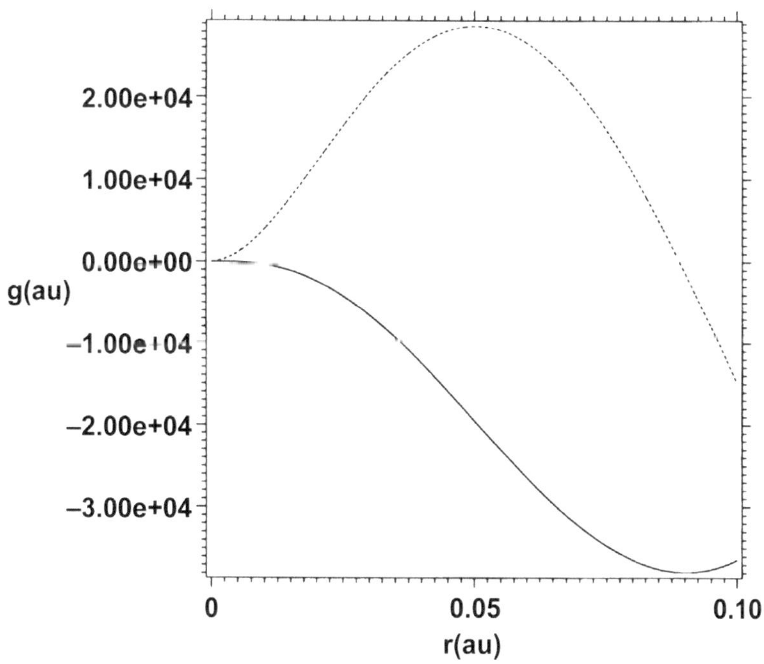

Fig. 6. Unnormalized wave function obtained from Eq. (23) by outward integration. Solid: real part. Dashed: imaginary part.

7. Acknowledgements

The author is grateful to T. Scott Carman for supporting this work. This work was performed under the auspices of the Lawrence Livermore National Security, LLC, (LLNS) under Contract No. DE-AC52-07NA27344.

8. References

[1] J. D. Bjorken and Sidney D. Drell, Relativistic Quantum Mechanics (McGraw Hill, New York, 1964), Chapter 2.
[2] C. G. Darwin, Proc. Roy. Soc. 118, 657 (1928).
[3] O. Laporte and G, Uhlenbeck, Phys. Rev. 37, 1380 (1931).
[4] R. Armour, Jr. Found. Phys. 34, 815 (2004) and references therein.
[5] B. Ritchie, Optics Communications 262, 229 (2006).
[6] H. Margenau, Phys. Rev. 46, 107 (1934).
[7] B. Ritchie, J. Mod. Optics, 55, 2903 (2008).
[8] R. Ritchie, Int. J. Quantum Chem. 111, 1 (2011).
[9] Richard Feynman, Quantum Electrodynamics (Benjamin, New York, 1962), p.37.
[10] For example Op. Cit. 1, p.
[11] M. D. Feit, J. A. Fleck, Jr. and A. Steiger, J. Comput. Phys. 47, 412 (1982).
[12] B. Ritchie, Phys. Rev. B 75, 052101 (2007).
[13] E. Schroedinger, Sitzungb. Preuss. Akad. Wiss. Ohys.-Math Kl, 24, 418 (1930).
[14] B. Ritchie, Optics Communications 280, 126 (2007).
[15] B. Ritchie, Optics Communications 282, 3286 (2009).
[16] B. Ritchie, Optics Communications 281, 3492 (2008).

4

Gouy Phase and Matter Waves

Irismar G. da Paz[1], Maria C. Nemes[2] and José G. P. de Faria[3]
[1]*Departamento de Física, Universidade Federal do Piauí e Universidade Federal de Minas Gerais*
[2]*Departamento de Física, Universidade Federal de Minas Gerais*
[3]*Departamento de Física e Matemática, Centro Federal de Educação Tecnológica de Minas Gerais*
Brazil

1. Introduction

Schrödinger conceived his wave equation having in mind de Broglie's famous relation from which we learnt to attribute complementary behavior to quantum objects depending on the experimental situation in question. He also thought of a wave in the sense of classical waves, like electromagnetic waves and others. However, the space-time asymmetry of the equation with governs quantum phenomena lead the scientific community to investigate the new physics this specific wave was about to unveil. It turns out that in certain experimental condition classical light has its behavior dictated by a bidimensional Schrödinger equation for a free particle. This fact is well known for several years (Yariv, 1991; Snyder & Love, 1991; Berman, 1997; Marte & Stenholm, 1997). For this special kind of waves it is possible to define the analog of a Hilbert space and operators which do not commute (as reviewed in section 2) in such a way that the mathematical analogy becomes perfect. A natural question emerging in this context, and the case of the present investigation is the following: how far, in the sense of leaning new physics, can we take this analogy ?

We have been able to show that the generalized uncertainty relation by Robertson and Schrödinger, naturally valid for paraxial waves, can shed new light on the physical context of a beautiful phenomenon, long discovered by Gouy (Gouy, 1890; 1891) which is an anomalous phase that light waves suffer in their passage by spatial confinement. This famous phase is directly related to the covariance between momentum and position and since for the "free particles" we are considering $\sigma_{xx}\sigma_{pp} - \sigma_{xp}^2 = constant$ we see that Gouy phase can be indirectly measured from the coordinate and momentum variances, quantities a lot easier to measure than covariance between x and p. On the other hand, as far as free atomic particles are concerned, experiments elaborated to test the uncertainty relation (Nairz et al., 2002) will reveal to us the matter wave equivalence of Gouy phase. Unfortunately the above quoted experiment was not designed to determine the phase and that is the reason why, so far, we have only an indirect evidence of the compatibility of theory and experiment. The last aim of our research is to try to encourage laboratories with facilities involving microwave cavities and atomic beams to perform an experiment to obtain the Gouy phase for matter waves.

We believe that Gouy phase for matter waves could have important applications in the field of quantum information. The transversal wavefunction of an atom in a beam state can be treated not only as a continuous variable system, but also as an infinite-dimensional discrete system.

The atomic wavefunction can be decomposed in Hermite-Gaussian or Laguerre-Gaussian modes in the same way as an optical beam (Saleh & Teich, 1991), which form an infinite discrete basis. This basis was used, for instance, to demonstrate entanglement in a two-photon system (Mair et al., 2001). However, it is essential for realizing quantum information tasks that we have the ability to transform the states from one mode to another, making rotations in the quantum state. This can be done using the Gouy phase, constructing mode converters in the same way as for light beams (Allen et al., 1992; Beijersbergen et al., 1993). In a recent paper is discussed how to improved electron microscopy of magnetic and biological specimens using a Laguere-Gauss beam of electron waves which contains a Gouy phase term (McMorran et al., 2011).

2. Analogy between paraxial equation and Schrödinger equation

One of the main differences in the dynamical behavior of electromagnetic and matter waves relies in their dispersion relations. Free electromagnetic wave packets in vacuum propagate without distortions while, *e.g.*, an initially narrow gaussian wave function of a free particle tends to increase its width indefinitely. However, the paraxial approximation to the propagation of a light wave in vacuum is formally identical to Schrödinger's equation. In this case they are bound to yield identical results.

We start our analysis by taking the simple route of a direct comparison between the Gaussian solutions of the paraxial wave equation and the two-dimensional Schrödinger equation.

Consider a stationary electric field in vacuum

$$E(\vec{r}) = A(\vec{r}) \exp(ikz). \tag{1}$$

The paraxial approximation consists in assuming that the complex envelope function $A(\vec{r})$ varies slowly with z such that $\partial^2 A/\partial z^2$ may be disregarded when compared to $k\partial A/\partial z$. In this condition, the approximate wave equation can be immediately obtained and reads (Saleh & Teich, 1991)

$$\left(\frac{\partial^2}{\partial x^2} + \frac{\partial^2}{\partial y^2} + i4\pi \frac{1}{\lambda_L} \frac{\partial}{\partial z} \right) A\left(x, y, z\right) = 0, \tag{2}$$

where λ_L is the light wavelength.

Consider now the two-dimensional Schrödinger equation for a free particle of mass m

$$\left(\frac{\partial^2}{\partial x^2} + \frac{\partial^2}{\partial y^2} + 2i \frac{m}{\hbar} \frac{\partial}{\partial t} \right) \psi(x, y, t) = 0. \tag{3}$$

Here, $\psi(x, y, t)$ stands for the wave function of the particle in time t. Assuming that the longitudinal momentum component p_z is well-defined (Viale et al., 2003), i.e., $\Delta p_z \ll p_z$, we can consider that the particle's movement in the z direction is classical and its velocity in this direction remains constant. In this case one can interpret the time variation Δt as proportional to Δz, according to the relation $t = z/v_z$. Now using the fact that $\lambda_P = h/p_z$ and substituting in Equation (3) we get

$$\left(\frac{\partial^2}{\partial x^2} + \frac{\partial^2}{\partial y^2} + i4\pi \frac{1}{\lambda_P} \frac{\partial}{\partial z} \right) \psi\left(x, y, t = z/v_z\right) = 0, \tag{4}$$

where λ_P is the wavelength of particle. As we can see the Equations (2) and (4) are formally identical.

The analogy between classical light waves and matter waves is more apparent if we use the formalism of operators in the classical approach introduced by Stoler (Stoler, 1981). In this formalism, the function $A(x,y,z)$ is represented by the ket vector $|A(z)\rangle$. If we take the inner product with the basis vectors $|x,y\rangle$, we obtain $A(x,y,z) = \langle x,y|A(z)\rangle$. The differential operators $-i(\partial/\partial x)$ and $-i(\partial/\partial y)$ acting on the space of functions containing $A(x,y,z)$ are represented in the space of abstract ket by the operators $\hat{k}_x$ and $\hat{k}_y$. The algebraic structure of operators $\hat{k}_x$, $\hat{k}_y$, $\hat{x}$ and $\hat{y}$ is specified by the following commutation relations

$$[\hat{x},\hat{k}_x] \equiv \hat{x}\hat{k}_x - \hat{k}_x\hat{x} = i, \quad [\hat{y},\hat{k}_y] = i, \quad [\hat{x},\hat{y}] = [\hat{x},\hat{k}_y] = [\hat{y},\hat{k}_x] = 0. \tag{5}$$

2.1 The generalized uncertainty relation for light waves

The analogy between the above equations in what concerns the uncertainty relation can be immediately constructed given the formal analogy between the equations.

Consider the plane wave expansion of the normalized wave $u(x,t)$ in one dimension (Jackson, 1999)

$$u(x,t) = \frac{1}{\sqrt{2\pi}} \int dk_x \; A(k_x) e^{i[k_x x - \omega(k_x)t]}. \tag{6}$$

The amplitudes $A(k_x)$ are determined by the Fourier transform of the $u(x,0)$ ($t = 0$ for simplicity)

$$A(k_x) = \frac{1}{\sqrt{2\pi}} \int dx \; u(x,0) e^{-ik_x x}. \tag{7}$$

The averages of functions $f(x,k_x)$ of x and k_x are evaluated as (Stoler, 1981)

$$\langle f(x,k_x)\rangle = \int dx \; u^*(x,0) f_s\left(x, -i\frac{\partial}{\partial x}\right) u(x,0), \tag{8}$$

in complete analogy with quantum mechanics. The function $f_s\left(x,-i\frac{\partial}{\partial x}\right)$ is obtained from $f(x,k_x)$ substituting the c-number variable k_x by the operator $-i\frac{\partial}{\partial x}$ followed by symmetric ordering. For example, if $f(x,k_x) = xk_x$, then $f_s(x,-i\frac{\partial}{\partial x}) = -\frac{i}{2}\left(x\frac{\partial}{\partial x} + \frac{\partial}{\partial x}x\right)$. Thus, we can write the variances

$$\sigma_{xx} = \langle \hat{x}^2\rangle - \langle \hat{x}\rangle^2, \tag{9}$$

$$\sigma_{k_x k_x} = \langle \hat{k}_x^2\rangle - \langle \hat{k}_x\rangle^2, \tag{10}$$

and the covariance

$$\sigma_{xk_x} = -\frac{i}{2} \int dx \; u^*(x) \left(x\frac{\partial}{\partial x} + \frac{\partial}{\partial x}x\right) u(x) - \langle \hat{x}\rangle\langle \hat{k}_x\rangle, \tag{11}$$

and get

$$\sigma_{xx}\sigma_{k_x k_x} - \sigma_{xk_x}^2 \geq \frac{1}{4}. \tag{12}$$

Equation (12) is the equivalent of generalized Schrödinger-Robertson uncertainty relation but for paraxial waves. It is also true in this context that the evolution given by Equation (2) preserves this quantity. This fact allows us to experimentally assess the covariance σ_{xk_x} by the measurements of σ_{xx} and $\sigma_{k_x k_x}$, which are quite simple to perform. Moreover, as we show next, σ_{xk_x} is directly related to the Rayleigh length and Gouy phase.

Next, we show one important result which is a consequence of this analogy - the Gouy phase for matter waves. The free time evolution of an initially Gaussian wave packet

$$\psi\left(x,y,0\right)=\left(\frac{1}{b_0\sqrt{\pi}}\right)\exp\left(-\frac{x^2+y^2}{2b_0^2}\right), \tag{13}$$

according to Schrödinger's equation is given by (da Paz, 2006)

$$\psi\left(x,y,t\right)=\left[\frac{1}{B\left(t\right)\sqrt{\pi}}\right]\exp\left(-\frac{x^2+y^2}{2B^2\left(t\right)}\right) \times \exp\left\{i\left[\frac{m\left(x^2+y^2\right)}{2\hbar R\left(t\right)}+\mu\left(t\right)\right]\right\}. \tag{14}$$

The comparison with the solution of the wave equation in the paraxial approximation with the same condition at $z=0$ yields

$$w\left(z\right)\longrightarrow B\left(t\right)=b_0\left[1+\left(\frac{t}{\tau_0}\right)^2\right]^{\frac{1}{2}}, \tag{15}$$

$$R\left(z\right)\longrightarrow R\left(t\right)=t\left[1+\left(\frac{\tau_0}{t}\right)^2\right], \tag{16}$$

$$\zeta\left(z\right)\longrightarrow\mu\left(t\right)=-\arctan\left(\frac{t}{\tau_0}\right), \tag{17}$$

and

$$z_0\longrightarrow\tau_0=\frac{mb_0^2}{\hbar}. \tag{18}$$

The parameter $B(t)$ ($w(z)$) is the width of the particle beam (of light beam), the parameter $R(t)$ ($R(z)$) is the radius of curvature of matter wavefronts (wavefront of light), $\mu(t)$ ($\zeta(z)$) is the Gouy phase for matter waves (for light waves). The parameter τ_0 is only related to the initial condition and is responsible for two regimes of growth of the beam width $B(t)$ (da Paz, 2006; Piza, 2001), in complete analogy with the Rayleigh length which separates the growth of the beam width $w(z)$ in two different regimes as is well known in optics (Saleh & Teich, 1991).
The above equations show that the matter wave propagating in time with fixed velocity in the propagation direction and the stationary electric field in the paraxial approximation are formally identical [if one replaces $t=z/v_z$ in the Equations (15–17)].
Next we show that $\mu(t)$ is directly related to the Schrödinger-Robertson generalized uncertainty relation. For quadratic unitary evolutions (as the free evolution in the present case) the determinant of the covariance matrix is time independent and for pure Gaussian states saturates to its minimum value,

$$\det\begin{pmatrix}\sigma_{xx} & \sigma_{xp}\\ \sigma_{xp} & \sigma_{pp}\end{pmatrix}=\frac{\hbar^2}{4} \tag{19}$$

where

$$\sigma_{xx}=\frac{B(t)^2}{2},\ \sigma_{pp}=\frac{\hbar^2}{2b_0^2}, \tag{20}$$

and

$$\sigma_{xp} = \frac{\hbar t}{2\tau_0} = -\frac{\hbar}{2}\tan 2\mu(t). \tag{21}$$

Since the covariance σ_{xp} is non-null if the Gaussian state exhibits squeezing (Souza et al., 2008), if one measures σ_{xp}, from the above relation it is possible to infer the Gouy phase for a matter wave which can be described by an evolving coherent wave packet. For light waves this is a simple task as can be seen below.

2.2 The Gouy phase for light waves

The generalized uncertainty relation for the Gaussian light field can be immediately obtained. Indeed the variances

$$\sigma_{xx} = \frac{w^2(z)}{4}, \tag{22}$$

$$\sigma_{k_x k_x} = \frac{k}{2z_0}, \tag{23}$$

$$\sigma_{xk_x} = \frac{z}{2z_0} = -\frac{1}{2}\tan 2\zeta(z), \tag{24}$$

satisfy the equality

$$\sigma_{xx}\sigma_{k_x k_x} - \sigma_{xk_x}^2 = \frac{1}{4}. \tag{25}$$

Analogue expressions can be found for the second moments of the y transverse component. The saturation at the value $1/4$ allows for the determination of the covariance σ_{xk_x}. From Equation (25) and using the expressions (22) and (23) we get

$$\sigma_{xk_x}(z) = \pm\frac{1}{2}\sqrt{\left[\frac{w(z)}{w_0}\right]^2 - 1}, \tag{26}$$

which is a function of z/z_0 just like expression (17) for the Gouy phase.
The connection between the Gouy phase and the covariance σ_{xk_x} is of purely kinematical nature. As pointed by Simon and Mukunda (Simon & Mukunda, 1993), the parameter space of the gaussian states has a hyperbolic geometry, and the Gouy phase has a geometrical interpretation related to this geometry.
Note that σ_{xk_x} can be positive or negative according to the Equation (26). However, the Equation (26) was deduced assuming that the focus of the beam is $z = 0$. If we shift the focus to any position z_c, as in the experiment, we must take this into account. The plus and minus sign in Equation (26) can be better understood if we look at the Equation (24)

$$\sigma_{xk_x} = \frac{z}{2z_0} \rightarrow \sigma_{xk_x} = \frac{z - z_c}{2z_0}, \tag{27}$$

which agrees with the experimental data as we show in what follows. Here we can see that for light waves propagating in the direction of focus ($z < z_c$) the covariance is negative, on the other hand, for light waves propagating after focus ($z > z_c$) the covariance is positive.
Now note Equation (26) suggests that by measuring the beam width $w(z)$ we can indirectly infer the value of σ_{xk_x} and thus the value of the Gouy phase by Equation (24). Next, we describe a simple experiment to measure $w(z)$. To experimentally obtain the beam width as a function of the propagation distance, we use the following experimental arrangement shown in Figure 1 (Laboratory of Quantum Optics at UFMG), where L_1 represents a divergent lens,

L_2 a convergent lens and D is a light detector. With this arrangement we can measure the width of the beam as a function of z. The width of the beam in position z is the width of the intensity curve, adjusted by a Gaussian function. In Figure 2, we show the width of the beam for different distances z, along with the corresponding result for σ_{xk_x}.

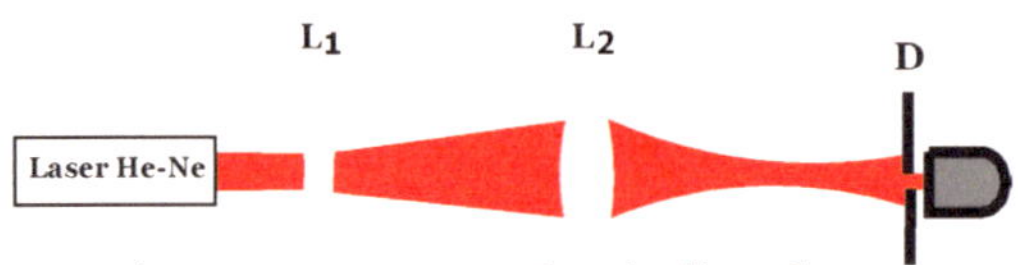

Fig. 1. Sketch of experimental arrangement used to indirectly measure the Gouy phase of a focused light beam.

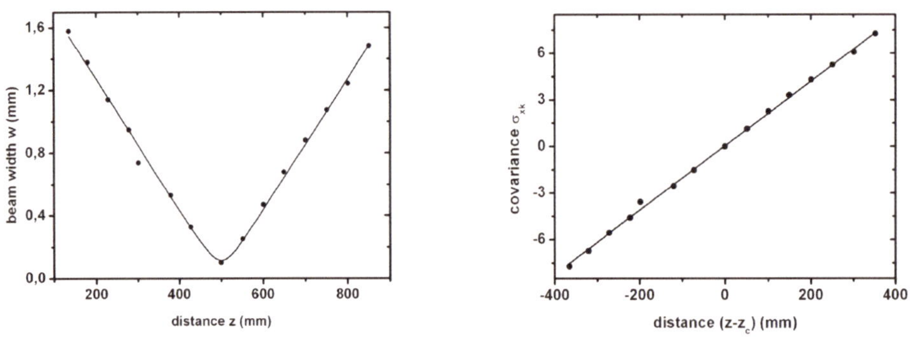

Fig. 2. On the left, the width of Gaussian beam $w(z)$ as a function of propagation direction z. Solid curve corresponds to the Equation (15) and the points were obtained of experiment. On the right, covariance σ_{xk_x} as a function of $z - z_c$. Solid curve corresponds to the Equation (27) and the points were obtained of experiment through the equation (26).

The determination of σ_{xk_x} or $w(z)$ allows us to determine $\xi(z)$ (see Figure 3).

3. Macromolecules diffraction and indirect evidence for the Gouy phase for matter waves

Recent experiments involving the diffraction of fullerene molecules and the uncertainty relation are shown to be quantitatively consistent with the existence of a Gouy phase for matter waves (da Paz et al., 2010). In Ref. (Nairz et al., 2002) an experimental investigation of the uncertainty relation in the diffraction of fullerene molecules is presented. In that experiment, a collimated molecular beam crosses a variable aperture slit and its width is measured as a function of the slit width. Before reaching the slit diffraction the molecular beam passes through a collimating slit whose width is fixed at $\sigma_0 = 10\ \mu m$, producing a correlated beam (see Figures 1 and 3 in Ref. (Nairz et al., 2002)).
The wave function of the fullerene molecules that leave the slit of width b_0, in the transverse direction, is given by

$$\psi_{k_x}(\bar{x},0) = \frac{1}{\sqrt{b_0\sqrt{\pi}}}\exp\left(-\frac{\bar{x}^2}{2b_0^2} + ik_x\bar{x}\right), \tag{28}$$

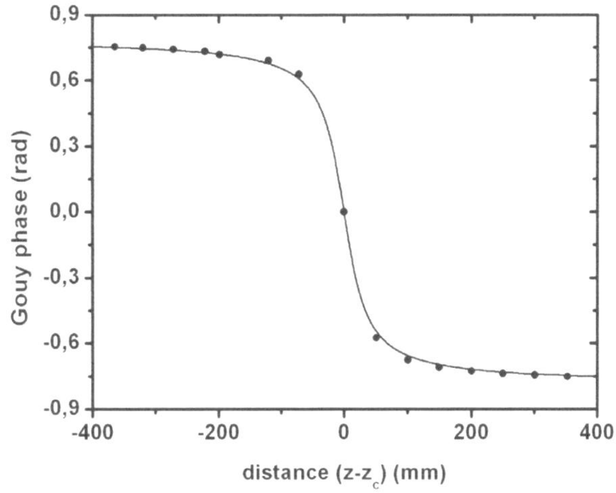

Fig. 3. Gouy phase for Gaussian light beam as a function of propagation direction $z - z_c$. Solid curve corresponds to the Equation (17) and the points were obtained of experiment through the Equation (24).

where k_x is the transverse wave number. The wave function on the screen is given by

$$\psi_{k_x}(x,t) = \int d\bar{x} G(x,t;\bar{x},0)\psi_{k_x}(\bar{x},0), \tag{29}$$

where

$$G(x,t;\bar{x},0) = \left(\frac{m}{2\pi i\hbar t}\right)^{\frac{1}{2}} \exp\left[\frac{im}{2\hbar t}(x-\bar{x})^2\right], \tag{30}$$

and $t = z/v_z$ is the propagation time from slit to detector, v_z is the most probable speed on the z direction. After some algebraic manipulations we obtain, for the normalized wave function at the detector, the following result

$$\psi_{k_x}(x,t) = \frac{1}{\sqrt{\sqrt{\pi}B(t)}} \exp\left[-\frac{1}{2B^2(t)}\left(x - \frac{b_0^2 t}{\tau_0}k_x\right)^2\right] \times \exp\left[\frac{i\mu(t)}{2} + \frac{im}{2\hbar R(t)}\left(x^2 - b_0^4 k_x^2 + \frac{2mb_0^4}{\hbar t}xk_x\right)\right], \tag{31}$$

where $B(t)$, $R(t)$ and $\mu(t)$ are given by the Equations (15), (16) and (17), respectively.
As discussed in Ref. (Viale et al., 2003), given the way the fullerene molecules are produced, it is reasonable to assume that the outgoing beam after the diffraction slit has a random transverse momentum. Due to the thermal production the beam contains different components k_x, although it has been collimated (Viale et al., 2003). The beam is an incoherent mixture of wave functions with wavenumber k_x randomly distributed according to probability distribution $\tilde{\mathbf{g}}^{(0)}(k_x)$. This distribution depends on the geometry of the collimator, secondary source, which reduces the beam width in the direction x. The index 0 represents the plane of the secondary source (the plane of the collimator), which means that the loss

of coherence of the beam is due to the production mechanism only. It is not physically reasonable to assume that a coherent wave packet leaves the diffraction slit due to the thermal production of the fullerene molecules as discussed above. Therefore, in order to introduce some incoherence along the spatial transverse direction, where the quantum effects occur, we use the formalism of density matrices (Gase, 1994; Ballentine, 1998; Scully & Zubairy, 1997; Fano, 1957). The density matrix of the beam at time t is given by

$$\rho(x,x',t) = \int dk_x \widetilde{\mathbf{g}}^{(0)}(k_x)\psi_{k_x}(x,t)\psi^*_{k_x}(x',t). \tag{32}$$

For simplicity, let us take a probability distribution of wave number k_x be a Gaussian function centered at $k_x = 0$ and width $\Delta k_x = \delta_{k_x}/\sqrt{2}$, i.e.,

$$\widetilde{\mathbf{g}}^{(0)}(k_x) = \frac{1}{\sqrt{\pi}\delta_{k_x}} \exp\left(-\frac{k_x^2}{\delta_{k_x}^2}\right). \tag{33}$$

This allows us to obtain for the density matrix Equation (32), the following result

$$\rho(x,x',t) = \frac{1}{\sqrt{\pi}\bar{B}(t)} \exp\left[-\frac{(x+x')^2 + \mathcal{M}_P^4(x-x')^2}{4\bar{B}^2(t)}\right] \exp\left[\frac{im}{2\hbar\bar{R}(t)}(x^2 - x'^2)\right], \tag{34}$$

where

$$\bar{B}(t) = b_0\left[1+\left(\frac{t}{\bar{\tau}_0}\right)^2\right]^{\frac{1}{2}}, \quad \bar{R}(t) = t\left[1+\left(\frac{\bar{\tau}_0}{t}\right)^2\right], \tag{35}$$

$$\bar{\tau}_0 = \mathcal{M}_P^{-2}\tau_0, \quad \mathcal{M}_P^2 = \sqrt{1+b_0^2\delta_{k_x}^2}. \tag{36}$$

We observe that the density matrix Equation (34) is a mixed state due to the incoherence of the source. The bar has been used to differentiate the parameters of the pure Gaussian state of matter waves of the respective parameters from a mixed Gaussian state. The quantity $\mathcal{M}_P^2$ is the quality factor of the particle beam. The quantity $\bar{\tau}_0$ is a generalization of the definition of time aging (Piza, 2001) (timescale) for partially coherent Gaussian state of matter waves. We see that this quantity is always smaller than the aging time of Gaussian pure states, τ_0, and in this case, a mixed Gaussian state will spread faster with time than the pure Gaussian states. In the coherent limit $\delta_{k_x} \to 0$ (ideal collimation), we obtain the parameters of pure Gaussian state, Equations (15), (16) and (17).
In the limit $t \to 0$ (the plane of source), we have

$$\rho^{(0)}(x,x') = \frac{1}{\sqrt{\pi}b_0} \exp\left[-\frac{(x^2+x'^2)}{2b_0^2}\right] \exp\left[-\frac{\delta_{k_x}^2}{4}(x-x')^2\right], \tag{37}$$

where the last exponential term of this equation make the role of the spectral degree of coherence defined in the theory of optical coherence (Mandel & Wolf, 1995). We see that the dependence of this term with the transverse position appears as the difference between the positions and, in this case, the source of fullerenes is a source of type Schell (Mandel & Wolf, 1995). Again, the source of fullerenes we refer to here is the collimation slit and not the oven. With the density matrix, we obtain the intensity at the detector by using $x = x'$ e $t = z/v_z$, i.e.,

$$I(x,t) = \rho(x,x,t) = \frac{1}{\sqrt{\pi}\bar{B}(t)} \exp\left[-\frac{x^2}{\bar{B}^2(t)}\right]. \tag{38}$$

Next, we calculate the new elements of the covariance matrix and obtain the following results

$$\sigma_{xx} = \langle \hat{x}^2 \rangle = \int dx x^2 \rho(x,x,t) = \frac{\bar{B}(t)}{2}, \tag{39}$$

$$\sigma_{pp} = \langle \hat{p}^2 \rangle = \int \left[\int dx \psi^*_{k_x}(x',t)(-\hbar^2 \frac{\partial^2}{\partial x^2}) \psi_{k_x}(x,t) \right]_{x=x'} \widetilde{\mathbf{g}}^{(0)}(k_x) dk_x = \frac{\hbar^2}{2} \frac{\mathcal{M}_P^4}{b_0^2}, \tag{40}$$

and

$$\sigma_{xp} = \langle \frac{\hat{x}\hat{p} + \hat{p}\hat{x}}{2} \rangle = \int \left[\int dx \psi^*_{k_x}(x',t) x(-i\hbar \frac{\partial}{\partial x}) \psi_{k_x}(x,t) \right]_{x=x'} \widetilde{\mathbf{g}}^{(0)}(k_x) dk_x = \frac{\hbar}{2} \mathcal{M}_P^2 \left(\frac{t}{\bar{\tau}_0} \right). \tag{41}$$

With these new elements, we obtain the following result for the determinant of covariance matrix

$$\det \begin{pmatrix} \sigma_{xx} & \sigma_{xp} \\ \sigma_{xp} & \sigma_{pp} \end{pmatrix} = \mathcal{M}_P^4 \frac{\hbar^2}{4}. \tag{42}$$

This result shows that the determinant of the covariance matrix remains time independent, but has a different value from $\frac{\hbar^2}{4}$, because now we have an incoherent state.
The experimental result for the width W_{FWHM} (full width at half maximum) at the detector, realized by the group of A. Zeilinger in Ref. (Nairz et al., 2002) is shown in Figure 4 and compared with our theoretical calculation, Equation (39) (where $W_{FWHM} = 2\sqrt{2\ln 2}\sigma_{xx}$). The points are experimental data extracted from Ref. (Nairz et al., 2002), the dashed curve is the beam width with incoherence effect and without convolution with the detector and the solid curve takes into account both effects. These curves show that to adjust the experimental points with theoretical model, we take into account the convolution with the detector and the partial coherence of the fullerenes source. To take into account the convolution with the detector, we use a detector width $FWHM$ of order of 12 μm, where we took as reference the value quoted in (Nairz et al., 2002). The parameter that measures the partial coherence in the transverse direction of the beam that best fits the experimental data is given by $\delta_{k_x} = 9.0 \times 10^6$ m^{-1}.
With this value of δ_{k_x} we calculate the initial transverse coherence length, i.e., $\ell_{0x} = \ell_x(t=0)$ and we obtain $\ell_{0x} = (\delta_{k_x}/\sqrt{2})^{-1} \approx 1.3 \times 10^{-7}$ m. As we do not take into account the coupling with the environment in our model, the initial coherence length remains constant in time, i.e., $\ell_x(t) = \ell_{0x}$. To compare the value of the coherence length with the value of the wavelength, we calculate λ_P through the equation $\lambda_P \approx \lambda_z = h/mv_z$ (where $v_z \approx 200$ m/s is the most probable speed) and we obtain $\lambda_P \approx 2.5$ pm. Thus, we have $\ell_{0x} \gg \lambda_P$, and the condition discussed in Ref. (Mandel & Wolf, 1995) for a locally coherent source is guaranteed. Because the source size is much larger than transverse coherence length, i.e., $\sigma_0 \gg \ell_{0x}$, the angle of beam divergence of fullerenes produced in the secondary source (collimation slit) is given by

$$\bar{\theta}_0 \approx \left(\frac{\lambda_P}{\pi} \right) \frac{1}{\ell_{0x}} = 6.1\ \mu rad, \tag{43}$$

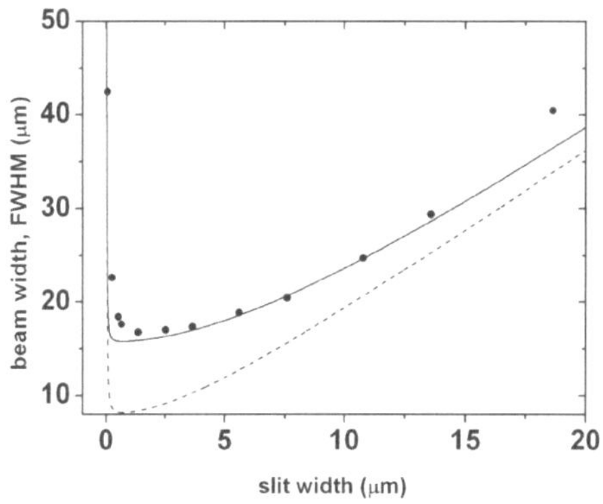

Fig. 4. Beam width of fullerene molecules C_{70} as a function of slit width. Solid and dashed curves correspond to our calculation, Equation (35), and the points are the experimental results obtained in Ref. (Nairz et al., 2002). Dashed curve corresponds to the incoherent case without convolution with the detector and solid curve corresponds to the case where both effects were taken into account. To adjust the theoretical calculation with the experimental data we use $\delta k_x = 9.0 \times 10^6\ \mathrm{m}^{-1}$ and $t = z/v_z = 6.65\ \mathrm{ms}$.

a value consistent with the experimental value quoted in Ref. (Nairz et al., 2000) ($2 \leq \theta \leq 10\ \mu rad$).
The range of wavelengths along the direction x is given by

$$\Delta\lambda_x = \frac{2\pi}{\Delta k_x} = 986\ \mathrm{nm}, \tag{44}$$

where $\Delta k_x = \delta_{k_x}/\sqrt{2} = 6.4 \times 10^6\ \mathrm{m}^{-1}$.
The value obtained for the range of wavelengths is the same order of magnitude of the transverse coherence length ℓ_{0x}, what justifies the existence of quantum effects along this direction. The component of the wave vector in the direction z has the value $k_z = mv_z/\hbar \approx 2.24 \times 10^{12}\ \mathrm{m}^{-1}$. The values found for k_z and Δk_x show that $k_z \gg \Delta k_x$ and thus, paraxial approximation is guaranteed for the partially coherent matter wave beam.

3.1 Covariance σ_{xp} and Gouy phase

In this section, we calculate the covariance between position and momentum and the Gouy phase for fullerenes molecules considering the free Schrödinger equation. We calculate the phase and show that it is also related to the covariance σ_{xp} as well as in the case of pure Gaussian states.
Starting from the determinant of the covariance matrix for mixed Gaussian state, Equation (42), we can express σ_{xp} in terms of the beam width, i.e.,

$$\sigma_{xp} = \frac{\hbar \mathcal{M}_P^2}{2}\left[\left(\frac{W_{FWHM}}{2\sqrt{\ln 2}b_0}\right)^2 - 1\right]^{\frac{1}{2}}, \tag{45}$$

where W_{FWHM} is measured in the laboratory. The curve for σ_{xp}, obtained with experimental data of the Ref. (Nairz et al., 2002) through the Equation (45), is showed in Figure 5 and compared with the theoretical value, Equation (41).

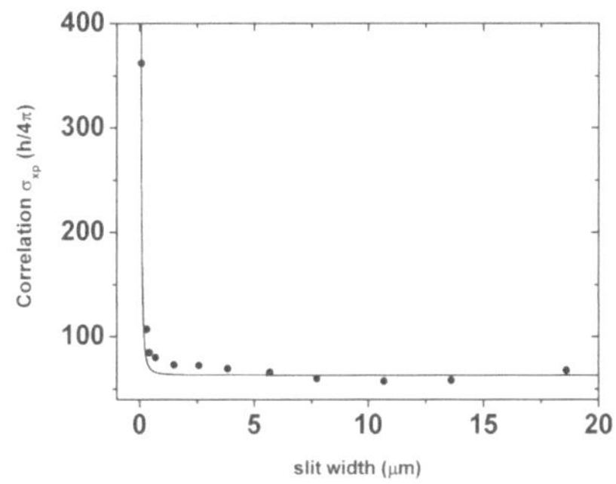

Fig. 5. Covariance σ_{xp} as a function of slit width. Solid curve corresponds to our calculation, Equation (41), and the points were obtained of experiment reported in Ref. (Nairz et al., 2002) through the Equation (45). The parameters are the same of Figure 4.

3.1.1 Gouy phase for a mixed Gaussian state

A more recent definition justifies the physical origin of the Gouy phase in terms of space enlargement, governed by the uncertainty relation, of a beam whose transverse field distribution is a Gaussian function (or arbitrary) (Feng & Winful, 2001). According to Equation (11) in Ref. (Feng & Winful, 2001) the Gouy phase $\mu(t)$ and the beam width $B(t)$ for a pure Gaussian state of matter waves are related by the expression

$$\mu(t) = -\frac{\hbar}{2m}\int^{t} \frac{dt}{B(t)^2}. \tag{46}$$

Here, we conjecture, based on the obtained results, that this definition holds for partially coherent Gaussian states since the spread of these states is also governed by the uncertainty relation. Thus, for a state given by Equation (34), the Gouy phase is

$$\mu(t) = -\frac{1}{2\mathcal{M}_p^2}\arctan\left(\frac{t}{\bar{\tau}_0}\right), \tag{47}$$

where the factor $\frac{1}{2}$ appears because we are working in one dimension. Note that, again $\mu(t)$ is related to σ_{xp} and is affected by the partial coherence of the initial wave packet, i.e.,

$$\mu(t) = -\frac{1}{2\mathcal{M}_p^2}\arctan\left(\frac{2\sigma_{xp}}{\hbar\mathcal{M}_p^2}\right). \tag{48}$$

In Figure 6, we show the phase extracted from Equation (48). As expected, the variation in phase is $\pi/4$, because we are dealing with a one-dimensional problem of diffraction and the propagation of the beam will be from $t = 0$ to $t = z/v_z$ (Feng & Winful, 2001). This result shows that the existence of a Gouy phase is compatible with the experimental data involving diffraction of fullerene molecules. It is an indirect evidence of the Gouy phase for matter waves (da Paz, 2011; da Paz et al., 2010).

4. Quantum lens and Gouy phase for matter waves

In the previous section, we have shown an indirect evidence for the Gouy phase for matter waves based on the analogy existent between the paraxial equation for wave optics and

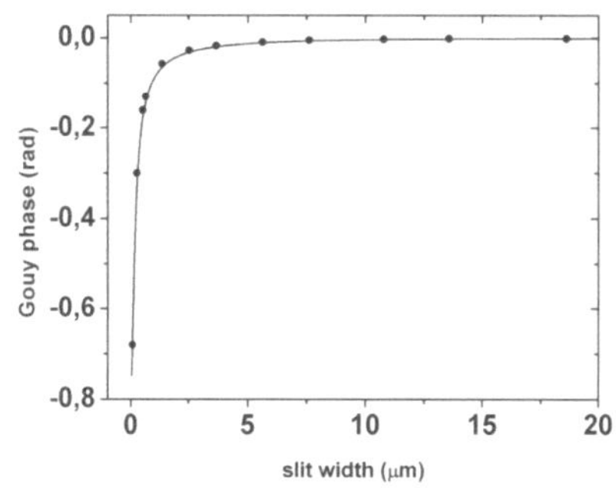

Fig. 6. Gouy phase as a function of slit width. Solid curve corresponds to our calculation, Equation (47), and the points were obtained of experiment reported in Ref. (Nairz et al., 2002). The parameters are the same of Figure 4.

Schrödinger equation for matter waves (da Paz, 2011; da Paz et al., 2010). Due to this formal similarity a question which arises naturally is if a similar phase anomaly may occur in the region around the focus of an atomic beam. In order to answer this question, in this section we present the evolution of an atomic beam described by a Gaussian wave packet interacting dispersively with a cavity field (da Paz, 2011; da Paz et al., 2007).
The model we use is the following (Averbukh etal., 1994; Rohwedder & Orszag, 1996; Schleich, 2001): consider two-level atoms moving along the Oz direction and that they penetrate a region where a stationary electromagnetic field is maintained. The region is the interval $z = -L_c$ until $z = 0$. The atomic linear moment in this direction is such that the de Broglie wavelength associated is much smaller than the wavelength of the electromagnetic field. We assume that the atom moves classically along direction Oz and the atomic transition of interest is detuned from the mode of the electromagnetic field (dispersive interaction). The Hamiltonian for this model is given by

$$\hat{H}_{AF} = \frac{\hat{p}_x^2}{2m} + g(\hat{x})\hat{a}^{\dagger}\hat{a} \tag{49}$$

where m is the atom mass, $\hat{p}_x$ and $\hat{x}$ are the linear momentum and position along the direction Ox, $\hat{a}^{\dagger}$ and $\hat{a}$ are the creation and destruction operators of a photon of the electromagnetic mode, respectively. The coupling between atom and field is given by the function $g(x) = \alpha E^2(x)$ where α is the atomic linear susceptibility, $\alpha = \frac{\wp^2}{\hbar\Delta}$, where $\wp^2$ is the square of the dipole moment and Δ is the detuning. $E(x)$ corresponds to the electric field amplitude in vacuum. The effective interaction time is $t_L = \frac{L_c}{v_z}$, where v_z is the longitudinal velocity of the atoms.
The dynamics of the closed system is governed by the Schrödinger equation

$$i\hbar\frac{d}{dt}|\Psi\rangle = \hat{H}_{AF}|\Psi\rangle. \tag{50}$$

At $t = 0$ the state of the system is given by a direct product of the state corresponding to the transversal component of the atom and a field state, $|\Psi_{cm}\rangle \otimes |\Psi_F\rangle$. The field state can be expanded in the eigenstates of the number operator $\hat{a}^{\dagger}\hat{a}$

$$|\Psi_F\rangle = \sum_n \omega_n|n\rangle, \quad \sum_n |\omega_n|^2 = 1. \tag{51}$$

When atom and field interact the atomic and field states get entangled. We can then write

$$|\Psi(t)\rangle = \sum_n \omega_n \int_{-\infty}^{+\infty} dx\, \psi_n(x,t)|x\rangle \otimes |n\rangle\,, \tag{52}$$

where

$$i\hbar\frac{\partial}{\partial t}\psi_n(x,t) = \left\{-\frac{\hbar}{2m}\nabla^2 + g(x)n\right\}\psi_n(x,t)\,, \tag{53}$$

or, if one defines

$$|\Psi_n(t)\rangle = \int_{-\infty}^{+\infty} dx\, \psi_n(x,t)|x\rangle\,, \tag{54}$$

the Equation (53) takes the form

$$i\hbar\frac{d}{dt}|\Psi_n(t)\rangle = \left[\frac{\hat{p}_x^2}{2m} + g(\hat{x})n\right]|\Psi_n(t)\rangle. \tag{55}$$

Next, we will use the harmonic approximation for $g(x)$ which is a fine approximation provided the dispersion of the wavepacket in the transverse direction b_0 is much smaller than the wavelength of the electromagnetic field mode λ (Schleich, 2001). Taking the main terms of the Taylor expansion of the function $g(x)$,

$$g(x) \approx g_0 - \frac{g_1^2}{2g_2} + \frac{1}{2}g_2\left(x - x_f\right)^2\,, \tag{56}$$

we get

$$\begin{aligned} i\hbar\frac{d}{dt}|\Psi_n(t)\rangle &= \left[\frac{\hat{p}_x^2}{2m} + \frac{1}{2}m\Omega_n^2(\hat{x} - x_f)^2\right]|\Psi_n(t)\rangle \\ &\equiv \hat{H}_n|\Psi_n(t)\rangle\,, \end{aligned} \tag{57}$$

where $x_f = -g_1/2g_2$ and $\Omega_n^2 = ng_2/m$. In order to obtain focalization of the atomic beam it is crucial that the initial state be compressed in momentum since this initial momentum compression is transferred dynamically to the x coordinate and a focus can be obtained (da Paz et al., 2007; Rohwedder & Orszag, 1996). In fact, the momentum compression is a necessary condition in optics to obtain a well defined focus (Saleh & Teich, 1991).

4.1 Time evolution

According to Bialynicki-Birula (Bialynicki-Birula, 1998), the general form of a Gaussian state in the position representation, is given by

$$\psi(x) = \left(\frac{u}{\pi}\right)^{\frac{1}{4}} \exp\left(-i\frac{\bar{x}\bar{p}}{2\hbar}\right)\exp\left[-\frac{(x-\bar{x})^2(u+iv)}{2} + i\frac{\bar{p}x}{\hbar}\right], \tag{58}$$

where $\bar{x}$ and $\bar{p}$ are the coordinates of "center of mass" of the distribution in phase space and u and v give the form of this distribution.

A dynamic governed by a Hamiltonian quadratic in position and momentum keep the Gaussian shape of a Gaussian initial state. This is the case of the problem treated here. The atomic motion can be divided into two stages: the first, the atom undergoes the action of a harmonic potential when it crosses the region of electromagnetic field while, in the second part, the atom evolves freely. In the two stages, the Hamiltonian governing the evolution are quadratic in atomic position and momentum [*cf.* Equation (57)]. Since the initial atomic state is Gaussian, we can consider that throughout evolution, such state will preserve the form given by Equation (58). In this case, the parameters $\bar{x}$, $\bar{p}$, u and v are functions of time, and their respective equations of motion can be derived from the Schrödinger equation.

Consider a particle of mass m moving under the action of a harmonic potential. The natural frequency of this movement is Ω_n. The Hamiltonian governing this dynamic is given by

$$\hat{H} = \frac{\hat{p}_x^2}{2m} + \frac{1}{2}m\Omega_n^2\hat{x}^2. \tag{59}$$

In position representation, the evolution of the state ψ of the particle is governed by the Schrödinger equation

$$i\hbar\frac{\partial}{\partial t}\psi(x,t) = \left[-\frac{\hbar^2}{2m}\frac{\partial^2}{\partial x^2} + \frac{1}{2}m\Omega_n^2 x^2\right]\psi(x,t). \tag{60}$$

Suppose that the initial state of the particle is Gaussian. We obtain the equations of motion for the parameters $\bar{x}$, $\bar{p}$, u and v by substituting the general form (58) in equation above, grouping the terms of same power in $(x-\bar{x})$, and then separating the real and imaginary parts. This procedure takes six equations for the four parameters mentioned. The system is therefore, "super-complete". Eliminating such redundancy, the equations of motion are the following

$$\dot{\bar{x}} = \frac{\bar{p}}{m}, \tag{61a}$$

$$\dot{\bar{p}} = -m\Omega_n^2\bar{x}, \tag{61b}$$

$$\dot{K} = i\frac{m\Omega_n^2}{\hbar} - i\frac{\hbar}{m}K^2, \tag{61c}$$

where we define $K = u + iv$. Here, the dots indicate time derivation. Note that the equations of motion for the coordinates of the centroid of the distribution are equivalent to the classical equations of movement to the position and momentum of a particle moving in a harmonic potential.

A important observation must be made here. One of the two equations removed is not consistent with the others in (61). This equation is the following:

$$\bar{p}\dot{\bar{x}} - \dot{\bar{p}}\bar{x} = \frac{\bar{p}^2}{m} + m\Omega_n^2\bar{x}^2 + \frac{\hbar^2}{m}u. \tag{62}$$

To see this, just replace the expressions (61a), (61b) in the above equation. We obtain $u = 0$, which makes no sense, since u represents the inverse square of the width of the Gaussian package. The only way to "dribble" this inconvenience is to redefine the general state as

$$\psi(x) = \left(\frac{u}{\pi}\right)^{\frac{1}{4}}\exp\left(-i\frac{\bar{x}\bar{p}}{2\hbar} + i\frac{\Phi}{2}\right)\exp\left[-\frac{(x-\bar{x})^2(u+iv)}{2} + i\frac{\bar{p}x}{\hbar}\right], \tag{63}$$

where Φ is a real function of time. This global phase, in general neglected (see, e.g., (Bialynicki-Birula, 1998; Piza, 2001)), ensures the consistency of the equations of motion because, in addition to Equations (61), we must have

$$\dot{\Phi} = -\frac{\hbar}{m}u. \tag{64}$$

$\Phi/2$ is known as *Gouy phase*. Equation (64) relates the Gouy phase with the inverse square of the beam width . The same result was obtained for light waves transversally confined in Ref. (Feng & Winful, 2001).

4.2 Focalization of the atomic beam

In Figure 7 we illustrated how the quantum lens work out. We consider that a initial Gaussian state compressed in the momentum (region I) penetrates in a region where a stationary electromagnetic field is maintained (region II). The atoms and the field inside the cavity interact dispersively. Dispersive coupling is actually necessary to produce a quantum lens, because the transitions cause aberration at the focus (Berman, 1997; Rohwedder & Orszag, 1996; Schleich, 2001). When the atomic beam leaves the region of the electromagnetic field, the atomic state evolves freely and the compression is transferred to the position (region III). Let us assume, as an initial atomic state, the compressed vacuum state

$$\langle x|\psi_n(t=0)\rangle = \psi_n(x,t=0) = \left(\frac{1}{b_0\sqrt{\pi}}\right)^{1/2} \exp\left(-\frac{x^2}{2b_0^2}\right), \tag{65}$$

where b_0 is the initial width of the packet and $b_0 > b_n = \sqrt{\hbar/(m\Omega_n)}$. For the parameters $\bar{x}$,

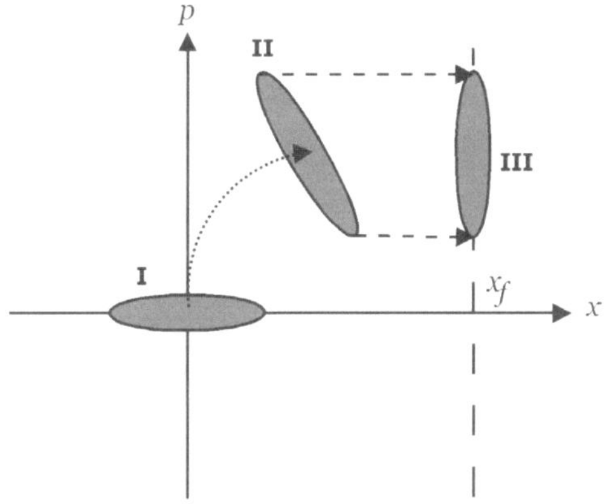

Fig. 7. Initial atomic compressed state in momentum . The evolution inside the cavity rotates the state and transfer the compression to the position.

$\bar{p}$, K and Φ, we get

$$\bar{x}(t<t_L) = -x_f \cos\Omega_n t, \tag{66}$$

$$\bar{p}(t<t_L) = m\Omega_n x_f \sin\Omega_n t, \tag{67}$$

and

$$K(t<t_L) = \left(\cos\Omega_n t + i\frac{b_n^2}{b_0^2}\sin\Omega_n t\right)^{-1}\left(\frac{1}{b_0^2}\cos\Omega_n t + i\frac{1}{b_n^2}\sin\Omega_n t\right), \tag{68}$$

for the initial conditions $\bar{x}_0 = -x_f, \bar{p}_0 = 0, u = b_0^{-2}$ and $v = 0$. Also, from Equation (68) we obtain

$$u(t<t_L) = \left[b_0^2\left(\cos^2\Omega_n t + \frac{b_n^4}{b_0^4}\sin^2\Omega_n t\right)\right]^{-1}. \tag{69}$$

Now u^{-1} is the width of the gaussian wavepacket squared. At this stage

$$\Phi(t < t_L) = -\frac{1}{\Omega_n \tau_n} \arctan\left[\frac{b_n^2}{b_0^2} \tan(\Omega_n t)\right]. \tag{70}$$

When the atomic beam leaves the region of the electromagnetic field, the atomic state evolves freely. The equations of motion can be obtained analogously and we get for $t > t_L$

$$\bar{x}(t > t_L) = -x_f \cos\phi_n + \Omega_n (t - t_L) x_f \sin\phi_n, \tag{71}$$

$$\bar{p}(t > t_L) = m\Omega_n x_f \sin\phi_n, \tag{72}$$

$$K(t > t_L) = \frac{\frac{b_n^2}{b_0^2}\cos\phi_n + i\sin\phi_n}{b_n^2\left[\cos\phi_n + i\frac{b_n^2}{b_0^2}\sin\phi_n + i\frac{t-t_L}{\tau_n}\left(\frac{b_n^2}{b_0^2}\cos\phi_n + i\sin\phi_n\right)\right]} \tag{73}$$

and

$$b_0^2 u(t > t_L) = \left[\left(\cos\phi_n - \frac{t - t_L}{\tau_n}\sin\phi_n\right)^2 + \frac{b_n^4}{b_0^4}\left(\sin\phi_n + \frac{t - t_L}{\tau_n}\cos\phi_n\right)^2\right]^{-1}, \tag{74}$$

where $\phi_n = \Omega_n t_L$ and $\tau_n = mb_n^2/\hbar$.

The focus will be located in the atomic beam region where the width of the wavepacket is minimal. In other words, when $u(t > t_L)$ be a maximum there will be the focus. This will happen when the function

$$D(t) = \left(\cos\phi_n - \frac{t - t_L}{\tau_n}\sin\phi_n\right)^2 + \frac{b_n^4}{b_0^4}\left(\sin\phi_n + \frac{t - t_L}{\tau_n}\cos\phi_n\right)^2 \tag{75}$$

attains its minimum value. The time for which its derivative vanishes is given by

$$t_f = \frac{z_f + L_c}{v_z} = t_L + \tau_n \frac{\left(1 - \frac{b_n^4}{b_0^4}\right)^2 \sin\phi_n \cos\phi_n}{\frac{b_n^4}{b_0^4}\cos^2\phi_n + \sin^2\phi_n}, \tag{76}$$

therefore the focus is located at

$$z_f = v_z \tau_n \frac{\left(1 - \frac{b_n^4}{b_0^4}\right)\tan\phi_n}{\frac{b_n^4}{b_0^4} + \tan^2\phi_n}. \tag{77}$$

The width of the Gaussian beam that passed through the lens, $B'(t) = 1/\sqrt{u(t)}$, can be written as

$$B'(t) = b_0' \left[1 + \left(\frac{t - t_f}{\tau_0'}\right)^2\right]^{\frac{1}{2}} \tag{78}$$

where we define

$$b_0' = Mb_0, \tag{79}$$

$$\tau_0' = M^2 \tau_0, \tag{80}$$

and

$$M = \frac{1}{\sqrt{\cos^2\phi_n + \frac{b_0^4}{b_n^4}\sin^2\phi_n}}. \tag{81}$$

The line was used here to differentiate the beam parameters after the focalization of their parameters before the focalization. We see that the waist of the beam is increased by factor M and the package time aging is increased by the M^2 (not confuse with the quality factor $\mathcal{M}_P$). In optics, the amount M is known as magnification factor (Saleh & Teich, 1991). If the state is not initially compressed, i.e., if $b_n = b_0$, does not exist focalization and in this case $b_0' = b_0$ and $\tau_0' = \tau_0$ as we can seen by the Equations (79), (80) and (81).
If we consider an interaction time of atoms with cavity field t_L very small, we have the so called thin lens regime. Because when the interaction time is very small, the movement of atoms along the transverse direction is also very small, i.e., the average transverse kinetic energy of atoms is much smaller than the average potential energy $\langle\hat{U}(x)\rangle$ produced by field, $\frac{\langle\hat{p}_x^2\rangle}{2m} \ll \langle\hat{U}(x)\rangle$ (Averbukh etal., 1994). The rotation angle of the atomic state caused by the interaction with the cavity field $\phi_n = \Omega_n t_L$ is directly proportional to the interaction time, thus, if t_L is too small, ϕ_n will also be very small. If we consider $\phi_n \ll 1$ and an initial atomic state compressed in the momentum with $b_n/b_0 \ll 1$, the expression for the focal distance, equation (77), acquires the simple form (Schleich, 2001)

$$z_f = \frac{mv_z^2}{ng_2L_c}. \tag{82}$$

4.3 Phase anomaly

If we integrate the equation of motion (64) for Φ considering the expression for $B'(t)$ given by the Equation (78), we obtain

$$\begin{aligned}\mu(t) = \frac{\Phi(t)}{2} &= -\frac{\hbar}{2m}\int_{t_f}^{t}\frac{dt}{B'^2(t)} \\ &= -\frac{1}{2}\arctan\left(\frac{t-t_f}{\tau_0'}\right).\end{aligned} \tag{83}$$

The integration interval is taken from t_f to t, because the Gouy phase is the phase of the Gaussian state relatives to the plane wave at the focus, i.e., at the focus the Gaussian state is in phase with the plane wave (Saleh & Teich, 1991; Boyd, 1980; Feng & Winful, 2001). At the focus, $\mu = 0$, as expected. Therefore, the Gouy phase of the atomic wave function undergoes a change of $\pi/2$ near the focus t_f. The fact that this variation is only $\pi/2$, in contrast with the value of π for the light, is due to the fact that the quantum lens focuses the atomic beam in the Ox direction, keeping the Oy direction unperturbed (i.e., the electromagnetic field acts as a cylindrical lens).

5. Experimental proposal

Consider a Rydberg atom with a level structure given in Figure 8 (left). Three Rydberg levels e, g, and i are taken into account. The transition between the states e and g is slightly detuned with a stationary microwave field stored in two separated cavities with frequency ω, C_1 and C_2, and completely detuned with the transition $g \to i$. These cavities are placed between two Ramsey zones, R_1 and R_2, where a microwave mode quasi-resonant with the atomic transition

$g \rightarrow i$ is stored (see Figure 8). If the electronic atomic state involve the levels i or g, the field in both Ramsey zones are adjusted to imprint a $\pi/2$ Rabi pulse on the internal state of the atom. Then, after the Ramsey zones, the electronic state changes as

$$|i\rangle = \frac{1}{\sqrt{2}}(|i\rangle + |g\rangle) \tag{84}$$

and

$$|g\rangle = \frac{1}{\sqrt{2}}(|i\rangle - |g\rangle). \tag{85}$$

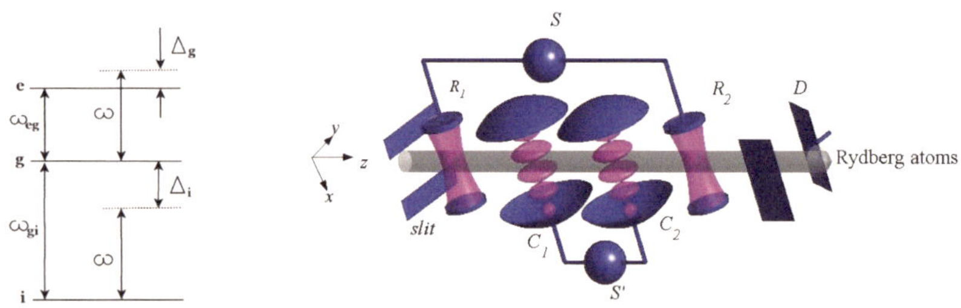

Fig. 8. On the left, atomic energy levels compared with the wavelength of the field inside the cavities C_1 and C_2. On the right, sketch of the experimental setup to measure the Gouy phase for matter waves. Rydberg atoms are sent one-by-one with well-defined velocity along the z-axis. A slit is used to collimate the atomic beam in the x-direction. The Ramsey zones R_1 and R_2 are two microwave cavities fed by a common source S, whereas C_1 and C_2 are two high-Q microwave cavities devised to work as thin lenses for the atomic beam. The field inside these cavities is supplied by common source S'. The state of each atom is detected by the detector D.

The experimental setup we propose to measure the Gouy phase shift of matter waves is depicted in Figure 8 (right). This proposal is based on the system of Ref. (Raimond et al., 2001). Rubidium atoms are excited by laser to a circular Rydberg state with principal quantum number 49 (Nussenzveig et al., 1993; Gallagher, 1994), that will be called state $|i\rangle$, and their velocity on the z direction is selected to a fixed value v_z. As it was stated before, we will consider a classical movement of the atoms in this direction, with the time component given by $t = z/v_z$. A slit is used to prepare a beam with small width in the x direction, but still without a significant divergence, such that the consideration that the atomic beam has a plane-wave behaviour is a good approximation.

If we disregard the cavities C_1 and C_2, the setup is that of an atomic Ramsey interferometer (Ramsey, 1985). The cavity R_1 has a field resonant or quasi resonant with the transition $|i\rangle \Leftrightarrow |g\rangle$ and results in a $\pi/2$ pulse on the atoms, that exit the cavity in the state $(|i\rangle + |g\rangle)/\sqrt{2}$ (Raimond et al., 2001; Ramsey, 1985; Kim et al., 1999; Gerry & Knight, 2005). After passing through the cavity R_1, the atoms propagate freely for a time t until the cavity R_2, that also makes a $\pi/2$ pulse on the atoms. Calling $\hbar\omega_g$ and $\hbar\omega_i$ the energy of the internal states $|g\rangle$ and $|i\rangle$ respectively, ω_r the frequency of the field in the cavities R_1 and R_2 and defining $\omega_{gi} \equiv \omega_g - \omega_i$, the probability that detector D measures each atom in the $|g\rangle$ state is (Raimond et al., 2001; Ramsey, 1985; Nogues et al., 1999)

$$P = \cos^2[(\omega_r - \omega_{gi})t] \, . \tag{86}$$

Upon slightly varying the frequency ω_r of the fields in cavities R_1 and R_2, the interference fringes can be seen (Raimond et al., 2001; Ramsey, 1985; Nogues et al., 1999).

5.1 Atom focalization by classical fields

The interaction between a two-level atom and a single mode of the electromagnetic field (EMF) is governed by the semiclassical hamiltonian

$$\hat{H}_{AF} = -\hat{d} \cdot \vec{E}(\hat{\vec{r}}, t). \tag{87}$$

$\hat{d} = \vec{\epsilon}_d \wp \hat{\sigma}_x^{eg}$ is the dipole moment operator, where $\vec{\epsilon}_d$ is the unitary vector along the direction of quantization, $\wp$ is the element of the transition matrix between the levels e (excited) and g (ground state), and $\hat{\sigma}_x^{eg} = |e\rangle\langle g| + |g\rangle\langle e|$. Assuming the longwave approximation, the electric field $\vec{E}$ is considered in the position $\vec{r}$ of the atomic center of mass ($\hat{\vec{r}}$ is the corresponding quantum operator). Here, $\vec{E}$ is treated classically.
Let us suppose that the atom interacts with a stationary electromagnetic wave kept in a cavity. Moreover, the atom moves along the Oz direction, while the stationary field is formed by two counterpropagating components along the Ox axis and linearly polarized in the direction Oy. We have

$$\vec{E}(\hat{\vec{r}}, t) = \vec{\epsilon}_y [E_0\, e^{i(k\hat{x} - \omega t)} + E_0\, e^{i(k\hat{x} + \omega t)} + h.c], \tag{88}$$

where E_0 is a complex amplitude. Thus, we can write

$$\vec{E}(\hat{\vec{r}}, t) = \vec{\epsilon}_y [2E_0\, e^{ik\hat{x}} \cos(\omega t) + h.c]. \tag{89}$$

Here, $k = \frac{2\pi}{\lambda}$, where λ is the wavelength of the EMF, and h.c. stands for hermitean conjugate. Without loss of generality, we can take E_0 real. Hence,

$$\vec{E}(\hat{\vec{r}}, t) = 4\vec{\epsilon}_y E_0 \cos(k\hat{x}) \cos(\omega t). \tag{90}$$

Assuming $\vec{\epsilon}_y = \vec{\epsilon}_d$, we have

$$\hat{H}_{AF} = -4pE_0 \cos(k\hat{x}) \cos(\omega t) \hat{\sigma}_x^{eg} \equiv \Omega_0 \cos(k\hat{x}) \cos(\omega t) \hat{\sigma}_x^{eg}, \tag{91}$$

where $|\Omega_0| = 4\wp E_0$ is the Rabi vacuum frequency.
The hamiltonian that governs the atomic dynamics during the interaction with the stationary field is given by

$$\hat{H} = \frac{\hat{p}_x^2}{2m} + \frac{\hbar\omega_{eg}}{2} \hat{\sigma}_z^{eg} + \hbar\Omega_0 \cos(k\hat{x}) \cos(\omega t) \hat{\sigma}_x^{eg}. \tag{92}$$

Here, $\hat{\sigma}_z^{eg} = |e\rangle\langle e| - |g\rangle\langle g|$, and m is the atomic mass. In the rotating wave approximation (RWA), we have

$$\hat{H} = \frac{\hat{p}_x^2}{2m} + \frac{\hbar\omega_{eg}}{2} \hat{\sigma}_z^{eg} + \frac{\hbar\Omega_0}{2} \cos(k\hat{x})(e^{-i\omega t}|e\rangle\langle g| + e^{i\omega t}|g\rangle\langle e|). \tag{93}$$

In order to remove the temporal dependence of $\hat{H}$, we define $|\tilde{\psi}(t)\rangle = \exp\left(i\frac{\omega t}{2}\hat{\sigma}_z^{eg}\right)|\psi(t)\rangle$. Then, the evolution of the state $|\tilde{\psi}(t)\rangle$ is governed by the equation

$$i\hbar \frac{d}{dt}|\tilde{\psi}\rangle = \hat{\tilde{H}}|\tilde{\psi}\rangle, \tag{94}$$

where

$$\hat{\tilde{H}} = \frac{\hat{p}_x^2}{2m} + \frac{\hbar\Delta}{2} \hat{\sigma}_z^{eg} + \frac{\hbar\Omega_0}{2} \cos(k\hat{x}) \hat{\sigma}_x^{eg}. \tag{95}$$

Here, we define $\Delta = \omega_{eg} - \omega$ as the detuning between the frequency of the atomic transition and the frequency of the field mode. In the limit of thin lens, the kinetic energy term can be neglected. So, we have

$$\hat{\hat{H}} = \frac{\hbar\Delta}{2}\hat{\sigma}_z^{eg} + \frac{\hbar\Omega_0}{2}\cos(k\hat{x})\hat{\sigma}_x^{eg}. \tag{96}$$

Let us define the operator $\hat{\Omega}_x = \Omega_0 \cos(k\hat{x})$. Consider the set composed by the states $|e,x\rangle = |e\rangle \otimes |x\rangle$ and $|g,x\rangle = |g\rangle \otimes |x\rangle$, where $|x\rangle$ is an eigenstate of the operator $\hat{x}$ with eigenvalue x. In the basis $\{|e,x\rangle, |g,x\rangle\}_{x\in\mathbb{R}}$, $\hat{\hat{H}}$ is represented by the matrix

$$\hat{\hat{H}}_x = \begin{pmatrix} \Delta & \Omega_x \\ \Omega_x & -\Delta \end{pmatrix}, \tag{97}$$

which is diagonalized by the eigenvectors

$$|+,x\rangle = \frac{\Omega_x}{\sqrt{(\Delta_x - \Delta)^2 + \Omega_x^2}}|e,x\rangle + \frac{\Delta_x - \Delta}{\sqrt{(\Delta_x - \Delta)^2 + \Omega_x^2}}|g,x\rangle, \tag{98}$$

$$|-,x\rangle = \frac{\Delta_x - \Delta}{\sqrt{(\Delta_x - \Delta)^2 + \Omega_x^2}}|e,x\rangle - \frac{\Omega_x}{\sqrt{(\Delta_x - \Delta)^2 + \Omega_x^2}}|g,x\rangle. \tag{99}$$

with the following eigenvalues

$$E_{\pm,x} = \pm\frac{\hbar}{2}\Delta_x = \pm\frac{\hbar}{2}\sqrt{\Delta^2 + \Omega_x^2}. \tag{100}$$

Here, $\Omega_x = \Omega_0 \cos(kx)$ is an eigenvalue of the operator $\hat{\Omega}_x$. In the dispersive limit, we have $\frac{|\Omega_x|}{\Delta} \ll 1$. In this limit, the eigenvectors and the eigenvalues given by the above equations can be approximated by

$$|+,x\rangle \to |e,x\rangle\,,\ |-,x\rangle \to |g,x\rangle, \tag{101}$$

$$E_{\pm,x} \to \pm\frac{\hbar}{2}\left(\Delta + \frac{\Omega_x^2}{2\Delta}\right). \tag{102}$$

Besides, assuming that the width of the atomic wavepacket is small compared with the wavelength of the stationary EMF (harmonic approximation), $\Omega_x^2 = \Omega_0^2\cos^2(kx)$ can be expanded to $\Omega_x^2 \approx \Omega_{min}^2 + \Omega_0^2 k^2 x^2$. The fact that we have chosen the point $\pi/2$ to do the expansion means that we are treating the case of blue detuning ($\Delta < 0$), i.e., the case in which the atoms will pass in the region of electric field node (Berman, 1997). Also we define $\Omega_x^2(\pi/2) \equiv \Omega_{min}^2$ to be different of zero, since the potential is produced by fields counterpropagating and it difficultly will be null for real cavities. Thus, taking the dispersive and harmonic approximations, we can define the effective hamiltonian

$$\hat{\hat{H}}_{eff} = \frac{\hbar}{2}\left(\Delta + \frac{\Omega_{min}^2}{2\Delta} + \frac{\Omega_0^2}{2\Delta}k^2\hat{x}^2\right)\hat{\sigma}_z^{eg}. \tag{103}$$

As discussed above, the devised experiment uses three atomic levels and the third level i possesses energy below of the energy of the levels g and e , as shown in Figure 8. The frequencies ω_{gi} and ω_{eg} satisfy $\omega_{gi} > \omega_{eg}$, thus the transition $g \to i$ is very far from the resonance with the stationary mode. In this case, adopting the same reasoning sketched above, the introduction of this level modifies the effective hamiltonian given in the Equation (103)

$$\hat{\tilde{H}}_{eff} = \frac{\hbar\Delta_i}{2}\hat{\sigma}_z^{gi} + \frac{\hbar}{2}\left(\Delta_g + \frac{\Omega_{min}^2}{2\Delta_g} + \frac{\Omega_0^2}{2\Delta_g}k^2\hat{x}^2\right)\hat{\sigma}_z^{eg}, \tag{104}$$

where $\Delta_i = \omega_{gi} - \omega$, $\Delta_g = \omega_{eg} - \omega$, and $\hat{\sigma}_z^{gi} = |g\rangle\langle g| - |i\rangle\langle i|$. Note that the effective coupling between the atom in the state i and the EMF is neglected in the above hamiltonian.
In order to discuss the focalization, let us consider the following initial state

$$|\tilde{\psi}(0)\rangle = \frac{1}{\sqrt{2}}(|i\rangle + |g\rangle) \otimes \psi_{cm}(x), \tag{105}$$

where $|\psi_{cm}\rangle$ stands for some state of the atomic center of mass coordinate. The atom prepared in this state interacts with the stationary field during a time interval t_L. After this interval, the state of the atom will be given by

$$|\tilde{\psi}(t_L)\rangle = \frac{1}{\sqrt{2}}[e^{i\phi_i}|i\rangle \otimes \psi_{cm}(x) + e^{-i(\phi_i-\phi_g)}|g\rangle \otimes \psi'_{cm}(x)], \tag{106}$$

where $\phi_i = \frac{\Delta_i}{2}t_L$, $\phi_g = \left(\frac{\Delta_g}{2} + \frac{\Omega_{min}^2}{4\Delta_g}\right)t_L$ are the phase shifts accumulated by the electronic levels during the interaction, and

$$\psi'_{cm}(x) = e^{i\frac{k^2\Omega_0^2 t_L}{4\Delta_g}x^2}\psi_{cm}(x) \tag{107}$$

is the evolved state of the center of mass of atomic beam composed of atoms in $|g\rangle$ state. As $\Delta_g < 0$, blue detuning, after to pass through the cavity satisfying the approaches that we use, the center of mass state gets a negative quadratic phase. An optical converging cylindrical lens with focal distance f puts a quadratic phase $-kx^2/(2f)$ on the electromagnetic beam (Saleh & Teich, 1991). By analogy, a thin lens for atoms should put a phase of the type $-k_P x^2/(2f_P)$ in the atomic beam, where $k_P = mv_z/\hbar$ is the atomic wave number and f_P the corresponding focal distance. If we compare this phase with the phase in the Equation (107), we get

$$f_P = \frac{2|\Delta_g|mv_z^2}{L_c\hbar k^2\Omega_0^2}. \tag{108}$$

This expression is the focal distance for a thin classical lens. Different from Equation (82) to thin quantum lens, this expression does not have a explicit dependence with photon number of the field mode.
The Rayleigh range z'_r and the beam waist w'_0 of the focused atomic beam also can be calculated using the analogy with the action of lenses in electromagnetic beams considering that the incident beam has plane wavefronts (Saleh & Teich, 1991)

$$z'_r = \left[\frac{1}{1+(z_r/f_P)^2}\right]z_r \ , \ w'_0 = \sqrt{\frac{1}{1+(z_r/f_P)^2}}\,w_0 \ , \tag{109}$$

where z_r and w_0 are the Rayleigh range and the beam waist of the incident beam, respectively, and f_P is the focal distance of the atomic lens.

5.2 Ramsey interferometry with focused atomic beam and Gouy phase

In order to experimentally observe this effect we propose an experiment with a focused Gaussian atomic beam. We will use a cylindrical focusing in the x direction, without changing the beam wavefunction in the y direction, what makes the total Gouy phase be $\pi/2$. By the analogy of the Schrödinger equation with the paraxial Helmholtz equation (Yariv, 1991; Snyder & Love, 1991; Berman, 1997; Marte & Stenholm, 1997; da Paz et al., 2007), we see that the cavities act on the $|g\rangle$ component of the atomic beam as cylindrical lenses with focal distance f_P. If we have $f_P = d/2$, where d is the distance between the cavities C_1 and C_2, the system will behave like the illustration in Figure 9. The cavity C_1 will transform the $|g\rangle$ component of the wavefunction in a converging beam with the waist on a distance $d/2$ (represented by solid lines). After its waist, the beam will diverge until the cavity C_2. The $|g\rangle$ component of the wavefunction on the position of cavity C_2 will have the same width and the opposite quadratic phase of the state $\psi'_{cm}(x)$ above, so the cavity C_2 will transform the divergent beam in a plane-wave beam again. The $|i\rangle$ component of the wavefunction, on the other hand, propagates as a plane-wave beam all the time (represented by dashed lines), as its interaction with the field of the cavities C_1 and C_2 is considered to be very small.

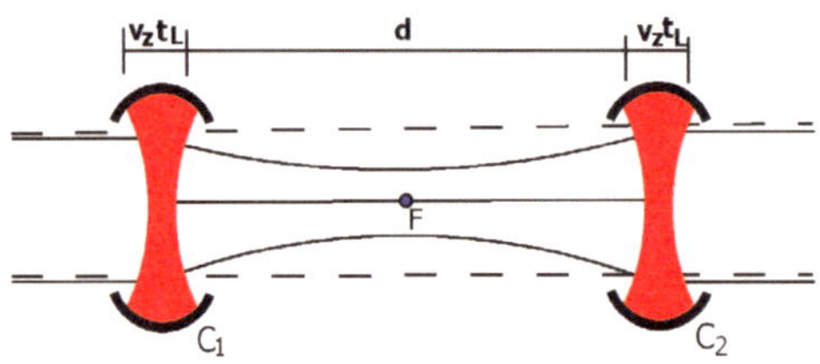

Fig. 9. Illustration of the operation of the cavities C_1 and C_2 as thin lenses for the atomic beam. The dashed lines represent the waist of the atomic beam if the cavities are empty. If a field is present, the solid lines represent the waist of a beam composed by atoms in the state $|g\rangle$. F denotes the focus region. On the other hand, if the beam is composed by atoms in the state $|i\rangle$, the waist does not change significatively.

By virtue of the $|g\rangle$ component acquires a $\pi/2$ Gouy phase due to the cylindrical focusing that is not shared by the $|i\rangle$ component, the interference pattern will be (da Paz et al., 2011)

$$P' = \cos^2[(\omega_r - \omega_{gi})t - \pi/2] \; . \tag{110}$$

The difference on the positions of the minimums and maximums of the patterns, one constructed when the field that forms the atomic lenses is present on the cavities C_1 and C_2 and other when the field is removed, should attest the existence of the Gouy phase for matter waves.

5.3 Experimental parameters and discussion

As experimental parameters, we propose the velocity of the atoms $v_z = 50$ m/s and a slit that generates an approximately Gaussian wavefunction for the atoms $\psi_0(x) \propto e^{-x^2/w_0^2}$ with $w_0 = 10\ \mu m$. The mass of Rubidium is $m = 1.44 \times 10^{-25}$ kg. With these parameters, the Rayleigh range of the atomic beam will be $z_r = k_P w_0^2/2 \simeq 3.5$ m, much larger than the length of the experimental apparatus, what justifies the plane-wave approximation. On the cavities C_1 and C_2, we consider an interaction time between the atoms and the atomic lenses $t_L = 0.2$ ms, that corresponds to a width $v_z t_L = 1$ cm for the field on the cavities. The

wavelength of the field of the cavities C_1 and C_2 must be $\lambda \simeq 5.8$ mm (Raimond et al., 2001), with frequency near but strongly detuned from the resonance of the transition $|g\rangle \Leftrightarrow |e\rangle$. The Rabi frequency is about $\Omega_0/(2\pi) = 47$ kHz (Raimond et al., 2001) and the detuning chosen is $\Delta_g/(2\pi) = -30$ MHz, what makes $\Delta_i/(2\pi) = +3.2$ GHz, such that with $\bar{n} = 3 \times 10^6$ photons, an effectively classical field, the focal distance for the atomic lenses is 10.5 cm for the $|g\rangle$ component and -11 m for the $|i\rangle$ component of the wavefunction. These parameters are consistent with a separation of $d = 21$ cm between C_1 and C_2. All the proposed parameters can be experimentally achieved (Raimond et al., 2001; Nogues et al., 1999; Gleyzes et al., 2007). Using the proposed parameters, we have $z_r' \simeq 3$ mm and $w_0' \simeq 0.3\ \mu m$. The fact that $d \gg z_r'$ justifies our consideration that the $|g\rangle$ component of the beam acquires a $\pi/2$ Gouy phase.
The interaction between the atomic beam and the field in the cavities C_1 and C_2 depends on the position x, according to Equation (103). If we do not want that photons be absorbed by the atoms, it is important that $\bar{n}\Omega_0^2 k^2 x^2/(\Delta_g^2) \ll 1$ for the entire beam (Scully & Zubairy, 1997). We have $\bar{n}\Omega_0^2 k^2 w_0^2/(\Delta_g^2) \simeq 8 \times 10^{-4}$ for the proposed parameters, where w_0 is the beam width, showing that the absorption of photons can be disregarded.
The phase difference between the electronic levels is given by

$$2\phi_i - \phi_g = \left(\Delta_{gi} - \frac{\Delta_g}{2} - \frac{\Omega_{min}^2}{4\Delta_g} \right) t_L, \tag{111}$$

where the last term is a dispersive phase that occurs because the intensity of the electric field is not exactly zero in the node $x = \pi/2$ for real cavities. In this case is important that the cavities C_1 and C_2 have a large quality factor Q. In fact, the ratio between the maximum and the minimum of intensity in a cavity should roughly be the quality factor Q. So the $|g\rangle$ component of the beam also acquires a phase $\bar{n}\Omega_0^2 t_L/(\Delta_g Q)$ on the passage in each cavity, and this phase will be added to the accumulated Gouy phase. If we want that this undesired phase be smaller than Gouy phase, we need $Q > 10^6$ for our proposed parameters. This can also be experimentally achieved (Raimond et al., 2001; Nogues et al., 1999; Gleyzes et al., 2007).

6. Conclusion

From the strict theoretical point of view we have used the formal analogy between matter and light waves to show that the well known Gouy phase in the context of classical optics, besides its geometrical character, reflects correlations of the same sort a free particle obeying a matter wave equation. Conversely we have seen that matter waves may also present the exact analogous to the Gouy phase of quantum optics and elaborated an experiment to measure it. We hope this work might encourage the groups with the appropriate facilities to realize the experiment and, who knows, find important applications for this matter phase. The verification of the Gouy phase in matter waves has the possibility to generate a great amount of development in atomic optics, in the same way as the electromagnetic counterpart Gouy phase had contributed to electromagnetic optics. For instance, it can be used to construct mode converters for atomic beams and trapped atoms, with potential applications in quantum information.

7. Acknowledgments

The authors thank Dr. P. L. Saldanha by fruitful discussions on experimental proposal to measure the Gouy phase. This work was in part supported by the Brazilian agencies CNPq, Capes and Fapemig.

8. References

Yariv, A. (1991). *Optical Electonics*, Saunders College Publishers, Philadelphia.

Snyder, A. W., and Love, J. D. (1991). *Waveguide Theory*, Chapman and Hall, London.

Berman, P. R. (1997). *Atom Interferometry*, Academic Press, San Diego.

Marte, M. A. M., and Stenholm, S. (1997). Paraxial light and atom optics: The optical Schrödinger equation and beyond, *Phys. Rev. A* 56, 2940.

Nairz, O., Arndt, M., and Zeilinger, A. (2002). Experimental verification of the Heisenberg uncertainty principle for fullerene molecules, *Phys. Rev. A* 65, 032109.

Saleh, B. E. A., and Teich, M. C. (1991). *Fundamentals of Photonics*, John Wiley et Sons, New York.

Gouy, C. R. (1890). *Acad. Sci. Paris* 110, 1251.

Gouy, C. R. (1891). Sur la propagation anomale des ondes, *Ann. Chim. Phip. Ser. 6* 24, 145.

Mair, A., Vaziri, A., Weihs, G., and Zeilinger, A. (2001). Entanglement of the orbital angular momentum states of photons, *Nature* 412, 313.

Allen, L., Beijersbergen, M. W., Spreeuw, R. J. C., and Woerdman, J. P. (1992). Orbital angular momentum of light and the transformation of Laguerre-Gaussian laser modes, *Phys. Rev. A* 45, 8185.

Beijersbergen, M. W., Allen, L., van der Veen, H. E. L. O., and Woerdman, J. P. (1993). Astigmatic laser mode converters and transfer of orbital angular momentum, *Opt. Comm.* 96, 123.

McMorran, B. J., Agrawal, A., Anderson, I. M., Herzing, A. A., Lezec, H. J., McClelland, J. J., and Unguris, J. (2011). Electron Vortex Beams with High Quanta of Orbital Angular Momentum, *Science* 331, 192.

Siegman, A. E. (1986). *Lasers*, University Science Books, Mill Valley.

Viale, A., Vicari, M., and Zanghì, N. (2003). Analysis of the loss of coherence in interferometry with macromolecules, *Phys. Rev. A* 68, 063610.

Jackson, J. D. (1999). *Classical Electrodynamics*, Jonh Wiley, New York.

Mandel, L., and Wolf, E. (1995). *Optical Coherence and Quantum Optics*, Cambridge, New York.

Gase, R. (1994). Methods of quantum mechanics applied to partially coherent light beams, *J. Opt. Soc. Am. A* 11, 2121.

Ballentine, L. E. (1998). *Quantum Mechanics*, World Scientific Publishing Co. Pte. Ltd, Singapore.

Scully, M. O., and Zubairy, M. S. (1997). *Quantum Optics*, Cambridge University Press, Cambridge.

Fano, U. (1957). Description of States in Quantum Mechanics by Density Matrix and Operator Techniques, *Rev. Mod. Phys.* 29, 74.

Stoler, D. (1981). Operator methods in physical optics, *J. Opt. Soc. Am.* 71, 334.

Simon, R., and Mukunda, N. (1993). Bargman Invariant and the Geometry of the Gouy Effect, *Phys. Rev. Lett.* 70, 880.

da Paz, I. G. (2006). *Matter Waves and Paraxial Propagation of Light*, Department of Physics, UFMG, MG, Brazil.

da Paz, I. G. (2011). *Gouy phase in matter waves: from pure and mixed Gaussian states*, Department of Physics, UFMG, MG, Brazil.

Souza, L. A. M., Nemes, M.C., Santos, M. F., and Peixoto de Faria, J. G. (2008). Quantifying the decay of quantum properties in single-mode states, *Opt. Comm.* 281, 4694.

Boyd, R. W. (1980). Intuitive explanation of the phase anomaly of focused light beams, *J. Opt. Soc. Am.* 70, 877.

Hariharan, P., and Robinson, P. A. (1996). The Gouy phase shift as a geometrical quantum effect, *J. Mod. Opt.* 43, 219.

Feng, S., Winful, H. G., and Hellwarth, R. W. (1998). Gouy shift and temporal reshaping of focused single-cycle electromagnetic pulses, *Opt. Lett.* 23, 385.

Feng, S., and Winful, H. G. (2001). Physical origin of the Gouy phase shift, *Opt. Lett.* 26, 485.

Yang, J., and Winful, H. G. (2006). Generakized eikonal treatment of the Gouy phase shift, *Opt. Lett.* 31, 104.

Holme, N. C. R., Daly, B. C., Myaing, M. T., and Norris, T. B. (2003). Gouy phase shift of single-cycle picosecond acoustic pulses, *Appl. Phys. Lett.* 83, 392.

Zhu, W., Agrawal, A., and Nahata, A. (2007). Direct measurement of the Gouy phase shift for surface plasmon-polaritons, *Opt. Express* 15, 9995.

Feurer, T., Stoyanov, N. S., Ward, D. W., and Nelson, K. A. (2002). Direct Visualization of the Gouy Phase by Focusing Phonon Polaritons, *Phys. Rev. Lett.* 88, 257.

Horvath, Z. L., and Bor, Z. (1999). Reshaping of femtosecond pulses by the Gouy phase shift, *Phys. Rev. E* 60, 2337.

Horvath, Z. L., Vinko, J., Bor, Z., and von der Linde, D. (1996). Acceleration of femtosecond pulses to superluminal velocities by Gouy phase shift, *Appl. Phys. B* 63, 481.

Gbur, G., Visser, T. D., and Wolf, E. (2002). Anomalous Behavior of Spectra near Phase singularities of Focused Waves, *Phys. Rev. Lett.* 88, 013901.

Nairz, O., Arndt, M., and Zeilinger, A. (2000). Experimental chanllenges in fullerene interferometry, *J. Modern Opt.* 47, 2811.

da Paz, I. G., Nemes, M. C., and Peixoto de Faria, J. G. (2007). Gouy phase and matter waves, *J. Phys.: Conference Series* 84, 012016.

da Paz, I. G., Monken, C. H., Padua, S., Nemes, M. C., and Peixoto de Faria, J. G. (2010). Indirect evidence for the Gouy phase for matter waves, *Phys. Lett. A* 374, 1660.

da Paz, I. G., Saldanha, P. L., Nemes, M. C., and Peixoto de Faria, J. G. (2011). Experimental proposal for measuring the Gouy phase for matter waves, *To be published* arXiv:1012.3910v1 [quant-ph].

Averbukh, I. S., Akulin V. M., and Schleich, W. P. (1994). Quantum Lens for Atomic Waves, *Phys. Rev. Lett.* 72, 437.

Rohwedder, B., and Orszag, M. (1996). Quantized light lenses for atoms: The perfect thick lens, *Phys. Rev. A* 54, 5076.

Schleich, W. P. (2001). *Quantum Optics in Phase Space*, Wiley-VCH, Berlin.

Bialynicki-Birula, I. (1998). Nonstandard Introduction to Squeezing of the Electromagnetic Field, *Acta Phys. Polonica B* 29, 3569.

Piza, A. F. R. T. (2001). *Mecânica Quântica*, Edusp, São Paulo.

Born, M., and Wolf, E. (1999). *Principles of Optics*, Cambridge University Press, Cambridge.

Ashkin, A. (1970). Acceleration and Trapping of Particles by Radiation Pressure, *Phys. Rev. Lett.* 24, 156.

Gordon, J. P., and Ashkin, A. (1980). Motion of atoms in a radiation trap, *Phys. Rev. A* 21, 1606.

Bjorkholm, J. E., Freeman, R. R., Ashkin, A., and Pearson, D. B. (1978). Observation of Focusing of Neutral Atoms by the Dipole Forces of Resonance-Radiation Pressure, *Phys. Rev. Lett.* 41, 1361.

Raimond, J. M., Brune, M., and Haroche, S. (2001). Colloquium: Manipulating quantum entanglement with atoms and photons in a cavity, *Rev. Mod. Phys.* 73, 565.

Nussenzveig, P., Bernardot, F., Brune, M., Hare, J., Raimond, J. M., Haroche, S., Gawlik, W. (1993). Preparation of high-principal-quantum-number "circular" states of rubidium, *Phys. Rev. A* 48, 3991.

Gallagher, T. F. (1994). *Rydberg Atoms*, Cambridge University Press, Cambridge.

Ramsey, N. F. (1985). *Molecular Beams*, Oxford University Press, New York.

Kim, J. I., Romero Fonseca, K. M., Horiguti, A. M., Davidovich, L., Nemes, M. C., and Toledo Piza, A. F. R. (1999). Classical behavior with small quantum numbers: the physics of Ramsey interferometry of Rydberg atoms, *Phys. Rev. Lett.* 82, 4737.

Gerry, C., and Knight, P. (2005). *Introductory Quantum Optics*, Cambridge University Press, Cambridge.

Nogues, G., Rauschenbeutel, A., Osnaghi, S., Brune, M., Raimond, J. M., and Haroche, S. (1999). Seeing a single photon without destroying it, *Nature* 400, 239.

Gleyzes, S., Kuhr, S., Guerlin, C., Bernu, J., Deleglise, S., Hoff, U. B., Brune, M., Raimond, J. M., and Haroche, S. (2007). Quantum jumps of light recording the birth and death of a photon in a cavity, *Nature* 446, 297.

Part 2

Methods of Computational Analysis

5

Simulation and Analysis of Transient Processes in Open Axially-symmetrical Structures: Method of Exact Absorbing Boundary Conditions

Olena Shafalyuk[1], Yuriy Sirenko[2] and Paul Smith[1]
[1]Macquarie University
[2]Usikov Institute of Radiophysics and Electronics
[1]Australia
[2]Ukraine

1. Introduction

Present day methodologies for mathematical simulation and computational experiment are generally implemented in electromagnetics through the solution of boundary-value (frequency domain) problems and initial boundary-value (time domain) problems for Maxwell's equations. Most of the results of this theory concerning open resonators have been obtained by the frequency-domain methods. At the same time, a rich variety of applied problems (analysis of complex electrodynamic structures for the devices of vacuum and solid-state electronics, model synthesis of open dispersive structures for resonant quasi-optics, antenna engineering, and high-power electronics, etc.) can be efficiently solved with the help of more universal time-domain algorithms.

The fact that frequency domain approaches are somewhat limited in such problems is the motivation for this study. Moreover, presently known remedies to the various theoretical difficulties in the theory of non-stationary electromagnetic fields are not always satisfactory for practitioners. Such remedies affect the quality of some model problems and limit the capability of time-domain methods for studying transient and stationary processes. One such difficulty is the appropriate and efficient truncation of the computational domain in so-called open problems, i.e. problems where the computational domain is infinite along one or more spatial coordinates. Also, a number of questions occur when solving far-field problems, and problems involving extended sources or sources located in the far-zone.

In the present work, we address these difficulties for the case of TE_{0n} - and TM_{0n} -waves in axially-symmetrical open compact resonators with waveguide feed lines. Sections 2 and 3 are devoted to problem definition. In Sections 4 and 5, we derive exact absorbing conditions for outgoing pulsed waves that enable the replacement of an open problem with an equivalent closed one. In Section 6, we obtain the analytical representation for operators that link the near- and far-field impulsive fields for compact axially-symmetrical structures and consider solutions that allow the use of extended or distant sources. In Section 7, we place some accessory results required for numerical implementation of the approach under

consideration. All analytical results are presented in a form that is suitable for using in the finite-difference method on a finite-sized grid and thus is amenable for software implementation. We develop here the approach initiated in the works by Maikov et al. (1986) and Sirenko et al. (2007) and based on the construction of the exact conditions allowing one to reduce an open problem to an equivalent closed one with a bounded domain of analysis. The derived closed problem can then be solved numerically using the standard finite-difference method (Taflove & Hagness, 2000).
In contrast to other well-known approximate methods involving truncation of the computational domain (using, for example, Absorbing Boundary Conditions or Perfectly Matched Layers), our constructed solution is exact, and may be computationally implemented in a way that avoids the problem of unpredictable behavior of computational errors for large observation times. The impact of this approach is most significant in cases of resonant wave scattering, where it results in reliable numerical data.

2. Formulation of the initial boundary-value problem

In Fig. 1, the cross-section of a model for an open axially-symmetrical ($\partial/\partial\phi \equiv 0$) resonant structure is shown, where $\{\rho,\phi,z\}$ are cylindrical and $\{\rho,\vartheta,\phi\}$ are spherical coordinates. By $\Sigma = \Sigma_\phi \times [0,2\pi]$ we denote perfectly conducting surfaces obtained by rotating the curve Σ_ϕ about the z-axis; $\Sigma^{\varepsilon,\sigma} = \Sigma_\phi^{\varepsilon,\sigma} \times [0,2\pi]$ is a similarly defined surface across which the relative

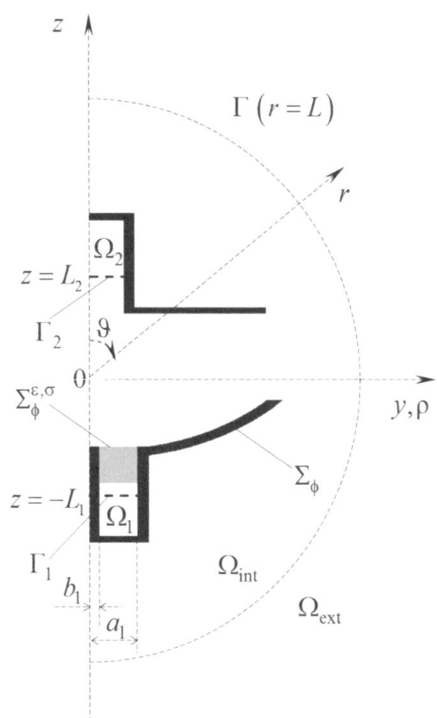

Fig. 1. Geometry of the problem in the half-plane $\phi = \pi/2$.

permittivity $\varepsilon(g)$ and specific conductivity $\sigma_0(g)=\eta_0^{-1}\sigma(g)$ change step-wise; these quantities are piecewise constant inside Ω_{int} and take free space values outside. Here, $g=\{\rho,z\}$; $\eta_0=(\mu_0/\varepsilon_0)^{1/2}$ is the impedance of free space; ε_0, and μ_0 are the electric and magnetic constants of vacuum.

The two-dimensional initial boundary-value problem describing the pulsed axially-symmetrical TE_{0n}- ($E_\rho=E_z=H_\phi\equiv 0$) and TM_{0n}- ($H_\rho=H_z=E_\phi\equiv 0$) wave distribution in open structures of this kind is given by

$$\begin{cases} \left[-\varepsilon(g)\dfrac{\partial^2}{\partial t^2}-\sigma(g)\dfrac{\partial}{\partial t}+\dfrac{\partial^2}{\partial z^2}+\dfrac{\partial}{\partial\rho}\left(\dfrac{1}{\rho}\dfrac{\partial}{\partial\rho}\rho\right)\right]U(g,t)=F(g,t), \quad t>0, \quad g\in\Omega \\ U(g,t)\big|_{t=0}=\varphi(g), \quad \dfrac{\partial}{\partial t}U(g,t)\Big|_{t=0}=\psi(g), \quad g=\{\rho,z\}\in\bar{\Omega} \\ E_{tg}(p,t)\big|_{p=\{\rho,\phi,z\}\in\Sigma}=0, \quad t\geq 0 \\ E_{tg}(p,t) \text{ and } H_{tg}(p,t) \text{ are continuous when crossing } \Sigma^{\varepsilon,\sigma} \\ U(0,z,t)=0, \quad |z|<\infty, \quad t\geq 0 \\ D_1\left[U(g,t)-U^{i(1)}(g,t)\right]\Big|_{g\in\Gamma_1}=0, \quad D_2\left[U(g,t)\right]\Big|_{g\in\Gamma_2}=0, \quad t\geq 0, \end{cases} \tag{1}$$

where $\vec{E}=\{E_\rho,E_\varphi,E_z\}$ and $\vec{H}=\{H_\rho,H_\varphi,H_z\}$ are the electric and magnetic field vectors; $U(g,t)=E_\phi(g,t)$ for TE_{0n}-waves and $U(g,t)=H_\phi(g,t)$ for TM_{0n}-waves (Sirenko et al., 2007). The SI system of units is used. The variable t which being the product of the real time by the velocity of light in free space has the dimension of length. The operators D_1, D_2 will be described in Section 2 and provide an ideal model for fields emitted and absorbed by the waveguides.

The domain of analysis Ω is the part of the half-plane $\phi=\pi/2$ bounded by the contours Σ_ϕ together with the artificial boundaries Γ_j (input and output ports) in the virtual waveguides Ω_j, $j=1,2$. The regions $\Omega_{\text{int}}=\{g=\{r,\vartheta\}\in\Omega:\ r<L\}$ and Ω_{ext} (free space), such that $\Omega=\Omega_{\text{int}}\cup\Omega_{\text{ext}}\cup\Gamma$, are separated by the virtual boundary $\Gamma=\{g=\{r,\vartheta\}\subset\Omega:\ r=L\}$.

The functions $F(g,t)$, $\varphi(g)$, $\psi(g)$, $\sigma(g)$, and $\varepsilon(g)-1$ which are finite in the closure $\bar{\Omega}$ of Ω are supposed to satisfy the hypotheses of the theorem on the unique solvability of problem (1) in the Sobolev space $\mathbf{W}_2^1(\Omega^T)$, $\Omega^T=\Omega\times(0;T)$ where $T<\infty$ is the observation time (Ladyzhenskaya, 1985). The 'current' and 'instantaneous' sources given by the functions $F(g,t)$ and $\varphi(g)$, $\psi(g)$ as well as all scattering elements given by the functions $\varepsilon(g)$, $\sigma(g)$ and by the contours Σ_ϕ and $\Sigma_\phi^{\varepsilon,\sigma}$ are located in the region Ω_{int}. In axially-symmetrical problems, at points g such that $\rho=0$, only H_z or E_z fields components are nonzero. Hence it follows that $U(0,z,t)=0$; $|z|<\infty$, $t\geq 0$ in (1).

3. Exact absorbing conditions for virtual boundaries in input-output waveguides

Equations

$$D_1\left[U(g,t)-U^{i(1)}(g,t)\right]\Big|_{g\in\Gamma_1}=0,\quad D_2\left[U(g,t)\right]\Big|_{g\in\Gamma_2}=0,\quad t\geq 0. \qquad (2)$$

in (1) give the exact absorbing conditions for the outgoing pulsed waves $U^{s(1)}(g,t)=U(g,t)-U^{i(1)}(g,t)$ and $U^{s(2)}(g,t)=U(g,t)$ traveling into the virtual waveguides Ω_1 and Ω_2, respectively (Sirenko et al., 2007). $U^{i(1)}(g,t)$ is the pulsed wave that excites the axially-symmetrical structure from the circular or coaxial circular waveguide Ω_1. It is assumed that by the time $t=0$ this wave has not yet reached the boundary Γ_1.

By using conditions (2), we simplify substantially the model simulating an actual electrodynamic structure: the Ω_j-domains are excluded from consideration while the operators D_j describe wave transformation on the boundaries Γ_j that separate regular feeding waveguides from the radiating unit. The operators D_j are constructed such that a wave incident on Γ_j from the region Ω_{int} passes into the virtual domain Ω_j as if into a regular waveguide – without deformations or reflections. In other words, it is absorbed completely by the boundary Γ_j. Therefore, we call the boundary conditions (2) as well as the other conditions of this kind 'exact absorbing conditions'.

In the book (Sirenko et al., 2007), one can find six possible versions of the operators D_j for virtual boundaries in the cross-sections of circular or coaxial-circular waveguides. We pick out two of them (one for the nonlocal conditions and one for the local conditions) and, taking into consideration the location of the boundaries Γ_j in our problem (in the plane $z=-L_1$ for the boundary Γ_1 and in the plane $z=L_2$ for Γ_2) as well as the traveling direction for the waves outgoing through these boundaries (towards $z=-\infty$ for Γ_1 and towards $z=\infty$ for Γ_2), write (2) in the form:

$$U^{s(1)}(\rho,-L_1,t)=\sum_n\left\{\int_0^t J_0\left[\lambda_{n1}(t-\tau)\right]\left[\int_{b_1}^{a_1}\frac{\partial U^{s(1)}(\tilde{\rho},z,\tau)}{\partial z}\bigg|_{z=-L_1}\mu_{n1}(\tilde{\rho})\tilde{\rho}\,d\tilde{\rho}\right]d\tau\right\}\mu_{n1}(\rho),\qquad (3)$$
$$b_1\leq\rho\leq a_1,\quad t\geq 0,$$

$$U(\rho,L_2,t)=-\sum_n\left\{\int_0^t J_0\left[\lambda_{n2}(t-\tau)\right]\left[\int_{b_2}^{a_2}\frac{\partial U(\tilde{\rho},z,\tau)}{\partial z}\bigg|_{z=L_2}\mu_{n2}(\tilde{\rho})\tilde{\rho}\,d\tilde{\rho}\right]d\tau\right\}\mu_{n2}(\rho),\qquad (4)$$
$$b_2\leq\rho\leq a_2,\quad t\geq 0$$

(nonlocal absorbing conditions) and

$$U^{s(1)}(\rho,-L_1,t)=\frac{2}{\pi}\int_0^{\pi/2}\frac{\partial W_1(\rho,t,\varphi)}{\partial t}d\varphi,\quad t\geq 0,\quad b_1\leq\rho\leq a_1$$
$$\begin{cases}\left[\dfrac{\partial^2}{\partial t^2}-\sin^2\varphi\dfrac{\partial}{\partial\rho}\dfrac{1}{\rho}\dfrac{\partial}{\partial\rho}\rho\right]W_1(\rho,t,\varphi)=\dfrac{\partial U^{s(1)}(\rho,z,t)}{\partial z}\bigg|_{z=-L_1}, & b_1\leq\rho\leq a_1,\quad t>0 \\ W_1(\rho,0,\varphi)=\dfrac{\partial W_1(\rho,t,\varphi)}{\partial t}\bigg|_{t=0}=0, & b_1\leq\rho\leq a_1,\end{cases}\qquad (5)$$

$$U(\rho,L_2,t)=\frac{2}{\pi}\int_0^{\pi/2}\frac{\partial W_2(\rho,t,\varphi)}{\partial t}d\varphi,\quad t\geq 0,\quad b_2\leq\rho\leq a_2$$

$$\begin{cases}\left[\dfrac{\partial^2}{\partial t^2}-\sin^2\varphi\dfrac{\partial}{\partial\rho}\dfrac{1}{\rho}\dfrac{\partial}{\partial\rho}\rho\right]W_2(\rho,t,\varphi)=-\left.\dfrac{\partial U(\rho,z,t)}{\partial z}\right|_{z=L_2}, & b_2\leq\rho\leq a_2,\quad t>0\\ W_2(\rho,0,\varphi)=\left.\dfrac{\partial W_2(\rho,t,\varphi)}{\partial t}\right|_{t=0}=0, & b_2\leq\rho\leq a_2\end{cases}\qquad(6)$$

(local absorbing conditions). The initial boundary-value problems involved in (5) and (6) with respect to the auxiliary functions $W_j(\rho,t,\varphi)$ must be supplemented with the following boundary conditions for all times $t\geq 0$:

$$\begin{cases}W_j(0,t,\varphi)=W_j(a_j,t,\varphi)=0 & (\mathrm{TE}_{0n}\text{-waves})\\ W_j(0,t,\varphi)=\left.\dfrac{\partial(\rho W_j(\rho,t,\varphi))}{\partial\rho}\right|_{\rho=a_j}=0 & (\mathrm{TM}_{0n}\text{-waves})\end{cases}\qquad(7)$$

(on the boundaries $\rho=0$ and $\rho=a_j$ of the region Ω_j for a circular waveguide) and

$$\begin{cases}W_j(b_j,t,\varphi)=W_j(a_j,t,\varphi)=0 & (\mathrm{TE}_{0n}\text{-waves})\\ \left.\dfrac{\partial(\rho W_j(\rho,t,\varphi))}{\partial\rho}\right|_{\rho=b_j}=\left.\dfrac{\partial(\rho W_j(\rho,t,\varphi))}{\partial\rho}\right|_{\rho=a_j}=0 & (\mathrm{TM}_{0n}\text{-waves})\end{cases}\qquad(8)$$

(on the boundaries $\rho=b_j$ and $\rho=a_j$ of the region Ω_j for a coaxial waveguide).
In (3) to (8) the following designations are used: $J_0(x)$ is the Bessel function, a_j and b_j are the radii of the waveguide Ω_j and of its inner conductor respectively (evidently, $b_j\neq 0$ if only Ω_j is a coaxial waveguide), $\{\mu_{nj}(\rho)\}$ and $\{\lambda_{nj}\}$ are the sets of transverse functions and transverse eigenvalues for the waveguide Ω_j.
Analytical representations for $\mu_{nj}(\rho)$ and λ_{nj} are well-known and for TE_{0n}-waves take the form:

$$\begin{cases}\mu_{nj}(\rho)=J_1(\lambda_{nj}\rho)\sqrt{2}\left[a_jJ_0(\lambda_{nj}a_j)\right]^{-1},\quad n=1,2,...,\\ \qquad\rho<a_j\quad\text{(circular waveguide)}\\ \lambda_{nj}>0\quad\text{are the roots of the equation}\quad J_1(\lambda a_j)=0,\end{cases}\qquad(9)$$

$$\begin{cases}\mu_{nj}(\rho)=G_1(\lambda_{nj},\rho)\sqrt{2}\left[a_j^2G_0^2(\lambda_{nj},a_j)-b_j^2G_0^2(\lambda_{nj},b_j)\right]^{-1/2},\quad n=1,2,...,\\ \qquad b_j<\rho<a_j\quad\text{(coaxial waveguide)}\\ \lambda_{nj}>0\quad\text{are the roots of the equation}\quad G_1(\lambda,a_j)=0\\ \qquad G_q(\lambda,\rho)=J_q(\lambda\rho)N_1(\lambda b_j)-N_q(\lambda\rho)J_1(\lambda b_j).\end{cases}\qquad(10)$$

For TM_{0n} -waves we have:

$$\begin{cases} \mu_{nj}(\rho) = J_1(\lambda_{nj}\rho)\sqrt{2}\left[a_j J_1(\lambda_{nj}a_j)\right]^{-1}, \quad n = 1,2,..., \\ \qquad \rho < a_j \quad \text{(circular waveguide)} \\ \lambda_{nj} > 0 \quad \text{are the roots of the equation} \quad J_0(\lambda a_j) = 0, \end{cases} \tag{11}$$

$$\begin{cases} \mu_{nj}(\rho) = \tilde{G}_1(\lambda_{nj},\rho)\sqrt{2}\left[a_j^2\tilde{G}_1^2(\lambda_{nj},a_j) - b_j^2\tilde{G}_1^2(\lambda_{nj},b_j)\right]^{-1/2}, \quad n = 1,2,... \\ \qquad \mu_0(\rho) = \left[\rho\sqrt{\ln(a_j/b_j)}\right]^{-1}, \quad b_j < \rho < a_j \quad \text{(coaxial waveguide)} \\ \lambda_{nj} > 0 \quad (n = 1,2,...) \quad \text{are the roots of the equation} \quad \tilde{G}_0(\lambda, b_j) = 0, \quad \lambda_{0j} = 0 \\ \qquad \tilde{G}_q(\lambda,\rho) = J_q(\lambda\rho)N_0(\lambda a_j) - N_q(\lambda\rho)J_0(\lambda a_j). \end{cases} \tag{12}$$

Here $N_q(x)$ are the Neumann functions. The basis functions $\mu_{nj}(\rho)$ satisfy boundary conditions at the ends of the appropriate intervals ($\rho < a_j$ or $b_j < \rho < a_j$) and the following equalities hold

$$\int_0^{a_j} \mu_{nj}(\rho)\mu_{mj}(\rho)\rho d\rho = \begin{cases} 0, & n \neq m \\ 1, & n = m \end{cases} \quad \text{or} \quad \int_{b_j}^{a_j} \mu_{nj}(\rho)\mu_{mj}(\rho)\rho d\rho = \begin{cases} 0, & n \neq m \\ 1, & n = m \end{cases} \tag{13}$$

in the circular or coaxial waveguide, respectively.

4. Exact radiation conditions for outgoing spherical waves and exact absorbing conditions for the artificial boundary in free space

When constructing the exact absorbing condition for the wave $U(g,t)$ crossing the artificial spherical boundary Γ, we will follow the sequence of transformations widely used in the theory of hyperbolic equations (e.g., Borisov, 1996) – incomplete separation of variables in initial boundary-value problems for telegraph or wave equations, integral transformations in the problems for one-dimensional Klein-Gordon equations, solution of the auxiliary boundary-value problems for ordinary differential equations, and inverse integral transforms.

In the domain $\Omega_{\text{ext}} = \Omega \setminus (\Omega_{\text{int}} \cup \Gamma)$, where the field $U(g,t)$ propagates freely up to infinity as $t \to \infty$, the 2-D initial boundary-value problem (1) in spherical coordinates takes the form

$$\begin{cases} \left[-\frac{\partial^2}{\partial t^2} + \frac{1}{r}\frac{\partial^2}{\partial r^2}r + \frac{1}{r^2}\frac{\partial}{\partial \vartheta}\left(\frac{1}{\sin\vartheta}\frac{\partial}{\partial \vartheta}\sin\vartheta\right)\right]U(g,t) = 0, \quad t > 0, \quad g = \{r,\vartheta\} \in \Omega_{\text{ext}} \\ U(g,t)\big|_{t=0} = 0, \quad \frac{\partial}{\partial t}U(g,t)\Big|_{t=0} = 0, \quad g \in \bar{\Omega}_{\text{ext}} \\ U(r,0,t) = U(r,\pi,t) = 0, \quad r \geq L, \quad t \geq 0. \end{cases} \tag{14}$$

Let us represent the solution $U(r,\vartheta,t)$ as $U(r,\vartheta,t)=u(r,t)\mu(\vartheta)$. Separation of variables in (14) results in a homogeneous Sturm-Liouville problem with respect to the function $\tilde{\mu}(\cos\vartheta)=\mu(\vartheta)$

$$\begin{cases} \left[\frac{d^2}{d\vartheta^2}+\operatorname{ctg}\vartheta\frac{d}{d\vartheta}-\frac{1}{\sin^2\vartheta}+\lambda^2\right]\tilde{\mu}(\cos\vartheta)=0, & 0<\vartheta<\pi \\ \tilde{\mu}(\cos\vartheta)\big|_{\vartheta=0,\pi}=0 \end{cases} \tag{15}$$

and the following initial boundary-value problem for $u(r,t)$:

$$\begin{cases} \left[-\frac{\partial^2}{\partial t^2}+\frac{\partial^2}{\partial r^2}-\frac{\lambda^2}{r^2}\right]r\,u(r,t)=0, & r\geq L, \quad t>0 \\ u(r,0)=\frac{\partial}{\partial t}u(r,t)\Big|_{t=0}=0, & r\geq L. \end{cases} \tag{16}$$

Let us solve the Sturm-Liouville problem (15) with respect to $\tilde{\mu}(\cos\vartheta)$ and λ. Change of variables $x=\cos\vartheta$, $\tilde{\mu}(x)=\tilde{\mu}(\cos\vartheta)$ yields the following boundary-value problem for $\tilde{\mu}(x)$:

$$\begin{cases} \left[\left(1-x^2\right)\frac{d^2}{dx^2}-2x\frac{d}{dx}+\left(\lambda^2-\frac{1}{1-x^2}\right)\right]\tilde{\mu}(x)=0, & |x|<1 \\ \tilde{\mu}(-1)=\tilde{\mu}(1)=0. \end{cases} \tag{17}$$

With $\lambda^2=\lambda_n^2=n(n+1)$ for each $n=1,2,3,...$ equation (17) has two nontrivial linearly independent solutions in the form of the associated Legendre functions $P_n^1(x)$ and $Q_n^1(x)$. Taking into account the behavior of these functions in the vicinity of their singular points $x=\pm 1$ (Bateman & Erdelyi, 1953), we obtain

$$\tilde{\mu}_n(\cos\vartheta)=\sqrt{(2n+1)/(2n(n+1))}P_n^1(\cos\vartheta). \tag{18}$$

Here $\{\tilde{\mu}_n(\cos\vartheta)\}_{n=1,2,...}$ is a complete orthonormal (with weight function $\sin\vartheta$) system of functions in the space $\mathbf{L}_2\left[(0<\vartheta<\pi)\right]$ and provides nontrivial solutions to (15). Therefore, the solution of initial boundary-value problem (14) can be represented as

$$U(r,\vartheta,t)=\sum_{n=1}^{\infty}u_n(r,t)\tilde{\mu}_n(\cos\vartheta), \quad u_n(r,t)=\int_0^{\pi}U(r,\vartheta,t)\tilde{\mu}_n(\cos\vartheta)\sin\vartheta d\vartheta, \tag{19}$$

where the space-time amplitudes $u_n(r,t)$ are the solutions to problems (16) for $\lambda^2=\lambda_n^2$. Our goal now is to derive the exact radiation conditions for space-time amplitudes $u_n(r,t)$ of the outgoing wave (19). By defining $w_n(r,t)=ru_n(r,t)$ and taking into account that $\lambda_n^2=n(n+1)$, we rewrite equation (16) as

$$\left[-\frac{\partial^2}{\partial t^2}+\frac{\partial^2}{\partial r^2}-\frac{n(n+1)}{r^2}\right]w_n(r,t)=0, \quad r\geq L, \quad t>0. \tag{20}$$

Now subject it to the integral transform

$$\tilde{f}(\omega)=\int_L^{\infty} f(r)Z_\gamma(\omega,r)dr, \quad \omega\geq 0, \tag{21}$$

where the kernel $Z_\gamma(\omega,r)=r^a\left[\alpha(\omega)J_\gamma(\omega r)+\beta(\omega)N_\gamma(\omega r)\right]$ satisfies the equation (Korn & Korn, 1961)

$$\left[\frac{\partial^2}{\partial r^2}+\frac{1-2a}{r}\frac{\partial}{\partial r}+\omega^2+\frac{a^2-\gamma^2}{r^2}\right]Z_\gamma(\omega,r)=0. \tag{22}$$

Here $\alpha(\omega)$, $\beta(\omega)$ are arbitrary functions independent of r, and a is a fixed real constant. Applying to (20) the transform (21) with $a=1/2$ and $\gamma=n+1/2$, we arrive at

$$\int_L^{\infty}\left[-\frac{\partial^2}{\partial t^2}-\omega^2\right]w_n(r,t)Z_\gamma(\omega,r)dr+Z_\gamma(\omega,r)\frac{\partial w_n(r,t)}{\partial r}\bigg|_L^{\infty}-w_n(r,t)\frac{\partial Z_\gamma(\omega,r)}{\partial r}\bigg|_L^{\infty}=0. \tag{23}$$

Since the 'signal' $w_n(r,t)$ propagates with a finite velocity, for any t we can always point a distance r such that the signal has not yet reached it, that is, for these t and r we have $w_n(r,t)\equiv 0$. Then we can rewrite equation (23) in the form

$$\int_L^{\infty}\left[-\frac{\partial^2}{\partial t^2}-\omega^2\right]w_n(r,t)Z_\gamma(\omega,r)dr-Z_\gamma(\omega,L)\frac{\partial w_n(r,t)}{\partial r}\bigg|_{r=L}+w_n(L,t)\frac{\partial Z_\gamma(\omega,r)}{\partial r}\bigg|_{r=L}=0. \tag{24}$$

From (24) the simple differential equation for the transforms $\tilde{w}_n(\omega,t)$ of the functions $w_n(r,t)$ follows:

$$\left[\frac{\partial^2}{\partial t^2}+\omega^2\right]\tilde{w}_n(\omega,t)=w_n(L,t)\frac{\partial Z_\gamma(\omega,r)}{\partial r}\bigg|_{r=L}-Z_\gamma(\omega,L)\frac{\partial w_n(r,t)}{\partial r}\bigg|_{r=L}. \tag{25}$$

In this equation, the values $\alpha(\omega)$ and $\beta(\omega)$ entering into $Z_\gamma(\omega,r)$ are not defined yet. With $\alpha(\omega)=1$ and $\beta(\omega)=0$, we have

$$Z_\gamma(\omega,r)=\sqrt{r}J_\gamma(\omega r) \tag{26}$$

and

$$\tilde{f}(\omega)=\int_L^{\infty} f(r)\sqrt{r}J_\gamma(\omega r)dr. \tag{27}$$

The last integral is the Hankel transform (Korn & Korn, 1961), which is inverse to itself, and

$$\begin{aligned} \tilde{f}(\omega)\sqrt{\omega} &= \int_0^\infty \left[f(r)\chi(r-L) \right] \sqrt{r\omega} J_\gamma(\omega r) dr, \\ f(r)\chi(r-L) &= \int_0^\infty \left[\tilde{f}(\omega)\sqrt{\omega} \right] \sqrt{r\omega} J_\gamma(\omega r) d\omega. \end{aligned} \tag{28}$$

By χ we denote the Heaviside step-function

$$\chi(r) = \begin{cases} 0 & \text{for} \quad r < 0 \\ 1 & \text{for} \quad r \geq 0 \end{cases}. \tag{29}$$

Taking into account (26), equation (25) can be rewritten in the form

$$\left[\frac{\partial^2}{\partial t^2} + \omega^2 \right] \tilde{w}_n(\omega,t) = g(\omega,t), \tag{30}$$

where

$$g(\omega,t) = w_n(L,t) \left[\frac{1}{2\sqrt{L}} J_\gamma(\omega L) + \omega\sqrt{L} J_\gamma'(\omega L) \right] - \sqrt{L} J_\gamma(\omega L) \left. \frac{\partial w_n(r,t)}{\partial r} \right|_{r=L}, \tag{31}$$

and the symbol $'$ denotes derivatives with respect to the whole argument ωL.
If G is a fundamental solution of the operator $B[G]$ (i.e., $B[G(t)] = \delta(t)$, where $\delta(t)$ is the Dirac delta function), then the solution to the equation $B[U(t)] = g(t)$ can be written as a convolution $U = (G * g)$ (Vladimirov, 1971). For $\left[\partial^2/\partial t^2 + \omega^2\right] G(t) = \delta(t)$ we have $G(t) = \chi(t)\omega^{-1}\sin\omega t$, and then

$$\begin{aligned} \tilde{w}_n(\omega,t) &= \int_0^\infty G(\omega,t-\tau) g(\omega,\tau) d\tau = \frac{1}{2\omega\sqrt{L}} J_\gamma(\omega L) \int_0^t w_n(L,\tau) \sin[\omega(t-\tau)] d\tau + \\ &+ \sqrt{L} J_\gamma'(\omega L) \int_0^t w_n(L,\tau) \sin[\omega(t-\tau)] d\tau - \frac{\sqrt{L}}{\omega} J_\gamma(\omega L) \int_0^t \left. \frac{\partial w_n(r,\tau)}{\partial r} \right|_{r=L} \sin[\omega(t-\tau)] d\tau. \end{aligned} \tag{32}$$

Applying the inverse transform (28) to equation (32), we can write

$$\begin{aligned} w_n(r,t)\chi(r-L) &= \int_0^\infty \tilde{w}_n(\omega,t) \omega \sqrt{r} J_\gamma(\omega r) d\omega = \\ &= \int_0^t \int_0^\infty J_\gamma(\omega r) J_\gamma(\omega L) \sin[\omega(t-\tau)] d\omega \sqrt{r} \left[\frac{1}{2\sqrt{L}} w_n(L,\tau) - \sqrt{L} \left. \frac{\partial w_n(r,\tau)}{\partial r} \right|_{r=L} \right] d\tau + \\ &+ \int_0^t \int_0^\infty \omega J_\gamma(\omega r) J_\gamma'(\omega L) \sin[\omega(t-\tau)] d\omega \sqrt{rL} w_n(L,\tau) d\tau. \end{aligned} \tag{33}$$

Let us denote

$$I_1 = \int_0^\infty J_\gamma(\omega r) J_\gamma(\omega L) \sin[\omega(t-\tau)] d\omega \quad \text{and} \quad I_2 = \int_0^\infty \omega J_\gamma(\omega r) J_\gamma'(\omega L) \sin[\omega(t-\tau)] d\omega. \tag{34}$$

Then from (Gradshteyn & Ryzhik, 2000) we have for $r > L > 0$

$$I_1 = \begin{cases} 0, & 0 < t-\tau < r-L \\ \dfrac{1}{2\sqrt{rL}} P_{\gamma-1/2}\left(\dfrac{r^2+L^2-(t-\tau)^2}{2rL}\right), & r-L < t-\tau < r+L \\ -\dfrac{\cos\gamma\pi}{\pi\sqrt{rL}} Q_{\gamma-1/2}\left(-\dfrac{r^2+L^2-(t-\tau)^2}{2rL}\right), & t-\tau > r+L, \end{cases} \tag{35}$$

where $P_\gamma(x)$ and $Q_\gamma(x)$ are the Legendre functions of the first and second kind, respectively. For $\gamma = n+1/2$, we can rewrite this formula as

$$. I_1 = \frac{1}{2\sqrt{rL}} P_n(q)\chi\left[(t-\tau)-(r-L)\right]\chi\left[(r+L)-(t-\tau)\right], \quad t-\tau > 0,. \tag{36}$$

where $q = \left[r^2 + L^2 - (t-\tau)^2\right]/2rL$ and $P_n(q)$ denotes a Legendre polynomial. Considering that

$$I_2 = \frac{\partial}{\partial L} I_1, \quad \frac{\partial P_n(q)}{\partial q} = \frac{1}{\sqrt{1-q^2}} P_n^1(q) \tag{37}$$

(Janke et al., 1960), and $d\chi(x)/dx = \delta(x)$, we can derive

$$\begin{aligned} I_2 = {} & \frac{1}{2\sqrt{rL}}\chi\left[(t-\tau)-(r-L)\right]\chi\left[(r+L)-(t-\tau)\right]\left[-\frac{P_n(q)}{2L} + \frac{P_n^1(q)}{\sqrt{1-q^2}}\left(\frac{1}{r}-\frac{q}{L}\right)\right] + \\ & + \frac{1}{2\sqrt{rL}} P_n(q)\left\{\delta(t-\tau-r+L) + \delta(r+L-t+\tau)\right\}. \end{aligned} \tag{38}$$

Finally, taking into account the relation $w_n(r,t) = r u_n(r,t)$, we have from (33)

$$\begin{aligned} u_n(r,t) = {} & \frac{L}{2r}\left\{\int_{t-(r+L)}^{t-(r-L)}\left[\left(\frac{L-rq}{rL\sqrt{1-q^2}} P_n^1(q) - \frac{1}{L}P_n(q)\right)u_n(L,\tau) - P_n(q)\left.\frac{\partial u_n(r,\tau)}{\partial r}\right|_{r=L}\right]d\tau + \right. \\ & \left. + u_n\left(L, t-(r-L)\right) + (-1)^n u_n\left(L, t-(r+L)\right)\right\}, \quad r > L. \end{aligned} \tag{39}$$

By using (19), we arrive at the desired radiation condition:

$$\begin{aligned} U(r,\vartheta,t) = {} & \frac{L}{2r}\sum_{n=1}^{\infty}\left\{\int_{t-(r+L)}^{t-(r-L)}\left[\left(\frac{L-rq}{rL\sqrt{1-q^2}} P_n^1(q) - \frac{1}{L}P_n(q)\right)\int_0^{\pi} U(L,\vartheta_1,\tau)\tilde{\mu}_n(\cos\vartheta_1)\sin\vartheta_1 d\vartheta_1 - \right.\right. \\ & \left.\left. - P_n(q)\int_0^{\pi}\left.\frac{\partial U(r,\vartheta_1,\tau)}{\partial r}\right|_{r=L}\tilde{\mu}_n(\cos\vartheta_1)\sin\vartheta_1 d\vartheta_1\right]d\tau + \right. \end{aligned} \tag{40}$$

$$+\int_0^{\pi}\left[U\left(L,\vartheta_1,t-(r-L)\right)+(-1)^n U\left(L,\vartheta_1,t-(r+L)\right)\right]\tilde{\mu}_n(\cos\vartheta_1)\sin\vartheta_1 d\vartheta_1\Bigg\}\times$$
$$\times\tilde{\mu}_n(\cos\vartheta),\quad r>L,\quad 0\le\vartheta\le\pi.$$

By passing to the limit $r\to L$ in (40), we obtain

$$\begin{aligned}U(L,\vartheta,t)=&\frac{1}{2}\sum_{n=1}^{\infty}\left\{\int_{t-2L}^{t}\left[\left(\frac{t-\tau}{L\sqrt{4L^2-(t-\tau)^2}}P_n^1\left(1-\frac{(t-\tau)^2}{2L^2}\right)-\frac{1}{L}P_n\left(1-\frac{(t-\tau)^2}{2L^2}\right)\right)\times\right.\right.\\&\times\int_0^{\pi}U(L,\vartheta_1,\tau)\,\tilde{\mu}_n(\cos\vartheta_1)\sin\vartheta_1 d\vartheta_1-\\&\left.-P_n\left(1-\frac{(t-\tau)^2}{2L^2}\right)\int_0^{\pi}\left.\frac{\partial U(r,\vartheta_1,\tau)}{\partial r}\right|_{r=L}\tilde{\mu}_n(\cos\vartheta_1)\sin\vartheta_1 d\vartheta_1\right]d\tau+\\&\left.+\int_0^{\pi}\left[U(L,\vartheta_1,t)+(-1)^n U(L,\vartheta_1,t-2L)\right]\tilde{\mu}_n(\cos\vartheta_1)\sin\vartheta_1 d\vartheta_1\right\}\tilde{\mu}_n(\cos\vartheta),\\&0\le\vartheta\le\pi.\end{aligned}\tag{41}$$

Formula (41) represents the exact absorbing condition on the artificial boundary Γ. This condition is spoken of as exact because any outgoing wave described by the initial problem (1) satisfies this condition. Every outgoing wave $U(g,t)$ passes through the boundary Γ without distortions, as if it is absorbed by the domain Ω_{ext} or its boundary Γ. That is why this condition is said to be absorbing.

5. On the equivalence of the initial problem and the problem with a bounded domain of analysis

We have constructed the following closed initial boundary-value problem

$$\begin{cases}\left[-\varepsilon(g)\dfrac{\partial^2}{\partial t^2}-\sigma(g)\dfrac{\partial}{\partial t}+\dfrac{\partial^2}{\partial z^2}+\dfrac{\partial}{\partial\rho}\left(\dfrac{1}{\rho}\dfrac{\partial}{\partial\rho}\rho\right)\right]U(g,t)=F(g,t),\quad t>0,\quad g\in\Omega_{\text{int}}\\ U(g,t)\big|_{t=0}=\varphi(g),\quad \dfrac{\partial}{\partial t}U(g,t)\Big|_{t=0}=\psi(g),\quad g\in\bar{\Omega}_{\text{int}}\\ E_{tg}(p,t)\big|_{p=\{\rho,\phi,z\}\in\Sigma}=0,\quad U(0,z,t)=0,\quad |z|\le L,\quad t\ge 0\\ E_{tg}(p,t)\ \text{and}\ H_{tg}(p,t)\ \text{are continuous when crossing}\ \Sigma^{\varepsilon,\sigma}\\ D_1\left[U(g,t)-U^{i(1)}(g,t)\right]\Big|_{g\in\Gamma_1}=0,\quad D_2\left[U(g,t)\right]\Big|_{g\in\Gamma_2}=0,\quad t\ge 0\\ D\left[U(g,t)\right]\Big|_{g\in\Gamma}=0,\end{cases}\tag{42}$$

where the operator D is given by (41). It is equivalent to the open initial problem (1). This statement can be proved by following the technique developed in (Ladyzhenskaya, 1985).

The initial and the modified problems are equivalent if and only if any solution of the initial problem is a solution to problem (42) and at the same time, any solution of the modified problem is the solution to problem (1). (In the Ω_{ext}-domain, the solution to the modified problem is constructed with the help of (40).) The solution of the initial problem is unique and it is evidently the solution to the modified problem according construction. In this case, if the solution of (42) is unique, it will be a solution to (1). Assume that problem (42) has two different solutions $U_1(g,t)$ and $U_2(g,t)$. Then the function $u(g,t)=U_1(g,t)-U_2(g,t)$ is also the generalized solution to (42) for $F(g,t)=U^{i(1)}(g,t)=\varphi(g)=\psi(g)\equiv 0$. This means that for any function $\gamma(g,t)\in \mathbf{W}_2^1(\Omega^T)$ that is zero at $t=T$, the following equality holds:

$$\int_{\Omega_{\text{int}}^T}\left[\varepsilon\frac{\partial u}{\partial t}\frac{\partial \gamma}{\partial t}-\left(\frac{1}{\rho^2}\frac{\partial}{\partial\rho}\rho u\right)\frac{\partial(\rho\gamma)}{\partial\rho}-\frac{\partial u}{\partial z}\frac{\partial\gamma}{\partial z}-\sigma\frac{\partial u}{\partial t}\gamma\right]dgdt+$$
$$+\int_{\Sigma_{\text{int}}^T}\left[\left(\frac{1}{\rho}\frac{\partial}{\partial\rho}\rho u\right)\gamma\cos(\vec{n},\vec{\rho})+\frac{\partial u}{\partial z}\gamma\cos(\vec{n},\vec{z})\right]dsdt=0. \tag{43}$$

Here, $\Omega_{\text{int}}^T=\Omega_{\text{int}}\times(0,T)$ and Σ_{int}^T are the space-time cylinder over the domain Ω_{int} and its lateral surface; $\cos(\vec{n},\vec{\rho})$ and $\cos(\vec{n},\vec{z})$ are the cosines of the angles between the outer normal $\vec{n}$ to the surface Σ_{int}^T and $\vec{\rho}$ - and $\vec{z}$ -axes, respectively; the element dg of the end surface of the cylinder Ω_{int}^T equals $\rho d\rho dz$.

By making the following suitable choice of function,

$$\gamma(g,t)=\begin{cases}\int_t^{\tau} u(g,\zeta)d\zeta & \text{for} \quad 0<t<\tau\\ 0 \quad \text{for} & \tau<t<T,\end{cases} \tag{44}$$

it is possible to show that every term in (43) is nonnegative (Mikhailov, 1976) and therefore $u(g,t)$ is equal to zero for all $g\in\Omega_{\text{int}}$ and $0<t<T$, which means that the solution to the problem (42) is unique. This proves the equivalency of the two problems.

6. Far-field zone problem. Extended and remote sources

As we have already mentioned, in contrast to approximate methods based on the use of the Absorbing Boundary Conditions or Perfectly Matched Layers, our approach to the effective truncation of the computational domain is rigorous, which is to say that the original open problem and the modified closed problem are equivalent. This allows one, in particular, to monitor a computational error and obtain reliable information about resonant wave scattering. It is noteworthy that within the limits of this rigorous approach we also obtain, without any additional effort, the solution to the far-field zone problem, namely, of finding the field $U(g,t)$ at arbitrary point in Ω_{ext} from the magnitudes of $U(g,t)$ on any arc $r=M\le L$, $0\le\vartheta\le\pi$, lying entirely in Ω_{int} and retaining all characteristics of the arc Γ. Thus in the case considered here, equation (39) defines the diagonal operator $S_{L\to r}(t)$ such that it operates on the space of amplitudes $u(r,t)=\{u_n(r,t)\}$ of the outgoing wave (19) according the rule

$$u(r,t)=S_{L\to r}(t)\left[u(L,\tau),\left.\frac{\partial u(\tilde{r},\tau)}{\partial\tilde{r}}\right|_{\tilde{r}=L}\right]; \quad r>L, \quad t\ge\tau\ge 0, \tag{45}$$

and allows one to follow all variations of these amplitudes in an arbitrary region of Ω_{ext}. The operator

$$U(r,\vartheta,t) = T_{L\to r}(t)\left[U(L,\tilde{\vartheta},\tau), \left.\frac{\partial U(\tilde{r},\tilde{\vartheta},\tau)}{\partial \tilde{r}}\right|_{\tilde{r}=L}\right]; \quad r > L, \quad 0 \le \vartheta \le \pi, \quad t \ge \tau \ge 0, \quad 0 \le \tilde{\vartheta} \le \pi, \tag{46}$$

given by (40), in turn, enables the variations of the field $U(g,t)$, $g \in \Omega_{\text{ext}}$, to be followed.
It is obvious that the efficiency of the numerical algorithm based on (42) reduces if the support of the function $F(g,t)$ and/or the functions $\varphi(g)$ and $\psi(g)$ is extended substantially or removed far from the region where the scatterers are located. The arising problem (the far-field zone problem or the problem of extended and remote sources) can be resolved by the following straightforward way.
Let us consider the problem

$$\begin{cases} \left[-\varepsilon(g)\dfrac{\partial^2}{\partial t^2} - \sigma(g)\dfrac{\partial}{\partial t} + \dfrac{\partial^2}{\partial z^2} + \dfrac{\partial}{\partial \rho}\left(\dfrac{1}{\rho}\dfrac{\partial}{\partial \rho}\rho\right)\right]U(g,t) = F(g,t) + \tilde{F}(g,t), \quad t > 0, \quad g \in \Omega \\ U(g,t)\big|_{t=0} = \varphi(g) + \tilde{\varphi}(g), \quad \dfrac{\partial}{\partial t}U(g,t)\Big|_{t=0} = \psi(g) + \tilde{\psi}(g), \quad g = \{\rho, z\} \in \bar{\Omega} \\ E_{tg}(p,t)\big|_{p=\{\rho,\phi,z\}\in\Sigma} = 0, \quad t \ge 0 \\ E_{tg}(p,t) \ \text{ and } \ H_{tg}(p,t) \ \text{ are continuous when crossing } \ \Sigma^{\varepsilon,\sigma} \\ U(0,z,t) = 0, \quad |z| < \infty, \quad t \ge 0 \\ D_1\left[U(g,t) - U^{i(1)}(g,t)\right]\Big|_{g\in\Gamma_1} = 0, \quad D_2\left[U(g,t)\right]\Big|_{g\in\Gamma_2} = 0, \quad t \ge 0, \end{cases} \tag{47}$$

which differs from the problem (1) only in that the sources $\tilde{F}(g,t)$ and $\tilde{\varphi}(g)$, $\tilde{\psi}(g)$ are located out of the domain Ω_{int} enveloping all the scatterers (Fig. 1). The supports of the functions $\tilde{F}(g,t)$, $\tilde{\varphi}(g)$, and $\tilde{\psi}(g)$ can be arbitrary large (and even unbounded) and are located in Ω_{ext} at any finite distance from the domain Ω_{int}.
Let the relevant sources generate a field $U^i(g,t)$ in the half plane $\Omega_0 = \{g : \rho > 0, |z| < \infty\}$. In other words, let the function $U^i(g,t)$ be a solution of the following Cauchy problem:

$$\begin{cases} \left[-\dfrac{\partial^2}{\partial t^2} + \dfrac{\partial^2}{\partial z^2} + \dfrac{\partial}{\partial \rho}\left(\dfrac{1}{\rho}\dfrac{\partial}{\partial \rho}\rho\right)\right]U^i(g,t) = \tilde{F}(g,t), \quad t > 0, \quad g \in \Omega_0 \\ U^i(g,t)\big|_{t=0} = \tilde{\varphi}(g), \quad \dfrac{\partial}{\partial t}U^i(g,t)\Big|_{t=0} = \tilde{\psi}(g), \quad g = \{\rho, z\} \in \bar{\Omega}_0 \\ U^i(0,z,t) = 0, \quad |z| < \infty, \quad t \ge 0. \end{cases} \tag{48}$$

It follows from (47), (48) that in the domain Ω_{ext} the function $U^s(g,t) = U(g,t) - U^i(g,t)$ satisfies the equations

$$\begin{cases} \left[\dfrac{\partial^2}{\partial t^2} + \dfrac{\partial^2}{\partial z^2} + \dfrac{\partial}{\partial \rho}\left(\dfrac{1}{\rho}\dfrac{\partial}{\partial \rho}\rho \right) \right] U^s(g,t) = 0, \quad t > 0, \quad g \in \Omega_{\text{ext}} \\ U^s(g,t)\Big|_{t=0} = 0, \quad \dfrac{\partial}{\partial t} U^s(g,t)\Big|_{t=0} = 0, \quad g = \{\rho, z\} \in \bar{\Omega}_{\text{ext}} \\ U^s(0,z,t) = 0, \quad |z| \ge L, \quad t \ge 0 \end{cases} \tag{49}$$

and determines there the pulsed electromagnetic wave crossing the artificial boundary Γ in one direction only, namely, from Ω_{int} into Ω_{ext}.

The problems (49) and (14) are qualitatively the same. Therefore, by repeating the transformations of Section 4, we obtain

$$\begin{aligned} U^s(L,\vartheta,t) = \frac{1}{2}\sum_{n=1}^{\infty}\Bigg\{ \int_{t-2L}^{t} \Bigg[\Bigg(\frac{t-\tau}{L\sqrt{4L^2-(t-\tau)^2}} P_n^1\left(1-\frac{(t-\tau)^2}{2L^2}\right) - \frac{1}{L}P_n\left(1-\frac{(t-\tau)^2}{2L^2}\right)\Bigg) \times \\ \times \int_0^{\pi} U^s(L,\vartheta_1,\tau)\tilde{\mu}_n(\cos\vartheta_1)\sin\vartheta_1 d\vartheta_1 - \\ - P_n\left(1-\frac{(t-\tau)^2}{2L^2}\right)\int_0^{\pi}\frac{\partial U^s(r,\vartheta_1,\tau)}{\partial r}\Bigg|_{r=L}\tilde{\mu}_n(\cos\vartheta_1)\sin\vartheta_1 d\vartheta_1\Bigg]d\tau + \\ + \int_0^{\pi}\left[U^s(L,\vartheta_1,t)+(-1)^n U^s(L,\vartheta_1,t-2L)\right]\tilde{\mu}_n(\cos\vartheta_1)\sin\vartheta_1 d\vartheta_1\Bigg\}\tilde{\mu}_n(\cos\vartheta), \quad 0\le\vartheta\le\pi, \end{aligned} \tag{50}$$

or, in the operator notations, $D\left[U(g,t)-U^i(g,t)\right]\Big|_{g\in\Gamma}=0$, – the exact absorbing condition allowing one to replace open problem (47) with the equivalent closed problem

$$\begin{cases} \left[-\varepsilon(g)\dfrac{\partial^2}{\partial t^2} - \sigma(g)\dfrac{\partial}{\partial t} + \dfrac{\partial^2}{\partial z^2} + \dfrac{\partial}{\partial \rho}\left(\dfrac{1}{\rho}\dfrac{\partial}{\partial \rho}\rho \right) \right] U(g,t) = F(g,t), \quad t > 0, \quad g \in \Omega_{\text{int}} \\ U(g,t)\Big|_{t=0} = \varphi(g), \quad \dfrac{\partial}{\partial t} U(g,t)\Big|_{t=0} = \psi(g), \quad g \in \bar{\Omega}_{\text{int}} \\ E_{tg}(p,t)\Big|_{p=\{\rho,\phi,z\}\in\Sigma} = 0, \quad U(0,z,t) = 0, \quad |z| \le L, \quad t \ge 0 \\ E_{tg}(p,t) \text{ and } H_{tg}(p,t) \text{ are continuous when crossing } \Sigma^{\varepsilon,\sigma} \\ D_1\left[U(g,t) - U^{i(1)}(g,t)\right]\Big|_{g\in\Gamma_1} = 0, \quad D_2\left[U(g,t)\right]\Big|_{g\in\Gamma_2} = 0, \quad t \ge 0 \\ D\left[U(g,t) - U^i(g,t)\right]\Big|_{g\in\Gamma} = 0. \end{cases} \tag{51}$$

7. Determination of the incident fields

To implement the algorithms based on the solution of the closed problems (42), (51), the values of the functions $U^{i(1)}(g,t)$ and $U^{i}(g,t)$ as well as their normal derivatives on the boundaries Γ_1 and Γ are required (see formulas (3), (5), (50)). Let us start from the function $U^{i(1)}(g,t)$. In the feeding waveguide Ω_1, the field $U^{i(1)}(g,t)$ incoming on the boundary Γ_1 can be represented (Sirenko *et al.*, 2007) as

$$U^{i(1)}(g,t)=\sum_n U_n^{i(1)}(g,t)=\sum_n v_{n1}(z,t)\,\mu_{n1}(\rho);\quad b_1<\rho<a_1,\quad z\leq -L_1. \tag{52}$$

Here (see also Section 3), $n=0,1,2,...$ only in the case of TM_{0n}-waves and only for a coaxial waveguide Ω_1. In all other cases $n=1,2,3,...$. On the boundary Γ_1, the wave $U^{i(1)}(g,t)$ can be given by a set of its amplitudes $\{v_{n1}(-L_1,t)\}_n$. The choice of the functions $v_{n1}(-L_1,t)$, which are nonzero on the finite interval $0<T_1\leq t\leq T_2<T$, is arbitrary to a large degree and depends generally upon the conditions of a numerical experiment. As for the set $\left\{\partial v_{n1}(z,t)/\partial z\big|_{z=-L_1}\right\}_n$, which determines the derivative of the function $U^{i(1)}(g,t)$ on Γ_1, it should be selected with consideration for the causality principle. Each pair $V_{n1}(\rho,t)=\left\{v_{n1}(-L_1,t)\mu_{n1}(\rho);\left(\partial v_{n1}(z,t)/\partial z\big|_{z=-L_1}\right)\mu_{n1}(\rho)\right\}$ is determined by the pulsed eigenmode $U_n^{i(1)}(g,t)$ propagating in the waveguide Ω_1 in the sense of increasing z. This condition is met if the functions $v_{n1}(-L_1,t)$ and $\partial v_{n1}(z,t)/\partial z\big|_{z=-L_1}$ are related by the following equation (Sirenko et al., 2007):

$$v_{n1}(-L_1,t)=-\int_0^t J_0\left[\lambda_{n1}(t-\tau)\right]\frac{\partial v_{n1}(z,\tau)}{\partial z}\bigg|_{z=-L_1}d\tau;\quad t\geq 0. \tag{53}$$

The function $U^{i}(g,t)$ generated by the sources $\tilde{F}(g,t)$, $\tilde{\varphi}(g)$, and $\tilde{\psi}(g)$ is the solution to the Cauchy problem (48). Let us separate the transverse variable ρ in this problem and represent its solution in the form (Korn & Korn, 1961):

$$U^{i}(\rho,z,t)=\int_0^\infty v_\lambda(z,t)J_1(\lambda\rho)d\lambda \tag{54}$$

$$\begin{aligned} v_\lambda(z,t)&=\int_0^\infty v_\mu(z,t)\delta(\mu-\lambda)d\mu-\lambda\int_0^\infty v_\mu(z,t)\left[\int_0^\infty J_1(\mu\rho)J_1(\lambda\rho)\rho d\rho\right]d\mu-\\ &=\lambda\int_0^\infty\left[\int_0^\infty v_\mu(z,t)J_1(\mu\rho)d\mu\right]J_1(\lambda\rho)\rho d\rho=\lambda\int_0^\infty U^{i}(\rho,z,t)J_1(\lambda\rho)\rho d\rho. \end{aligned} \tag{55}$$

In order to find the functions $v_\lambda(z,t)$, one has to invert the following Cauchy problems for one-dimensional Klein-Gordon equations:

$$\begin{cases}\left[-\dfrac{\partial^2}{\partial t^2}+\dfrac{\partial^2}{\partial z^2}-\lambda^2\right]v_\lambda(z,t)=F_\lambda(z,t), & t>0,\quad |z|<\infty\\ v_\lambda(z,t)\big|_{t=0}=\varphi_\lambda(z),\quad \dfrac{\partial}{\partial t}v_\lambda(z,t)\bigg|_{t=0}=\psi_\lambda(z), & |z|<\infty.\end{cases} \tag{56}$$

Here, $F_\lambda(z,t)$, $\varphi_\lambda(z)$ и $\psi_\lambda(z,t)$ are the amplitude coefficients in the integral presentations (54) for the functions $\tilde{F}(g,t)$, $\tilde{\varphi}(g)$, and $\tilde{\psi}(g)$.
Now, by extending the functions $F_\lambda(z,t)$ and $v_\lambda(z,t)$ with zero on the interval $t<0$, we pass on to a generalized version of problems (56) (Vladimirov, 1971)

$$B(\lambda)[v_\lambda] \equiv \left[-\frac{\partial^2}{\partial t^2}+\frac{\partial^2}{\partial z^2}-\lambda^2\right]v_\lambda(z,t)=F_\lambda(z,t)-\delta^{(1)}(t)\varphi_\lambda(z)-\delta(t)\psi_\lambda(z)=f_\lambda(z,t), \quad |t|<\infty, \quad |z|<\infty \tag{57}$$

($\delta^{(1)}(t)$ is the generalized derivative of the function $\delta(t)$). Their solutions can be written by using the fundamental solution $G(z,t,\lambda)=(-1/2)\chi(t-|z|)J_0\left[\lambda\left(t^2-z^2\right)^{1/2}\right]$ of the operator $B(\lambda)$ as follows:

$$v_\lambda(z,t)=\left[G(z,t,\lambda)*f_\lambda(z,t)\right]=\int_{-\infty}^{\infty}\int_{0}^{\infty}G(z-\tilde{z},t-\tau,\lambda)f_\lambda(\tilde{z},\tau)d\tau\, d\tilde{z}. \tag{58}$$

Equations (54) and (58) completely determine the desired function $U^i(g,t)$.

8. Conclusion

In this paper, a problem of efficient truncation of the computational domain in finite-difference methods is discussed for axially-symmetrical open electrodynamic structures. The original problem describing electromagnetic wave scattering on a compact axially-symmetric structure with feeding waveguides is an initial boundary-value problem formulated in an unbounded domain. The exact absorbing conditions have been derived for a spherical artificial boundary enveloping all sources and scatterers in order to truncate the computational domain and replace the original open problem by an equivalent closed one. The constructed solution has been generalized to the case of extended and remote field sources. The analytical representation for the operators converting the near-zone fields into the far-zone fields has been also derived.
We would like to make the following observation about our approach.

- In our description, the waveguide Ω_1 serves as a feeding waveguide. However, both of the waveguides can be feeding or serve to withdraw the energy; also both of them may be absent in the structure.
- The choice of the parameters $\alpha(\omega)$ and $\beta(\omega)$ determining $Z_\gamma(\omega,r)$ (see Section 4) affects substantially the final analytical expression for the exact absorbing condition on the spherical boundary Γ. When constructing boundary conditions (41), (50), we assumed that $\alpha(\omega)=1$ and $\beta(\omega)=0$. In (Sirenko et al., 2007), for a similar situation, the exact absorbing conditions for outgoing pulsed waves were constructed with the assumption that $\alpha(\omega)=-N_\gamma(\omega L)$ and $\beta(\omega)=J_\gamma(\omega L)$. With such $\alpha(\omega)$ and $\beta(\omega)$, equation (21) is the Weber-Orr transform (Bateman & Erdelyi, 1953). However, the final formulas corresponding to (39), (40) for this case turn into identities as $r \to L$, which present a considerable challenge for using them as absorbing conditions. In addition, the analytical expressions with the use of Weber-Orr transform are rather complicated to implement numerically.

- The function $U^i(g,t)$ (see Section 7) can be found in spherical coordinates as well. In this situation, we arrive (see Section 4) at the expansions like (19) with the amplitude coefficients $v_n(r,t)$ determined by the Cauchy problems

$$\begin{cases} \left[-\frac{\partial^2}{\partial t^2} + \frac{\partial^2}{\partial r^2} - \frac{n(n+1)}{r^2} \right] r v_n(r,t) = F_n(r,t), & r \geq 0, \quad t > 0 \\ v_n(r,0) = \varphi_n(r), \quad \frac{\partial}{\partial t} v_n(r,t)\Big|_{t=0} = \psi_n(r), & r \geq 0 \end{cases}, \tag{59}$$

  where $F_n(r,t)$, $\varphi_n(r)$, and $\psi_n(r)$ are the amplitude coefficients for the functions $\tilde{F}(g,t)$, $\tilde{\varphi}(g)$, and $\tilde{\psi}(g)$.
- The standard discretization of the closed problems (42), (51) by the finite difference method using a uniform rectangular mesh attached to coordinates $g = \{\rho, z\}$ leads to explicit computational schemes with uniquely defined mesh functions $U(j,k,m) = U(\rho_j, z_k, t_m)$. The approximation error is $O(\bar{h}^2)$, where $\bar{h}$ is the mesh width in spatial coordinates, $\bar{l} = \bar{h}/2$ for $\theta = \max_{g \in \Omega_{\text{int}}} [\varepsilon(g)] < 2$ or $\bar{l} < \bar{h}/2$ for $\theta \geq 2$ is the mesh width in time variable t; $\rho_j = j\bar{h}$, $z_k = k\bar{h}$, and $t_m = m\bar{l}$. The range of the integers $j = 0,1,...,J$, $k = 0,1,...K$, and $m = 0,1,...M$ depends both on the size of the Ω_{int} domains and on the length of the interval $[0,T]$ of the observation time t. The condition providing uniform boundedness of the approximate solutions $U(j,k,m)$ with decreasing $\bar{h}$ and $\bar{l}$ is met (see, for example, formula (1.50) in (Sirenko *et al.*, 2007)). Hence the finite-difference computational schemes are stable, and the mesh functions $U(j,k,m)$ converge to the solutions $U(\rho_j, z_k, t_m)$ of the original problems (42), (51).

As opposed to the well-known approximate boundary conditions standardly utilized by finite-difference methods, the conditions derived in this paper are exact by construction and do not introduce an additional error into the finite-difference algorithm. This advantage is especially valuable in resonant situations, where numerical simulation requires large running time and the computational errors may grow unpredictably if an open problem is replaced by an insufficiently accurate closed problem.

9. References

Bateman, H. & Erdelyi, A. (1953). *Higher transcendental functions*, McGraw-Hill, ISBN 048644614X, New York

Borisov, V.V. (1996). *Electromagnetic fields of transient currents*, St. Petersburg Univ. Press, ISBN 5-288-01256-3, St. Petersburg

Gradshteyn, I.S. & Ryzhik, I.M. (2000). *Table of Integrals, Series, and Products*, ISBN 0-12-294757-6, Academic Press, San Diego, London

Janke, E., Emde, F., & Lösch, F. (1960). *Tables of Higher Functions*, McGraw-Hill, ISBN 0070322457, New York

Korn, G.A. & Korn, T.M. (1961). *Mathematical Handbook for Scientists and Engineers*, Mc Graw-Hill, ISBN 0070353700, New York

Ladyzhenskaya, O.A. (1985). *The boundary value problems of mathematical physics*, Springer-Verlag, ISBN 3540909893, New York

Maikov, A.R., Sveshnikov, A.G., & Yakunin, S.A. (1986). Difference scheme for the Maxwell transient equations in waveguide systems. *Journal of Computational Mathematics and Mathematical Physics*, Vol. 26, pp. 851–863, ISSN 0965-5425

Mikhailov, V.P. (1976). *Partial Differential Equations*. Nauka, ISBN 20203-032-13-76, Moscow

Sirenko, Y.K., Strom, S., & Yashina, N.P. (2007). *Modeling and Analysis of Transient Processes in Open Resonant Structures. New Methods and Techniques*, Springer, ISBN-10 0-387-30878-4, New York

Taflove, A. & Hagness, S.C. (2000). *Computational Electrodynamics: the Finite-Difference Time-Domain Method*, Artech House, ISBN 0-89006-792-9, Boston

Vladimirov, V.S. (1971). *Equations of mathematical physics*, Dekker, ISBN 978-0824717131, New York

6

Fractional Operators Approach and Fractional Boundary Conditions

Eldar Veliev[1], Turab Ahmedov[2] and Maksym Ivakhnychenko[3]
[1]*State Research and Design Institute of Basis Chemistry*
[2]*Institute of Mathematics NAS of Azerbaijan*
[3]*Institute of Radiophysics and Electronics NAS of Ukraine*
[1,3]*Ukraine*
[2]*Azerbaijan*

1. Introduction

Tools of fractional calculus including fractional operators and transforms have been utilized in physics by many authors (Hilfer, 2000). Fractional operators defined as fractionalizations of some commonly used operators allow describing of intermediate states. For example, fractional derivatives and integrals (Oldham & Spanier, 1974; Samko et al., 1993) are generalizations of derivative and integral. Fractional curl operator defined in (Engheta, 1998) is a fractionalized analogue of conventional curl operator used in many equations of mathematical physics. A fractionalized operator generalizes the original operator. The idea to use fractional operators in electromagnetic problems was formulated by N. Engheta (Engheta, 2000) and named "fractional paradigm in electromagnetic theory".

Our purpose is to find possible applications of the use of fractional operators in the problems of electromagnetic wave diffraction. In this paper two-dimensional problems of diffraction by infinitely thin surfaces are considered: a strip, a half-plane and a strip resonator (Fig.1). Assume that an incident field is an E-polarized plane wave, described by the function

$$\vec{E}^i = \vec{z}E_z^i(x,y) = \vec{z}e^{-ik(x\cos\theta + y\sin\theta)}, \quad (1)$$

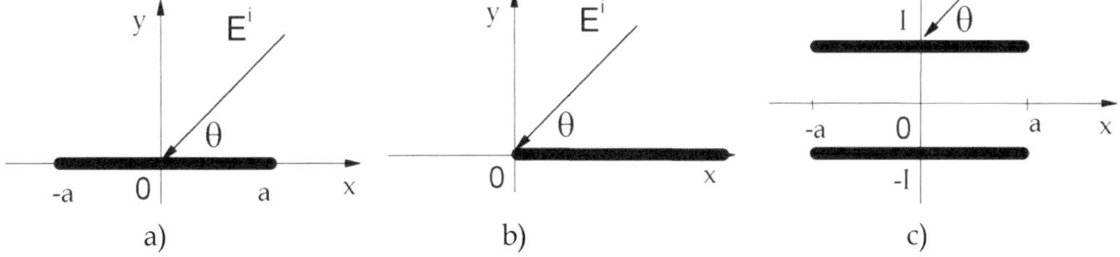

Fig. 1. Geometry of the diffraction problems: a) strip, b) half-plane, c) two parallel strips.

where θ is the incidence angle, $k=\frac{2\pi}{\lambda}$ is the wavenumber. Here, the time dependence is assumed to be $e^{-i\omega t}$ and omitted throughout the paper. There are three structures considered in this paper:

- a strip located in the plane $y=0$ ($x\in[-a,a]$) infinite along the axis z (Fig. 1a);
- a half-plane ($y=0$, $x\geq 0$) (Fig. 1b);
- two parallel strips infinite along the axis z (a strip resonator). The first strip is located at $y=l$, $x\in[-a,a]$, and the second one is at $y=-l$, $x\in[-a,a]$ (Fig. 1c).

One may ask what new features are that the fractional operators can bring to the theory of diffraction. The concept of intermediate states, obtained with the aid of fractional derivatives and integrals, yields to various generalizations of commonly used models in electrodynamics such as:

- Intermediate waves. For instance, intermediate waves between plane and cylindrical waves (Engheta, 1996, 1999) can be obtained using fractional integral of scalar Green's function:

$$G^{\alpha}(x,y;k)\equiv\frac{1}{2}\left({}_{-\infty}D_{y}^{-\alpha}G_2(x,y;k)-{}_{-\infty}D_{-y}^{-\alpha}G_2(x,y;k)\right),\ 0\leq\alpha\leq 1,$$

where G_2 is two-dimensional Green's function of the free space. G^{α} describes an intermediate case between one- and two-dimensional Green's functions and have the following behavior in the far-zone (Engheta, 1999):

$$G^{\alpha}\sim\frac{i}{4\pi}\cos\left(\frac{\pi\alpha}{2}\right)(k\sin|\varphi|)^{-\alpha}\sqrt{\frac{2\pi}{k\rho}}e^{ik\rho-i\pi/4}+\frac{i}{2k^{\alpha}}\Gamma(\alpha)\frac{e^{ik|x|}}{k|y|^{1-\alpha}},\quad k\rho=k\sqrt{x^2+y^2}\to\infty,\ \varphi\neq 0.$$

This function consists of two waves: a cylindrical wave and a non-uniform plane wave propagating in the x direction and behaving with y as $|y|^{\alpha-1}$.

- Fractional Green's function G^{α} defined as a fractional derivative (integral) of the ordinary Green's function of the free space - $G^{\alpha}\equiv{}_{-\infty}D_{ky}^{\alpha}G$. α denotes the fractional order and varies from 0 to 1 ($0\leq\alpha\leq 1$). In two-dimensional case G^{α} is expressed as

$$G^{\alpha}(x-x',y-y')=-\frac{i}{4}D_{ky}^{\alpha}H_0^{(1)}\left(k\sqrt{(x-x')^2+(y-y')^2}\right). \tag{2}$$

- Fractional Green's theorem which involves fractional derivatives of ordinary Green's function and fractional derivatives of the considered function on a boundary of a domain (Veliev & Engheta, 2003). The corresponding equations will be presented later in this paper.
- Fractional boundary conditions (FBC) defined via fractional derivatives of the tangential electric field components $U(x,y)$. For an infinitely thin boundary S located in the plane $y=d$, FBC are defined as

$$D_y^{\alpha}U(x,y)|_{y\in S}=0,\ y\to\pm d.$$

The order of the fractional derivative α is assumed to be between 0 and 1. Fractional derivative D^{α} is applied along the direction normal to the surface S. Fractional boundary conditions describe an intermediate boundary between the perfect electric conductor (PEC) and the perfect magnetic conductor (PMC), obtained from FBC if the fractional order equals to 0 and 1, respectively.
We will use the symbol $D_y^{\alpha} f$ to denote operator of fractional derivative or integral ${}_{-\infty}D_y^{\alpha} f$, which is defined by the integral of Riemann-Liouville on semi-infinite interval (Samko et al., 1993):

$$({}_{\infty}D_x^{\alpha} f)(x) = \frac{1}{\Gamma(1-\alpha)} \frac{d}{dx} \int_{-\infty}^{x} \frac{f(t)dt}{(x-t)^{\alpha}}, \quad 0 < \alpha < 1,$$

where $\Gamma(1-\alpha)$ is Gamma function.
This paper is devoted to the problems of diffraction by a strip, a strip resonator and a half-plane characterized with fractional boundary conditions with $0 \le \alpha \le 1$ expressed as

$$D_{ky}^{\alpha} E_z(x,y) = 0, \; y \to \pm 0, \quad x \in L,$$

where $L = (-a,a)$ for a strip and $L = (0,\infty)$ for a half-plane. For convenience, fractional derivative is applied with respect to dimensionless variable ky. The function $E_z(x,y)$ denotes z-component of the total electric field, $E_z(x,y) = E_z^i + E_z^s$, that is the sum of the incident plane wave $E_z^i(x,y)$ and the scattered wave $E_z^s(x,y)$.
In case of a strip resonator we have two equations to impose fractional boundary conditions:

$$D_{ky}^{\alpha} E_z(x,y) = 0, \; y \to l \pm 0, \quad x \in (-a,a),$$

$$D_{ky}^{\alpha} E_z(x,y) = 0, \; y \to -l \pm 0, \quad x \in (-a,a).$$

From the one hand, introduction of new boundary conditions should describe a new physical boundary world, and from the other hand they must allow to build an effective computational algorithm to solve the stated problems with a desired accuracy. Simple mathematical description of the scattering properties of surfaces is a common problem in modeling in diffraction theory.
One of the well-studied boundaries, which can be treated as an intermediate state between PEC and PMC, is an impedance boundary defined by the equation

$$\vec{n} \times \vec{E}(\vec{r}) = \eta \vec{n} \times (\vec{n} \times \vec{H}(\vec{r})), \; \vec{r} \to S,$$

where $\vec{n}$ is the normal to the surface S. The value of the impedance η varies from 0 for PEC to $i\infty$ for PMC.
There are many papers devoted to diffraction by impedance boundaries. Impedance boundary conditions (IBC) have been used for the modeling of the scattering properties of good conductors, gratings, etc. In each case there are formulas to define the value of the impedance as a function of material parameters. IBC are approximate BC and therefore they have limitations in usage and cannot describe all diversity of boundaries.

Further approximation of IBC can be made with the aid of derivatives of higher but integer orders or generalized boundary conditions (Hope & Rahmat-Samii, 1995; Senior & Volakis, 1995). A general methodology to obtain exact IBC of higher order in spectral domain is presented in (Hope & Rahmat-Samii, 1995), where flat covers (and also surfaces with curvature) consisting of homogeneous materials with an arbitrary (linear, bi-anisotropic) constitutive equations. It is possible to obtain exact IBC in the spectral domain that can be often done in an analytical form very often. However, it is not always possible to get IBC in the spatial domain in an exact form. That is why it is necessary to approximate IBC in the spectral domain in order to apply inverse Fourier transform.
Another boundary condition that generalizes the perfect boundaries like PEC and PMC was introduced in (Lindell & Sihvola, 2005a). The corresponding surface was named perfect electromagnetic conductor (PEMC) and the mentioned condition is defined as

$$\vec{H} + M\vec{E} = 0 .$$

For $M = 0$, PEMC defines a PEC boundary and for $M = \infty$ we get a PMC. The physical model of PEMC boundary was proposed in (Lindell & Sihvola, 2005b) where it was shown that the PEMC condition can simulate reflection from an anisotropic layer for the normal incidence of the plane wave. Diffraction by a PEMC boundary has not been considered yet. Further generalization of PEMC can be made using concept of the generalized soft-and-hard surface (GSHS) (Haninnen et al., 2006):

$$\vec{a} \cdot \vec{E} = 0 , \; \vec{b} \cdot \vec{H} = 0 ,$$

where $\vec{a}$, $\vec{b}$ are complex vectors that satisfy equations $\vec{n} \cdot \vec{a} = \vec{n} \cdot \vec{b} = 0$ and $\vec{a} \cdot \vec{b} = 1$. GSHS can transform an incident plane wave with any given polarization into any other polarization of the reflected plane wave if the vectors $\vec{a}$, $\vec{b}$ are chosen appropriately (Haninnen et al., 2006).
Fractional boundary conditions (FBC) can be compared with impedance boundary conditions (IBC). First of all FBC are intermediate between PEC and PMC as well as IBC. The value of fractional order $\alpha = 0$ ($\alpha = 1$) corresponds to the value of impedance $\eta = 0$ ($\eta = i\infty$), respectively. For other values of $0 < \alpha < 1$ the deeper analysis is needed.
Physical analysis of the strip with FBC shows that the induced surface currents behave similarly to the currents on an impedance strip. Due to specific properties the strip with FBC is compared with the well-known impedance strip. It can be shown that for a wide range of input parameters the "fractional strip" behaves similarly to the impedance strip if the fractional order is chosen appropriately (Veliev et al., 2008b). The proposed method used for a "fractional strip" has some advantages over the known methods applied to the analysis of the wave scattering by an impedance strip.
The purpose of this work is to build an effective analytic-numerical method to solve two-dimensional diffraction problems for the boundaries described by fractional boundary conditions with $\alpha \in [0,1]$. The method will be applied to two canonical scattering objects: a strip and a half plane. The method is based on presenting the scattered field via fractional Green's function,

$$E_Z^S(x,y) \equiv \int_L f^{1-\alpha}(x')G^{\alpha}(x-x',y)dx' ,$$

where $f^{1-\alpha}(x)$ is the unknown function and $G^{\alpha}(x-x',y)=-\frac{i}{4}D_{ky}^{\alpha}H_0^{(1)}(k\sqrt{(x-x')^2+y^2})$ is the fractional derivative of the Green's function defined by equation (2). This presentation leads to the following dual integral equations (DIE) with respect to the Fourier transform $F^{1-\alpha}(q)=\int_L f^{1-\alpha}(\xi)e^{-ikq\xi}d\xi$ of the function $f^{1-\alpha}(x)$

$$\begin{cases} \int_{-\infty}^{\infty} F^{1-\alpha}(q)e^{ikd_L\xi q}(1-q^2)^{\alpha-1/2}dq = -4\pi e^{i\pi/2(1-\alpha)}\sin^{\alpha}\theta e^{-ikd_L\xi\cos\theta}, & \xi \in L, \\ \int_{-\infty}^{\infty} F^{1-\alpha}(q)e^{ikd_L\xi q}dq = 0, & \xi \notin L, \end{cases}$$

where $d_L = a$ for $L=(-a,a)$, $d_L = 1$ for $L=(0,\infty)$.
In the case of a strip resonator, we obtain more complicated set of integral equations which will be presented later in this paper.
The method generalizes the known method used for the PEC and PMC strip and half plane. As will be shown later, this method allows obtaining a solution for the value $\alpha = 0.5$ in the explicit analytical form. For other values of $\alpha \in [0,1]$ the scattering problems are reduced to solving of the infinite systems of linear algebraic equations (SLAE). In order to discretize the DIE the function $f^{1-\alpha}(x)$ is represented as a series in terms of orthogonal polynomials: Gegenbauer polynomials for the strip and Laguerre polynomials for the half-plane. These representations result in a special kind of the edge conditions for the fractional current density function $f^{1-\alpha}(x)$. The physical characteristics of the considered scattering objects can be found with any desired accuracy by solving SLAE.

2. Diffraction by a strip with fractional boundary conditions

Assume that an E-polarized plane wave is characterized with the function $\vec{E}^i = \vec{z}E_z^i(x,y) = \vec{z}e^{-ik(x\cos\theta + y\sin\theta)}$. The total field $\vec{E} = \vec{z}E_z(x,y)$ must satisfy fractional boundary conditions

$$D_{ky}^{\alpha}E_z(x,y) = 0, \quad y \to \pm 0, \quad x \in L, \tag{3}$$

where $L=(-a,a)$ for a strip. For convenience, fractional derivative ${}_{-\infty}D_{ky}^{\alpha}$ is applied with respect to a dimensionless variable ky. The function $E_z(x,y)$ denotes the z-component of the total electric field $E_z(x,y)=E_z^i + E_z^s$ that is the sum of the incident plane wave $E_z^i(x,y)$ and the scattered field $E_z^s(x,y)$. Solution to the diffraction by the screen $S=\{(x,y): y=0, -a<x<a\}$ is to be sought under the following conditions:

- The total field $\vec{E}$ must satisfy the Helmholtz equation everywhere outside the screen

$$\left(\frac{\partial^2}{\partial x^2}+\frac{\partial^2}{\partial y^2}+k^2\right)E_z(x,y)=0. \tag{4}$$

- The scattered field $E_z^s(x,y)$ must satisfy Sommerfeld radiation condition at the infinity

$$\lim_{r\to\infty}\sqrt{r}\left(\frac{\partial E_z^s}{\partial r}-iE_z^s\right)=0,\ r=\sqrt{x^2+y^2} \tag{5}$$

- The total field $\vec{E}$ must satisfy the edge condition, i.e. the finiteness of energy in every local area near the edges of the screen (Honl et al., 1961).
- The total field $E_z(x,y)$ must satisfy the boundary conditions (3).

The method is based on representation of the scattered field with the aid of the fractional derivative of the Green's function:

$$E_z^s(x,y)\equiv\int_L f^{1-\alpha}(x')G^{\alpha}(x-x',y)dx'. \tag{6}$$

In (6), the function $f^{1-\alpha}(x)$ is the unknown function called the density of the fractional potential, and G^{α} is the fractional derivative of two-dimensional the Green's function of the free space defined by equation (2).

For the limit cases of the fractional order with $\alpha=0$ and $\alpha=1$ representation (6) corresponds to the single-layer and double-layer potentials commonly used to present the scattered fields in diffraction problems:

$$E_z^s(x,y)=\begin{cases}-\frac{i}{4}\int_{-a}^{a}f'(x')H_0^{(1)}(k\sqrt{(x-x')^2+y^2})dx', & \alpha=0\\ -\frac{i}{4}\int_{-a}^{a}f(x')\frac{\partial}{\partial y}H_0^{(1)}(k\sqrt{(x-x')^2+y^2})dx', & \alpha=1\end{cases}$$

More general representations (6) can be derived from the fractional Green's theorem (Veliev & Engheta, 2003) which generalizes the ordinary Green's theorem.

2.1 Fractional Green's theorem

Consider a function $\psi(\vec{r})$, which satisfies inhomogeneous scalar Helmholtz equation with the source density given by the function $\rho(\vec{r})$:

$$\Delta\psi(\vec{r})+k^2\psi(\vec{r})=-4\pi\rho(\vec{r}). \tag{7}$$

Besides, define $G(\vec{r},\vec{r}_0)$ as the Green's function of the Helmholtz equation:

$$\Delta G(\vec{r},\vec{r}_0)+k^2G(\vec{r},\vec{r}_0)=-4\pi\delta(\vec{r}-\vec{r}_0). \tag{8}$$

Here, $\delta(\vec{r}-\vec{r}_0)$ is the three-dimensional Dirac delta function, $\vec{r}$ and $\vec{r}_0$ are the position vectors for the observation and source points, respectively, $\Delta=\frac{\partial^2}{\partial x^2}+\frac{\partial^2}{\partial y^2}+\frac{\partial^2}{\partial z^2}$ is the Laplacian, and k is a scalar constant. After applying fractional derivatives to equations (7) and (8) with respect to the x variable, multiplying the first equation with ${}_{-\infty}D_x^{\nu}G(\vec{r},\vec{r}_0)$, and the second with ${}_{-\infty}D_x^{\mu}\psi(\vec{r})$, subtracting one from another, integrating this over all source coordinates x_0, y_0, z_0 inside S, and finally using the Green's theorem, we obtain the following representation:

$$ {}_{-\infty}D_x^{\beta}\psi(\vec{r}) = \begin{cases} \int\limits_V {}_{-\infty}D_{x_0}^{\beta-\nu}\rho(\vec{r}_0)\bullet {}_{-\infty}D_{x_0}^{\nu}G(\vec{r},\vec{r}_0)dv_0 + \\ +\frac{1}{4\pi}\oint\limits_S [{}_{-\infty}D_{x_0}^{\nu}G(\vec{r},\vec{r}_0)\bullet\nabla_0 {}_{-\infty}D_{x_0}^{\beta-\nu}\psi(\vec{r}_0) - {}_{-\infty}D_{x_0}^{\beta-\nu}\psi(\vec{r}_0)\bullet\nabla_0 {}_{-\infty}D_{x_0}^{\nu}G(\vec{r},\vec{r}_0)]\bullet ds_0, & \vec{r}\in V \\ 0, & \vec{r}\notin V \end{cases} \quad (9) $$

where $\mu+\nu=\beta$. Operator ∇_0 denotes the operator of gradient in respect of variable $\vec{r}_0(x_0, y_0, z_0)$. Here it was used the property of the fractional derivative of the Dirac delta function:

$$ \int\limits_V F(\vec{r}_0)\, {}_{-\infty}D_{x_0}^{\nu}\delta(\vec{r}_0-\vec{r})dv_0 = {}_{-\infty}D_x^{\nu}F(\vec{r}), \quad (10) $$

We use the uniform symbol ${}_{-\infty}D_x^{\alpha}$ (or D_x^{α}) to denote both fractional derivatives and fractional integrals, and it defines a fractional derivative for $0<\alpha<1$ and a fractional integral for $\alpha<0$.

Equation (9) is a generalization of well-known Green's theorem for the case of fractional derivatives.

Consider some important particular cases, which can be obtained from (9).

In the case of excitation in a free space so that the volume V is the whole space, the surface integrals in (9) vanish, and we have:

$$ {}_{-\infty}D_x^{\beta}\psi(\vec{r}) = \int\limits_V {}_{-\infty}D_{x_0}^{\beta-\nu}\rho(\vec{r}_0)\bullet {}_{-\infty}D_{x_0}^{\nu}G(\vec{r},\vec{r}_0)dv_0 . \quad (11) $$

Originally function $\psi(\vec{r})$ characterizes the field excited by the source with the volume density $\rho(\vec{r})$. From the other hand, for $\beta=0$ representation (11) means that the field $\psi(\vec{r})$ is expressed through the distribution of fractional sources with density $D^{-\nu}\rho(r_0)$ inside the volume V and by using fractional integral of conventional Green's function $D^{\nu}G(r_0, r)$. Assuming $\rho(\vec{r})=0$, we can obtain some other important representations:

$$ {}_{-\infty}D_x^{\beta}\psi(\vec{r}) = \begin{cases} \frac{1}{4\pi}\oint\limits_S [{}_{-\infty}D_{x_0}^{\beta}G(\vec{r},\vec{r}_0)\bullet\nabla_0\psi(\vec{r}_0) - \psi(\vec{r}_0)\bullet\nabla_0 {}_{-\infty}D_{x_0}^{\beta}G(\vec{r},\vec{r}_0)]ds_0, & \text{if } \nu=\beta, \mu=0 \\ \frac{1}{4\pi}\oint\limits_S [G(\vec{r},\vec{r}_0)\bullet\nabla_0 {}_{-\infty}D_{x_0}^{\beta}\psi(\vec{r}_0) - {}_{-\infty}D_{x_0}^{\beta}\psi(\vec{r}_0)\bullet\nabla_0 G(\vec{r},\vec{r}_0)]ds_0, & \text{if } \nu=0 \end{cases} \quad (12) $$

From this representation we see that the fractional derivative of function $\psi(\vec{r})$ is expressed either via the value of the function and its first derivative at the boundary and the fractional derivatives of Green's function, or by the fractional derivatives of the function at the boundary and the usual Green's function.

If $\nu = -\mu$, i.e. $\beta = 0$, we obtain a representation for the function $\psi(\vec{r})$ itself:

$$\psi(\vec{r}) = \frac{1}{4\pi}\oint_S [{}_{-\infty}D_{x_0}^{-\mu}G(\vec{r},\vec{r}_0)\bullet\nabla_{0\,-\infty}D_{x_0}^{\mu}\psi(\vec{r}_0) - {}_{-\infty}D_{x_0}^{\mu}\psi(\vec{r}_0)\bullet\nabla_{0\,-\infty}D_{x_0}^{-\mu}G(\vec{r},\vec{r}_0)]ds_0 \,. \qquad (13)$$

This expression means that the function $\psi(\vec{r})$ is represented through its fractional derivatives at the boundary and the fractional derivatives of Green's function. The equation (13) can be useful in scattering problems. If we have boundary conditions for the function $\psi(\vec{r})$ on the surface S as $\nabla_{0\,-\infty}D_{x_0}^{\mu}\psi(\vec{r}_0)|_{\vec{r}_0\in S}=0$ (or ${}_{-\infty}D_{x_0}^{\mu}\psi(\vec{r}_0)|_{\vec{r}_0\in S}=0$) then one of the surface integrals in (13) vanishes and we get a simple presentation for $\psi(\vec{r})$. This fact will be used to present the scattered field in all diffraction problems considered in this paper (6). Equations (12), (13) generalize the Huygens principle in such a sense that the fractional derivative of the function $\psi(\vec{r})$, which characterizes a wave process, is presented as a superposition of waves radiated by elementary "fractional" sources distributed on the given surface. "Fractional" potentials, $\oint_S {}_{-\infty}D_{x_0}^{\beta-\nu}\psi(\vec{r}_0)\bullet\nabla_{0\,-\infty}D_{x_0}^{\nu}G(\vec{r},\vec{r}_0)\bullet ds_0$, $\oint_S {}_{-\infty}D_{x_0}^{\nu}G(\vec{r},\vec{r}_0)\bullet\nabla_{0\,-\infty}D_{x_0}^{\beta-\nu}\psi(\vec{r}_0)\bullet ds_0$, can be treated as a generalization of well-known single and double layer potentials.

2.2 Solution to integral equations

Substituting the expression (6) for $E_z(x,y)$ into fractional boundary conditions (3) we get the equation

$$\lim_{y\to 0} D_{ky}^{\alpha}\int_L f^{1-\alpha}(x')G^{\alpha}(x-x',y)dx' = -\lim_{y\to 0} D_{ky}^{\alpha}E_z^i(x,y)\,, \qquad (14)$$

It is convenient to use the Fourier transform of the fractional potential density $f^{1-\alpha}(x)$

$$F^{1-\alpha}(q) \equiv \int_{-\infty}^{\infty}\tilde{f}^{1-\alpha}(\xi)e^{-ikq\xi}d\xi = a\int_{-1}^{1}f^{1-\alpha}(a\xi)e^{-ikq\xi}d\xi\,,$$

where a new function $\tilde{f}^{1-\alpha}(\xi)$ is introduced:

$$\tilde{f}^{1-\alpha}(\xi) \equiv af^{1-\alpha}(a\xi)\,,\quad |\xi|<1\,,$$

$$\tilde{f}^{1-\alpha}(\xi) \equiv 0\,,\quad |\xi|\geq 1\,.$$

Then the scattered field is expressed via the Fourier transform $F^{1-\alpha}(q)$ as

$$E_z^s(x,y) = -i\frac{e^{\pm i\pi\alpha/2}}{4\pi}\int_{-\infty}^{\infty}F^{1-\alpha}(q)e^{ik(xq+|y|\sqrt{1-q^2})}(1-q^2)^{(\alpha-1)/2}dq\,, \qquad (15)$$

where the upper (lower) sign is chosen for y>0 (y<0). Here, in (15), the following representation for the fractional Green's function was used:

$$G^{\alpha}(x-x',y) = -iD_{ky}^{\alpha}H_0^{(1)}(k\sqrt{(x-x')^2+y^2}) = -i\frac{e^{sign(y)i\pi\alpha/2}}{4\pi}\int_{-\infty}^{\infty}e^{ik((x-x')q+|y|\sqrt{1-q^2})}(1-q^2)^{(\alpha-1)/2}dq \qquad (16)$$

It can be shown that the equation (14) can be reduced to dual integral equations (DIE)

$$\begin{cases} \int_{-\infty}^{\infty} F^{1-\alpha}(q) e^{ika\xi q}(1-q^2)^{\alpha-1/2} dq = -4\pi e^{i\pi/2(1-\alpha)} \sin^{\alpha}\theta e^{-ikd_L \xi\cos\theta}, & |\xi|<1, \\ \int_{-\infty}^{\infty} F^{1-\alpha}(q) e^{ika\xi q} dq = 0, & |\xi|>1, \end{cases} \quad (17)$$

For the limit cases of the fractional order $\alpha = 0$ and $\alpha = 1$ the equations (17) are reduced to the well known integral equations used for PEC and PMC strips (Honl et al., 1961; Veliev & Veremey, 1993; Veliev & Shestopalov, 1988; Uflyand, 1977), respectively. In this paper the method to solve DIE (17) is proposed for arbitrary value of $\alpha \in [0,1]$.
DIE (17) can be solved analytically for one special case of $\alpha = 0.5$. In this case we get the solutions for any value of k as

$$f^{0.5}(x) = -2ik\sin^{1/2}\theta e^{-ikx\cos\theta + i\pi/4}, \quad (18)$$

$$F^{0.5}(q) = -4ie^{i\pi/4}\sin^{1/2}\theta \frac{\sin ka(q+\cos\theta)}{q+\cos\theta}. \quad (19)$$

In the case of arbitrary α the solutions can be obtained numerically. First, we modify the equations (17). After multiplying by $e^{-ika\tau\xi}$ and integrating in ξ from -1 to 1, the first equation in (17) can be rewritten in the following form:

$$\int_{-\infty}^{\infty} F^{1-\alpha}(q) \frac{\sin ka(q-\tau)}{q-\tau}(1-q^2)^{\alpha-1/2} dq = -4\pi e^{i\pi/2(1-\alpha)} \sin^{\alpha}\theta \frac{\sin ka(\tau+\cos\theta)}{\tau+\cos\theta}. \quad (20)$$

In order to discretize this equation, we present the unknown function $\tilde{f}^{1-\alpha}(\xi)$ as a uniformly convergent series in terms of the orthogonal polynomials with corresponding weight functions which allow satisfying the edge conditions:

$$\tilde{f}^{1-\alpha}(\xi) = \left(1-\xi^2\right)^{\alpha-1/2} \sum_{n=0}^{\infty} f_n^{\alpha} \frac{1}{\alpha} C_n^{\alpha}(\xi), \quad (21)$$

where $C_n^{\alpha}(x)$ are the Gegenbauer polynomials and f_n^{α} are the unknown coefficients. Gegenbauer polynomials can be treated as intermediate polynomials between Chebyshev polynomials of the first and second kind:

$$\lim_{\alpha\to 0} \frac{C_n^{\alpha}(\xi)}{\alpha} = \begin{cases} \frac{2}{n} T_n(\xi), & n \neq 0 \\ 1, & n = 0 \end{cases}, \quad \lim_{\alpha\to 1} \frac{C_n^{\alpha}(\xi)}{\alpha} = C_n^1(\xi) = U_n(\xi).$$

The Fourier transform $F^{1-\alpha}(q)$ is expressed as the series

$$F^{1-\alpha}(q) = \frac{2\pi}{\Gamma(\alpha+1)} \sum_{n=0}^{\infty} (-i)^n \frac{\Gamma(n+2\alpha)}{\Gamma(n+1)} \frac{J_{n+\alpha}(kaq)}{(2kaq)^{\alpha}} f_n^{\alpha}, \quad (22)$$

where $J_{n+\alpha}(kaq)$ is the Bessel function.

It must be noted that the edge conditions are chosen in the following form

$$\tilde{f}^{1-\alpha}(\xi) = O\left((1-\xi^2)^{\alpha-1/2}\right), \ \xi \to \pm 1 . \tag{23}$$

For special cases of $\alpha = 0$ and $\alpha = 1$ the edge conditions have the form as

$$\tilde{f}^{1-\alpha}(\xi) = \begin{cases} O\left((1-\xi^2)^{-1/2}\right), & \alpha = 0 \\ O\left((1-\xi^2)^{1/2}\right), & \alpha = 1 \end{cases}, \quad \xi \to \pm 1 \tag{24}$$

These are well-known Meixner edge conditions in diffraction problems (Honl et al., 1961). Substituting (22) into (17) and taking into account the properties of discontinuous integrals of Weber-Shafheitlin (Bateman & Erdelyi, 1953) and the following formula (Prudnikov et al., 1986)

$$\frac{1}{\pi}\int_{-\infty}^{\infty} \frac{J_{n+\nu}(\varepsilon q)}{q^{\nu}} \frac{\sin \varepsilon (q-\beta)}{q-\beta} dq = \frac{J_{n+\nu}(\varepsilon \beta)}{\beta^{\nu}} , \tag{25}$$

one can show that the homogenous equation in the set (17) is satisfied identically.
The first equation of (17) written in the form (20) can be reduced to an infinite system of linear algebraic equations (SLAE) with respect to the unknown coefficients f_n^{α} :

$$\sum_{n=0}^{\infty} (-i)^n \frac{\Gamma(n+2\alpha)}{\Gamma(n+1)} C_{mn}^{\alpha} f_n^{\alpha} = B_m^{\alpha}, \quad m = 0,1,2,..,\infty \tag{26}$$

where the matrix coefficients are expressed as

$$C_{mn}^{\alpha} = \int_{-\infty}^{\infty} J_{n+\alpha}(kaq) J_{m+\alpha}(kaq) \frac{(1-q^2)^{\alpha-1/2}}{q^{2\alpha}} dq ,$$

$$B_m^{\alpha} = -2\Gamma(\alpha+1)(2ka)^{\alpha} e^{i\pi/2(1-\alpha)} \sin^{\alpha}\theta \frac{J_{m+\alpha}(ka\cos\theta)}{(\cos\theta)^{\alpha}} .$$

It can be shown that the SLAE (26) can be reduced to SLAE of the Fredholm type of the second kind (Veliev et al., 2008a). Then the coefficients f_n^{α} can be found with any desired accuracy (within the machine precision) using the truncation of SLAE. The fractional density $f^{1-\alpha}(x)$ is computed by using (21) and the scattered field (6) and other physical characteristics can be obtained as series in terms of the found coefficients f_n^{α} .
In order to solve the diffraction problem on a plane screen with fractional boundary conditions and obtain a convenient SLAE we applied several techniques. First of all, the fractional Green's theorem presented above allowed searching the unknown scattered field as a potential with the fractional Green's function. The order of the fractional Green's function is defined from the fractional order of the boundary conditions. In general, the fractional derivative of Green's function may have a complicated form, but we used the Fourier transform where application of the fractional derivative maps to a simple multiplication by $(iq)^{\alpha}$. Finally, utilization of the orthogonal Gegenbauer polynomials along with the specific form of the edge conditions

allowed to reduce integral equations to SLAE in a convenient form. One can compare the method presented for fractional boundary conditions with the known methods applied to solve diffraction by an impedance strip. The impedance strip requires to consider two unknown densities in presentation of the scattered field as a sum of single- and double-layer potentials. The usage of two unknown functions leads to more complicated SLAE in spite of the SLAE obtained for fractional boundary conditions.

2.3 Physical characteristics

We consider such electrodynamic characteristics of the scattered field as the radiation pattern (RP), monostatic radar cross-section (MRCS) and surface current densities depending on the coefficients f_n^α. The scattered field $E_z^s(x,y)$ in the far-zone $kr \to \infty$ in the cylindrical coordinate system (r,ϕ), $x = r\cos\varphi, y = r\sin\varphi$, is expressed as

$$E_z^s(r,\varphi) = \frac{i}{4\pi}(\pm i)^\alpha \int_{-\infty}^{+\infty} F^{1-\alpha}(\cos\beta) e^{ikr\cos(\varphi\pm\beta)} \sin^\alpha \beta \, d\beta ,$$

where the upper sign is chosen for $\varphi \in [0,\pi]$, and the lower one when $\varphi \in [\pi, 2\pi]$. Using the stationary phase method for $kr \to \infty$ we present $E_z^s(x,y)$ as

$$E_z^s(x,y) \approx A(kr)\Phi^\alpha(\varphi), \; kr \to \infty, \tag{27}$$

where

$$A(kr) = \sqrt{\frac{2}{\pi kr}} e^{ikr - i\pi/4}, \; \Phi^\alpha(\varphi) = -\frac{i}{4}(\pm i)^\alpha F^{1-\alpha}(\cos\varphi)\sin^\alpha\varphi .$$

The function $\Phi^\alpha(\varphi)$ describes RP and can be expressed via the coefficients f_n^α as

$$\Phi^\alpha(\varphi) = \frac{\pi i (\pm i)^\alpha}{2\Gamma(\alpha+1)} \tan^\alpha\varphi \sum_{n=0}^{\infty} (-i)^n f_n^\alpha \frac{\Gamma(n+2\alpha)}{\Gamma(n+1)} \frac{J_{n+\alpha}(ka\cos\varphi)}{(2ka)^\alpha} .$$

In physical optics (PO) approximation ($ka \gg 1$) $\Phi^\alpha(\varphi)$ has a simpler form. Using the following formula

$$\lim_{ka\to\infty} \frac{\sin ka(\alpha-\beta)}{\alpha-\beta} = \pi\delta(\alpha-\beta), \tag{28}$$

in IE (20) we get the following expressions for $F^\alpha(q)$ and $\Phi^\alpha(\varphi)$:

$$F^{1-\alpha}(q) \approx -4i^\alpha \frac{\sin^{1-\alpha}\theta}{(1-q^2)^{(1-2\alpha)/2}} \frac{\sin ka(q-\cos\theta)}{q-\cos\theta} ,$$

$$\Phi^\alpha(\varphi) \approx (\mp 1)^\alpha \sin\varphi \left(\frac{\sin\theta}{\sin\varphi}\right)^\alpha \frac{\sin ka(\cos\varphi+\cos\theta)}{\cos\varphi+\cos\theta} .$$

In the special case of $\alpha = 0.5$ and arbitrary value of ka we get an analytical expression for the RP

$$\Phi^{0.5}(\varphi) = (\mp 1)^{1/2} \sqrt{\sin\varphi \sin\theta} \frac{\sin ka(\cos\varphi + \cos\theta)}{\cos\varphi + \cos\theta}.$$

Bi-static radar cross section (BRCS) is expressed from RP $\Phi(\varphi)$ as $\frac{\sigma_{2d}}{\lambda}(\varphi) = \frac{2}{\pi}|\Phi(\varphi)|^2$. MRCS σ_{2D}^{mono} is defined as $\sigma_{2D}^{mono} = \frac{\sigma_{2d}}{\lambda}(\theta) = \frac{2}{\pi}|\Phi(\theta)|^2$.

We have the following representations in PO approximation

$$\frac{\sigma_{2d}}{\lambda} = \frac{2}{\pi}\sin^2\varphi \left(\frac{\sin\theta}{\sin\varphi}\right)^{2\alpha} \left\{\frac{\sin ka(\cos\varphi + \cos\theta)}{\cos\varphi + \cos\theta}\right\}^2, \quad ka \gg 1,$$

$$\sigma_{2D}^{mono} = \frac{2}{\pi}\sin^2\theta \left\{\frac{\sin ka(2\cos\theta)}{2\cos\theta}\right\}^2, \quad ka \gg 1.$$

It must be noted that the density function $f^{1-\alpha}(x)$ in the integral (6) does not describe the density of physical surface currents on the strip for $0 < \alpha < 1$. The function $f^{1-\alpha}(x)$ is defined as the discontinuity of fractional derivatives of E-field at the plane $y = 0$:

$$f^{1-\alpha}(x) = {}_{-\infty}D_{ky}^{1-\alpha} E_z(x,y)|_{y=+0} - {}_{-\infty}D_{ky}^{1-\alpha} E_z(x,y)|_{y=-0}, \; x \in (-a,a). \tag{29}$$

For the limit cases of $\alpha = 0$ and $\alpha = 1$ the equation (29) is reduced to well-known presentations for electric and magnetic surface currents, respectively, i.e.

$$f^{1-\alpha}(x) = \begin{cases} \frac{\partial E_z(x,y)}{\partial y}|_{y=+0} - \frac{\partial E_z(x,y)}{\partial y}|_{y=-0} = H_x(x,+0) - H_x(x,-0), & \alpha = 0 \\ \\ E_z(x,+0) - E_z(x,-0), & \alpha = 1 \end{cases}$$

In order to obtain physical surface currents from $f^{1-\alpha}(x)$ we have to apply additional integration. In case of E-polarized incident plane wave we have the following induced currents on a strip: electric current $\vec{j}^{\alpha(e)} = \vec{z} j_z^{\alpha(e)}$ and magnetic current $\vec{j}^{\alpha(m)} = \vec{x} j_x^{\alpha(m)}$ expressed from $f^{1-\alpha}(x)$ as

$$j_z^{\alpha(e)}(x) = -2i\cos\left(\frac{\pi\alpha}{2}\right)\frac{i}{4\pi}\int_{-\infty}^{+\infty} F^{1-\alpha}(q) e^{ikax}(1-q^2)^{\alpha/2} dq,$$

$$j_x^{\alpha(m)}(x) = -2\sin\left(\frac{\pi\alpha}{2}\right)\frac{i}{4\pi}\int_{-\infty}^{+\infty} F^{1-\alpha}(q) e^{ikax}(1-q^2)^{\alpha/2-1/2} dq.$$

The detailed analysis of the scattering properties of the strip with fractional boundary conditions one can find in papers (Veliev et al., 2008a; Veliev et al., 2008b).

2.4 H-polarization

In the case of the H-polarized incident plane wave $\vec{H}^i\left(0,0,H_z^i\right)$, where $H_z^i(x,y)=e^{-ik(x\cos\theta+y\sin\theta)}$, the method proposed above can be applied as well. We define fractional boundary conditions as

$$D_{ky}^{1-\alpha}H_z(x,y)\big|_{y\to\pm0}=D_{ky}^{1-\alpha}\left[H_z^i(x,y)+H_z^s(x,y)\right]\big|_{y\to\pm0}=0\,,\quad x\in(-a,a).$$

The case of $\alpha=0$ corresponds to diffraction of the H-polarized plane wave on a PEC strip, while the case of $\alpha=1$ describes diffraction of the H-polarized plane wave on a PMC strip. As before, we represent the scattered field via the fractional Green's function

$$H_z^s(x,y)\equiv\int_{-a}^{a}f^\alpha(x')G^{1-\alpha}(x-x',y)dx'\,.$$

After substituting (18) into fractional boundary conditions (19) we get the equation

$$\lim_{y\to0}D_{ky}^{1-\alpha}\int_{-a}^{a}f^\alpha(x')G^{1-\alpha}(x-x',y)dx'=-\lim_{y\to0}D_{ky}^{1-\alpha}H_z^i(x,y)\,.$$

This equation can be solved by repeating all steps of the E-polarization case after changing α to $1-\alpha$.

3. Diffraction by a half-plane with fractional boundary conditions

Another problem studied in this paper is the diffraction by a half-plane with fractional boundary conditions. The method introduced to solve the dual integral equation (DIE) for a finite object (a strip) will be modified to solve DIE for semi-infinite scatterers such as half-plane. There are many papers devoted to the classical problem of diffraction by a half-plane. The method to solve the scattering problem for a perfectly conducting half-plane is presented in (Honl et al., 1961). Usually, it is solved using Wiener-Hopf method. The first application of the method to a PEC half-plane can be referred to the papers of Copson (Copson, 1946) and independently to papers of Carlson and Heins (Carlson & Heins, 1947). In 1952 Senior first applied Wiener-Hopf method to the diffraction by an impedance half-plane (Senior, 1952) and later oblique incidence was considered (Senior, 1959). Diffraction by a resistive and conductive half-plane and also by various types of junctions is analyzed in details in (Senior & Volakis, 1995). We propose a new approach for the rigorous analysis of the considered problem which generalizes the results of (Veliev, 1999) obtained for the PEC boundaries and includes them as special cases.

Let an E-polarized plane wave $E_z^i(x,y)=e^{-ik(x\cos\theta+y\sin\theta)}$ (1) be scattered by a half-plane ($y=0$, $x>0$). The total field $E_z=E_z^i+E_z^s$ must satisfy fractional boundary conditions

$$D_{ky}^{\alpha}E_z(x,y)=0\,,\ y\to\pm0\,,x>0\,, \tag{30}$$

and Meixner's edge conditions must be satisfied for $x\to0$.

Following the idea used for the analysis of diffraction by a strip we represent the scattered field using the fractional Green's function

$$E_z^s(x,y) \equiv \int_0^{\infty} f^{1-\alpha}(x') G^{\alpha}(x-x',y)\,dx', \tag{31}$$

where $f^{1-\alpha}(x)$ is the unknown function, G^{α} is the fractional Green's function (2).
After substituting the representation (31) into fractional boundary conditions (30) we get the equation

$$\frac{-i}{4}\lim_{y\to 0} D_{ky}^{2\alpha}\int_0^{\infty} f^{1-\alpha}(x') H_0^{(1)}\left(k\sqrt{(x-x')^2+y^2}\right)dx' = -\lim_{y\to 0} D_{ky}^{\alpha} E_z^i(x,y),\ x>0. \tag{32}$$

The Fourier transform of $f^{1-\alpha}(x)$ is defined as

$$F^{1-\alpha}(q) = \int_{-\infty}^{\infty} \tilde{f}^{1-\alpha}(\xi) e^{-ikq\xi} d\xi = \int_0^{\infty} f^{1-\alpha}(x) e^{-ikqx} dx,$$

where $\tilde{f}^{1-\alpha}(\xi) \equiv f^{1-\alpha}(\xi)$ for $\xi>0$ and $\tilde{f}^{1-\alpha}(\xi) \equiv 0$ for $\xi<0$.
Then the scattered field will be expressed via the Fourier transform $F^{1-\alpha}(q)$ as

$$E_z^s(x,y) = -i\frac{e^{\pm i\pi\alpha/2}}{4\pi}\int_{-\infty}^{\infty} F^{1-\alpha}(q) e^{ik(xq+|y|\sqrt{1-q^2})}(1-q^2)^{(\alpha-1)/2} dq\,. \tag{33}$$

Using the Fourier transform the equation (32) is reduced to the DIE with respect to $F^{1-\alpha}(q)$:

$$\begin{cases} \int_{-\infty}^{\infty} F^{1-\alpha}(q) e^{ik\xi q}\left(1-q^2\right)^{\alpha-1/2} dq = -4\pi e^{i\pi/2(1-\alpha)}\sin^{\alpha}\theta e^{-ik\xi\cos\theta},\ \xi>0, \\ \int_{-\infty}^{\infty} F^{1-\alpha}(q) e^{ik\xi q} dq = 0, \quad \xi<0. \end{cases} \tag{34}$$

The kernels in integrals (34) are similar to the ones in DIE (17) obtained for a strip if the constant d_L is equal to 1 ($L=(0,\infty)$ in the case of a half-plane).
For the limit cases of the fractional order $\alpha=0$ and $\alpha=1$ these equations are reduced to well known integral equations used for the PEC and PMC half-planes (Veliev, 1999), respectively. In this paper the method to solve DIE (5) is proposed for arbitrary values of $\alpha\in[0,1]$.
DIE allows an analytical solution in the special case of $\alpha=0.5$ in the same manner as for a strip with fractional boundary conditions. Indeed, for $\alpha=0.5$ we obtain the solution for any value of k as

$$F^{0.5}(q) = -2\sin^{1/2}\theta\, e^{i\pi/4}\frac{\pi}{k}\delta(q+\cos\theta),$$

$$f^{0.5}(x) = -2\sin^{1/2}\theta e^{i\pi/4} e^{-ikx\cos\theta}.$$

The scattered field can be found in the following form:

$$E_z^s(x,y) = \frac{i}{2k} e^{\pm i\pi\alpha/2} e^{i\pi/4} \sin^{\alpha-1/2}\theta e^{ik(-\cos\theta x + |y|\sin\theta)}, \qquad \alpha = 0.5, \ \text{for } y > 0 \ \ (y < 0).$$

In the general case of $0 < \alpha < 1$ the equations (34) can be reduced to SLAE. To do this we represent the unknown function $\tilde{f}^{1-\alpha}(\xi)$ as a series in terms of the Laguerre polynomials with coefficients f_n^α:

$$\tilde{f}^{1-\alpha}(x) = e^{-x} x^{\alpha-1/2} \sum_{n=0}^{\infty} f_n^\alpha L_n^{\alpha-1/2}(2x). \qquad (35)$$

Laguerre polynomials are orthogonal polynomials on the interval $L = (0,\infty)$ with the appropriate weight functions used in (35) . It can be shown from (35) that $\tilde{f}^{1-\alpha}(\xi)$ satisfies the following edge condition:

$$\tilde{f}^{1-\alpha}(\xi) = O\left(\xi^{\alpha-1/2}\right), \ \xi \to 0. \qquad (36)$$

For the special cases of $\alpha = 0$ and $\alpha = 1$, the edge conditions are reduced to the well-known equations (Honl et al., 1961) used for a perfectly conducting half-plane.
After substituting (35) into the first equation of (34) we get an integral equation (IE)

$$\sum_{n=0}^{\infty} f_n^\alpha \int_{-\infty}^{\infty} \left[\int_0^{\infty} e^{-t} t^{\alpha-1/2} L_n^{\alpha-1/2}(2t) e^{-ikqt} dt \right] \times e^{ik\xi q} \left(1-q^2\right)^{\alpha-1/2} dq = R(\xi), \qquad (37)$$

where $R(\xi) = -4\pi e^{i\pi/2(1-\alpha)} \sin^\alpha\theta e^{-ik\xi\cos\theta}$ is known.
Using the representation for Fourier transform of Laguerre polynomials (Prudnikov et al., 1986) we can evaluate the integral over dt as

$$\int_0^\infty e^{-t} t^{\alpha-1/2} L_n^{\alpha-1/2}(2t) e^{-ikqt} dt = \int_0^\infty e^{-t(1+ikq)} t^{\alpha-1/2} L_n^{\alpha-1/2}(2t) dt = \frac{\Gamma(n+\alpha+1/2)}{\Gamma(n+1)} \frac{(ikq-1)^n}{(ikq+1)^{n+\alpha+1/2}}$$

After some transformations IE (37) is reduced to

$$\sum_{n=0}^{\infty} f_n^\alpha \frac{\Gamma(n+\alpha+1/2)}{\Gamma(n+1)} \int_{-\infty}^{\infty} \frac{(ikq-1)^n}{(ikq+1)^{n+\alpha+1/2}} \left(1-q^2\right)^{\alpha-1/2} e^{ik\xi q} dq = R(\xi), \ \xi > 0. \qquad (38)$$

Then we integrate both sides of equation (38) with appropriate weight functions, as $\int_0^\infty (\cdot) e^{-\xi} \xi^{\alpha-1/2} L_m^{\alpha-1/2}(2\xi) d\xi$. Using orthogonality of Laguerre polynomials we get the following SLAE:

$$\sum_{n=0}^{\infty} f_n^\alpha C_{mn}^\alpha = B_m^\alpha, \quad m = 0,1,2,\ldots,\infty,$$

with matrix coefficients

$$C_{mn}^{\alpha} = \frac{\Gamma(n+\alpha+1/2)}{\Gamma(n+1)} \int_{-\infty}^{\infty} \frac{(ikq+1)^{m-n-\alpha-1/2}}{(ikq-1)^{n-m-\alpha-1/2}} \left(1-q^2\right)^{\alpha-1/2} dq ,$$

$$B_m^{\alpha} = 4\pi e^{-i\pi/2\alpha} \frac{|\sin\theta|^{\alpha} (1-ik\cos\theta)^m}{(1+ik\cos\theta)^{m+\alpha+1/2}} .$$

It can be shown that the coefficients f_n^{α} can be found with any desired accuracy by using the truncation of SLAE. Then the function $\tilde{f}^{1-\alpha}(x)$ is found from (35) that allows obtaining the scattered field (33).

4. Diffraction by two parallel strips with fractional boundary conditions

The proposed method to solve diffraction problems on surfaces described by fractional boundary conditions can be applied to more complicated structures. The interest to such structures is related to the resonance properties of scattering if the distance between the strips varies. Two strips of the width *2a* infinite along the axis z are located in the planes $y = l$ and $y = -l$. Let the E-polarized plane wave $E_z^i(x,y) = e^{-ik(x\cos\theta + y\sin\theta)}$ (1) be the incident field. The total field $E_z = E_z^i + E_z^s$ satisfies fractional boundary conditions on each strip:

$$D_{ky}^{\alpha} E_z(x,y) = 0 , \quad y \to \pm l \pm 0 , \quad x \in (-a,a) , \tag{39}$$

and Meixner's edge conditions must be satisfied on the edges of both strips ($y = \pm l$, $x \to \pm a$).

The scattered field $E_z^s(x,y)$ consists of two parts

$$E_z^s(x,y) \equiv E_z^{1s}(x,y) + E_z^{2s}(x,y) ,$$

where

$$E_z^{js}(x,y_j) \equiv \int_{-a}^{a} f_j^{1-\alpha}(x') G^{\alpha}(x-x',y_j) dx' , \ j = 1,2 . \tag{40}$$

Here, G^{α} is the fractional Green's function defined in (2). $y_{1,2}$ are the coordinates in the corresponding coordinate systems related to each strip,

$$y_1 = y - l , \ x_1 = x ,$$

$$y_2 = y + l , \ x_2 = x .$$

Using Fourier transforms, defined as

$$F_j^{1-\alpha}(q) \equiv \int_{-\infty}^{\infty} \tilde{f}_j^{1-\alpha}(\xi) e^{-ikq\xi} d\xi = a \int_{-1}^{1} f_j^{1-\alpha}(a\xi) e^{-ikq\xi} d\xi ,$$

$$\tilde{f}_j^{1-\alpha}(\xi) \equiv a f_j^{1-\alpha}(a\xi) , \ j = 1,2 ,$$

the scattered field is expressed as

$$E_z^{1s}(x,y) = -i\frac{e^{\pm i\pi\alpha/2}}{4\pi}\int_{-\infty}^{\infty} F_1^{1-\alpha}(q)e^{ik[xq+|y-l|\sqrt{1-q^2}]}(1-q^2)^{(\alpha-1)/2}dq\,,\ y>l\ (y<l), \tag{41}$$

$$E_z^{2s}(x,y) = -i\frac{e^{\pm i\pi\alpha/2}}{4\pi}\int_{-\infty}^{\infty} F_2^{1-\alpha}(q)e^{ik[xq+|y+l|\sqrt{1-q^2}]}(1-q^2)^{(\alpha-1)/2}dq\,,\ y>-l\ (y<-l). \tag{42}$$

Fractional boundary conditions (30) correspond to two equations

$$D_{ky}^{\alpha}E_z(x,y)=0\,,\ y\to l\pm 0\,,\ x\in(-a,a)\,. \tag{43}$$

$$D_{ky}^{\alpha}E_z(x,y)=0\,,\ y\to -l\pm 0\,,\ x\in(-a,a)\,. \tag{44}$$

After substituting expressions (41) and (42) into the equations (43) and (44) we obtain

$$\begin{aligned}&\int_{-\infty}^{\infty} F_1^{1-\alpha}(q)e^{ikxq}(1-q^2)^{\alpha-1/2}dq = -4\pi i e^{i\pi\alpha/2}\sin^{\alpha}\theta e^{-ik(x\cos\theta+l\sin\theta)} - \\ &-\int_{-\infty}^{\infty} F_2^{1-\alpha}(q)e^{ik[xq+2l\sqrt{1-q^2}]}(1-q^2)^{\alpha-1/2}dq\end{aligned}\,, \tag{45}$$

$$\begin{aligned}&\int_{-\infty}^{\infty} F_2^{1-\alpha}(q)e^{ikxq}(1-q^2)^{\alpha-1/2}dq = -4\pi i e^{i\pi/2\alpha}\sin^{\alpha}\theta e^{-ik(x\cos\theta-l\sin\theta)} - \\ &-\int_{-\infty}^{\infty} F_1^{1-\alpha}(q)e^{ik[xq+2l\sqrt{1-q^2}]}(1-q^2)^{\alpha-1/2}dq\end{aligned}\,. \tag{46}$$

Multiplying both equations with $e^{-ikx\tau}$ and integrating them in ζ on the interval $[-a,a]$, the system (45), (46) leads to

$$\begin{cases}\int_{-\infty}^{\infty} F_1^{1-\alpha}(q)\dfrac{\sin ka(q-\tau)}{q-\tau}(1-q^2)^{\alpha-1/2}dq = -4\pi i e^{i\pi\alpha/2}\sin^{\alpha}\theta\dfrac{\sin ka(\tau+\cos\theta)}{\tau+\cos\theta}e^{-ikl\sin\theta} - \\ \int_{-\infty}^{\infty} F_2^{1-\alpha}(q)\dfrac{\sin ka(q-\tau)}{q-\tau}e^{i2kl\sqrt{1-q^2}}(1-q^2)^{\alpha-1/2}dq \\ \\ \int_{-\infty}^{\infty} F_2^{1-\alpha}(q)\dfrac{\sin ka(q-\tau)}{q-\tau}(1-q^2)^{\alpha-1/2}dq = -4\pi i e^{i\pi\alpha/2}\sin^{\alpha}\theta\dfrac{\sin ka(\tau+\cos\theta)}{\tau+\cos\theta}e^{ikl\sin\theta} - \\ -\int_{-\infty}^{\infty} F_1^{1-\alpha}(q)\dfrac{\sin ka(q-\tau)}{q-\tau}e^{i2kl\sqrt{1-q^2}}(1-q^2)^{\alpha-1/2}dq\end{cases} \tag{47}$$

Similarly to the method described for the diffraction by one strip, the set (47) can be reduced to a SLAE by presenting the unknown functions $f_j^{1-\alpha}(x)$ as a series in terms of the orthogonal polynomials. We represent the unknown functions $\tilde{f}_j^{1-\alpha}(\xi)$ as series in terms of the Gegenbauer polynomials:

$$\tilde{f}_j^{1-\alpha}(\xi) = \left(1-\xi^2\right)^{\alpha-1/2}\sum_{n=0}^{\infty} f_n^{j,\alpha}\frac{1}{\alpha}C_n^{\alpha}\left(\xi\right),\ j=1,2\,.$$

For the Fourier transforms $F_j^{1-\alpha}(q)$ we have the representations (22). Substituting the representations for $F_j^{1-\alpha}(q)$ into the (47), using the formula (25), then integrating $\int_{-\infty}^{\infty}(.)\frac{J_{m+\alpha}(ka\tau)}{m^{\alpha}}d\tau$ for $m=0,1,2,..$, we obtain the following SLAE:

$$\begin{cases}\sum_{n=0}^{\infty}(-i)^n\frac{\Gamma(n+2\alpha)}{\Gamma(n+1)}C_{mn}^{11,\alpha}f_n^{1,\alpha}+\sum_{n=0}^{\infty}(-i)^n\frac{\Gamma(n+2\alpha)}{\Gamma(n+1)}C_{mn}^{12,\alpha}f_n^{2,\alpha}=B_m^{1,\alpha}\\ \sum_{n=0}^{\infty}(-i)^n\frac{\Gamma(n+2\alpha)}{\Gamma(n+1)}C_{mn}^{21,\alpha}f_n^{1,\alpha}+\sum_{n=0}^{\infty}(-i)^n\frac{\Gamma(n+2\alpha)}{\Gamma(n+1)}C_{mn}^{22,\alpha}f_n^{2,\alpha}=B_m^{2,\alpha}\end{cases},\quad m=0,1,2,..$$

where the matrix coefficients are defined as

$$C_{mn}^{11,\alpha}=C_{mn}^{22,\alpha}=\int_{-\infty}^{\infty}\frac{J_{m+\alpha}(ka\tau)}{\tau^{\alpha}}\frac{J_{n+\alpha}(ka\tau)}{\tau^{\alpha}}(1-\tau^2)^{\alpha-1/2}d\tau,$$

$$C_{mn}^{12,\alpha}=C_{mn}^{21,\alpha}=\int_{-\infty}^{\infty}\frac{J_{m+\alpha}(ka\tau)}{\tau^{\alpha}}\frac{J_{n+\alpha}(ka\tau)}{\tau^{\alpha}}e^{i2kl\sqrt{1-\tau^2}}(1-\tau^2)^{\alpha-1/2}d\tau,$$

$$B_m^{1,\alpha}=e^{-2ikl\sin\theta}B_m^{2,\alpha}=-2ie^{i\pi\alpha/2}\Gamma(\alpha+1)\sin^{\alpha}\theta e^{-ikl\sin\theta}(2ka)^{\alpha}\frac{J_{m+\alpha}(ka\cos\theta)}{(\cos\theta)^{\alpha}}.$$

Consider the case of the physical optics approximation, where $ka\gg 1$. In this case we can obtain the solution of (47) in the explicit form. Indeed, using the formula (28) we get

$$\begin{cases}\pi F_1^{1-\alpha}(\tau)(1-\tau^2)^{\alpha-1/2}=\\ =-4\pi ie^{i\pi\alpha/2}\sin^{\alpha}\theta\frac{\sin ka(\tau+\cos\theta)}{\tau+\cos\theta}e^{-ikl\sin\theta}-\pi F_2^{1-\alpha}(\tau)e^{i2kl\sqrt{1-\tau^2}}(1-\tau^2)^{\alpha-1/2}\\ \pi F_2^{1-\alpha}(\tau)(1-\tau^2)^{\alpha-1/2}=\\ =-4\pi ie^{i\pi\alpha/2}\sin^{\alpha}\theta\frac{\sin ka(\tau+\cos\theta)}{\tau+\cos\theta}e^{ikl\sin\theta}-\pi F_1^{1-\alpha}(\tau)e^{i2kl\sqrt{1-\tau^2}}(1-\tau^2)^{\alpha-1/2}\end{cases}\quad(48)$$

Finally, we obtain the solution as

$$\begin{cases}F_1^{1-\alpha}(\tau)=4ie^{i\pi\alpha/2}\sin^{\alpha}\theta\frac{\sin ka(\tau+\cos\theta)}{\tau+\cos\theta}\frac{1}{(1-\tau^2)^{\alpha-1/2}}\frac{(e^{ikl\sin\theta}e^{i2kl\sqrt{1-\tau^2}}-e^{-ikl\sin\theta})}{(1-e^{i4kl\sqrt{1-\tau^2}})}\\ F_2^{1-\alpha}(\tau)=4ie^{i\pi\alpha/2}\sin^{\alpha}\theta\frac{\sin ka(\tau+\cos\theta)}{\tau+\cos\theta}\frac{1}{(1-\tau^2)^{\alpha-1/2}}\frac{(e^{-ikl\sin\theta}e^{i2kl\sqrt{1-\tau^2}}-e^{ikl\sin\theta})}{(1-e^{i4kl\sqrt{1-\tau^2}})}\end{cases}\quad(49)$$

Having expressions for $F_j^{1-\alpha}(q)$ we can obtain the physical characteristics. The radiation pattern of the scattered field in the far zone (27) is expressed as

$$\Phi^{\alpha}(\varphi) = \Phi_1^{\alpha}(\varphi) + \Phi_2^{\alpha}(\varphi),$$

where

$$\Phi_1^{\alpha}(\varphi) = -\frac{i}{4} e^{\pm i\pi/2\alpha} F_1^{1-\alpha}(\cos\varphi) \sin^{\alpha}\varphi e^{-ikl\cos\varphi},$$

$$\Phi_2^{\alpha}(\varphi) = -\frac{i}{4} e^{\pm i\pi/2\alpha} F_2^{1-\alpha}(\cos\varphi) \sin^{\alpha}\varphi e^{ikl\cos\varphi}.$$

5. Conclusion

The problems of diffraction by flat screens characterized by the fractional boundary conditions have been considered. Fractional boundary conditions involve fractional derivative of tangential field components. The order of fractional derivative is chosen between 0 and 1. Fractional boundary conditions can be treated as intermediate case between well known boundary conditions for the perfect electric conductor (PEC) and perfect magnetic conductor (PMC). A method to solve two-dimensional problems of scattering of the E-polarized plane wave by a strip and a half-plane with fractional boundary conditions has been proposed. The considered problems have been reduced to dual integral equations discretized using orthogonal polynomials. The method allowed obtaining the physical characteristics with a desired accuracy. One important feature of the considered integral equations has been noted: these equations can be solved analytically for one special value of the fractional order equal to 0.5 for any value of frequency. In that case the solution to diffraction problem has an analytical form. The developed method has been also applied to the analysis of a more complicated structure: two parallel strips. Introducing of fractional derivative in boundary conditions and the developed method of solving such diffraction problems can be a promising technique in modeling of scattering properties of complicated surfaces when the order of fractional derivative is defined from physical parameters of a surface.

6. References

Bateman, H. & Erdelyi, A. (1953). *Higher Transcendental Functions, Volume 2*, McGraw-Hill, New York

Carlson J.F. & Heins A.E. (1947). The reflection of an electromagnetic plane wave by an infinite set of plates. *Quart. Appl. Math.*, Vol.4, pp. 313-329

Copson E.T. (1946). On an integral equation arising in the theory of diffraction, *Quart. J. Math.*, Vol.17, pp. 19-34

Engheta, N. (1996). Use of Fractional Integration to Propose Some 'Fractional' Solutions for the Scalar Helmholtz Equation. A chapter in Progress in Electromagnetics Research (PIER), Monograph Series, Chapter 5, Vol.12, Jin A. Kong, ed.EMW Pub., Cambridge, MA, pp. 107-132

Engheta, N. (1998). Fractional curl operator in electromagnetic. *Microwave and Optical Technology Letters*, Vol.17, No.2, pp. 86-91

Engheta, N. (1999). Phase and amplitude of fractional-order intermediate wave, Microwave and optical technology letters, Vol.21, No.5

Engheta, N. (2000). Fractional Paradigm in Electromagnetic Theory, *a chapter in IEEE Press*, chapter 12, pp.523-553

Hanninen, I.; Lindell, I.V. & Sihvola, A.H. (2006). Realization of Generalized Soft-and-Hard Boundary, *Progress In Electromagnetics Research*, PIER 64, pp. 317-333

Hilfer, R. (1999). *Applications of Fractional Calculus in Physics*, World Scientific Publishing, ISBN 981-0234-57-0, Singapore

Honl, H., A.; Maue, W. & Westpfahl, K. (1961). *Theorie der Beugung*, Springer-Verlag, Berlin

Hope D. J. & Rahmat-Samii Y. (1995). *Impedance boundary conditions in electromagnetic*, Taylor and Francis, Washington, USA

Lindell I.V. & Sihvola A.H. (2005). Transformation method for Problems Involving Perfect Electromagnetic Conductor (PEMC) Structures. *IEEE Trans. Antennas Propag.*, Vol.53, pp. 3005-3011

Lindell I.V. & Sihvola A.H. (2005). Realization of the PEMC Boundary. *IEEE Trans. Antennas Propag.*, Vol.53, pp. 3012-3018

Oldham, K.B. & Spanier, J. (1974). *The Fractional Calculus: Integrations and Differentiations of Arbitrary Order*, Academic Press, New York

Prudnikov, H.P.; Brychkov, Y.H. & Marichev, O.I. (1986). *Special Functions, Integrals and Series, Volume 2*, Gordon and Breach Science Publishers

Samko, S.G.; Kilbas, A.A. & Marichev, O.I. (1993), *Fractional Integrals and Derivatives, Theory and Applications*, Gordon and Breach Science Publ., Langhorne

Senior, T.B.A. (1952). Diffraction by a semi-infinite metallic sheet. *Proc. Roy. Soc. London, Seria A*, 213, pp. 436-458.

Senior, T.B.A. (1959). Diffraction by an imperfectly conducting half plane at oblique incidence. *Appl. Sci. Res., B8*, pp. 35-61

Senior, T.B. & Volakis, J.L. (1995). *Approximate Boundary Conditions in Electromagnetics*, IEE, London

Uflyand, Y.S. (1977). The method of dual equations in problems of mathematical physics [in russian]. *Nauka*, Leningrad

Veliev, E.I. & Shestopalov, V.P. (1988). A general method of solving dual integral equations. *Sov. Physics Dokl.*, Vol.33, No.6, pp. 411–413

Veliev, E.I. & Veremey, V.V. (1993). Numerical-analytical approach for the solution to the wave scattering by polygonal cylinders and flat strip structures. *Analytical and Numerical Methods in Electromagnetic Wave Theory*, M. Hashimoto, M. Idemen, and O. A. Tretyakov (eds.), Chap. 10, Science House, Tokyo

Veliev, E.I. (1999). Plane wave diffraction by a half-plane: a new analytical approach. *Journal of electromagnetic waves and applications*, Vol.13, No.10, pp. 1439-1453

Veliev, E.I. & Engheta, N. (2003). Generalization of Green's Theorem with Fractional Differintegration, *IEEE AP-S International Symposium & USNC/URSI National Radio Science Meeting*

Veliev, E.I.; Ivakhnychenko, M.V. & Ahmedov, T.M. (2008). Fractional boundary conditions in plane waves diffraction on a strip. *Progress In Electromagnetics Research*, Vol.79, pp. 443–462

Veliev, E.I.; Ivakhnychenko, M.V. & Ahmedov, T.M. (2008). Scattering properties of the strip with fractional boundary conditions and comparison with the impedance strip. *Progress In Electromagnetics Research C*, Vol.2, pp. 189-205

Part 3

Electromagnetic Wave Propagation and Scattering

7

Atmospheric Refraction and Propagation in Lower Troposphere

Martin Grabner and Vaclav Kvicera
Czech Metrology Institute
Czech Republic

1. Introduction

Influence of atmospheric refraction on the propagation of electromagnetic waves has been studied from the beginnings of radio wave technology (Kerr, 1987). It has been proved that the path bending of electromagnetic waves due to inhomogeneous spatial distribution of the refractive index of air causes adverse effects such as multipath fading and interference, attenuation due to diffraction on the terrain obstacles or so called radio holes (Lavergnat & Sylvain, 2000). These effects significantly impair radio communication, navigation and radar systems. Atmospheric refractivity is dependent on physical parameters of air such as pressure, temperature and water content. It varies in space and time due the physical processes in atmosphere that are often difficult to describe in a deterministic way and have to be, to some extent, considered as random with its probabilistic characteristics.

Current research of refractivity effects utilizes both the experimental results obtained from in situ measurements of atmospheric refractivity and the computational methods to simulate the refractivity related propagation effects. The two following areas are mainly addressed. First, a more complete statistical description of refractivity distribution is sought using the finer space and time scales in order to get data not only for typical current applications such as radio path planning, but also to describe adverse propagation in detail. For example, multipath propagation can be caused by atmospheric layers of width of several meters. During severe multipath propagation conditions, received signal changes on time scales of minutes or seconds. Therefore, for example, the vertical profiles of meteorological parameters measured every 6 hours by radiosondes are not sufficient for all modelling purposes. The second main topic of an ongoing research is a development and application of inverse propagation methods that are intended to obtain refractivity fields from electromagnetic measurements.

In the chapter, recent experimental and modelling results are presented that are related to atmospheric refractivity effects on the propagation of microwaves in the lowest troposphere. The chapter is organized as follows. Basic facts about atmospheric refractivity are introduced in the Section 2. The current experimental measurement of the vertical distribution of refractivity is described in the Section 3. Long term statistics of atmospheric refractivity parameters are presented in the Section 4. Finally, the methods of propagation modelling of EM waves in the lowest troposphere with inhomogeneous refractivity are discussed in the Section 5.

2. Atmospheric refractivity

2.1 Physical parameters of air and refractivity formula

The refractive index of air n is related to the dielectric constants of the gas constituents of an air mixture. Its numerical value is only slightly larger than one. Therefore, a more convenient atmospheric refractivity N (N-units) is usually introduced as:

$$N = (n-1)\times 10^6 \tag{1}$$

It can be simply demonstrated, based on the Debye theory of polar molecules, that refractivity can be calculated from pressure p (hPa) and temperature T (K) as (Brussaard, 1996):

$$N = \frac{77.6}{T}\left(p + 4810\frac{e}{T}\right) \tag{2}$$

where e (hPa) stands for a water vapour pressure that is related to the relative humidity H (%) by a relation:

$$H = 100\, e/e_s(t) \tag{3}$$

where e_s (hPa) is a saturation vapour pressure. The saturation pressure e_s depends on temperature t (°C) according to the following empirical equation:

$$e_s(t) = a\exp\left(bt/(t+c)\right) \tag{4}$$

where for the saturation vapour above liquid water $a = 6.1121$ hPa, $b = 17.502$ and $c = 240.97$ °C and above ice $a = 6.1115$ hPa, $b = 22.452$ and $c = 272.55$ °C.

It is seen in Fig.1a where the dependence of the refractivity on temperature and relative humidity is depicted that refractivity generally increases with humidity. Its dependence on temperature is not generally monotonic however. For humidity values larger than about 40%, refractivity also increases with temperature.

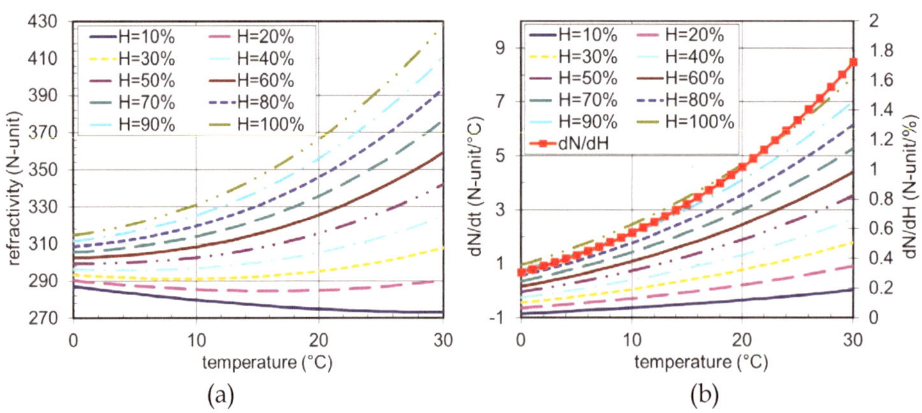

Fig. 1. The radio refractivity dependence on temperature and relative humidity of air for pressure $p = 1000$ hPa (a), refractivity sensitivity dependence on temperature and relative humidity of air (b).

The *sensitivity* of refractivity on temperature and relative humidity of air is shown in Fig. 1b. For t = 10°C (cca average near ground temperature in the Czech Republic), H = 70% (cca average near ground relative humidity) and p = 1000 hPa, the sensitivities are dN/dt = 1.43 N-unit/°C, dN/dH = 0.57 N-unit/% and dN/dp = 0.27 N-unit/hPa. The refractivity variation is usually most significantly influenced by the changes of relative humidity as a water vapour content often changes rapidly (both in space and time) and it is least sensitive to pressure variation. However a decrease in pressure with altitude is mainly responsible for a standard vertical gradient of the atmospheric refractivity.
During standard atmospheric conditions, the temperature and pressure are decreasing with the height above the ground with lapse rates of about 6 °C/km and 125 hPa/km (near ground gradients). Assuming that relative humidity is approximately constant with height, a standard value of the lapse rate of refractivity with a height h can be obtained using pressure and temperature sensitivities and their standard lapse rates. Such an estimated standard vertical gradient of refractivity is about $dN/dh \approx$ -42 N-units/km. It will be seen that such value is very close to the observed long term median of the vertical gradient of refractivity.

2.2 EM wave propagation basics

Ray approximation of EM wave propagation is convenient to see the basic propagation characteristics in real atmosphere. The ray equation can be written in a vector form as:

$$\frac{\mathrm{d}}{\mathrm{d}s}\left(n\frac{\mathrm{d}\mathbf{r}}{\mathrm{d}s}\right) = \nabla n \tag{5}$$

where a position vector $\mathbf{r}$ is associated with each point along a ray and s is the curvilinear abscissa along this ray. Since the atmosphere is dominantly *horizontally stratified*, the gradient ∇n has its main component in vertical direction. Considering nearly horizontal propagation, the refractive index close to one and only vertical component of the gradient ∇n, one can derive from (5) that the *inverse* of the radius of ray curvature, ρ, is approximately equal to the negative height derivative of the refractive index, $-dn/dh$. Using the conservation of a relative curvature: $1/R - 1/\rho$ = const. = $1/R_{ef} - 1/\infty$ one can transform the curvilinear ray to a straight line propagating above an Earth surface with the effective Earth radius R_{ef} given by:

$$R_{ef} = R\Big/\left(1-\frac{R}{\rho}\right) = R\Big/\left(1+R\frac{\mathrm{d}N}{\mathrm{d}h}10^{-6}\right) \tag{6}$$

where R stands for the Earth radius and dN/dh denotes a vertical gradient of refractivity.
Three typical propagation conditions are observed depending on the numerical value of the gradient. If $dN/dh \approx$ -40 N-units/km, than from (6): $R_{ef} \approx 4/3\,R$ and standard atmospheric conditions take place. The standard value of the vertical refractivity gradient is approximately equal to the long term median of the gradient observed in mild climate areas. The median gradients observed in other climate regions may be slightly different, see the world maps of refractivity statistics in (Rec. ITU-R P.453-9, 2009).
Sub-refractive atmospheric conditions occur when the refractivity gradient has a significantly larger value, super-refractive conditions occur when the refractivity gradient is well below the standard value of -40 N-units/km. During sub-refractive atmospheric conditions, the effective

Earth radius R_{ef} decreases, terrain obstacles are relatively higher and the received signal may by attenuated due to diffraction loss appearing if the obstacle interfere more than 60% of the radius of the 1st Fresnel ellipsoid on the line between the transmitter and receiver. During super-refractive conditions, on the other hand, the effective Earth radius is lower than the Earth radius R or it is even negative when $dN/dh < -157$ N-units/km. It means a radio path is more "open" in the sense that terrain obstacles are relatively lower. Super-refractive conditions are often associated with multipath propagation when the received signal fluctuates due to constructive and destructive interference of EM waves coming to the receiver antenna with different phase shifts or time delays.

In principle, the EM wave propagation characteristics during clear-air conditions are straightforwardly determined by the state of atmospheric refractivity. Nevertheless, atmospheric refractivity varies in time and space more or less randomly and full details of it are out of reach in practice. Therefore the *statistics* of atmospheric refractivity and related propagation effects are of main interest. The statistical data important for the design of terrestrial radio systems have to be obtained from the experiments, an example of which is described further.

3. Measurement of refractivity and propagation

3.1 Measurement setup

A propagation experiment focussed on the atmospheric refractivity related effects has been carried out in the Czech Republic since November 2007. First, the combined experiment consists of the measurement of a received power level fluctuations on the microwave terrestrial path operating in the 10.7 GHz band with 5 receiving antennas located in different heights above the ground. Second, atmospheric refractivity is determined in the several heights (19 heights from May, 2010) at the receiver site from pressure, temperature and relative humidity that are simultaneously measured by a meteo-sensors located on the 150 meters tall mast. Refractivity is calculated using (2) - (4). Figure 2a shows the terrain profile of the microwave path.

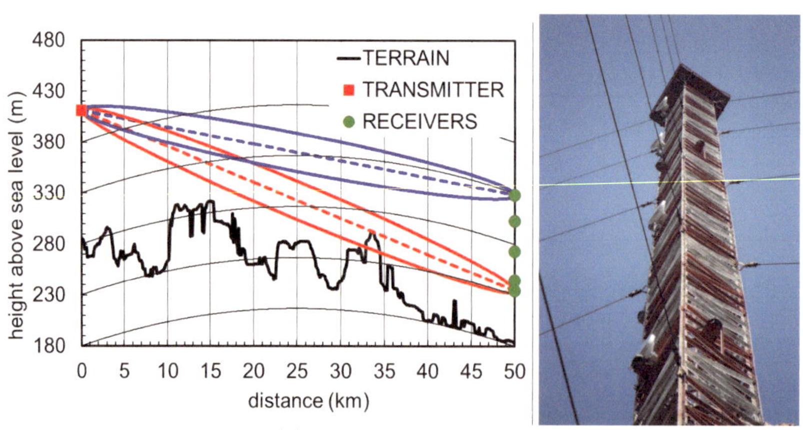

(a) (b)

Fig. 2. (a) The terrain profile of an experimental microwave path, TV Tower Prague – Podebrady mast, with the first Fresnel ellipsoids of the lowest and the highest paths for $k = R_{ef}/R = 4/3$, (b) the parabolic receiver antennas placed on the 150 m high mast (Podebrady site).

The distance between the transmitter and receivers is 49.8 km. It can be seen in Fig. 2a a terrain obstacle located about 33 km from the transmitter site. The height of the obstacle is such that about 0% of the first Fresnel ellipsoid radius of the lowest path (between the transmitter antenna and the lowest receiver antenna) is free. It follows that under standard atmospheric conditions ($k = R_{ef}/R = 4/3$) the lowest path is attenuated due to the diffraction loss of about 6 dB. Tables 1a and 1b show the parameters of the measurement setup.

Heights of meteorological sensors	5.1 m, 27.6 m, 50.3 m, 75.9 m, 98.3 m, 123.9 m, 19 sensors approx. every 7 m (from May 2010)
Pressure sensor height	1.4 m
Temperature/humidity sensor	Vaisala HMP45D, accuracy ±0.2°C, ±2% rel. hum.
Pressure sensor	Vaisala PTB100A, accuracy ±0.2 hPa

Table 1a. The parameters of a measurement system (meteorology).

TX tower ground altitude	258.4 m above sea level
TX antenna height	126.3 m
Frequency	10.671 GHz
Polarization	Horizontal
TX output power	20.0 dBm
Path length	49.82 km
Parabolic antennas	diameter 0.65 m, gain 33.6 dBi
RX dynamical range	> 40 dB
RX tower ground altitude	188.0 m above sea level
RX antennas heights	51.5 m, 61.1 m, 90.0 m, 119.9 m, 145.5 m
Est. uncertainty of received level	±1 dB

Table 1b. The parameters of a measurement system (radio, TX = transmitter, RX = receiver).

3.2 Examples of refractivity effects

In order to get a better insight into atmospheric refractivity impairments occurring in real atmosphere, several examples of measured vertical profiles of temperature, relative humidity, modified refractivity and of received signal levels are given. The modified refractivity M is calculated from refractivity N as:

$$M(h) = N(h) + 157h \tag{7}$$

where h(km) stands for the height above the ground. The reason of using M instead of N here is to clearly point out the possible ducting conditions ($\mathrm{d}N/\mathrm{d}h < -157$ N-units/km) when $\mathrm{d}M/\mathrm{d}h < 0$ M-units/km.

Figure 3 shows the example of radio-meteorological data obtained during a very calm day in autumn 2010. The relative received signal levels measured at 51.5 m (floor 0), 90.0 m (floor 2) and at 145.5 m (floor 4) are depicted. The lowest path (floor 0) is attenuated of about 6 dB due to diffraction on a path obstacle. The situation is atypical since the received signal level is very steady and does not fluctuate practically. The vertical gradient of modified refractivity has approximately the same value (≈ 110 M-units/km or -47 N-units/km) during the whole day, the propagation conditions correspond to standard atmosphere.

A more typical example of measured data is shown in Fig. 4. Temperature and relative humidity change appreciably with height and in time. Specifically, temperature inversion is seen before 4:00 and after 20:00, the standard gradient takes place in the middle of the day. The received signal level recorded on the lowest path shows a typical enhancement at the beginning and at the end of the day which is caused by super-refractive propagation conditions. On the other hand the signal received at the higher antennas fluctuates mildly around 0 dB with more pronounced variations of the signal in the morning and at night.

Sub-refractive propagation conditions were observed between 2:00 and 4:00 on 14 October 2010 as shown in Fig. 5. One can see that increased attenuation due to diffraction on the path obstacle appears on the lowest path (floor 0) at that time. This well corresponds with the sub-refractive gradient of modified refractivity observed; see the lower value of dM/dh near the ground between 2:00 and 4:00 which is caused by strong temperature inversion together with no compensating humidity effect. The received signal measured on the higher antennas that are not affected by diffraction stays around the nominal value with some smaller fluctuations probably due to multipath and focusing/defocusing effects.

A typical example of multipath propagation is shown in Fig. 6. In the middle of the day from about 7:00 to 18:00, the received signal is steady at all heights and the atmosphere seems to be well mixed. On the other hand, multipath propagation occurring in the morning and at night is characterized by relatively fast fluctuations of the received signal. It is seen that all the receivers are impaired in the particular multipath events. Deep fading (attenuation > 20 dB) is quite regularly changing place with significant enhancement of the received signal level.

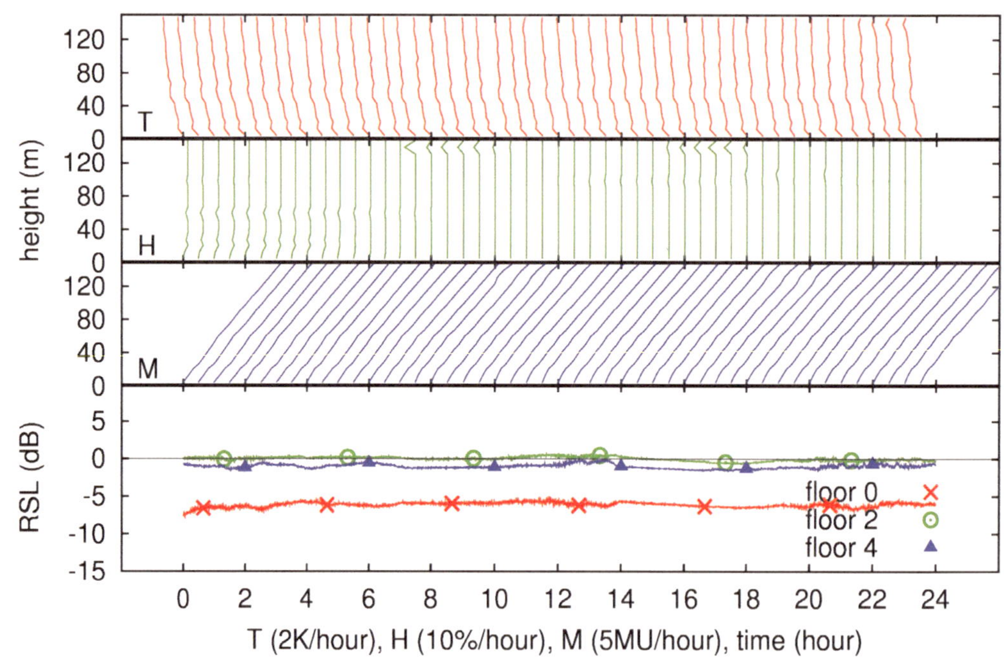

Fig. 3. The vertical profiles of temperature T, relative humidity H, modified refractivity M and received signal levels relative to free-space level observed on 17 November 2010

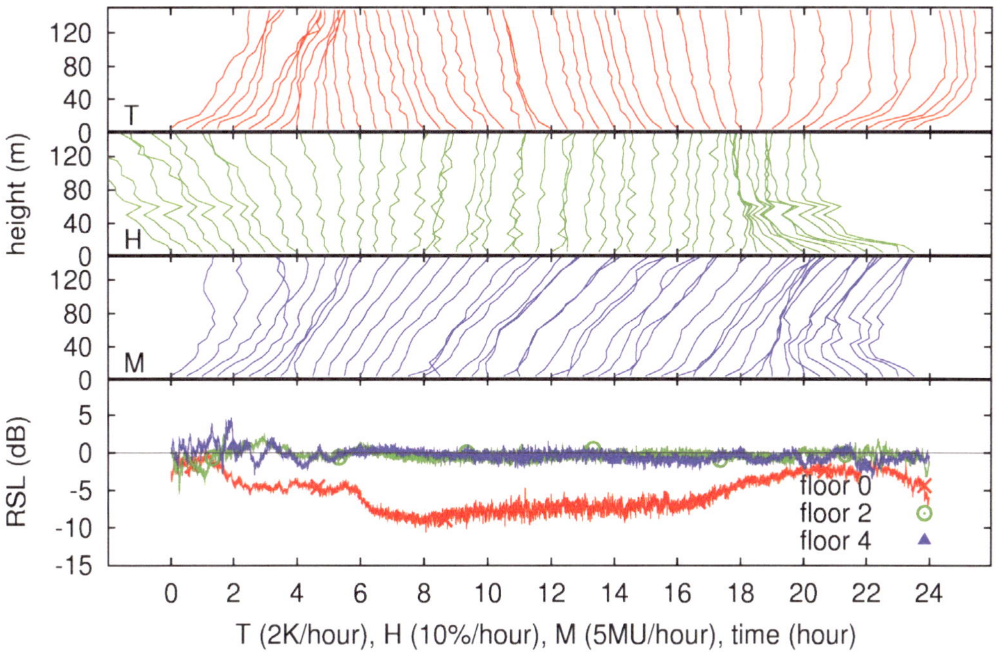

Fig. 4. The vertical profiles of temperature *T*, relative humidity *H*, modified refractivity *M* and received signal levels relative to free-space level observed on 26 June 2010

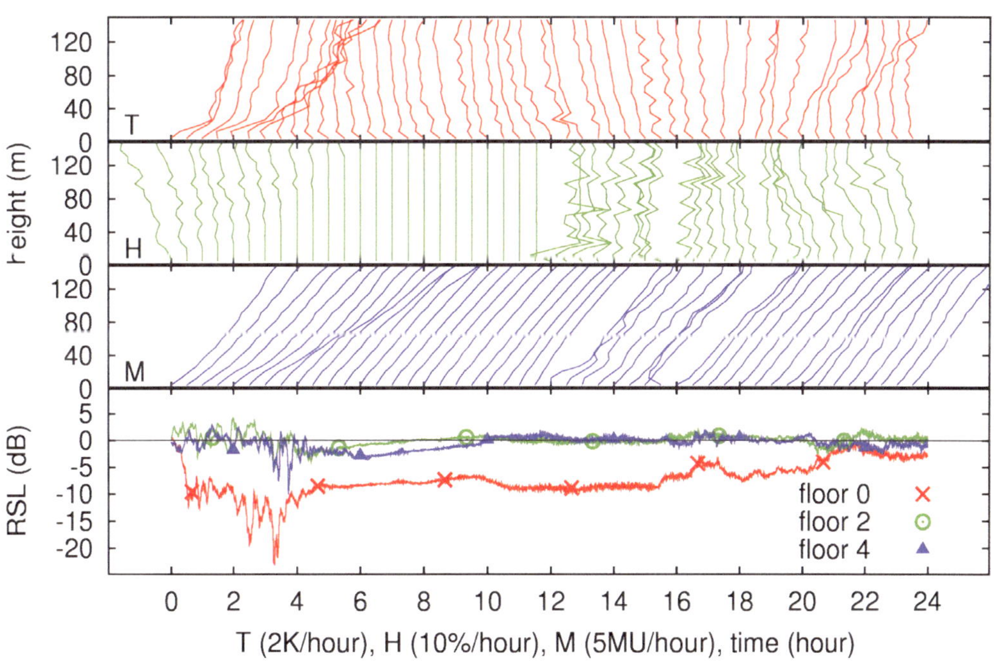

Fig. 5. The vertical profiles of temperature *T*, relative humidity *H*, modified refractivity *M* and received signal levels relative to free-space level observed on 14 October 2010

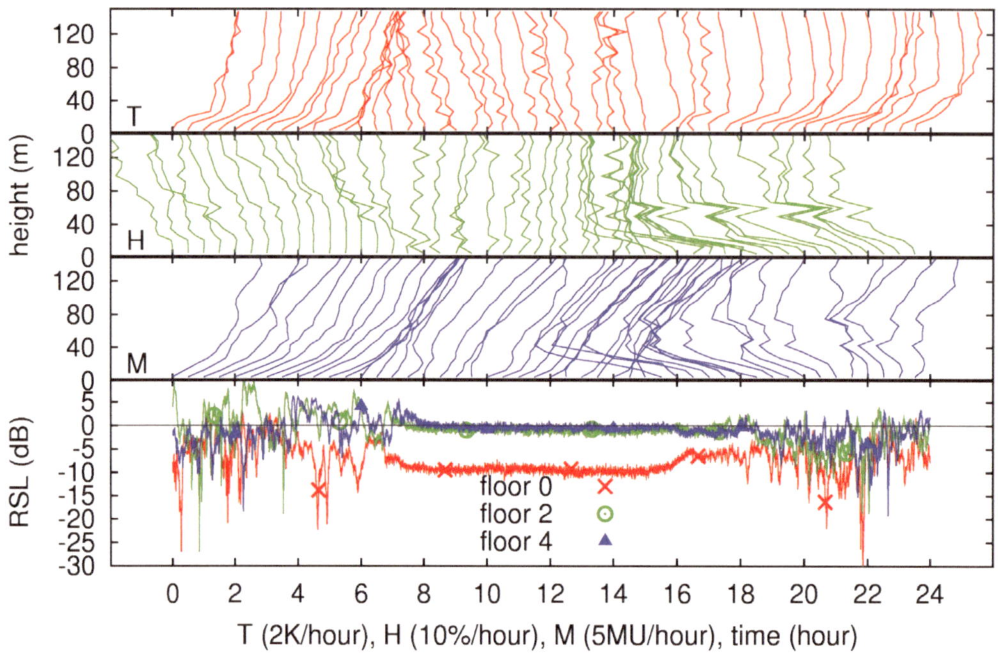

Fig. 6. The vertical profiles of temperature *T*, relative humidity *H*, modified refractivity *M* and received signal levels relative to free-space level observed on 12 September 2010

4. Refractivity statistics

As already mentioned, the physical processes in troposphere are complex enough to allow only statistical description of spatial and temporal characteristics of atmospheric refractivity. Nevertheless the statistics of important refractivity parameters such as an average vertical gradient are extremely useful in practical design of terrestrial radio paths when the long term statistics of the received signal have to be estimated, see (Rec. ITU-R P.530-12, 2009).

4.1 Average vertical gradient of refractivity

The prevailing vertical gradient of refractivity can be regarded as the single most important characteristics of atmospheric refractivity. According to (6), it is related to the effective Earth radius discussed above and it specifically determines the influence of terrain obstacles on terrestrial radio propagation paths. The examples of measured vertical profiles presented in the previous section show that the near-ground refractivity profile evolution is complex enough to not be described by only a single value of the gradient. The question arises what should be considered as a prevailing vertical gradient at a particular time. The gradient value is usually obtained from the refractivity difference at fixed heights, e.g. at 0 and 65 meters above the ground (Rec. ITU-R P.453-9, 2009). If more accurate data is available, the prevailing vertical gradient of refractivity can be calculated using a linear regression approach.

Two year data (2008-2009) of measured vertical profiles were analysed by means of linear regression of refractivity in the heights (0 - 120 m) and the statistics of the vertical gradient so obtained were calculated. The results are in Fig. 7a where the annual cumulative distribution functions of the gradient are depicted. The quantiles provided by ITU-R datasets are also shown for comparison. It is clear that extreme gradients are less probable in reality than predicted by ITU-R. Linear regression tends to filter out the extreme gradients (otherwise obtained from two-point measurements) which do not fully represent the vertical distribution as a whole.

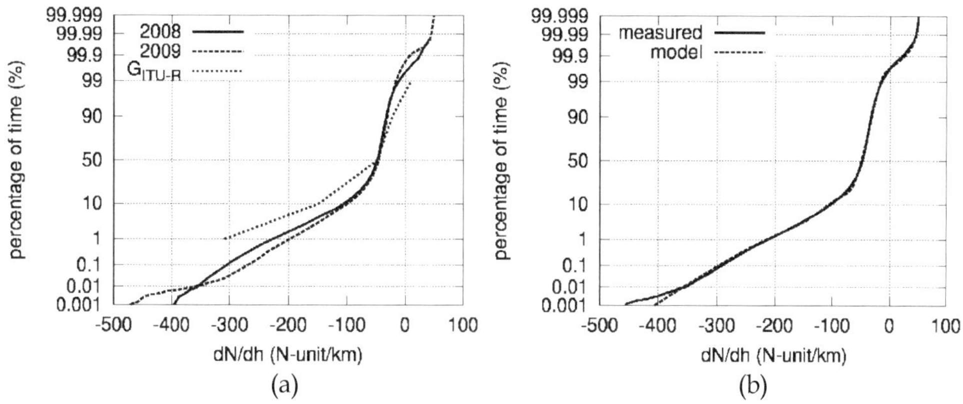

Fig. 7. Annual cumulative distributions of the vertical gradient of atmospheric refractivity obtained in 2008, 2009 (a), cumulative distribution obtained from the whole season (2 years) and fitted model (b).

Taking into account the importance of the gradient statistics for the design of terrestrial radio path, it seems desirable to have a suitable model. Several models of the gradient statistics were proposed, see (Brussaard, 1996), that can be fitted to measured data. Since they are often discontinuous in the probability density, they can be thought to be little unnatural. One can see in Fig. 7b where the two-year cumulative distribution is shown that the distribution consists of three parts: the part around the standard (median) gradient and two other parts - tails. Therefore the following model of the probability density $f(x)$ and of the cumulative distribution function $F(x)$ is proposed:

$$f(x)=\sum_{i=1}^{3} p_i \frac{1}{\sigma\sqrt{2\pi}} \exp\left[\frac{(x-\mu_i)^2}{2\sigma_i^2}\right]; \qquad \sum_{i=1}^{3} p_i = 1 \tag{8}$$

$$F(x)=\sum_{i=1}^{3} p_i \frac{1}{2}\left[1+\operatorname{erf}\left(\frac{x-\mu_i}{\sigma_i\sqrt{2}}\right)\right] \tag{9}$$

where the p_i, μ_i and σ_i are the relative probabilities, the mean values and the standard deviations of the Gaussian distributions forming the three parts of the whole distribution. Fitted model parameters (see Fig. 7b) are summarized in Table 2.

i	p_i	μ_i	σ_i
1	0.086	-128.0	75.1
2	0.793	-46.1	11.8
3	0.121	-99.6	24.8

Table 2. Vertical refractivity gradient distribution parameters

4.2 Ducting layers

Although the ducting layers appearing in the first several tens or hundreds meters above the ground have significant impact on the propagation of EM waves on nearly horizontal paths, surprisingly little is known about their occurrence probabilities or about their spatial/temporal properties (Ikegami et al., 1966). This is true especially in the lowest troposphere where the usual radio-sounding data suffers from insufficient spatial and also time resolution. In the following, the parameters of ducting layers observed during the experiment are analysed by means of the modified Webster duct model.

An analytic approach to the modelling of refractivity profiles was proposed in (Webster, 1982). The refractivity profile with the height h (m) was to be approximated by the formula similar to the following modified model:

$$N(h) = N_0 + G_N h + \frac{dN}{2} \tanh \frac{2.96(h - h_0)}{dh} \tag{10}$$

where the refractivity N_0 (N-units), the gradient G_N (N-units/m), the duct depth dN (N-units), the duct height h_0 (m) and the duct width dh (m) are model parameters. A hyperbolic tangent is used in (10) instead of arctangent in the original Webster model because the "tanh" function converges faster to a constant value for increasing arguments than the "arctan" does. As a consequence, there is a sharper transition between the layer and the ambient gradient in the modified model and so the duct width values dh are more clearly recognizable in profiles. Figure 8 shows the meaning of the model parameters by an example where the modified refractivity profile is also included. It is seen from (7) and (10) that the model for modified refractivity profiles differs only in the value of the gradient: $G = G_N + 0.157$ (N-units/m).

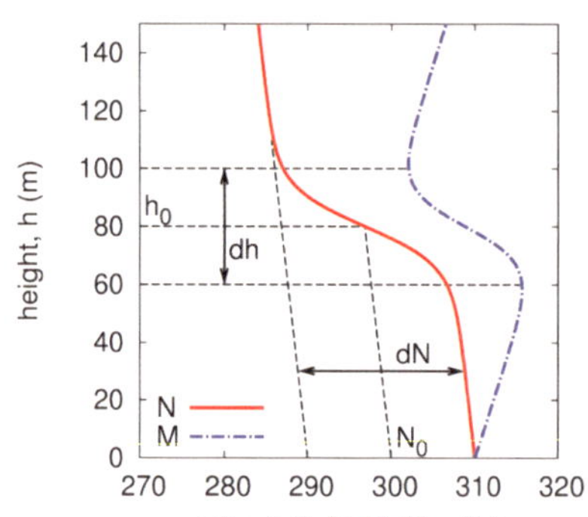

Fig. 8. Duct model parameter definition with the values of parameters: N_0 = 300 N-units, G_N = -40 N-units/km, dN = -20 N-units, h_0 = 80 m, dh = 40 m.

The above model was fitted to the refractivity profiles measured in between May and November 2010. More than $3 \cdot 10^5$ profiles were analysed and related model parameters were obtained. Figure 9 shows two examples of 1-hour measured data and fitted models. Significant dynamics is clearly seen in the evolving elevated ducting layers. It is also clear from the examples in Fig. 9 that the model is not able to capture all the fine details of measured profiles but it serves very well to describe the most important features relevant for radio propagation studies. Sometimes, the part or the whole ducting layer is located above the measurement range and so it is out of reach of modelling despite its effect on the propagation might be serious. This should be kept in mind while studying the statistical results presented below.

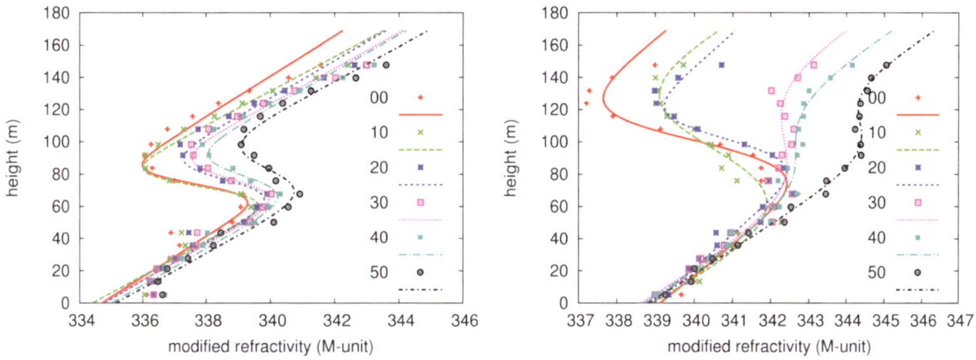

(a) (b)

Fig. 9. The examples of time evolution of elevated ducting layers observed on the 1st of August 2010 at 00:00-00:50 (a) and on the 14th of July 2010 at 22:00-22:50 (b), measured data with points, fitted profiles with lines.

Figure 10 shows the empirical cumulative distributions of duct model parameters obtained from the fitting procedure. The medians (50% of time) of duct parameters can be read as $N_0 = 320$ N-units, $G = 116$ N-units/km, $dN = -2.2$ N-units, $h_0 = 61$ m, $dh = 73$ m. The probability distributions of N_0 and G are almost symmetric around the median. On the other hand, the depth dN and width dh distributions are clearly asymmetric showing that the smaller negative values of the depth and the smaller values of width are observed more frequently. Almost linear cumulative distribution of the duct height h_0 between 50 and 100 m above the ground suggests that there is no preferred duct height here.

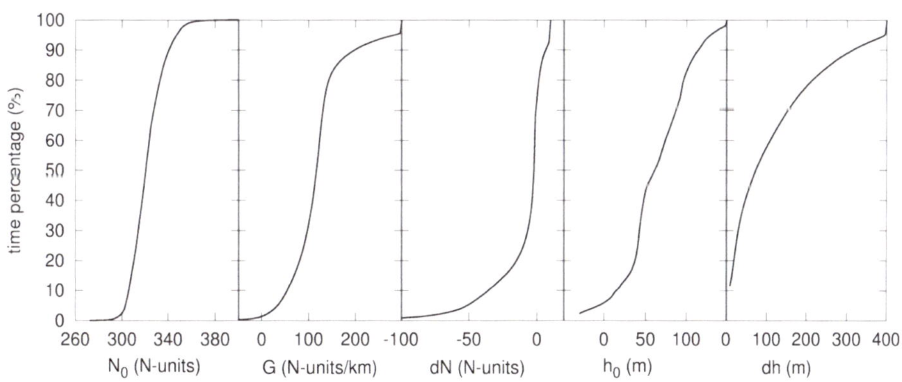

Fig. 10. The cumulative distribution functions of duct parameters obtained from measured profiles of atmospheric refractivity at Podebrady, 05/2010 - 11/2010.

Important interrelations between duct parameters are revealed by empirical joint probability density functions (PDF) presented in Fig. 11 - 15. The 2D maps show the logarithm of joint PDFs of all combinations of 5 parameters of the duct model (10). In these plots, dark areas mean the high probability values and light areas mean the low probability values. It is generally observed that there are certain preferred areas in the parameter space where the combinations of duct parameters usually fall in. For example, it is seen in Fig. 13a that the absolute value of the negative duct depth is likely to increase with the increasing gradient G. On the other hand, there are empty areas in the parameter space where the combinations of parameters are not likely to appear. One may find this information helpful when analysing terrestrial propagation using random ducts generated by the Monte Carlo method.

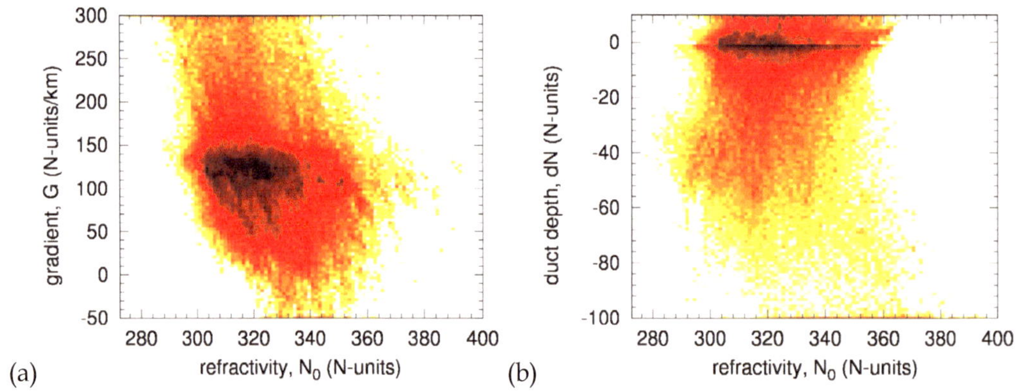

Fig. 11. The logarithm of the joint probability density function of duct parameters, obtained from measured profiles of atmospheric refractivity at Podebrady, 05/2010 - 11/2010.

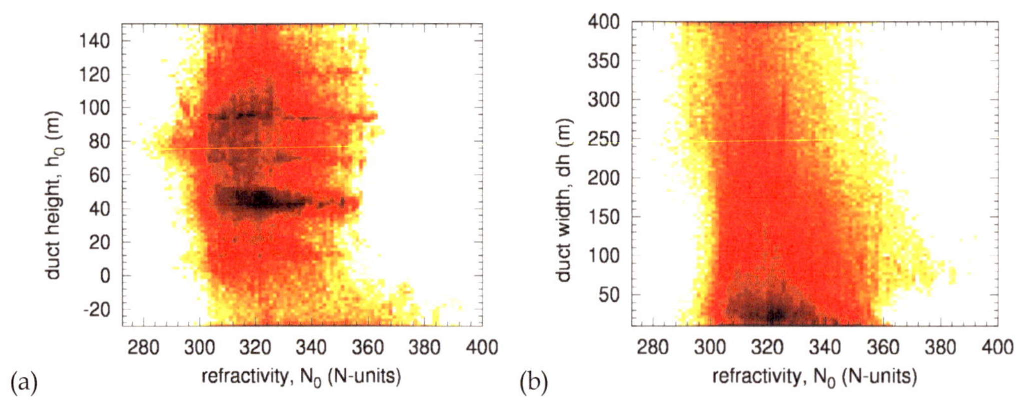

Fig. 12. The logarithm of the joint probability density function of duct parameters, obtained from measured profiles of atmospheric refractivity at Podebrady, 05/2010 - 11/2010.

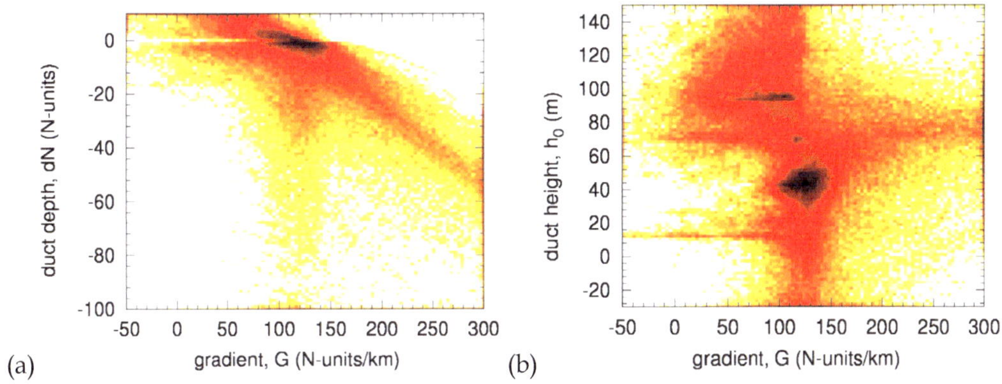

Fig. 13. The logarithm of the joint probability density function of duct parameters, obtained from measured profiles of atmospheric refractivity at Podebrady, 05/2010 - 11/2010.

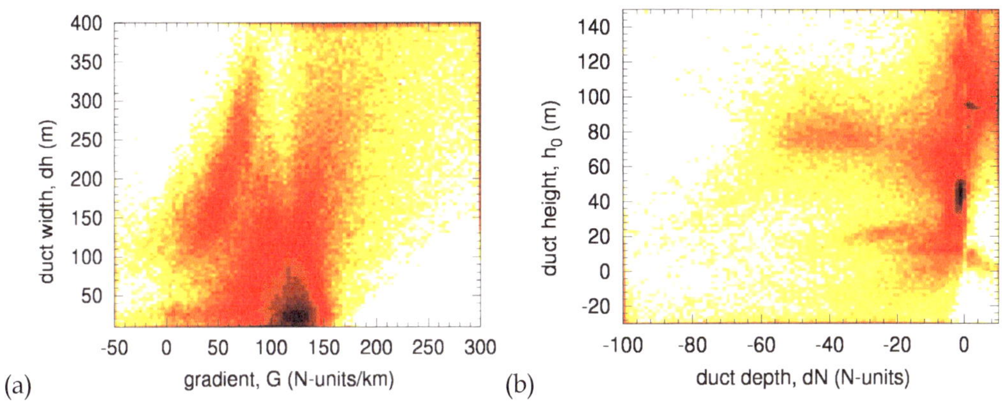

Fig. 14. The logarithm of the joint probability density function of duct parameters, obtained from measured profiles of atmospheric refractivity at Podebrady, 05/2010 - 11/2010.

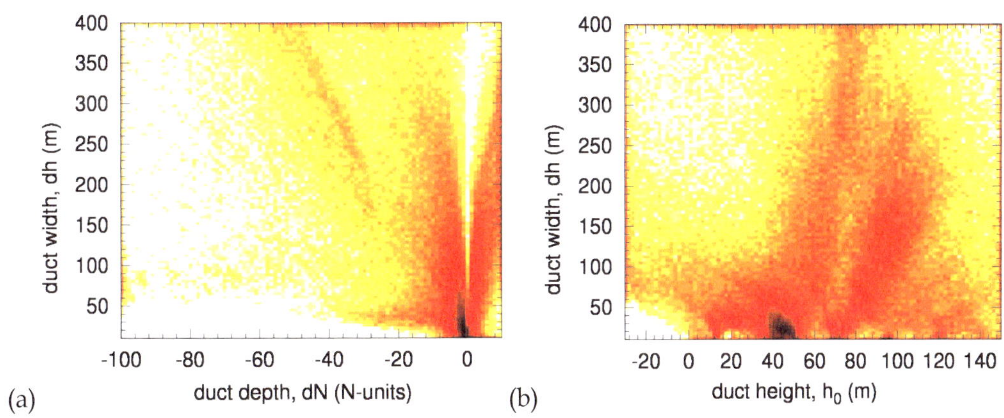

Fig. 15. The logarithm of the joint probability density function of duct parameters, obtained from measured profiles of atmospheric refractivity at Podebrady, 05/2010 - 11/2010.

5. Modelling of EM waves in the troposphere

Several numerical methods have been used in order to assess the effects of atmospheric refractivity on the propagation of electromagnetic waves in the troposphere. They can be roughly divided into two categories - ray tracing methods based on geometrical optics and full-wave methods. The ray tracing methods numerically solve the ray equation (5) in order to get the ray trajectories of the electromagnetic wave within inhomogeneous refractivity medium. The ray tracing provides a useful qualitative insight into refraction phenomena such as bending of electromagnetic waves. Its utilization for quantitative modelling is limited to conditions where the electromagnetic waves of sufficiently large frequency may be approximated by rays. Geometrical optics description is known to fail at focal points and caustics where the full-wave methods provide more accurate results.

The full-wave numerical methods solve the wave equation that is a partial differential equation. Among time domain techniques, finite difference time domain (FDTD) based approaches were proposed (Akleman & Sevgi, 2000) that implement sliding rectangular window where 2D FDTD algorithm is applied. Nevertheless, tropospheric propagation simulation in frequency domain is more often. In particular , there is a computationally efficient approach based on the paraxial approximation of Helmholtz wave equation, so called Parabolic Equation Method (PEM), which is the most often used full-wave method in tropospheric propagation.

5.1 Split step parabolic equation method

We start the brief summary of PEM (Levy, 2000) with the scalar wave equation for an electric or magnetic field component ψ:

$$\nabla^2\psi + k^2 n^2 \psi = 0 \tag{11}$$

where $k = 2\pi/\lambda$ is the wave number in the vacuum and $n(r,\theta,\varphi)$ is the refractive index. Spherical coordinates with the origin at the center of the Earth are used here. Further, we assume the azimuthal symmetry of the field, $\psi(r,\theta,\varphi) = \psi(r,\theta)$, and express the wave equation in cylindrical coordinates:

$$\frac{\partial^2\psi}{\partial z^2} + \frac{\partial^2\psi}{\partial x^2} + \frac{1}{x}\frac{\partial\psi}{\partial x} + k^2 m^2(x,z)\psi = 0 \tag{12}$$

where:

$$m(x,z) = n(x,z) + z / R \tag{13}$$

is the modified refractive index which takes account of the Earth's radius R and where $x = r\theta$ is a horizontal range and $z = r - R$ refers to an altitude over the Earth's surface. We are interested in the variations of the field on scales larger than a wavelength. For near horizontal propagation we can separate "phase" and "amplitude" functions by the substitution of:

$$\psi(x,z) = u(x,z)\frac{e^{jkx}}{\sqrt{x}} \tag{14}$$

in equation (12) to obtain:

$$\frac{\partial^2 u}{\partial z^2} + \frac{\partial^2 u}{\partial x^2} + 2\mathrm{j}k\frac{\partial u}{\partial x} + k^2\left(m^2 - 1 + \frac{1}{(2kx)^2}\right)u = 0 \tag{15}$$

Paraxial approximation is made now. The field $u(x,z)$ depends only little on z, because main dependence of $\psi(x,z)$ is covered in the exp(jkx) factor in (14). Then it is assumed that:

$$\left|\frac{\partial^2 u}{\partial x^2}\right| << 2k\left|\frac{\partial u}{\partial x}\right| \tag{16}$$

and the $1/(2kx)^2$ term can be removed from (15) since $kx >> 1$ when the field is calculated far enough from a source. We obtain the following parabolic equation:

$$\frac{\partial^2 u}{\partial z^2} + 2\mathrm{j}k\frac{\partial u}{\partial x} + k^2(m^2(x,z) - 1)u = 0 \tag{17}$$

An elliptic wave equation is therefore simplified to a parabolic equation where near horizontal propagation is assumed. This equation can be solved by the efficient iterative methods such as the Fourier split-step method. Let us assume the modified refractivity m is constant. Then we can apply Fourier transform on the equation (17) to get:

$$-p^2 U + 2\mathrm{j}k\frac{\partial U}{\partial x} + k^2(m^2 - 1)U = 0 \tag{18}$$

where Fourier transform is defined as:

$$U \equiv U(x,p) = \mathrm{F}\{u(x,z)\} = \int_{-\infty}^{\infty} u(x,z)\mathrm{e}^{-\mathrm{j}pz}\mathrm{d}z \tag{19}$$

From (18), we obtain:

$$\frac{\partial U(x,p)}{\partial x} = \left(\frac{p^2 - k^2(m^2 - 1)}{2\mathrm{j}k}\right)U(x,p) \tag{20}$$

$$U(x,p) = \mathrm{e}^{-\mathrm{j}x(p^2/(2k))} \cdot \mathrm{e}^{\mathrm{j}x(k(m^2-1)/2)} \tag{21}$$

and we get the formula for step-by-step solution:

$$U(x + \Delta x, p) = \left(\mathrm{e}^{-\mathrm{j}\Delta x(p^2/(2k))} \cdot \mathrm{e}^{\mathrm{j}\Delta x(k(m^2-1)/2)}\right)U(x,p) \tag{22}$$

The field in the next layer $u(x+\Delta x,z)$ is computed using the field in the previous layer $u(x,z)$:

$$u(x + \Delta x, z) = \mathrm{e}^{\mathrm{j}\Delta x(k(m^2-1)/2)} \cdot \mathrm{F}^{-1}\left\{ U(x,p)\mathrm{e}^{-\mathrm{j}\Delta x(p^2/(2k))}\right\} \tag{23}$$

Fourier transformation is applied in z-direction and the variable p represents the "spatial frequency" (wave number) of this direction: $p = k_z = k\sin(\xi)$ and ξ is the angle of propagation.

The assumption that m is constant is not fulfilled, but equation (23) is used anyway. The resulting error is proportional to Δx and to horizontal and vertical gradients of refractivity. In practice, the value of Δx can be of several hundred wavelengths.

5.2 Application example and comparison with measured data

The parabolic equation method outlined above has been applied frequently to investigate the propagation characteristics on terrestrial (and also on Earth - space) paths under the influence of different refractivity conditions (Barrios, 1992, 1994; Levy, 2000) including the ducting layers described in the section 4.2. Users agree the method gives reliable results provided all the relevant details of terrain profile and of refractivity distribution are known and modelled correctly. This is however not always the case in practice. It is believed that the modelling results have to be compared with real world data whenever possible in order to validate the method under different propagation conditions and to know more about the expected errors due to incomplete knowledge of propagation medium.

Let us illustrate the particular example of conditions where the parabolic equation method performs successfully regardless the fact that refractivity profile along the propagation path is only roughly estimated. Figures 16a and 16b show the results of PEM propagation simulation performed using refractivity gradients measured during the 4th of November, 2008 at the receiver site. Sub-refractive conditions that occurred early morning caused a significant diffraction fading of more than 20 dB on the two lowest paths see Fig. 16b. On the other hand, the higher paths (receiving antennas located at 90 m and above) were not affected by diffraction effects.

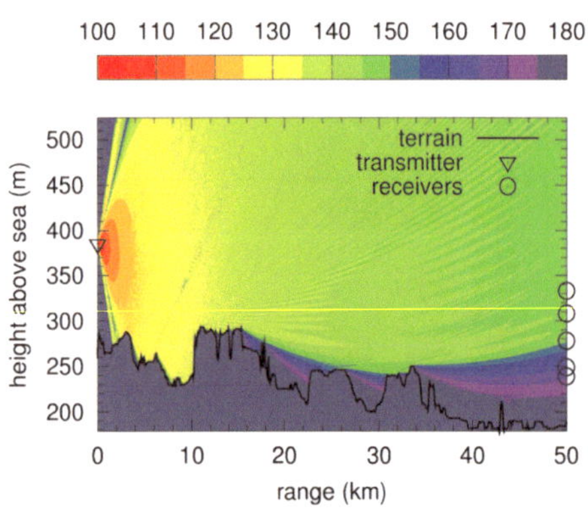

(a)

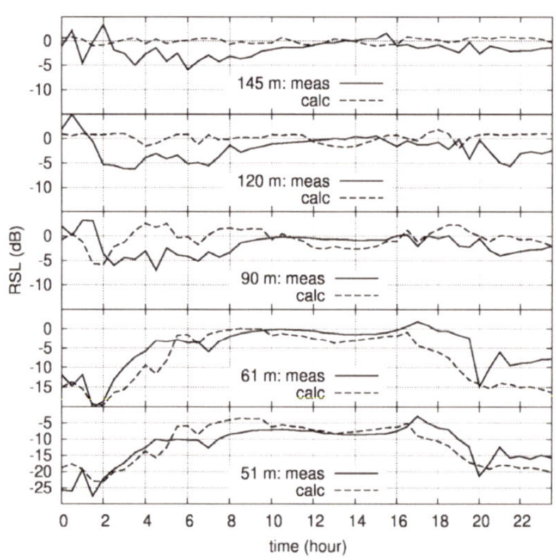

(b)

Fig. 16. Spatial distribution of received power loss during sub-refractive condition on the path TV Tower Prague - Podebrady calculated by PEM (a), received signal levels measured in 5 receivers located in different heights and received signal levels modelled by PEM using time dependent vertical gradient of refractivity (b).

The results shown in Fig. 16b confirm that a very good agreement between PEM simulation and measurement can be achieved if the diffraction fading due to sub-refractive conditions (see time about 2:00) is the most important effect influencing the received power. It suggests that sub-refractive gradients are likely to be approximately the same along the whole propagation path and the approximation of horizontally independent refractivity, which is usually applied in PEM, is reasonable in this case. On the other hand, similar conclusion cannot be reached when multipath propagation occurs because only slight change in a refractivity profile along the propagation path may vary the received power distribution profoundly. These facts have to be kept in mind when the simulation results are interpreted.

6. Conclusion

Some results of the ongoing studies focussed on the propagation impairments of the atmospheric refractivity in the lowest troposphere were presented. Concurrent measurements of the vertical distribution of atmospheric refractivity together with the multi-receiver microwave propagation experiment were described. A new statistical model of vertical refractivity gradient was introduced. The unique joint statistics of ducting layers parameters were presented. The application of parabolic equation method was demonstrated on the example of a diffraction fading event. Simulated and measured time series were compared. A good agreement between simulation and measured data has been witnessed.

Future works in the area of the atmospheric refractivity related propagation effects should, for example, investigate the relations between the time evolution of duct parameters and multipath propagation characteristics, which is the area where only little is known at this moment.

7. Acknowledgment

This research was financially supported by the Czech Science Foundation under the project no. P102/10/1901.

8. References

Akleman, F.; Sevgi, L. (2000). A novel time-domain wave propagator. *IEEE Transactions on Antennas and Propagation,* Vol. 48, No. 5, (May 2000), pp. (839-841), 0018-926X

Barrios, A. (1992). Parabolic equation modelling in horizontally inhomogeneous environments. *IEEE Transactions on Antennas and Propagation,* Vol. 40, No. 7, (July 1992), pp. (791-797), 0018-926X

Barrios, A. (1994). A terrain parabolic equation model for propagation in the troposphere. *IEEE Transactions on Antennas and Propagation,* Vol. 42, No. 1, (January 1994), pp. (90-98), 0018-926X

Brussaard, G. (Ed.). (1996). *Handbook on Radiometeorology,* International Telecommunication Union, ISBN 92-61-06241-5, Geneva, Switzerland.

Ikegami, F.; Haga, M.; Fukuda, T.; Yoshida, H. (1966). Experimental studies on atmospheric ducts and microwave fading. *Review of the Electrical Communication Laboratory,* Vol. 14, No. 7-8, (July-August 1966), pp. (505-533)

Kerr, D. E. (Ed.). (1987). *Propagation of Short Radio Waves,* Peter Peregrinus Ltd./IEE, ISBN 0-86341-099-5, London, UK.

Lavergnat, J.; Sylvain, M. (2000). *Radio Wave Propagation: Principles and Techniques,* John Wiley & Sons., ISBN 0-471-49027-X, Chichester, UK.

Levy, M. (2000). *Parabolic Equation Methods for Electromagnetic Wave Propagation,* The Institution of Electrical Engineers, ISBN 0-85296-764-0, London, UK.

Rec. ITU-R P.453-9 (2009). The radio refractive index: its formula and refractivity data, *ITU-R Recommendations and Reports*, ITU, Geneva, Switzerland, 2009.

Rec. ITU-R P.530-12 (2009). Propagation data and prediction methods required for the design of terrestrial line-of-sight systems, *ITU-R Recommendations and Reports*, ITU, Geneva, Switzerland, 2009.

Webster, A. R. (1982). Raypath parameters in tropospheric multipath propagation. *IEEE Transactions on Antennas and Propagation,* Vol. 30, No. 4, (July 1982), pp. (796-800), 0018-926X

8

Atmospheric Attenuation due to Humidity

Milda Tamošiūnaitė[1], Mindaugas Žilinskas[1,2],
Milda Tamošiūnienė[3] and Stasys Tamošiūnas[1,4]
[1]*Vilnius University, Faculty of Physics*
[2]*The Communications Regulatory Authority of the Republic of Lithuania, Department of Radio Communication*
[3] *Center for Physical Sciences and Technology, Semiconductor Physics Institute*
[4]*Vilnius University, Institute of Applied Research*
Lithuania

1. Introduction

Humidity remains in the atmosphere even on bright days. Water of all three states can be found naturally in the atmosphere: liquid (rain, fog, and clouds), solid (snowflakes, ice crystals), and gas (water vapour). Water in any state is an obstacle in the link of the electromagnetic wave. When the wave passes through the water particles, a part of its energy is absorbed and a part is scattered. Therefore the electromagnetic wave is attenuated. Prediction of the influence of these factors is very important in radio system design. Attenuation due to rain, fog, and clouds can lead to the perturbations of the wireless, mobile, satellite and other communications. Another problem is the refractive index of the atmosphere, which affects the curvature of the electromagnetic wave path and gives some insight into the fading phenomenon. The anomalous electromagnetic wave propagation can cause disturbances to radar work, because variation of the refractive index of the atmosphere can induce loss of radar coverage. Accurate prediction of losses due to these factors can ensure a reliability of the radio system, decrease an equipment cost, furthermore, the radio systems can become less injurious to health of people.

When there are no possibilities to gather data for calculations of the specific attenuation due to rain, clouds and fog, and atmospheric refractive index, the values recommended by the International Communication Union's Radiocommunication sector (ITU-R) can be used. But the recommended values are not always exact. In design of the radio links, the most desirable operating frequencies are below 10 GHz, because in such cases atmospheric absorption and rainfall loss may generally be neglected (Freeman, 2007). However, in most countries, the frequency-band below 10 GHz is highly congested. In addition, high frequencies provide larger bandwidth, narrower beam width, good resolution and smaller component size (Bhattacharyya et al., 2000). Therefore, the operating frequencies of 10 GHz and above are often used in design of radio systems. The higher the operating frequency, the greater attenuation due to hydrometeors (rain, cloud, fog, snow, and etc.) is observed (Tamošiūnaitė et al., 2010a).

In (Ishimaru, 1978), it was mentioned that the electromagnetic wave attenuation due to snow is less than attenuation due to rain, and that the attenuation due to dry snow may be neglected in microwave band. However, the attenuation due to wet snow is higher. Some results of attenuation due to hail are presented in (Ishimaru, 1978). In this chapter, our attention would be concentrated on the attenuation due to rain, clouds, and fog. The variation of the radio refractivity will be the object of our investigation presented there as well.

2. Attenuation due to rain

The electromagnetic wave attenuation due to rain (the rain attenuation) is one of the most noticeable components of excess losses, especially at frequencies of 10 GHz and above (Freeman, 2007). The methods of prediction of the rain attenuation can be grouped into two groups: the physical (exact) models and the empirical models. The physical models attempt to reproduce the physical behaviour involved in the attenuation processes while the empirical methodologies are based on measurement databases from stations in different climatic zones within a given region. The empirical methods are used widely and frequently with the best success (Emiliani et al., 2004). Two main causes of attenuation are scattering and absorption. When the wavelength is large compared to the size of raindrop, scattering is predominant. Conversely, when the wavelength is small compared to the raindrop's size, attenuation due to absorption is predominant (Ivanovs & Serdega 2006). Water molecules are dipoles. The raindrop's dipoles have the same time variation as the electromagnetic waves and therefore act as an antenna, which re-radiates the electromagnetic wave energy. Hence, a raindrop becomes an "antenna" with low directivity. Consequently, some energy is reradiated in arbitrary directions giving a net loss of energy in the direction towards the receiver (Ivanovs & Serdega 2006). Water is a loss-making dielectric medium. The relative dielectric constant of water is high, compared to the dielectric constant of the surrounding air. It depends on temperature and the operating frequency of the radio system. The specific heat of the water is high. Therefore, water absorbs a large amount of warmth, while warms itself. The surface tension of water is high. This is the reason why the molecules of water are holding together. One of the problems in prediction of electromagnetic wave power losses is description of shape of the raindrop. It depends on the size of droplet. It is known, that only very small droplets are like spheres. Such droplets form in clouds, as water vapour condenses on the nuclei of condensation. Further, these droplets grow by coalescence. Shape of the raindrops, that are larger than 1 mm in diameter, is no more spherical. They are not tear-shaped, as it commonly presented in pictures. The shape of falling large raindrops is more like a hamburger shape. Therefore, horizontally polarized waves suffer greater attenuation than vertically polarized waves (Freeman, 2007).
As mentioned above, the water molecules are polar ones. Those molecules rotate in such way that positive part of one molecule would be as near as possible to the negative part of another molecule. Therefore, molecules are rotating, hammering one on another and heating (Tamošiūnaitė et al., 2010a). The water molecule also rotates when a negative charge is brought near to it. The fields of electromagnetic wave vary up as time goes and force the water molecules to rotate respectively to the variation of fields.

2.1 Specific rain attenuation

One of the most widely used rain attenuation prediction methods is an empirical relationship between the specific rain attenuation α [dB·km^{-1}] and the rain rate R [mm·h^{-1}] (Freeman, 2007, Rec. ITU-R P.838-3, 2005):

$$\alpha = aR^b \tag{1}$$

where a and b are functions of operating frequency f and rain temperature t; the value of R [mm·h^{-1}] is for an exceedance of 0.01% of the time for point rainfall rates with an integration time of one minute. The coefficients a and b (coefficients a_h and b_h to be used for horizontal polarized waves; coefficients a_v and b_v to be used for vertical polarized waves) are presented in (Freeman, 2007; Recommendation ITU-R P. 838-3, 2005).

2.1.1 Rain rate

In determination of the rain attenuation, the main parameter is rain rate R, which is expressed in [mm·h^{-1}]. Gauges at the surface measure the accumulation of rain-water (flux) in a known time interval and report the result as a rain rate (accumulation per unit time) averaged over some measurement or aggregation interval (Crane, 1996). The rain rate can be described as the thickness of the precipitation layer, which felled down over the time period of one hour in the case when the precipitation is not evaporated, not soaked into the soil, and is not blown away by the wind (Tamošiūnaitė et al., 2010a). The evaluation of R-value is the first step in the rain attenuation prediction. The rain attenuation depends on the meteorological conditions in the considered localities. This is the reason to analyze the rain attenuation in particular locations (eg. country, city, climatic region).

First attempts to predict the rain attenuation under Baltic region climate conditions are described in (Tamošiūnas et al., 2005, 2006; Ivanovs & Serdega, 2006; Zilinskas et al., 2006, 2008). It was mentioned in (Ivanovs & Serdega, 2006), that rain events produce unavailability of microwave link, which sometimes lead operators to economical losses or even license loosing.

The significant differences in annual, seasonal, monthly, and daily amounts of rainfall are observed in localities of Lithuania. The noticeable local differences of rainfall amounts are characteristic of Lithuania as well. The precipitation amount is probably the most changeable meteorological index on Lithuania's territory. It varies from 901 mm in Šilalė district to 520 mm in Pakruojis district (Bukantis, 2001). No month of a year could be described as "an average month" in Lithuania. This is the reason to revise the suitability of the models that derived under climatic conditions other than Lithuanian ones. The models using only annual amount of rainfall was analyzed in (Tamošiūnas et al., 2005). Considering the peculiarities of Lithuania's climate, the change in (Chebil et al., 1999) model was made. This new model for the electromagnetic wave attenuation due to rain medium in atmosphere for the first time has been presented in (Tamošiūnas et al., 2006). Calculation of radio wave attenuation due to rain using annual precipitation and heavy rainfall data is described in (Zilinskas et al., 2006). The heavy rainfall events and showers with thunderstorms occur during the warm season (from May to September) in Lithuania.

2.1.2 Integration time

As was mentioned above, the R-values are expressed in [mm·h^{-1}]. However, time intervals between the readings of rainfall amount in many cases must be much shorter. Those intervals are called the integration time τ. In (Ivanovs & Serdega, 2006; Tamošiūnas et al, 2007; Tamošiūnaitė et al., 2010a) it was mentioned, that the period of time between the readings of the rainfall amount values is a very important parameter, because it can significantly change the R-value. High R-values "hides" when τ is long.

Consider an example. There were raining. The duration of the rain was 5 minutes. The total amount of the precipitation was 5 mm. It did not rain during remaining 55 minutes of one hour. Thereby, if we would count the average *R*-value for that hour ($\tau = 60$ min.), it would be equal to 5 mm·h^{-1}. But if we would count the *R*-value for every minute of that hour, we would find that *R*-values are much higher. Consider that in every of those 5 rainy minutes the amount of the precipitation was 1 mm. Consequently, for each of those 5 minutes the *R*-value would be 60 mm·h^{-1}. That is why the average *R*-values are unreliable.
In Lithuania, the τ values must be as small as possible (Tamošiūnaitė et al., 2010a).

2.1.3 "One-minute" rain rate

Almost all rain attenuation methods require "one-minute" rain rate value. The "one-minute" rain rate value $R_{(1\text{ min.})}$ is expressed in [mm·h^{-1}]. $R_{(1\text{ min.})}$-value can be defined as the *R*-value for 0.01% of time of the year, obtained using the rainfall amount value, which was measured in $\tau = 1$ min and multiplied by 60 (Karasawa & Matsudo, 1991).
However, in many instances data collection is oriented toward agricultural and hydrological purposes, for which annual, monthly, daily, and less commonly, 3- and 6-hourly totals are collected. Therefore the models for conversion of $R_{(\tau\text{ min.})}$-values into $R_{(1\text{ min.})}$-values are used. A review of models for estimation of 1 min rainfall rates for microwave attenuation calculations are presented in (Tattelman & Grantham, 1985).
One of such conversion models was presented in (Moupfouma &Martin, 1995):

$$R_{(1\text{ min.})} = \left(R_{(\tau\text{ min.})}\right)^d \quad (2)$$

$$d = 0.987\tau^{0.061} \quad (3)$$

where $R_{(1\text{ min.})}$ is the "one-minute" rain rate value, $R_{(\tau\text{ min.})}$ is the rain rate value measured in τ minutes ($\tau \geq 1$ min.).
In (Zilinskas et al, 2008) another model (4) for calculation of the $R_{(1\text{ min.})}$-value was presented. That model was derived on the basis of model presented in (Rice & Holmberg, 1973) in accordance with the peculiarities of Lithuanian climate.

$$R_{(1\text{ min.})} = \frac{\ln\left(0.0144\dfrac{M_{V-IX}}{t}\right)}{0.03} \quad (4)$$

where M_{V-IX} is amount of rainfall which precipitated in May-September, *t* is the number of hours in a year when the value of rain rate could be equal or exceed the $R_{(1\text{ min.})}$-value.
According to data that was collected in Lithuanian weather stations and (4) formula, the average $R_{(1\text{ min.})}$-value for Lithuanian territory was calculated. That value is 60.23 mm·h^{-1}. This value is double the value, which is suggested by ITU-R (Tamošiūnaitė et al., 2010a).
According to (1) formula, the values of coefficients *a* and *b* (presented in Freeman, 2007), and the value of $R_{(1\text{ min.})}$ = 60.23 mm·h^{-1}, the dependency of the average specific electromagnetic wave attenuation due to rain, α, on the operating frequency *f* was estimated. The results are shown in Fig. 1.

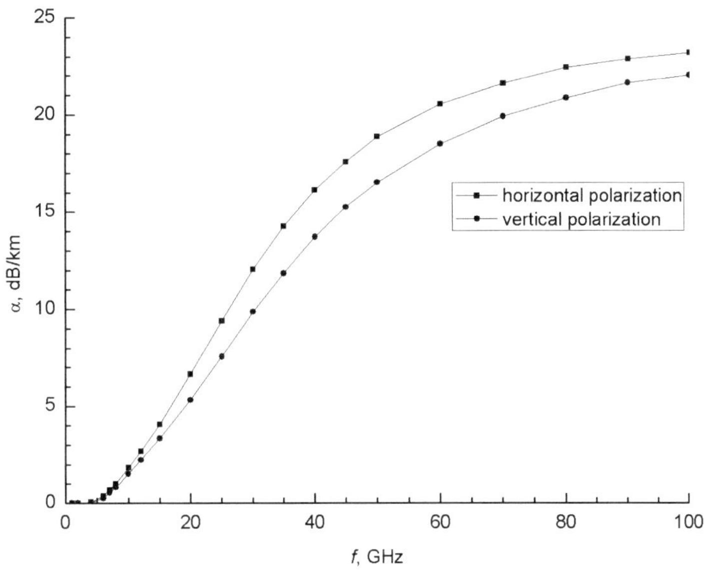

Fig. 1. The dependency of the average specific electromagnetic wave attenuation due to rain α on the operating frequency f, in Lithuania.

2.1.4 Worst month statistics

The "Worst-month" model was proposed by ITU-R in (Rec. ITU-R P.481-4, 2005). This model is a supplement of the "One-minute" models, which were explained above. In "One-minute" models a lot of precipitation data must be collected and calculated. Furthermore, majority of those models are appropriate only in cases when the reliability of the radio wave system must be equal 99.99%. The main advantage of the "Worst-month" model is that only the worst-month statistics must be collected. Furthermore, the "Worst-month" model is appropriate in cases when the required reliability of the radio system is other than 99.99%.

The worst-month is the month (or 30 days period) from a year (or twelve consecutive calendar months), during which the threshold is exceeded for the longest time. This month is not necessarily the same month in different year. The fraction of time when the threshold value of rain rate (so, and rain attenuation value) was exceeded is identical to probability that the threshold value of rain rate would be exceeded (Crane, 1996).

The average annual worst-month time percentage of excess, p_m, is proportional to the average annual time percentage of excess, p, in such relation:

$$p_m = Qp \tag{5}$$

where Q is the conversion factor; p_m [%] and p [%] must refer to the same threshold levels (the same rain rate value).

The conversion factor Q is a two-parameters (Q_1, β) function of p. In most cases a high reliability of the radio system is required ($p \le 3$ %). Then Q can be expressed as (Rec. ITU-R P.481-4, 2005):

$$Q = Q_1 p^{-\beta} \tag{6}$$

For global planning purposes the following values of the parameters Q_1 and β may be used: $Q_1 = 2.85$ and $\beta = 0.13$ (Rec. ITU-R P.481-4, 2005).

For global rain rate applications, the following values for the parameters Q_1 and β should be used: $Q_1 = 2.82$ and $\beta = 0.15$, for tropical, subtropical and temperate climate regions with frequent rain; $Q_1 = 4.48$ and $\beta = 0.11$, for dry temperate, polar and desert regions. Yet ITU-R recommends that more precise values of Q_1 and β should be used where possible.

Since

$$p = \frac{p_{\mathrm{m}}}{Q} \tag{7}$$

and (6), consequently:

$$p = \frac{1}{Q_1} p_{\mathrm{m}}^{\frac{1}{1-\beta}} \tag{8}$$

Mark $\frac{1}{Q_1} = q$ and $\frac{1}{1-\beta} = \xi$, then:

$$p = q p_{\mathrm{m}}^{\xi} \tag{9}$$

According to (2), (3) and annual data, the relation between p and $R_{(1\,\mathrm{min.})}$ can be found. This relation could be compared to the relation calculated according to (8) and ITU-R suggested Q_1 and β values. According to Lithuanian climate, the values $Q_1 = 2.82$ and $\beta = 0.15$ should be appropriate.

For example, we evaluated the "Worst-month" model in Vilnius, the capital of Lithuania. The results are shown in Fig. 2. As can be seen, the values $Q_1 = 2.82$ and $\beta = 0.15$ are

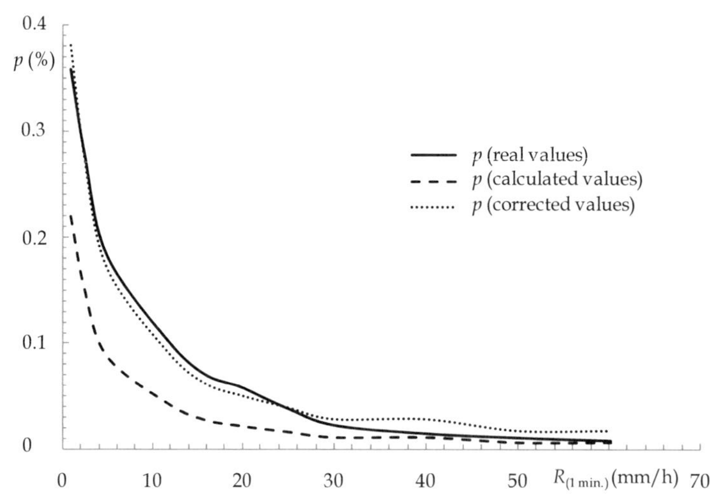

Fig. 2. The correlation between the real, calculated and corrected values of p (in Vilnius).

appropriate only in cases when $R_{(1\ \text{min.})} > 38$ mm·h^{-1}. When $R_{(1\ \text{min.})} \leq 38$ mm·h^{-1}, the calculated values are apparently distant from the real values. Therefore, the values of Q_1 and β must be corrected. The best correlation is when in (6) there are $q = 0.5$ and $\xi = 1.03$. Consequently, the corrected Q_1 and β values should be $Q_1 = 2$ and $\beta = 0.03$. But still, as can be seen in Fig. 2, the corrected values are only correct when $R_{(1\ \text{min.})} \leq 30$ mm·h^{-1}. Furthermore, when $R_{(1\ \text{min.})} > 34$ mm·h^{-1}, the values $Q_1 = 2.82$ and $\beta = 0.15$ are more proper than $Q_1 = 2$ and $\beta = 0.03$. As a result, in cases when $R_{(1\ \text{min.})} \leq 34$, the values $Q_1 = 2$ and $\beta = 0.03$ should be used, and in cases when $R_{(1\ \text{min.})} > 34$ mm·h^{-1}, the ITU-R suggested values $Q_1 = 2.82$ and $\beta = 0.15$ may be used.

3. Attenuation due to clouds

The effect of rain attenuation is greater than that of clouds in many cases, but clouds occur more often than rain. In clouds, water droplets are generally less than 0.01 cm in diameter (Freeman, 2007). In (Altshuler & Mart, 1989), it was mentioned that cloud attenuation was primarily due to absorption by the cloud droplets, and scattering losses were secondary. With increase in operating frequency the attenuation due to clouds also increases, but as the temperature of the clouds decreases the attenuation value increases (Sarkar et. al., 2005).

On average, the clouds cover more than 50% of the territory of Lithuania. According to the data of its weather stations, November and December are the cloudiest months. The clearest sky is in May and June. There are about 100 overcast days in the year.

3.1 Liquid water content

The liquid water content M is one of the most important parameters of the clouds. M describes the mass of water drops in the volume units of the cloud. It has been mentioned in (Freeman, 2007) that the specific cloud attenuation α_C [dB/km] is a function of the liquid water content M [g/m^3], the frequency f, and the temperature within the cloud T. The measurements of M at a point in space or averaged over a radio wave path are very complicated. Direct methods for measuring M consists of extracting a known volume through a cotton pad or of rotating cups in an impeller apparatus, both to be weighed; also, resistance changes can be measured with a hot wire probe attached to an aircraft flying through clouds (Liebe et al., 1989). The liquid water content in the cloud varies in a wide range. In most of the cloud attenuation models, it is required to know the value of M.

The climate conditions (humidity, temperature, etc.) and cloud morphology are different over various localities of several regions; accordingly, the liquid water contents differ within the clouds as well. This factor must be considered when analyzing rain attenuation and cloud attenuation. Our first attempt to determine the specific cloud attenuation under the Lithuanian climatic conditions is presented in (Tamošiūnaitė et al., 2008; Zilinskas et al., 2008). The humid weather predominates over the year in Lithuania.

3.2 Calculation of the specific cloud attenuation

The specific cloud attenuation is a function of clouds' liquid water content and a coefficient, which is a function of frequency and temperature. In this case, the main problem is the value of clouds' water content, because the direct measurements at a point in space are problematic. In cases when such data is unavailable, models that require only the meteorological parameters, measured at ground level, can be used. These models are based on the fact that the condensation is possible when the water vapour concentration exceeds the saturation density at the temperature, which is prevailing at that height. The water vapour density can be estimated from the humidity measurements carried out at ground level. The cloud's water content value can be estimated as the difference between water vapour concentration and saturation density at cloud temperature. The specific cloud attenuation is, unlike the case of rain, independent of drop-size distribution (Freeman, 2007). Several cloud attenuation models were developed. In (Freeman, 2007), the specific cloud attenuation was expressed as the function of liquid water content M:

$$\alpha_C = K_C M \tag{10}$$

where K_C is the attenuation constant.
The attenuation constant K_C is the function of frequency f and temperature T. The values of K_C for pure water droplets are presented in (Freeman, 2007). The values of K_C for salt-water droplets (over the sea and ocean surfaces) are higher. The necessity to know M value is limiting the direct use of relationship (10).
Often there are no possibility to measure the liquid water content and temperature within the clouds. In such cases the methods that require only meteorological parameters measured at the ground level may be used. The basic idea of such models (Dintelmann &Ortgies, 1989) is that the water vapour in the atmosphere would lead to the formation of clouds whenever there would be a possibility for condensation at some height h above ground level. There is also mentioned that the condensation is possible when the water vapour density ρ exceeds the saturation density ρ_s at temperature T prevailing at that height. It is assumed that the water vapour density ρ can be estimated from humidity measurements carried out at ground level.
The height at which cloud exists is very important for accurate determination of results of attenuation due to clouds (Sarkar et al., 2005). It was assumed in (Ito, 1989, as cited in Dintelmann &Ortgies, 1989) that clouds are created starting in the vicinity of the height h, and h [km] follows ground temperature T_0 [K] as:

$$h = 0.89 + 0.165(T_0 - 273)\,. \tag{11}$$

Relation (11) is based on analysis of temperature profiles in rain and on the Aerological Data of Japan and we have specified the applicability of this relation in the territory of Lithuania.
The condensed water content M is estimated as the difference between ρ and saturation density ρ_s at cloud temperature (Dintelmann & Ortgies, 1989):

$$M = \rho - \rho_s \tag{12}$$

where ρ_s [g/m^3] is the saturated vapour density.
It is assumed that clouds are formed when $M > 0$. As mentioned above, the determination of the water content value M is complicated. Its values differ in each group of the clouds (the clouds are grouped according to their shape, height, and structure). In our calculations, the main problem was determination of M. According to (Dintelmann & Ortgies, 1989), the values of water vapour density ρ at the height h can be estimated from the equation of state, assuming an adiabatic process:

$$\rho = \frac{\rho_0 T_0}{T}\left(1 - \frac{\kappa - 1}{\kappa} \cdot \frac{\mu g h}{R T_0}\right)^{\frac{\kappa}{\kappa - 1}} \tag{13}$$

where ρ_0 is the water vapour density at the ground level, T_0 is the ground level temperature, T is the absolute temperature in the vicinity of h, denotes the specific heat ratio which is 4/3 for the water vapour molecule, μ is the water molar mass, g is the acceleration due to gravity, h is the height, and R is the fundamental gas constant. The values of ρ_0 can be determined by using known relations (Freeman, 2007).

We assume that the clouds are created starting in the vicinity of the height h. We determine the values of h by using relation (11) or the data of the dew point temperature, temperature at the ground level, and the temperature gradient of 6.5°C/km (Rec. ITU-R P. P.835-3, 2004). The values of h obtained here we compared to the cloud base height values measured at the weather stations (see Table 1). The analysis of the cloud cover over the localities of Lithuania data shows that the relationship (11) can be used only in the cases when the middle or high clouds are formed over those localities.

T_0 [K]	Cloud base height (data of weather stations)	Cloud base height (equation 11)
280.1	0.6-1.0	2.06
280.1	2.0-2.5	2.06
280.4	2.0-2.5	2.11
281.5	2.0-2.5	2.29
281.6	2.0-2.5	2.31
282.6	2.0-2.5	2.47
284.4	2.0-2.5	2.77

Table 1. Temperature at the ground level and the values of the cloud base heights (data of weather station) in Vilnius in April 2007, as well as the height h determined using equation (8) (Tamošiūnaitė et al., 2008).

4. Attenuation due to fog

The influence of the fog on the attenuation of the electromagnetic waves can to lead to the perturbation of the wireless communication. In (Chen et al., 2004), it was mentioned that fog may be one of dominant factors in determination of the reliability of millimeter wave systems, especially in coastal areas, where dense moist fog with high liquid water content happen frequently. Fog results from the condensation of atmospheric water vapour into water droplets that remain suspended in air (Freeman, 2007). Moist fog frequently appears over the localities of Lithuania (Tamosiunas et al., 2009). There are several meteorological mechanisms for determination whether fog will form and of degree of its intensity. The physical mechanism of the formation of the fog can be reduced to three processes: cooling, moistening, and vertical mixing of air parcels with different temperatures and humidity (Duynkerke et al., 1991). All three processes can occur, although one meteorological mechanism may dominate. This circumstance leads to the different types of the fog. In (Galati et. al., 2006), the fog is classified in four types: strong advection fog, light advection fog, strong radiation fog, and light radiation fog.

The calculation methods for determination of fog attenuation are used in many cases. The propagation properties for microwave and millimeter-wave frequencies at the foggy air conditions were examined in (Liebe et. al, 1989). The values of the specific attenuation were derived from a complex refractivity based on the Rayleigh absorption approximation of Mie's scattering theory. In (Liebe et. al, 1989), the particle mass content and permittivity, which depends on the frequency and the temperature, were key variables. Attenuation due to fog is a complex function of the particle size distribution, density, extent, index of refraction, and wavelength (Altshuler, 1984). Normalized fog attenuation directly, given only the wavelength and fog temperature is presented in (Altshuler, 1984):

$$A = -1.347 + 0.0372\lambda + \frac{18}{\lambda} - 0.022T_0 \tag{14}$$

where A is attenuation in [(dB/km)/(g/m³)], λ is wavelength in [mm], t is temperature in [°C]; the relation (14) is valid only for 3 mm< λ <3 cm and -8°C< T < 25°C.
It was mentioned in (Altshuler, 1984], that the total fog attenuation could be obtained by multiplying the normalized attenuation by the fog density in [g/m³] and the fog extent in [km]. In (Zhao &Wu, 2000), it was mentioned that fog is often characterized by the visibility and the visibility is defined as the greatest distance at which it is just possible for an observer to see a prominent dark object against the sky at the horizon.
Attenuation due to fog can be expressed in terms of the water content M, and the microstructure of the fog can be ignored (Galati et al., 2000). In (Altshuler, 1984), the empirical formula for fog visibility as a function of fog density was derived:

$$V = 0.024M^{-0.65} \tag{15}$$

where V is the visibility in [km] and M is the liquid water content in [g/m³].
It was mentioned in (Altshuler, 1984), that the empirical formula (15) is valid for drop diameter between 0.3 μm and 10 μm. For the case of dense haze or other special type fogs, it is recommended to replace the coefficient 0.024 with 0.017 (Altshuler, 1984). If the visibility data are available, but the fog density data are not available, the following expression may be used (Altshuler, 1984):

$$M = \left(\frac{0.024}{V}\right)^{1.54} \tag{16}$$

In (Chen et al., 2004; Galati et al., 2006; Recommendation ITU-R PN 840-4, 2009), based on the Rayleigh approximation, the specific attenuation due to the fog α_{fog} has been written as:

$$\alpha_{fog} = KM \text{ [dB/km]}, \tag{17}$$

where K is specific attenuation coefficient.

$$K = 6.0826 \cdot 10^{-4} f^{-1.8963} \theta^{7.8087 - 0.01565 f - 3.0730 \cdot 10^{-4} f^2} \tag{18}$$

where θ =300/T, f is frequency, and T is temperature [K].

V, km	M, g/m^3
0.1	0.111
0.2	0.038
0.3	0.020
0.5	0.010
1.0	0.003

Table 2. The values of visibility *V* measured in the localities of Lithuania and the values of fog water content *M* (Tamosiunas et al., 2009).

The values of the visibility measured in the localities of Lithuania and the values of fog water content *M* determined using (16) are presented in Table 2. The highest value of the specific fog attenuation determined using *M*-data presented in Table 2 was 0.59 dB/km.
In (Naveen Kumar Chaudhary et al., 2011), it was concluded, that the link reliability can be improved by increasing the transmission power or using high gain directional antennas in the cases when the foggy conditions occur and the visibility is less than 500 meters. For the same value of visibility, the fog attenuation decreases when the temperature increases (Naveen Kumar Chaudhary et al., 2011).

5. Radio refractive index and its variability

The atmospheric refractive index is the ratio of the velocity of propagating electromagnetic wave in free space and its velocity in a specific medium (Freeman, 2007). The value of the atmosphere's refractive index is very close to the unit. Furthermore, changes of the refractive index value are very small in time and space. In the aim to make those changes more noticeable, the term of refractivity is used. It is a function of temperature, atmospheric pressure and partial vapour pressure. The value of the refractivity is about million times greater than the value of refractive index.
In design of the radio communication networks, it is important to know the atmospheric radio refractive index. The path of a radio ray becomes curved when the radio wave propagates through the Earth's atmosphere due to the variations in the atmospheric refractivity index along its trajectory (Freeman, 2007). Refractivity of the atmosphere affects not only the curvature of the radio ray path but also gives some insight into the fading phenomenon. The anomalous electromagnetic wave propagation can be a problem for radars because the variation of the refractive index can induce loss of radar coverage (Norland, 2006). In practice, the propagation conditions are more complicated in comparison with the conditions predictable in design of radio system in most cases.
The anomalous propagation is due to the variations of the humidity, temperature and pressure at the atmosphere that cause variations in the refractive index (Norland, 2006). The climatic conditions are very changeable and unstable in Lithuania (Pankauskas & Bukantis, 2006). The territory of Lithuania belongs to the area where there is the excess of moisture. The relative humidity is about 70% in spring and summer while in winter it is as high as 85 - 90% (Bagdonas & Karalevičienė, 1987). Lithuanian climate is also characterized by large temperature fluctuations. Difference between the warmest and coldest months is 21.8°C (Pankauskas & Bukantis, 2006). It was noted in (Priestley & Hill, 1985; Kablak, 2007) that even small changes of temperature, humidity and partial water vapour pressure lead to changes in the atmospheric refractive index. In (Zilinskas et al., 2008), the measurements of these meteorological parameters were analyzed in the different time of year and different

time of day. The values of the refractive index have been determined by using measured meteorological data. In (Žilinskas et al., 2010), it was mentioned that seasonal variation of refractivity gradient could cause microwave systems unavailability.

5.1 Calculation of radio refractivity

As mentioned above, the value of the radio refractive index, n, is very close to the unit and changes in this value are very small in the time and in the space. With the aim to make those changes more noticeable, the term of radio refractivity, N, is used (Freeman, 2007; Rec. ITU-R P. 453-9, 2003):

$$N = (n-1)\cdot 10^6 . \tag{20}$$

According to the recommendation of ITU -R (Rec. ITU-R P. 453-9, 2003):

$$N = \frac{77.6}{T}\left(p + 4810\frac{e}{T}\right) \tag{21}$$

where T [K] is a temperature; p [hPa] is the atmospheric pressure; e [hPa] is partial water vapour pressure. The refractivity is expressed in N - units.

It was mentioned in (Freeman, 2007; Rec. ITU-R P. 453-9, 2003), that expression (21) may be used for all radio frequencies; for frequencies up to 100 GHz, the error is less than 0.5%. There are two terms (the "dry term" and the "wet term") in relationship (21).

The values of the refractivity N in Lithuania were determined by using (21). The data of temperature, humidity, and atmospheric pressure were taken from a meteorological data website (http://rp5.ru).

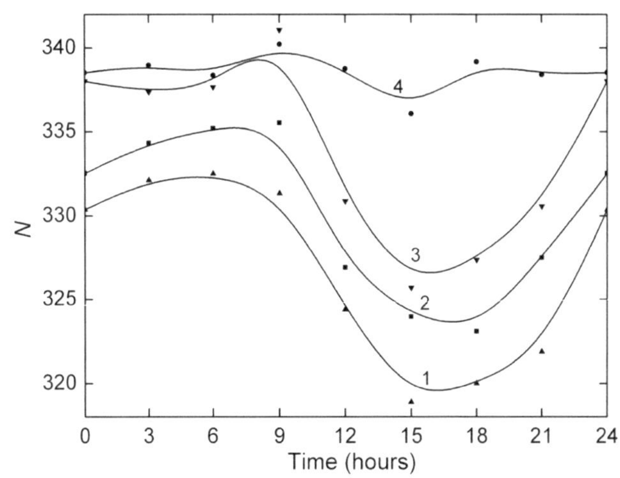

Fig. 3. The dependences of average N- values on the time of day in cities of Lithuania: Vilnius (curve 1), Mažeikiai (curve 2), Kaunas (curve 3), and Klaipėda (curve 4) in July 2008 (Valma, et al., 2010).

The dependences of average N-values on the time of day in cities of Lithuania are presented in Fig. 3. As can be seen, the behaviours of those dependences at the diurnal time are similar in all localities that are situated in the Continental part of Lithuania (Vilnius, Kaunas and

Mažeikiai) and slightly different in Seacoast (Klaipėda). The climate of Klaipėda is moderate and warm (Pankauskas &Bukantis, 2006; Zilinskas et al., 2008). The climate of Continental part of Lithuania is typical climate of the middle part of the Eastern Europe. This may explain the difference between the daily variations of N in Klaipėda and in other localities analyzed here. In Lithuania, the highest N-values were in July.

6. Conclusions

The main models for calculation of electromagnetic wave attenuation due to atmosphere humidity were revised. In Lithuania, when the reliability of the radio system of 99,99% is required, the $R_{(1\ \text{min.})}$-value is $R_{(1\ \text{min.})} = 60.23\ \text{mm/h}$. It is twice the ITU-R recommended value. The dependency of the average specific electromagnetic wave attenuation due to rain on the operating frequency (0-100 GHz) was determined. The attenuation of horizontally polarized electromagnetic waves is greater than the attenuation of vertically polarized electromagnetic waves. In cases when the required reliability of the radio system is other than 99,99%, the "Worst-month" model can be used. However, for small $R_{(1\ \text{min.})}$-values the parameters of that model should be corrected. In Vilnius, the city of Lithuania, when $R_{(1\ \text{min.})} > 34\ \text{mm/h}$, ITU-R recommended values $Q_1 = 2.82$ and $\beta = 0.15$ could be used. In cases when $R_{(1\ \text{min.})} \leq 34\ \text{mm/h}$, the corrected values $Q_1 = 2$ and $\beta = 0.03$ are more appropriate.

The main problem of models for calculation of electromagnetic wave attenuation due to clouds and fog is the required value of liquid water content. In Lithuania it is impossible to gather such meteorological information. Therefore, models excluding or calculating the liquid water content were revised. The variations of the atmospheric humidity, temperature and pressure can cause the fluctuations of the atmospheric refractive index. In Lithuania, the atmosphere refractive index fluctuates most in July. The variations of N in diurnal time are similar in all localities that are situated in the Continental part of Lithuania and slightly different in Seacoast.

7. References

Altshuler, E. E. A simple expression for estimating attenuation by fog at millimeter wavelengths. *IEEE Transactions on Antennas and Propagation*, Vol.32, No.7, (July 1984), pp.757-758, ISSN 0018-926X

Altshuler, E. E. & Marr, R. A. Cloud attenuation at millimeter wavelengths. *IEEE Transactions on Antennas and Propagation*, Vol.37, No.11, (November 1989), pp.1473-1479, ISSN 0018-926X

Bagdonas, A. & Karalevičienė, R. (1987). *The Reference Book of Agrometeorologist,* Mokslas, Lithuania (in Lithuanian).

Bhattacharyya, S., Dan, M. & Sen, A. K. Modelling of drop size distribution of rain from rain rate and attenuation measurements at millimetre and optical wavelengths. *International Journal of Infrared and Millimeter Waves,* Vol. 21, No. 12, (2000), pp. 2065-2075, ISSN 1572-9559

Bukantis, A. Climate fluctuations in Lithuania against a background of global warming. *Acta Zool. Lituanica*, Vol. 11, No. 2 (2001), pp. 113.120, ISSN 1392-1657

Characteristics of precipitation for propagation modeling. *Recommendation ITU-R* P.837-4, 2003

Chebil J., & Raihman T. A. Development of the one-minute Rain Rate Contour Maps for Microwave Applications in Malaysia Peninsula. *Electronics Letters*, Vol.35, No.20 (1999)pp. 1772-1774 ISSN 0013-5194

Chen, H., Dai, J. & Liu, Y. Effect of fog and clouds on the image quality in millimeter communications. *International Journal of Infrared and Millimeter Waves,* Vol.25, No.5, (May 2004), pp. 749-757, ISSN 1572-9559

Conversion of annual statistics to worst-month statistics. *Recommendation ITU-R P.481-4,* 2005

Crane R. K. *Electromagnetic Wave Propagation Through Rain,* 1996, John Wiley&Sons, Inc., New York, Chichester, Weinheim, Brisbane, Singapore, Toronto, ISBN 10 0471613762

Dintelmann, F., & Ortgies, G. Semiempirical model for cloud attenuation prediction. *Electronics Letters,* Vol.25, No.22, (October 1982), pp.1487-1488, ISSN 0013-5194

Dissanayake, A., Allnutt, J. & Haidara, F. A prediction model that combines rain attenuation and other propagation impairments along Earth-satellite paths. *IEEE Transactions on Antennas and Propagation,* Vol.45, No.10, (October 1997), pp.1546-1558, ISSN 0018-926X

Emiliani, L. D., Agudelo, J., Gutierrez, E., Restrepo, J. & Fradique-Mendez, C. Development of rain -attenuation and rain-rate maps for satellite system design in the Ku and Ka bands in Colombia. *IEEE Transactions on Antennas and Propagation Magazine,* Vol. 46. No. 6, (2004) pp. 54-68, ISSN 1045-9243

Freeman, R. L. (2007). *Radio System Design for Telecommunications, Third Edition,* John Wiley&Sons, ISBN: 978-0-471-75713-9, New York

Galati, G., Dalmasso, I., Pavan. G., & Brogi, G. (2006). Fog detection using airport radar. *Proceedings of IRS 2006 International Radar Symposium,* pp. 209-212, *Krakow, Poland May 24-26, 2006,*

Ishimaru, A. (1978). *Wave propagation and Scattering in Random Media,* Academic Press, ISBN 10 0123747015, New York, San Francisco, London

Ito, S. Dependence of 0° C isotherm heoght on temperature at ground level in rain. *IEICE Transactions,* Vol. E72, No.2 (February 1989), pp. 98-100

Ivanovs, G., & Serdega, D. Rain intensity influence on to microwave line payback terms. *Electronics and Electrical Engineering,* No.6 (70), (2006), pp. 60-64, ISSN 1392 - 1215

Kablak, N. I. Refractive index and atmospheric correction to the distance to the Earth's artificial satellites, *Kinematics and Physics of Celestial Bodies,* Vol. 23, No.2 (2007), pp.84-88, ISSN 1934-8401

Karasawa, Y. & Matsudo, T. One-minute rain rate distributions in Japan derived from A Me DAS one-hour rain rate data, *IEEE Transactions on Geoscience and Remote Sensing,* Vol.29, No. 6, (1991), pp. 890-894, ISSN 0196-2892

Liebe, H. J., Manabe, T., & Hufford, G. A., Millimeter-wave attenuation and delay rates due to fog/cloud conditions. *IEEE Transactions on Antennas Propagation,* Vol.37, No.12, (December 1989), pp.1617-1623, ISSN 10018-926X

Moupfouma F., & Martin, L. Modelling of the rainfall rate cumulative distribution for the design of sattelite and terrestrial communication systems. *International Journal of*

Satellite Communications and Networking, Vol. 3, No.2, (March/April 1995), pp. 105-115, ISSN 1542-0973

Naveen Kumar Chaudhary, Trivedi, D. K., &Roopam Gupta. Radio link reliability in Indian semi-desert terrain under foggy conditions. *International Journal of Latest Trends in Computing*, Vol.2, No.1, (March 2011), pp.47-50, E-ISSN 2045-5364

Norland, N. (2006). Temporal Variation of the Refractive Index in Coastal waters. *Proceedings of IRS 2006, International Radar Symposium*, pp. 221-224, *Krakow, Poland, 24 - 26 May 2006*, ISBN 83-7207-621-9

Pankauskas, M., & Bukantis, A. The dynamics of the Baltic Sea Region climate humidity. *Annales Geographicae*, Vol.39, No.1 (2006), pp. 5-14, ISSN 1822-6701

Priestley J. T.& Hill R. J. Measuring High-Frequency Refractive Index in the Surface Layer. *Journal of Atmospheric and Oceanic Technology*, Vol. 2, No. 2. (1985), pp. 233-251, ISSN 0739-0572

Reference Standard Atmospheres. *Draft Revision to Recommendation ITU-R P.835-3*, 2004

Rice, P., & Holmberg, N. Cumulative time statistics of surface-point rainfall rates. *IEEE Transactions on Communications*, Vol.21, No.10, (October 1973), pp. 1131-1136, ISSN 0090-6778

Sarkar, S. K., Kumar, A., Ahmad, I. & Gupta, M. M. Cloud morphology over three Indian tropical stations for Earth space communication. *International Journal of Infrared and Millimeter Waves*, Vol.27, No.7, (July 2006), pp. 1005-1017, ISSN 1572-9559

Specific attenuation model for rain use in prediction methods, *Recommendation ITU-R P. 838-3*, 2005

Tamošiūnaitė, M., Tamošiūnas, S., Tamošiūnienė M., & Žilinskas, M. Influence of clouds on attenuation of electromagnetic waves. *Lithuanian Journal of Physics*, Vol.48, No.1, (2008) pp. 65-72, ISSN 1648-8504

Tamošiūnaitė, M., Tamošiūnienė, M., Gruodis, A., & Tamošiūnas, S. Prediction of electromagnetic wave attenuation due to water in atmosphere. 1. Attenuation due to rain. *Innovative Infotechnologies for Science, Business and Education*, Vol. 2, No. 9, (2010), pp. 3-10, ISSN 2029-1035

Tamošiūnaitė, M., Tamošiūnas, S., Daukšas, V., Tamošiūnienė, M., & Žilinskas, M. Prediction of electromagnetic waves attenuation due to rain in the localities of Lithuania. *Electronics and Electrical Engineering*, No. 9 (105), (2010), pp. 9-12, ISSN 1392-1215

Tamošiūnas S., Žilinskas M., Nekrošius A., & Tamošiūnienė M. Calculation of radio signal attenuation using local precipitation data. *Lithuanian Journal of .Physics*, Vol. 45, No. 5, (2005) pp. 353-357, ISSN 1648-8504

Tamošiūnas, S., Žilinskas, M., Šileika, M., & Tamošiūnienė, M. Revised model of attenuation of electromagnetic waves due to rain. *Lithuanian Journal of Physics*, Vol.46, No.4, (2006) pp. 433-436, ISSN 1648-8504

Tamosiunas, S., Tamosiunaite, M., Zilinskas, M., Tamosiuniene, M. The influence of fog on the propagation of the electromagnetic waves under Lithuanian climate conditions. *PIERS Online*, Vol. 5, No. 6, (2009), pp. 576-580, 321-325, ISSN 1931-7360

The radio refractive index: its formula and refractivity data. *Recommendation ITU-R P.453-9*, 2003

Valma, E., Tamošiūnaitė, M., Tamošiūnas, S., Tamošiūnienė, M., & Žilinskas, M. Determination of radio refractive index using meteorological data. *Electronics and Electrical Engineering,* No. 10 (106), (2010), pp. 125-128, ISSN 1392-1215

Zhao, Z., Wu, Z. Millimeter- wave attenuation due to fog and clouds. *International Journal of Infrared and Millimeter Waves,* Vol.21, No.10, (2000), pp. 1607-1615, ISSN 1572-9559

Zilinskas, M., Tamosiunas, S., & Tamosiuniene, M. (2006). Calculation of radio signal attenuation using annual precipitation and heavy rainfall data. *Proceedings of EMC 2006 18 th International Wroclaw Symposium and Exhibition on Electromagnetic Compatibility*, pp. 490-493, ISBN 83 7025 947 X, Wroclaw, Poland, June 28-30, 2006

Zilinskas, M., Tamosiunaite, M., Tamosiunas, S., & Tamosiuniene, M. The influence of the climatic peculiarities on the electromagnetic waves attenuation in the Baltic Sea region. *PIERS Online*, Vol. 4, No. 3, (2008), 321-325, ISSN 1931-7360

Zilinskas, M., Tamosiunaite, M., Tamosiunas, S., Tamosiuniene M. (2009). Calculation of electromagnetic waves attenuation due to rain for various percentages of time. *Proceedings of PIERS 2009 Progress in Electromagnetics Research Symposium,* pp. 541-545, Beijing, China, March 23-27 2009, ISSN 1559 9450, ISBN 978 1 934142 08 0

9

Effects of Interaction of Electromagnetic Waves in Complex Particles

Ludmilla Kolokolova[1], Elena Petrova[2] and Hiroshi Kimura[3]
[1]*University of Maryland, College Park,*
[2]*Space Research Institute, Moscow,*
[3]*Center for Planetary Science, Kobe*
[1]*USA,*
[2]*Russia,*
[3]*Japan*

1. Introduction

The majority of natural materials (rocks, soil, wood, etc.) are inhomogeneous and have a complex structure. Very often they are conglomerates or aggregates, i.e. made of small grains stuck together. This is especially typical for planetary aerosols and all types of cosmic dust (interstellar, circumstellar, interplanetary, cometary, etc.). Cosmic dust, specifically, cometary will be the main test object for this paper. This is related to the fact that cosmic dust is usually studied through remote sensing, specifically through the study of electromagnetic waves it scatters and emits. Due to this, the field of light scattering by cosmic dust has always been at the frontier of the study of interaction of electromagnetic waves with non-spherical and inhomogeneous particles. It has inspired publication of the scholarly books by van de Hulst (1957), Schuerman (1980), Kokhanovsky (2001), Hovenier et al. (2004), Voshchinnikov (2004), Borghese et al. (2010), and Mishchenko et al. (2000, 2002, 2010) and numerous book chapters, e.g., Mukai (1989), Lien (1991), Gustafson (1999), Gustafson et al. (2001), Kolokolova et al. (2004a, b).

To consider the scattering of electromagnetic waves by an object of complex structure, we will determine this object as a configuration of discrete finite constituents. They will be called inclusions in the case of inhomogeneous particles, or monomers in the case when they are constituent particles of an aggregate. Their volume is large enough that we may ignore their atomic structure and characterize their material by a specified complex refractive index, $m=n+i\kappa$, whose real part is responsible for the refraction and imaginary part for the absorption of the light by the material. The surrounding medium is assumed to be homogeneous, linear, isotropic, and, in the case of aggregates, non-absorbing. Although we discuss some approximations, our consideration is based on the Maxwell equations fully describing the interaction of the electromagnetic radiation with the material. The non-linear optical effects, non-elastic scattering, quickly-changing illumination and morphology of the scattering object are beyond the scope of our study.

As mentioned above, our test example will be cosmic dust that typically can be presented as aggregates of submicron monomers. In the optical wavelengths they are good representatives of inhomogeneous particles with inclusions of size comparable with the wavelength, more exactly of size parameter $x=2\pi a/\lambda> 1$ where a is the radius of the monomers and λ is the wavelength. The main light scattering characteristics that we use in our consideration are intensity (the first Stokes parameter, I) and linear polarization, P. The latter we describe as $P=Q/I$ where Q is the second Stokes parameter; $P>0$ when the scattering plane is perpendicular to the polarization plane and $P<0$ when the scattering plane coincides with the polarization plane. We ignore the third Stokes parameter U since in the vast majority of the observational data the third Stokes parameter is equal to zero. We mainly consider how electromagnetic scattering affects phase curves, i.e. dependences of I and P on the phase angle, α, i.e. the angle source-scatterer-observer. It is related to the scattering angle as 180°- α. The phase curves typical for cosmic dust are presented in Fig. 1.1. Their major features that we will discuss later are forward and back scattering enhancements in the intensity phase curve and negative polarization at small phase angles. In Section 5 we also briefly consider spectral dependence of the intensity and polarization and circular polarization defined as V/I where V is the fourth Stokes parameter. All the ideas considered below can be easily extended to the case of other complex particles or media with inhomogeneities characterized by $x>1$.

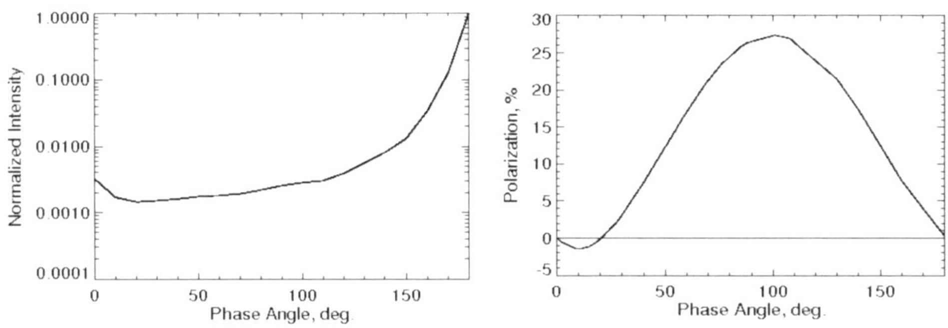

Fig. 1.1. Typical phase curves of intensity (left) and polarization (right) for cosmic dust. Intensity is normalized to the value at 180°. Notice forward and back scattering enhancements in the intensity curve and a negative polarization branch in the polarization curve at small phase angles.

In Sections 2-4 we consider main interactions between constituents of a complex particle and describe the conditions and consequences of these interactions. The focus of our consideration is how the electromagnetic interactions change as the constituents (e.g. monomers in aggregates) become more closely packed. In Section 5 we discuss the results of rigorous computer simulations of the electromagnetic interactions. The simulations are illustrated by the results of computer modeling of light scattering by aggregates. For the modeling, we use the T-matrix approach for clusters of spheres by Mackowski & Mishchenko (1996) that, being a rigorous solution of the Maxwell equations, allows us to account for all physical phenomena that occur at the light scattering by aggregates of small particles, including far-field and near-field effects, and diffuse and coherent scattering.

2. Electrostatic approximation: Effective medium theories

An extreme case of electromagnetic interaction between constituents of a complex particle occurs when this interaction can be considered in the electrostatic approximation. This consideration works when a complex particle can be represented as a matrix material that contains inclusions and both the size of the inclusions and distances between them are much smaller than the wavelength. This approach implies that the inhomogeneous particle is much larger than the inclusions and can be considered as a medium. Such a medium can be presented as homogeneous and characterized by some "effective" refractive index whose value can be found if refractive indexes of the matrix and inclusion materials are known. Such an approach to the complex particles (or media) is called mixing rules or effective medium theories. After the effective refractive index is found, it can be used to model the material of the particle whose size and shape correspond to the macroscopic particle and then consider scattering of radiation by such a macroscopic particle as if it is homogeneous. Numerous mixing rules have been developed for a variety of inclusion types (non-Rayleigh, non-spherical, layered, anisotropic, chiral) and their distribution within the medium including aligned inclusions and fractal structures (see, e.g., Bohren & Huffman, 1983; Sihvola, 1999; Choy, 1999; Chylek et al., 2000). However, still the most popular remain the simplest Maxwell Garnett (1904) and Bruggeman (1935) mixing rules. The Maxwell Garnett rule represents the medium as inclusions embedded into the matrix material and the result depends upon which material is chosen as the matrix. The Bruggeman rule was obtained for a conglomerate of particles made of materials with the refractive indexes of inclusions and matrix embedded into the material with the effective refractive index. This formula is symmetric with respect to the interchange of materials and can be easily generalized for the n-component medium.

As we mentioned above, the derivation of the mixing rules is based on an assumption that the external field is an electrostatic one, which requires the inclusions to be much smaller than the wavelength of electromagnetic wave. More exactly, the criterion of the validity of effective medium theories is $xRe(m) \ll 1$ (Chylek et al., 2000) where x is the size parameter of inclusions and $Re(m)$ is the real part of the refractive index for the matrix material. Comparison of effective medium theories with more rigorous calculations, e.g. those that use Discrete Dipole Approximation, DDA (Lumme & Rahola, 1994; Wolff et al., 1998; Voshchinnikov et al., 2007; Shen et al., 2008), and experiments (Kolokolova & Gustafson, 2001) show that even for $xRe(m) \sim 1$ effective medium theories provide reasonable results. The best accuracy can be obtained for cross sections and the worst for polarization, especially at phase angles smaller than 50° and larger than 120°.

There were a number of attempts to consider heterogeneous grains using effective-medium theories, particularly to treat cosmic aggregates as a mixture of constituent particles (inclusions) and voids (matrix material) (e.g. Greenberg & Hage, 1990; Mukai et al., 1992; Li & Greenberg, 1998b; Voshchinnikov et al., 2005, 2006). In the visual these aggregates with the monomer size parameter of $x > 1$ are, most likely, out of the range of the validity of the effective medium theories. However, for the thermal infrared, cosmic aggregates can be treated with the effective medium theories if they are sufficiently large; remember, that the macroscopic particle should be large enough to allow considering it as a medium.

If the distance between inclusions becomes larger than the wavelength, the electrostatic approximation should be replaced by the far-field light scattering (see Section 3). If the inclusions or monomers in aggregates become comparable or larger than the wavelength i.e.

the criterion $xRe(m)<1$ is violated, cooperative effects in electromagnetic interaction between the inhomogeneities become dominating. To account for them one needs to consider rigorously the interaction of electromagnetic waves that occurs in such complex objects counting on the near-field effects (Section 4).

3. Far-field light scattering

The fundamental solution of the Maxwell equations as a harmonic plane wave describes the energy transfer from one point to another. The plane electromagnetic wave propagates in the infinite nonabsorbing medium with no change in intensity and polarization state. The presence of a finite scattering object results in modification of the field of the incident wave; this modification is called the electromagnetic scattering.

If the scattering object (e.g., particle) is located from the observer at such a distance that the scattered field becomes a simple spherical wave with amplitude decreasing in inverse proportion to the distance to the scattering object, the equations describing the scattered field become much simpler. This is the so-called far-field approximation. There are several criteria of this approximation (e.g., Mishchenko et al., 2006, Ch. 3.2): $2\pi(R-a)/\lambda >>1$, $R>>a$, and $R>>\pi a^2/\lambda$, where R is the distance between the object and the observer and a is the radius of the object. The first relation means that the distance from any point inside the object to the observer must be much larger than the wavelength. Then, the field produced by any differential volume of the object (the so-called partial field) becomes an outgoing spherical wave. The second relation requires the observer to be at a distance from the object much larger than the object size. Then, the spherical partial waves coming to the observer propagate almost in the same direction. The third relation can be interpreted as a requirement that the observer is sufficiently far from the scatterer so that the constant-phase surfaces of the waves generated by differential volumes of the scattering object locally coincide in the observation point and form an outgoing spherical wave.

If the scattering object is an ensemble of particles, it is convenient to present the total scattered field as a vector superposition of the fields scattered by individual particles and, thus, to introduce the concept of multiple scattering. It is worth noting that at multiple scattering the mutual electromagnetic excitations occur simultaneously and are not temporally discrete and ordered events (Mishchenko et al., 2010). However, the concept of multiple scattering is a useful mathematical abstraction facilitating, in particular, the derivation of such important theories as the microphysical theories of radiative transfer and coherent backscattering (see below).

In some cases the scattering by a complex object can be considered in the far-field approximation that substantially simplifies the equations that describe the scattering. The conditions for this are the following: (1) the constituent scatterers of the complex object are far from each other to allow each constituent to be in the far-field zone of the others, and (2) the observer is located in the far fields of all of the constituent scatterers. Natural examples of such objects are atmospheric clouds and aerosols.

3.1 Diffuse light scattering

The properties of the light that is scattered by an ensemble of scatterers (e.g., small particles) only once are fully determined by the properties of the constituents. If the particles are much smaller than the wavelength, they scatter light in the Rayleigh regime and produce symmetric photometric phase function with the minimum at 90° and also symmetric, bell-shaped, polarization phase function with the maximum at 90°. For larger particles, the phase curves demonstrate a resonant structure with several, or even numerous, minima and maxima in both intensity and polarization depending on the size parameter of particles and the refractive index. Nowadays, the single scattering properties can be reliably calculated for particles of various types (e.g., Mishchenko et al., 2002).

If a complex object can be presented as a cluster of sparsely distributed particles, i.e. the far-field requirements are satisfied, the intensity of light scattered by the object is proportional to the number of constituents, *N*. While the number *N* and the packing density are increasing, the effects of mutual shadowing, multiple scattering, interference, and the interaction in the near field may destroy this dependence.

The evolution of the scattering characteristics of a cluster of separated particles with increasing number of the constituent partciels can be illustrated with the results of model calculations preformed with the T-matrix method for randomly oriented clusters of spheres (Mackowski & Mishchenko, 1996). We consider a restricted spherical volume and randomly fill it with small non-intersecting identical spheres (in the same manner as Mishchenko, 2008; Mishchenko et al., 2009a, b; Petrova & Tishkovets, 2011; see example in Fig.3.1). In Fig. 3.1 we show the absolute values of intensity and the degree of linear polarization calculated for a single small nonabsorbing spherical particle and the volume containing different number of such particles. There we define the intensity as $F^{11}Q_{sca}X_v$, where F^{11} is the first element of the scattering matrix normalized in such a way that this quantity integrated over all phase angles is equal to unity, Q_{sca} is the scattering efficiency of the cluster, and X_v is the size parameter of the cluster calculated from the volume of the constituents as $x_1N^{1/3}$.

When the number of particles in the cluster grows, the amplitude of the bell-shaped branches of polarization decreases, and the curves of intensity in the phase interval from 20° to 150° become flatter. If the phase curves for individual particles contained substantial interference features typical for relatively large spheres (larger than the particles considered in the example in Fig. 3.1), these features would be continuously smoothed with increasing packing density (see, e.g., Mishchenko, 2008). Such a smoothing can be interpreted as a result of the increasing contribution of multiple scattering, when many scattering events force light to "forget" the initial direction and to contribute equally to all exit directions. This also causes the depolarization effect, i.e. the light multiply scattered by an ensemble of particles is characterized by smaller values of polarization than the polarization of the light scattered by an individual particle of the ensemble. This happens since the position of the scattering plane changes at each consequent scattering, thus changing the polarization plane of the scattered light. Multiple changes that resulted from multiple scattering by randomly distributed particles randomize the polarization plane and, thus, lower the polarization of the resultant light. It is remarkable that diffuse multiple scattering is unable to change the state of polarization. As a result of this, the polarization always changes its sign at the same phase angle as for an individual particle no matter how many particles are in the cluster (Fig 3.1).

Since the behavior of the diffuse multiple scattering in the sparse media is rather well investigated in the framework of the radiative transfer theory, here we only recall the main properties of the scattered electromagnetic radiation. It increases, when either the particle size, or the number of particles in the medium, or the real part of the refractive index, or the packing density grow. If the imaginary part of the refractive index increases, the contribution of the radiation scattered twice predominates. The latter is partially polarized and can strongly depend on phase angle. For densely packed clusters or media, a study of the scattering based on the diffuse scattering is not relevant as it lacks consideration of such effects as shadowing and near-field interaction (see Section 4).

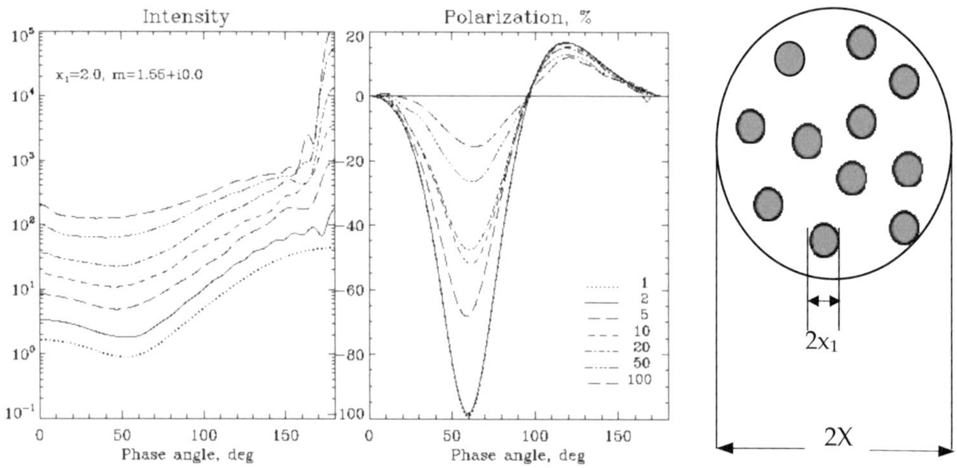

Fig. 3.1. The intensity and polarization of light scattered by a single spherical particle (dotted curve) and clusters of such particles contained in the volume of the size parameter X=20. The values of the size parameter x_1 and the refractive index m of the constituent particles and the number of particles in the volume are listed in the figure. The packing density of the cluster (defined as $\rho = N x_1^3/X^3$) grows from 0.1% to 10% (for N=1 and 100, respectively). An example of the cluster is shown on the right.

Numerous computations have shown that the light-scattering characteristics of aggregates substantially differ from those of a cluster of separated monomers and change if the structure and porosity of the aggregates change (West, 1991; Lumme & Rahola, 1994; Kimura, 2001; Kimura et al., 2003, 2006; Mann et al., 2004; Petrova et al., 2004; Tishkovets et al., 2004; Mishchenko & Liu, 2007; Mishchenko et al., 2007; 2009a; 2009b; Zubko et al., 2008; Okada & Kokhanovsky, 2009; and references therein). These changes cannot result from the diffuse multiple scattering between the aggregate monomers, which can only suppress the resonant features typical for the phase function of constituents and depolarize the scattered light. The specific shape of the phase curves shown in Fig. 1.1 is caused by more complex cooperative effects.

A striking feature in the intensity phase curve in Fig 1.1 is a strong increase of the intensity as the phase angles become larger than 160°. Development of such an increase with increasing number of the particles in the volume is evident in the plots shown in the left panel of Fig. 3.1. This strong forward scattering enhancement is caused by constructive interference of light scattered by the particles in the exact forward direction. In this direction, the waves scattered once by all the particles are of the same phase (if the particles are identical) irrespective of the particle positions (see Bohren &Huffman, 1983; Section 3.3). The oscillating behavior of the intensity curves in the forward scattering domain also points to the interference nature of this feature. In the absence of multiple scattering, this interference would result in an increase of intensity by a factor of $N(N-1)$ as compared to the scattering by a single particle or by a factor of N^2, if the non-coherent single scattered components are taken into account. Such an increase is really observed, when the packing density is small. However its development slows down with increasing packing density and practically stops, when the packing density exceeds approximately 15%. Such a behavior results from the fact that the incident light exciting a particle gets attenuated by its neighbors. This effect finally leads to the exponential extinction of light considered in the radiative transfer theory. The polarization caused by the single scattering interference in the forward scattering region is the same as that for the constituents, if they are identical.

One more interesting feature starts to develop in the intensity phase curve when the number of particles in the volume grows. This is the enhancement toward zero phase angle, which becomes noticeable at N=50 at phase angles smaller than 15°. It is accompanied by a change in the polarization state at small phase angles. These features are a typical manifestation of the coherent-backscattering (or weak-localization) effect, which is considered in the next section.

3.2 Coherent backscattering effect

The enhancement of intensity that started to emerge in the backscattering domain (Fig. 3.1), when the packing density approached 5%, is a frequent feature of the phase curves of many scattering objects observed in laboratory (particulate samples) or in nature (regolith surfaces). This is the so-called brightness opposition effect. Explanation of its origin is illustrated in Fig. 3.2a (see Mishchenko et al., 2006 and references therein). The conjugate waves scattered along the same sequence of particles in the medium but in opposite directions interfere, and the result depends on the respective phase differences. For any observational direction far from the exact backscattering, the average effect of interference is negligible, since the particle positions are random. However, at exactly the backscattering direction, the phase difference is always zero and, consequently, the interference is always constructive, which causes the intensity enhancement to the opposition. This effect is called coherent backscattering.

Interference in the backscattering direction may manifest itself in one more effect: it may lead to appearance of a branch of negative polarization at small phase angles (the so-called polarization opposition effect). This effect is schematically explained in Fig. 3.2b (also see Shkuratov, 1989; Muinonen, 1990; Shkuratov et al., 1994; Mishchenko, 2008). Particles 1-4 are in the plane perpendicular to the direction of the incident nonpolarized light. The particles 1 and 2 are in the scattering plane, while particles 3 and 4 are in the perpendicular plane. Let us assume that the particles are small relatively to the wavelength. Then they scatter light in the Rayleigh regime; the radiation scattered by such a Rayleigh particle is positively polarized for all phase angles. For the light scattered by the pair of particles 1-2, the resultant polarization keeps the polarization plane of the single scattering, i.e. it stays positive. However, the light scattering by the pair 3-4 occurs in the plane perpendicular to the resultant scattering plane; this makes the light scattered by this pair polarized in the scattering plane, i.e. negatively. The phase difference between the waves passing through particles 3 and 4 in opposite directions is always zero, while for particles 1 and 2 such phase difference is zero only at exactly the backscattering direction and quickly changes with changing the phase angle. Consequently, the conditions for negative polarization of the scattered light are on average more favorable in a wider range of phase angles than those for positive polarization. This forms a branch of negative polarization with the minimum at a phase angle whose value is comparable with the width of the brightness peak of coherent backscattering. Since only definite configurations of particles contribute to this effect, polarization opposition effect is less strong than the opposition effect in intensity.

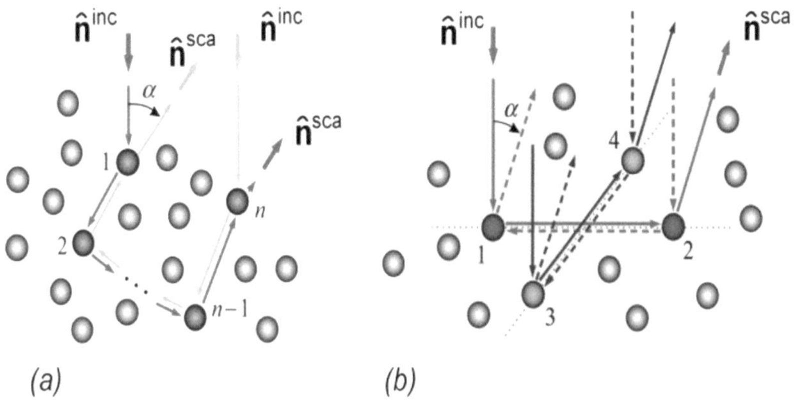

Fig. 3.2. Schematic explanation of the coherent backscattering effect (from Mishchenko, 2009a, b).

An example of such a behavior is shown in Fig. 3.3. It is seen that the formation of the intensity enhancement at small phase angles is accompanied by development of a negative polarization branch as the number of particles in the ensemble grows. Notice that the effect results from the fact that the polarization of the single-scattered light is positive. If the polarization of the single scattered light is negative, the interference results in positive polarization. If the polarization of singly scattered light changes its sign at a specific scattering angle, the interference leads to a complex angular dependence of polarization for the ensemble of scatterers as seen in Fig. 3.1.

In the interference presentation of the brightness and polarization opposition effects it was clearly assumed that the scatterers are in the far-field zones of each other, since some phase and polarization are attributed to the wave scattered by one particle and exiting the other one. However, recently it has been demonstrated that the conclusion on the interference nature of the opposition effects remains also valid for more closely packed media. In Fig. 3.4 we present some results obtained by Mishchenko et al. (2009a, b). They examined the influence of the packing density on the opposition phenomena in order to determine the range of applicability of the low-packing density concept of the coherent backscattering theory to densely packed media. As in the previous example, the ensemble of varying packing density was enclosed in a spherical volume of size parameter X (shown on the right of Fig. 3.4). When the number of particles in the volume of X=40 grows (N=500 corresponds to the packing density ρ=6.25%), the opposition peak grows, and the branch of negative polarization becomes deeper (Fig. 3.4 a-b). At the same time, the angular width of the opposition peak (determined as the angular position of the point, where the curve changes its slope) and the angular position of the polarization minimum are almost the same and remain constant with increasing number of particles. However, as the packing density grows (in Fig 3.4c this was achieved by decreasing the volume X) the shape of the negative branch transforms. To some value of the packing density, it is asymmetric, and its minimum is shifted to opposition as predicted by the theory of coherent backscattering (Mishchenko et al., 2006 and references therein). When the packing density grows up to substantial values (Fig. 3.4c, N= 300 that correspond to ρ = 30%), the effects related to the interaction of particles in the near field become noticeable. They manifest themselves in the transformation of the shape of the negative branch and its widening, which we discuss in Section 4.

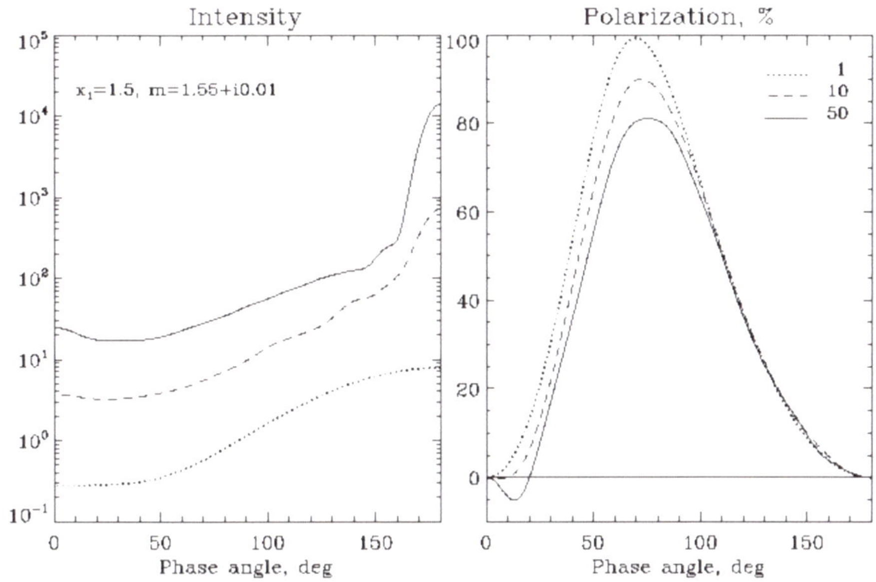

Fig. 3.3. Same as Fig. 3.1, but X=15, x_1=1.5, and m=1.55+i0.01. The numbers of particles in the volume are listed in the right top corner of the polarization plot.

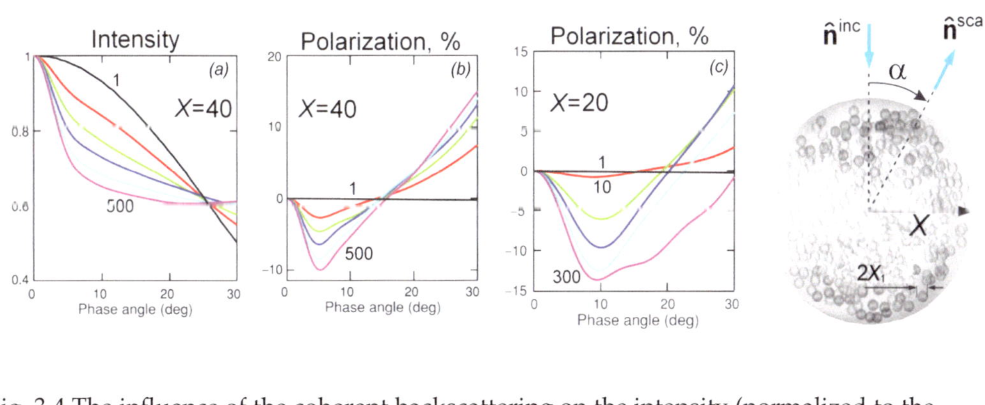

Fig. 3.4 The influence of the coherent backscattering on the intensity (normalized to the value at zero phase angle) and polarization of light scattered by ensembles of nonabsorbing spherical monomers of x_1=2 and m=1.32. Note that such individual monomers have polarization equal to zero in the backscattering domain and positive for the other phase angles. The size parameter of the volume X and the smallest and largest numbers of particles are shown in the plots. The geometry of the scattering ensemble is shown on the right. Adapted from Mischchenko et al. (2009b).

3.2.1 Some experimental facts

The above described opposition phenomena - a nonlinear enhancement of brightness to opposition and a negative branch of linear polarization of the scattered light – have been observed for cosmic dust in a variety of environments (debris disks, comets, Saturn's rings, asteroids and satellites of planets) as well as for laboratory particulate samples. Numerous experimental studies showed that the characteristics of these effects and their phase profiles are undoubtedly connected with absorption and microphysical structure of the scattering objects. In particular, it was found that a very sharp narrow brightness peak and an asymmetric branch of negative polarization with the minimum close to zero phase angle (less than 2°) are typical of bright and porous objects (see, e.g., the review by Rosenbush et al., 2002). These strongly expressed manifestations of the coherent backscattering mechanism appear due to a rather large free path of light in such a sparse particulate medium as regolith. Since the width of the coherent peak in intensity is inversely proportional to the free path, for extremely sparse media like atmospheric clouds this peak should be very narrow and cannot be observed. This peak also cannot be observed for the media that have a small restricted volume like small aggregates, especially if they are absorbing (e.g., Etemad et al., 1987). The absence of very sharp opposition features in aggregates and other individual particles of complex structure was confirmed by both observations of the cosmic dust and laboratory measurements (e.g., Levasseur-Regourd & Hadamcik, 2003; Shkuratov et al., 2004). This effect is also seen in Figs. 3.3-3.4 when the number of monomers in aggregates is small. These particles demonstrate a moderate increase of brightness to opposition and the branch of negative polarization with a shape close to symmetric.

Astronomical observations also revealed that dark or densely packed media demonstrate wider, if any, peaks of brightness near opposition and more symmetric branches of negative polarization (e.g., Shkuratov et al., 2002; Belskaya et al., 2005). This contradicts to the theory of coherent backscattering, which predicts that the opposition effects in brightness and polarization have the same cause and should appear simultaneously. Moreover, since only certain particle configurations contribute to polarization opposition effect, it might be less pronounced than brightness opposition effect. The shadow hiding, which is usually invoked to explain the widening of the opposition brightness peaks in dark surfaces (Lumme & Bowell, 1981), cannot induce such a significant negative polarization of the scattered light (e.g., Shkuratov & Grynko, 2005). Accurate consideration of the electromagnetic field in the particle vicinity, accounting for the presence of neighbor particles in the densely packed scattering clusters allows revealing one more scattering effect – the influence of the near field, which is considered in the next section.

4. Near-field effects

In the case of compact aggregates/media the electromagnetic interaction becomes even more complex, because the electromagnetic field in the close vicinity of the scattering particle is inhomogeneous due to the lag of the wave within the particle with respect to the incident wave. This effect is mostly expressed if the scatterer is comparable in size to the wavelength. Direct calculations using the Lorentz-Mie theory for spherical particles show that the constant phase surface of the total field is funnel shaped in the particle vicinity (Fig. 4.1a). Consequently, the field inhomogeneity near the particle causes a rotation of the total field vector relatively to the incident field vector. This results in the formation of a Z-component of the total field that lies in the scattering plane and, consequently, reduces the scattered intensity in the back and forward scattering regions and increases the negative polarization (Tishkovets, 1998; Tishkovets et al., 1999; 2004a, b; Petrova et al., 2007).

To illustrate the influence of the field inhomogeneity in the vicinity of a particle, let us consider Rayleigh test particles located on a constant phase surface near a larger particle in its inhomogeneous zone (Fig. 4.1b). First, assume that the incident field is polarized in the scattering plane (as shown in Fig. 4.1a). If the test particles were far from each other and from other particles, i.e., in the homogeneous field, their dipole moments would be parallel to the x_0 axis. In this case, the intensity of the light scattered by all four test particles-dipoles would concentrate in the direction $\alpha = 0°$ and 180° and would be zero in the direction $\alpha = 90°$. If the test particles are, however, in the inhomogeneous zone near a wavelength-sized particle, the dipole moments induced in particles *1* and *3* have a nonzero component in the direction of wave propagation, i.e., along the z_0 axis. This results in decreasing intensity of the scattered light in the direction $\alpha = 0°$ and 180°, whereas the intensity in the direction $\alpha = 90°$ becomes nonzero. In both cases, the scattered wave is polarized the same way as the incident one, i.e. in the scattering plane (negatively). Now assume that the incident wave is polarized perpendicular to the scattering plane. Then particles *1* and *3* produce the radiation that is polarized perpendicular to the scattering plane and does not depend on phase angle. The radiation scattered by particles *2* and *4* has a component parallel to the z_0 axis (i.e., polarized in the scattering plane) that depends on α. As a result, the intensity again decreases in the directions $\alpha = 0°$ and 180° and increases in side directions, and polarization gets a negative component. So, at any polarization of the incident wave, the field inhomogeneity in the vicinity of the scattering particle induces a rotation of the field vector and leads to appearance of Z-component of the total field, which affects the angular distribution of the scattered intensity and causes negative polarization (for more details, see Tishkovets, 1998; Tishkovets et al., 1999; 2004a, b; Petrova et al., 2007).

One more type of interaction of particles in the near field is the mutual shielding of particles (Tishkovets, 2008; Petrova et al., 2009). The scheme with the test dipoles (Fig. 4.1b) helps to estimate qualitatively the result of the shielding. For the sake of simplicity, let us assume that at a given polarization of the incident radiation, the dipole moment of particle *1* is oriented exactly opposite to the k_{sc} vector. In this case, particle *1* does not radiate in the k_{sc} direction. It does not matter whether we take the shielding into account or not. When the incident radiation is polarized in the y_0z_0 plane, in the case of ignoring the shielding, particle *1* would radiate like particle *3* or like all the particles in the homogeneous field. However, when the large particle shields particle *1*, the latter does not radiate in the α direction, i.e., its positive polarization does not contribute to the scattered light. Thus, the shielding diminishes the positively polarized scattered radiation and diminishes the intensity in the α direction. However, in the backscattering direction, dipole *1* contributes to the scattered radiation, which induces an increase in the intensity with respect to that in the α direction. Contrary to the field inhomogeneity in the near zone, which is most noticeable for the wavelength-sized particles, the mutual shielding effect is independent of the size of the particles located in the near field.

Under the above described conditions the wave coming from one particle to another is not spherical, and the single-scattering characteristics of individual monomers, such as their phase matrix, are not applicable. In other words, in densely packed systems the scatterers become highly dependent. The influence of the interaction in the near field on intensity and polarization of the scattered light can be easily demonstrated by the models, where the near-field contribution is ignored in the calculations of the light-scattering characteristics. The example presented in Fig. 4.2 clearly shows that the interaction in the near field substantially diminishes the backscattering peak in intensity induced by the coherent backscattering effect and changes the shape of the negative polarization branch.

Contrary to the coherent backscattering mechanism, the near-field effects work in a wide angular range. In the backscattering domain they distort the manifestations of the coherent backscattering. Their influence on polarization is rather complex and significantly depends on the size parameter of monomers, their packing density, and the refractive index. For example, with increasing packing density (i.e., when the near-field effects manifest themselves more clearly), the negative branch becomes deeper and wider if the aggregate is composed of larger monomers, while it may become shallower for smaller constituents. The modeling experiments with particles of different properties show that the most permanent and noticeable manifestation of the near-field effects in polarization is the shift of the polarization minimum out of opposition (Petrova et al., 2007; 2009). In other words, while the coherent backscattering mechanism forms the negative branch with the minimum near zero phase angle, the interaction in the near field causes the shift of the polarization minimum to larger phase angles and makes the negative polarization branch more symmetric.

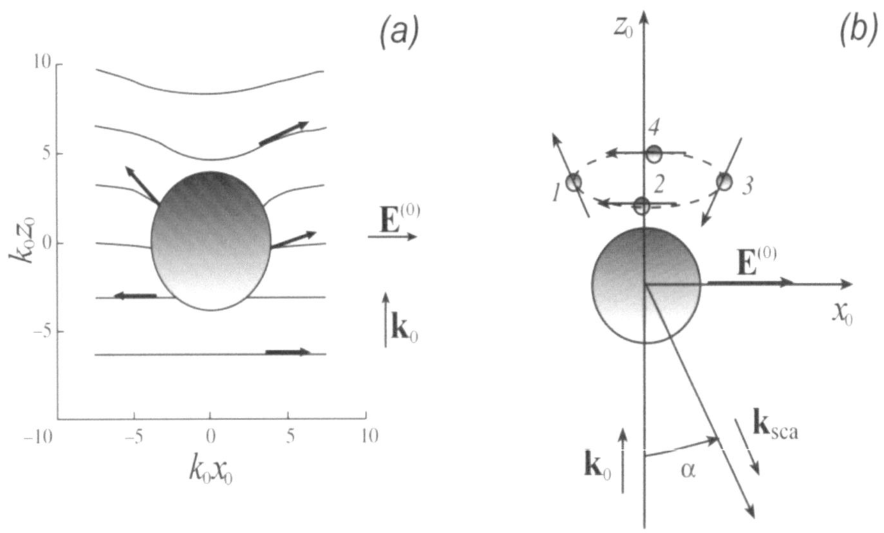

Fig. 4.1. (a) The scheme shows the constant phase surfaces and directions of electric field vectors (sum of the incident and scattered waves) in the close vicinity of a particle with x=4.0 and m = 1.32 + i0.05. The incident wave propagates along the wave vector k_0 and is polarized in the x_0z_0 plane. Adapted from Tishkovets et al. (2004a). (b) The scheme for the scattering of inhomogeneous waves by the Rayleigh test particles 1 - 4. Particles *1* and *3* are in the x_0z_0 plane, while particles *2* and *4* are in the y_0z_0 plane. The incident wave propagates along the z_0 axis and is polarized in the x_0z_0 plane. The scattered wave propagates to the direction of the phase angle α. The vectors at the Rayleigh particles show the directions of the induced dipole moments. Adapted from Petrova et al. (2009).

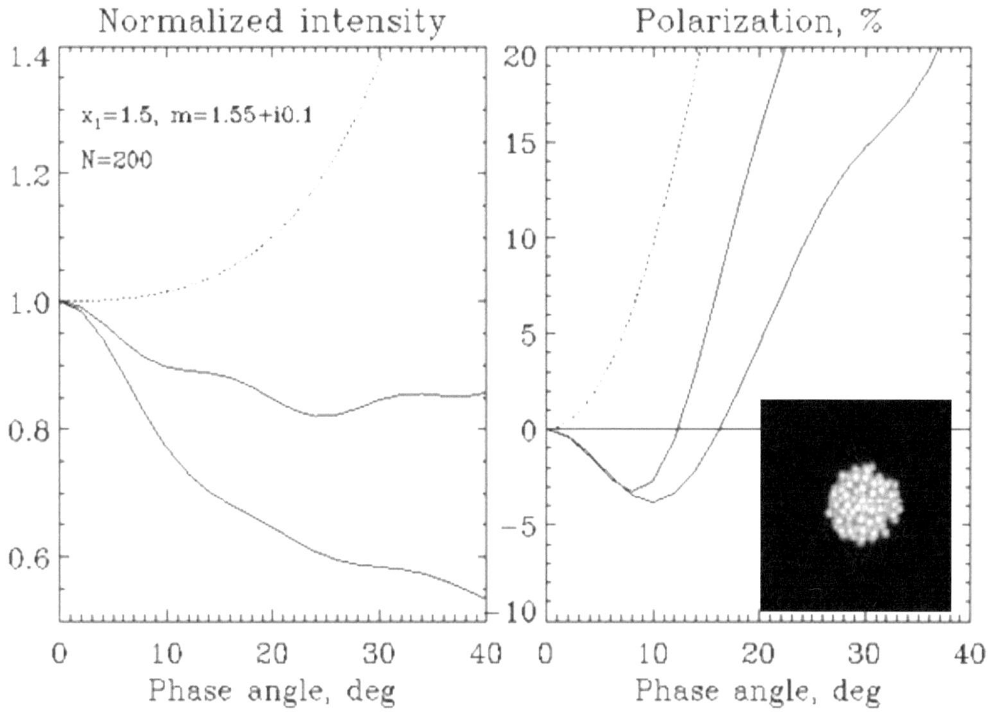

Fig. 4.2. The influence of the interaction in the near field on the intensity (normalized to the value at zero phase angle) and polarization in the backscattering domain for the compact cluster shown in the insert. Thick and thin curves present the models calculated with and without the near-field effects respectively. Dashed curves show intensity and polarization for the individual monomer. The parameters x_1, m, and N are shown in the figure. The data for the figures were kindly provided by V.P. Tishkovets.

Due to their nature, the manifestations of the near-field effects can be more easily observed in absorbing aggregates when the packing density exceeds 10-15%. One of such examples is shown in Figs. 4.3 for the whole range of phase angles and separately for the backscattering domain. For rather small number of monomers, the conditions for diffuse scattering and coherent backscattering are applied. With increasing number of monomers, the forward-scattering peak develops, the intensity profile becomes flatter, and the polarization maximum gets depressed. Then the opposition peak in intensity grows, and the negative branch of polarization appears. However, the opposition features do not develop as quickly as in nonabsorbing aggregates (compare Fig. 3.4), because the free paths become somewhat shorter when absorbing monomers are added into the volume. Partly due to this effect, partly due to the interaction in the near field - which becomes more important with increasing packing density - the polarization minimum moves out of opposition. Further increase of the packing density makes the near-field effects even more decisive. We see that the opposition peak stops to grow, while the negative branch continues to develop; it becomes wider and deeper (the curves for N=200).

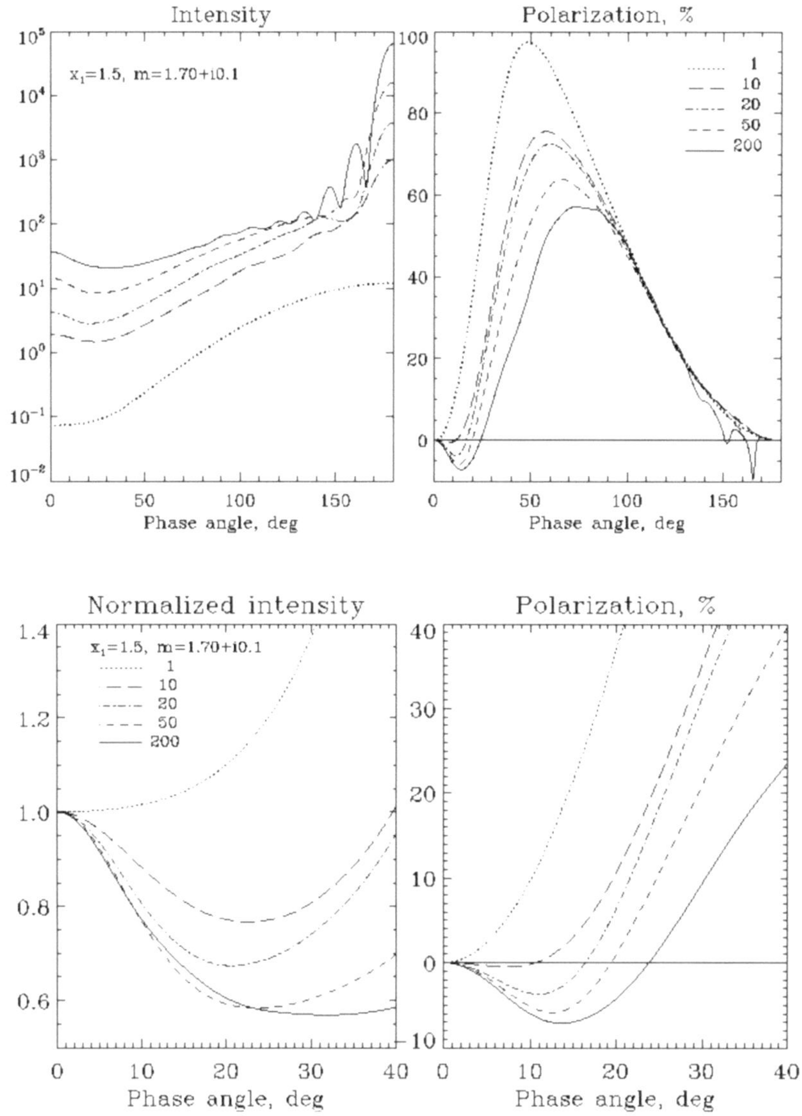

Fig. 4.3. Top panel: same as Fig. 3.1-3.3, but for the parameters listed in the plot. The packing density varied from 0.1% to 20% (*N* changes from 1 to 200, respectively). Bottom panel: larger scale for the backscattering domain; the intensity is normalized to the value at zero phase angle.

5. Modeling of light scattering by aggregates

In this section we explore how the considered above phenomena associated with electromagnetic interaction between constituents in a complex medium affect the angular and spectral dependence of intensity and linear polarization of the scattered radiation. We show how these results can be applied to the study of cosmic dust and other types of complex particles. We also briefly consider how the cooperative effects affect circular polarization of aggregates that contain optically active materials, e.g. complex organics of biological origin.

To model electromagnetic scattering by complex dispersed systems, several methods are now available. They are based on the numerically exact solutions of the Maxwell equations. One of them, the so-called superposition T-matrix method (Mishchenko et al., 2002; Mackowski & Mishchenko, 1996), was used to obtain the intensity and linear polarization of clusters of particles discussed above. Since these computations are time and resource consuming, they cannot be presently fulfilled for very large clusters/layers of particles, such as regolith. Nevertheless, they allow us to obtain the scattering characteristics of aggregates of a restricted number of monomers that are typical for cosmic dust, and to study the dependence of the light-scattering characteristics on the size of monomers, their packing density and refractive index.

5.1 Dependence of light scattering characteristics on the physical properties of aggregates

Exploring the light scattering characteristics of aggregates, we continue to focus on the dependence of intensity and linear polarization on phase angle, i.e. photometric and polarimetric phase curves. Our goal is to find out how the phase curves depend on such characteristics of aggregates as the size and composition of the monomers, their number and arrangement in the aggregate. In the previous sections we were mainly interested in the models of complex objects that allowed us to better see specific physical phenomena such as coherent backscattering or near-field effects. This section is directed to provide a basis for the interpretation of experimental data, specifically the observations of cosmic dust. This is why in this section we use more realistic models of natural aggregates, namely the aggregates grown under ballistic process (Meakin et al. 1984). There are commonly used two types of such aggregates: Ballistic Particle-Cluster Aggregate (BPCA) that grows at collision of individual monomers with the aggregate and Ballistic Cluster-Cluster Aggregate (BCCA) that grows at collision of clusters of monomers. Examples of such aggregates are shown in Fig. 5.1. Notice that BPCAs are usually more compact than BCCAs. The packing density of ballistic aggregates is defined as the ratio of the volume taken by their monomers to the total volume of the aggregate which is the volume of a sphere of the characteristic radius A calculated as $A^2 - 5/3 \sum_{i,j=1,..N} (r_i - r_j)^2/(2N^2)$ (Kozasa et al., 1992) where r_i is location of the center of the ith monomer and the total number of the monomers is N. Packing density depends on the number of monomers; as this number increases, it decreases significantly for BCCAs and slightly for BPCAs (Kolokolova et al., 2007).

The results of the modeling of the light scattering characteristics of BCCA and BPCA at some refractive indexes and monomer size are shown in Figs. 5.2 -5.3; for more results see LISA database at https://www.cps-jp.org/~lisa/. There instead of intensity I we use albedo, a characteristic that is usually used in astronomical observations to describe the reflectivity of an object. In the case of aggregates, albedo is defined as $(I/I_0)*(\pi/G)$ where I_0 is the intensity of the incident light and G is the aggregate geometric cross section (Hanner et al., 1981; Kimura et al., 2003). We show spectral dependence of albedo and polarization in two filters: 450 nm (blue filter) and 600 nm (red filter). Following astronomical definitions, if albedo or polarization have larger values in the red filter we say that they have a red color and if the values are larger in the blue filter we say that they have a blue color.

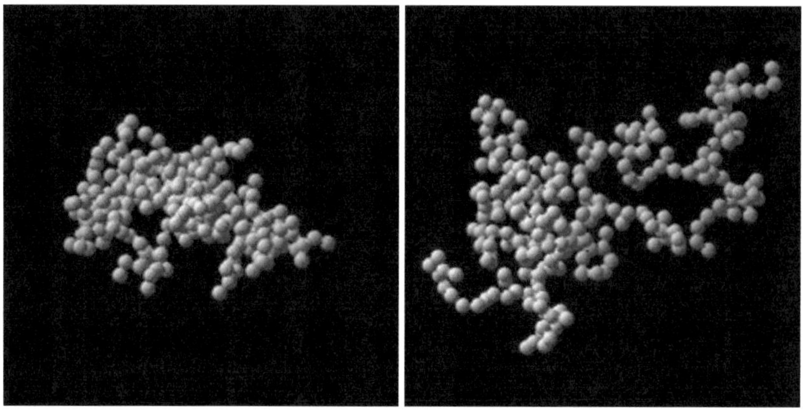

Fig. 5.1. Samples of BPCA (left) and BCCA (right) aggregates. These aggregates were used in Kimura et al. (2003, 2006) computations to model light-scattering characteristics of cometary dust.

First, notice in Fig. 5.2-5.3 the features of the modeled phase curves described in the previous sections, namely: (1) strong forward scattering resulted from the interference of the light single-scattered by individual monomers; (2) rather low values of the maximum polarization that manifests depolarizing effects of the diffuse scattering and influence of the near-field effects; (3) some, although small, backscattering enhancement; and (4) rather small but symmetric branch of negative polarization at small phase angles. The last two features indicate a serious influence of the near-field effects. This is not surprising as the monomers in aggregates touch each other, i.e. they do are located in the inhomogeneous field produced by their neighbors. As it was shown in Section 4, the near-field effects affect the shape of the intensity curve and result in a more pronounced and symmetric negative polarization branch and in diminished values of the positive polarization.

The figures also show a difference between the plots obtained for aggregates of different physical properties. The most influential parameter seems to be the monomer size whose variations change the shape of the polarization phase curve and the dependence of the albedo on the wavelength. The real part of the refractive index mostly affects the maximum polarization. The imaginary part of the refractive index affects the spectral dependence of photometric phase curve and the values of albedo but does not much affect polarization. Notice also that the more compact BPCAs depolarize the light more strongly than the more porous BCCAs, although their other characteristics are rather similar.

Although the curves in Figs 5.2-5.3 resemble the typical observational curves shown in Fig.1.1, they have some characteristics that are not typical for cometary dust. Observational facts summarized in Kolokolova et al. (2004a, b) indicate that comets usually have red photometric and polarimetric colors, i.e. their albedo and polarization have larger values at longer wavelengths. Unlike the observational data, the results of the modeling shown in Figs. 5.2-5.3 always demonstrate predominantly blue photometric color. In the case of the monomers of radius 120 nm and the refractive index equal to $1.4+i0.01$, the results of the modeling also demonstrate blue polarimetric color for some phase angles. Also, in the majority of plots, the value of albedo at zero phase angle is higher than the one observed, which is equal to 3 - 5% (Hanner, 2003).

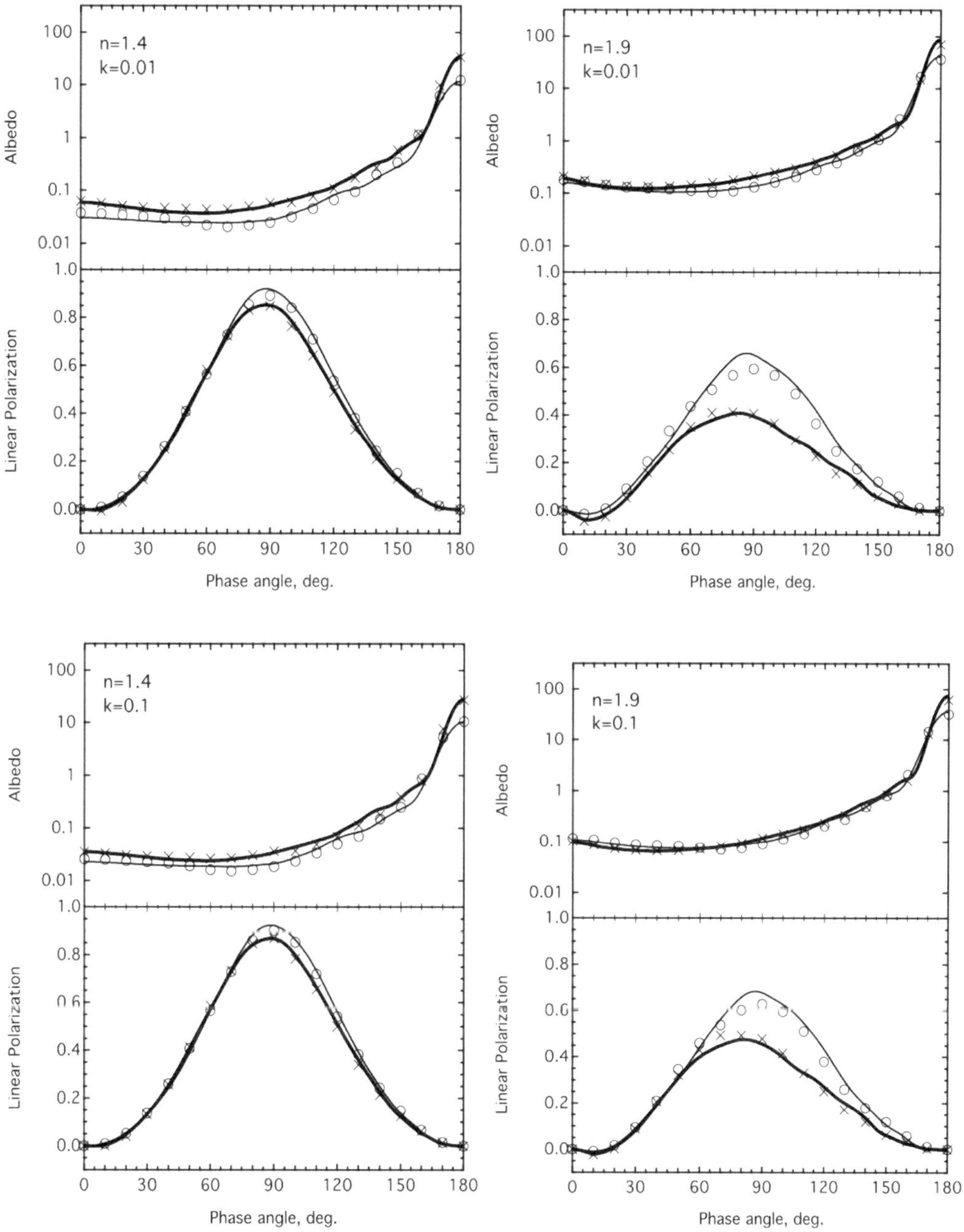

Fig. 5.2. Albedo (in %) and polarization as functions of phase angle for aggregates of monomer radius equal to 90 nm. Real part of the refractive index, *n*, and imaginary part of the refractive index, κ, are shown in the top left corner of each figure. Results for the wavelength 450 nm are shown by thick line (BCCA) and crosses (BPCA) and for 600 by thin line (BCCA) and circles (BPCA). All aggregates consisted of 128 monomers.

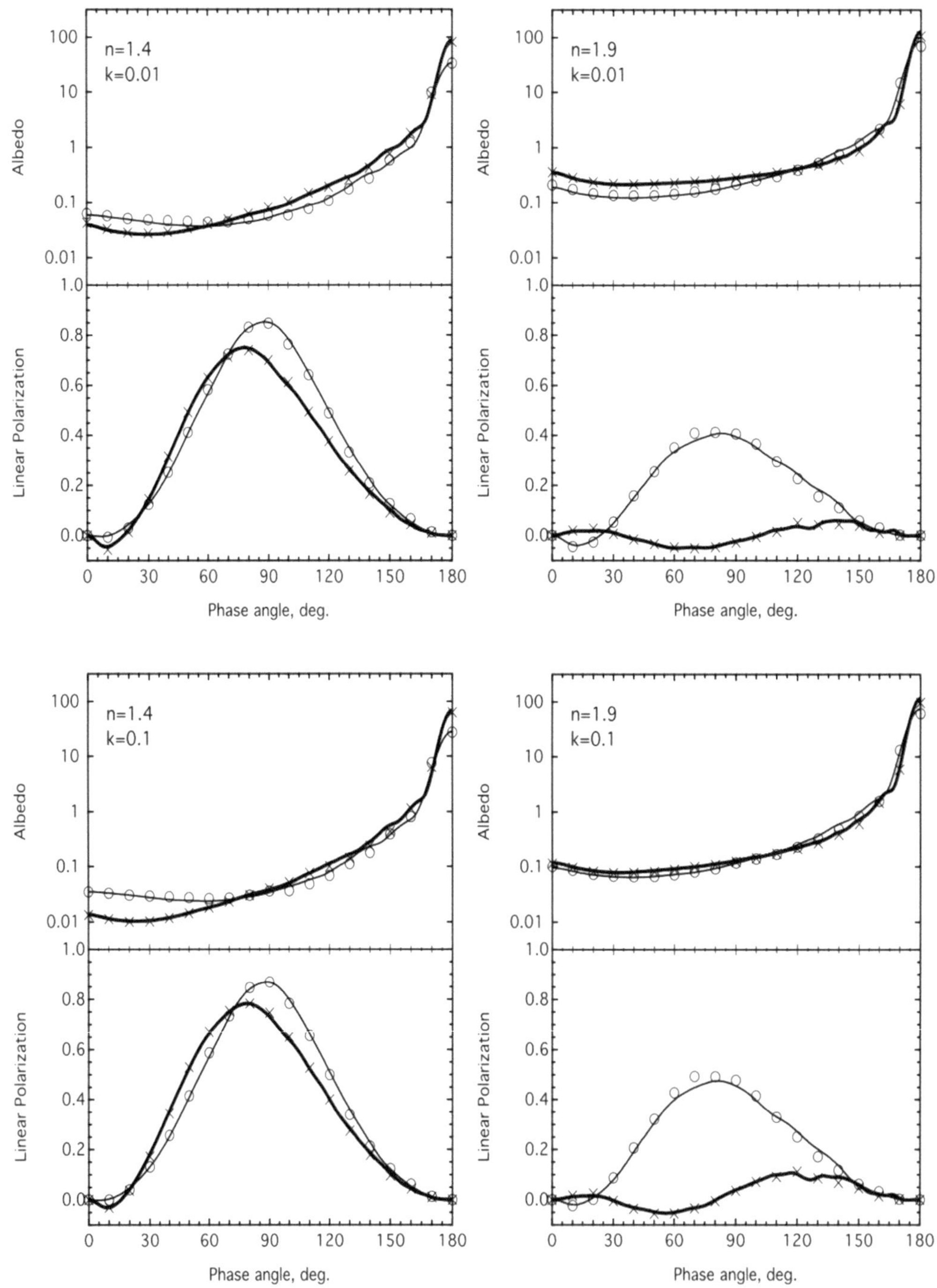

Fig. 5.3. The same as Fig. 5.2 but for monomers of radius 120 nm.

Our computations, summarized in Kimura et al. (2003, 2006) provided characteristics of the aggregates that satisfy the observational data for cometary dust. The best fit was achieved for the monomers of radius 100 nm and the refractive index that was determined based on *in situ* study of comet Halley, which is equal to *1.88+i0.47* for the wavelength λ=450nm and to *1.98+i0.48* for λ=600nm. It appears that for such a dark material a crucial characteristic is the number of monomers in the aggregate. Fig. 5.4 shows that increasing the number of monomers in the aggregate results in a more pronounced negative polarization branch and in a stronger depolarization of the positive polarization. This allows us to suggest that in the case of aggregates of thousands of monomers it is possible to reach the observable values of negative (~0.015) and positive (~0.3) polarization.

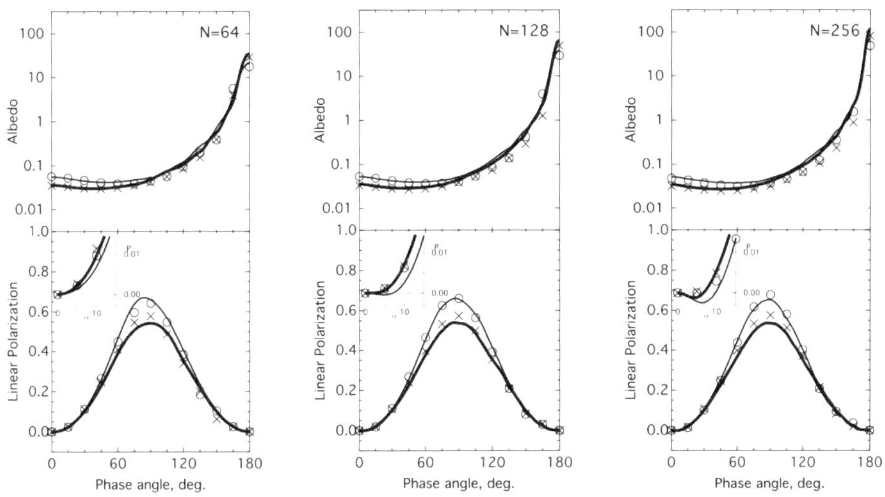

Fig. 5.4. Albedo (in %) and polarization as functions of phase angle depending on the aggregate size (number of monomers in the aggregate). The monomer radius is equal to 100 nm. The refractive index was taken as typical for cometary dust (based on *in situ* data for comet Halley) and is equal to *1.88+i0.47* for the wavelength λ=450nm and *1.98+i0.48* for λ=600nm. The number of monomers in the aggregate is 64 (left panel), 128 (middle panel), 256 (right panel). Development of the negative polarization is shown in the inserts. Notice also a decrease of the polarization maximum more pronounced for the shorter wavelength. The figure was adapted from Mann et al. (2004).

Figs. 5.2-5.4 also demonstrate that the polarimetric color is often less red in the case of more compact BPCAs. We explain this by a stronger depolarization of light in the case of more compact aggregates. Such a depolarization is even more evident from Figs. 3.3 and 4.3 where aggregates with higher packing density (more particles in the volume) always demonstrate smaller polarization maximum. Depolarization of light with increasing packing density is consistent with increasing electromagnetic interaction between the monomers resulted from both diffuse multiple scattering and near-field effects as considered in Sections 3-4

Kolokolova & Kimura (2010) showed that a measure of the depolarization can be the number of monomers covered by a single wavelength; the more monomers the wavelength covers, the more depolarized is the scattered light. It is clear that a single wavelength covers more monomers in the case of more compact aggregates. It also covers more monomers if the wavelength is longer. Thus, we can expect the scattered light to be more depolarized at longer wavelengths and the color of polarization should be blue. Blue polarimetric color is frequently observed. For example, it is typical for asteroid surfaces and interplanetary dust. However, as we already mentioned, cometary dust has a red polarimetric color. In our opinion, this is good evidence that cometary aggregates are highly porous. For porous aggregates, an increase in the wavelength may not increase the number of monomers covered by a single wavelength. Then the polarimetric color is defined by properties of individual monomers. Specifically, the monomer size parameter decreases with increasing wavelength that moves it closer to the Rayleigh regime of scattering characterized by higher polarization, thus, resulting in the red color of polarization.

An interesting observational result was reported by Kiselev et al. (2008) who summarized the observational data of spectral behavior of comet polarization and showed that cometary dust is characterized by a red polarimetric color in the visible (wavelengths of 400-800nm) but it changes to a blue polarimetric color in the near infrared (wavelengths of 1000-3000nm). They also showed that some comets exhibit a blue polarimetric color even in the visible. These observations can be interpreted based on the dependence of electromagnetic interaction on the number of monomers covered by a single wavelength. Fig. 5.5 illustrates our point. One can see there that in the case of a porous aggregate a small change in the wavelength does not change the number of particles it covers. However, at longer wavelength even in porous aggregates the number of monomers covered by a single

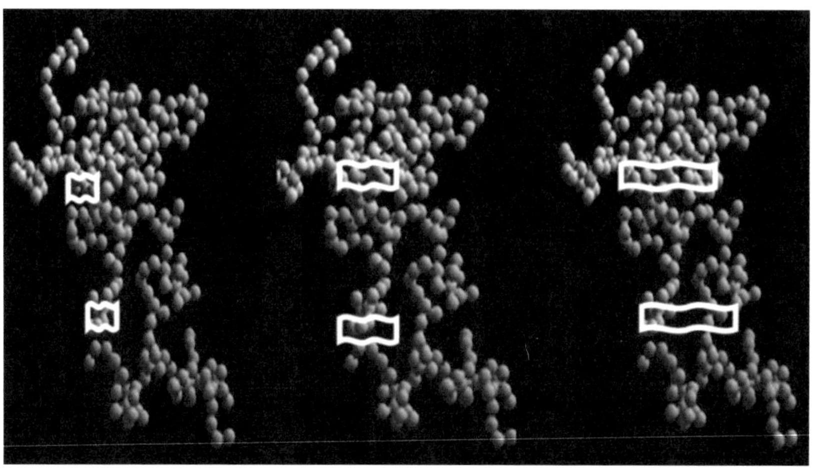

Fig. 5.5. Illustration of the effect of increasing wavelength on the light scattering by an aggregate. In a compact aggregate (top part of the aggregate) the longer the wavelength the more monomers it covers, so the interaction between the monomers becomes stronger, and the light becomes more depolarized. This results in a decrease of polarization with wavelength, i.e. blue color of polarization. For a porous aggregate (bottom part of the aggregate), the number of monomers covered by a single wavelength does not change much as the wavelength increases, i.e. the change in the interaction between the monomers cannot overpower the change in the monomer size parameter, and so the polarization color stays red. However, as the wavelength reaches some critical value, the number of covered monomers in the porous aggregate changes significantly(as shown in the right-hand aggregate) and interaction becomes the main factor that defines the polarization color which then becomes blue.

wavelength increases causing depolarization of the scattered light. This explains the change in the observed polarimetric color as the observations move to the near infrared. In the case of more compact aggregates, even a slight change in wavelength increases the number of covered monomers resulting in blue polarimetric color even in the visible. Thus, it is likely that the dust in the comets with blue polarimetric color, as well as asteroidal and interplanetary dust, is characterized by more compact particles. The wavelength where polarimetric color changes from red to blue may be used to determine the porosity of aggregate particles.

5.2 Spectral manifestation of coherent backscattering

In Section 3.2 we discussed how coherent backscattering affects intensity and polarization phase curves producing there brightness and polarization opposition effects. Recently it has been found that coherent backscattering also manifests itself in spectral data. It affects the depth of the absorption bands and makes it dependent on the phase angle. The physics of this is clear: since coherent backscattering produces brightness opposition effect of different steepness at different absorptions, the steepness of the opposition effect is different within and outside of the absorption bands and, thus, the absorption bands should have different depth and, most likely, shape at different phase angles. This fact was confirmed at observations of Saturn's satellites. Their spectra have distinct ice absorption bands in the near infrared and these bands do change with phase angle (Fig. 5.6). Although this effect has been studied so far for regolith surfaces it should also exist for any medium whose light scattering is affected by coherent backscattering.

We modeled spectral manifestation of the coherent backscattering using the T-matrix code and presenting the surface of Saturn's satellites as a large icy aggregate similar to those described in Sections 3 and shown in Fig. 3.4. Fig 5.7 presents the results of our simulations of the ice absorption band at 2.8 μm at different size of monomers and packing density of the aggregate. One can see that the simulations correctly reproduce the observed tendencies. More so, the variations in the rate of the change of the absorption band depth and shape promise that the study of the spectra at several phase angles can serve as a new remote sensing tool to reveal properties of monomers and their arrangement in aggregates.

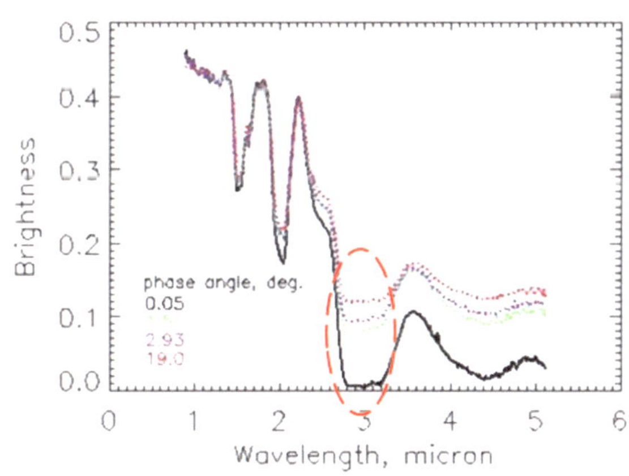

Fig. 5.6. Spectrum of Saturn's icy satellite Rhea at a variety of phase angles (from Kolokolova et al., 2010). It is clearly seen that the depth of the absorption bands varies with phase angle as it should be in accordance with the coherent backscattering. The red dashed ellipse shows the band whose modeling is presented in Fig. 5.7.

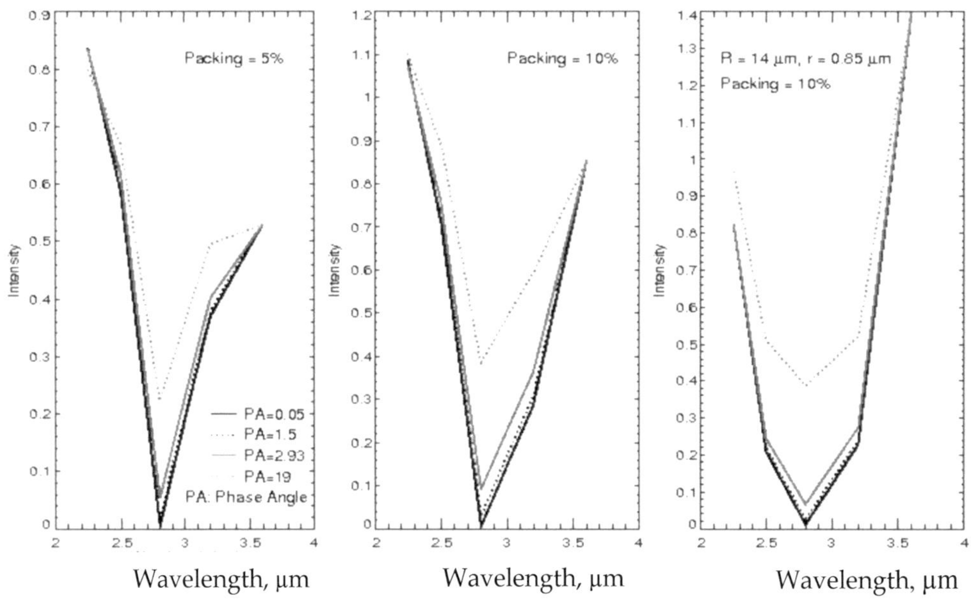

Fig. 5.7. Simulations of the phase angle variations in the spectra of icy aggregates. Different phase angles (PA) are indicated in the left panel. The left panel is for the monomer of radius 1.0 μm and packing 5%, the middle panel is for the same monomers but different packing, 10%, and the right panel is for the same packing as the middle one but for smaller monomers, r= 0.85 μm. In all cases the overall size of the aggregate is 14 μm. Adapted from Kolokolova et al. (2011a).

5.3 Circular polarization of the light scattered by aggregates

Circular polarization was observed in the light scattered by the dust in comets (Rosenbush et al., 2007) and molecular clouds (Hough et al., 2001). It is well known that circular polarization manifests violation of mirror symmetry in the medium. Van de Hulst (1957) showed (see his Section 5.22) that circular polarization arises when the medium has unequal number of left-handed and right-handed identical but mirror asymmetric particles. This immediately shows that if we consider light scattering by a single aggregate, let say BPCA or BCCA, then even in the case of random orientation of this aggregate its circular polarization does not vanish as the majority of ballistic aggregates are asymmetric (Fig. 5.1). This was repeatedly shown by computer simulations of light scattering by aggregates (Kolokolova et al., 2006; Guirado et al., 2007). However, ensembles of natural aggregates, such as cosmic dust, usually do not have domination of particles of a specific handiness. So, in the case when some ensemble of natural aggregates demonstrates circular polarization, it has another violation of mirror symmetry than that resulted from the asymmetric arrangement of the monomers in the aggregates.

One of the most common violations is alignment of elongated particles (e.g., in magnetic field). This is a very common situation for cosmic dust and numerous papers on alignment of aggregates and their circular polarization have been published (see reviews by Lazarian, 2007; 2009 and reference therein). One more opportunity for mirror asymmetry of aggregates is optical activity of their material. Optical activity is typical for organics of

biological origin due to the homochirality of their molecules (i.e. domination of left handed amino acids and right handed sugars). Recently the T-matrix code by Mackowski & Mishchenko (1996) has been updated to allow accounting for the optical activity of the monomer material (Mackowski et al., 2011). Below we show some results of the computer modeling based on this code.
To avoid the influence of mirror asymmetry of the aggregate itself, described above, we performed the calculations for a completely symmetric aggregate like a cube of spheres or 3D-cross. The optical activity was described by a complex parameter $\beta=\beta_r+i\beta_i$ that demonstrated the difference in the complex refractive index for the light with left-handed and right-handed polarization; here βr described the circular birefringence of the material and β_i described its circular dichroism. The code correctly predicted the equal but opposite sign of the circular polarization in the case of aggregates of the opposite sign of β. The modeling by Kolokolova et al. (2011b) showed that the circular polarization quickly increased with increasing optical activity, size of monomers, and especially size of the aggregate. An interesting result was a strong dependence of the circular polarization on the packing density of the aggregates. Fig. 5.8 shows that the circular polarization is much larger and increases more quickly with the size of aggregate in the case when the aggregate is more compact. This probably demonstrates an increasing influence of the diffuse multiple scattering as the aggregate becomes larger or more compact, and more monomers are involved in the light scattering.

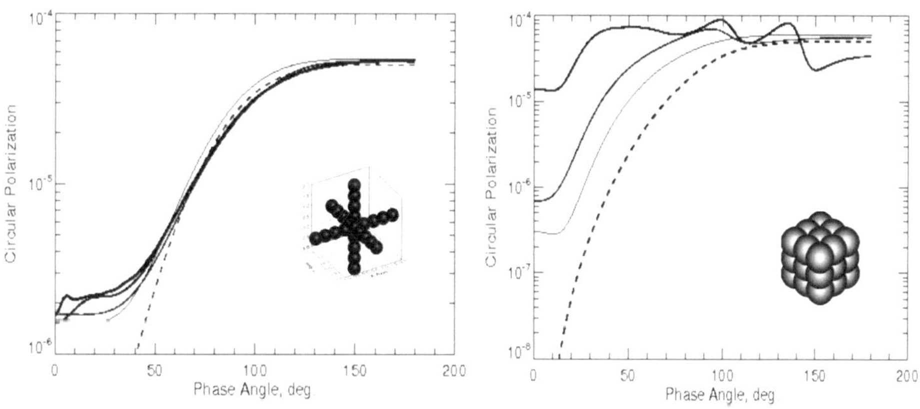

Fig. 5.8. Dependence of absolute values of circular polarization on the size of a 3D-cross aggregate (left) and cubic aggregate (right). The radius of the monomers is 50 nm; the wavelength is 650 nm. The dashed line is for a single monomer; solid lines are for the aggregates of 9, 125, and 343 monomers (thickness of the line increases with the number of monomers). In the simulations we used $m=1.55002+i0.0006002$ *and* $\beta=7.034*10^{-6}-i*0.8692*10^{-8}$ which were estimated based on the measured excess of left-handed amino acids in some meteorites (Pizzarello & Cronin, 2000; Pizzarello & Cooper, 2001).

It is evident that diffuse multiple scattering can affect circular polarization because at each consequent scattering on an optically active monomer circular polarization should increase. This effect is opposite to the depolarization of linearly polarized light in a result of multiple scattering. Linear polarization depends on the plane in which the scattering happens, and at multiple scattering this plane changes randomly thereby randomizing the resultant polarization (see Section 3.1). Orientation of the scattering plane does not affect circular polarization, and its formation is determined only by the fact that the light repeatedly interacts with optically-active scatterers. Since the cubic aggregate shown in Fig. 5.8 represents the case of a densely packed aggregate, we expect that its light scattering is also affected by near-field effects. How near-field effects influence circular polarization is a topic of a separate study that still needs to be done.

6. Conclusions

We have briefly described a progress recently made in the understanding and modeling of a variety of physical effects associated with electromagnetic interaction between constituent scatterers in a complex object such as an inhomogeneous particle or an aggregate of small monomers. Our test objects were aggregates as a common example of natural particles. In the case when such aggregates are made of particles much smaller than wavelength, effective medium theories can be applied to study their light scattering. However, natural, especially cosmic, particles are aggregates of monomers larger than wavelength when observed in the visible spectral range. Their light scattering requires a more sophisticated approach. We showed that with increasing packing density of aggregates interaction of their monomers becomes more complex and involves diffuse multiple scattering, coherent scattering, and, at even larger packing densities, near-field effects. The diffuse multiple scattering simplifies dependencies of intensity and polarization on phase angle reducing the resonant oscillations typical for single scattering by particles of size larger than wavelength. In its turn, coherent scattering complicates the phase curves adding brightness and polarization opposition feature in the backscattering domain. Development of these features becomes even more complex when the packing density increases and near-field effects become not negligible. The near-field effects affect all phase angles, changing value and location of both the polarization minimum and maximum as well as behavior of the intensity. The correct accounting for all these effects is possible by using rigorous solutions of the Maxwell equations for complex objects. In the case of aggregates, such a solution is provided by the superposition T-matrix approach (Mackowski & Mishchenko, 1996). We use this approach to simulate properties of large aggregates. This allows us not only to study all types of interaction separately and find conditions for their realization, but also to interpret the observational data for cosmic dust. The T-matrix modeling provides: (1) explanation of specifics of phase dependencies of intensity and polarization for cometary and other cosmic dust; (2) explanation of spectral dependence of polarization for comets and asteroids and its variations with wavelength; (3) explanation of variations in depth of spectral bands observed for Saturn's satellites; (4) study of circular polarization of light scattered by objects of biological interest. This modeling also allows us to reveal the characteristics of dust particles in a variety of natural environments thereby validating it as a powerful tool for remote sensing applications.

7. Acknowledgement

This work was supported by a grant of NASA Astrobiology Program and by a grant of Japanese Society for the Promotion of Science.

8. References

Belskaya, I. N., Shkuratov, Yu. G, Efimov, Yu. S., Shakhovskoj, N.M., Gil-Hutton, R., Cellino, A., et al. (2005) The F-type asteroids with small inversion angles of polarization. *Icarus,* 178, pp. 213-221.

Bohren, C., & Huffman, D. (1983) *Absorption and Scattering of Light by Small Particles,* J. Wiley & Sons, NY.

Borghese, F., Denti, P., & Saija, R. (2010) *Scattering from Model Nonspherical Particles: Theory and Applications to Environmental Physics Physics of Earth and Space Environments),* Springer, Berlin.

Bruggeman, D. A. G. (1935) Berechnung verschiedener physikalischer Konstanten von heterogenenSubstanzen. I. Dielektrizitatskonstanten and Leitfahigkeiten der Mischkorper aus isotropen Substanzen. *Annalen der Physik,* 24, pp. 636–664.

Choy, T. C. (1999) *Effective medium theory: principles and applications,* Clarendon Press, Oxford University Press, Oxford England, New York.

Chylek, P, Videen, G., Geldart, D., Dobbie, J., & Tso, H. W. (2000) Effective medium approximations for heterogeneous particles. In: *Light scattering by nonspherical particles* (Mishchenko, M., Hovenier, J., & Travis, L., Eds.), pp. 274-308, Academic Press, NY.

Etemad, S., Thompson, R., Andrejco, W. J., John, S., & MacKintosh, F.C. (1987) Weak localization of photons: termination of coherent random walks by absorption and confined geometry. *Phys. Rev. Lett.,* 59, pp. 1420-1423.

Greenberg, J. M., & Hage, J. I. (1990) From interstellar dust to comets - A unification of observational constraints, *Astrophys. J., Part 1,* 361, pp. 260-274.

Guirado, D., Hovenier, J. W., & Moreno, F. (2007) Circular polarization of light scattered by asymmetrical particles, *J. Quant. Spectr. Radiat. Transfer,* 106, pp. 63-73.

Gustafson, B. Å. S. (1999) Scattering by complex systems I: Methods. In: *Formation and Evolution of Solids in Space* (Greenberg, J. M. & Li, A., Eds.), pp. 535-549, Kluwer Academic Publishers, Dordrecht.

Gustafson, B. Å. S., Greenberg, J. M., Kolokolova, L., et al. (2001) Interactions with Electromagnetic Radiation: Theory and Laboratory Simulations, In: I*nterplanetary Dust,* pp. 509 -523, Springer-Verlag.

Hanner, M. S. (2003) The scattering properties of cometary dust, J. *Quant. Spectrosc. Radiat. Transfer,* 79-80, pp. 164-173

Hanner, M. S., Giese, R. H., Weiss, K., & Zerull, R. (1981) On the definition of albedo and application to irregular particles, *Astron. Astrophys.* 104, pp. 42-46.

Hough, J. H., Bailey, J., Chrysostomou, A., Gledhill, T., Lucas, O., Tamura, M., Clark, S., Yates, J., & Menard, F. (2001) Circular polarization in star-forming regions: possible implications for homochirality, *Advances in Space Research,* 27, pp. 313-322.

Hovenier, J. W., Van der Mee, C., & Domke, H. (2004) *Transfer of polarized light in planetary atmospheres: basic concepts and practical methods,* Kluwer Academic Publishers, Dordrecht.

van de Hulst, H. C., (1957) *Electromagnetic scattering by small particles,* Dover Publ. Inc., New York.

Kimura, H. (2001) Light-scattering properties of fractal aggregates: numerical calculations by superposition technique and the discrete-dipole approximation. *J Quant. Spectrosc. Radiat. Transfer,* 70, 581-594.

Kimura, H., Kolokolova, L., & Mann, I., (2003) Optical properties of cometary dust: Constraints from numerical studies on light scattering by aggregate particles, *Astron.Astrophys.*, 407, L5-L9.
Kimura, H., Kolokolova, L., & Mann, I. (2006) Light scattering by cometary dust numerically simulated with aggregate particles consisting of identical spheres, *Astron.Astrophys.*, 449, pp. 1243 - 1254.
Kiselev, N., Rosenbush, V., Kolokolova, L., & Antonyuk, K. (2008) The anomalous spectral dependence of polarization in comets, *J. Quant. Spectrosc. Radiat. Transfer,* 109, pp. 1384-1389.
Kokhanocsky, A. (2001) *Optics of Light Scattering Media: Problems and Solutions,* Springer, Berlin.
Kolokolova, L., & Gustafson, B. Å. S. (2001) Scattering by inhomogeneous particles: Microwave analog experiment comparison to effective medium theories. *J. Quant. Spectrosc. Radiat. Transfer,* 70, pp. 611-625.
Kolokolova, L., Hanner, M., Levasseur-Regourd, A.-Ch., & Gustafson, B. Å. S. (2004a) Physical properties of cometary dust from light scattering emission, In *Comets II,* (Festou, M. C., Keller, H. U., & Weaver, H. A., Eds.), pp. 577 – 604, Univ.of Arizona Press, Tucson.
Kolokolova, L., Kimura, H., & Mann, I. (2004b) Characterization of dust particles using photopolarimetric data: Example of cometary dust, In: *Photopolarimetry in Remote Sensing* (Videen, G., Yatskiv, Ya., & Mishchenko, M., Eds.), pp. 431 - 454, Kluwer Acad. Publ., Dordrecht-London.
Kolokolova, L., Kimura, H., Ziegler, K., & Mann, I. (2006) Light-scattering properties of random-oriented aggregates: do they represent the properties of an ensemble of aggregates? *J. Quant. Spectr. Radiat. Transfer,* 100, pp. 199-206.
Kolokolova, L., Kimura, H., Kiselev, N., & Rosenbush, V. (2007) Two different evolutionary types of comets proved by polarimetric and infrared properties of their dust. *Astron. Astrophys.*, 463, pp.1189-1196.
Kolokolova, L., Buratti, B., & Tishkovets, V. (2010) Impact of coherent backscattering on the spectra of icy satellites of Saturn the implications of its effects for remote sensing, *Astrophys J. Let.*, 711, L71-L74.
Kolokolova, L., & Kimura, H. (2010) Effects of Electromagnetic Interaction in the Polarization of Light Scattered by Cometary and Other Types of Cosmic Dust, *Astron.Astrophys.*, 513, id.A40.
Kolokolova, L., Liu, L., Buratti, B., & Mishchenko, M.I. (2011a) Modeling variations in the near-infrared spectra produced by the coherent backscattering effect, *J. Quant. Spectrosc. Radiat. Transfer,* DOI: 10.1016/j.jqsrt.2011.03.010.
Kolokolova, L., Sparks, W., & Mackowski, D. (2011b) Astrobiological remote sensing with circular polarization. In: *Polarimetric Detection, Characterization, and Remote Sensing* (Mishchenko, M. I., Yatskiv, Ya. S., Videen, G., & Rosenbush, V. K., Eds.), Springer, Berlin, in press.
Kozasa, T., Blum, J., & Mukai, T. (1992) Optical properties of dust aggregates. I - Wavelength dependence, *Astron. Astrophys.,* 263, pp. 423-432.
Lazarian, A. (2007) Tracing magnetic fields with aligned grains, *J. Quant. Spectr. Radiat. Transfer,* 106, pp. 225-256

Lazarian, A. (2009) Quantitative Theory of Grain Alignment: Probing Grain Environment and Grain Composition, In: *Cosmic Dust - Near and Far* (Henning, Th., Grün, E., & Steinacker, J., Eds.), pp. 482-493, Astromomical Society of the Pacific, San Francisco.

Levasseur-Regourd, A.-Ch., & Hadamcik, E. (2003) Light scattering by irregular dust particles in the solar system: observations and interpretation by laboratory measurements. *J. Quant. Spectrosc. Radiat. Transfer*, 79–80, pp. 903–910.

Li, A., & Greenberg, J. M. (1998) From interstellar dust to comets: infrared emission from comet Hale-Bopp C/1995 O1), *Astrophys.J. Let.*, 498, L83-L87.

Lien, D. J. (1991) Optical properties of cometary dust, In: *Comets in the post-Halley era. Vol. 2* (Newburn R. L., Neugebauer, M., & K. Rahe, Eds.), pp. 1005-1041, Kluwer Acad. Publishers, Dordrecht.

Lumme, K., & Bowell, E. (1981) Radiative transfer in the surfaces of atmosphereless bodies. I. Theory. *Astron. J.* 86, pp. 1694-1704.

Lumme, K., & Rahola, J. (1994) Light scattering by porous dust particle in the discrete-dipole approximation. *Astrophys. J.*, 425, pp. 653-667.

Mackowski, D.W., & Mishchenko, M.I. (1996) Calculation of the T matrix and the scattering matrix for ensembles of spheres. *J. Opt. Soc. Am. A.*, 13, pp. 2266-2278.

Mackowski, D., Kolokolova, L., & Sparks, W. (2011) T-matrix approach to calculating circular polarization of aggregates made of optically active materials, *J. Quant. Spectrosc. Radiat. Transfer,* DOI: 10.1016/j.jqsrt.2011.02.003.

Mann, I., Kimura, H., & Kolokolova, L. (2004) A comprehensive model to describe light scattering properties of cometary dust, *J. Quant. Spectrosc. Radiat. Transfer,* 89, pp. 291-301.

Maxwell Garnett, J. C. (1904) Colours in metal glasses and in metallic films. *Philos Trans Roy Soc A,* 203, pp.385–420.

Meakin, P. (1984) Effects of cluster trajectories on cluster-cluster aggregation: a comparison of linear and Brownian trajectories in two- and three-dimensional simulations. *Phys. Rev. A,* 29, pp. 997–999.

Mishchenko, M. I. (2008) Multiple scattering, radiative transfer, and weak localization in discrete random media: unified microphysical approach. *Rev. Geophys.* 46, RG2003.

Mishchenko, M. I., Hovenier, J.W., & Travis, L.D. Eds.(2000) *Light Scattering by Nonspherical Particles. Theory, Measurements, and Applications.* Academic Press, London.

Mishchenko, M. I., Travis, L. D., & Lacis, A.A. (2002) *Scattering, Absorption, and Emission of Light by Small Particles.* Cambridge Univ. Press, Cambridge.

Mishchenko, M. I., Travis, L.D., & Lacis, A. A. (2006) Multiple scattering of light by particles: Radiative transfer and coherent backscattering. Cambridge Univ. Press, Cambridge.

Mishchenko, M. I., & Liu, L. (2007) Weak localization of electromagnetic waves by densely packed many-particle groups: exact 3D results. *J Quant Spectrosc Radiat Transfer,* 106, pp.616-621.

Mishchenko, M. I., Liu, L., Mackowski, D. W., Cairns, B., & Videen, G. (2007) Multiple scattering by random particulate media: exact 3D results. *Opt. Express,* 15, pp. 2822-2836.

Mishchenko, M. I., Dlugach, J. M., Liu, L., Rosenbush, V. K., Kiselev, N. N., & Shkuratov, Yu. G. (2009a) Direct solutions of the Maxwell equations explain opposition phenomena observed for high-albedo solar system objects. *Astrophys. J. Let.*, 705, pp. L118-L22.

Mishchenko, M. I., Dlugach, J. M., & Liu, L. (2009b) Azimuthal asymmetry of the coherent backscattering cone: Theoretical results. *Physical Review A*, 80, pp. 053824 - 053832

Mishchenko, M. I., Rosenbush, V. K., Kiselev, N. N., Lupishko, D. F., Tishkovets, V. P., Kaydash, V. G., Belskaya, I. N., Efimov, Yu. S., & Shakhovskoy, N. M. (2010) *Polarimetric Remote Sensing of Solar System Objects*, Akademperiodyka, Kyiv.

Muinonen, K. (1990) *Light scattering by inhomogeneous media: backward enhancement and reversal of linear polarization.* PhD dissertation, University of Helsinki, Finland.

Mukai, T. (1989) Cometary dust and interplanetary particles. In: *Evolution of Interstellar Dust and Related Topics* (Bonetti, A., Greenberg J. M., & Aiello, S., Eds.), pp. 397-445, North-Holland, Amsterdam.

Mukai, T., Ishimoto, H., Kozasa, T., Blum, J., & Greenberg, J. M. (1992) Radiation pressure forces of fluffy porous grains, *Astron. Astrophys.*, 262, pp. 315-320.

Okada, Y., & Kokhanovsky, A. A. (2009) Light scattering and absorption by densely packed groups of spherical particles. *J Quant Spectrosc Radiat Transfer*, 110, pp. 902-917.

Petrova, E. V., Jockers, K., & Kiselev, N. N. (2000) Light Scattering by Aggregates with Sizes Comparable to the Wavelength: An Application to Cometary Dust. *Icarus*, 148, pp. 526-536.

Petrova, E. V., Tishkovets, V. P., & Jockers, K. (2004) Polarization of light scattered by Solar system bodies and the aggregate model of dust particles. *Solar System Res.*, 38, pp. 309-324.

Petrova, E. V., Tishkovets, V. P., & Jockers, K. (2007) Modeling of opposition effects with ensembles of clusters: Interplay of various scattering mechanisms, *Icarus*, 188, pp. 233-245.

Petrova, E. V., Tishkovets, V. P., & Jockers, K. (2009) Interaction of particles in the near field and opposition effects in regolith-like surfaces, *Sol. System Res.*, 43, pp. 100-115.

Petrova, E.V., & Tishkovets, V.P. (2011) Light scattering by aggregates of varying porosity and the opposition phenomena observed in the low-albedo particulate media. *J Quant Spectrosc Radiat Transfer*, DOI: 10.1016/j.jqsrt.2011.01.011.

Pizzarello, S., & Cronin, G. R. (2000) Non-racemic amino acids in the Murray and Murchison meteorites, *Geochimica et Cosmochimica Acta*, 64, pp. 329-338.

Pizzarello, S., & Cooper, G. (2001) Molecular and chiral analyses of some protein amino acid derivatives in the Murchison and Murray meteorite, *Meteoritics & Planetary Science*, 36, pp. 897-909.

Rosenbush, V., Kiselev, N., Avramchuk, V., & Mishchenko, M. (2002) Photometric and polarimetric opposition phenomena exhibited by Solar system bodies. In: *Optics of Cosmic Dust* (Videen, G., & Kocifaj, M., Eds.) pp. 191-224 Kluwer Academic Publishers, Dordrecht.

Rosenbush, V., Kolokolova, L., Lazarian, A., Shakhovskoy, N., & Kiselev, N. (2007) Circular polarization in comets: Observations of Comet C/1999 S4 LINEAR) and tentative interpretation, *Icarus*, 186, pp. 317-330.

Schuerman, D. W. (1980) *Light scattering by irregularly shaped particles.* New York, Plenum Press.

Shen, Y., Draine, B. T., & Johnson, E. T. (2008) Modeling Porous Dust Grains with Ballistic Aggregates. I. Geometry and Optical Properties, *Astrophys. J.*, 689, pp. 260-275.

Shkuratov, Yu. G. (1989) New mechanism of the negative polarization of light scattered by atmosphereless cosmic bodies. *Astron Vestnik,* 23, pp. 176–180 [in Russian].

Shkuratov, Yu. G., Muinonen, K., Bowell, E., Lumme, K., Peltoniemi, J. I., Kreslavsky, V.F., Stankevich, D.G., Tishkovetz, V.P., Opanasenko, N.V., & Melkumova, L.Y. (1994) A critical review of theoretical models of negatively polarized light scattered by atmosphereless Solar system bodies. *Earth, Moon, & Planets*, 65, pp. 210-246.

Shkuratov, Yu., Ovcharenko, A., Zubko, E., Miloslavskaya, O., Muinonen. K,. Piironen, J., et al. (2002) The opposition effect and negative polarization of structural analog for planetary regoliths. *Icarus*, 159, pp. 396-416.

Shkuratov, Yu., Ovcharenko, A., Zubko, E., Volten, H., Munoz, O., & Videen, G. (2004) The negative polarization of light scattered from particulate surfaces and of independently scattering particles. *J. Quant. Spectrosc. Rad. Transfer,* 88, pp. 267-284.

Shkuratov, Yu. G., & Grynko, Ye. S. (2005) Light scattering by media composed of semitransparent particles of different shapes in ray optics approximation: consequences for spectroscopy, photometry, and polarimetry of planetary regoliths. *Icarus*, 173, pp.16-28.

Sihvola, A., (1999) *Electromagnetic Mixing Formulae and Applications,* IEE Publ., London.

Tishkovets, V. P. (1998) Backscattering of light by close-packed system of particles. *Opt. Spectrosc.*, 85, pp. 212–217.

Tishkovets, V. P., Shkuratov, Yu. G., & Litvinov, P. V. (1999) Comparison of collective effects at scattering by randomly oriented cluster of spherical particles. *J. Quant. Spectrosc. Radiat. Transfer,* 61, pp. 767–773.

Tishkovets, V., Litvinov, P., Petrova, E., Jockers, K., & Mishchenko, M. (2004a) Backscattering effects for discrete random media. In:. *Photopolarimetry in remote sensing* (Videen, G., Yatskiv, Ya. S., & Mishchenko, M. I., Eds), pp. 221-242, Kluwer Acad. Publ., Dordrecht.

Tishkovets, V. P., Petrova, E. V., & Jockers, K. (2004b) Optical properties of aggregate particles comparable in size to the wavelength, *J. Quant. Spectrosc. Radiat. Transfer,* 86, pp. 241-265.

Tishkovets, V. P. (2008) Light scattering by closely packed clusters: shielding of particles by each other in the near field. *J Quant Spectrosc Radiat Transfer,* 109, pp. 2665-2672.

Voshchinnikov, N. V. (2004) *Optics of cosmic dust,* Cambridge, UK: Cambridge Scientific Publishers.

Voshchinnikov, N. V., Il'in, V. B., & Henning, Th. (2005) Modelling the optical properties of composite and porous interstellar grains, *Astron. Astrophys.*, 429, pp.371-381

Voshchinnikov, N. V., Il'in, V. B., Henning, Th., & Dubkova, D. N. (2006) Dust extinction and absorption: the challenge of porous grains, *Astron. Astrophys.*, 445, pp.167-177.

Voshchinnikov, N., V., Videen, G., & Henning, Th. (2007) Effective medium theories for irregular fluffy structures: aggregation of small particles, *Applied Optics IP,* 46, pp.4065-4072.

West, R. A. (1991) Optical properties of aggregate particles whose outer diameter is comparable to the wavelength. *Appl. Opt.,* 30, pp. 5316–5324.

Wolff, J., Clayton, G. C., & Gibson, S. J. (1998) Modeling composite and fluffy grains. II. Porosity and phase function. *Astrophys. J.*, 503, pp. 815–830.

Zubko, E., Shkuratov, Yu., Mishchenko, M., & Videen, G. (2008) Light scattering in a finite multi-particle system. *J Quant Spectrosc Radiat Transfer*, 109, pp. 2195–2206.

10

Models for Scattering from Rough Surfaces

F. Ticconi[1], L. Pulvirenti[2] and N. Pierdicca[2]
[1]*Dept. of Information Engineering and Computer Science, University of Trento,*
[2]*Dept. Information Engineering, Electronics and Telecommunications - "Sapienza University of Rome,*
Italy

1. Introduction

Models for scattering of electromagnetic waves from random rough surfaces have been developed during the last two centuries and the scientific interest in the problem remains strong also today due to the importance of this phenomenon in diverse areas of science, such as measurements in optics, geophysics, communications and remote sensing of the Earth.
Such models can be categorised into empirical models, analytical models and a combination of the two. Though very simple, empirical models are greatly dependent on the experimental conditions. In spite of their complexity, only theoretical models can yield a significant understanding of the interaction between the electromagnetic waves and the Earth's surface, although an exact solution of equations governing this interaction may not always be available and approximate methods have to be used. The semi-empirical models, which are based on both physical considerations and experimental observations, can be set between these two kinds of models and can be easily inverted. In this survey, we will focus on the analytical models and we study more in detail the *Kirchhoff Approximation* (KA), the *Small Perturbation Method* (SPM) and the *Integral Equation Method* (IEM). The Kirchhoff Approximation and the Small Perturbation Methods represent early approaches to scattering which are still much used, whereas the Integral Equation Method represents a newer approach which has a larger domain of validity. These methods have been found to be the most common in the literature and many of the other methods are based or have much in common with these approaches. In section 2, we begin by giving a brief presentation of the scattering problem and introduce some concepts and results from the theory of electromagnetic fields which are often used in this context. We will also define the bistatic scattering coefficient, due to the importance of this type of measurement in many remote sensing applications, and in particular in the retrieval of soil moisture content. In section 3, we give a brief presentation on the Kirchhoff Approximation and its close variants, the *Physical Optics* (PO) and the *Geometrical Optics* (GO). In section 4, we give a brief presentation of the Small Perturbation Method and in section 5 we will present the Integral Equation Model.

2. Some concepts of the electromagnetic theory and surface parameters

In this section we will give a brief presentation of some concepts on theories of electromagnetism and statistical characterisation of surfaces, which are often used for modelling scattering of electromagnetic waves from random rough surfaces. We will also define the bistatic scattering coefficient due to the importance of this type of measurement in many remote sensing applications.

2.1 The Maxwell's equations and the wave equation

The basic laws of the electromagnetism are given by the Maxwell's equations which, for linear, homogeneous, isotropic, stationary and not dispersive media, can be written as (Balanis, 1989):

$$\nabla \times \mathbf{E} = -\frac{\partial \mathbf{B}}{\partial t} \tag{2.1.1}$$

$$\nabla \times \mathbf{H} = \frac{\partial \mathbf{D}}{\partial t} + \mathbf{J}_c + \mathbf{J}_i \tag{2.1.2}$$

$$\nabla \cdot \mathbf{D} = \rho \tag{2.1.3}$$

$$\nabla \cdot \mathbf{B} = 0 \tag{2.1.4}$$

where **E** is the electric field vector, **D** is the electric flux density, **H** is the magnetic field vector, **B** is the magnetic flux density, **J** is the conduction electric current density, $\mathbf{J}_i$ is the impressed electric current density and ρ is the electric charge density. Maxwell's equations together with the boundary conditions, give a complete description of the field vectors at any points (including discontinuities) and at any time. In rough surface scattering, the surface enters in the boundary conditions (see equations (2.2.1)-(2.2.4)), which have to be also supplied at infinity.
If we consider *time-harmonic* variation of the electromagnetic field, the instantaneous field vectors can be related to their complex forms. Thus the Maxwell's equations can be written in a much simpler form:

$$\nabla \times \mathbf{E} = -j\omega\mu\mathbf{H} \tag{2.1.5}$$

$$\nabla \times \mathbf{H} = (\sigma + j\omega\varepsilon)\mathbf{E} + \mathbf{J}_i \equiv j\omega\varepsilon_c\mathbf{E} + \mathbf{J}_i \tag{2.1.6}$$

$$\nabla \cdot \varepsilon\mathbf{E} = \rho \tag{2.1.7}$$

$$\nabla \cdot \mu\mathbf{H} = 0 \tag{2.1.8}$$

where we assumed the region characterised by permeability μ, permittivity ε and conductivity σ (lossy medium). To obtain the governing equation for the electric field, we take the curl of (2.1.5) and then replace (2.1.6). Thus,

$$\nabla \times \nabla \times \mathbf{E} + \omega^2\mu\varepsilon_c\mathbf{E} = j\omega\mu\mathbf{J}_i \tag{2.1.9}$$

which is known as the inhomogeneous Helmholtz vector wave equation. In a free-source region, $\nabla \cdot \mathbf{E} = 0$ and (2.1.9) simplifies to:

$$\nabla^2 \mathbf{E} + \omega^2 \mu \varepsilon_c \mathbf{E} = 0 \qquad (2.1.10)$$

In rectangular coordinates, a simple solution to (2.1.10) has the form:

$$\mathbf{E}(\mathbf{r}) = \mathbf{E}_0 e^{-j\mathbf{k}\cdot\mathbf{r}} \qquad (2.1.11)$$

where $\mathbf{E}_0$ is a constant complex vector which determines the polarisation characteristics and the complex propagation vector, $\mathbf{k}$, is defined as:

$$\mathbf{k} = \hat{\mathbf{x}} k_x + \hat{\mathbf{y}} k_y + \hat{\mathbf{z}} k_z \qquad (2.1.12)$$

with the components satisfying

$$k_x^2 + k_y^2 + k_z^2 = \omega^2 \mu \varepsilon_c = k^2 \qquad (2.1.13)$$

Equation (2.1.11) represents a plane wave and k is the propagation constant. Most analytical methods for scattering from rough surfaces assume this kind of incident wave, which if linearly polarised can be rewritten as:

$$\mathbf{E}^i(\mathbf{r}) = \hat{\mathbf{p}} E_0 e^{-j\mathbf{k}_i\cdot\mathbf{r}} = \hat{\mathbf{p}} E^i \qquad (2.1.14)$$

where $\mathbf{k}_i = \hat{\mathbf{k}}_i k$, $\hat{\mathbf{p}}$ is the unit polarisation vector and E_0 is the amplitude. The associated magnetic field is given by:

$$\mathbf{H}^i(\mathbf{r}) = \hat{\mathbf{k}}_i \times \mathbf{E}^i(\mathbf{r}) / \eta \qquad (2.1.15)$$

where $\eta = \sqrt{\mu/\varepsilon_c}$ is the wave impendence in the medium.

2.2 Integral theorems and other results used in scattering models

We will present some results for electromagnetic fields which are often used as a starting point in the analytical models for scattering from rough surfaces. These equations are approximated and simplified using different methods and assumptions in the analytical solutions for scattering from rough surfaces. We will not show how the equations in this section are derived, but derivation can be found in the references.

Consider an electromagnetic plane wave incident on a rough surface as shown in figure 2.2.1.

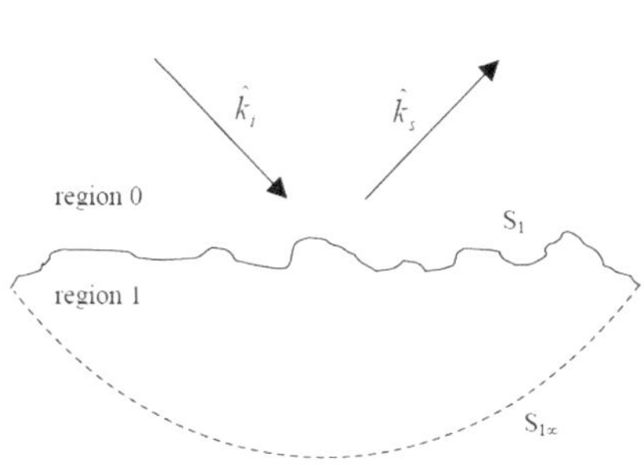

Fig. 2.2.1. Scattering of electromagnetic field on surface separating two media.

Across any surface interface, the electromagnetic field should satisfy continuity conditions given by (Balanis, 1989):

$$\hat{\mathbf{n}} \times (\mathbf{E} - \mathbf{E}_1) = 0 \tag{2.2.1}$$

$$\hat{\mathbf{n}} \times (\mathbf{H} - \mathbf{H}_1) = \mathbf{J}_s \tag{2.2.2}$$

$$\hat{\mathbf{n}} \cdot (\varepsilon \mathbf{E} - \varepsilon_1 \mathbf{E}_1) = \rho_s \tag{2.2.3}$$

$$\hat{\mathbf{n}} \cdot (\varepsilon \mathbf{H} - \varepsilon_1 \mathbf{H}_1) = 0 \tag{2.2.4}$$

where $\hat{\mathbf{n}}$ is the unit normal vector of the rough surface (pointing in the region 0). The electric surface current density, $\mathbf{J}_s$, and the charge surface density, ρ_s, at the rough interface are zero unless the scattering surface (or one of the media) is a perfect conductor.
Using the fact that the fields satisfy the Helmholtz wave equation (2.1.9), it can be shown that in the region 0, the electromagnetic fields **E** and **H**, satisfy Huygens' principle and the radiation boundary condition at infinity and **E** is given by (Ulaby et al, 1982; Tsang et al, 2000):

$$\mathbf{E}(\mathbf{r}) = \mathbf{E}^i(\mathbf{r}) + \int_{S_1} \left\{ -j\omega\mu \overline{\overline{\mathbf{G}}}(\mathbf{r},\mathbf{r}') \cdot \hat{\mathbf{n}}' \times \mathbf{H}(\mathbf{r}') - \nabla \times \overline{\overline{\mathbf{G}}}(\mathbf{r},\mathbf{r}') \cdot \hat{\mathbf{n}}' \times \mathbf{E}(\mathbf{r}') \right\} ds' \tag{2.2.5}$$

where $\overline{\overline{\mathbf{G}}}$ is the dyadic Green function (to the vector Helmholtz equation) which is represented by:

$$\overline{\overline{\mathbf{G}}}(\mathbf{r},\mathbf{r}') = \left(\overline{\overline{\mathbf{I}}} + \frac{\nabla\nabla}{k^2} \right) g(\mathbf{r},\mathbf{r}') \tag{2.2.6}$$

Here $\overline{\overline{\mathbf{I}}}$ is the unit dyadic and $g(\mathbf{r},\mathbf{r}')$ is the Green function that satisfies the scalar wave equation. It assumes the following expression:

$$g(\mathbf{r},\mathbf{r}') = \frac{e^{-jk|\mathbf{r}-\mathbf{r}'|}}{4\pi|\mathbf{r}-\mathbf{r}'|} \tag{2.2.7}$$

In (2.2.5) the first term on the right-hand side represents the field generated by a current source in an *unbounded* medium with permittivity ε and permeability μ and corresponds to the incident field. Hence, the electromagnetic field in the region 0 is expressed as the sum of two contributions: one is given by the incident field $\mathbf{E}^i(\mathbf{r})$; the other contribution is given by the surface integrals that involve the tangential components $\mathbf{E}_t$ and $\mathbf{H}_t$ of the fields at the boundary S_1 (note that $\hat{\mathbf{n}}' \times \mathbf{E} = \hat{\mathbf{n}}' \times \mathbf{E}_t$ and $\hat{\mathbf{n}}' \times \mathbf{H} = \hat{\mathbf{n}}' \times \mathbf{H}_t$) and represents the scattered field due to the presence of surface.
The equation (2.2.5) constitutes the mathematical basis of Huygens' principle in vector form. According to this principle, the electromagnetic field in a source-free region $(\mathbf{J} = 0)$ is uniquely determined once its tangential components are assigned on the boundary of the region. However, since in the region 0, the existence of the impressed current **J** has been

assumed, the total electric field can be expressed as the sum of two terms, the incident and scattering ones:

$$\mathbf{E}(\mathbf{r}) = \mathbf{E}^i(\mathbf{r}) + \mathbf{E}^s(\mathbf{r}) \tag{2.2.8}$$

Thus, the scattered field can be written as:

$$\mathbf{E}^s(\mathbf{r}) = \int_{S_1} \left\{ -j\omega\mu \overline{\overline{\mathbf{G}}}(\mathbf{r},\mathbf{r}')\cdot \hat{\mathbf{n}}' \times \mathbf{H}(\mathbf{r}') - \nabla \times \overline{\overline{\mathbf{G}}}(\mathbf{r},\mathbf{r}')\cdot \hat{\mathbf{n}}' \times \mathbf{E}(\mathbf{r}') \right\} ds' \tag{2.2.9}$$

If the observation point is in the far field region, the Green function in (2.2.9) can be simplified and the scattering field can be written as (Ulaby et al,1982; Tsang et al, 2000):

$$\mathbf{E}^s(\mathbf{r}) = K\mathbf{r} \times \int_{S_1} \left[\left(\hat{\mathbf{n}}' \times \mathbf{E}(\mathbf{r}') \right) - \eta \hat{\mathbf{r}} \times \left(\hat{\mathbf{n}}' \times \mathbf{H}(\mathbf{r}') \right) \right] e^{j\mathbf{k}\cdot\mathbf{r}'} ds' \tag{2.2.10}$$

where $K = -jke^{-jkr}/4\pi r$ and $\hat{\mathbf{r}}$ is the unit vector pointing in the direction of observation. The tangential surface fields $\hat{\mathbf{n}} \times \mathbf{E}$ and $\hat{\mathbf{n}} \times \mathbf{H}$ can be also expressed as (Poggio & Miller, 1973):

$$\hat{\mathbf{n}} \times \mathbf{E} = 2\hat{\mathbf{n}} \times \mathbf{E}^i - \frac{2}{4\pi}\hat{n} \times \int \varepsilon \, ds' \tag{2.2.11}$$

$$\hat{\mathbf{n}} \times \mathbf{H} = 2\hat{\mathbf{n}} \times \mathbf{H}^i + \frac{2}{4\pi}\hat{\mathbf{n}} \times \int \mathcal{H} ds' \tag{2.2.12}$$

and

$$\hat{\mathbf{n}} \times \mathbf{E}_t = -\frac{2}{4\pi}\hat{\mathbf{n}}_t \times \int \varepsilon_t \, ds' \tag{2.2.13}$$

$$\hat{\mathbf{n}} \times \mathbf{H}_t = \frac{2}{4\pi}\hat{\mathbf{n}}_t \times \int \mathcal{H}_t \, ds' \tag{2.2.14}$$

where

$$\varepsilon = jk\eta\left(\hat{\mathbf{n}}' \times \mathbf{H}'\right)G_1 - \left(\hat{\mathbf{n}}' \times \mathbf{E}'\right) \times \nabla' G_1 - \left(\hat{\mathbf{n}}' \cdot \mathbf{E}'\right)\nabla' G_1 \tag{2.2.15}$$

$$\mathcal{H} = \frac{jk}{\eta}\left(\hat{\mathbf{n}}' \times \mathbf{E}'\right)G_1 - \left(\hat{\mathbf{n}}' \times \mathbf{H}'\right) \times \nabla' G_1 - \left(\hat{\mathbf{n}}' \cdot \mathbf{H}'\right)\nabla' G_1 \tag{2.2.16}$$

$$\varepsilon_t = -\left[jk_2\eta_2\left(\hat{\mathbf{n}}' \times \mathbf{H}'\right)G_2 - \left(\hat{\mathbf{n}}' \times \mathbf{E}'\right) \times \nabla' G_2 - \left(\hat{\mathbf{n}}' \cdot \mathbf{E}'\right)\nabla' G_2\left(1/\varepsilon_r\right) \right] \tag{2.2.17}$$

$$\mathcal{H}_t = -\left[\frac{jk_2}{\eta_2}\left(\hat{\mathbf{n}}' \times \mathbf{E}'\right)G_2 - \left(\hat{\mathbf{n}}' \times \mathbf{H}'\right) \times \nabla' G_2 - \left(\hat{\mathbf{n}}' \cdot \mathbf{H}'\right)\nabla' G_2\left(1/\mu_r\right) \right] \tag{2.2.18}$$

and $\hat{\mathbf{n}}$, $\hat{\mathbf{n}}'$, $\hat{\mathbf{n}}_t$, $\hat{\mathbf{n}}_t'$ are the unit normal vectors to the surface and $\hat{\mathbf{n}}_t = -\hat{\mathbf{n}}$, $\hat{\mathbf{n}}_t' = -\hat{\mathbf{n}}'$, $\hat{\mathbf{n}} \times \mathbf{E}$ and $\hat{\mathbf{n}} \times \mathbf{H}$ are the total tangential fields on the rough surface in the medium above the separating interface; G_1 and G_2 are the Green's functions in medium above and below the interface, respectively, and $\varepsilon_r = \varepsilon_2/\varepsilon_1$, $\mu_r = \mu_2/\mu_1$, $\eta_2 = \sqrt{\mu_2/\varepsilon_2}$ and $k_2 = \omega\sqrt{\mu_2\varepsilon_2}$.

2.3 The nature of surface scattering

When an electromagnetic wave impinges the surface boundary between two semi-infinitive media, the scattering process takes place only at the surface boundary if the two media can be assumed homogeneous. Under such supposition, the problem at issue is indicated as *surface scattering* problem. On the other hand, if the lower medium is inhomogeneous or is a mixture of materials of different dielectric properties, then a portion of the transmitted wave scattered backward by the inhomogeneities may cross the boundary surface into the upper medium. In this case scattering takes place within the volume of the lower medium and it is referred to as *volume scattering*. In most cases both the scattering processes are involved, although only one of them can be dominant. In the case of bare soil, which will be assumed to be a homogeneous body, surface scattering is the only process taken into consideration.

When the surface boundary separating the two semi-infinitive media is perfectly smooth the reflection is in the specular direction and is described by the Fresnel reflection laws. On the other hand, when the surface boundary becomes rough, the incident wave is partly reflected in the specular direction and partly scattered in all directions. Qualitatively, the relationship between surface roughness and surface scattering can be illustrated through the example shown in Figure 2.3.1. For the specular surface, the angular radiation pattern of the reflected wave is a delta function centred about the specular direction as shown in Figure 2.3.1 (a). For the slightly rough surface (Figure 2.3.1 (b)), the angular radiation pattern consists of two components: a reflected component and a scattered component. The reflected component is again in the specular direction, but the magnitude of its power is smaller than that for smooth surface. This specular component is often referred to as the *coherent* scattering component. The scattered component, also known as the *diffuse* or *incoherent* component, consists of power scattered in all directions, but its magnitude is smaller than that of the coherent component. As the surface becomes rougher, the coherent component becomes negligible.

Note that the specular component represents also the mean scattered field (in statistical sense), whereas the diffuse component has a stochastic behaviour, associated to the randomness of the surface roughness.

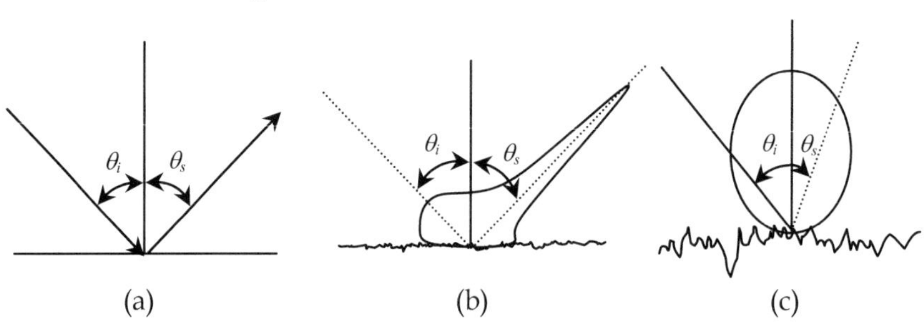

Fig. 2.3.1. Relative contributions of coherent and diffuse scattering components for different surface-roughness conditions: (a) specular, (b) slightly rough, (c) very rough.

2.3.1 Characterisation of soil roughness

A rough surface can be described by a height function $\varsigma = z(x,y)$. There are basically two categories of methods which are being used to measure surface roughness. The roughness can be carried out by means of various experimental approaches able to reproduce the surface profile by using contact or laser probes, or it can be estimated using some theory which relates scattering measurements to surface roughness. In general, the study of scattering in remote sensing is performed by using random rough surface models, where the elevation of surface, with respect to some mean surface, is assumed to be an ergodic[1], and hence stationary[2], random process with a Gaussian height distribution.

Accordingly, the *degree of roughness*, or simply the *roughness*, of a random surface is characterised in terms of statistical parameters that are measured in units of wavelength. For this reason, a given surface that may "appear" very rough to an optical wave, may "appear" very smooth to a microwave.

The two fundamental parameters commonly used are the standard deviation of the surface height variation (or rms height) and the surface correlation length. Such parameters describe the statistical variation of the random component of surface height relative to a reference surface, that may be the unperturbed surface of a period pattern, as in the case of a row-tilled soil surface (Figure 2.3.1.1. (a)), or may be the mean plane surface if only random variations exist (Figure 2.3.1.1 (b)).

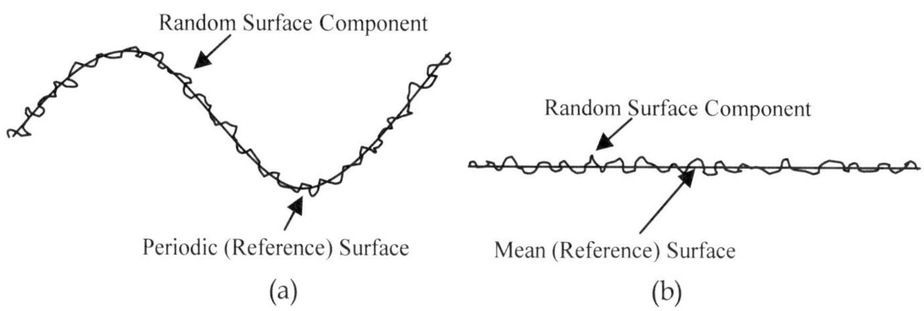

Fig. 2.3.1.1. Two configurations of height variations: (a) random height variations superimposed to a periodic surface; (b) random variations superimposed to a flat surface.

Let $z(x)$ be a representative realisation of the ergodic and stationary process that describes a generic rough surface in a one-dimensional case. The mean value, which throughout this chapter will be denoted by angular brackets $\langle ... \rangle$, is equal to the spatial average over a statistically representative segment of the surface, of dimensions L_x, centred at the origin:

$$\bar{z} = \frac{1}{L_x} \int_{-L_x/2}^{L_x/2} z(x) dx = \langle z(x) \rangle \qquad (2.3.1.1)$$

As it can be noted from the above definition, for a stationary surface the average does not depend on x. The second moment is:

$$\overline{z^2} = \frac{1}{L_x} \int_{-L_x/2}^{L_x/2} z^2(x) dx = \langle z^2(x) \rangle \qquad (2.3.1.2)$$

[1] A process is ergodic when one realisation is representative of all the process, i.e. the statistical averages over an extracted random variable may be replaced by spatial averages over a single realisation.

[2] The stationarity implies that all the statistically properties of a random process are invariant under the translation of spatial coordinates.

Using the above expressions, the standard deviation of the surface height, σ, is therefore defined as:

$$\sigma = \left[\overline{z^2} - \left(\overline{z} \right)^2 \right]^{1/2} = \left\langle \left[z(x) - \overline{z} \right]^2 \right\rangle^{1/2} \tag{2.3.1.3}$$

Such quantities characterise the dispersion of the surface height relative to the reference plane. Taking into account the stationary properties of the process and considering its mean value null, the variance, σ^2, is coincident with the second moment and does not depend on x. The autocorrelation function of the height random process $z(x)$ is given by:

$$R_z(\tau) = \frac{1}{L_x} \int_{-L_x/2}^{L_x/2} z(x) z(x+\tau) dx = \langle z(x) z(x+\tau) \rangle \tag{2.3.1.4}$$

The normalised autocorrelation function (ACF), better known as the correlation coefficient, assumes for a process with zero mean value the following expression:

$$\rho(\tau) = \frac{\int_{-L_x/2}^{L_x/2} z(x) z(x+\tau) dx}{\int_{-L_x/2}^{L_x/2} z^2(x) dx} = \frac{R_z(\tau)}{R_z(0)} = \frac{R_z(\tau)}{\sigma^2} \tag{2.3.1.5}$$

It is a measure of the similarity between the height z at point x and at point distant τ from x. It has the following properties:

$$|\rho(\tau)| \le \rho(0) = 1;$$

$$\lim_{\tau \to \infty} \rho(\tau) = 0.$$

The spectral density or power spectrum is defined, for an ergodic random process, as the Fourier transform of the autocorrelation function $R_z(x)$:

$$\tilde{W}(k_x) = \int_{-\infty}^{\infty} R_z(x)\, e^{jk_x x} dx \tag{2.3.1.6}$$

where k_x is the Fourier transform variable.

However, taking into account the equation (2.3.1.5), it is common practice in characterising the random surface to define the power spectrum of the normalised autocorrelation function:

$$W(k_x) = \int_{-\infty}^{\infty} \rho(x)\, e^{jk_x x} dx \tag{2.3.1.7}$$

The Gaussian distribution plays a central role in modelling scattering from random rough surfaces because it is encountered under a great number of different conditions and because Gaussian variates have the unique property that the random process is entirely determined by the height probability distribution and autocorrelation. All higher order correlations can be expressed in terms of the (second order) autocorrelation function, which simplifies modelling the surface scattering process. A simple and often used form for the autocorrelation is the Gaussian function but other forms have also been studied (Saillard & Sentenac, 2001).

The roughness spectrum at the n'th power of the autocorrelation function, $W^{(n)}$, which often enters into closed form solutions of the scattering problem, is given by the Fourier transform:

$$W^{(n)}(k_x) = \int_{-\infty}^{\infty} \rho^n(x)\, e^{jk_x x} dx \tag{2.3.1.8}$$

The consideration of a realistic autocorrelation function is in fact a relevant problem for a better modelling of the soil scattering. Some often used forms (see for instance (Fung, 1994)) of the autocorrelation function are the Gaussian correlation function, the exponential correlation function, combinations of the Gaussian and exponential functions and the so called 1.5-power correlation function. For all of these, the roughness spectrum at the n'th power can be evaluated analytically (see (Fung, 1994)). For instance, for an isotropically rough surface, the normalised Gaussian autocorrelation in a single dimension assumes the following expression:

$$\rho(x) = \exp\left(-\frac{x^2}{l^2}\right) \tag{2.3.1.9}$$

where l is the *correlation length*. Such surface parameter is defined as the displacement x for which $\rho(x)$ is equal to $1/e$

$$\rho(l) = 1/e \tag{2.3.1.10}$$

The correlation length of a surface provides a reference for estimating the statistical independence of two points on the surface; if the two points are separated by a horizontal distance greater than l, then their heights may be considered to be (approximately) statistically independent of one another. In the extreme case of a perfectly smooth (specular) surface, every point on the surface is correlated with every other point with a correlation coefficient of unity. Hence, $l = \infty$ in this case.
Referring to equation (2.3.1.9), the n'th power roughness spectrum is equal to:

$$W^{(n)}(k_x) \equiv \int \rho^n(x) e^{jk_x x} dx = l\sqrt{\frac{\pi}{n}} e^{-\frac{k_x^2 l^2}{4n}} \tag{2.3.1.11}$$

Beside the height random function $z(x)$, the slope function is another important characterisation of the rough surface. It is defined as:

$$Z_x = \lim_{\Delta x \to 0} \frac{z(x+\Delta x) - z(x)}{\Delta x} \tag{2.3.1.12}$$

Considering the stationary random process $z(x)$ as normally distributed with zero mean and variance σ^2, being Z_x the first derivative, its distribution is again normal with zero mean and variance related to the second derivative of the autocorrelation function of $z(x)$ at the origin (Beckman & Spizzichino, 1963):

$$\sigma_s^2 = \langle Z_x^2 \rangle = -\sigma^2 \rho''(0) \quad (2.3.1.13)$$

The rms slope is subsequently indicated as m:

$$m = \left(-\sigma^2 \rho''(0)\right)^{1/2} \quad (2.3.1.14)$$

When the normalised autocorrelation function is Gaussian (equation (2.3.1.9)), the rms slope is equal to:

$$m = \sqrt{2}\frac{\sigma}{l} \quad (2.3.1.15)$$

2.4 Bistatic scattering coefficient

A quantity often used in models and measurements of scattering in the microwave region is the *bistatic scattering coefficient* $\sigma_{q,p}^o(\theta_i,\phi_i,\theta_s,\phi_s)$. It describes the target's scattering properties at a given frequency, polarisation, incidence and observing directions, being independent on the specific measurement system used. It is possible to define $\sigma_{q,p}^o$ directly in terms of the incident and scattering field E_p^i and E_q^s as follows (Ulaby et al, 1982):

$$\sigma_{q,p}^o(\theta_i,\phi_i,\theta_s,\phi_s) = \frac{4\pi R^2 \left|E_p^s\right|^2}{A_0 \left|E_q^i\right|^2} \quad (2.4.1)$$

where the ensemble average must be considered in case the scattered field is the fluctuating zero mean component (i.e., the diffuse or incoherent component mentioned before) generated by a natural target or random rough surface. Such equation shows $\sigma_{q,p}^o$ as the ratio of the total power scattered by an equivalent isotropic scatterer in direction (θ_s,ϕ_s) to the product of the incident power density in direction (θ_i,ϕ_i) and the illuminated area.

The *backscattering coefficient* $\sigma_{q,p}^o(\theta_i)$ is a special case of $\sigma_{q,p}^o(\theta_i,\phi_i,\theta_s,\phi_s)$; it is defined for $\theta_s = \theta_i$ and $\phi_s = \phi_i \pm \pi$ (Figure 2.4.1), which corresponds to the incident and scattered direction being the same except for a reversal in sense.

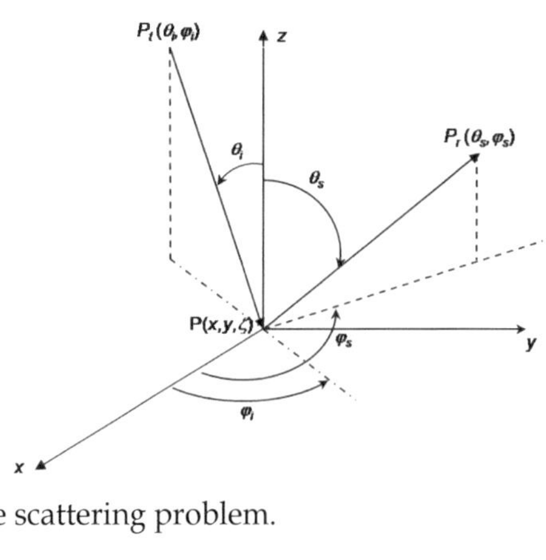

Fig. 2.4.1. Geometry of the scattering problem.

3. The Kirchhoff approximation

In this section we shall consider the Kirchhoff (also sometimes referred to as the tangent plane approximation) approach to describe the scattering from rough surfaces, which was one of the first methods applied. We will consider surfaces with random surface profiles (i.e. not period surfaces) and within the context of the vector theory we will discuss the Kirchhoff Approximation. We will consider here the case of scattering from 2-dimensional dielectric surfaces. We will present results for the case of a surface which can be characterised as a Gaussian random process. We will also mention some extensions of the Kirchhoff approximation and will give references to further reading about the Kirchhoff approach. The reference list is by no means complete, since the literature on the Kirchhoff approximation is vast. A good representation of the Kirchhoff method can be found for instance in (Tsang et al, 2000, Tsang & Kong, 2001, Ulaby et al, 1982).

3.1 Formulation of the scattering problem

The geometry of the scattering problem we consider is shown in figure 2.4.1. We consider a monochromatic, linearly polarised incident plane wave with electric and magnetic field given by the equations (2.1.14) and (2.1.15), respectively.
It can be shown, similarly to equation (2.2.10), that the far zone scattering field, E_{qp}^{s}, can be written in terms of the tangential surface fields in the medium above the separating surface as (Stratton-Chu integral) (Ulaby et al, 1982):

$$E_{qp}^{s} = K\int\left\{\hat{\mathbf{q}}\cdot\left[\hat{\mathbf{k}}_s\times\left(\hat{\mathbf{n}}\times\mathbf{E}_p\right)+\eta\left(\hat{\mathbf{n}}\times\mathbf{H}_p\right)\right]\right\} e^{j(\mathbf{k}_s\cdot\mathbf{r})}ds \tag{3.1.1}$$

where

$$\mathbf{k}_s = k\hat{\mathbf{k}}_s = k\left(\sin\theta_s\cos\phi_s\hat{\mathbf{x}}+\sin\theta_s\sin\phi_s\hat{\mathbf{y}}+\cos\theta_s\hat{\mathbf{z}}\right) = k_{sx}\hat{\mathbf{x}}+k_{sy}\hat{\mathbf{y}}+k_{sz}\hat{\mathbf{z}} \tag{3.1.2}$$

What needs to be calculated are the tangential surface fields in equation (3.1.1). In equations (2.2.11) - (2.2.12) and (2.2.15) - (2.2.16) we presented integral equations for the tangential surface fields in the medium above the scattering dielectric surface. It should be noted that these expressions are exact. However, they cannot in general be solved analytically and therefore approximations have to be introduced. Below we will show that by introducing an approximation called the *tangent plane approximation* (or the *Kirchhoff approximation*), closed analytical solutions can be obtained to the scattering problem.

3.2 The tangent plane approximation and the Kirchhoff fields

In the Kirchhoff approach, the total fields at any point of the surface (i.e., the incident plus the scattered one, to be considered inside the integral (3.1.1)) are approximated by the fields that would be present on an infinitely extended tangent plane at that particular point on the surface. The reflection is therefore considered to be locally specular. It is due to this fact that the Kirchhoff approximation is also referred to as the *tangent plane approximation*. The Kirchhoff approach requires to be valid that every point on the surface has a large radius of curvature relative to the wavelength of the incident field.
Thus, under the tangent-plane approximation, the total field at a point on the surface is assumed equal to the incident field plus the field reflected by an infinite plane tangent to the point. Hence, the tangential surface fields are (Ulaby et al, 1982):

$$\hat{\mathbf{n}} \times \mathbf{E} = \left(\hat{\mathbf{n}} \times \mathbf{E}\right)_k = \hat{\mathbf{n}} \times \left(\mathbf{E}^i + \mathbf{E}^r\right) \tag{3.2.1}$$

$$\hat{\mathbf{n}} \times \mathbf{H} = \left(\hat{\mathbf{n}} \times \mathbf{H}\right)_k = \hat{\mathbf{n}} \times \left(\mathbf{H}^i + \mathbf{H}^r\right) \tag{3.2.2}$$

Here the subscript k stands for the Kirchhoff approximation.
The way to proceed from here, in most presentations of the Kirchhoff method, consists in expressing the tangential fields under the Kirchhoff approximation in terms of the incident electric field components and the local Fresnel reflection coefficients, which depend on the local angles of incidence. This results in the following expressions:

$$\hat{\mathbf{n}} \times \mathbf{E} = \left[(1+R_h)\left(\hat{\mathbf{p}} \cdot \hat{\mathbf{t}}\right)\left(\hat{\mathbf{n}} \times \hat{\mathbf{t}}\right) - (1-R_v)\left(\hat{\mathbf{n}} \cdot \hat{\mathbf{k}}_i\right)\left(\hat{\mathbf{p}} \cdot \hat{\mathbf{d}}\right)\hat{\mathbf{t}}\right] E_0 \, e^{-jk\hat{\mathbf{k}}_i \cdot \mathbf{r}} \tag{3.2.3}$$

$$\eta\left(\hat{\mathbf{n}} \times \mathbf{H}\right) = -\left[(1-R_h)\left(\hat{\mathbf{n}} \cdot \hat{\mathbf{k}}_i\right)\left(\hat{\mathbf{p}} \cdot \hat{\mathbf{t}}\right)\hat{\mathbf{t}} + (1+R_v)\left(\hat{\mathbf{p}} \cdot \hat{\mathbf{d}}\right)\left(\hat{\mathbf{n}} \times \hat{\mathbf{t}}\right)\right] E_0 \, e^{-jk\hat{\mathbf{k}}_i \cdot \mathbf{r}} \tag{3.2.4}$$

where the unit vectors $\hat{\mathbf{t}}$, $\hat{\mathbf{d}}$, $\hat{\mathbf{k}}_i$ define the local reference coordinate system (see (Fung, 1994)) and $\hat{\mathbf{n}}$ is the unit normal vector to the interface in the above medium. R_v and R_h are the Fresnel reflection coefficients for vertical and horizontal polarisation respectively.
Upon substituting (3.2.3) and (3.2.4) in (3.1.1), the scattered field is:

$$E_{qp}^s = K \int \left\{ \hat{\mathbf{q}} \cdot \left[\hat{\mathbf{k}}_s \times \left(\hat{\mathbf{n}} \times \mathbf{E}_p\right) + \eta\left(\hat{\mathbf{n}} \times \mathbf{H}_p\right)\right]\right\} e^{j(\mathbf{k}_s - \mathbf{k}_i)\cdot \mathbf{r}} ds \tag{3.2.5}$$

where the phase factor, $\exp\left(-jk\hat{\mathbf{k}}_i \cdot \mathbf{r}\right)$, of the incident wave has been pointed out from the equations (3.2.3) and (3.2.4). Such equation represents the scattered field formulated under the tangent-plane, or Kirchhoff approximation. As it stands the expression is a complicated function of the surface function and its partial derivatives. No analytic solution has been obtained from (3.2.5) without additional simplifying assumptions. Here we will show the results presented in (Ulaby et al, 1982): for surface with large (with respect to wavelength) standard deviation of surface heights, for which the stationary-phase approximation (Geometric Optics, GO) will be used, and for surfaces with small slopes and a medium or small standard deviation of surface heights, for which a scalar approximation (Physical Optics, PO) will be used.

3.2.1 The scattered field under the stationary-phase approximation (Geometric Optic, GO)

Under the stationary-phase approximation the local tangent plane on a surface point can be considered infinitely wide and, as consequence, the angular re-irradiation pattern originating from that specific point can be represented by a delta function centred in the specular direction. This means that scattering can occur only along directions for which there are specular points on the surface. Hence local diffraction effects are excluded. The approximating relations are obtained from the phase Q of (3.2.5), that is:

$$Q = k\left(\hat{\mathbf{k}}_s - \hat{\mathbf{k}}_i\right) \cdot \mathbf{r} \equiv \mathbf{q} \cdot \mathbf{r} = q_x x + q_y y + q_z z \tag{3.2.1.1}$$

where

$$\hat{\mathbf{k}}_s = \hat{\mathbf{x}}\sin\theta_s\cos\phi_s + \hat{\mathbf{y}}\sin\theta_s\sin\phi_s + \hat{\mathbf{z}}\cos\theta_s \quad (3.2.1.2)$$

$$\hat{\mathbf{k}}_i = \hat{\mathbf{x}}\sin\theta_i\cos\phi_i + \hat{\mathbf{y}}\sin\theta_i\sin\phi_i - \hat{\mathbf{z}}\cos\theta_i \quad (3.2.1.3)$$

$$q_x = k\left(\sin\theta_s\cos\phi_s - \sin\theta_i\cos\phi_i\right) \quad (3.2.1.4)$$

$$q_y = k\left(\sin\theta_s\sin\phi_s - \sin\theta_i\sin\phi_i\right) \quad (3.2.1.5)$$

$$q_z = k\left(\cos\theta_s + \cos\theta_i\right) \quad (3.2.1.6)$$

The phase Q is said to be stationary at a point if its rate of change is zero at the point, that is:

$$\frac{\partial Q}{\partial x} = 0 = q_x + q_z\frac{\partial z}{\partial x}$$

$$\frac{\partial Q}{\partial y} = 0 = q_y + q_z\frac{\partial z}{\partial y}$$

Hence, the partial derivatives of the surface slopes can be replaced by the components of the phase as:

$$Z_x = \frac{\partial z}{\partial x} = -\frac{q_x}{q_z} \quad (3.2.1.7)$$

$$Z_y = \frac{\partial z}{\partial y} = -\frac{q_y}{q_z} \quad (3.2.1.8)$$

Since, the local unit vector $\hat{\mathbf{n}}$ is a function of the surface derivatives:

$$\hat{\mathbf{n}} = \frac{-Z_x\hat{\mathbf{x}} - Z_y\hat{\mathbf{y}} + \hat{\mathbf{z}}}{\sqrt{1 + Z_x^2 + Z_y^2}} \quad (3.1.1.9)$$

the use of (3.2.1.7) and (3.2.1.8) makes $\hat{\mathbf{n}}\times\mathbf{E}$ and $\hat{\mathbf{n}}\times\mathbf{H}$ independent on the integration variables. Thus, the expression for $\mathbf{E}^s$ can be rewritten as:

$$\mathbf{E}^s = K\hat{\mathbf{k}}_s \times\left[\left(\hat{\mathbf{n}}\times\mathbf{E}\right) - \eta\hat{\mathbf{k}}_s\times\left(\hat{\mathbf{n}}\times\mathbf{H}\right)\right] I_1 \quad (3.2.1.10)$$

where

$$I_1 = \int e^{jk\left(\hat{\mathbf{k}}_s - \hat{\mathbf{k}}_i\right)\cdot\mathbf{r}} ds \quad (3.2.1.11)$$

The scattering field corresponding to transmission of $\hat{\mathbf{p}}$ polarisation and reception of $\hat{\mathbf{q}}$ polarisation can be written as (Ulaby et al, 1982):

$$E_{qp}^s = \hat{\mathbf{q}} \cdot \mathbf{E}^s \equiv K I_1 E_0 U_{qp} \tag{3.2.1.12}$$

where

$$U_{qp} = \frac{1}{E_0} \hat{\mathbf{q}} \cdot \hat{\mathbf{k}}_s \times \left[\left(\hat{\mathbf{n}} \times \mathbf{E} \right) - \eta \hat{\mathbf{k}}_s \times \left(\hat{\mathbf{n}} \times \mathbf{H} \right) \right] \tag{3.2.1.13}$$

To compute the scattering coefficient, defined in (2.4.1), for different polarisation states, it is necessary to calculate the ensemble average of $|I_1|^2$:

$$\left\langle |I_1|^2 \right\rangle = \iint \left\langle e^{jk\left(\hat{\mathbf{k}}_s - \hat{\mathbf{k}}_i\right)\cdot(\mathbf{r}-\mathbf{r}')} \right\rangle ds ds' \tag{3.2.1.14}$$

By assuming the surface roughness as a stationary and isotropic Gaussian random process, with zero mean, variance σ^2, and correlation coefficient ρ, and in the assumption that the standard deviation of surface heights is large (that is, $(q_z\sigma)^2$ large) the integral can be solved. The result is (Ulaby et al, 1982):

$$\left\langle |I_1|^2 \right\rangle = \frac{2\pi A_0 q^2}{q_z^4 \sigma^2 |\rho''(0)|} \exp\left[-\frac{q_x^2 + q_y^2}{2q_z^2 \sigma^2 |\rho''(0)|} \right] \tag{3.2.1.15}$$

where the illuminated area A_0 is $(2L)^2$, $\rho''(0)$ is the second derivatives of ρ evaluated at the origin and $\sigma^2 |\rho''(0)|$ corresponds to the mean-squared slope of the surface (Ulaby et al, 1982) (Section 2.3.1).
Upon substituting (3.2.1.15) into the product in the scattered-field expression, it follows:

$$\left\langle E_{qp}^s E_{qp}^{s\,*} \right\rangle = \left| K E_0 U_{qp} \right|^2 \left\langle |I_1|^2 \right\rangle \tag{3.2.1.16}$$

Substituting (3.2.1.16) in the definition of the scattering coefficient given by equation (2.4.1), it assumes the following expression:

$$\sigma_{qp}^o = \frac{\left(kq \left| U_{qp} \right|^2 \right)}{2q_z^4 \sigma^2 |\rho''(0)|} \exp\left[-\frac{q_x^2 + q_y^2}{2q_z^2 \sigma^2 |\rho''(0)|} \right] \tag{3.2.1.17}$$

In the derivation of σ_{qp}^o, the effects of shadowing and multiple scattering have been ignored. It is important to underline that (3.2.1.17) is valid only for surface with sufficiently large standard deviation of surface heights. Under such assumption, that is $(q_z\sigma)^2$ large, the scattering is purely incoherent. As $(q_z\sigma)^2$ decreases, some scattered energy begins to appear in the coherent component. To examine such situation, a different approximation to the tangential fields is needed to permit small $(q_z\sigma)^2$. This is discussed in the next section.

3.2.2 The scattered field under the scalar approximation (Physical Optics, PO)

A different Kirchhoff approach is the Physical Optics solution to (3.1.1). The Physical Optics approach involves the integration of the Kirchhoff scattered field over the entire rough surface, not just the portions of surface which contribute specularly to the scattered direction. Unlike the Geometric Optics solution, the Physical Optics solution predicts a coherent component.

The power in the incoherent reflected field can be found by expanding the Stratton-Chu equation in a Taylor series in surface slope distribution. In (Ulaby et al, 1982) the Physical Optics solution is called scalar approximation because slopes are ignored in the surface coordinate system, leading to a decoupling of polarisation in the vector scattering equations. Accordingly, the basic scattered-field expression can be rewritten in the form:

$$E_{qp}^{s} = KE_0 \int \overline{U}_{qp} \exp\left[jk\left(\hat{\mathbf{k}}_s - \hat{\mathbf{k}}_i\right)\cdot \mathbf{r} \right] ds \tag{3.2.2.1}$$

where $\overline{U}_{qp}$ are given in (Ulaby et al, 1982). To find $\left\langle E_{qp}^{s} E_{qp}^{s\,*} \right\rangle$ for the scattering-coefficient computation, the following integral needs to be computed:

$$I = \iint \left\langle \overline{U}_{qp} \overline{U}_{qp}^{*} \exp\left[jk\left(\hat{\mathbf{k}}_s - \hat{\mathbf{k}}_i\right)\cdot \left(\mathbf{r} - \mathbf{r}'\right) \right] \right\rangle ds ds' \tag{3.2.2.2}$$

Since all $\overline{U}_{qp}$ are expressed in a Taylor series in surface slope distribution, Z_x and Z_y:

$$\overline{U}_{qp} = a_0 + a_1 Z_x + a_2 Z_y \tag{3.2.2.3}$$

where a_i are polarisation-dependent coefficients, the product $\overline{U}_{qp}\overline{U}_{qp}^{*}$ can be written up to the first order in slope as:

$$\overline{U}_{qp}\overline{U}_{qp}^{*} \approx a_0 a_0^{*} + a_0 a_1^{*} Z_x + a_0^{*} a_1 Z_x + a_0 a_2^{*} Z_y + a_0^{*} a_2 Z_y \tag{3.2.2.4}$$

Since $\left(q_z \sigma\right)^2$ is no longer required to be large and assuming the size of the illuminated area equal to $2L \times 2L$, the ensemble average of the first term in (3.2.2.4) can be expressed as (for more details see (Ulaby et al, 1982))

$$I_0 = \left|a_0\right|^2 e^{-q_z^2\sigma^2} \sum_{n=0}^{\infty} \frac{\left(q_z^2\sigma^2\right)^n}{n!} \int_{-2L}^{2L} \int_{-2L}^{2L} \rho^n \left(2L - |u|\right)\left(2L - |v|\right) e^{jq_x u + jq_y v} du dv \tag{3.2.2.5}$$

where the $n = 0$ term corresponds to coherent scattering. It can be shown that this coherent-scattering coefficient can be expressed as:

$$\sigma_{qp}^{o\,c} = \pi k^2 \left|a_0\right|^2 \delta\left(q_x\right)\delta\left(q_y\right) e^{-q_z^2\sigma^2} \tag{3.2.2.6}$$

which shows that coherent scattering is important only when $q_z\sigma$ is small. The rest of the series in (3.2.2.5) represents incoherent scattering. The integral I_0 for $n \geq 1$ can be rewritten in the following manner pointing out the illuminated area $A_0 = \left(2L\right)^2$:

$$I_0 = |a_0|^2 e^{-q_z^2\sigma^2} A_0 \sum_{n=1}^{\infty} \frac{\left(q_z^2\sigma^2\right)^n}{n!} \int_{-\infty}^{\infty}\int_{-\infty}^{\infty} \rho^n e^{jq_x u + jq_y v} du dv \qquad (3.2.2.7)$$

For an isotropically rough surface with correlation length l and Gaussian normalised autocorrelation function, $\rho = \exp\left[-\xi^2/l^2\right]$, the integral (3.2.2.7) can be shown to be:

$$\int_{-\infty}^{\infty}\int_{-\infty}^{\infty} e^{-n\xi^2/l^2 + jq_x u + jq_y v} du dv = \frac{\pi l^2}{n} e^{-\frac{\left(q_x^2+q_y^2\right)l^2}{4n}} \qquad (3.1.2.8)$$

It is clear that different solutions may be obtained for the integral if the normalised surface autocorrelation function is assumed to take some other functional forms. Upon substituting (3.2.2.7) and (3.2.2.8) into the factor $\left\langle E_{qp}^s E_{qp}^{s\,*}\right\rangle$, the scattering coefficient for the incoherent part of the $|a_0|^2$ term has the following expression:

$$\sigma_{qp}^{o\ inc} = \left(|a_0| kl/2\right)^2 e^{-q_z^2\sigma^2} \sum_{n=1}^{\infty} \frac{\left(q_z^2\sigma^2\right)^n}{n!n} e^{-\frac{\left(q_x^2+q_y^2\right)l^2}{4n}} \qquad (3.2.2.9)$$

If the normalised surface autocorrelation is not known, $\sigma_{qp}^{o\ inc}$ can be written as:

$$\sigma_{qp}^{o\ inc} = \frac{k^2|a_0|^2}{4\pi} e^{-q_z^2\sigma^2} \sum_{n=1}^{\infty} \frac{\left(q_z^2\sigma^2\right)^n}{n!} \int_{-\infty}^{\infty}\int_{-\infty}^{\infty} \rho^n e^{jq_x u + jq_y v} du dv \qquad (3.2.2.10)$$

An additional contribution to the total scattering coefficient comes from the slope terms in (3.2.2.4). It can be computed taking into account in the ensemble average $\left\langle E_{qp}^s E_{qp}^{s\,*}\right\rangle$ the integrals of the slope terms in the x- and y-direction. The results of such integrals for a Gaussian normalised autocorrelation function are reported in (Ulaby et al, 1982). Also the expressions of the polarisation-dependent coefficients a_i can be found in the same reference. However, the expressions of the coefficient a_0 for each polarisation are reported below for the two particular cases of backscattering and scattering in the specular direction.
In the backscattering:

HH polarisation:	$a_0 = 2R_h(\theta_i)\cos\theta_i$
VH polarisation:	$a_0 = 0$
VV polarisation:	$a_0 = -2R_v(\theta_i)\cos\theta_i$
HV polarisation:	$a_0 = 0$

Conversely, in the specular direction case:

HH polarisation:	$a_0 = -2R_h(\theta_i)\cos\theta_i$
VH polarisation:	$a_0 = 0$
VV polarisation:	$a_0 = 2R_v(\theta_i)\cos\theta_i$
HV polarisation:	$a_0 = 0$

The quantity $q_{x,y,z}$ are defined in the previous section.

3.3 On the range of validity of the Kirchhoff method and shadowing effects

The basic assumption of the Kirchhoff method is that plane-boundary reflection occurs at every point on the surface. Thus, when statistical surfaces are considered, their horizontal-scale roughness, the correlation length l, must be larger than the electromagnetic wavelength, while their vertical-scale roughness, the standard deviation σ of surface heights, must be small enough so that the average radius of curvature is larger than the electromagnetic wavelength. Mathematically, for stationary isotropic Gaussian surface the above-stated restriction are (Ulaby et al, 1982):

$$kl > 6 \tag{3.3.1}$$

$$\sigma < \frac{l^2}{2.76\lambda} \tag{3.3.2}$$

where k is the wave number and λ is the electromagnetic wavelength. Note that the surface standard deviation should be small relative to the correlation length, but it can be comparable to or even larger than the electromagnetic wavelength. This means that large standard deviations can be tolerated if the correlation length is large enough to preserve an acceptable average radius of curvature. The conditions reported above are for the Kirchhoff approximation. The scattering models described in section 3.2.1 and 3.2.2 require additional approximations reported in the following table:

Validity limits of Kirchhoff Approximation (KA) (Gaussian surface)	
$l^2 > 2.76\sigma\lambda$ and	$kl > 6$
Stationary Phase Aproximation (GO)	Scalar Approximation (PO)
$k\sigma > 2$	$k\sigma < 1$ and $rms_{slope} < 0.25$

Table 3.3.1. Validity of GO and PO for stationary isotropic Gaussian surfaces with standard deviation σ and correlation length l.

3.4 Some concluding remarks on the Kirchhoff method

As was mentioned in the previous paragraph, the Kirchhoff method does neither in itself account for shadowing and nor does it (in the form described here) account for multiple scattering on the surface. Due to the lack of these two effects energy conservation is not satisfied. However, in (Ulaby et al, 1982) this conservation is demonstrating with the inclusion of these two effects.

In the literature, the surface height distribution is in most cases assumed to be Gaussian. The reason for this is, as mentioned previously, that the surface roughness rms height and the autocorrelation function entirely determine the random process, and therefore the bistatic scattering coefficient can be expressed in terms of these two quantities.

The Kirchhoff method has been applied to surfaces described by fractal geometry. As an example we can mention that in (Franceschetti et al, 1999) a fractional Brownian motion model was used for modelling the scattering from natural rough surface. In combination with the Kirchhoff method an analytical solution for the bistatic scattering coefficient was obtained.

4. The small perturbation method

The Small Perturbation Method (SPM) belongs to a large family of perturbation expansion solutions to the wave equation. The approach is based on formulating the scattering as a partial differential equation boundary value problem. The basic idea is to find a solution in terms of plane waves that matches the surface boundary conditions, which state that the tangential component of the field must be continuous across the boundary. The surface fields are expanded in a perturbation series with respect to surface height, e.g., $\mathbf{E} = \mathbf{E}_0 + \mathbf{E}_1 + \ldots$. In the expansion $\mathbf{E}_0$ would be the surface field if the surface was flat. The philosophy behind this approach is that small effective surface currents on a mean surface replace the role of a small-scale roughness. So this method applies to surfaces with small surface height variations and small surface slopes compared with the wavelength but independently of the radius of curvature of the surface. Therefore, the surface needs no longer to be approximated by planes. The small-scale roughness is expanded in a Fourier series and the contribution to the field is therefore analysed in terms of different wavelength components.

Here we will report only the expressions of the bistatic scattering coefficient. A more detailed description of their computation process can be found in (Ulaby et al, 1982).

4.1 A small presentation of the SPM

The zero order solution of the SPM is the same as for a plane interface, while the first order solution gives the incoherent scattered field due to single scattering. For the latter case, the bistatic scattering coefficient for either a horizontally or vertically polarised incident wave is (Ulaby et al, 1982):

$$\sigma_{qp}^{o} = 8\left|k^2\sigma\cos\theta_i\cos\theta_s\alpha_{qp}\right|^2 W\left(k_x + k\sin\theta_i, k_y\right) \tag{4.1.1}$$

where

$$k_x = -k\sin\theta_s\cos\phi_s$$

$$k_y = -k\sin\theta_s\sin\phi_s$$

$$W\left(k_x, k_y\right) = \frac{1}{2\pi}\int_{-\infty}^{\infty}\int_{-\infty}^{\infty}\rho(u,v)e^{-jk_xu-jk_yv}dudv$$

σ and $\rho(u,v)$ are, respectively, the variance of surface heights and the surface correlation coefficient; α_{qp} are coefficients that depend on polarisation, incidence and scattering angle, and on complex relative dielectric constant ε_c of the homogeneous medium below the interface. The detailed expressions of α_{qp} are reported in (Ulaby et al, 1982).

4.2 Some remarks on the region of validity of the SPM

The Small Perturbation Method is applied to surfaces with a surface height standard deviation much less than the incident wavelength (5 percent or less) and an average surface slope comparable to or less than the surface standard deviation times the wave number. For a surface with Gaussian correlation function, such two conditions can be expressed analytically as follows, but they should be viewed only as a guideline for applying the SPM scattering model:

$$k\sigma < 0.3$$

$$\sqrt{2}\,\sigma/l < 0.3$$

The SPM has been compared to more accurate numerical simulations in (Thorsos & Jackson, 1989; 1991) for one-dimensional rough surfaces with a Gaussian roughness spectrum. Under these conditions the authors show that the first-order SPM gives accurate results for $k\sigma << 1$ and $kl \approx 1$. The results also show that for $k\sigma << 1$ and $kl > 6$, the sum of the first three orders of the SPM is required to obtain accurate results.
It has been argued that the SPM does account for multiple scattering up to the order of the perturbative expansion. This means that the first order perturbative solution does not account for multiple scattering but that some multiple scattering effects can be observed in the higher order solutions.

Validity limits of Small Perturbation Method (SPM) (Gaussian surface)		
$k\sigma < 0.3$	and	$rms_{slope} < 0.3$

Table 4.2.1. Validity of SPM for stationary isotropic Gaussian surfaces with standard deviation σ and root mean square slope rms_{slope}.

5. The Integral Equation Method (IEM)

A relatively new method for calculating scattering of electromagnetic waves from rough surfaces is the Integral Equation Method (IEM). The IEM has been used extensively in the microwave region in recent years and it has proved to provide good predictions for a wide range of surface profiles. The method can be viewed as an extension of the Kirchhoff method and the Small Perturbation Method since it has been shown to reproduce results of these two methods in appropriate limits. The IEM is a relatively complicated method in its general form (including multiple scattering) and it is beyond the scope of the present overview to give a full presentation of the method. A more detailed presentation of the IEM can be found in (Fung, 1994).

5.1 On the formulation of the IEM

The starting point of the IEM is the Stratton-Chu integral for the scattered field, equation (3.1.1). The tangential surface fields which enter the Stratton-Chu integral are given in equations (2.2.11) - (2.2.12) and (2.2.15) - (2.2.16). In the Kirchhoff approach, the tangential fields are approximated using the tangent plane approximation, replacing the complete tangential surface fields with the Kirchhoff tangential surface fields of equations (3.2.1) and (3.2.2). It is clear that the Kirchhoff tangential surface fields cannot provide alone a good estimate of the surface fields since the integral form in equations (2.2.11) - (2.2.12) are not accounted for in the Kirchhoff approach. In the IEM, a complementary term is included in equations (3.2.1) and (3.2.2) to correct for this:

$$\hat{\mathbf{n}}\times\mathbf{E}=\left(\hat{\mathbf{n}}\times\mathbf{E}\right)_k+\left(\hat{\mathbf{n}}\times\mathbf{E}\right)_c \tag{5.1.1}$$

$$\hat{\mathbf{n}}\times\mathbf{H}=\left(\hat{\mathbf{n}}\times\mathbf{H}\right)_k+\left(\hat{\mathbf{n}}\times\mathbf{H}\right)_c \tag{5.1.2}$$

In these equations, the first terms on the right hand side are the tangential fields under Kirchhoff approximation and the complementary fields are given by:

$$\left(\hat{\mathbf{n}}\times\mathbf{E}\right)_c=\hat{\mathbf{n}}\times\left(\mathbf{E}^i-\mathbf{E}^r\right)-\frac{2}{4\pi}\hat{\mathbf{n}}\times\int\varepsilon ds' \tag{5.1.3}$$

$$\left(\hat{\mathbf{n}}\times\mathbf{H}\right)_c=\hat{\mathbf{n}}\times\left(\mathbf{H}^i-\mathbf{H}^r\right)+\frac{2}{4\pi}\hat{\mathbf{n}}\times\int\mathcal{H} ds' \tag{5.1.4}$$

$\mathbf{E}^r$ and $\mathbf{H}^r$ being the reflected electric and magnetic fields propagating along the reflected direction. To use (5.1.1) and (5.1.2) for estimating the tangential field, both the Kirchhoff field and the complementary field need to be expressed in terms of the incident field components and the surface reflectivity properties. Using the local coordinate system defined by the unit vectors $\hat{\mathbf{t}}$, $\hat{\mathbf{d}}$, $\hat{\mathbf{k}}_i$ (for their expressions refer to (Fung, 1994)), the incident electric and magnetic field can be expressed into locally horizontally and vertically polarised components. Accordingly, after some manipulations (see (Fung, 1994) for more details), the Kirchhoff and complementary tangential fields can be rewritten as:

$$\left(\hat{\mathbf{n}}\times\mathbf{E}\right)_k=\hat{\mathbf{n}}\times\left[\left(1+R_h\right)\left(\hat{\mathbf{p}}\cdot\hat{\mathbf{t}}\right)\hat{\mathbf{t}}+\left(1-R_v\right)\left(\hat{\mathbf{p}}\cdot\hat{\mathbf{d}}\right)\hat{\mathbf{d}}\right]E^i \tag{5.1.5}$$

$$\eta\left(\hat{\mathbf{n}}\times\mathbf{H}\right)_k=\hat{\mathbf{n}}\times\left[\left(1-R_h\right)\left(\hat{\mathbf{p}}\cdot\hat{\mathbf{t}}\right)\hat{\mathbf{d}}+\left(1+R_v\right)\left(\hat{\mathbf{p}}\cdot\hat{\mathbf{d}}\right)\hat{\mathbf{t}}\right]E^i \tag{5.1.6}$$

$$\begin{aligned}\left(\hat{\mathbf{n}}\times\mathbf{E}\right)_c=&-\frac{1}{4\pi}\left(\hat{\mathbf{n}}\times\hat{\mathbf{t}}\right)\left\{\hat{\mathbf{n}}\times\hat{\mathbf{t}}\cdot\hat{\mathbf{n}}\times\int\left[\left(1+R_h\right)\varepsilon+\left(1-R_h\right)\varepsilon_t\right]ds'\right\}\\&-\frac{1}{4\pi}\hat{\mathbf{t}}\left\{\hat{\mathbf{t}}\cdot\hat{\mathbf{n}}\times\int\left[\left(1-R_v\right)\varepsilon+\left(1+R_v\right)\varepsilon_t\right]ds'\right\}\end{aligned} \tag{5.1.7}$$

$$\begin{aligned}\left(\hat{\mathbf{n}}\times\mathbf{H}\right)_c=&\frac{1}{4\pi}\left(\hat{\mathbf{n}}\times\hat{\mathbf{t}}\right)\left\{\hat{\mathbf{n}}\times\hat{\mathbf{t}}\cdot\hat{\mathbf{n}}\times\int\left[\left(1+R_v\right)\mathcal{H}+\left(1-R_v\right)\mathcal{H}_t\right]ds'\right\}\\&+\frac{1}{4\pi}\hat{\mathbf{t}}\left\{\hat{\mathbf{t}}\cdot\hat{\mathbf{n}}\times\int\left[\left(1-R_h\right)\mathcal{H}+\left(1+R_h\right)\mathcal{H}_t\right]ds'\right\}\end{aligned} \tag{5.1.8}$$

It can be noted that, while (5.1.5) and (5.1.6) are expressed in terms of known quantities, that is the incident electric or magnetic fields, the local Fresnel reflection coefficient and the local incident angle, (5.1.7) and (5.1.8) are integral equations. In order to obtain estimates of (5.1.7) and (5.1.8), IEM substitutes the unknown expressions of the tangential fields in the right-hand side of (5.1.7) and (5.1.8), that is the $\left(\hat{\mathbf{n}}'\times\mathbf{E}'\right)$ and $\left(\hat{\mathbf{n}}'\times\mathbf{H}'\right)$ terms which appear in ε, ε_t, $\mathcal{H}$ and $\mathcal{H}_t$, with the Kirchhoff tangential fields, $\left(\hat{\mathbf{n}}'\times\mathbf{E}'\right)_k$ and $\left(\hat{\mathbf{n}}'\times\mathbf{H}'\right)_k$, respectively. This is the fundamental approximation adopted by IEM model. However, even with this simplification the obtained integral expressions remain too complex for practical use.

Much simpler approximate expressions of the tangential Kirchhoff and complementary fields can be obtained differentiating them for each linear incident and scattered polarisation. The resulting approximated equations (electric and magnetic surface field equations for horizontal, vertical and cross polarisation) can be found in (Fung, 1994).
Then, the simplified tangential surface fields can be inserted in the Stratton-Chu integral. The far field scattered from the rough surface can be expressed as a combination of the Kirchhoff and the complementary term:

$$E_{qp}^{s} = E_{qp}^{k} + E_{qp}^{c} \tag{5.1.9}$$

where

$$E_{qp}^{k} = CE_0 \int f_{qp}\, e^{j(\mathbf{k}_s - \mathbf{k}_i)\cdot \mathbf{r}} dxdy \tag{5.1.10}$$

and

$$E_{qp}^{c} = \frac{C}{8\pi^2} \int \tilde{F}_{qp} e^{j\mathbf{k}_s \cdot \mathbf{r}} dxdy = \frac{CE_0}{8\pi^2} \int F_{qp}\, e^{ju(x-x')+jv(y-y')+j\mathbf{k}_s\cdot\mathbf{r}-j\mathbf{k}_i\cdot\mathbf{r}'} dxdydudvdx'dy' \tag{5.1.11}$$

The quantities f_{qp} and $\tilde{F}_{qp}$, respectively the Kirchhoff and complementary field coefficients, that appear in the above equations are defined as follows:

$$f_{qp} = \left[\hat{\mathbf{q}} \times \mathbf{k}_s \cdot \left(\hat{\mathbf{n}} \times \mathbf{E}_p\right)_k + \eta \hat{\mathbf{q}} \cdot \left(\hat{\mathbf{n}} \times \mathbf{H}_p\right)_k \right] D_1 \Big/ E_i \tag{5.1.12}$$

$$\tilde{F}_{qp} = 8\pi^2 \left[\hat{\mathbf{q}} \times \mathbf{k}_s \cdot \left(\hat{\mathbf{n}} \times \mathbf{E}_p\right)_c + \eta \hat{\mathbf{q}} \cdot \left(\hat{\mathbf{n}} \times \mathbf{H}_p\right)_c \right] D_1 \tag{5.1.13}$$

where $D_1 = \sqrt{1 + Z_x^2 + Z_y^2}$ and E^i is the complex amplitude of the incident electric field.
In general, both f_{qp} and $\tilde{F}_{qp}$ are dimensionless, complicated expressions and depended on spatial variables. Therefore several approximations are made to make these functions independent of spatial variables (Fung, 1994).
In particular, the f_{qp} coefficients depend on the Fresnel reflection coefficients, and hence on the local angle, and on the slope terms, Z_x and Z_y. The first dependency is removed by approximating the local incidence angle in the Fresnel reflection coefficients by the *incident angle*, θ_i, for surface with *small scale roughness* and by the *specular angle*, θ_{sp}, $\cos\theta_{sp} = -\hat{\mathbf{n}} \times \hat{\mathbf{k}}_i$, for surface with *large scale roughness*. The rule that defines the bound between the two regions is reported here assuming a Gaussian autocorrelation function:

$$R_{v,h} = \begin{cases} R_{v,h}(\theta_i) & k^2\sigma l < 1.2\sqrt{\varepsilon_r} \\ R_{v,h}(\theta_{spec}) & kl > 5 \end{cases} \tag{5.1.14}$$

In order to remove the dependence on the slope terms, the integral (5.1.10) is solved by parts and the edge terms were discarded.
To obtain the expressions of the complementary coefficients $\tilde{F}_{qp}$, the computation is rather lengthy and complicated. When the equations (2.2.15) - (2.2.18) are substituted in the

approximated expressions of tangential complementary fields, the spectral representations of Green's function and of its gradient are introduced, *assuming however the same Green's functions for both the medium*:

$$G = -\frac{1}{2\pi}\int \frac{j}{q} e^{ju(x-x')+jv(y-y')-jq|z-z'|} du dv \tag{5.1.15}$$

$$\nabla' G = -\frac{1}{2\pi}\int \frac{\mathbf{g}}{q} e^{ju(x-x')+jv(y-y')-jq|z-z'|} du dv \tag{5.1.16}$$

$\mathbf{g} = u\hat{\mathbf{x}} + v\hat{\mathbf{y}} \mp q\hat{\mathbf{z}}$ and $q = \sqrt{k^2 - u^2 - v^2}$ are the propagation vector and its z-component of the generic plane wave that appears in the plane waves expansion of the field, whereas z and z' are the random variables representing the surface heights at different locations on the random surface. In (Fung, 1994), the $|z - z'|$ terms and the term with the $\mp$ are dropped in the equations (5.1.15) and (5.1.16) in order to simplify the calculation. However, in an improved version of the IEM (see (Chen et al, 2000)) these terms are kept in the analysis. In addition, as was the case for the Kirchhoff coefficients, f_{qp}, the dependence through the slope terms is removed by integrating by parts and discarding the edge terms. Instead, as regard the Fresnel reflection coefficients, the local angle is always replaced by incident angle (Fung, 1994; Wu et al, 2001).
Moreover, it is important to underline that the tangential and normal field components that appear in the expressions of the $\tilde{F}_{qp}$ coefficients through equations (2.2.15) - (2.2.18) can be approximated by the tangential Kirchhoff fields. The complimentary field coefficients F_{qp} that appear in the right term of the equation (5.1.11) are obtained from the definition of the $\tilde{F}_{qp}$ after the Green's function and its gradient are replaced by the *simplified* spectral representation, above mentioned, and after the phase factor of the Green function and u, v, x', y' integrations are factored out. The expressions of such coefficients together with the expressions of the Kirchhoff ones are reported in (Brogioni et al, 2010).
Once the field coefficients, f_{qp} and F_{qp}, are made independent of spatial variables, it is possible to provide the expression of the incoherent scattered power:

$$\begin{aligned}\left\langle \left|E_{qp}^{s}\right|^{2}\right\rangle - \left|\left\langle E_{qp}^{s}\right\rangle\right|^{2} &= \left\langle E_{qp}^{s} E_{qp}^{s\,*}\right\rangle - \left\langle E_{qp}^{s}\right\rangle \left\langle E_{qp}^{s}\right\rangle^{*} = \\ &= \left\langle E_{qp}^{k} E_{qp}^{k\,*}\right\rangle - \left\langle E_{qp}^{k}\right\rangle \left\langle E_{qp}^{k}\right\rangle^{*} + \\ &+ 2\,\mathrm{Re}\left[\left\langle E_{qp}^{c} E_{qp}^{k\,*}\right\rangle - \left\langle E_{qp}^{c}\right\rangle \left\langle E_{qp}^{k}\right\rangle^{*}\right] + \left\langle E_{qp}^{c} E_{qp}^{c\,*}\right\rangle - \left\langle E_{qp}^{c}\right\rangle \left\langle E_{qp}^{c}\right\rangle^{*}\end{aligned} \tag{5.1.17}$$

and from this the bistatic scattering coefficient:

$$\sigma_{qp}^{o} = \sigma_{qp}^{k} + \sigma_{qp}^{kc} + \sigma_{qp}^{c} \tag{5.1.18}$$

From the above expression it follows that the scattering coefficient is given by the sum of three terms: the Kirchhoff, the complementary and the cross term. The first is originated by Kirchhoff fields, the second by the interaction between Kirchhoff and complementary fields, whereas the last is due only to complementary fields.

To carry out the average operation an assumption about the type of surface height distribution is necessary. In order to simplify the calculation of the incoherent power terms the rough surface is assumed characterised by a Gaussian height distribution. Accordingly, the terms in (5.1.18) assume the following expressions, reported in (Fung, 1994):

$$\sigma_{qp}^{k} = \frac{k^2}{4\pi}\left| f_{qp}\right|^2 e^{-\sigma^2 (k_{sz}+k_z)^2} \int\left\{\exp\left[\sigma^2 (k_{sz}+k_z)^2 \rho(\xi,\zeta)\right]-1\right\}\exp\left[j(k_{sx}-k_x)\xi + j\left(k_{sy}-k_y\right)\zeta\right]d\xi\, d\zeta \tag{5.1.19}$$

$$\begin{aligned}\sigma_{qp}^{kc} = \frac{k^2}{16\pi^3}\mathrm{Re}\Big\{ &\int\left(F_{qp} f_{qp}^{*}\right)\exp\left[-\sigma^2\left(k_{sz}^2+k_z^2+k_z k_{sz}\right)\right] \\ &\iint\exp\left[-\sigma^2 k_z k_{sz}\rho(\xi-\xi',\zeta-\zeta')\right]\Big\{\exp\left[\sigma^2 k_{sz}(k_{sz}+k_z)\rho(\xi,\zeta)+\right. \\ &\left.\sigma^2 k_z (k_{sz}+k_z)\rho(\xi',\zeta')\right]-1\Big\}\exp\left[jk_{sz}\xi + jk_{sy}\zeta + j\nu(\zeta-\zeta')+ju(\xi-\xi')\right] \\ &\exp\left[-jk_x\xi' - jk_y\zeta'\right]d\xi\, d\zeta\, d\xi'\, d\zeta'\, du\, dv\Big\}\end{aligned} \tag{5.1.20}$$

$$\begin{aligned}\sigma_{qp}^{c} = \left|\frac{k}{16\pi^{2.5}}\right|^2 \Big\{ &\iint\left(F_{qp}F_{qp}^{*}\right)\exp\left[-\sigma^2\left(k_{sz}^2+k_z^2\right)\right] \\ &\exp\left[-\sigma^2 k_z k_{sz}\rho(\tau+\xi-\xi',\kappa+\zeta-\zeta')\right]\Big(\exp\left[\sigma^2 k_{sz}^2\rho(\xi,\zeta)\right] \\ &\exp\left\{\sigma^2 k_z\rho(\xi',\zeta')+\sigma^2 k_z k_{sz}\left[\rho(\xi+\tau,\zeta+\kappa)+\rho(\xi'-\tau,\zeta'-\kappa)\right]\right\}-1\Big) \\ &\exp\left\{j\left[(k_{sx}+u)\xi+\left(k_{sy}+v\right)\zeta-(k_x+u)\xi'-\left(k_y+v\right)\zeta'\right]\right\} \\ &\exp\left\{j\left[(u-u')\tau+(v-v')\kappa\right]\right\}d\xi\, d\zeta\, d\xi'\, d\zeta'\, d\tau\, d\kappa\, du\, dv\, du'\, dv'\end{aligned} \tag{5.1.21}$$

The above expressions consist of multiple integrals which are too complex and hence not practical to use. In order to evaluate these integrals, the model is approximated in two different forms depending upon whether the surface height is moderate or large in terms of the incident wavelength ($k\sigma$). The first case is referred to as low frequency approximation, whilst the other is referred to as high frequency approximation. An indicative threshold value of $k\sigma < 2$ is reported in (Fung, 1994). The detailed expressions of σ_{qp}^{k}, σ_{qp}^{kc}, σ_{qp}^{c} valid separately when $k\sigma < 2$ and for large $k\sigma$ are given in (Fung, 1994) and are not reported here. For both the approximations, in the expression of the bistatic scattering coefficient two types of terms can be distinguished: one representing single-scattering and the other representing multiple-scattering. The latter may be viewed as a correction to the single term for both the high- and the low-frequency regions. This division is important to identify weather single or multiple scattering is significant for applications. For completeness we report here the total single scattering coefficient obtained by selecting the single scattering contributions in the expressions of σ_{qp}^{k}, σ_{qp}^{kc}, σ_{qp}^{c} valid when $k\sigma < 2$ (for the detailed explanation refer to (Fung, 1994)):

$$\sigma^{o}{}_{qp}=\frac{k^2}{2}e^{-\sigma^2(k_z^2+k_{sz}^2)}\sum_{n=1}^{\infty}\frac{\sigma^{2n}}{n!}\left|(k_z+k_{sz})^n\ f_{qp}\ e^{-\sigma^2 k_z k_{sz}}+\frac{(k_{sz})^n F_{qp}(-k_x,-k_y)+(k_z)^n F_{qp}(-k_{sx},-k_{sy})}{2}\right|^2 W^{(n)}(k_{sx}-k_x,k_{sy}-k_y)$$

(5.1.22)

6. Conclusions

We have presented the results from a literature search of models for scattering of electromagnetic waves from random rough surfaces. In particular we have focused on the calculation of the *bistatic scattering coefficient* in three different classes of methods: the *Kirchhoff Approximation*, the *Method of Small Perturbation* and the *Integral Equation Method.* Of these, the first two, are amongst the early approaches which however are still much used. The latter is an example of more recent approaches which have been developed as an attempt to extend the validity of the former methods.

7. References

Balanis C. A. (1989). *Advanced Engineering Electromagnetism*, John Wiley and Sons Inc.

Beckman P., Spizzichino A. (1963). *The Scattering of Electromagnetic Waves from Rough Surfaces*, Pergamon press, Oxford [etc.].

Brogioni M., Macelloni G., Paloscia S., Pampaloni P., Pettinato S., Pierdicca N, Ticconi F. (2010). "Sensitivity of Bistatic Scattering to Soil Moisture and Surface Roughness of Bare Soils", *International Journal of Remote Sensing*, vol. 31, no. 15, pp. 4227-4255.

Chen K.S., Wu T.D., Tsay M.K., Fung A.K. (2000). "A Note on the Multiple Scattering in an IEM Model", *IEEE Transactions on Geoscience and Remote Sensing*, vol. 38, no. 1, pp. 249-256.

Franceschetti G., Iodice A., Migliaccio M., Riccio D. (1999), "Scattering from Natural Rough Surfaces Modeled by Fractional Brownian Motion Two Dimensional Processes", *Proc. SPIE, Remote Sensing of the Ocean and Sea Ice*, vol. 4172, pp. 169-176.

Fung A. K. (1994). *Microwave Scattering and Emission Models and Their Application*, Boston, MA, Arthech House.

Poggio A. J., Miller E. K. (1973). "Integral Equation solution of Three Dimensional Scattering Problems", *Computed Techniques for Electromagnetics*, Pergamon, New York.

Saillard M., Sentenac A. (2001). "Rigorous Solutions for Electromagnetic Scattering from Rough Surfaces", *Waves in Random Media*, vol. 11, pp. 103-137.

Tsang L., Kong J. A., Ding K.-H. (2000). *Electromagnetic Waves, theories and applications*, Jon Wiley and Sons, New York.

Tsang L., Kong J. A. (2001). *Scattering of Electromagnetic Waves, Advanced Topics*, Jon Wiley and Sons, New York.

Thorsos E. I., Jackson D. (1989). "The Validity of the Perturbation Approximation for Rough Surface Scattering using a Gaussian Roughness Spectrum", *Journal Acoust. Soc. Am.*, vol. 86 (1), pp. 261-277.

Thorsos E. I., Jackson D. (1991). "Studies of Scattering Theory using Numerical Methods", *Waves in Random Media*, vol. 3, pp. 165-190.

Ulaby F. T., Moore R. K., Fung A. K. (1982). *Microwave remote sensing: active and passive*, Addison-Wesley Publishing Company, London [etc.].

Wu T-D., Chen K. S., Shi J., Fung A. K. (2001). "A Transition Model for the Reflection Coefficient in Surface Scattering", *IEEE Transactions on Geoscience and Remote Sensing*, vol. 39, no. 9, pp. 2040-2050.

11

Electromagnetic Wave Propagation in Circular Tunnels

Osama M. Abo-Seida
Mathematics Department, Faculty of Science, Kafr El-Sheikh University, Egypt

1. Introduction

The problem of guided wave propagation in a wave guide with imperfectly reflecting boundaries arises in several applications, such as propagation in mine tunnels and in screened surface wave guides. In recent years, many studies have been carried out on the propagation characteristic of radio waves in tunnels in UHF band [1-2]. From the theoretical point of view, tunnels can be regarded as hollow waveguides surrounded by a lossy dielectric medium, such as concrete, ground, and so on. The problem of radio communication in tunnels has found solutions using leaky transmission lines as supports for propagation of transverse electromagnetic modes [3-4]. These modes are characterized by the fact that there is no cutoff frequency, and by an attenuation which increases with increasing frequency. However, when the frequency is high enough, natural propagation modes, which are transverse electric or transverse magnetic, can appear and interfere with the transmission line supported transverse electromagnetic modes [5].

Natural propagation can be helpful to solve some specific problems, such as radio communications in mines with rooms and pillars which cannot be solved easily with the help of transmission lines. Natural propagation modes are also useful for short range communications, for example, in some road tunnels. For distances shorter than 200m, these modes can be more suitable the transmission line supported modes [6].

In order to maximize the performance attainable feeding in the interior of tunnels with antenna systems, it is important to estimate the value and the attenuation of the electromagnetic strength inside the tunnel. Investigations concerning radiowave propagation in railways tunnels have been performed [7] and the results confirm the existence of a waveguide effect strongly related to the antenna positions. Recently, Abo-Seida et al. [8] and Abo-Seida [9] studied the electromagnetic field due to vertical magnetic dipole buried in stratified media. Also, Abo-Seida [10] computed the attenuation below and above the cutoff frequency in a rectangular tunnel.

It might be thought that waveguides with circular cross sections would be preferred to guides with rectangular cross sections, just as circular pipes are commonly used for carrying water and fluids in preference to rectangular pipes. However, circular waveguides have the disadvantage that there is only a very narrow range between the cutoff wavelength of the dominant mode and the cutoff wavelength of the next higher mode. As with rectangular guides, the modes may be classified as transverse electric (TE) or transverse magnetic (TM), according to whether it is the electric or magnetic lines of force that lie in planes perpendicular to the axis of the guide. The different modes are designated by a double subscript system analogous to that for rectangular guides.

2. Geometrical structure and basic equations

The concrete tunnel is practically of finite thickness. At low frequencies, below cutoff frequencies, the skin depth may exceed this thickness and thus the field penetrates to the outside. However, our aim is to check the possibilities of radio communication in tunnels and to replace the tunnel by a circular waveguide. It is possible to make this replacement in practical problems involving conducting walls.

A uniform waveguide of circular cross section is most conveniently described by polar coordinate system (r, ϕ, z), and divide the possible solutions for circular guides into transverse magnetic and transverse electric waves. For the TM waves H_z is identically zero and the wave equation for E_z is used. Also, TE waves E_z is identically zero and H_z is used. The modes are further labeled by a two-dimensional order number (m, n).

For a circular waveguide with a radius r the critical frequencies are given by

$$f_{cmn} = \begin{cases} \dfrac{CY_{mn}}{2r} \,, & TM_{mn} \text{ modes} \\[2ex] \dfrac{CY'_{mn}}{2r} \,, & TE_{mn} \text{ modes} \end{cases} \tag{1}$$

where $m \geq 0$, Y_{mn} and Y'_{mn} are the nth zero of Bessel function J_m and of its derivative, respectively, and $C = 3 \times 10^8 \, ms^{-1}$ is the speed of light.

The cutoff wavelengths are given for a circular waveguide by [11]

$$\lambda_{cmn} = \begin{cases} \dfrac{2\pi}{Y_{mn}} r \,, & TM_{mn} \text{ modes} \\[2ex] \dfrac{2\pi}{Y'_{mn}} r \,, & TE_{mn} \text{ modes} \end{cases} \tag{2}$$

where m, n are equal to $1,2,3,........$ for the TM_{mn} modes, and equal to $0,1,2,3,........$ for the TE_{mn} modes.

Below the lowest cutoff frequency, propagation is not possible. The attenuation α is independent of the electrical properties of the wall and have the form equation

$$\alpha = \frac{2\pi}{\lambda_c} \sqrt{1 - \left(\frac{\lambda_c}{\lambda}\right)^2} = \frac{2 \times 3.14 \times 8.69}{\lambda_c} \sqrt{1 - \left(\frac{\lambda_c}{\lambda}\right)^2}$$

$$= \frac{54.6}{\lambda_c} \sqrt{1 - \left(\frac{\lambda_c}{\lambda}\right)^2} \text{ dB/ unit length} \tag{3}$$

where λ_c is the longest cutoff wavelength of the waveguide and the value 8.69 is the decibels of attenuation per unit length.

Above its cutoff frequency, the attenuation α for the TM_{mn} modes is given by [11]

$$\alpha = 8.69 \frac{R}{\xi r} \frac{1}{\sqrt{1-(\frac{\lambda}{\lambda_{mn}})^2}} \text{ dB/m} \tag{4}$$

and for the TE_{mn} modes, this attenuation α is given by

$$\alpha = 8.69 \frac{R}{\xi r} \left[\frac{m^2}{Y'^2_{mn} - m^2} + (\frac{\lambda}{\lambda_{mn}})^2 \right] \frac{1}{\sqrt{1-(\frac{\lambda}{\lambda_{mn}})^2}} \text{ dB/m} \tag{5}$$

where ξ is the intrinsic impedance of the propagation medium and equal to $(\mu/\varepsilon)^{\frac{1}{2}}$, $R = 10.88 \times 10^{-3} \sqrt{(10^7/\sigma)(1/\lambda)}$ ohms, σ is the conductivity of the guide walls in mho/m, μ is the permeability of the propagation medium in henry/m, ε is the permittivity of the propagation medium in farad/m.

3. Determination of the attenuation constant

We consider the circular -waveguide model of the tunnel. In practice, the concrete tunnel wall is of finite thickness. This tunnel is considered to have a radius $r = 4.08$ m. Its conductivity is taken as 10^{-1} and 10^{-2} mho/m as in [10].
The cutoff frequency is 20 MHz corresponding to approximately a tunnel shaped as a circular cylinder, inside it a wave is propagating in the
TE_{11} mode. Taking account of the conductivity of the walls, the attenuation constants of the TE_{11} and TE_{01} modes have been determined and the results plotted in Figs (1) and (2), respectively, for both values of σ and for frequencies in the range 20-300 MHz.
As in the case of TE_{mn} modes, the attenuation of the electromagnetic waves is calculated for both the TM_{11} and TM_{01} modes. The obtained results are plotted in Figs (3) and (4), respectively, as a function of the frequency.

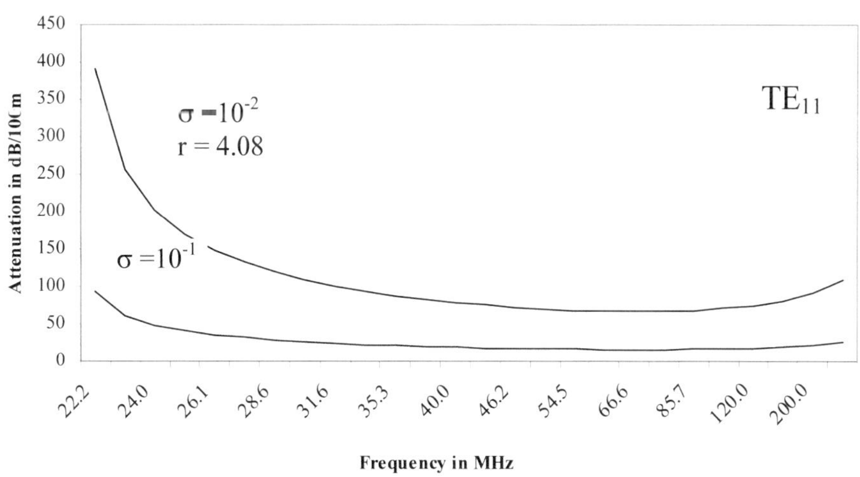

Fig. 1. Frequency and attenuation in a circular tunnel

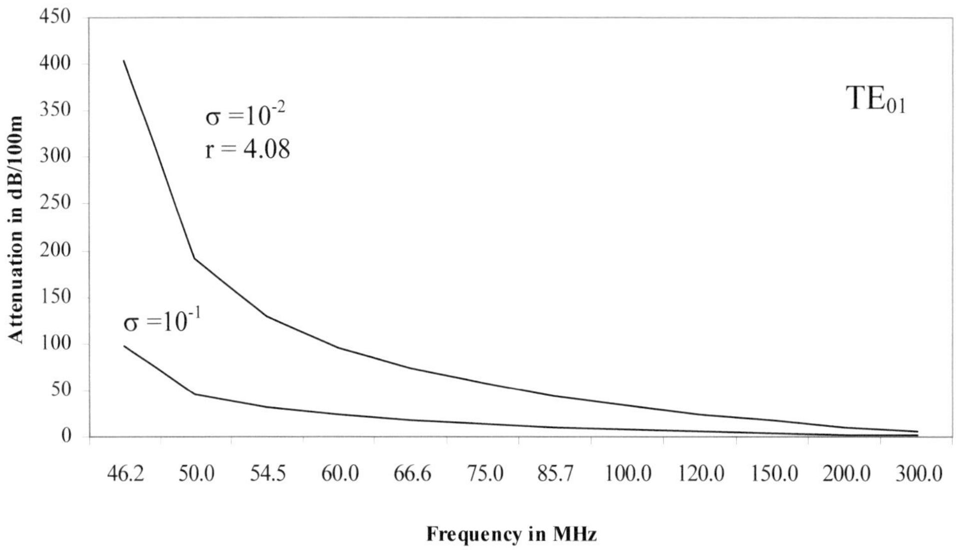

Fig. 2. Frequency and attenuation in a circular tunnel

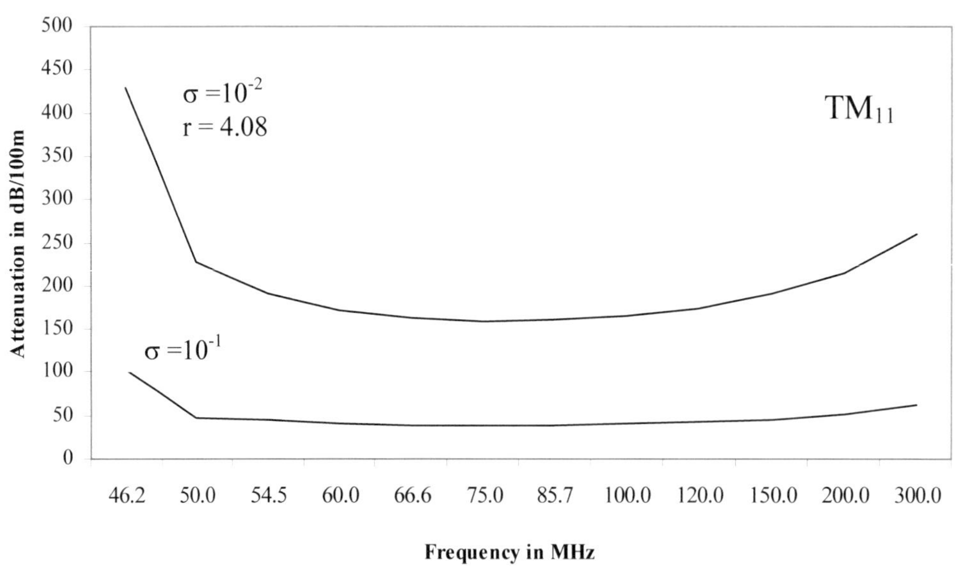

Fig. 3. Frequency and attenuation in a circular tunnel

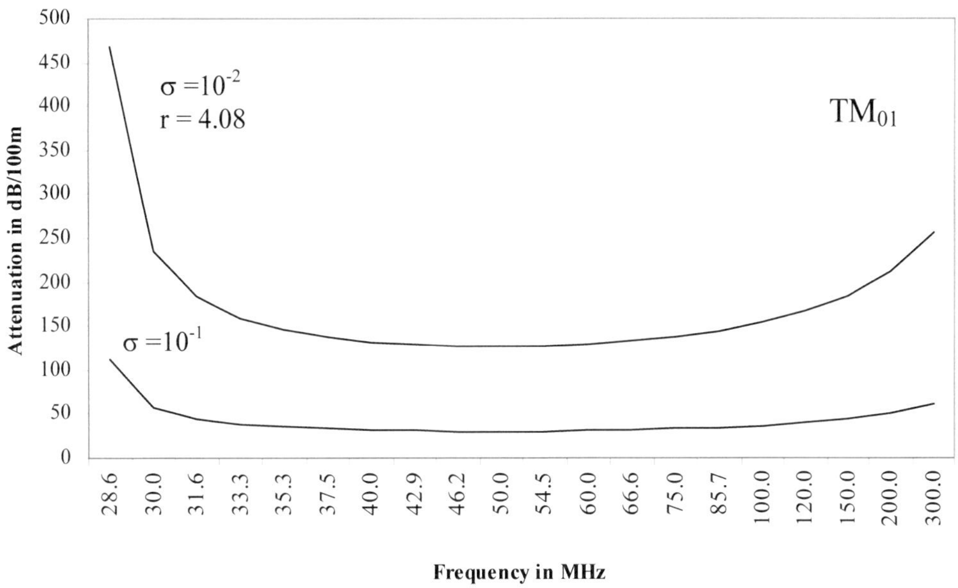

Fig. 4. Frequency and attenuation in a circular tunnel

4. Conclusion

The propagation of electromagnetic waves in a circular tunnel have been performed and the results confirm the existence of a waveguide effect strongly related to the antenna positions. The cutoff frequency was calculated and three different types of frequency ranges were characterized. The numerical results presented here indicate that the different values of m and n modes.

This work enable us to distinguish three different ranges of frequency, characterized by three different propagation mechanisms.

5. References

Emslie, A.; Lagace, R. & Strong, P.(1975). Theory of the propagation of UHF radio waves in cool mine tunnels, *IEEE Trans. Antennas Propagat.* AP-23, pp. 192-205.

Delogne, P. (1982). *Leaky feeders and subsurface radio communication*, Peter Peregrinus, Stevegage, U.K.

Delogne,P. & Safak, M. (1975). Electromagnetic Theory of the leaky, Coaxial Cable, *The Radio and Electronic Engineer*, vol. 45, pp. 233-240.

Deryck, L. (1975). Control of mode conversions on bifilar line in tunnels, *The Radio and Electronic Engineer,* vol. 45, pp. 241-247.

Deryck, L. (1978). Natural propagation of electromagnetic waves in tunnels, *IEEE Transactions on Vehicular Technology*, vol. VT-27, pp. 145-150.

Damoso, E. & Padova, S. (1976). Propagation and radiation of VHF radio signals in motorway tunnels, *IEEE Trans. on Vehicular Technology*, vol. VT-25, pp. 39-45.

Mandich, D. & Carpine, E. An experimental study of wave-guiding effect on radiowave propagation in arched railways tunnels at 450 and 900 MHz , *Proceedings of the International Conference on Electromagnetic in Advanced Applications (ICEAA 95)*, pp. 69-72, Sept. 12-15, Politecnico di Torino, Italy, 1995.

Abo-Seida, O., Bishay, S. & El-Morabie, K. (2006). *IEEE Trans. Geosci. Remote Sensing* 44 (8) pp. 2135-2142.

Abo-Seida, O. (2006). *Journal of Electromagnetic Waves and Applications,* 20 (6), pp.707-715.

Abo-Seida, O. (2002). Propagation of electromagnetic waves in a rectangular tunnel , *ACES Journal*, vol. 17, pp. 170-175.

Marcuvitz, N. (1951). *Waveguide Handbook* , Mc Graw-Hill, New York and London, pp. 55-72.

Part 4

Analysis and Applications of Periodic Structures and Waveguide Components

12

Propagation of Electromagnetic Waves in Thin Dielectric and Metallic Films

Luc Lévesque
Royal Military College of Canada
Canada

1. Introduction

Matrix formalism is a very systematic method to find the reflectance or transmittance in a stratified medium consisting of a pile of thin homogeneous films. Fitting experimental values of reflectance curve to expressions obtained from the matrix formalism method is an efficient method to estimate the refractive index (n) of a dielectric and/or the real and imaginary parts of a metal permittivity (ε). In the next section, the method of matrix formalism is briefly reviewed with some examples to show how it can be applied in curve fitting to determine refractive indices or the metal permittivity. Applications to more complex structures such as planar waveguides and periodic grating are presented in sections 2 and 3, respectively.

1.1 Matrix formalism for the transverse electric and magnetic waves in stratified thin films

Maxwell equations will be applied at each interface between two homogeneous media to find the characteristic matrix defining a thin film. Let us consider figure 1 for a transverse electric (TE) wave with the E-field vector perpendicular to the plane of incidence for one thin homogeneous film

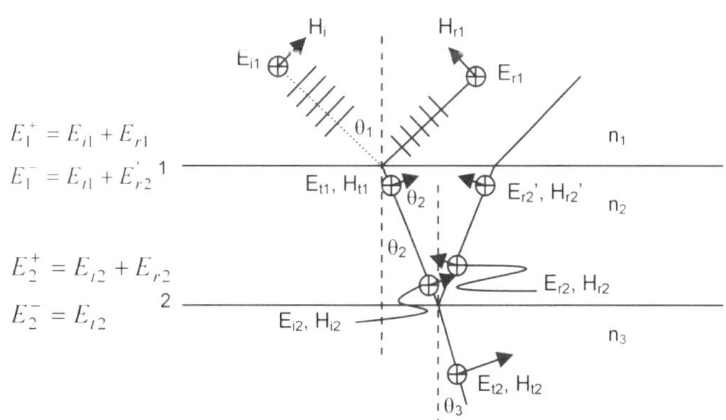

$$H_{1x}^+ = \sqrt{\frac{\varepsilon_o}{\mu_o}}\, n_1 (E_{i1} - E_{r1}) \cos\theta_1$$

$$H_{1x}^- = \sqrt{\frac{\varepsilon_o}{\mu_o}}\, n_2 (E_{t1} - E_{r2}') \cos\theta_2$$

$$H_{2x}^+ = \sqrt{\frac{\varepsilon_o}{\mu_o}}\, n_2 (E_{i2} - E_{r2}) \cos\theta_2$$

$$H_{2x}^- = \sqrt{\frac{\varepsilon_o}{\mu_o}}\, n_3 E_{t1} \cos\theta_3$$

Fig. 1. Electric field (E) and magnetic field (H) in each medium of refractive index n_1, n_2 and n_3.

In figure 1, the H-field is related to the E-field using:

$$H_{i,r,t} = \sqrt{\frac{\varepsilon_o}{\mu_o}} n_{i,r,t} E_{i,r,t} \tag{1}$$

where ε_o and μ_o are referred to as the electric permittivity and the magnetic permeability, respectively. Letters i , r and t stand for incident, reflected and transmitted rays, respectively and the homogeneous medium is identified using numbers 1, 2 or 3.
As both the E and H fields are continuous at boundary 1, one may write E_1 and H_1 as:

$$E_1 = E_1^- = E_{t1} + E'_{r2} \tag{2}$$

$$H_{1x} = H_{1x}^- = n_2 \sqrt{\frac{\varepsilon_o}{\mu_o}} (E_{t1} - E'_{r2}) \cos\theta_2 = Y_2 (E_{t1} - E'_{r2}) \tag{3}$$

where
$$Y_2 = n_2 \sqrt{\frac{\varepsilon_o}{\mu_o}} \cos\theta_2 \tag{4}$$

The system of equations (2) and (3) can be written under the matrix form as:

$$\begin{bmatrix} E_1 \\ H_{1x} \end{bmatrix} = \begin{bmatrix} 1 & 1 \\ Y_2 & -Y_2 \end{bmatrix} \begin{bmatrix} E_{t1} \\ E'_{r2} \end{bmatrix} \tag{5}$$

At interface 2, we merely write

$$E_2 = E_2^+ = E_{i2} + E_{r2} \tag{6}$$

By making use of the E-field amplitude phase shift, it can be shown that E_{i2} and E'_{r2} can be expressed as:

$$E_{i2} = E_{t1} e^{-jk_2 h_2} \tag{7}$$

and

$$E'_{r2} = E_{r2} e^{-jk_2 h_2} \tag{8}$$

respectively, with

$$h_2 = d_2 \cos\theta_2 \tag{9}$$

where d_2 is the thickness of the homogenous thin film and θ_{i2} is the angle defined as shown in figure 1. k_2 is the wave-vector in the thin homogeneous film (medium 2) , which is given as

$$k_2 = \frac{2\pi}{\lambda} n_2 \tag{10}$$

where λ is the wavelength of the monochromatic incident light when propagating in a vacuum.

Equations (7) and (8) are used to express the tangential component of the H-field vector at interface 2 as:

$$H_{2x} = Y_2(E_{t1}e^{-jk_2h_2} - E'_{r2}e^{jk_2h_2}) \tag{11}$$

Using Equations (7) and (8), Equations (6) and (11) are expressed under the matrix form and by matrix inversion one can show that:

$$\begin{bmatrix} E_{t1} \\ E'_{r2} \end{bmatrix} = \begin{bmatrix} e^{jk_2h_2}/2 & e^{jk_2h_2}/2Y_2 \\ e^{-jk_2h_2}/2 & -e^{-jk_2h_2}/2Y_2 \end{bmatrix} \begin{bmatrix} E_2 \\ H_{2x} \end{bmatrix} \tag{12}$$

Lastly, substituting Equation (12) into Equation (5) the E and H field components at interface 1 are related to those at interface 2 by:

$$\begin{bmatrix} E_1 \\ H_{1x} \end{bmatrix} = \begin{bmatrix} \cos(k_2h_2) & j\sin(k_2h_2)/Y_2 \\ jY_2\sin(k_2h_2) & \cos(k_2h_2) \end{bmatrix} \begin{bmatrix} E_2 \\ H_{2x} \end{bmatrix} = M_2 \begin{bmatrix} E_2 \\ H_{2x} \end{bmatrix} \tag{13}$$

The 2x2 matrix in equation (13) is the characteristic matrix (M_2) of the homogenous thin film. Note that M_2 is unimodular as its determinant is equal to 1. Assuming another film lying just underneath the thin film shown in figure 1, from Equation (13) we imply that field components E and H at interface 2 will be related to those at interface 3 by the matrix equation:

$$\begin{bmatrix} E_2 \\ H_{2x} \end{bmatrix} = \begin{bmatrix} \cos(k_3h_3) & j\sin(k_3h_3)/Y_3 \\ jY_3\sin(k_3h_3) & \cos(k_3h_3) \end{bmatrix} \begin{bmatrix} E_3 \\ H_{3x} \end{bmatrix} = M_3 \begin{bmatrix} E_3 \\ H_{3x} \end{bmatrix} \tag{14}$$

Substituting Equation (14) into Equation (13) one finds:

$$\begin{bmatrix} E_1 \\ H_{1x} \end{bmatrix} = M_2M_3 \begin{bmatrix} E_3 \\ H_{3x} \end{bmatrix} \tag{15}$$

By applying this method repeatedly for a stratified system of N thin homogeneous thin films we can write:

$$\begin{bmatrix} E_1 \\ H_{1x} \end{bmatrix} = M_2M_3 ... M_N \begin{bmatrix} E_N \\ E_{Nx} \end{bmatrix} = M \begin{bmatrix} E_N \\ E_{Nx} \end{bmatrix} \tag{16}$$

where $Y_l = \sqrt{\frac{\varepsilon_o}{\mu_0}} n_l \cos\theta_l$, $k_l = \frac{2\pi}{\lambda} n_l$ and $h_l = d_l \cos\theta_l$ for interfaces l = 2,3, ... , N. (Born & Wolf, 1980) show that the reflection and transmission coefficient amplitudes for a system of N-1 layers (l = 2 to N) lying on a substrate of refractive index n_s can be expressed from the matrix entries of the system matrix M as:

$$r_{\perp} = \frac{E_{i1}}{E_{r1}} = \frac{Y_1m_{11} + Y_oY_sm_{12} - m_{21} - Y_sm_{22}}{Y_1m_{11} + Y_oY_sm_{12} + m_{21} + Y_sm_{22}} \tag{17}$$

where $r_{\perp}$ is referred to as the reflection coefficient for the TE wave. Admittances Y_1 and Y_s for the incident medium and the substrate hosting the system of N-1 homogeneous thin films are given by:

$$Y_1 = \sqrt{\frac{\varepsilon_o}{\mu_o}} n_1 \cos\theta_1 \tag{18}$$

and

$$Y_s = \sqrt{\frac{\varepsilon_o}{\mu_o}} n_s \cos\theta_N \tag{19}$$

For the case where the H-field is perpendicular to the plane of incidence (TM wave), the impedances Y_1, Y_l *and* Y_s must be replaced by Z_1 , Z_l and Z_s , which are given by

$$Z_1 = \sqrt{\frac{\mu_o}{\varepsilon_o}} \frac{\cos\theta_1}{n_1} \tag{20}$$

$$Z_l = \sqrt{\frac{\mu_o}{\varepsilon_o}} \frac{\cos\theta_l}{n_l} \quad \text{for } l = 2,3,\ldots, \text{N-1} \tag{21}$$

and

$$Z_s = \sqrt{\frac{\mu_o}{\varepsilon_o}} \frac{\cos\theta_N}{n_s} \tag{22}$$

1.2 Examples with dielectrics and metal thin films with some experimental results

Expressions derived in the previous section can be applied to find the reflectance curve of thin dielectric or metal films. They can be applied to fit experimental reflectivity data points to determine refractive indices of a dielectric film or metal film relative permittivity and even their thickness. Before we illustrate how it is used, let us apply Equation (17) for the simple case of Fresnel reflection coefficient amplitude for an interface between two semi-infinite media.

1.2.1 Interface between two semi-infinite media (Fresnel reflection coefficient)

This situation can be mimicked by setting $d_2 = 0$ into Equation (13). In other words, interfaces 1 and 2 in Figure 1 collapse into one single interface separating two semi-infinite media of refractive index n_1 and n_2.

Characteristic matrix in Equation (13) can be used to find the matrix system for two semi-infinite media. Setting for $d_2 = 0$, the matrix system for the two semi-infinite media becomes the identity matrix as h_2 equals 0. This means that $m_{11} = m_{22} = 1$ and $m_{12} = m_{21} = 0$. Substituting the matrix entries into Equation (17) one obtains:

$$r_{\perp} = \frac{Y_1 - Y_s}{Y_1 + Y_s} = \frac{n_1\cos\theta_1 - n_2\cos\theta_2}{n_1\cos\theta_1 + n_2\cos\theta_2} \tag{23}$$

In the previous equation we use $n_s = n_2$ and $\theta_N = \theta_2$ for this single interface system. For the TM wave, it can be shown that:

$$r_{||} = \frac{Z_1 - Z_s}{Z_1 + Z_s} = \frac{n_2 \cos\theta_1 - n_1 \cos\theta_2}{n_2 \cos\theta_1 + n_1 \cos\theta_2} \tag{24}$$

We then retrieve the results for the Fresnel reflection coefficients. Results for the transmission coefficient amplitude (t) can be obtained in the same manner.

1.2.2 Reflectance curve for a thin metallic film of silver or gold (surface plasmons)

A matrix approach is used to compute the reflectance of a thin film coupled to the hypotenuse of a right angle prism. The system shown in Figure 2 can be modeled by using three characteristic matrices for the matching fluid, the glass slide, the metal film and then accounting for the various Fresnel reflection losses at both the entrance and output face of the prism.

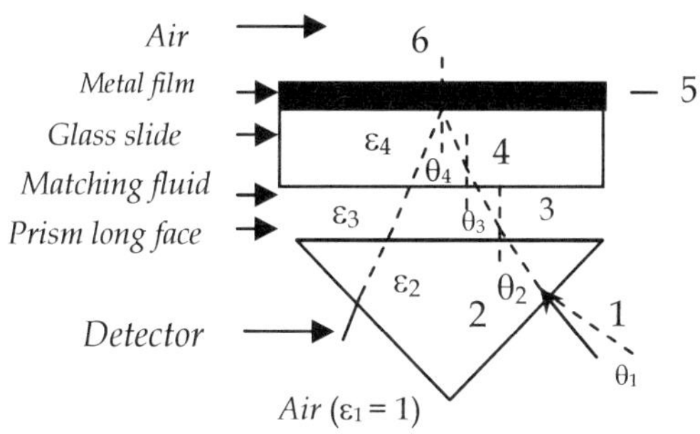

Fig. 2. Path of a laser beam propagating through all interfaces bounded by two given media. For a one way trip the media are (1) air, (2) glass, (3) matching fluid (greatly exaggerated), (4) glass (slide), (5) metal film (Au of Ag) and (6) air.

(Lévesque, 2011) expressed the characteristic matrix M of the sub-system of three layers in Figure 2 as

$$M = M_3 M_4 M_5 \tag{25}$$

where M_3, M_4 and M_5 are the characteristic matrices for the index matching fluid layer, the glass slide and the metal thin film, respectively. Each of these matrices is given by

$$M_i = \begin{bmatrix} \cos(\beta_i) & \frac{-i}{q_i}\sin(\beta_i) \\ -iq_i \sin(\beta_i) & \cos(\beta_i) \end{bmatrix} \tag{26}$$

where i =3, 4 or 5. β_i and q_i for p-polarized light are expressed as

$$\beta_i = \frac{2\pi}{\lambda} n_i d_i \cos\theta_i \tag{27}$$

$$q_i = \sqrt{\frac{\mu_o}{\varepsilon_o}}\sqrt{\frac{1}{\varepsilon_i}}\cos\theta_i = \sqrt{\frac{\mu_o}{\varepsilon_o}}q_i^{'} \tag{28}$$

respectively, where ε_I (= n_i^2) is the relative permittivity of the material. Using Snell's law, note that

$$\cos\theta_i = \left(1 - \frac{\varepsilon_2 \sin^2\theta_2}{\varepsilon_i}\right)^{1/2} \tag{29}$$

We are assuming all media to be non-magnetic and d_3, d_4 and d_5 are the thicknesses of the matching fluid, glass slide and the metal film, respectively. ε_3, ε_4 and ε_5 (= $\varepsilon'_5 + i\varepsilon_5''$) are the relative permittivity for the matching fluid, the glass slide and the metal film, respectively. ε_5' and ε_5'' are respectively, the real and imaginary parts of the metal film relative permittivity. By taking into account the Fresnel reflection losses F_1 at the input and output faces of the glass prism, the reflectance for the p-polarized light R_{Det} is given by:

$$R_{Det} = F_1 \left| \frac{(m_{11} + m_{12}q_6^{'})q_1^{'} - (m_{21} + m_{22}q_6^{'})}{(m_{11} + m_{12}q_6^{'})q_1^{'} + (m_{21} + m_{22}q_6^{'})} \right|^2 \tag{30}$$

where m_{ij} are the entries of matrix M and F_1 is given by

$$F_1 = \left[\frac{4n_2 \cos\theta_1 \cos\theta_2}{(\cos\theta_2 + n_2\cos\theta_1)^2} \right]^2 \tag{31}$$

In previous equation n_2 is the refractive index of the prism. Investigations on optical reflectivity were done on glass slides which were sputtered with gold or silver. These glass slides were pressed against a right angle prism long face and a physical contact was then established with a refractive index matching fluid. The prism is positioned on a rotary stage and a detector is measuring the signal of the reflected beam after minute prism rotations of roughly 0.03°. The p-polarized light at λ = 632.8 nm is incident from one side of a glass prism and reflects upon thin metal films as shown in figure 3. As exp (-jωt) was assumed in previous sections, all complex permittivity ε must be expressed as $\varepsilon = \varepsilon' + j\varepsilon''$.

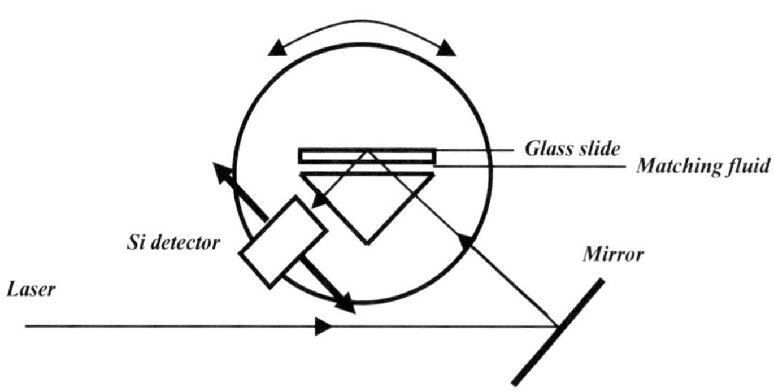

Fig. 3. Experimental set-up to obtain reflectivity data points.

If no film is coating the glass slide, a very sharp increase in reflectivity is expected when θ_4 approaches the critical angle. This sudden increase would occur at $\theta_c = \sin^{-1}(1/n_2) \sim 41.3^o$. The main feature of the sharp increase in the reflectivity curve is still obvious in the case of a metalized film. This is so as the penetration of the evanescent field is large enough to feel the presence of air bounding the thin metal film. As silver or gold relative permittivity (optical constant) is complex, $\cos\theta_5$ becomes complex in general and as a result θ_5 is not represented in Fig. 2. This means physically that the field penetrates into the metal film and decays exponentially through the film thickness. At an optimum thickness, the evanescent field excites charge oscillations collectively at the metal-film-air surface (c.f.fig.2), which is often used to probe the metal surface. This phenomenon known as Surface Plasmon Resonance (Raether, 1988; Robertson & Fullerton, 1989; Welford, 1991) is occurring at an angle of θ_2 that is a few degrees greater than θ_c. For a He-Ne laser beam at λ = 632.8 nm, that is incident from the prism's side (c.f.fig.2) and then reflecting on silver or gold metal films, surface plasmons (SP) are excited at θ_2 near 43° and 44°, respectively. At these angles, the incident light wave vector matches that of the SP wave vector. At this matching condition, the incident energy delivered by the laser beam excites SP and as a result of energy conservation the reflected beam reaches a very low value. At an optimal thickness, the reflectance curve displays a very sharp reflectivity dip. Figure 4 shows the sudden increase at the critical angle followed by a sharp dip in the reflectance curve in the case of a gold film of various thicknesses, which is overlaying the glass slide.

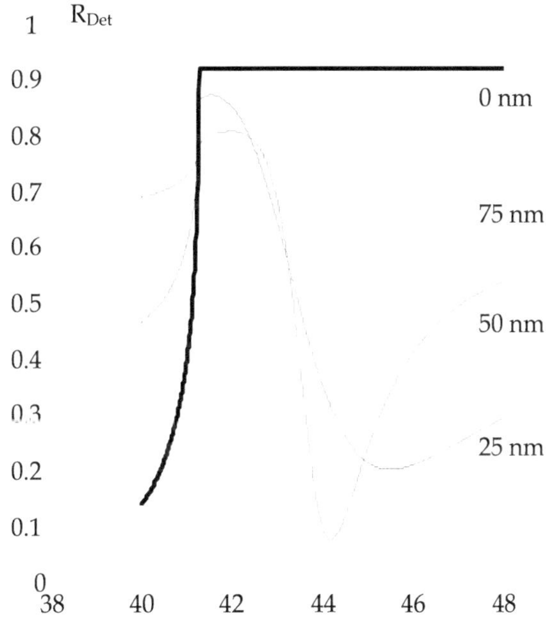

Fig. 4. Reflectance curves for gold films of various thicknesses d_5 obtained from Eq.(30). We used d_3 = 10000 nm and d_4 = 1000000 nm (1mm), n_2 =1.515, n_3 =1.51, n_4 =1.515 and ε_5 = -11.3+3j.

Reflectance curves for gold films sputtered on glass slides show a sudden rise at the critical angle θ_c followed by a sharp drop reaching a minimum near 44°. For all film thicknesses, a sudden rise occurs at the critical angle. Note that the reflectance curve for a bare glass slide

(d_5 = 0 nm) is also shown in figure 4. At smaller thicknesses, the electromagnetic field is less confined within the metallic film and does penetrate much more into the air. The penetration depth of the electromagnetic field just before reaching the critical angle ($\theta <$ 41.3°) is indicated by a lower reflectance as d_5 gets closer to zero, as shown in Fig.4. The reflectivity drop beyond θ_c is known as Surface Plasmon Resonance (SPR). SPR is discussed extensively in the literature and is also used in many applications. Good fitting of both regions displaying large optical intensity change is also useful in chemical sensing devices. As a result fitting of both regions is attempted using the exact function curve without any approximations given by Eq. (30). Eq. (30) is only valid for incident plane wave. Therefore, the reflectivity data points were obtained for a very well-collimated incident laser beam. A beam that is slightly converging would cause more discrepancy between the curve produced from Eq. (30) and the reflectivity data points. Although the Fresnel loss at the transparent matching fluid-glass slide interface is very small, it was taken into account in Eq. (30), using d_3=10 000nm (10 μm) in matrix M_3. The theoretical reflectance curve is not affected much by the matching fluid thickness d_3. It was found that d_3 exceeding 50 μm produces larger oscillations in the reflectance curve predicted by Eq.(30). As the oscillations are not noticeable amongst the experimental data points, the value of d_3 = 10 μm was deemed to be reasonable. A function curve from Eq. (30) is generated by changing three output parameters ε_5', ε_5'' and d_5. The sum of the squared differences (SSQ) between R_{Det} and the experimental data points R_i is calculated. The best fit is determined when the SSQ is reaching a minimum. The SSQ is defined as:

$$SSQ = \sum_{i=1}^{N} (R_i - R_{Det})^2 \tag{32}$$

where i is a subscript for each of the N data points from the data acquisition. Each sample was placed on a rotary stage as shown in Figure 3 and a moving Si-pin diode is rotating to track down the reflected beam to measure a DC signal as a function of θ_2. The reflectivity data points and typical fits are shown in Figure 5.

In the fit in Fig. 5a, we used $n_2 = n_4$ = 1.515, n_3 = 1.47(glycerol) for red light, d_3 = 10 000 nm, ε_5 =-11.55+3.132j and d_5 =43.34 nm.

In the fit in Fig. 5b, we used $n_2 = n_4$ = 1.515, n_3 = 1.47(glycerol) for red light, d_3 = 10 000 nm, ε_5 = -10.38+2.22j and d_5 = 53.8 nm. The three output parameters (ε_5', ε_5'' and d_5) minimizing the SSQ determine the best fit. Plotting the SSQ in 3D as a function of ε_5' and d_5 at ε_5'' = 3.132 shows there is indeed a minimum in the SSQ for the fit shown in Fig.5a. Figure 6 shows a 3D plot of the SSQ near the output parameters that produced (Lévesque, 2011) the best fit in Fig. 5a. 3D plots at values slightly different from ε_5'' = 3.132 yield larger values for the minimum.

2. Wave propagation in a dielectric waveguide

In this section, we apply the matrix formalism to a dielectric waveguide. We will describe how the reflectance curve changes for a system such as the one depicted in Fig. 2 if a dielectric film is overlaying the metal film. It will be shown that waveguide modes can be excited in a dielectric thin film overlaying a metal such as silver or gold and that waveguide modes supported by the dielectric film depend upon its thickness.

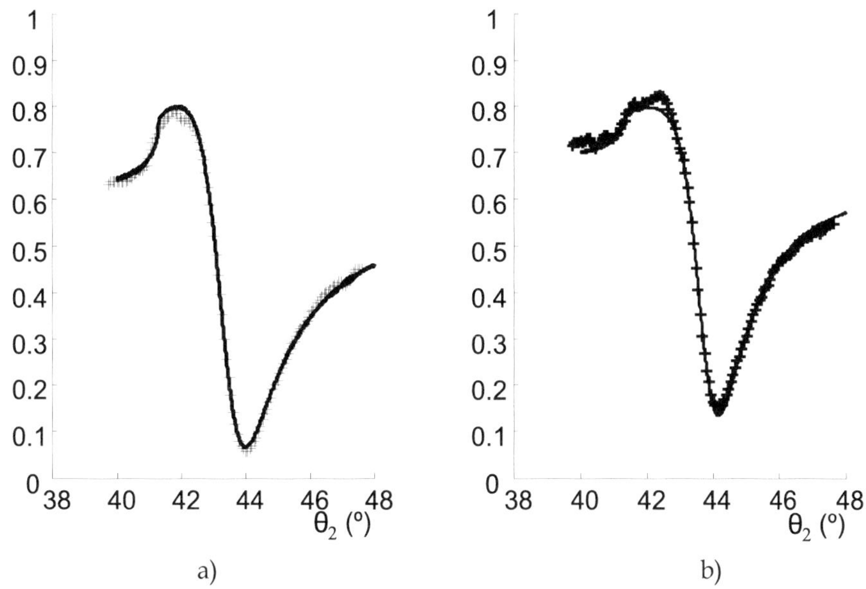

Fig. 5. Reflectivity data points (+) and a fit (solid line) produced from Eq.(30) for two different gold films.

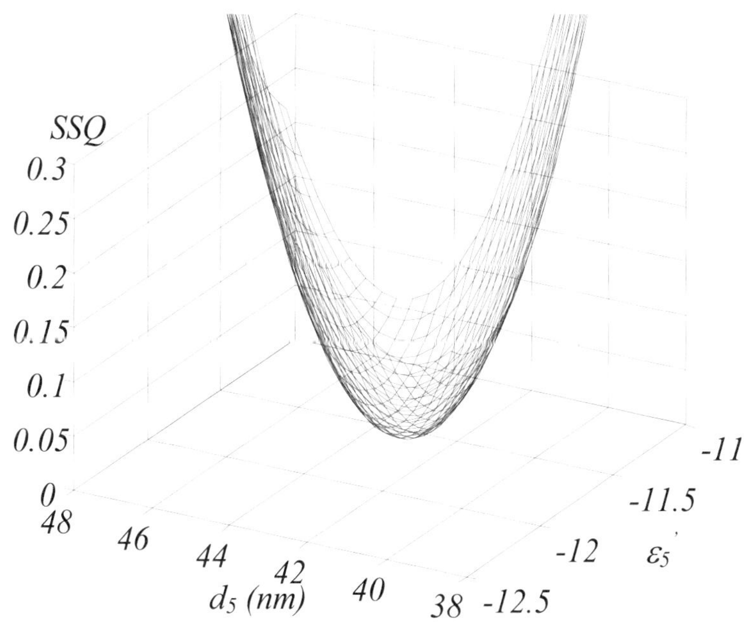

Fig. 6. 3D plot of SSQ as a function of two output parameters at a given value of ε_5''(= 3.132). We assumed the glycerol layer (d_3) to be 10 μm and the thickness of the glass slide is 1 mm (d_4). The SSQ reaches a minimum of 0.01298 for ε_5' =-11.55, ε_5'' = 3.132 and d_5 =43.34 nm.

2.1 Wave propagation in dielectric films

Let us consider a dielectric film of thickness d_6 overlaying the metal film in Figure 2. We will be assuming that the top surface of the overlaying dielectric is bounded by the semi-infinite air medium. The characteristic matrix M for the sub-system of four layers can be expressed as:

$$M = M_3 M_4 M_5 M_6 \tag{33}$$

where M_3, M_4, M_5 and M_6 are the characteristic matrices for the index matching fluid layer, the glass slide, the metal thin film and the thin dielectric film, respectively. Each matrix in Eq.(33) is given by Eq.(26) for i = 3,4,5 and 6 and the reflectance for the p-polarized wave is given by Eq.(30). For this four layer system, q_6' in Eq.(30) should be replaced by q_7' (air) and m_{11}, m_{12}, m_{21} and m_{22} are the entries of the system matrix given by Eq.(33). The expression for q_6 is given by Eq.(28) and is used in the computation of M_6 for the dielectric film characteristic matrix.

2.1.1 Computation of reflectance with a thin dielectric film and experimental results

Eq. (30) can be used with the minor modifications discussed in section 2.1 to find the reflectance of the system in Fig. 2 with an extra dielectric film processed on the metal film. The dielectric film can support waveguide modes if the laser beam is directed at very precise incident angle θ_2. Let us consider a transparent polymer film with a real permittivity $\varepsilon_6 = 2.30$ processed on a silver film. The computation is done for a silver film that is 50 nm thick. Silver permittivity is assumed to be $\varepsilon_5 = -18.0 + 0.6i$ and the prism refractive index to be 2.15 (ZrO_2) for He-Ne laser at $\lambda = 632.8$ nm. We also assume that the metal film is directly coated on the prism long face and as a result we set $d_3 = d_4 = 0$. In other words M_3 and M_4 are expressed by identity matrices. Figure 7 is showing the reflectance curve for a dielectric film of different thickness that is overlaying the silver film coated on the high refractive index ZrO_2 prism.

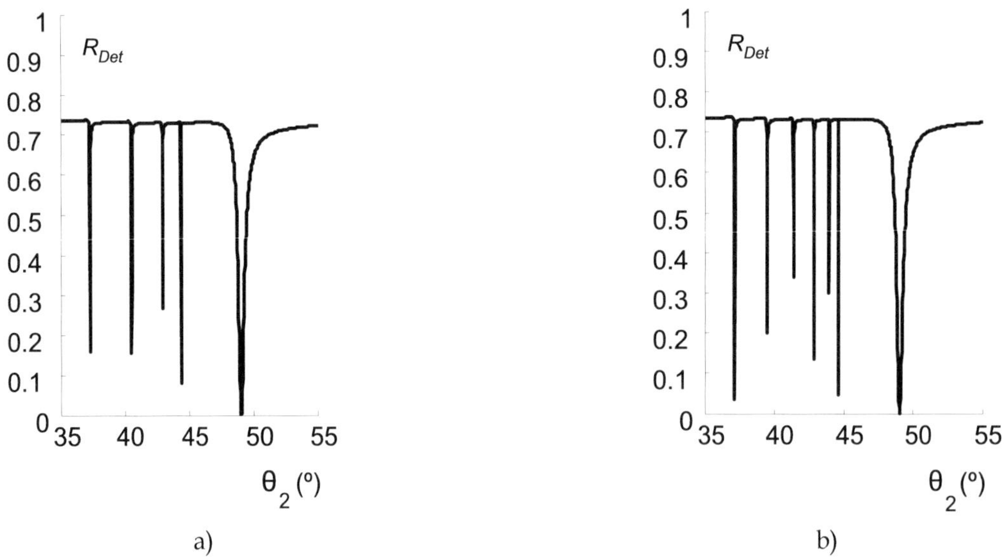

Fig. 7. a) Reflectance curve for a lossless dielectric film of 1.7 μm overlaying a thin silver film
b) Reflectance curve for a lossless dielectric film of 2.5 μm overlaying a thin silver film

In figure 7a, a series of very sharp reflectivity drops occur in the reflectance curve for θ_2 within the range 35°-45°. These sharp reflectivity drops with small full width at half maximum (FWHM) are waveguide modes supported by the dielectric film. The last reflectivity dip with a larger FWHM near $\theta_2 \sim 50°$ is due to surface plasmon resonance (SPR) and is mostly depending upon the metal film properties and its thickness as discussed in section 1.2.2. A thicker dielectric film (c.f. fig. 7b) can support more waveguide modes and as a result the number of sharp reflectivity dip for θ_2 within the range 35° to 45° is expected to be greater. Note that the FWHM of the SPR dip remains at the same position as the metal film thickness was not changed. These waveguide modes do not propagate a very large distance as light is slightly attenuated when reflecting at the metal-dielectric film interface. Therefore, at precise angle θ_2 the incident light is probing the dielectric film locally before being reflected by the thin metal film. Nevertheless, the laser beam is simultaneously probing the metal and the dielectric films because it creates SPR on the thin metal film and waveguide modes are being supported by the dielectric film. In practice, dielectric films are not lossless (Podgorsek & Franke, 2002) and their permittivity should be expressed using a small imaginary part. Let us assume that each dielectric films in Fig. 7 have a permittivity of $\varepsilon_6 = 2.30 + 0.005j$.

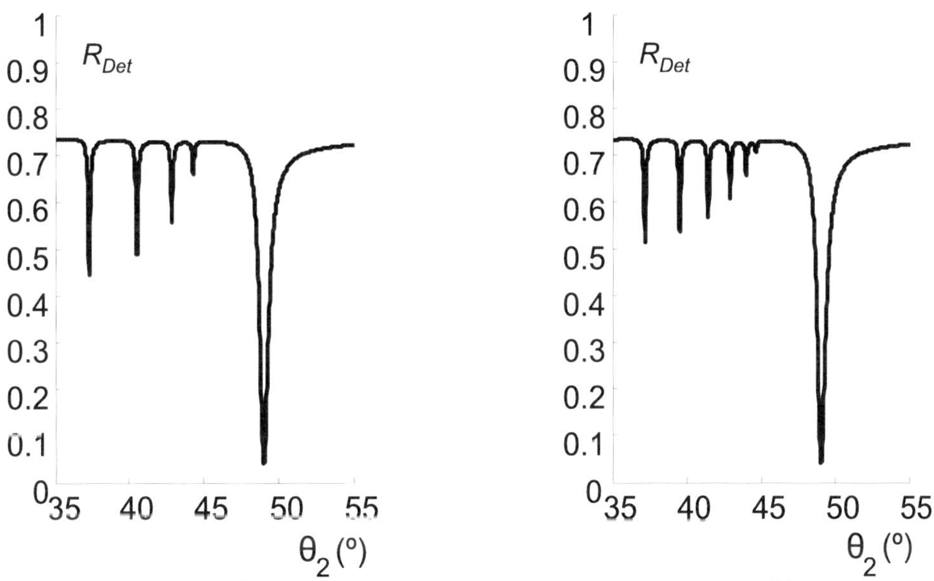

Fig. 8. a) Reflectance curve for a dielectric film ($\varepsilon_6 = 2.30 + 0.005j$, $d_6 = 1.7$ μm) overlaying the metal film. b) Reflectance curve for a dielectric film ($\varepsilon_6 = 2.30 + 0.005j$, $d_6 = 2.5$ μm) overlaying the metal film.

Note from figures 7 and 8 that the waveguide mode dips are greatly attenuated when a small imaginary part is assumed in the dielectric film permittivity. The dips at larger angles (near 45°) are getting smaller as the propagation distance into in the dielectric film is larger as θ_2 increases. Note that the SPR dip is not much affected by the imaginary part of ε_6. Essentially, the whole 4-layer system of prism material-silver film- dielectric film-air can be mounted on a rotary stage and the angle θ_2 can be varied using a set-up similar to that shown in figure 3. As it is difficult to obtain a large dynamic range in the measurements of reflectivity data points, scans must be done successively to cover a long range of incident angle. Figure 9 shows reflectance curves for a transparent layer of polyimide processed directly on silver films. Ranges of incident angle θ_2 where no noticeable change in reflectivity were observed are not shown. Only the dips in the reflectivity data points are fitted by Eq. (30).

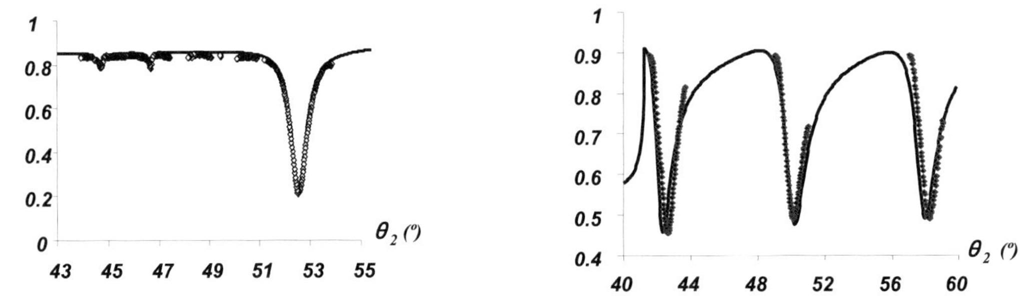

a) b)

Fig. 9. Reflectivity data points of the 4-layer system and fits (solid lines) from Eq. (30) for a) ZrO_2 prism-Ag-polyimide film b) glass prism-Ag-polyimide film

Using a method based on the optimization of the sum of square (SSQ) as presented in section 1.2.2, thicknesses and the complex permittivities of both films can be estimated. Values obtained from the minimization of the SSQ are given in table 1. Uncertainties are estimated from a method described by (Lévesque et al., 1994).

Prism	*ZrO2*	*Glass*
ε_5' (Ag)	-17.41±0.10	-18.1 ±0.1
ε_5'' (Ag)	0.2 ±0.1	1.36 ±0.08
d_5	615 ±15Å	146 ±5Å
ε_6' (Pi)	2.495±0.001	2.230±0.002
ε_6'' (Pi)	0.011±0.003	0.0017±0.0002
d_6	1.488±0.004μm	1.723±0.003μm

Table 1. Thicknesses and permittivities of the silver (Ag) and polyimide (*Pi*) films.

3. Diffraction efficiency (DE) in dielectric periodic grating structures

Abrupt changes in reflectivity or transmission were first observed in gratings as early as 1902 (Wood, 1902). These so-called anomalies in diffraction efficiency (DE) occurring over an angle range or a wavelength spectrum are very different from the normally smooth diffraction curves. These abrupt changes in DE led researchers to design and investigate resonant filters for applications in many devices including gratings.

Rigorous coupled wave analysis (RCWA) has been used extensively (Moharam et al., 1995; Lalanne & Morris, 1996; Lenaerts et al., 2005) to calculate diffraction efficiencies (DE) in waveguide structures. The application of RCWA to resonant-grating systems has been investigated mostly for both the *TE* and *TM* polarization. In this section, the basic binary dielectric rectangular-groove grating is treated with careful considerations on the computation of DE. The results obtained for binary dielectric rectangular-groove grating are also applied to metallic grating. Introduction to photonic bandgap systems are discussed and some examples are presented at the end of this section.

3.1 Theory of coupled wave analysis

As the numerical *RCWA* method is introduced extensively in the literature, only the basics equations will be presented in this section. Computation will be done for the *TM* wave on ridge binary grating bounded by two semi-infinite dielectric media of real permittivities ε_1 and ε_3. The type of structures presented in this section is depicted in figure 10.

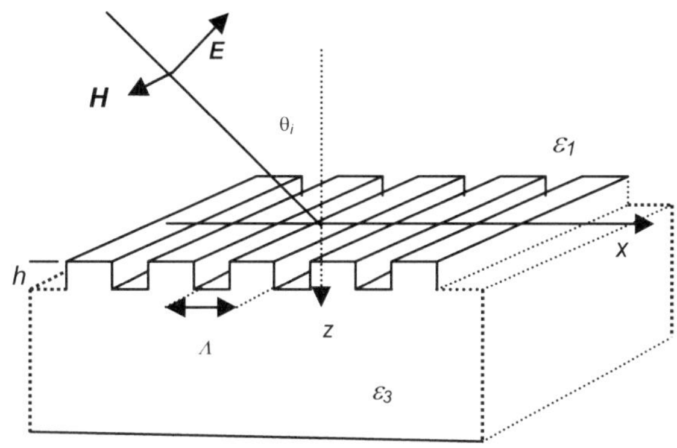

Fig. 10. Basic structure of the binary rectangular-groove grating bounded by two semi-infinite dielectrics.

The relative permittivity $\varepsilon(x)$ of the modulated region shown in figure 10 is varying periodically along the x-direction and is defined as:

$$\varepsilon(x) = \sum_s \varepsilon_s \exp(-2\pi jsx / \Lambda) \tag{34}$$

where ε_s is the sth Fourier component of the relative permittivity in the grating region ($0< z <h$), which can be complex in the case of metallic gratings. The incident normalized magnetic field that is normal to the plane of incidence (*cf. fig.10*) is given by:

$$H_{inc,y} = \exp[-jk_o n_1(\sin\theta_i x + \cos\theta_i z)] \tag{35}$$

where $k_o = 2\pi/\lambda$. θ_i is the incident angle with respect to the z-axis as shown in figure 10. The normalized solutions in regions 1 ($z < 0$) and 3 ($z > h$) are expressed as:

$$H_{1,y} = H_{inc,y} + \sum_i R_i \exp[-j(k_{xi}x - k_{1,zi}z)] \tag{36}$$

$$H_{3,y} = \sum_i T_i \exp[-j(k_{xi}x + k_{3,zi}(z-h))] \tag{37}$$

where k_{xi} is defined by the Floquet condition, i.e.,

$$k_{xi} = k_o(n_1 \sin\theta_i - i(\lambda / \Lambda)) \quad (38).$$

In previous equations, Λ is the grating spacing, n_1 ($\equiv \sqrt{\varepsilon_1}$) is the refractive index of medium 1 and

$$k_{l,zi} = \begin{Bmatrix} k_o\left[n_l^2 - \left(\dfrac{k_{xi}}{k_o}\right)^2\right]^{1/2} & k_o n_l > k_{xi} \\ -jk_o\left[\left(\dfrac{k_{xi}}{k_o}\right)^2 - n_l^2\right]^{1/2} & k_{xi} > k_o n_l \end{Bmatrix} \quad (39)$$

with l = 1,3. n_3 ($\equiv \sqrt{\varepsilon_3}$) is the refractive index of medium 3.

R_i and T_i are the normalized electric-field amplitudes of the ith diffracted wave in media 1 and 3, respectively. In the grating region ($0 < z < h$) the tangential magnetic (y-component) and electric (x-component) fields of the *TM* wave may be expressed as a Fourier expansion:

$$\begin{aligned} H_y &= \sum_i U_{yi}(z)\exp(-jk_{xi}x) \\ E_x &= j\left(\frac{\mu_o}{\varepsilon_o}\right)^{1/2} \sum_i S_{xi}(z)\exp(-jk_{xi}x) \end{aligned} \quad (40)$$

where $U_{yi}(z)$ and $S_{xi}(z)$ are the normalized amplitudes of the ith space-harmonic which satisfy Maxwell's equations, i.e.,

$$\begin{aligned} \frac{\partial E_x}{\partial z} &= -j\omega\mu_o H_y + \frac{\partial E_z}{\partial x} \\ \frac{1}{\varepsilon(x)}\frac{\partial H_y}{\partial x} &= j\omega\varepsilon_o E_z \end{aligned} \quad (41)$$

where a temporal dependence of exp (jωt) is assumed ($j^2 = -1$) and ω is the angular optical frequency. ε_o and μ_o are respectively the permittivity and permeability of free space. As the exp (jωt) is used, all complex permittivity must be expressed under $\varepsilon = \varepsilon' - j\varepsilon''$.

Substituting the set of equations (40) into Maxwell's equations and eliminating E_z, the coupled-wave equations can be expressed in the matrix form as:

$$\begin{bmatrix} \partial U_y / \partial z' \\ \partial S_x / \partial z' \end{bmatrix} = \begin{bmatrix} 0 & E \\ B & 0 \end{bmatrix} \begin{bmatrix} U_y \\ S_x \end{bmatrix} \quad (42)$$

where z' equals $k_o z$.

Previous equations under the matrix form can be reduced to

$$\left[\partial^2 U_y / \partial z'^2\right] = [EB]\left[U_y\right] \quad (43)$$

where $B = K_xE^{-1}K_x - I$. E is the matrix formed by the permittivity elements, K_x is a diagonal matrix, with their diagonal entries being equal to k_{xm} / k_o and I is the identity matrix. The solutions of Eq. (43) and the set of Eq. (42) for the space harmonics of the tangential magnetic and electric fields in the grating region are expressed as:

$$\begin{aligned} U_{yi}(z) &= \sum_{m=1}^{n} w_{i,m}(c_m^+ \exp[-jk_o q_m z] + c_m^- \exp[jk_o q_m (z-h)]) \\ S_{xi}(z) &= \sum_{m=1}^{n} v_{i,m}(-c_m^+ \exp[-jk_o q_m z] + c_m^- \exp[jk_o q_m (z-h)]) \end{aligned} \tag{44}$$

where, $w_{,i,,m}$ and q_m are the elements of the eigenvector matrix W and the positive square root of the eigenvalues of matrix **G** (=-**EB**), respectively. The quantities c_m^+ and c_m^- are unknown constants (vectors) to be determined from the boundary conditions. The amplitudes of the diffracted fields R_i and T_i are calculated by matching the tangential electric and magnetic field components at the two boundaries. Using Eqs. (35) , (36), (44) and the previously defined matrices, the boundary conditions at the input boundary (z = 0) are:

$$\delta_{i,0} + R_i = Wc^+ + WXc^- \tag{45}$$

and

$$\frac{j\cos\theta}{n_1}\delta_{i,0} - jZ_1 R_i = Vc^+ - VXc^- \tag{46}$$

where X and Z_1 are diagonal matrices with diagonal elements $\exp(-jk_oq_mh)$ and $k_{1zi}/(n_1^2\,k_o)$, respectively. c^+ and c^- are vectors of the diffracted amplitude in the ith order. From (42) and (44), it can be shown that

$$V = jE^{-1}WQ \tag{47}$$

where $v_{m,l}$ are the elements of the product matrix with Q being a diagonal matrix with diagonal entries q_l.
At z = h, the boundary conditions are:

$$WXc^+ + Wc^- = T_i \tag{48}$$

and

$$VXc^+ - Vc^- = jZ_3T_i \tag{49}$$

where Z_3 is the diagonal matrix with diagonal elements $k_{3zi}/\ (n_3^2\ k_o)$. Multiplying each member of Eq. (48) by $-jZ_3$ and using Eq. (49) to eliminate T_i vectors c^- and c^+ are related by:

$$c^- = (jZ_3W + V)^{-1}(-jZ_3W + V)Xc^+ \tag{50}$$

Multiplying each member of Eq. (45) by jZ_1 and using Eq. (46) to eliminate R_i a numerical computation can be found for c^+ by making use of Eq.(50), that is:

$$c^{+} = C^{-1}(\frac{j\cos\theta}{n_1} + jZ_1)\delta_{i,o} \quad (51)$$

where

$$C = [(jZ_1W + V) + (jZ_1W - V)X(jZ_3W + V)^{-1}(-jZ_3W + V)X] \quad (52)$$

Note in Eq. (51) that $\delta_{i,0}$ is a column vector. In the case of a solution truncated to the first negative and positive orders,

$$\delta_{i,o} = \begin{bmatrix} 0 \\ 1 \\ 0 \end{bmatrix} \quad (53)$$

assuming the incident wave to be a plane wave. In this particular case

$$(\frac{j\cos\theta}{n_1} + jZ_1)\delta_{i,o} = \begin{bmatrix} 0 \\ j\frac{\cos\theta}{n1} + jZ_1(2,2) \\ 0 \end{bmatrix} \quad (54)$$

where $Z_1(2,2)$ is the element on line 2 and column 2 of matrix Z_1. Finally, the vector on the right-hand side of Eq.(54) is applied to the inverse matrix of C to find the column vector for the diffracted amplitude c^+ from Eq. (51). Then c^- is found from Eq. (50) and the normalized electric field amplitudes for R_i and T_i can be found from Eqs. (48) and (49).
Substituting Eq. (34) and Eq.(44) into Maxwell's equations and eliminating E_z , it can be shown that

$$\frac{\partial S_{xm}}{\partial z'} = j(\sum_p \frac{k_{xm}}{k_o}\varepsilon_{i-p}^{-1}\frac{k_{xm}}{k_o}U_{yp} - U_{yi}) \quad (55)$$

Eq. (55) is one of the two coupled-wave equations involving the inverse permittivity for the case of *TM* polarization only. In the conventional formulation (Wang et al., 1990; Magnusson & Wang, 1992; Tibuleac & Magnusson, 1997) the term ε_{i-p}^{-1} is treated by taking the inverse of the matrix *E* defined by the permittivity components (Moharam & Gaylord, 1981), with the *i*, *p* elements being equal to $\varepsilon_{(i-p)}$. In the reformulation of the eigenvalue problem (Lalanne & Morris, 1996), the term ε_{i-p}^{-1} is considered in a different manner by forming a matrix *A* of the inverse-permittivity coefficient harmonics for the two regions inside the modulated region. Fourier expansion in Eq.(34) is modified to:

$$1/\varepsilon(x) = \sum_s (1/\varepsilon)_s \exp(-2\pi jsx / \Lambda) \quad (56)$$

where $(1/\varepsilon)_s$ is the sth Fourier component of the relative permittivity in the grating region.
Since the coupled-wave equations do not involve the inverse of the permittivity in the coupled-wave equations for the TE wave, matrix **A** is not needed in numerical computations

and the eigenvalue problem is greatly simplified in this case. As a result, solutions for the TE wave are more stable in metallic lamellar gratings.

Only the DE in reflection and transmission for zeroth order are computed in the examples that will be discussed throughout this section. The diffraction efficiencies in both reflection (DE_R) and transmission (DE_T) are defined as:

$$DE_R = R_0 R_0^* \operatorname{Re}(k_{1,z0} / (k_o n_1 \cos\theta_i)) \tag{57}$$

and

$$DE_T = T_0 T_0^* \operatorname{Re}(\frac{k_{3,z0}}{n_3^2}) / (\frac{k_o \cos\theta}{n_1}) \tag{58}$$

3.1.1 Examples with binary dielectric periodic gratings

Let us consider a binary rectangular-groove grating with real permittivity ε_L and ε_H as shown in figure 10. In the case of notch filters the higher permittivity value ε_H ($\Lambda / 2 < x < \Lambda$) is greater than ε_L ($0 < x < \Lambda / 2$). Figure 11 shows the numerical computation for *DE* from the RCWA formulation for the *TM* wave when only three orders (m = -1, 0, 1) are retained in the computation.

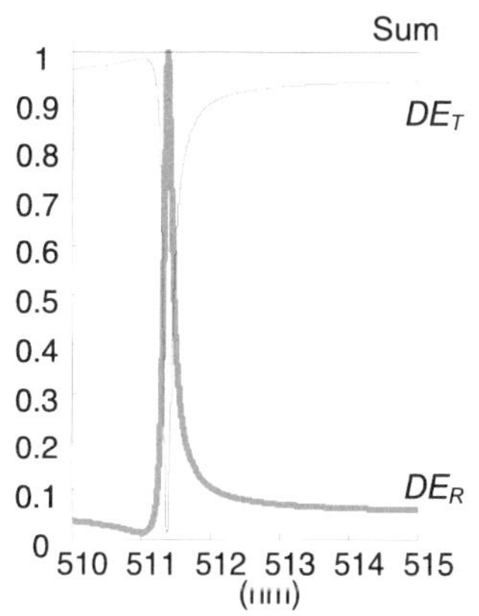

Fig. 11. DE_R and DE_T for a binary dielectric periodic grating for ε_L =4.00, ε_H = 4.41, Λ =314 nm, n_1 =1.00 (air), n_3 =1.52 (glass) and h = 134 nm.

From the principle of energy conservation, the sum of DE_R and DE_T must be equal to unity. This principle is useful to decide if the number of orders retained in the computation is sufficient. As no deviation from unity is seen in the sum of DE_R and DE_T in figure 11, three orders is deemed to be enough to describe the diffraction efficiencies within this narrow wavelength spectrum. At a wavelength of roughly 511.3 nm all the optical energy is reflected back in the opposite direction from that of the incident light. As a result, DE_T is reaching a zero value as destructive interferences occur within the grating at this precise wavelength value of 511.3 nm.

3.1.2 Examples with metallic periodic gratings

The theory presented in section 3.1 can be applied to metallic periodic gratings. For the TM wave many terms need to be retained in the calculation to reach convergence (Li &

Haggans, 1993). For the sake of saving time, a fairly accurate computation is reached after retaining ten orders when using the reformulated eigenvalue problem (Lalanne & Morris, 1996). Figure 12 shows DE_R for a metallic periodic grating using a 3D plot. Metallic periodic grating are used to excite surface plasmons (SP) to improve Surface-enhanced-Raman-Scattering (SERS) sensor performances (Sheng et al., 1982). At a given wavelength λ the reflectivity of the metallic grating should be symmetric with the incident angle θ. If a reflectivity drop occurs due to SP at θ, the metallic periodic grating should display a similar drop at $-\theta$. Note that two minima occur on either side of normal incidence ($\theta = 0^o$) and one single minimum is displayed at normal incidence for $\lambda \sim 630$ nm. Basically each minimum in DE_R forms two valleys which crisscross at normal incidence and $\lambda \sim 630$ nm. This point will become important in the next section where photonic band gap is discussed.

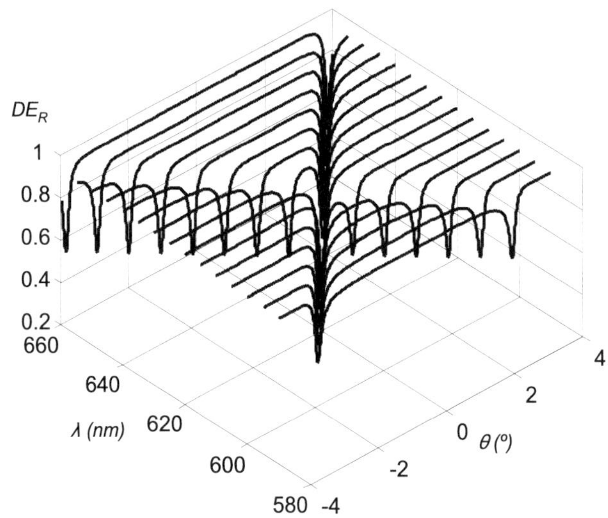

Fig. 12. 3D plot of DE_R for a periodic metallic grating. In the calculation, we used $n_1 = 1$, $\varepsilon_3 = \varepsilon_2 = -17.75 - 0.7j$, $\Lambda = 600$nm, $\varepsilon_L = -17.75 - 0.7j$, $\varepsilon_H = 1$, and h = 10.5 nm.

3.2 Photonic bandgap in metallic periodic gratings

Resonant surface plasmon (SP) coupling involving metallic periodic gratings has been extensively studied over the past years and more recently in work looking at photonic devices (Park et al., 2003; Barnes et al. 2003; Ebbesen et al., 1998; Ye & Zhang, 2004) surface-enhanced Raman scattering (Sheng et al., 1982) and photonic bandgaps (Barnes et al., 1996). Corrugated surfaces are commonly produced by direct exposure of a photoresist film to a holographic interference pattern. There is some experimental evidence that owing to nonlinear response of the photoresist, this technique leads to the presence of higher harmonics in addition to the fundamental pattern that is inscribed (Gallatin, 1987; Pai & Awada, 1991). The higher harmonics can then influence the propagation of the SP on the metallic periodic grating and, in particular, can generate a bandgap in the plasmon dispersion curve.

3.2.1 Generating a photonic bandgap with two metallic periodic gratings

Let us consider two metallic sinusoidal gratings with vectors K1 ($= 2\pi/\Lambda1$) and K2 ($=2\pi/\Lambda2$) inscribed at the same location on the film surface. One grating acts as a coupler that allows light

to generate SPs while the second grating creates a bandgap in the dispersion curve for the SP propagation. Herein, we consider the case $K_2 = 2K_1$. More complicated cases such as $K_2 < 2K_1$ have also been investigated and may be found in the literature (Lévesque & Rochon , 2005).
The SP dispersion curve for a uniform silver or gold film in the absence of a gap is shown in Figure 13a and is described by:

$$k_{SP} = \frac{\omega}{c}\left(\frac{\varepsilon_m \varepsilon_d}{\varepsilon_m + \varepsilon_d}\right)^{1/2} \tag{59}$$

where k_{SP} is the wave vector of the SP modes coupled at the surface and ε_m and ε_d are the permittivities of the metal and dielectric material (air). The dispersion line for light incident at an angle θ and scattered by a vector K_1 is given by:

$$k_{light} = \frac{\omega}{c}\sin\theta \pm m_1 K_1 \tag{60}$$

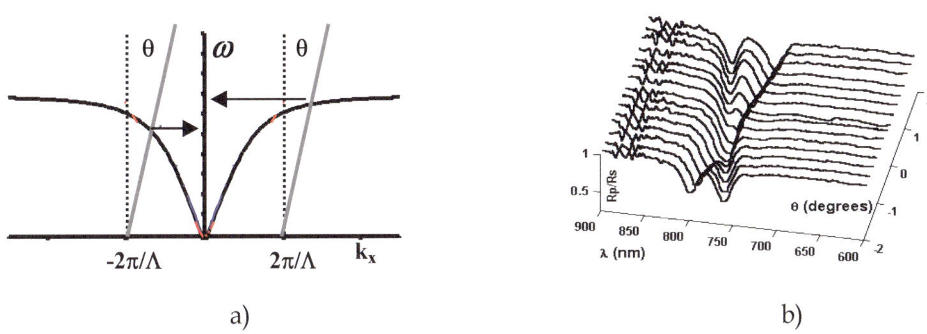

Fig. 13. a) SP dispersion curves for one periodic grating of Bragg vector K_1 b) Normalized reflectance (R_p/R_s) curve for a single metallic grating with $\Lambda \sim 755$ nm.

Note from Figure 13a that at normal incidence ($\theta = 0^o$) SPs will be excited at a single wavelength from a loss or gain of light momentum by the grating Bragg vector $K_1 = 2\pi/\Lambda$. Scattering of incident light from the metallic grating at a given incident angle can fulfill the phase-matching condition ($k_{SP} = k_{light}$) for SP excitation. As θ increases from zero SPs can be generated if light scatters by a Bragg vector $\pm K_1$, i.e., two valleys will form for $\theta > 0^o$ and $\theta < 0^o$ as shown in Figure 13b. Experimentally, a fairly sharp dip in the reflectivity curve (R_p) was observed for the p-polarized light but not for the s-polarized light (R_s). To emphasize the SP contribution, the R_p reflectance curves were normalized to R_s in the range 600-900 nm spectral range. Curves shown in Figure 13b were predicted by DE_R computations shown in Figure 12.
The SP dispersion curves for the doubly corrugated surfaces and light lines are shown in Figure 14a.
It can be seen from Figure 14a) that two SPs can be generated at normal incidence as a bandgap is being created by the grating of Bragg vector K_2. As a result, SPs can be generated at $\omega = \omega_1$ and $\omega = \omega_2$. This means the band will open as shown in Figure 14b, where two minima are shown in the experimental data points for all incident angles (Lévesque & Rochon , 2005). Each of these minima corresponds to SP excitation at the air-metal grating surface when light is scattered by Bragg vector $\pm K_1$.

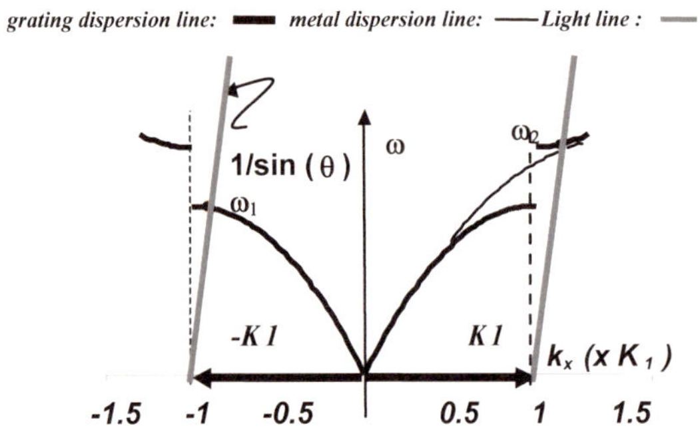

a)

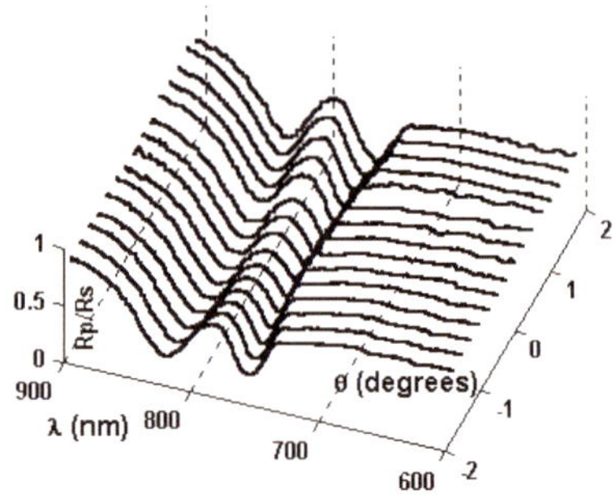

b)

Fig. 14. a) SP dispersion curves for two superimposed periodic grating of Bragg vector K_1 and K_2 b) Normalized reflectance curve for a doubly corrugated surface with $\Lambda_1 \sim 755$ nm and $\Lambda_2 \sim 375$ nm.

3.2.2 Processing of the single and double metallic corrugated surfaces

Surface corrugations with selected pitches (Bragg vectors) can be produced on azopolymer films by direct exposure of an interference pattern from two coherent light beams at $\lambda = 532$ nm, as shown in Figure 15a. The two desired spacing are obtained by adjusting the angle φ (c. f fig.15a) between the writing beams, and their depth is determined by their exposure time. The films under investigation have two superimposed sinusoidal gratings with vectors K_1 and K_2. These azopolymer films were prepared on glass slides and then coated with a 50

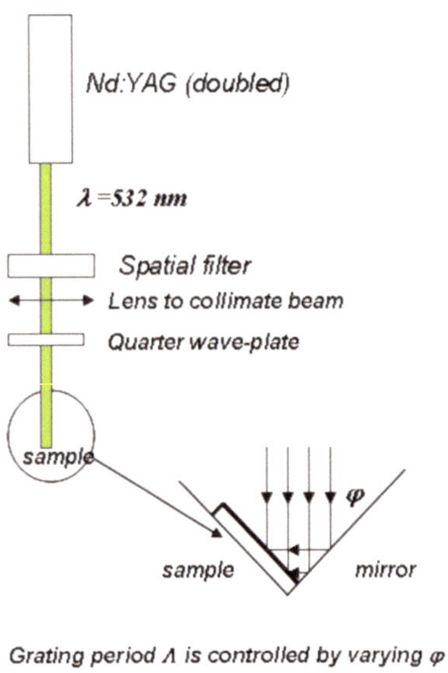

a)

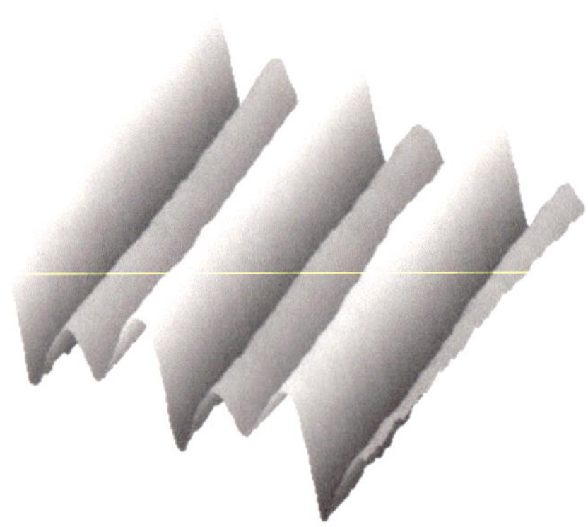

b)

Fig. 15. a) Experimental set-up to produce corrugated metallic grating. b) Atomic force microscope image of a double metallic grating. Pitches here are 700 and 375 nm with their respective depths of 19± 1 nm and 7.0±0.5 nm.

nm thick gold film by sputter. The surface profile s(x) shown in the atomic force microscope image in Figure 15b can be represented as

$$s(x) = h_1 \sin(K_1 x) + h_2 \sin(K_2 x + \phi_2) \quad (61)$$

where x is the spatial coordinate, h_1 and h_2 are the amplitudes of the two harmonic components, and ϕ_2 is their relative phase.

4. Conclusion

The matrix formalism was shown to be efficient to predict the reflectance curves of both uniform films and periodic corrugated surfaces. It was shown in this chapter that the reflectance derived from the matrix formalism allows precise determinations of refractive indices and thickness when it is fitted to experimental data points. The principle to determine a good fit from the minimization of the sum of squares was presented in some details. The application of the sum of squares in more complex structures involving transparent overlaying films were also introduced along with waveguide modes. It was also shown that the matrix formalism can be used in numerical techniques and can be applied to periodic gratings to predict diffraction efficiencies. Systems of metallic periodic gratings were discussed and it was shown that photonic bandgap can be produced by superposition of two inscribed corrugated surface on an azopolymer film. The modulated films were made by holographic technique to write surface relief structures. One grating is written to have a spacing vector K_2 to generate a bandgap in the SP dispersion curve. A second grating with grating spacing vector K_1 is superimposed and allows the coupling of the incident light to generate the SP itself.

5. References

Barnes, W.L., Priest, T.W., Kitson, S.C. & Sambles, J.R. (1996). Physical origin of photonic energy gaps in the propagation of surface plasmons on gratings, *Phys Rev B* , Vol. 54, No. 9 (1 September 1996), pp. 6227-6244, ISSN 1098-0121

Barnes, W.L., Dereux, A. & Ebbesen, T.W. (2003). Surface plasmon sub wavelength optics, *Nature*, Vol. 424, No. 6950, (14 August 2003), pp. 824-830, ISSN 0028-0836.

Born, M. & Wolf, E. (1980). *Principles of Optics* (6th Ed.), Cambridge University Press, ISBN 0 521 63921 2, Cambridge, UK.

Ebbesen, T.W., Lezec, H.J., Ghaemi, H.F., Thio, T. & Wolff, P.A. (1998). Extraordinary optical transmission through sub wavelength hole arrays, *Nature* , Vol. 391, No. 6668, (February 12 1998), pp. 667-669, ISSN 0028-0836.

Gallatin, G.M. (1987). Properties and applications of layered grating resonances, in Applications and Theory of Periodic Structures, Diffraction Gratings, and Moiré Phenomena III, *Proc. SPIE,* Vol. 815, pp.158-167, ISBN 9780892528509.

Lalanne, P. & Morris, G.M. (1996). Highly improved convergence of the coupled-wave method for TM polarization, *J Opt Soc Am A,* Vol. 13, No.4, (April 1st 1996), pp. 779-784, ISSN 1084-7529.

Lenaerts, C., Michel, F., Tilkens, B., Lion, Y. & Renotte, Y. (2005). High transmission efficiency for surface plasmon resonance by use of a dielectric grating, *Appl Opt,* Vol. 44, No.28, (October 1st 2005), pp. 6017-6022, ISSN 1559-128X.

Lévesque, L., Paton B.E. & Payne S.H. (1994). Precise thickness and refractive index determination of polyimide films using attenuated total reflection, *Appl Opt,* Vol. 33, No.34, (Dec. 1st, 1994), pp. 8036-8040, ISSN 1559-128X.

Lévesque, L. & Rochon, P. (2005). Surface plasmon photonic bandgap in azopolymer gratings sputtered with gold, *J Opt Soc Am A*, Vol. 22, No. 11, (November 1st 2005), pp. 2564-2568, ISSN 1084-7529.

Lévesque, L. (2011). Determination of thickness and optical constants of metal films from an extended ATR spectrum by using a statistical method, *Optics & Laser technology*, Vol. 43, No. 1, (February 2011), pp. 14-19, ISSN 0030-3992.

Li, L. & Haggans, C.W. (1993). Convergence of the coupled-wave method for metallic lamellar diffraction gratings, *J Opt Soc Am A*, Vol. 10, No. 6, (June 1st 1993) , pp. 1184-1189, ISSN 1084-7529.

Magnusson, R. & Wang, S.S. (1992). New principle for optical filters, *Appl Phys Lett*, Vol. 61, No. 9, (August 31 1992), pp. 1022-1024, ISSN 003-6951.

Moharam, M.G. & Gaylord, T.K. (1981). Rigorous coupled-wave analysis of planar-grating diffraction, *J Opt Soc Am* (1917-1983), Vol. 71, No. 7, (July 1st 1981), pp. 811-818, ISSN 0030-3941.

Moharam, M.G., Grann, E.B., Pommet D.A. & Gaylord T.K. (1995). Formulation for stable and efficient implementation of the rigorous coupled-wave analysis of binary gratings, *J Opt Soc Am A*, Vol. 12, No.12, (May 1st 1995), pp. 1068-1076, ISSN 1084-7529.

Pai, D.M. & Awada, K.A. (1991). Analysis of dielectric gratings of arbitrary profiles and thicknesses, *J Opt Soc Am A*, Vol. 8, No. pp.755-762, ISSN 1084-7529.

Park, S., Lee, G., Song, S.H., Oh, C.H. & Kim, P.S. (2003). Resonant coupling of surface plasmons to radiation modes by use of dielectric gratings, *Opt Lett* , Vol. 28,No. 20, (October 15th 2003), pp. 1870-1872, ISSN 0146-9592.

Raether, H. (1988). *Surface Plasmons on Smooth and Rough Surfaces and on Gratings*, Springler-Verlag, ISBN 3-540-17363-3, Berlin, Germany.

Roberston, WM. & Fullerton E. (1989). Reexamination of surface-plasma-wave technique for determining the dielectric constant and thickness of metal films, *J Opt Soc Am B*, Vol. 6, No. 8, (August 1st 1989), pp. 1584-1589, ISSN 0740-3224.

Sheng, P., Stepleman, R.S. and Sanda, P.N. (1982). Exact eigenfunctions for square-wave gratings: Application to diffraction and surface-plasmon calculations, *Phys Rev B*, Vol. 26, No. 6, (15 September 1982) pp. 2907-2916, ISSN 1098-0121.

Tibuleac, S. & Magnusson, R. (1997). Reflection and transmission guided-mode resonance filters, *J Opt Soc Am A*, Vol. 14, No. 7, (July 1st, 1997) , pp. 1617-1626, ISSN 1084-7529.

Wang, S.S, Magnusson, R., Bagby, J.S., & Moharam, M.G. (1990). Guided-mode resonances in planar dielectric-layer diffraction gratings, *J Opt Soc Am A*, Vol. 7, No.8, (August 1st, 1990), pp.1470-1474, ISSN 1084-7529.

Welford, K. (1991). Surface plasmon-polaritons and their uses, *Opt Quant Elec,* Vol. 23, No. 1, (January 1991), pp.1-27, ISSN 0036-8075.

Wood, R.W. (1902). Remarkable spectrum from a diffraction grating, Philos Mag , Vol. 4, No. 21, (September 1902), pp. 396-402, ISSN 1941-5990.

Ye, Y.H. & Zhang, J.Y. (2004) . Middle-infrared transmission enhancement through periodically perforated metal films, *Appl Phys Lett* , Vol. 84, No.16, (April 19th,2004), pp. 2977-2979 , ISSN 003-6951.

13

Quasi-optical Systems Based on Periodic Structures

Gennadij Vorobjov[1], Yulya Shulga[1] and Vitaliy Zhurbenko[2]
[1]*Sumy State University, Ukraine*
[2] *Technical University of Denmark,*
[1]*Ukraine*
[2]*Denmark*

1. Introduction

Open resonators and open waveguides are widely used in millimeter and submillimeter wave electronics because they provide lower loss and higher Q-factor in comparison to the standard closed structures [Valitov et al., 1969; Shestopalov, 1985; Weinstein, 1966, 1995] . Examples of high performance measurement equipment employing open resonators (based on spherical or semispherical mirrors) include resonant wave meters, reference oscillators, systems for measurement of intrinsic electromagnetic properties of dielectric materials, and others [Valitov et al, 1969; Milovanov and Sobenin, 1980; Valitov and Makarenko, 1984]. Semispherical and spherocylindrical open resonators in combination with reflective diffraction gratings are used in various diffraction radiation oscillators [Shestopalov, 1976, 1985, 1991] providing higher frequency stability and output power in comparison to the standard devices such as traveling-wave tubes, klystrons, and magnetrons. Open resonators with echelette-type corner mirrors have been chosen as the basis for highly efficient Gunn and IMPATT diode oscillators. Quasi-optical resonators of such devices adopt reactive reflection and transmission-type schemes [Sukhoruchko et al., 2003]. Open resonators has found a wide practical application in relativistic electronics. Several types of oscillators and amplifiers have been created on their basis [Balakirev et al., 1993]. It has been demonstrated by [Weinstein and Solntsev, 1973] that Smith-Purcell effect (diffraction radiation) can be used to build an amplifier based on an open waveguide.
The constantly growing interest in the implementation of millimeter and submillimeter wave radiation in different areas of science and technology puts forward demands for components with high performance and flexible functionality. One of the most promising strategies for the development of such components is to modify their electromagnetic structure in order to increase operating frequency band and improve efficiency of interaction between the electron beam and electromagnetic wave. Following this strategy, several new approaches have been proposed based on modification of open coupled electromagnetic structures such as coupled open resonators [Shestopalov, 1991], open waveguides [Weinstein, 1995; Weinstein and Solntsev, 1973], open resonators with dispersion elements [Marshall et al., 1998], as well as the metal-dielectric structures [Shestopalov, 1991] which are particularly useful for electromagnetic wave excitation

employing Cherenkov effect. Unfortunately, the practical realization of the proposed structures is a rather difficult task because of complicated electromagnetic analysis and a lack of systematic approach.
The objective of this chapter is to perform a comparative analysis of classical quasi-optical structures and their new modifications. The strategies for further development of these structures will be discussed based on the performed analysis.
The chapter starts with a description of basic properties of a classical regular open resonator as a basis for new modified millimeter and submillimeter wave coupled resonant structures. The properties of open resonators and open waveguides based on periodic metal and metal-dielectric discontinuities excited by both the electron beam and the surface wave of the dielectric waveguide are considered.

2. The coupled quasi-optical systems based on open resonators

This section is dedicated to the analysis of simple (regular) resonant systems and coupled quasi-optical systems based on periodic metal and metal-dielectric structures such as open resonators with diffraction grating, coupled open resonators and resonators with layered metal-dielectric structures.

2.1 The main properties of classical quasi-optical resonators

A classical quasi-optical resonator consists of two-mirrors. In the simplest case considered here, the open resonator contains two opposing flat infinitely thin parallel aligned disks. This system of mirrors is referred as plane-parallel resonator and known from optics as the main part of Fabry-Perot interferometer.
The plane-parallel resonators exhibit a number of valuable properties: sparse spectrum of resonant frequencies, homogeneous field along the symmetry axis of the resonator and the wavelength in the resonator is slightly different from the wavelength in the free space.
While simple, this arrangement is rarely used in practice due to the difficulty of alignment, comparatively large size, and insufficient mode separation. Therefore the resonators based on the reflectors with quadratic phase correction are more promising in the millimeter and submillimeter wave range. These type of resonators are referred as confocal resonators and contain spherical mirrors. These resonators exhibit a better spectral resolution in comparison to the plane-parallel resonators. Besides, confocal resonators are less sensitive to misalignment. The resonator with spherical reflectors typically exhibits lower power loss per one propagation in comparison to the open resonator with plane mirrors having the same aperture. The other important advantage is the large separation between the fundamental and the higher order modes TEM_{mnq}, where $m, n = 0, 1, 2, ...$ is the number of half-waves in transverse direction and q is the longitudinal index which corresponds to the number of half-waves in the direction of propagation. For the resonator with spherical mirrors the resonance distances or the resonance wavelengths of the oscillation modes should comply with the following condition:

$$\frac{2H}{\lambda} = q + \frac{1}{\pi}(m + 2n + 1)\arccos\sqrt{g_1 g_2}, \tag{1}$$

where H is the distance between the mirrors; λ is the wavelength in the open resonator; $g_1 = 1 - \frac{H}{R_1}$; $g_2 = 1 - \frac{H}{R_2}$; R_1, R_2 are the curvature radii of the mirrors.

Limiting the size of resonator's apertures results in radiation loss and has negligible effect on the field distribution in the open resonator. Therefore the field must be concentrated close to the center of the mirror in order to reduce the losses. This, in turn, restricts the choice of ratio between the radius of the curved mirrors and the distance between them. In order to construct resonators with the field concentrated close to the center of the mirror, the distance between the mirrors must be selected within the following intervals:

$$0 < g_1 g_2 < 1. \tag{2}$$

This expression is known as the condition of stability of the resonator with quadratic correction; g_1, g_2 are the parameters that depend on geometry of the resonator.

The behavior of oscillations in plane-parallel and spherical-mirror resonators is quite different. The field distribution in the plane-parallel resonator mostly depends on the dimensions of the plane plates, while field distribution in the resonator with spherical mirrors is mostly determined by their radius and the ratio of the distance between mirrors and the radius, $\frac{H}{R}$.

The semi-spherical resonators which consist of a plane and a spherical mirror have also received a great deal of interest in microwave and millimeter-wave applications. It is known that the fundamental modes of the semi-spherical resonator are represented by the azimuthal oscillations TEM_{m0q}. If the field spot on the plane mirror is considerably smaller than its diameter then the semi-spherical resonators can be substituted by the equivalent resonators with two spherical mirrors having doubled distance between them. The distribution of amplitudes in both cases is identical to a high degree of accuracy. The Q-factor of the semi-spherical resonator depends on diffraction losses at the edges of the plane and the spherical mirrors, ohmic losses in the mirrors, the coupling losses and the losses related to attenuation in the medium.

2.2 Resonators with periodic metal grating

The plane-parallel mirror of the semi-spherical resonator can be substituted by a diffraction grating as it is shown in Fig. 1. Such an electrodynamic structure is often used in diffraction radiation oscillators - orotrons [Shestopalov, 1976, 1991; Marshall et al., 1998; Ginzburg et al., 2000; Bratman et al., 2002; Rusin et al., 2002].

The orotron's operation principle is based on the diffraction radiation effect caused by the electron beam propagating above the diffraction grating of the open resonator. The electron beam interacts with the incident field diffracted from the grating which results in oscillation and amplification of the electromagnetic signal. Therefore, the orotron's output characteristics are strictly defined by the properties of the implemented open resonator. The periodic structure in the open resonator of the orotron considerably changes the characteristics of the previously described classical resonant quasi-optical structures. The substitution of the plane mirror by a diffraction grating considerably increases the total loss resulting in the Q-factor degradation by almost four times. The decrease of the Q-factor occurs as the result of additional losses, which are originated from a power leakage of the waveguide waves propagating along the grooves to the edges of the mirror where the reflection coefficient is not equal to one.

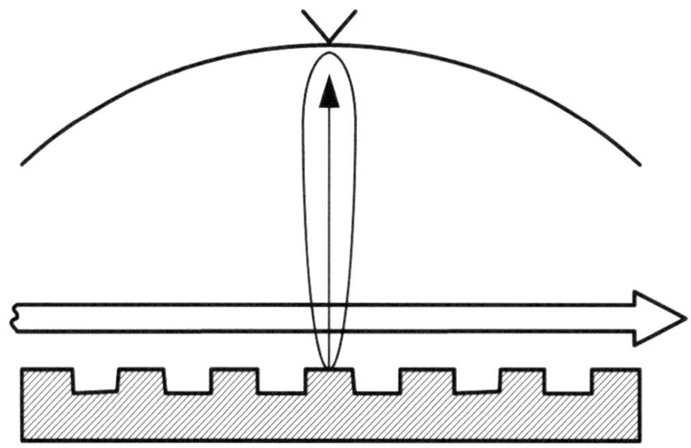

Fig. 1. Semispherical open resonator with diffraction grating

To overcome this drawback, a semi-spherical resonator where only the central part of the plane mirror was covered with the diffraction grating, has been proposed [Shestopalov, 1976, 1991]. This resonator has a wider distance between the oscillation frequencies. The achieved radiation loss depends on the parameters and the position of the grating. The width of the grating defines the number of the oscillation modes excited in the open resonator and the frequency of the higher order resonances. Losses in the open resonator are greatly dependent on the ratio between the period of the grating and the wavelength. The maximally achieved Q-factor of the resonator also greatly depends on the groove depths of the reflective grating oscillations could be varied by several times.

The fundamental mode of the semi-spherical resonator with a local diffraction grating is TEM_{20q}. The research in [Shestopalov, 1976, 1991] proved that the perturbation caused by the grating is insignificant in such a system if the minimum of the field distribution is above the boundary between the grating and the mirror. This is the case when the width of the diffraction grating is larger or equal than the width of the main lobe in TEM_{20q} oscillation mode.

Corner-echelette open resonators are widely used for realization of semiconductor sources in the microwave and millimeter-wave range. For example, modifications of quasi-optical reflection and transmission-type solid-state pump oscillators with spherical-corner-echelette open resonator have been shown in [Belous et al., 2003]. As shown in [Sukhoruchko et al., 2003], the corner-echelette resonator has the following properties: the degree of sparseness of the spectrum is lower than for the resonator with plane echelette mirror; however, the spectrum contains the oscillation modes with extremely high Q-factor, which are known as the quasi-fundamental oscillation modes; the field of the quasi-fundamental oscillation modes is concentrated around the axis of the open resonator resulting in a larger power density in comparison to the fundamental and higher order oscillation modes; the field distribution close to the surface of the corner-echelette mirror transforms and near the center of the resonator becomes similar to the field in a rectangular waveguide; corner-echelette mirror can be considered as a multi-step impedance transformer.

2.3 Coupled open resonators

The work by [Shestopalov, 1991] is dedicated to the diffraction radiation devices employing coupled open resonators. The coupled resonators have an advantage of providing a wider operating frequency range in comparison to the single resonator structures. The coupling between open resonators can be realized either by means of the field diffracted at the edges of the mirrors using series positioning of the resonators (Fig. 2a) or the field diffracted on a metal-strip grating using parallel connection of open resonators (Fig. 2b) with respect to the axis of the distributed excitation source. In the electron devices, the electron beam is such a source. In case of experimental modeling of diffraction radiation it is the surface wave of the single-mode dielectric waveguide.

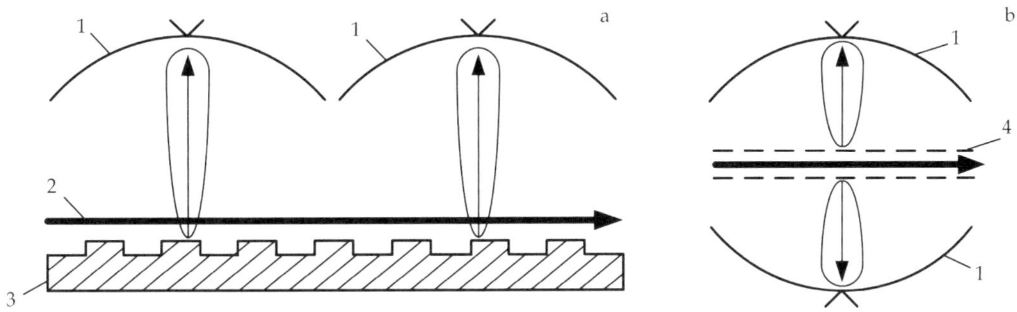

Fig. 2. Electrodynamic systems based on coupled open resonators: a - series connection of open resonators; b - parallel connection of open resonators

The system of series open resonators is, in the case shown in Fig. 2a, consists of two semi-spherical resonators with the common plane mirror realized as a reflective diffraction grating. In the parallel coupling case (Fig. 2b), a two-layer metal-strip diffraction grating is placed between the spherical mirrors.

Systems of coupled open resonators consist of spherical mirrors **1** with the radius R=60 mm and aperture A=55 mm reduced to 35 mm along the axis of the dielectric waveguide **2**. The lower plane mirror **3** of the system shown in Fig. 2a is either a reflective or semitransparent diffraction grating and serves as a common mirror for the first and the second open resonator. In the system with parallel open resonators, plane mirrors **4** with semitransparent diffraction gratings in their central sections were placed between spherical mirrors **1**.

Parameters of the gratings are chosen to ensure the operation at a frequency f_0 = 46 GHz. These gratings transform the surface wave of the dielectric waveguide into a free space wave propagating normal to the surface of the grating [Shestopalov, 1976]. The energy is coupled out from the system through the coupling slots in the spherical mirrors. The signals are then fed to a detector and measured using a standard measurement equipment [Shestopalov, 1976, 1991].

The described coupled resonators have been analyzed with regard to their spectra and resonance characteristics of oscillation. The measured characteristics of the equivalent single hemispherical and spherical open resonators have been used as a reference.

Fig. 3 shows the resonant frequencies versus the distance between the mirrors (H) in the system with coupling through the diffraction field (Fig. 2a) and in a reference hemispherical open resonator. The data presented in Fig. 3 characterizes the capability of the considered resonance system to support a limited number of TEM_{mnq} oscillation modes.

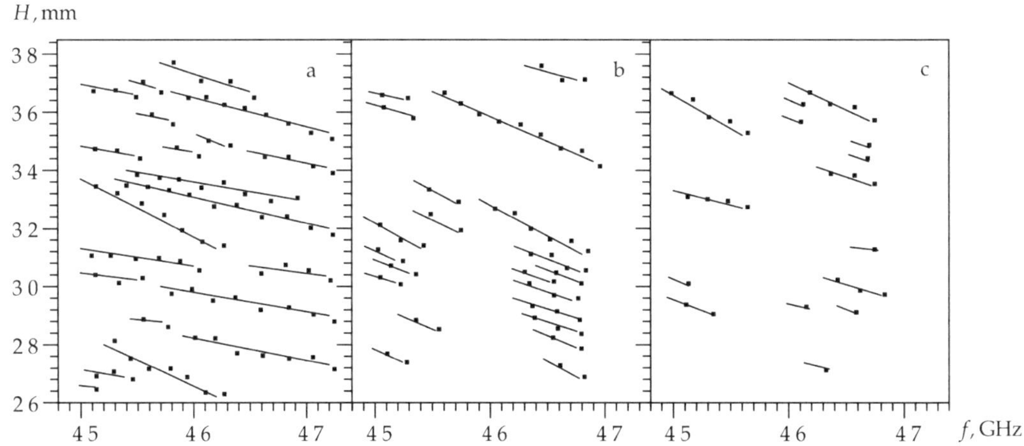

Fig. 3. Spectra of resonant frequencies of (a) a hemispherical resonator and (b, c) a diffraction-coupled resonators with (b) reflective and (c) strip diffraction gratings.

The data in Fig. 3a shows that for the hemispherical open resonator, the fundamental TEM_{00q} modes exist in the entire frequency range $f = 45 - 47\,GHz$ while changing the distance between the mirrors. The implementation of a dispersive element such as the reflective diffraction grating in the open resonator allows for the modes with transverse indices m and n. The TEM_{20q} oscillation mode is usually a fundamental mode for such resonators [Shestopalov, 1991]. In addition to the fundamental modes, depending on the parameters of the open resonator and the diffraction grating, the other types of higher order oscillations (e.g., TEM_{02q}) occur influencing the coupling between the two open resonators through the diffraction fields.

Figure 3b shows the resonances of two coupled open resonators tuned to a frequency f=46 GHz. As can be seen from these spectra, the second hemispherical open resonator is excited at the edge points of the frequency band in the interval H=27-33 mm. There are no oscillations around the resonance frequency of the open resonator, which is due to the minimum amplitude of the diffraction field in a case when the diffraction-grating–dielectric-waveguide system emits radiation along the normal. Detuning from the frequency f_0 in the interval $\Delta f \approx \pm 1\,GHz$ leads to the deviation of the main lobe direction from the normal, which increases the intensity of the diffraction field and, consequently, leads to the excitation of the second resonator at the edges of the frequency range. As the distance H increases, the coupling between the resonators becomes stronger reaching its maximum magnitude when the distances between the mirrors are equal to each other. In this case, oscillations in the second open resonator arise even at a frequency $f \approx 46\,GHz$.

Coupled open resonators with a strip grating at the center of the common plane mirror (Fig. 2b) exhibit similar properties. The decrease in the number of oscillation modes in such a system (Fig. 3c) is due to the selective properties of the employed diffraction grating [Shestopalov, 1991]: the intensity of radiation emitted from the volume of the open resonator to free space through the diffraction grating reaches its maximum at $H \approx (\lambda/4)(2N+1)$, while the accumulation of energy inside the volume of the open resonator appears at values $H \approx (\lambda N)/2$, where λ is the radiated wavelength, N = 1, 2, ... The coupling in open resonators reaches its maximum when the distances between the resonators are approximately equal to each other, i.e., when the resonators are tuned to close frequencies. The typical response of the previously described coupled open resonators is presented in Fig. 4. Here P/P_{max} is the power in the open resonators normalized to the maximum power P_{max}. The resonance curve of a hemispherical open resonator is shown for comparison (curve 1) in the same figure. As can be seen from the presented data, the transmission band of coupled open resonators measured at the level of $0,5P_{\text{max}}$ increases by a factor of nearly two, resulting in $\Delta f \approx 250\,MHz$. The resonance curves corresponding to coupled open resonators with reflective and strip diffraction gratings virtually coincide with each other under these conditions, which indicates the existence of efficient coupling in these systems through the diffraction of the fields at the periphery of the mirrors.

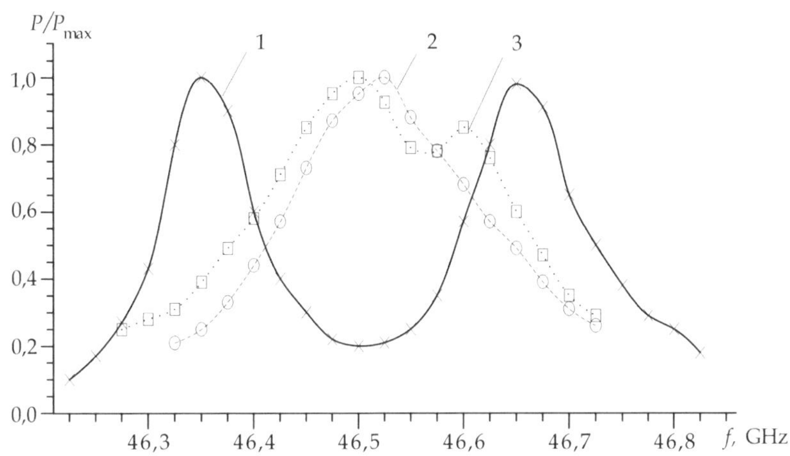

Fig. 4. Response of (1) a hemispherical open resonator and (2, 3) coupled open resonators with (2) metal-strip and (3) reflective diffraction gratings

The open resonator with spherical mirrors, which is a basis for the second scheme of coupled resonators (Fig. 2b) supports similar to the case of the hemispherical open resonators fundamental TEM_{00q} modes. This follows from the analysis of the achieved resonance frequencies. The field distribution in an open resonator with spherical mirrors is the same as in the hemispherical open resonator [Shestopalov, 1976]. However, the distance between the resonance frequencies in the open resonator with spherical mirrors is two times smaller than in a hemispherical open resonator. Inserting an additional plane mirror with a strip diffraction grating in a spherical open resonator will result in the spectrum of the coupled system similar to the spectrum of the hemispherical open resonator (Fig. 3a). The metal-strip diffraction gratings couple two hemi-spherical open resonators simultaneously filtering out the angular spectrum of plane waves excited in the system. Consequently, the variation of the position of these diffraction gratings in the volume of the resonator with respect to the spherical mirrors changes the spatial distribution of the fields corresponding to the oscillation modes excited in the considered system of coupled open resonators. Similar to the hemispherical open resonator with a reflective diffraction grating, TEM_{20q} and TEM_{02q} oscillation modes, as well as the higher order oscillation modes arise due to introducing a coupling element such as a double-layer diffraction grating.

The measured data for the resonance curves of coupled open resonators indicates that the achieved bandwidth of the system becomes much broader when the open resonators are tuned to close frequencies rather than in the case when the resonators are coupled through the diffracted fields. Fig. 5 presents the response of the open resonators coupled through a strip diffraction grating and for the open resonator with spherical mirrors. The achieved bandwidth of the coupled system was observed within the range $f = 44,5 \div 49,5\,GHz$ for equal distances of spherical mirrors from the planes of the coupling element with a total distance between the spherical mirrors equal to H=31 mm. The achieved bandwidth measured at the $0,5P_{\max}$ power level is equal to $\Delta f \approx 1,3\,GHz$. The narrowing of the transmission band of coupled open resonators observed in the higher frequency band (f=48,5 GHz) is due to the deviation of the radiation pattern for the diffraction grating-dielectric waveguide system from the normal and, consequently, the decrease in the coupling coefficient between the resonators.

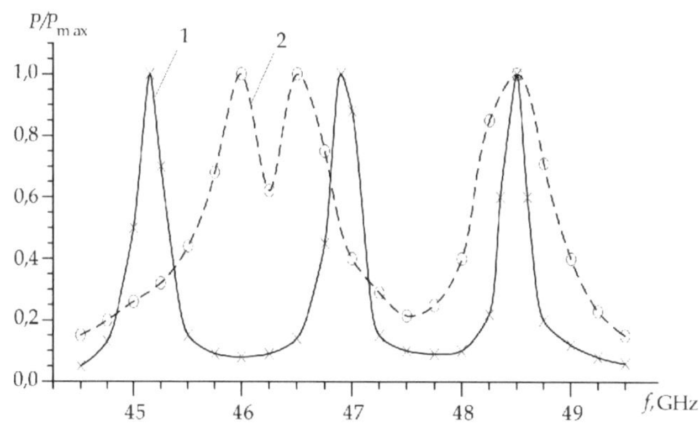

Fig. 5. Response of (1) the spherical open resonator and (2) the system of resonators coupled through semitransparent diffraction gratings.

Analysis of the achieved bandwidth Δf for the single resonator and coupled systems shows that the maximum bandwidth in systems with comparable H can be achieved when two open resonators are coupled through a strip diffraction grating. The bandwidth of the system with parallel open resonators is almost five times wider than the bandwidth of the system with series open resonators. It should be noted that the Q-factor of the coupled open resonators is of the same order as the Q-factor of the single open resonators. Therefore the open resonators coupled through the strip diffraction grating are preferable for systems requiring wideband operation. Such resonators also provide a reduced size of the system along the electron beam propagation axis.

2.4 Open resonators with metal-dielectric structures

Coupled systems based on open resonators and open waveguides with metal-dielectric structures allow to realize different modes of energy transformation depending on parameters of the electromagnetic system [Shestopalov, 1991].

The simplest open resonator employing a metal-dielectric slab is shown in Fig. 6a. It consists of a metal plane and a dielectric slab with a planar metallic diffraction grating on its surface.

ε is the permittivity of the dielectric. The source of electromagnetic energy is distributed along the grating. It excites various spatial harmonics of Cherenkov diffraction radiation of order $n = 0, \pm 1, \pm 2, \ldots$ and the power density S_n, which depends on the parameters of the structure. Fig. 6a demonstrates the excitation of Cherenkov ($S_{0\varepsilon}$) and minus first diffraction ($S_{-1\varepsilon}$) harmonics in the dielectric as well as minus first diffraction harmonic (S_{-1v}) in open volume, which can be reflected back by a metal plane and fed to the metal-dielectric channel. A number of numerical and experimental methods for simulation of different excitation modes of Cherenkov diffraction radiation has been developed [Vorobyov et al., 1997, 2007]. They allow to determine the quantitative relation between the power densities of spatial harmonics in the structure as well as to optimize and tune their parameters.

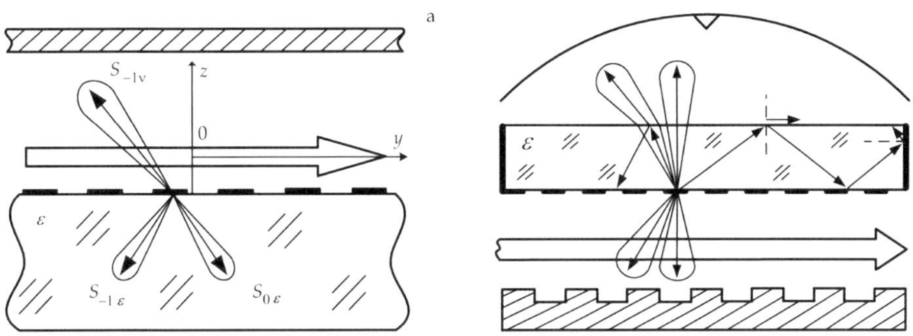

Fig. 6. Quasi-optical resonators based on metal-dielectric slabs

A more complicated case of the open resonator with a metal-dielectric structure is shown in Fig. 6b. The resonator consists of a spherical mirror, a plane mirror such as a reflecting diffraction grating, and a layered metal-dielectric structure between the two mirrors. Such an electromagnetic structure is often used in Cherenkov diffraction oscillators. Fig. 6b demonstrates possible modes of Cherenkov diffraction radiation excited by a source of electromagnetic energy distributed between the metal-dielectric grating and plane mirror.

The metal-dielectric slab (Fig. 6b) of the open resonator introduces qualitatively new electromagnetic properties in such a system. It is possible to attenuate the power in the open resonator, increase the amplitude of the oscillating wave and the value of Q-factor as well as to improve selectivity by choosing parameters of the metal-dielectric slab.

3. Coupled open waveguides employing periodic structures

This section describes the main properties of quasi-optical open waveguides with periodic metal-dielectric structures excited by distributed sources of electromagnetic energy such as electron beam or surface wave of a dielectric waveguide. Such structures are promising for the design of low-voltage amplifiers based on Smith-Purcell effect [Weinstein and Solntsev, 1973; Smith and Parcell, 1953] and other microwave and millimeter wave electron devices [Joe et al., 1994, 1997].

3.1 Amplifiers based on Smith-Purcell effect

Fig. 7 shows the structure of the amplifier using a planar layered metal-dielectric stack and based on Smith-Purcell effect. The open waveguide of the considered system consists of the periodic rectangular grating structure **1** with the period of 2*l*, width 2*d* and grating depth of *h*; the planar layered metal-dielectric structure **2** with the thickness $\Delta = H - s$ which is positioned in parallel to the grating at a distance s. A non-relativistic sheet electron beam **3** with the finite thickness (r-b) is propagating along the axis 0y at a distance *b* above the grating. The entire structure is considered to be infinite within the plane x0y.

The electromagnetic problem is solved using the method of partial domains. The field in each domain is determined from the Maxwell equations, equation for the electron beam propagation, and corresponding boundary conditions. In order to obtain the dispersion equation, we have to perform the following operations: determine the linear approximation of the equation for the variable component of the convectional current intensity and the field in the beam, and transform it into a homogeneous form; determine the electromagnetic field in the interaction region in hot (with electron beam) and cold (without electron beam) regime.

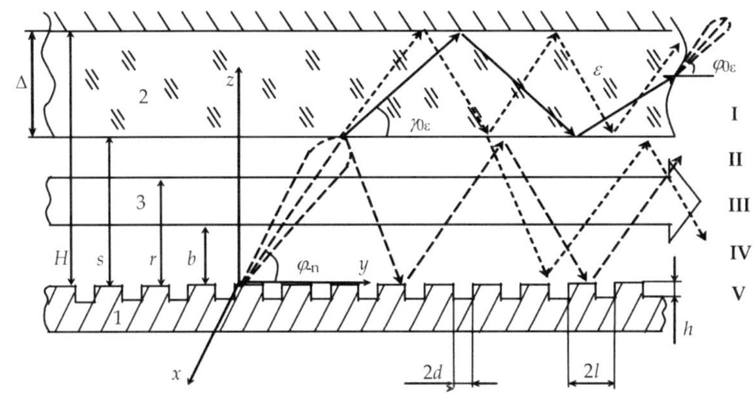

Fig. 7. The amplifier with the metal-dielectric layer based on the Smith-Purcell effect: 1 - periodic metal structure, 2 - planar layered metal-dielectric structure, 3 - electron beam

To this end, the electric field E, the beam velocity v_0, and the charge density ρ_0 are expressed as a sum of constant and small harmonically time-dependent variable quantities [Shmatko, 2008]. The charge density constant ρ_0 of the beam is considered to be compensated by external sources. The solution of the problem in such a way leads to the following dispersion:

$$1+\frac{k}{ld}\tan kh\sum_{n=-\infty}^{\infty}\frac{\sin^2\alpha_n d}{\alpha_n^2}\left[\frac{\Gamma_n\left(\cos\left(\xi_n\Gamma_n(b-r)\right)-\sin\left(\xi_n\Gamma_n(b-r)\right)\Pi\right)}{\Pi+\sin\left(\xi_n\Gamma_n(b-r)\right)}\right]=0\,, \tag{3}$$

where $\Pi = \dfrac{\Gamma_n\cos\left(\xi_n(r-s)\right)\dfrac{\sigma_n}{\varepsilon\xi_n}\tan\left(\varsigma_n(s-H)\right)+\sin\left(\xi_n(r-s)\right)}{\cos\left(\xi_n(r-s)\right)-\dfrac{\sigma_n}{\varepsilon\xi_n}\tan\left(\varsigma_n(s-H)\right)\sin\left(\xi_n(r-s)\right)}$; $k=\dfrac{\omega}{c}$ - is the wave number; ω - is the frequency; c - is the speed of light; ε - is the relative permittivity of the dielectric layer; $\alpha_n = \alpha_0 + \dfrac{\pi n}{l}$ - is the propagation constant of the electromagnetic wave propagating along the axis 0y; $n = 0, \pm 1, \pm 2$ - is the spatial harmonic number; $\xi_n = \sqrt{\kappa^2 - \alpha_n^2}$, $\sigma_n = \sqrt{\varepsilon\kappa^2 - \alpha_n^2}$ - are the transverse wave numbers; $\kappa = \frac{kl}{\pi}$; $\Gamma_n = \sqrt{1 - \frac{\omega_p^2}{(\omega - v_0\alpha_n)^2 - \omega_p^2}}$; $\omega_p = \sqrt{\frac{e\rho_0}{m_e\varepsilon_0}}$ - is the plasma frequency of the electron beam (e and m_e - are the charge and the mass of the electron, ε_0 - is the electric constant).

The following analysis of equation (3) concerns defining the propagation constant α_n, which is generally a complex number. The imaginary part $-i\alpha_n$ is responsible for the solutions increasing along the axis 0y and specifies the electromagnetic wave amplification in the system. The analysis of the given dispersion equation would be difficult without simplifications. If we separate this equation in three terms, which correspond to zeroth order harmonic, first harmonic, and the sum of the rest harmonics, and then use an approximation of the maximum interaction between the electron beam the fields of the slow wave structures (b = 0, s = r), it is possible to reformulate (3) into the following:

$$\frac{\cot(\pi\kappa\delta)}{\kappa\theta} = \frac{1}{\sqrt{(\mu+1)^2 - \kappa^2}}\left(\frac{\Gamma_1(1+\Gamma_1 F\mathrm{M})}{\Gamma_1 F + \mathrm{M}} - 1\right) - \left(2\ln\left(\sin\left(\frac{\pi\theta}{2}\right)\right) + \frac{\varepsilon\cot\left(\pi\chi\sqrt{\varepsilon\kappa^2 - \mu^2}\right)}{\sqrt{\varepsilon\kappa^2 - \mu^2}}\right), \quad (4)$$

where

$\Gamma_1 = \sqrt{1 - \frac{g^2\kappa^2}{[\kappa - \beta(\mu+1)]^2 - g^2\kappa^2}}$; $\mathrm{M} = \tanh\left(\pi\zeta\Gamma_1\sqrt{(\mu+1)^2 - \kappa^2}\right)$; $F = \frac{\sqrt{\varepsilon\kappa^2 - (\mu+1)^2}}{\varepsilon\sqrt{\kappa^2 - (\mu+1)^2}}$; $g = \frac{\omega_p}{\omega}$ is the spatial charge parameter, in practice $g \le 0{,}01$; $\beta = \frac{v_0}{c}$ - is the relative electron beam velocity, $\mu = \kappa\hat{\alpha}_n - 1$; $\hat{\alpha}_n = \frac{\alpha_n}{k}$ - is the unitless propagation constant; $\gamma = \frac{h}{l}$, $\zeta = \frac{b}{l}$, $\theta = \frac{d}{l}$, $\chi = \frac{H}{l}$ - are the unitless geometrical parameters of the system.

The solution to the dispersion equation is found by using Newton iterative method for the range of electron velocities $\beta = 0{,}05 \div 0{,}2$ and various values of the electromagnetic parameters of the system κ, χ, ζ and $\varepsilon = 1 \div 210$. The lower ε limit corresponds to the case where there is no dielectric between the grating and the metal mirror; the upper limit corresponds to the case when it is possible to excite Cherenkov radiation at non-relativistic electron beam velocities ($\beta \approx 0{,}07$).

The numerical analysis of the dispersion equation for an open waveguide with no dielectric layer shows that there are two direct waves and two waves traveling in the backward direction exist in the system without the presence of an electron beam. They have different wave numbers and corresponding phase velocities. In addition to the previously described electromagnetic waves in the open waveguide, there are also two electron beam waves: a fast space-charge wave and a slow space-charge wave. All four electromagnetic waves in the system while synchronized with the electron beam spatial waves have regions with a positive amplitude growth that allows for signal amplification and realization of the following regimes varying parameter β: surface wave mode with the maximum amplitude exists when synchronized with the slow space-charge wave; the volume waves of the diffraction radiation at the angle less than $\pi/2$ with respect to the grating plane which transfer power from the beam to the field by means of the slow and fast space-charge waves. It must be noted that the regime employing fast space-charge wave is observed starting from the relative beam thickness $\zeta = 0{,}6$, and the regime employing slow space-charge wave is observed starting from $\zeta = 0{,}02$. Depending on the excitation region the maximum amplitude of the amplified signal is observed at $\zeta = 0{,}4 \div 0{,}6$ that corresponds to the electron beam thickness $r \approx 0{,}1$ mm, which is typical for the microwave tubes. The further increase in r has no influence on the amplitude of the amplified signal but results in generation of a discrete set of radiated electromagnetic waves the number of which depends on thickness of the electron beam and the frequency. This effect can be physically explained by the dispersion properties of the electron beam and partial reflection of the electromagnetic waves from its boundaries (equivalent of the low reflection coefficient resonator).

Introducing a dielectric layer with small values $\varepsilon \approx 3$ will result in changing the phase velocities of the electromagnetic waves in the open resonator and synchronization with the electron beam. This also leads to the generation of additional waves with the parameters close to those for the case with $\varepsilon = 1$. Transverse wave numbers σ_n, ξ_n (3) determine the wave modes classification in the open waveguide: volume waves propagating between the periodic structure and the metal plane; the volume waves in the dielectric layer; the surface waves above the periodic structure.

The increase in permittivity ε of the dielectric layer leads to accumulation of power from volume waves that is caused by the improvement of its resonance properties due to reflection from the boundaries of the dielectric. Fig. 8 represents graphically the solution of the dispersion equation with regard to the real (Re μ) and imaginary (Im μ) parts of the gain factor versus the parameter β for $\varepsilon = 50, \kappa = 0{,}083, \chi = 10$. The presented curves allow to analyze the propagation properties for fourteen wave modes in the open waveguide, two of which, 7 and 8, are the fast and slow space-charge waves of the electron beam respectively. The waves from 1 to 6 are propagating in the same direction as the electron beam propagation, while the waves 9 and 10 are propagating in the opposite direction. The parameters of the mentioned waves satisfy the condition for their propagation in the dielectric layer. Consequently, the increase of the number of wave modes in the open waveguide results in distribution of the energy between them and leads to a decrease of the amplitude growth factors for some waves as compared to the case when $\varepsilon = 1$.

Parameter χ (the distance between the mirrors of the open waveguide normalized to the grating period) has a significant influence on the propagation properties of the modes and defines the field distribution between the mirrors of the open waveguide. Fig. 9 graphically represents the solution of the transcendent equation (4) with regard to the real and imaginary parts of the gain factor versus β for different values of the parameter χ and the direct volume wave of the periodic structure ($\varepsilon = 1$). The presented data illustrates that changing the distance from $\chi = 8$ to $\chi = 14$ leads to a considerable decrease in the amplitude of the signal. The maximum amplitude of the signal is observed at the radiation along the normal direction ($\chi = 8$, $\mu \approx 0$) while the minimum is observed in a tangential direction ($\chi \approx 14$, $\mu \approx 0{,}073$). However, it is not possible to ensure the excitation of the traveling wave mode along the axis of the open waveguide for radiation in the normal direction. In practice, this would result in a feedback and instability. This operation mode is similar to the operation of the microwave tubes such as orotron and diffraction radiation oscillator [Shestopalov, 1991].

It should also be noted that increasing the distance between the mirrors results in increase of a number of surface waves and decrease of the gain factor for the volume waves. In the extreme case when the values $\chi \to \infty$, the volume waves transfer into surface waves and the system is similar to the traditional devices such as the backward-wave oscillator and the traveling-wave tube.

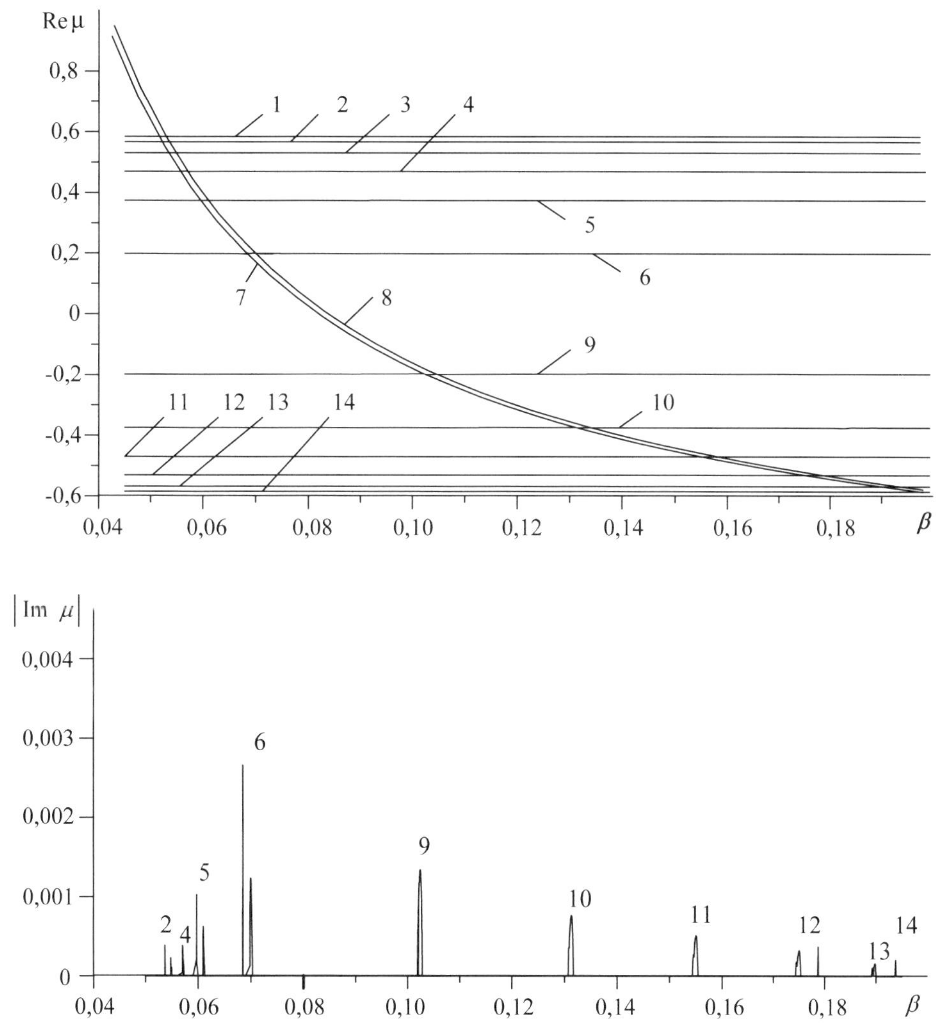

Fig. 8. Solutions of the dispersion equation (4) for $\varepsilon = 50$

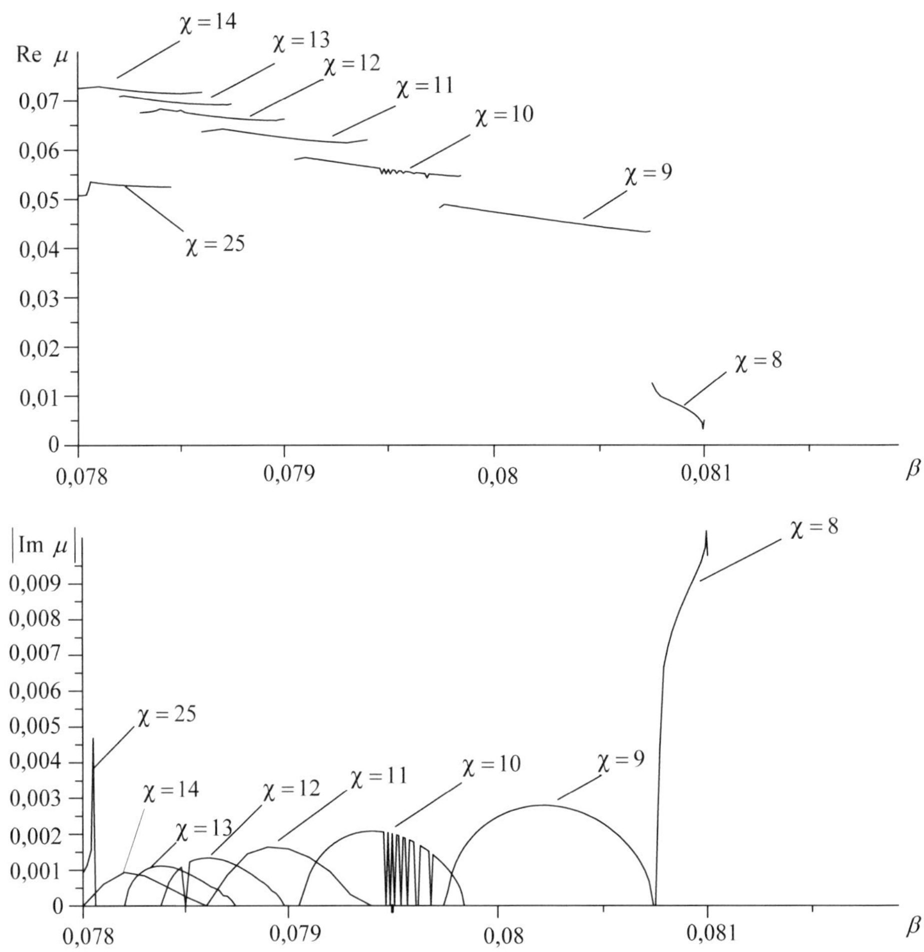

Fig. 9. Influence of the parameter χ on the solutions of the dispersion equation (4) for $\varepsilon = 1$

3.2 Experimental modeling of coupled open waveguides

The experimental modeling is one of the most efficient methods for solving problems of diffraction electronics. The radiation of the electron beam is simulated by a surface wave in the planar dielectric waveguide placed above the diffraction grating. The modeling techniques have been sufficiently developed and summarized in the literature [Shestopalov, 1976, 1985, 1991]. Nevertheless, each structure has its own specific features which have to be taken into account while developing and realizing the experimental setup.

There are three components in the previously described electromagnetic system which can be considered separately during the experimental modeling of the wave processes in amplifiers based on Smith-Purcell effect. They determine the general electromagnetic properties of the open waveguide. These components are the dielectric waveguide which feeds the surface wave into the system; diffraction grating which transforms the surface wave from the dielectric waveguide into the volume wave; the planar layered metal-dielectric structure which serves for both a transformation of the surface wave into the volume wave for the dielectric layer and reflection of the radiation arriving from the diffraction grating - dielectric waveguide interface. Compared to the system without the metal-dielectric layer, the wave processes in the open waveguide with the metal-dielectric stack are more complicated in comparison to the systems without such a stack due to the presence and superposition of different waves such as the volume wave incident to the layered metal-dielectric structure from the diffraction-grating-dielectric-waveguide interface and the waves propagating in the dielectric.

The parameters of the diffraction-grating-dielectric-waveguide system are chosen to satisfy the condition of the volume wave existence in the open waveguide [Shestopalov, 1991]:

$$\varphi_{-1} = \arccos\left(1/\beta_w - n/k\right), \tag{5}$$

where φ_{-1} - is the radiation angle, $\beta_w = v_w/c$ - is the relative velocity of the wave in the waveguide, v_w - is the phase velocity, $k = l/\lambda$ - is the wave number, λ - is the wave length.

The period of the diffraction grating has been chosen such that the main lobe of the radiation pattern ($n=-1$) is at an angle $\varphi = 70°$ for the wavelength of 9 mm and the parameter $\beta_w \approx 0{,}9$ which corresponds to the material of the dielectric waveguide implemented in the experiment (polystyrene waveguide with a cross-section $7{,}2\times3{,}4\ mm^2$). The depth of the grating slots was chosen to minimize the influence of their resonance properties on the radiation characteristics. The waveguide length L is 150 mm, that satisfies the requirement $L/\lambda \geq 10$. This ensured the excitation and propagation of electromagnetic wave along the open waveguide axis.

The distance between the dielectric waveguide and the surface of the diffraction grating, a, is a very important parameter for the optimization of the system. The diffraction of the surface waves on the diffraction grating is nontrivial in this case because the value a is chosen to be smaller than the wavelength. However, a strong coupling between the waveguide and the diffraction grating effects the field distribution in the waveguide and, consequently, the propagation constant β_w. The strong coupling results in interference between the wave propagating along the waveguide and the wave being scattered by the diffraction grating. Such an interference might result in additional propagation modes in the waveguide and, consequently, in the parasitic spatial harmonics [Shestopalov, 1991].

The behavior of the planar metal-dielectric structure of the open waveguide is similar to the behavior of the shielded planar dielectric waveguide. In order to analyze the physical phenomena of the electromagnetic wave excitation in the layered metal-dielectric structure, the electromagnetic field can be represented as a composition of plane electromagnetic waves. Based on this, the metal-dielectric structure can support two types of waves: the one excited by the diffraction grating-dielectric waveguide interface (these waves not necessarily undergo total internal reflection in the dielectric for certain angles φ_{-n}); the second type of waves is excited by a guided surface electromagnetic wave in the dielectric waveguide and is totally reflected from the boundaries of the layered metal-dielectric structure (the wave satisfies the following condition $\cos\gamma_{0\varepsilon} = cv_w/\sqrt{\varepsilon}$)(see Fig. 7). The second wave allows to model Cherenkov radiation. However, this concept of wave decomposition does not consider the multimode nature of the metal-dielectric wave-guiding structure. The modes exist due to the finite layer thickness Δ comparable to the wavelength. The metal layer on the side wall of the dielectric does not prevent the wave propagation but results in increase of the effective thickness of the layer and number of the higher order modes in the metal-dielectric structure.

The experiments were performed in the frequency range from 30 GHz to 37 GHz within the interval $\Delta \approx \lambda - 4\lambda$ and using a dielectric with permittivity $\varepsilon = 2$.

Fig. 10 shows of the normalized radiation pattern in the open waveguide at the center frequency $f = 33{,}4\ GHz$. The diagrams in Fig. 10a depicts the radiation from the end of the metal-dielectric structure in the mode of Cherenkov radiation for the phase velocities satisfying the condition $\varepsilon\beta_w^2 > 1$ for guiding electromagnetic wave on the homogeneous surface of the dielectric. Propagation of the most portion of power in the surrounding environment is typical for the dielectric layer with the thickness less than the wavelength (Figure 10a - curve 1). This holds when the single-mode condition satisfies the condition of synchronization between the phase velocities of waves in dielectric and wave in the surrounding environment. The dielectric layer is actually operates as an antenna, which radiates the power in the direction close to the axis y. The observed asymmetry in the patterns is caused by the technical difficulties to measure the radiation at angles $\varphi_{0\varepsilon} \approx 0 - 10°$. The side lobes are caused by the mismatch with the open area, multiple reflections from the measurement setup, and by a power leakage from the dielectric-waveguide-to-metallic-waveguide transitions. The observed peaks in the radiation pattern are due to the strong coupling between the dielectric waveguide and the dielectric layer at the center and critical frequencies.

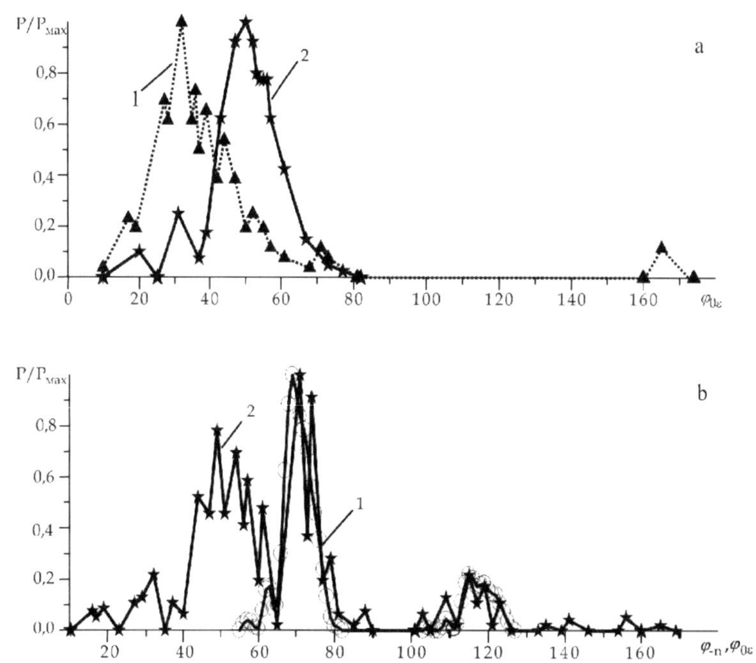

Fig. 10. Radiation patterns of the open waveguide components: a - dielectric layer - dielectric waveguide ($\Delta \approx \lambda$ - curve 1, $\Delta \approx 4\lambda$ - curve 2); b - diffraction-grating-dielectric-waveguide (curve 1), diffraction-grating-dielectric-waveguide-dielectric-layer system (curve 2)

For dielectric layers with $\Delta > \lambda$, the wave is totally reflected from the boundaries and a significant portion of the power is concentrated in the dielectric. The direction of radiation from the end changes to a higher angle (Fig. 10a - graph 2) and approach the calculated values determined from the geometrical optics ($\varphi_{0\varepsilon} \approx 62^\circ$ at $\gamma_{0\varepsilon} \approx 39^\circ$, Fig. 7).
Fig. 10b (curve 1) demonstrates the patterns of the diffraction-grating-dielectric-waveguide radiating system. It is clear from the presented data that the main radiation maximum is in agreement with the calculated value of $\varphi_{-n} = 70^\circ$. At such an angle, the beam for $\varepsilon = 2$, which incidents side wall of the dielectric layer, is slightly refracted and leaves the dielectric from the opposite side at an angle, which is approximately equal to the angle of radiation. This fact is illustrated in Figure 10b for the diffraction-grating-dielectric-waveguide-dielectric-layer system for $\Delta \approx 4\lambda$ (graph 2).
Covering the dielectric layer with a metal (Fig. 7) results in the fact that the radiation arriving from the diffraction-grating-dielectric-waveguide system will be reflected and fed into the open waveguide volume exciting the wave along its axis. Correspondingly, there are two volume waves propagating in the system: the wave in the layered metal-dielectric structure and the wave in the volume of the open waveguide. These waves are coupled to each other by means of the surface wave of the common radiation source - the dielectric waveguide. The existence of the forward and backward coupled waves in the open waveguide might result in parasitic resonances during the modeling. The wave numbers are complex if there is a coupling between the direct and the backward waves. This indicates the excitation of complex decaying waves. The waves are synchronized and the power of the forward wave is pumped into the backward wave and vice versa. Such a power exchange is performed along the significant propagation distance if the coupling is weak. The propagation becomes impossible and the transmission line turns into a sort of a resonator for certain frequencies. In such a system the waveguide characteristics such as the standing wave ratio (SWR) and the transmission coefficient ($K_{tr} = P_{output}/P_{input}$, where P_{output} and P_{input} are the power values at the dielectric waveguide output and input respectively) become fundamental. The waveguide characteristics of the dielectric-waveguide-dielectric-layer system (curve 1), dielectric-waveguide-diffraction-grating-dielectric-layer system (curve 2) and the open waveguide system in general (curve 3) are represented in Fig. 11 for $\Delta \approx \lambda$. The presented data indicates that the SWR of the open waveguide elements and the system in general are within the interval 1,05 ÷ 1,4. These reflections are due to the out of band mismatch of the dielectric-waveguide-metallic-waveguide transitions. The achieved SWR is considerably different from SWR for the open waveguide with no dielectric layer which is approximately 2,0 (curve 4) due to the resonance nature of the system. Substantial changes in the behavior of the K_{tr} versus frequency are also observed. Curves 1 and 2 indicate an efficient transformation of surface waves into the volume waves, while graph 3 indicates the presence of the coupled waves in the system and it is substantially different from the behavior of K_{tr} for the open waveguide with no layered metal-dielectric structure in it (curve 4). It can be assumed that for $\Delta \approx \lambda$ a large amount of power escapes from the dielectric and propagates in the open waveguide. The observed maxima and minima of the spectrum of K_{tr} can be explained by the fact that the waves propagating in the open waveguide are combined in- and out of phase.
The increase in the thickness of the dielectric layer results in the fact that the most amount of power is concentrated in the dielectric which leads to decrease in coupling between the layered metal-dielectric structure and the dielectric waveguide, and, in general, increase in K_{tr} for the open waveguide components (Fig. 12, curves 1 and 2) at $\Delta \approx 4\lambda$.

At the same time, the behavior of the transmission coefficient in the considered frequency band indicates the decrease in the coupling between the waves propagating in the open waveguide (Fig. 12, curve 3).

The analysis described above for the characteristics of the open waveguide and its components indicates that it is possible to control the electromagnetic processes in the system by varying the thickness of the dielectric layer: adjust the coupling between the radiation of the dielectric waveguide and the waves propagating in the open waveguide. The increase in coupling is useful for enhancing the efficiency of the interaction between the electron beam and the open waveguide fields in the amplifier applications. The decrease in the coupling is interesting for realization of power decoupling from the open waveguide through the dielectric layer.

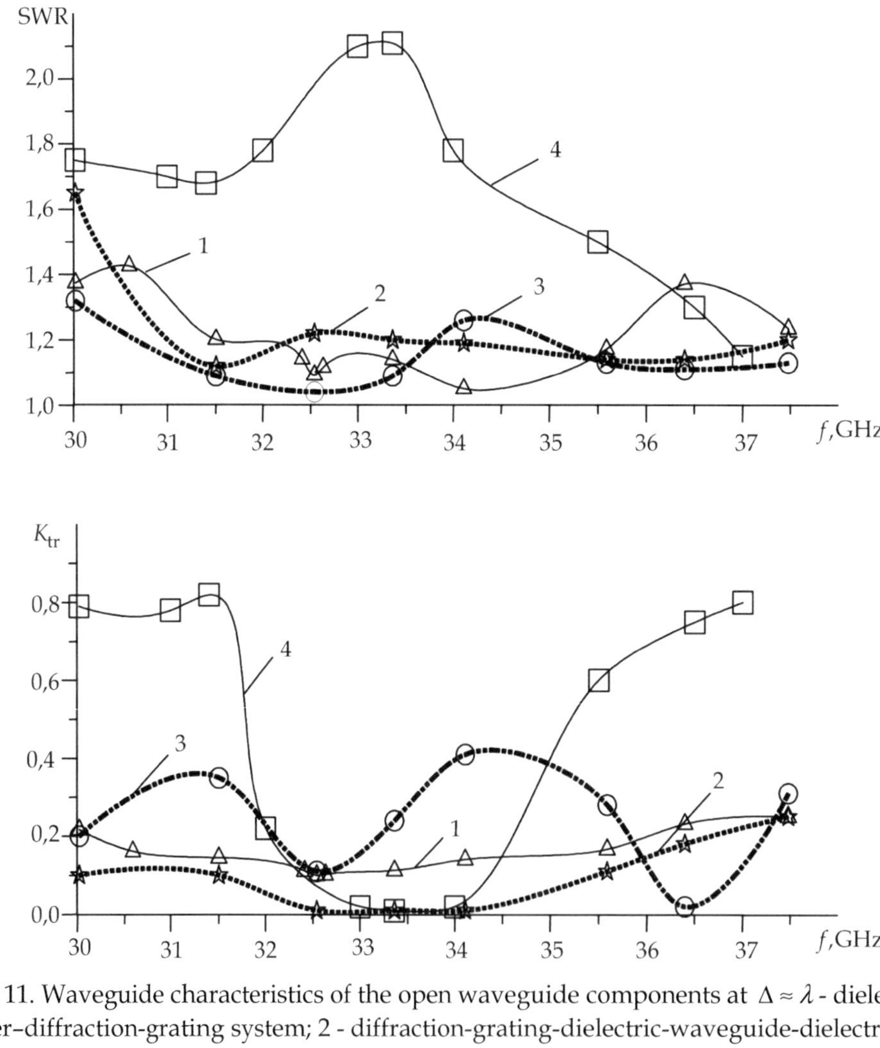

Fig. 11. Waveguide characteristics of the open waveguide components at $\Delta \approx \lambda$ - dielectric-layer–diffraction-grating system; 2 - diffraction-grating-dielectric-waveguide-dielectric-layer system; 3 - open waveguide with the dielectric layer; 4 - open waveguide without the dielectric layer

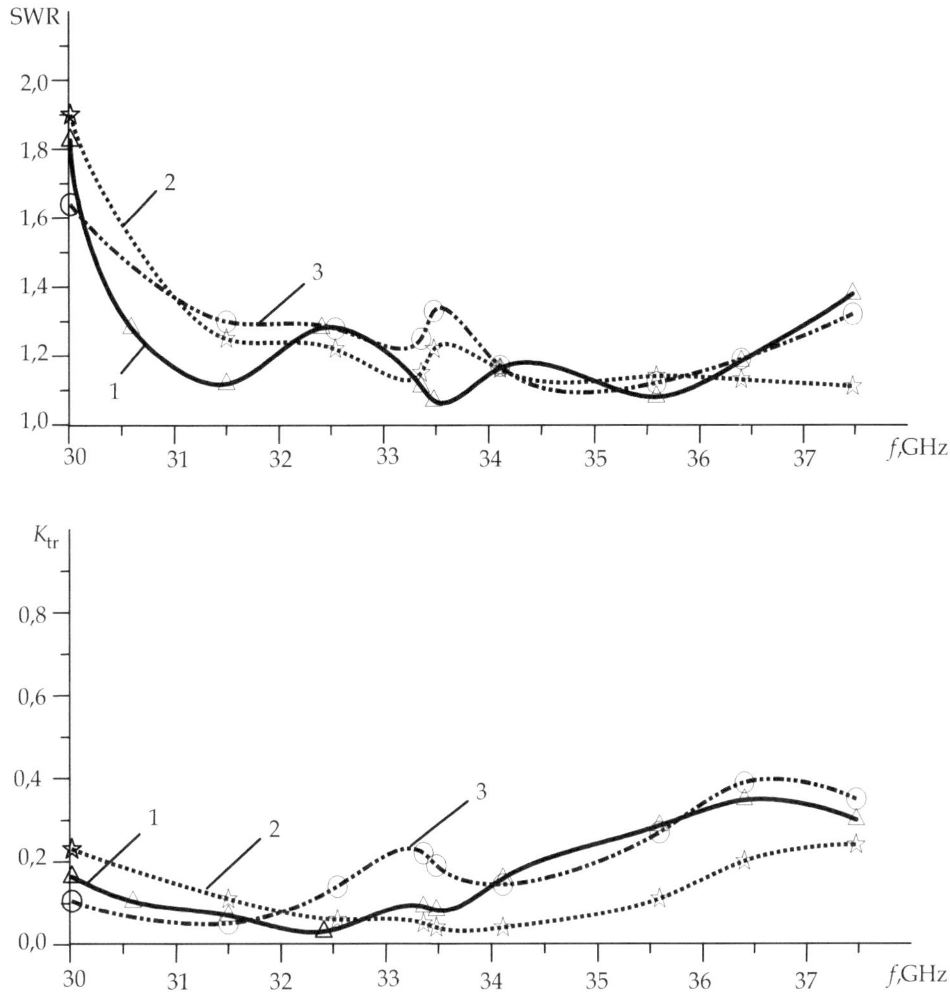

Fig. 12. Characteristics of the open waveguide components at $\Delta \approx 4\lambda$: 1 - dielectric-layer–dielectric-waveguide system; 2 - diffraction-grating-dielectric waveguide-dielectric-layer system; 3 - open wavegude with the dielectric layer

4. The implementation of coupled quasi-optical systems in vacuum electron devices

A two-stage diffraction radiation oscillator has been realized using the structure shown in Fig. 2a in the frequency range $f = 43 \div 98\,GHz$. The system consists of two short-focus spherical mirrors [Shestopalov, 1991] and the common cylindrical mirror with a diffraction grating along its longitudinal axis. The electron beam generated by the electron gun and focused by the static magnetic field propagates above the diffraction grating exciting electromagnetic oscillations in the coupled open resonators. In case of weak coupling between the open resonators, the device operates as a multifrequency oscillator at specific frequencies. In case of optimal coupling, the device operates as a broadband diffraction radiation oscillator with coupled resonators. The operating frequency band in this case is more than 1,5 times wider compared to the single resonator diffraction radiation oscillator. The device operates as an amplifier if the microwave signal is applied to the input of the first (with respect to the gun) resonator and the beam current J is less than the starting current J_n. These regimes have been tested in the millimeter wave range ($f = 43 \div 98\,GHz$). Figure 13 shows the data when the device operates as an oscillator in case of optimal coupling between the open resonators. The power of such a diffraction radiation oscillator at $f_0 = 84\,GHz$ was measured to be 0,4 W with the beam current $J = 1{,}5 J_n \left(J_n \approx 30\,mA\right)$. The range of electron frequency tuning at these conditions was 1,5 times wider than in the case of a single-resonator oscillator, which is comparable with the results obtained by the previously described modeling (Fig. 4). A similar behavior has been also observed in the regime of amplification at $J \approx 0{,}8 \div 0{,}9 J_n$, which confirms the possibility to build a regenerative amplifier based on coupled open resonators with a broader transmission band than just using a single-resonator amplifier [Shestopalov, 1991].

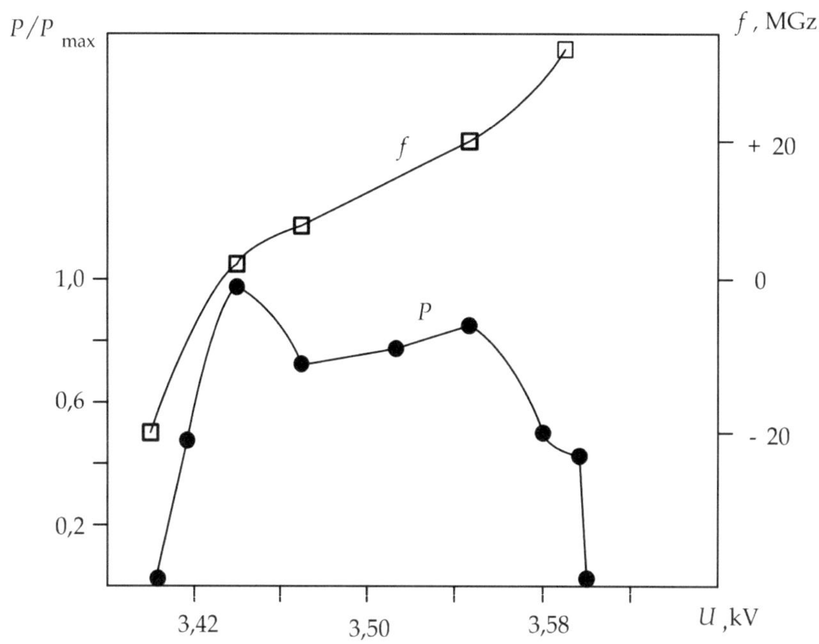

Fig. 13. The bandwidth and a tuning range of the diffraction radiation oscillator based on two coupled resonators

Figure 14 presents the diagrams of a vacuum electron devices with open resonators connected in series with respect to the axis of the electron beam. An orotron shown in Fig. 14a consists of two coupled open resonators **1**. Each of these resonators consists of two mirrors **2** and **3**. Energy is coupled out through a waveguide in mirror **2**. Mirror **3** has a parabolic cylinder shape. Metal-strip diffraction gratings **4** located in the center of the adjacent parabolic mirrors **3** are made of metal bars. The electron gun **5** generates a focused electron beam **6** and is placed between the parabolic mirrors **3**. A collector **7** is positioned at the end of the interaction region.

The operation of the orotron can be described in the following way: the electron gun generates a focused electron beam which than experiences a bunching within the small interaction length due to the spatial charge in the interaction zone formed by the open resonators and gratings. The diffraction radiation is produced in the open resonators as electrons propagate through the gap between the diffraction gratings. The electrons are than striking a collector at the other end of the interaction region. The orotron operates as an oscillator if the electron beam current is much higher than the starting current. The orotron operates as an amplifier if the condition of self-excitation is not satisfied and a signal from an external microwave source is fed to the input of one of the resonators. It should also be noted that the orotron may function as a frequency multiplier if using two coupled open resonators. This device is a low-power oscillator. The increase of the electron beam current density is limited due to overheating of the strip diffraction grating.

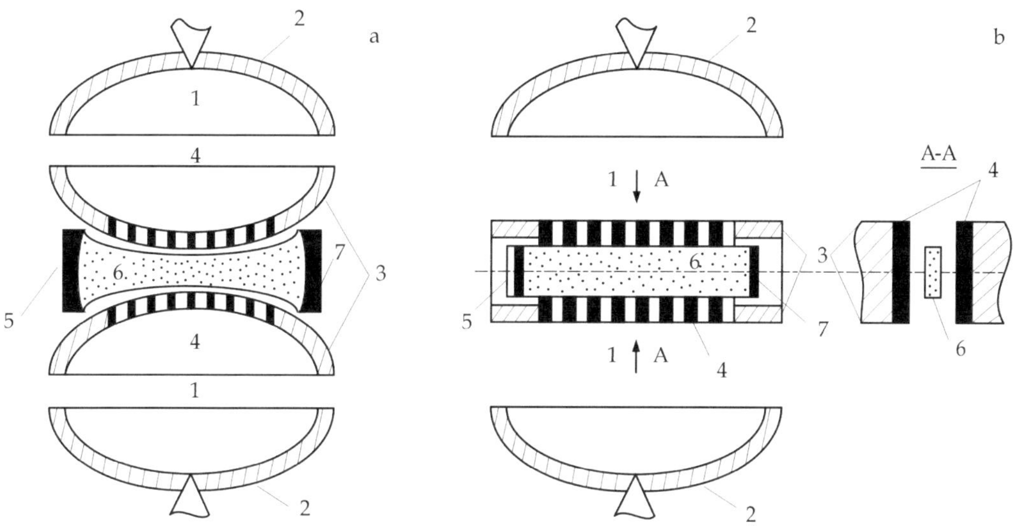

Fig. 14. Vacuum electron devices based on parallel connection of open resonators: a - an orotron with coupling through the strip diffraction gratings and b - diffraction radiation oscillator with coupling through the reflective diffraction gratings

A higher power level can be achieved in diffraction radiation oscillators based on coupled open resonators schematically shown in Fig. 14b. The design and the principle of operation of such a device are similar to the design and the principle of operation of the previously described orotron. The coupling of resonators **1** is achieved through the slots in the identical reflective diffraction gratings **4** placed in the center of the adjacent mirrors **3** and perpendicularly oriented with regard to the planes of these mirrors. The electron beam is focused with a magnetic field. The use of bulky gratings attached to the mirrors simplifies the temperature dissipation and, consequently, allows for higher electron beam currents. Furthermore, one of the resonators in such a system may be realized with an option for mechanical tuning where a moving short-circuit plunger located on the opposite side of the coupling slot. Figure 15 shows the oscillation bandwidth and frequency tuning characteristic for different distances h of the plunger for the case when the open resonator is centered at $f_0 = 36\ GHz$.

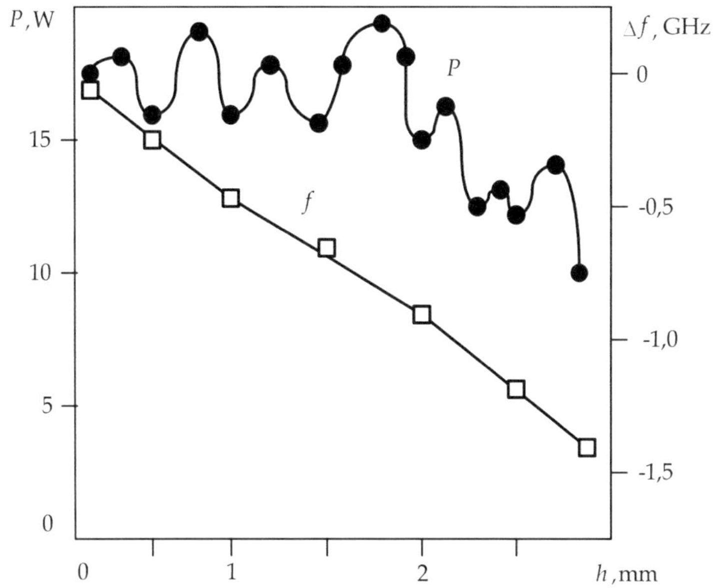

Fig. 15. The output power and the frequency tuning range of the diffraction radiation oscillator with a tunable resonator coupled to the open resonator

The presented data shows that, one can smoothly tune the oscillation frequency within a sufficiently broad frequency range by mechanically tuning the volume resonator with a fixed value of H for the mirrors in the open resonator. The variation of the output power in the considered frequency band does not exceed 3 dB. This characteristic of the considered device indicates the possibility for improving the vibration stability of the system in comparison to the vibration stability of systems with mechanical tuning of mirrors. The grating-coupled open resonators could also be used to build reflection type diffraction radiation oscillators [Shestopalov, 1991]. In this case, the collector should be replaced by an electron reflector, producing a backward electron beam. Such devices exhibit low starting currents and able to operate in the regime of stochastic oscillations [Korneenkov et al., 1982]. The wide functionality of open resonators with layered metal-dielectric structures allowed to build several types of diffraction based devices with complex resonant structures such as Cherenkov diffraction oscillator and Cherenkov backward-wave tube. Fig. 16 shows the example of Cherenkov backward-wave tube and Cherenkov diffraction oscillator.

The electron beam **1** of the backward-wave tube is generated by the electron gun **2**. The beam propagates through the channel **3** formed by the adjacent surfaces of the resonator **4** to the slow-wave structure **5**. The electron beam interacts with the field of the slow-wave structure **5** resulting in modulation of charge density. Simultaneously, Cherenkov radiation occurs when the electrons velocity exceed the phase velocity of the electromagnetic wave in the dielectric. The radiation is directed into the dielectric. The resonator **4** has a field distribution allowing a feedback (solid lines with arrows). Oscillations occur in the resonator effectively extracting power from the modulated electron beam via the strip grating **6** when the frequency is synchronized with the eigen frequency of the resonator. The power is coupled from the resonator **4** via the waveguide **7** with $\varepsilon_1 > \varepsilon$. The synchronization between the electron beam and the wave in the dielectric is achieved by choosing the proper value for ε and adjusting the accelerating voltage for the electron beam.

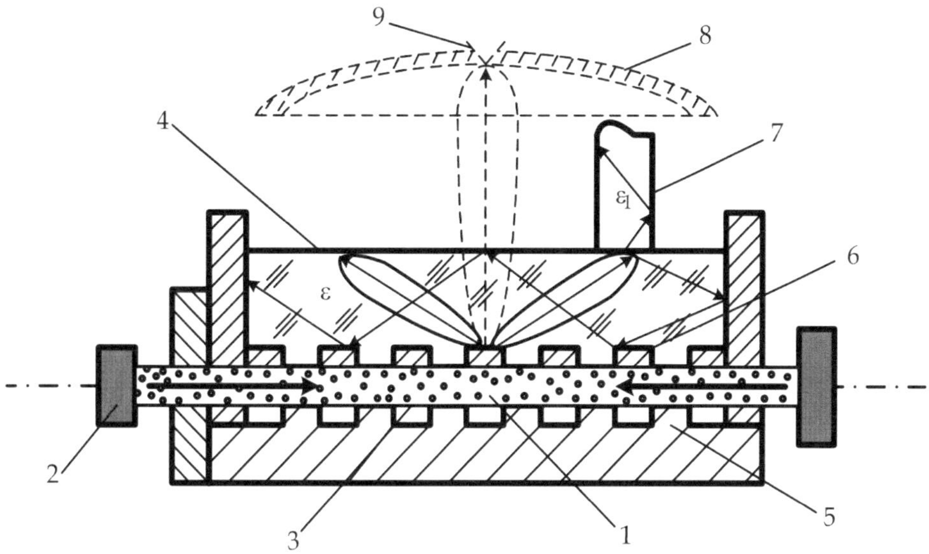

Fig. 16. Realization of Cherenkov backward-wave tube and Cherenkov diffraction oscillator

The similar electron optics is used for excitation of Cherenkov diffraction radiation. The slow-wave structure (diffraction grating) **5** is positioned in the central part of the fixed mirror. The moving mirror **8** with a coupling slot **9** is used for coupling the power out of the device. In contrast to the backward-wave oscillator, the geometrical parameters of the gratings **5**, **6** are optimized for efficient excitation of radiation in the normal direction with respect to the axis of the electron gun **2** (dotted oscillation pattern in Fig. 16) and for maximum power density of Cherenkov radiation within the dielectric resonator **4**.
Recently, significant attention is drawn to amplifiers based on Smith-Purcell effect, which has been described in section 3. An amplifier employing sheet electron beam is shown in Figure 17.

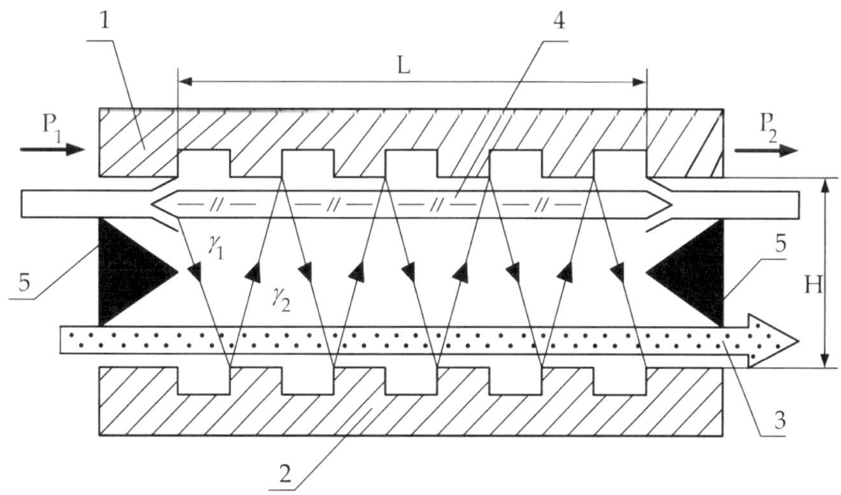

Fig. 17. Travelling wave tube based on the Smith-Purcell effect

The open waveguide with length L is formed by the surfaces of parallel passive **1** and active **2** mirrors realized as reflecting diffraction gratings with the periods l_1 and l_2 and a distance H between them. The sheet electron beam **3** propagates above the surface of the active mirror **2**. The dielectric waveguide **4** is placed close to the surface of the passive mirror **1**, and the matched absorption loads **5** are positioned at the ends of the waveguide. The periods l_1 and l_2 of the diffraction gratings comply with the relations that follow from the conditions of the in-phase mode of radiation (shown with arrows) from the active and the passive mirrors of the open waveguide:

$$l_1 = \frac{\lambda}{1+\tilde{\varepsilon}_w\left(\sqrt{\varepsilon}-1\right)-\cos\gamma_2}; \; l_2 = \frac{\lambda}{5/\sqrt{KU_0}-\cos\gamma_2}; \tag{6}$$

$$arctg\frac{2H}{L}+\left(\frac{65\lambda}{L}\right)^{\circ} \le \gamma_2 < 90°; L \ge 10\lambda,$$

where $\tilde{\varepsilon}_w = \frac{c^2}{v_w^2}$ - is the effective permittivity of the waveguide; U_0 - is the accelerating voltage of the electron beam, V; K=505 1/V.

The range of angle γ_2 and the length L of the waveguide are chosen to minimize diffraction loss into the free space. A high-frequency signal of power P_1 with a wavelength λ is fed to the dielectric waveguide **4**. The transformation of the surface wave into the volume wave radiated in the direction of angle $\gamma_1 = \arccos\left(c/v_w + \lambda/l_1\right)$ occurs on the diffraction grating of the passive mirror **1**. The non-reflected portion of the volume wave excites the spectrum of the spatial harmonics having different phase velocities when the volume wave of the transformed input signal incidents the grating of the active mirror **2**. The electron beam velocity v_e synchronizes with one of the surface waves which results in bunching of electrons radiating at a frequency of input signal in the direction of angle $\gamma_2 = \arccos\left(c/v_e + \lambda/l_2\right)$. The reverse transformation of the volume wave into the surface wave, which is followed by a radiation into the open waveguide, occurs at the grating of the passive mirror. The signal P_2 is amplified in the case of the in-phase radiation from the mirrors. The periodic re-emission results in increase of amplitude of the volume wave propagated along the open waveguide and the amplitude of the surface wave propagating in the same direction along the dielectric waveguide which is used to couple the amplified signal P_2 out to the load. The matched loads **5** and the dielectric waveguide **4** decrease the probability of regeneration effects that might occur in the amplifier both due to the reflections from the open waveguide ends and due to the parasitic oscillations due to multiple reflections between the active and the passive mirrors in direction of angles γ_1 and γ_2 close to $\pi/2$.

The prototype of the suggested travelling wave tube has been realized in the V band. The open waveguide was formed by two cylinder-shaped mirrors (the passive mirror with the curve radius $R_{curv} = 20\ mm$ and the active mirror with $R_{curv} = 110\ mm$). The grating periods l_1 and l_2 were chosen according to (6) resulting in $\gamma_1 = \gamma_2$. High-frequency signal was fed into the amplifier from the resonance carcinotron in the frequency band $f = 68 \div 72\ GHz$

using a quartz dielectric waveguide and a sheet beam with a cross-section $5 \times 0{,}2\, mm^2$. The electron beam was propagating along the active mirror with accelerating voltages in the range $U_0 = 2200 \div 2500\, V$. The system was built in the vacuum shell between the poles of the electromagnet that limited the open waveguide length to L=40 mm and allowed to ensure about a double transformation of the surface wave into the diffraction radiation. The achieved experimental results show that amplification of rather broadband signals (up to 2 GHz) along with increase in gain is possible if increase the beam current. At the same time, the limited length of the open waveguide did not allow a sufficient number of transformations of the surface waves into the volume waves, which limited the increase of the gain K.

A further improvement of the output parameters of the amplifier could be achieved by increasing the interaction region and the electron beam current. This could be achieved, for instance, by means of using axial-symmetric electromagnetic systems and a better electron focusing optics.

5. Conclusion

The chapter provides a summary of results on both the classical quasi-optical systems forming a basis for development of new modifications of oscillation systems of the microwave and millimeter-wave band devices and more advanced coupled electromagnetic systems with complex periodic structures such as coupled open resonators, open resonators and waveguides with layered metal-dielectric structures. It is demonstrated that the coupled open resonators exhibit wider frequency tuning range while preserving high values of Q-factor. Coupled systems such as open resonators and waveguides with layered metal-dielectric structures have qualitatively new properties: by varying the parameters of metal-dielectric structure one could achieve attenuation or amplification of the oscillations and their selection. New modifications of Cherenkov traveling wave tube such as Cherenkov diffraction oscillator and amplifier based on the Smith-Purcell effect are suggested and realized.

6. References

Balakirev, V.A., Karbushev, N.I., Ostrovsky, A.O., and Tkach, Yu.V. (1993). *Theory of Cherenkov amplifiers and oscillators based on relativistic interaction of oscillation bundles*, Naukova Dumka, Kyiv.

Belous, O.I., Fisun, A.I., and Sukhoruchko, O.N. (2003). Synthesis of basic components of a low-noise input circuit for millimeter wavelength. *Telecommunication and Radio Engineering*, Vol. 59, No 1-2, p.p. 111-118.

Bratman, V.L., Dumesl, B.S., and Fedotov, A.E. (2002). Broadband orotron operation at millimeter and sub-millimeter waves. *International Journal of Infrared and Millimeter Waves*, Vol. 23, No 11, p.p. 1595-1601.

Ginzburg, N.S., Zavolsky, N.A., and Zapevalov, V.Ye. (2000). Non-stationary processes in orotron with diffraction output of oscillation. *JTP*, Vol. 70, No 4, p.p. 99-104.

Joe, J., Scharer, J., Booske, J.H., and Mevey, B. (1994). Wave dispersion and growth analysis of low-voltage Cherenkov amplifiers. *Phys. Plasmas*, Vol. 1, No 1, p.p. 176-188.

Joe, J., Louis, L.J., Booske J.H., and Basten, M.A. (1997). Experimental and theoretical investigations of a rectangular grating structure for low-voltage traveling-wave tube amplifiers. *Phys. Plasmas*, Vol. 4, No 7, p.p. 2707-2715.

Korneenkov, V.K., Miroshnichenko, V.S., and Tsvyk A.I. (1982). About excitation of accidental oscillations in diffraction radiation oscillator-free electron laser. *Report. AN. USSR, Ser.A*, No 5, p.p. 59-61.

Marshall, E.M., Philips, P.M., and Walsh, J.E. (1998). Planar orotron experiments in millimeter wavelength band. *IEEE Transactions on Plasma Science*, Vol. 16, No 2, p.p. 199-205.

Milovanov, O.S., and Sobenin, N.P. (1980). *Microwave equipment*, Atomizdat, Moscow.

Rusin, F.S., Bratman, V.L., and Fedotov, A.E. (2002). Orotron: perspectives of advancing into submillimeter wavelength band. *Vacuum microwave electronics: Trans. Reviews*, p.p. 121-124.

Shestopalov, V.P. (1976). *Diffraction electronics*, Kharkiv University Publishers, Kharkiv.

Shestopalov, V.P. (1985). *Physical basis for millimeter- and submillimeter-wave equipment*, Naukova Dumka, Vol 1 (*Open-type structures*), Vol. 2 (*Sources, element basis - Radio systems*), Kyiv.

Shestopalov, V.P. (1991). *Diffraction radiation oscillators*, Naukova Dumka, Kyiv.

Shmatko, A.A (2008). *An electron-wave systems of millimeter wave range*, Vol. 1, KNU V.N. Karazina, Kharkov.

Smith, S.I., and Parcell, E.M. (1953).Visible light from localized surface charges moving across a grating. *Phys. Rev.*,Vol. 92, No 4, p.p. 1069-1073.

Sukhoruchko, O.N., Tkachenko, V.I., and Fisun, A.I. (2003). Modelling of elements of the air intake low-noise duct with parametric signal amplification. *Applied Radioelectronics*, Vol. 2, p.p. 163-167.

Valitov, R.A., Dyubko, S.F., Kamyshan, V.V et al. (1969). *Submillimeter-wave equipment*, Sov. Radio, Moscow.

Valitov, R.A., and Makarenko, B.I. (1984). *Measurements upon millimeter and submillimeter waves: Methods and equipment*, Radio i Svyaz, Moscow.

Vorobyov G.S. Pushkarev K.A.and Tsvyk A.I. (1997). Numerical analysis of shielding properties of diffraction grating excited by electron beam radiation on metal-dielectric structures. *Radiotehnika I Elektronika*, Vol.42, №2, p.p. 738-740.

Vorobyov G.S., Petrovsky M.V., Zhurba V.O., Ruban A.I., Belous O.I., and Fisun A.I. (2007). Perspectives of application of new modifications of resonant quasi-optical structures in EHF equipment and electronics. *Telecommunications and Radio Engineering*, Vol.66, Issue 20, p.p. 1839-1862.

Weinstein, L.A. (1966). *Open resonators and open waveguides*, Sov. Radio, Moscow.

Weinstein, L.A., and Solntsev, V.A. (1973). *Lectures on microwave electronics*, Sov. Radio, Moscow.

Weinstein, L.A. (1995). *Theory of diffraction. Microwave electronics*, Radio i Svyaz, Moscow.

14

Waveguide Mode Converters

Yoshihiro Kokubo
Graduate School of Engineering, University of Hyogo
Japan

1. Introduction

Metallic waveguides have major advantages, such as low propagation loss and high power transmission in the microwave frequency range. However, one disadvantage is that the usable frequency range is restricted to $f_c < f < 2f_c$, because the TE_{20} mode is possible in a frequency region higher than $2f_c$ for rectangular metallic waveguides. A ridge waveguide (Cohn, 1947) (or double-ridge waveguide) has an advantage in that it can spread the propagating frequency range as a result of reduction in the cutoff frequency for the TE_{10} mode. However, one disadvantage is that the attenuation constant becomes large.

Power sources, such as watt class IMPATT diodes or Gunn diodes, are readily available, and for high frequency use, power sources are sometimes combined, due to their low power rating. However, power combiners consisting of cavity resonators usually have narrow bandwidths (For example, Matsumura et al., 1987). Power dividers and power combiners may be easily setup using mode converters. For example, a TE_{10}–TE_{30} mode converter easily offers a three-port power divider, and a three-way power combiner can be composed by reversal. A power combiner is useful for application to Gunn diodes in a waveguide array (Bae et al., 2000), because it converts the TE_{30} mode to the TE_{10} mode.

2. Design method of the mode converters

We have reported that single-mode propagation is available for a metallic waveguide with dielectric rods arrayed at the center of the waveguide in the frequency under twice the cutoff frequency region using the TE_{10} mode, and in the frequency over twice the cutoff frequency region using the TE_{20} mode, because of restrictions of the TE_{10} mode (Kokubo, 2007; Kokubo & Kawai, 2009). However, a TE_{20}-like mode, which is propagated in the second band, is an odd mode, and generation systems for odd modes have seldom been reported. In this investigation, a mode converter is proposed which passes through the TE_{10} mode for the low frequency range and efficiently converts TE_{10} to TE_{20} mode for the high frequency range.

2.1 Design method of the TE_{10}-to-TE_{20} mode converter

The frequency eigenvalues of a conventional metallic waveguide in a given k wavevector are shown in Fig. 1. In this figure, the wavevector k and frequency ω are normalized using the width of the waveguide w. The electromagnetic wave propagates the TE_{10} mode only for $0.5<\omega w/2\pi c<1$, and can propagate TE_{10} and TE_{20} modes for $1<\omega w/2\pi c<1.5$. If these two modes are excited by only the TE_{10} mode, the group velocity of TE_{10} (A) must be changed to that of TE_{20} (B) for $1<\omega w/2\pi c<1.5$. On the other hand, the group velocity (C) is not changed for $0.5<\omega w/2\pi c<1$, because this remains in the TE_{10} mode. If the distribution of the transverse electromagnetic field is gradually changed from TE_{10} to TE_{20}, and group velocity (A) is also gradually changed to (B), then the reflection may be reduced for $1<\omega w/2\pi c<1.5$. On the other hand, if the group velocity (C) is not significantly changed, the reflection may also be suppressed for $0.5<\omega w/2\pi c<1$. Since the mode profile gradually shifts from TE_{10} to TE_{20}, the dielectric rods are replaced from near the sidewall to the center of the waveguide. In other words, the basic setup is shown in Fig. 2.

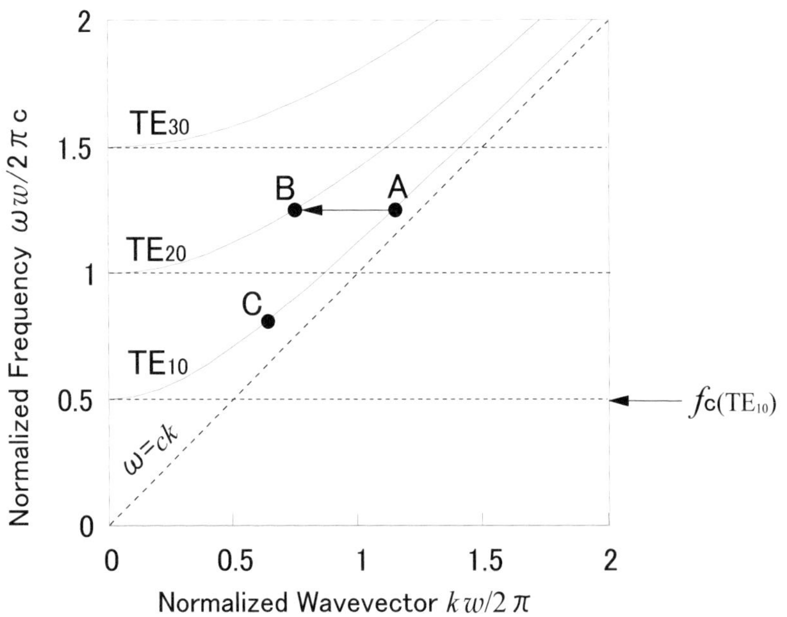

Fig. 1. The frequency eigenvalues of a conventional metallic waveguide in a given k wavevector.

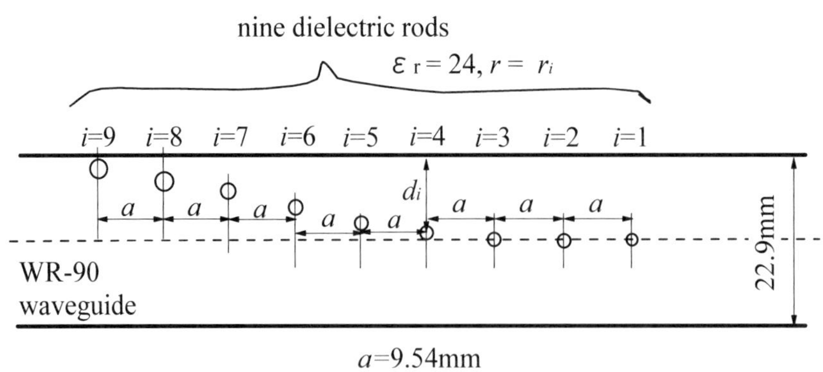

Fig. 2. The proposed structure of the TE_{10} to TE_{20} mode converter.

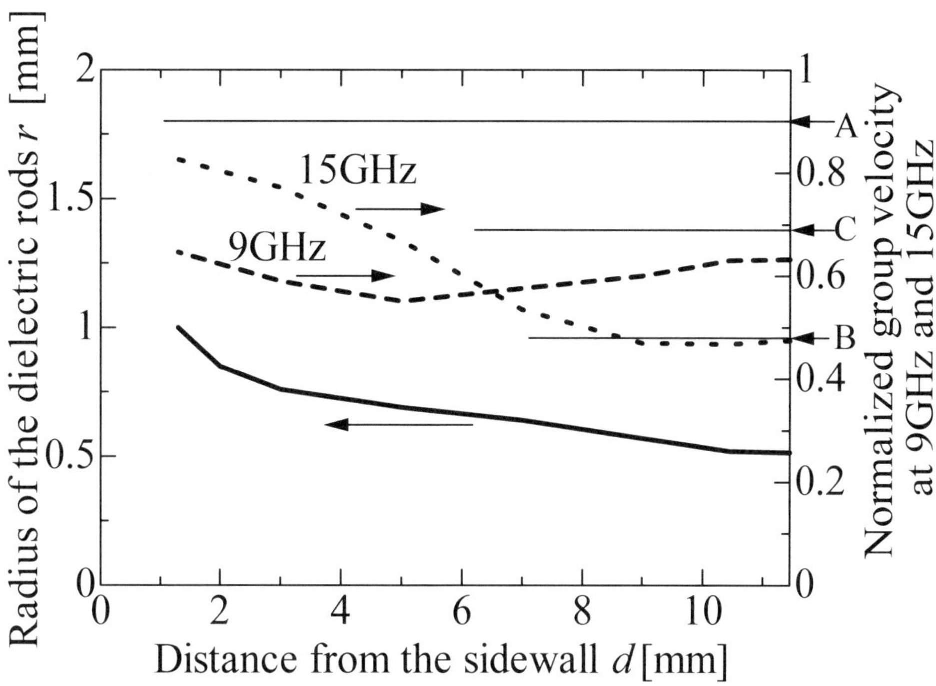

Fig. 3. The group velocity in a metallic waveguide with a periodic array of dielectric rods for various distances from the sidewall, d, and various radii of the rods, r, at 15 and 9 GHz. (Kokubo, 2010)

The group velocity is given by $v_g = \frac{1}{\left(\frac{dk}{d\omega}\right)}$. However, it is not simple to determine the group velocity in the waveguide shown in Fig. 2. The propagation modes in a waveguide having in-line dielectric rods with period a are calculated using a supercell approach (Benisty, 1996) by application of appropriate periodic Bloch conditions at the boundary of the unit cell (Boroditsky et al.; Kokubo & Kawai, 2008). When the location of the dielectric rods is fixed at a distance d from the sidewall, the group velocity v_g at both of the first and the second bands is changed by varying the radius r. However, the group velocities are also changed at the same time and cannot be changed individually.

If the group velocity is normalized using light velocity in a vacuum, v_g/c is the same as the gradient of the characteristic curve. Therefore, when d and r are fixed to certain values, v_g/c is calculated for the periodic structure of the dielectric rods at a specific frequency. If group velocity (A) is gradually changed to (B) for $1<\omega w/2\pi c<1.5$ when d is varied, and group velocity (C) is not changed for $0.5<\omega w/2\pi c<1$, then one unit of each pair of d and r connects to its respective pair to form a structure shown in Fig. 2.

The metallic waveguide is assumed to be a WR-90 (22.9×10.2 mm, cutoff frequency $f_c \approx$ 6.55GHz) and period a is fixed at 9.54 mm. Fig. 3 shows a sample of calculated results of normalized velocity along the axis of the waveguide at 15 GHz and 9 GHz for dielectric rods ($LaAlO_3$: ε_r= 24, radius r [mm]) aligned at a distance from the sidewall d [mm] (Kokubo, 2010). It is desirable that the normalized velocity (A) (TE_{10}: v_g/c =0.900) monotonically decreases to (B) (TE_{20}: v_g/c =0.487) at 15 GHz and normalized velocity (C) (TE_{10}: v_g/c =0.686) is not changed at 9 GHz. However, at 15 GHz, such a condition is not found around d=10 mm, because the placement of the dielectric rod at the center of the waveguide is the same as that where the electric field becomes a minimum. On the other hand, placement of the dielectric rod near the sidewall of the waveguide is the same as that where the electric field becomes a maximum. At the transition region, around d=10 mm, the characteristics are complex. The design takes priority, in order to not vary the group velocity at 9 GHz. Since the group velocity must become slow with dielectric material at 9 GHz and becomes slowest at d=5 mm, the design takes priority at 15 GHz for mode conversion, because both 15 GHz and 9 GHz conditions cannot be satisfied at the same time. The final design of the mode converter is shown in Fig. 2. Three dielectric rods are located at the center of the waveguide. Two of these have r = 0.515 mm and the remainder have a half cross-section radius of 0.36 mm in order to decrease electromagnetic reflection. Nine dielectric rods are placed from the center of the waveguide to near the sidewall with increasing radius of the rods r and with constant a = 9.54 mm. Table 1 shows the relation between the distance d and radius r (Kokubo, 2010). The S parameters between the input port (Port 1) and output. port (Port 2) are calculated using the HFSS software by Ansys Inc. and the results are shown as solid lines in Figs. 4(a) and (b). The electromagnetic waves pass through as the TE_{10} mode for 7-11.2 GHz and are converted to the TE_{20} mode for 14.1-16.1 GHz under a condition of over 95% efficiency. However, as the TE_{20} mode is not sufficiently small under -15 dB, optimization of the design is necessary for reduction of reflections. Though reflection as TE_{20} is not small at high frequency, if a coaxial-waveguide converter is used for the introduction of electromagnetic waves to a waveguide, odd mode may not have a strong influence on even symmetry structure.

Rod Number i	Distance d_i from the Sidewall [mm]	Radius r_i of the dielectric rod [mm]
1	11.45	0.36
2	11.45	0.515
3	11.45	0.515
4	10.45	0.52
5	9	0.57
6	7	0.64
7	5	0.69
8	3	0.76
9	1.3	1.0

Table 1. Location and radii of the dielectric rods

2.2 Simple fabrication method

For the fabrication of a mode converter, such as the Type A illustrated in Fig. 5(a), it is necessary to locate the dielectric rods in the waveguide without a gap at top and bottom.
Such a structure may be difficult to fabricate. As a solution, holes with diameters 0.2 mm larger than the rods were fabricated at the top of the waveguide and the dielectric rods were inserted (Type B, Fig. 5(b)). The S parameters were calculated using the HFSS software and the results are shown as dotted lines in Figs. 4(a) and (b). The results for these different structural conditions (solid lines and dotted lines) are almost same.

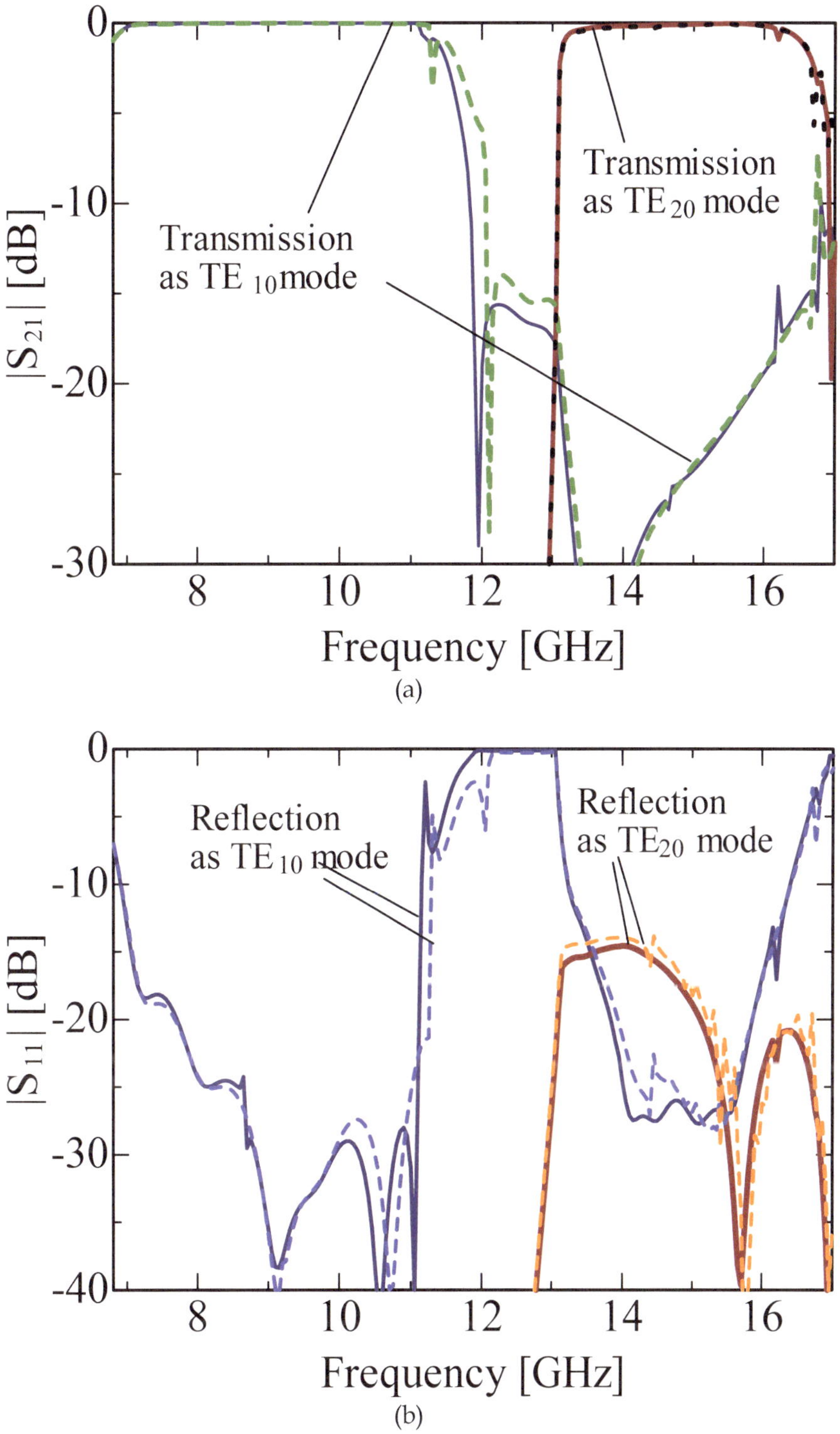

Fig. 4. S parameter for the mode converter; (a) $|S_{21}|$ and (b) $|S_{11}|$.(Kokubo, 2010)

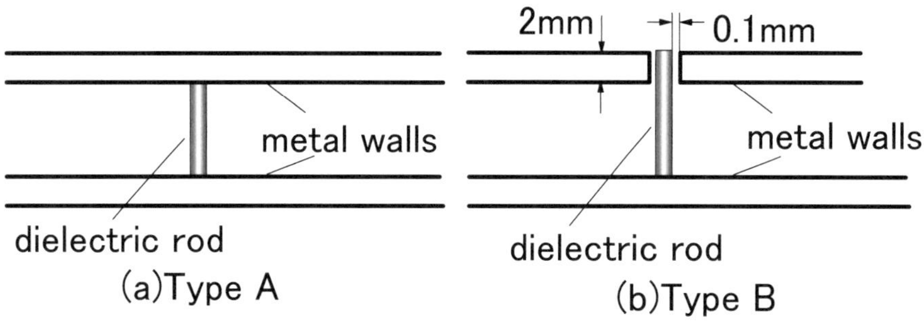

Fig. 5. (a) Dielectric rod located in a waveguide without gaps at top and bottom.
(b) Dielectric rod inserted in a hole made at the top of the waveguide with a diameter 0.2 mm larger than the diameter of the rod. (Kokubo, 2010)

2.3 Reduction of reflection for TE_{20} mode

As shown in Fig. 4(b), reflection as TE_{20} mode is not small enough. This reason is dielectric rods are asymmetric arrangement for electromagnetic wave. Fig. 6 shows an improved structure of the TE_{10} to TE_{20} mode converter.

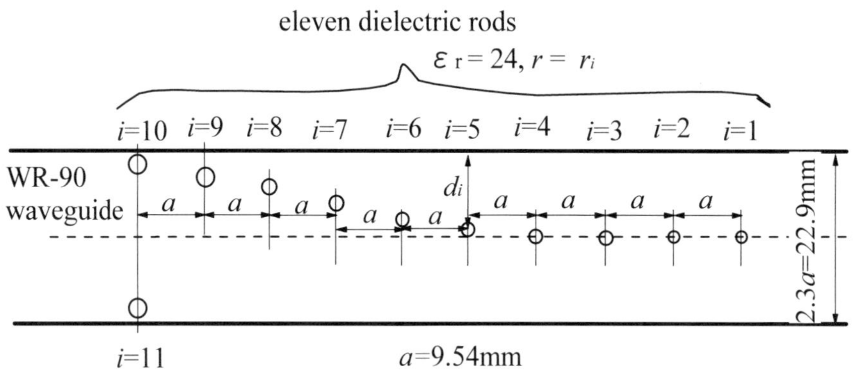

Fig. 6. An improved structure of the TE_{10} to TE_{20} mode converter.

Rod Number i	Distance d_i from the Sidewall [mm]	Radius r_i of the dielectric rod [mm]
1	11.45	0.38
2	11.45	0.55
3	11.45	0.55
4	11.45	0.55
5	10.45	0.515
6	9	0.57
7	7	0.64
8	5	0.69
9	3	0.76
10	1.25	0.92
11	21.65	0.96

Table 2. Location and radii of the dielectric rods

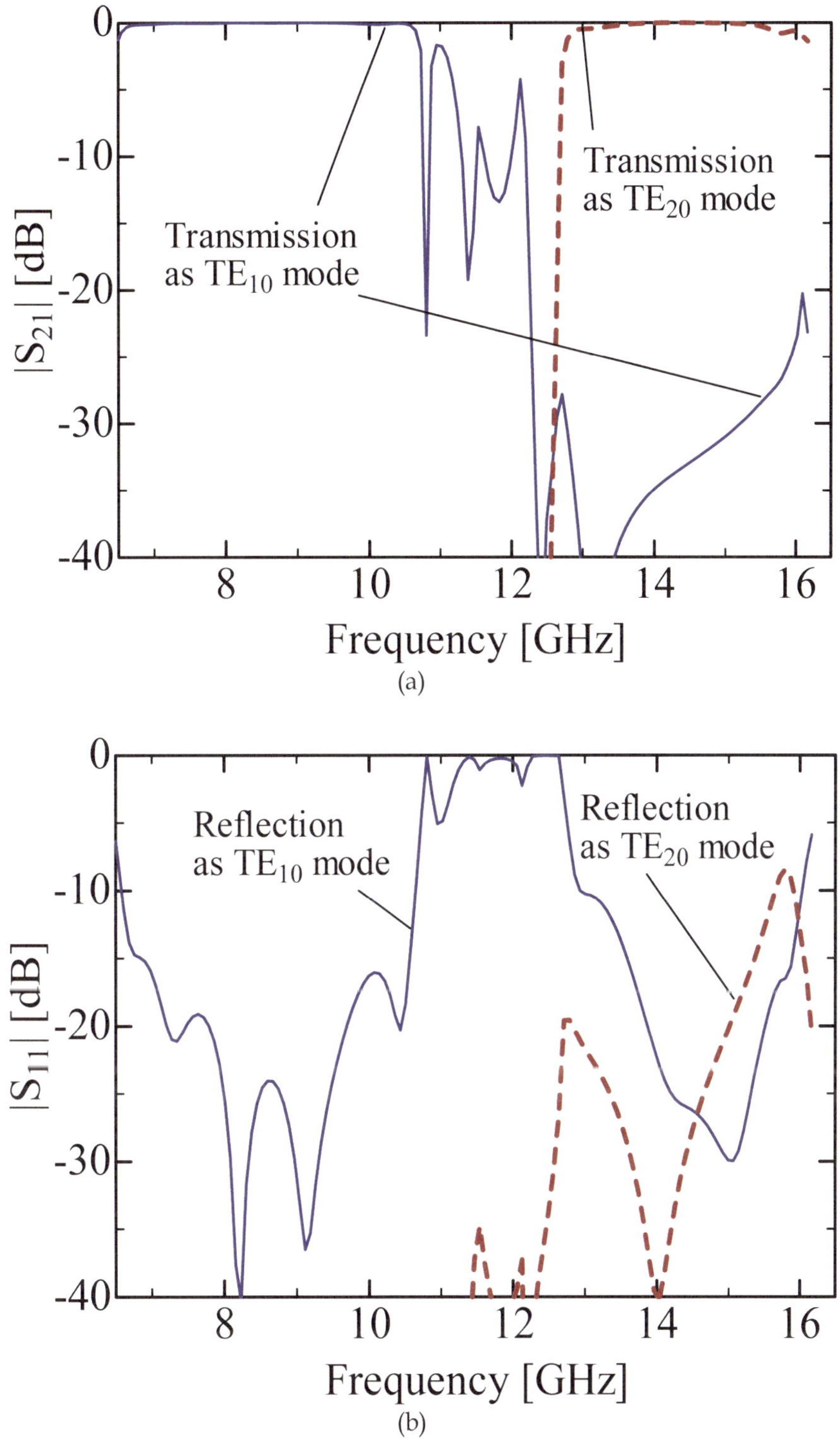

Fig. 7. S parameter for the mode converter; (a) $|S_{21}|$ and (b) $|S_{11}|$.

2.4 Design method of the TE_{10}-to-TE_{40} mode converter

A TE_{10} to TE_{40} mode converter can be considered by combination of TE_{10} to TE_{20} mode converters. Another structure of the TE_{10}-to-TE_{20} mode converter is proposed and is shown in Fig. 8. The locations of the dielectric rods are indicated in Table 1. The structure of the proposed TE_{10}-to-TE_{40} mode converter, which is composed of three TE_{10}-to-TE_{20} mode converters, is shown in Fig. 9. The S parameters between the input port (port 1) and output port (port 2) calculated using HFSS are shown in Figs. 10(a), (b) and (c).

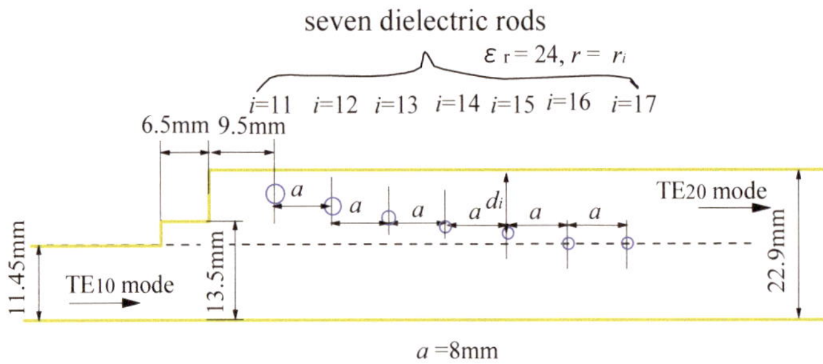

Fig. 8. Structure of the proposed TE_{10}-to-TE_{20} mode converter.

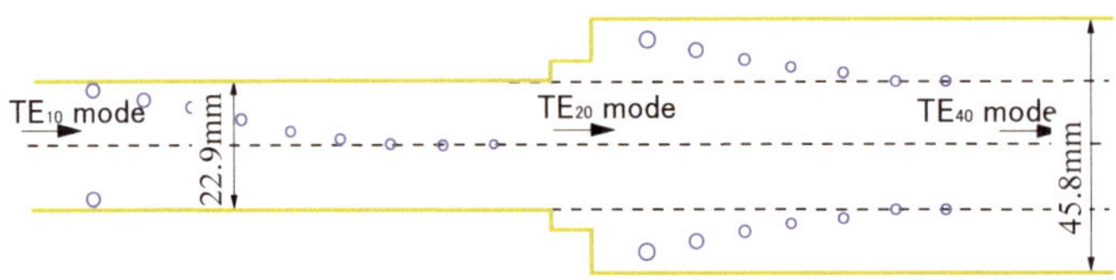

Fig. 9. Structure of the proposed TE_{10}-to-TE_{40} mode converter. (Kokubo, 2011a)

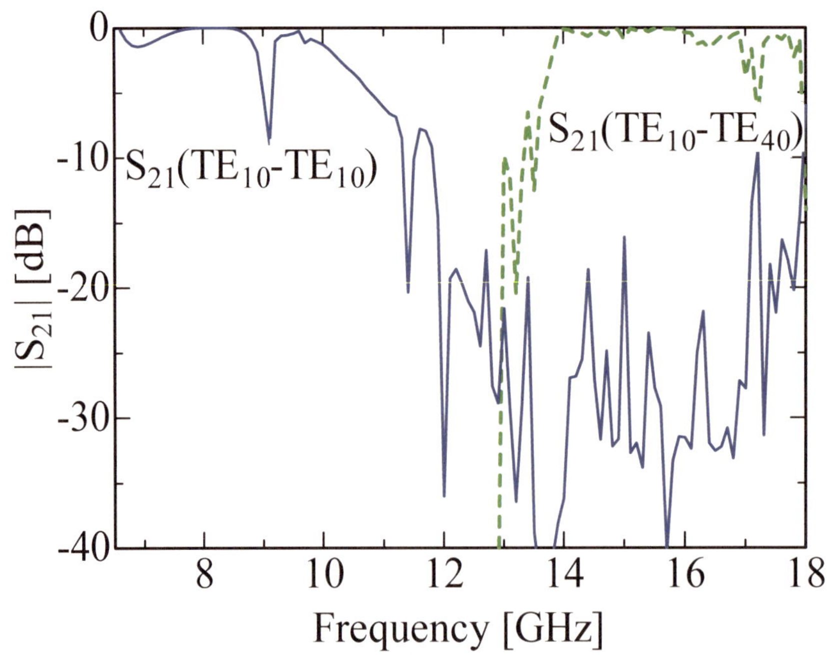

(a)

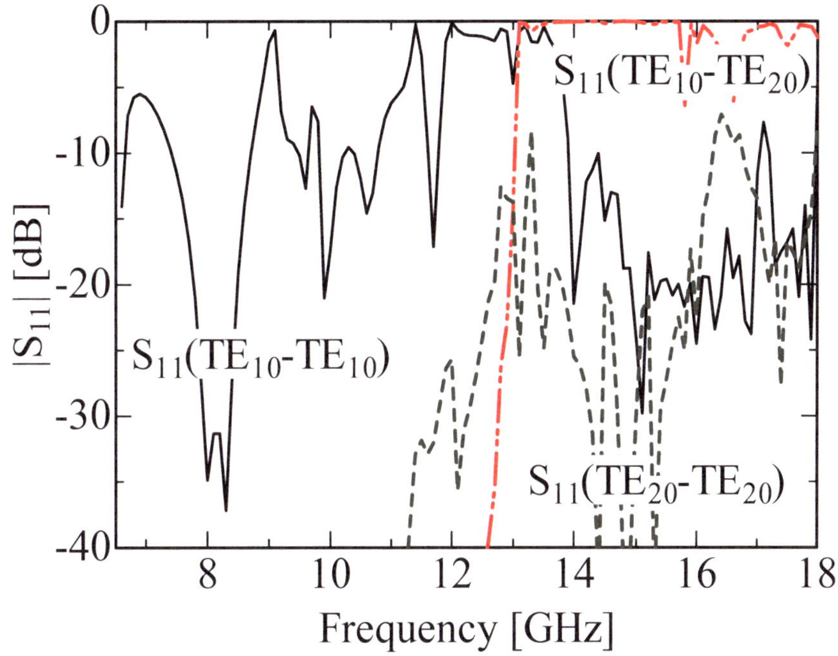

(b)

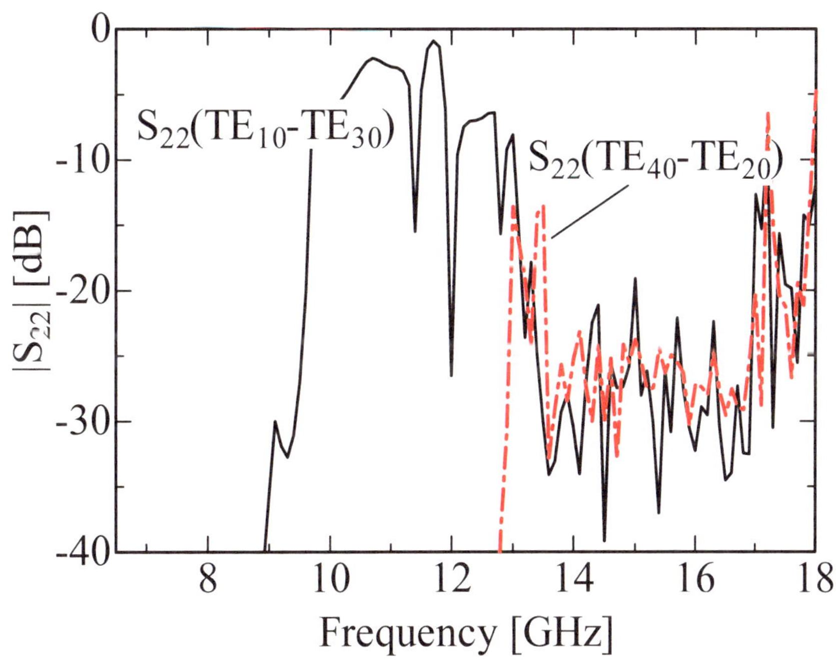

(c)

Fig. 10. S parameters for the TE_{10}-to-TE_{40} mode converter. (a) $|S_{21}|$, (b) $|S_{11}|$, and (c) $|S_{22}|$. (Kokubo, 2011a)

2.5 Design method of the TE_{30}-to-TE_{10} mode converter

A structure that contains two arrays of dielectric rods can convert the TE_{30} mode into the TE_{10} mode (Kokubo, 2009). The TE_{30} mode electromagnetic waves in this type of waveguide are converted to the TE_{10} mode for 7.1–8.9 GHz with over 95% efficiency. However, this structure cannot pass through electromagnetic waves around 2.5–3 GHz without high reflection even if the waveguide is straight. Therefore, a mode converter is proposed that passes the TE_{10} mode at low frequencies and efficiently converts the TE_{30} mode into the TE_{10} mode at high frequencies.

A metallic waveguide that contains two in-line dielectric rods can propagate single modes in two frequency regions (Shibano et al., 2006). The propagation modes in a waveguide with two in-line dielectric rods with period a are calculated using a supercell approach (Benisty, 1996) by applying appropriate periodic Bloch conditions at the boundary of the unit cell (Boroditsky et al., 1998; Kokubo & Kawai, 2008).

The metallic waveguide width is assumed to be w_1 = 21.4 mm, which is 3 times wider than the WR-28 waveguide (7.11× 3.56 mm; $f_c \approx$ 21.1 GHz), and period a is fixed at 5 mm. Fig. 11 shows a sample of the calculated normalized velocity along the waveguide axis at 9 and 27 GHz for dielectric rods ($LaAlO_3$: ε_r = 24; radius r [mm]) aligned at a distance w_2 [mm] between two arrays. The waveguide must be designed so as not to vary the group velocity (C) (TE_{10}: v_g/c = 0.627) at 9 GHz. After calculating v_g for pairs of w_{2i} and r_i, if group velocity (A) (TE_{30}: v_g/c = 0.627) is gradually changed to (B) (TE_{10}: v_g/c = 0.966) at 27GHz, then each pair of w_{2i} and r_i are combined (see Fig. 12).

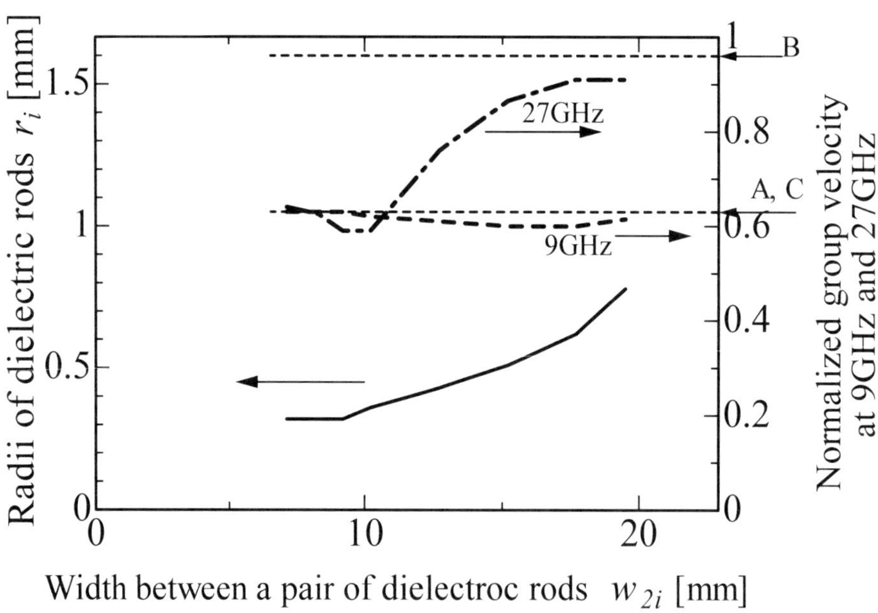

Fig. 11. Group velocity (dotted lines) in a metallic waveguide containing a periodic array of dielectric rods with various distances w_{2i} between the rods and corresponding radii r_i (solid line) of the rods, at 9 and 27 GHz. (Kokubo, 2011b)

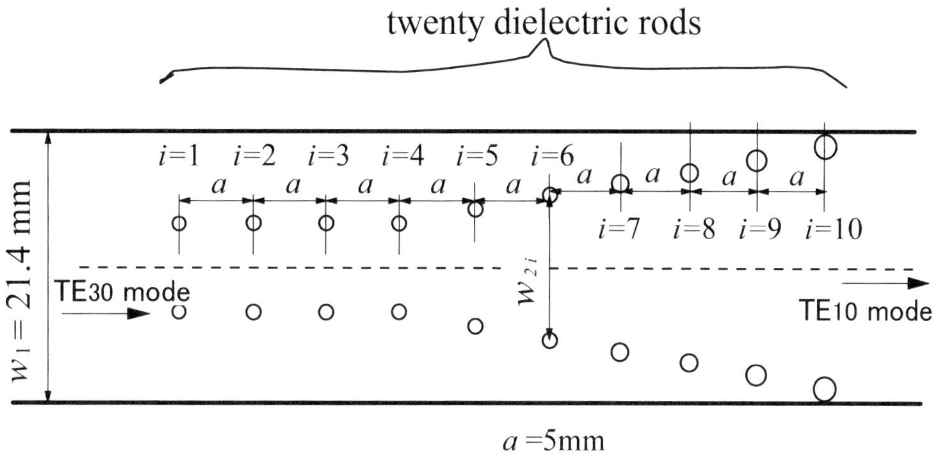

Fig. 12. Proposed structure of the TE_{30}-to-TE_{10} mode converter. (Kokubo, 2011b)

Rod Number i	Distance w_{2i} between a pair of the rods [mm]	Radius r_i of the dielectric rod [mm]
1	7.13	0.22
2	7.13	0.32
3	7.13	0.32
4	7.13	0.32
5	8.3	0.32
6	10.2	0.36
7	12.7	0.43
8	15.2	0.51
9	17.7	0.62
10	19.5	0.78

Table 3. Location and radius of the dielectric rods

The first pair of rods has r_1 = 0.22 mm to reduce electromagnetic reflection at low frequencies. If the first pair of rods is absent, then reflections will be above −10 dB at low frequencies. Twenty dielectric rods are placed from one third of the width of the waveguide to near the sidewall with increasing rod radius r_i and constant a (= 5 mm). Table 3 shows the relation between the distance w_{2i} and radius r_i. The S parameters between the input port (port 1) and output port (port 2) are calculated using HFSS software by Ansys Inc., and the results are shown in Figs. 13(a-c). Electromagnetic waves propagate as the TE_{10} mode for 8.2–14.8 GHz and the TE_{30} mode is converted into the TE_{10} mode for 22.2–28.4 GHz at an efficiency of over 95%.

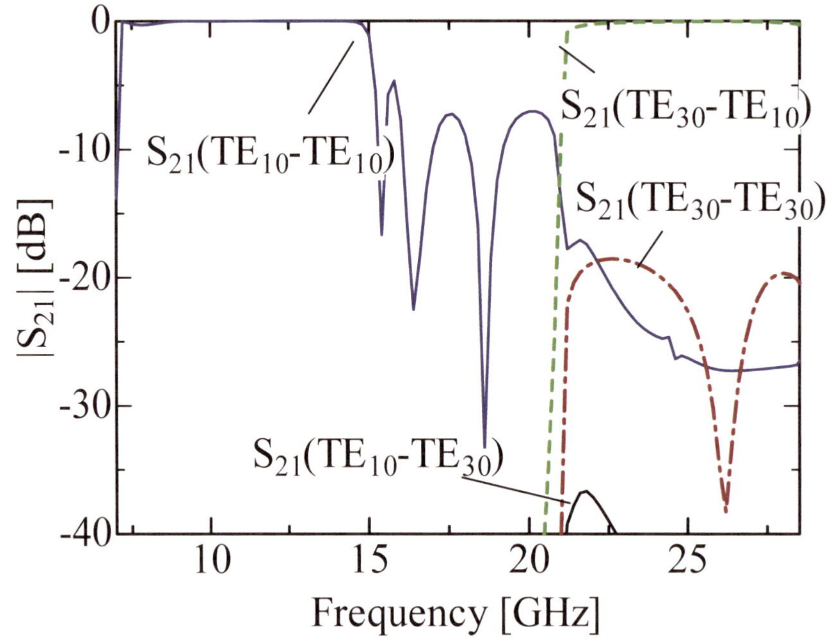

0
-10
-20
-30
-40
10
15
20
25
Frequency [GHz]
$|S_{21}|$ [dB]
$S_{21}(TE_{10}-TE_{10})$
$S_{21}(TE_{30}-TE_{10})$
$S_{21}(TE_{30}-TE_{30})$
$S_{21}(TE_{10}-TE_{30})$

(a)

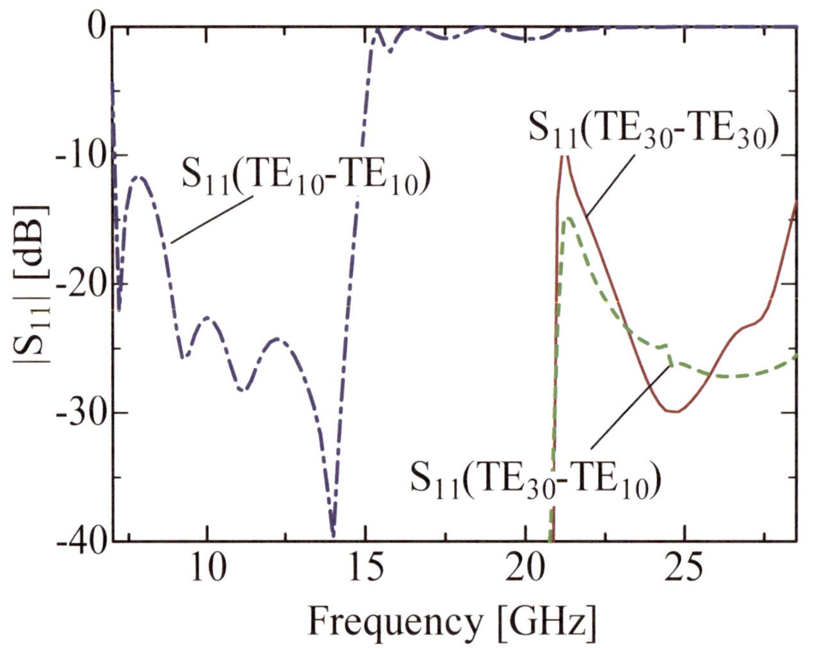

0
-10
-20
-30
-40
10
15
20
25
Frequency [GHz]
$|S_{11}|$ [dB]
$S_{11}(TE_{10}-TE_{10})$
$S_{11}(TE_{30}-TE_{30})$
$S_{11}(TE_{30}-TE_{10})$

(b)

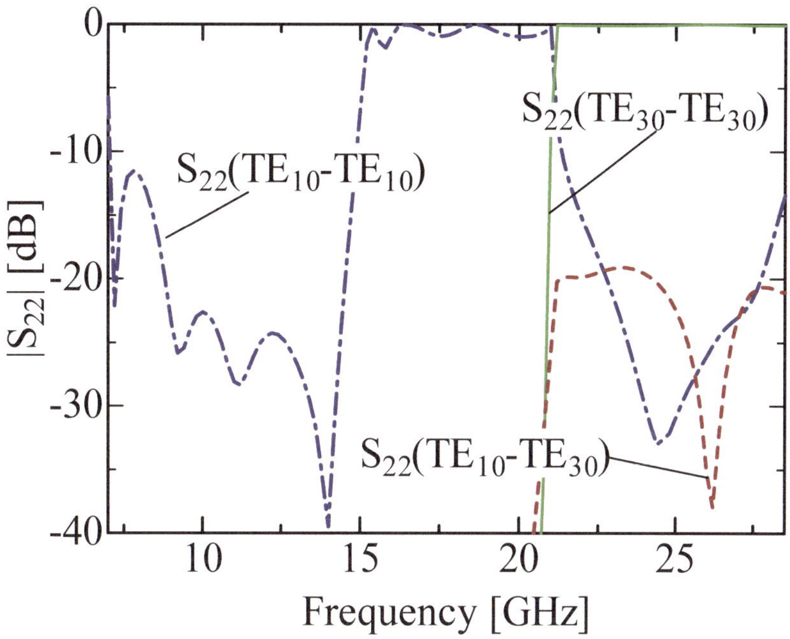

(c)

Fig. 13. S parameters for the mode converter; (a) $|S_{21}|$, (b) $|S_{11}|$, and (c) $|S_{22}|$. (Kokubo, 2011b)

3. Conclusion

We have previously reported that single-mode propagation is available for a metallic waveguide with dielectric rods arrayed at the center of a waveguide using the TE_{10} mode, and the TE_{20} mode. However, a TE_{20}-like mode, which is propagated in the second band, is an odd mode, and generation is not easy. In this investigation, a mode converter is proposed which passes through the TE_{10} mode for the low frequency range and converts TE_{10} to the TE_{20} mode for the high frequency range by small variation of the group velocity. It was shown that the electromagnetic waves pass through as the TE_{10} mode for 7-11.2 GHz and are converted to the TE_{20} mode for 14.1-16.1 GHz under a condition of over 95% efficiency.
It was shown that electromagnetic waves propagate as the TE_{10} mode around 8 GHz and that the TE_{40} mode is converted into the TE_{10} mode around 16 GHz.

4. Acknowledgment

This work was supported by JSPS KAKENHI 22560336.

5. References

Bae, J.; Fujita, M. & Mizuno, K. (2000). W-Band Overmoded-Waveguide Oscillator with Gunn Diodes, *2000 Asia-Pacific Microwave Conference*, (Dec. 2000) pp.1093-1096, ISBN 0-7803-6435-X.

Benisty, H. (1996). Modal analysis of optical guides with two-dimensional photonic band-gap boundaries, *Journal of Applied Physics*, vol.79, no.10, (Oct. 1996) pp.7483-7492, ISSN 0021-8979.

Boroditsky, M.; Coccioli, R. & Yablonovitch, E. (1998). Analysis of photonic crystals for light emitting diodes using the finite difference time domain technique, *Proceedings of SPIE*, vol.3283, (1998), pp.184-190, ISSN 0277-786X.

Cohn, S., B. (1947). Properties of Ridge Wave Guide, *Proceedings of the IRE*, Vol.35, (Aug. 1947) pp. 783-788, ISSN 0096-8390.

Kokubo, Y. (2007). Wide band metallic waveguide with in-line dielectric rods, *IEICE Transactions on Electronics*, vol. J90-C, no. 9, (Sep. 2007) pp.642-643, ISSN 1345-2827 (Japanese edition).

Kokubo, Y. & Kawai, T. (2008). Wide band metallic waveguide with asymmetric in-line dielectric rods, *IEICE Transactions on Electronics*, vol.E-91-C, no. 12, (2008)pp.1966-1968, ISSN 0916-8524.

Kokubo, Y. & Kawai, T. (2009). 90-degree H-plane bent waveguide using dielectric rods, *Microwave and Optical Technology Letters*. vol.51, no.9, (Sep. 2009) pp.2015-2017, ISSN 0895-2477.

Kokubo, Y. (2009). Rectangular TE_{30} to TE_{10} mode converter, *IEICE Transactions on Electronics*, vol. E92-C, no. 8, (Aug. 2009) pp.1087-1090, ISSN 0916-8524.

Kokubo, Y. (2010). Frequency Range Dependent TE_{10} to TE_{20} Mode Converter, *Microwave and Optical Technology Letters*, vol. 52, no. 1, (Jan. 2010) pp. 169-171, ISSN 0895-2477.

Kokubo, Y. (2011a). TE_{10}-to-TE_{40} Metallic Waveguide Mode Converter, *European Microwave Conference (EuMC2011)*, (submitted).

Kokubo, Y. (2011b). Frequency-Dependent Rectangular TE_{30}-to-TE_{10} Mode Converter, *IEICE Transactions on Electronics*, (submitted).

Matsumura, H.; & Mizuno, H. (1987). Design of Microwave Combiner with Waveguide Ports, *IEICE Transactions on Electronics*, vol.J70-C, no.9, (Sep. 1987) pp.1279-1285, ISSN 1345-2827(Japanese edition).

Shibano, T.; Maki, D. & Kokubo, Y. (2006). Dual Band Metallic Waveguide with Dual in-line Dielectric Rods, *IEICE Transactions on Electronics*, vol.J89-C, No.10, (Oct. 2006) pp.743-744, ISSN 1345-2827 (Japanese Edition); Correction and supplement, ibid, Vol.J90-C, No.3, (Mar. 2007) p.298, ISSN 1345-2827. (Japanese Edition).

Part 5

Electromagnetic Material Analysis and Characterization

15

Resonance Properties of Scattering and Generation of Waves on Cubically Polarisable Dielectric Layers

Lutz Angermann[1] and Vasyl V. Yatsyk[2]
[1]*Institut für Mathematik, Technische Universität Clausthal, Erzstraße 1, Clausthal-Zellerfeld*
[2]*Usikov Institute of Radiophysics and Electronics, National Academy of Sciences of Ukraine, Kharkov*
[1]*Germany*
[2]*Ukraine*

1. Introduction

In this paper we investigate the problem of scattering and generation of waves on an isotropic, non-magnetic, linearly polarised (E-polarisation), non-linear, layered, cubically polarisable, dielectric structure, which is excited by a packet of plane waves, in the range of resonant frequencies. We consider wave packets consisting of both strong electromagnetic fields at the excitation frequency, leading to the generation of waves, and of weak fields at the multiple frequencies, which do not lead to the generation of harmonics but influence on the process of scattering and generation of waves. The analysis of the quasi-homogeneous electromagnetic fields of the non-linear dielectric layered structure made it possible to derive a condition of phase synchronism of waves. If the classical formulation of the mathematical model is supplemented by this condition of phase synchronism, we arrive at a rigorous formulation of a system of boundary-value problems with respect to the components of the scattered and generated fields (see Angermann & Yatsyk (2011)). This system is transformed to equivalent systems of non-linear problems, namely a system of one-dimensional non-linear Fredholm integral equations of the second kind and a system of non-linear boundary-value problems of Sturm-Liouville type. The numerical algorithms of the solution of the non-linear problems are based on iterative procedures which require the solution of a linearised system in each step. In this way the approximate solution of the non-linear problems is described by means of solutions of linearised problems with an induced dielectric permeability. The analytical continuation of these problems into the region of complex values of the frequency parameter allows us to switch to the analysis of spectral problems. The corresponding eigen-frequencies form a discrete, countable set of points, with the only possible accumulation point at infinity, and lie on a complex two-sheeted Riemann surface. In the frequency domain, the resonant scattering and generation properties of non-linear structures are determined by the proximity of the excitation frequencies of the non-linear structures to the complex eigen-frequencies of the corresponding homogeneous linear spectral problems with the induced non-linear dielectric permeability of the medium.

Results of calculations of characteristics of the scattered field of a plane wave are presented, taking into account the third harmonic generated by non-linear cubically polarisable layers with both negative as well as positive values of the cubic susceptibility of the medium.
Within the framework of the closed system, which is given by a system of self-consistent boundary-value problems, we show the following. The variation of the imaginary part of the permittivity of the layer at the excitation frequency can take both positive and negative values along the height of the non-linear layer. It is induced by the non-linear part of the permittivity and is caused by the loss of energy in the non-linear medium (at the frequency of the incident field), which is spent for the generation of the electromagnetic field of the third harmonic (at the triple frequency). The magnitude of this variation is determined by the amplitude and phase characteristics of the fields which are scattered and generated by the non-linear layer.
It is shown that layers with negative and positive values of the coefficient of cubic susceptibility of the non-linear medium have fundamentally different scattering and generation properties in the range of resonance. For instance, in the case of negative values of the susceptibility, a decanalisation of the electromagnetic field can be detected. With the increase of intensity of the incident field, the maximal variations of the reflection and transmission coefficients can be observed in the vicinity of the normal incidence of the plane wave. A previously transparent structure becomes semi-transparent, and the reflection and transmission coefficients become comparable. For the layer considered here, the maximal portion of the total energy generated in the third harmonic is observed in the direction normal to the structure and amounts to 3.9% of the total dissipated energy. For a layer with a positive value of the susceptibility an effect of energy canalisation is observed. The increase of intensity of the incident field leads to an increase of the angle of transparancy which increasingly deviates from the direction normal to the layer and which is responsible for a reflection coefficient close to zero. In this case, the maximal portion of energy generated in the third harmonic is observed near the angle of transparency of the non-linear layer. In the numerical experiments there have been reached intensities of the excitation field of the layer such that the relative portion of the total energy generated in the third harmonic is 36%. The paper also presents results of numerical calculations that describe properties of the non-linear dielectric permeabilities of the layers as well as their scattering and generation characteristics. The tests are illustrated by figures showing the dependence on the amplitudes and the angles of incidence of the plane wave for layers with negative and positive values of the coefficient of the cubic susceptibility of the non-linear medium.

2. Maxwell equations and wave propagation in non-linear media with cubic polarisability

In this section we give a short overview on the derivation of the mathematical model. A more detailed explanation can be found in Angermann & Yatsyk (2011). Electrodynamical wave phenomena in charge- and current-free media can be described by the Maxwell equations

$$\nabla \times \mathbf{E} = -\frac{1}{c}\frac{\partial \mathbf{B}}{\partial t}, \quad \nabla \times \mathbf{H} = \frac{1}{c}\frac{\partial \mathbf{D}}{\partial t}, \quad \nabla \cdot \mathbf{D} = 0, \quad \nabla \cdot \mathbf{B} = 0. \tag{1}$$

Here $\mathbf{E} = \mathbf{E}(\mathbf{r}, t)$, $\mathbf{H} = \mathbf{H}(\mathbf{r}, t)$, $\mathbf{D} = \mathbf{D}(\mathbf{r}, t)$ and $\mathbf{B} = \mathbf{B}(\mathbf{r}, t)$ denote the vectors of electric and magnetic field intensities, electric displacement, and magnetic induction, respectively, and $(\mathbf{r}, t) \in \mathbb{R}^3 \times (0, \infty)$. The symbol ∇ represents the formal vector of partial derivatives w.r.t. the spatial variables, i.e. $\nabla := (\partial/\partial x, \partial/\partial y, \partial/\partial z)^\top$, where the symbol $\cdot^\top$ denotes the transposition. In addition, the system (1) is completed by the material equations

$$\mathbf{D} = \mathbf{E} + 4\pi\mathbf{P}, \qquad \mathbf{B} = \mathbf{H} + 4\pi\mathbf{M}, \tag{2}$$

where $\mathbf{P}$ and $\mathbf{M}$ are the vectors of the polarisation and magnetic moment, respectively.
In the present paper, the non-linear medium under consideration is located in an infinite plate of thickness $4\pi\delta$, where $\delta > 0$ is a given parameter: $\{\mathbf{r} = (x, y, z)^\top \in \mathbb{R}^3 : \ |z| \leq 2\pi\delta\}$.
As in the books Akhmediev & Ankevich (2003), Kivshar & Agrawal (2005) and Miloslavsky (2008), the investigations will be restricted to non-linear media without dispersion (cf. Agranovich & Ginzburg (1966)), i.e. we use the following expansion of the polarisation vector in terms of the electric field intensity:

$$\mathbf{P} = \boldsymbol{\chi}^{(1)}\mathbf{E} + (\boldsymbol{\chi}^{(2)}\mathbf{E})\mathbf{E} + ((\boldsymbol{\chi}^{(3)}\mathbf{E})\mathbf{E})\mathbf{E} + \dots, \tag{3}$$

where $\boldsymbol{\chi}^{(1)}$, $\boldsymbol{\chi}^{(2)}$, $\boldsymbol{\chi}^{(3)}$ are the media susceptibility tensors of rank one, two and three, with components $\{\chi_{ij}^{(1)}\}_{i,j=1}^3$, $\{\chi_{ijk}^{(2)}\}_{i,j,k=1}^3$ and $\{\chi_{ijkl}^{(3)}\}_{i,j,k,l=1}^3$, respectively (see Butcher (1965)). In the case of media which are invariant under the operations of inversion, reflection and rotation, in particular of isotropic media, the quadratic term disappears.
It is convenient to split $\mathbf{P}$ into its linear and non-linear parts as $\mathbf{P} = \mathbf{P}^{(L)} + \mathbf{P}^{(NL)} := \boldsymbol{\chi}^{(1)}\mathbf{E} + \mathbf{P}^{(NL)}$. Similarly, with $\boldsymbol{\varepsilon} := \mathbf{I} + 4\pi\boldsymbol{\chi}^{(1)}$ and $\mathbf{D}^{(L)} := \boldsymbol{\varepsilon}\mathbf{E}$, where $\mathbf{I}$ denotes the identity in $\mathbb{C}^3$, the displacement field in (2) can be decomposed as

$$\mathbf{D} = \mathbf{D}^{(L)} + 4\pi\mathbf{P}^{(NL)}. \tag{4}$$

$\boldsymbol{\varepsilon}$ is the linear term of the permittivity tensor. Furthermore we assume that the media are non-magnetic, i.e $\mathbf{M} = \mathbf{0}$, so that

$$\mathbf{B} = \mathbf{H} \tag{5}$$

by (2). Resolving the equations (1), (4) and (5) w.r.t. $\mathbf{H}$, a single vector-valued equation results:

$$\nabla^2\mathbf{E} - \nabla(\nabla \cdot \mathbf{E}) - \frac{1}{c^2}\frac{\partial^2}{\partial t^2}\mathbf{D}^{(L)} - \frac{4\pi}{c^2}\frac{\partial^2}{\partial t^2}\mathbf{P}^{(NL)} = 0. \tag{6}$$

In addition, if the media under consideration are isotropic and transversely inhomogeneous w.r.t. z, i.e. $\boldsymbol{\varepsilon} = \varepsilon^{(L)}\mathbf{I}$ with a scalar, possibly complex-valued function $\varepsilon^{(L)} = \varepsilon^{(L)}(z)$, if the wave is linearly E-polarised, i.e. $\mathbf{E} = (E_1, 0, 0)^\top$, $\mathbf{H} = (0, H_2, H_3)^\top$, and if the electric field $\mathbf{E}$ is homogeneous w.r.t. the coordinate x, i.e. $\mathbf{E}(\mathbf{r}, t) = (E_1(t; y, z), 0, 0)^\top$, then equation (6) simplifies to

$$\nabla^2\mathbf{E} - \frac{\varepsilon^{(L)}}{c^2}\frac{\partial^2}{\partial t^2}\mathbf{E} - \frac{4\pi}{c^2}\frac{\partial^2}{\partial t^2}\mathbf{P}^{(NL)} = 0, \tag{7}$$

where ∇^2 reduces to the Laplacian w.r.t. y and z, i.e. $\nabla^2 := \partial^2/\partial y^2 + \partial^2/\partial z^2$.
The stationary problem of the diffraction of a plane electromagnetic wave (with oscillation frequency $\omega > 0$) on a transversely inhomogeneous, isotropic, non-magnetic, linearly polarised, dielectric layer filled with a cubically polarisable medium (see Fig. 1) is studied in the frequency domain (i.e. in the space of the Fourier amplitudes of the electromagnetic field), taking into account the multiple frequencies $s\omega$, $s \in \mathbb{N}$, of the excitation frequency generated by non-linear structure, where a time-dependence of the form $\exp(-is\omega t)$ is assumed. The transition between the time domain and the frequency domain is performed by means of

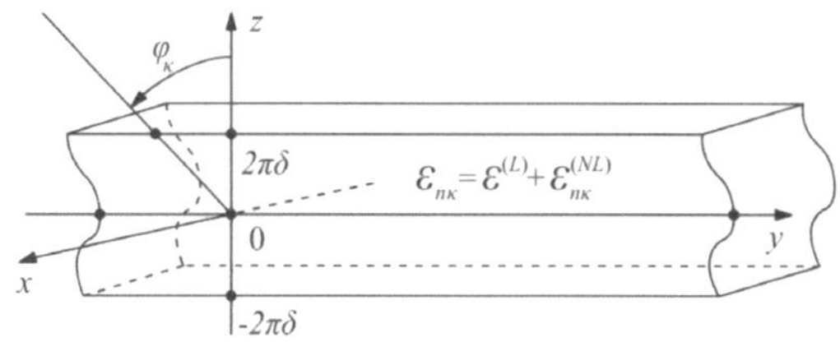

Fig. 1. The non-linear dielectric layered structure

direct and inverse Fourier transforms:

$$\hat{\mathbf{F}}(\mathbf{r},\hat{\omega}) = \int_{\mathbb{R}} \mathbf{F}(\mathbf{r},t)e^{i\hat{\omega}t}dt, \qquad \mathbf{F}(\mathbf{r},t) = \frac{1}{2\pi}\int_{\mathbb{R}} \hat{\mathbf{F}}(\mathbf{r},\hat{\omega})e^{-i\hat{\omega}t}d\hat{\omega},$$

where $\mathbf{F}$ is one of the vector fields $\mathbf{E}$ or $\mathbf{P}^{(NL)}$. Applying formally the Fourier transform to equation (7), we obtain the following representation in the frequency domain:

$$\nabla^2\hat{\mathbf{E}}(\mathbf{r},\hat{\omega}) + \frac{\varepsilon^{(L)}\hat{\omega}^2}{c^2}\hat{\mathbf{E}}(\mathbf{r},\hat{\omega}) + \frac{4\pi\hat{\omega}^2}{c^2}\hat{\mathbf{P}}^{(NL)}(\mathbf{r},\hat{\omega}) = 0. \tag{8}$$

A stationary (i.e. $\sim \exp(-i\hat{\omega}t)$) electromagnetic wave propagating in a weakly non-linear dielectric structure gives rise to a field containing all frequency harmonics (see Sukhorukov (1988), Vinogradova et al. (1990)). Therefore, the quantities describing the electromagnetic field in the time domain subject to equation (7) can be represented by means of the Fourier series

$$\mathbf{F}(\mathbf{r},t) = \frac{1}{2}\sum_{n\in\mathbb{Z}} \mathbf{F}(\mathbf{r},n\omega)e^{-in\omega t}, \qquad \mathbf{F} \in \{\mathbf{E},\mathbf{P}^{(NL)}\}. \tag{9}$$

Applying to (9) the Fourier transform, we obtain

$$\hat{\mathbf{F}}(\mathbf{r},\hat{\omega}) = \frac{1}{2}\int_{\mathbb{R}}\sum_{n\in\mathbb{Z}} \mathbf{F}(\mathbf{r},n\omega)e^{-in\omega t}e^{i\hat{\omega}t}dt = \frac{\sqrt{2\pi}}{2}\mathbf{F}(\mathbf{r},n\omega)\delta(\hat{\omega},n\omega), \qquad \mathbf{F} \in \{\mathbf{E},\mathbf{P}^{(NL)}\}, \tag{10}$$

where $\delta(s,s_0) := \frac{1}{\sqrt{2\pi}}\int_{\mathbb{R}} e^{i(s-s_0)t}dt$ is the Dirac delta-function located at $s = s_0$.
Substituting (10) into (8), we obtain an infinite system of coupled equations w.r.t. the Fourier amplitudes of the electric field intensities of the non-linear structure in the frequency domain:

$$\nabla^2\mathbf{E}(\mathbf{r},s\omega) + \frac{\varepsilon^{(L)}(s\omega)^2}{c^2}\mathbf{E}(\mathbf{r},s\omega) + \frac{4\pi(s\omega)^2}{c^2}\mathbf{P}^{(NL)}(\mathbf{r},s\omega) = 0, \quad s \in \mathbb{Z}. \tag{11}$$

In the case of a three-component E-polarised electromagnetic field

$$\mathbf{E}(\mathbf{r},s\omega) = (E_1(s\omega;y,z),0,0)^\top, \quad \mathbf{H}(\mathbf{r},s\omega) = (0;H_2(s\omega;y,z),H_3(s\omega;y,z))^\top, \tag{12}$$

the system (11) reduces to a system of scalar equations w.r.t. E_1:

$$\nabla^2 E_1(\mathbf{r},s\omega) + \frac{\varepsilon^{(L)}(s\omega)^2}{c^2}E_1(\mathbf{r},s\omega) + \frac{4\pi(s\omega)^2}{c^2}P_1^{(NL)}(\mathbf{r},s\omega) = 0, \quad s \in \mathbb{N}. \tag{13}$$

In writing equation (13), the property $E_1(\mathbf{r}; j\omega) = \overline{E}_1(\mathbf{r}; -j\omega)$ of the Fourier coefficients and the lack of influence of the static electric field $E_1(\mathbf{r}, s\omega)|_{s=0} = 0$ on the non-linear structure were taken into consideration.

We assume that the main contribution to the non-linearity is introduced by the term $\mathbf{P}^{(NL)}(\mathbf{r}, s\omega)$ (cf. Yatsyk (2007), Shestopalov & Yatsyk (2007), Kravchenko & Yatsyk (2007), Angermann & Yatsyk (2008), Yatsyk (2006), Schürmann et al. (2001), Smirnov et al. (2005), Serov et al. (2004)), and we take only the lowest-order terms in the Taylor series expansion of the non-linear part $\mathbf{P}^{(NL)}(\mathbf{r}, s\omega) = \left(P_1^{(NL)}(\mathbf{r}, s\omega), 0, 0\right)^\top$ of the polarisation vector in the vicinity of the zero value of the electric field intensity, cf. (3). In this case, the only non-trivial component of the polarisation vector is determined by susceptibility tensor of the third order $\chi^{(3)}$. In the time domain, this component can be represented in the form (cf. (3) and (9)):

$$P_1^{(NL)}(\mathbf{r}, t) \doteq \frac{1}{8} \sum_{\left\{\substack{n,m,p,s \in \mathbb{Z}\setminus\{0\} \\ n+m+p=s}\right.} \chi_{1111}^{(3)}(s\omega; n\omega, m\omega, p\omega) E_1(\mathbf{r}, n\omega) E_1(\mathbf{r}, m\omega) E_1(\mathbf{r}, p\omega) e^{-is\omega t}, \tag{14}$$

where the symbol $\doteq$ means that higher-order terms are neglected. Applying to (14) the Fourier transform w.r.t. time, we obtain an expansion in the frequency domain:

$$\begin{aligned} P_1^{(NL)}(\mathbf{r}, s\omega) &= \frac{1}{4} \sum_{j \in \mathbb{N}} 3\chi_{1111}^{(3)}(s\omega; j\omega, -j\omega, s\omega) |E_1(\mathbf{r}, j\omega)|^2 E_1(\mathbf{r}, s\omega) \\ &+ \frac{1}{4} \sum_{\left\{\substack{n,m,p \in \mathbb{Z}\setminus\{0\} \\ n \neq -m,\ p=s \\ m \neq -p,\ n=s \\ n \neq -p,\ m=s \\ n+m+p=s}\right.} \chi_{1111}^{(3)}(s\omega; n\omega, m\omega, p\omega) E_1(\mathbf{r}, n\omega) E_1(\mathbf{r}, m\omega) E_1(\mathbf{r}, p\omega). \end{aligned} \tag{15}$$

We see that, under the above assumptions, the electromagnetic waves in a non-linear medium with a cubic polarisability are described by an infinite system (13)&(15) of coupled non-linear equations (Yatsyk (2007), Shestopalov & Yatsyk (2007), Kravchenko & Yatsyk (2007), Angermann & Yatsyk (2010)). In what follows we will consider the equations in the frequency space taking into account the relation $\kappa = \frac{\omega}{c}$.

Here we study non-linear effects involving the waves at the first three frequency components of E_1 only. That is, we further neglect the influence of harmonics of order higher than 3. Then it is possible to restrict the examination of the system (13)&(15) to three equations, and also to leave particular terms in the representation of the polarisation coefficients. Taking into account only the non-trivial terms in the expansion of the polarisation coefficients and using the so-called *Kleinman's rule* (i.e. the equality of all the coefficients $\chi_{1111}^{(3)}$ at the multiple frequencies, Kleinman (1962), Miloslavsky (2008)), we arrive at the following system:

$$\begin{cases} \nabla^2 E_1(\mathbf{r}, \kappa) + \varepsilon^{(L)} \kappa^2 E_1(\mathbf{r}, \kappa) + 4\pi\kappa^2 \left(P_1^{(PSM)}(\mathbf{r}, \kappa) + P_1^{(GC)}(\mathbf{r}, \kappa)\right) = -4\pi\kappa^2 P_1^{(G)}(\mathbf{r}, \kappa), \\ \nabla^2 E_1(\mathbf{r}, 2\kappa) + \varepsilon^{(L)} (2\kappa)^2 E_1(\mathbf{r}, 2\kappa) + 4\pi(2\kappa)^2 \left(P_1^{(PSM)}(\mathbf{r}, 2\kappa) + P_1^{(GC)}(\mathbf{r}, 2\kappa)\right) = 0, \\ \nabla^2 E_1(\mathbf{r}, 3\kappa) + \varepsilon^{(L)} (3\kappa)^2 E_1(\mathbf{r}, 3\kappa) + 4\pi(3\kappa)^2 P_1^{(PSM)}(\mathbf{r}, 3\kappa) = -4\pi(3\kappa)^2 P_1^{(G)}(\mathbf{r}, 3\kappa), \end{cases} \tag{16}$$

where

$$P_1^{(PSM)}(\mathbf{r},n\kappa) := \frac{3}{4}\chi_{1111}^{(3)}(|E_1(\mathbf{r},\kappa)|^2 + |E_1(\mathbf{r},2\kappa)|^2 + |E_1(\mathbf{r},3\kappa)|^2)E_1(\mathbf{r},n\kappa), \quad n = 1,2,3,$$

$$P_1^{(GC)}(\mathbf{r},\kappa) := \frac{3}{4}\chi_{1111}^{(3)}\frac{[\overline{E}_1(\mathbf{r},\kappa)]^2}{E_1(\mathbf{r},\kappa)}E_1(\mathbf{r},3\kappa)E_1(\mathbf{r},\kappa), \quad P_1^{(G)}(\mathbf{r},\kappa) := \frac{3}{4}\chi_{1111}^{(3)}E_1^2(\mathbf{r},2\kappa)\overline{E}_1(\mathbf{r},3\kappa),$$

$$P_1^{(GC)}(\mathbf{r},2\kappa) := \frac{3}{4}\chi_{1111}^{(3)}\frac{\overline{E}_1(\mathbf{r},2\kappa)}{E_1(\mathbf{r},2\kappa)}E_1(\mathbf{r},\kappa)E_1(\mathbf{r},3\kappa)E_1(\mathbf{r},2\kappa),$$

$$P_1^{(G)}(\mathbf{r},3\kappa) := \frac{3}{4}\chi_{1111}^{(3)}\left\{\frac{1}{3}E_1^3(\mathbf{r},\kappa) + E_1^2(\mathbf{r},2\kappa)\overline{E}_1(\mathbf{r},\kappa)\right\}.$$

The terms $P_1^{(PSM)}(\mathbf{r},n\omega)$ are usually called *phase self-modulation (PSM)* terms (cf. Akhmediev & Ankevich (2003)). The permittivity of the non-linear medium filling a layer (see Fig. 1) can be represented as

$$\varepsilon_{n\kappa} = \varepsilon^{(L)} + \varepsilon_{n\kappa}^{(NL)} \quad \text{for} \quad |z| \le 2\pi\delta. \tag{17}$$

Outside the layer, i.e. for $|z| > 2\pi\delta$, $\varepsilon_{n\kappa} = 1$. The linear and non-linear terms of the permittivity of the layer are given by the coefficients at $(n\kappa)^2E_1(\mathbf{r},n\kappa)$ in the second and third addends in each of the equations of the system, respectively. Thus $\varepsilon^{(L)} = D_1^{(L)}(\mathbf{r},n\kappa)/E_1(\mathbf{r},n\kappa) = 1 + 4\pi\chi_{11}^{(1)}$, where the representations for the linear part of the complex components of the electric displacement $D_1^{(L)}(\mathbf{r},n\kappa) = E_1(\mathbf{r},n\kappa) + 4\pi P_1^{(L)}(\mathbf{r},n\kappa) = \varepsilon^{(L)}E_1(\mathbf{r},n\kappa)$ and the polarisation $P_1^{(L)}(\mathbf{r},n\kappa) = \chi_{11}^{(1)}E_1(\mathbf{r},n\kappa)$ are taken into account. Similarly, the third term of each equation of the system makes it possible to write the non-linear component of $\varepsilon_{n\kappa}$ in the form

$$\begin{aligned}\varepsilon_{n\kappa}^{(NL)} &= \alpha(z)\big[|E_1(\mathbf{r},\kappa)|^2 + |E_1(\mathbf{r},2\kappa)|^2 + |E_1(\mathbf{r},3\kappa)|^2 \\ &\quad + \delta_{n1}\frac{[\overline{E}_1(\mathbf{r},\kappa)]^2}{E_1(\mathbf{r},\kappa)}E_1(\mathbf{r},3\kappa) + \delta_{n2}\frac{\overline{E}_1(\mathbf{r},2\kappa)}{E_1(\mathbf{r},2\kappa)}E_1(\mathbf{r},\kappa)E_1(\mathbf{r},3\kappa)\big],\end{aligned} \tag{18}$$

where $\alpha(z) := 3\pi\chi_{1111}^{(3)}(z)$ is the so-called *function of cubic susceptibility.* For transversely inhomogeneous media (a layer or a layered structure), the linear part $\varepsilon^{(L)} = \varepsilon^{(L)}(z) = 1 + 4\pi\chi_{11}^{(1)}(z)$ of the permittivity is described by a piecewise smooth or even a piecewise constant function. Similarly, the function of the cubic susceptibility $\alpha = \alpha(z)$ is also a piecewise smooth or even a piecewise constant function. This assumption allows us to investigate the diffraction characteristics of a non-linear layer and of a layered structure (consisting of a finite number of non-linear dielectric layers) within one and the same mathematical model.

Here and in what follows we use the following notation: $(\mathbf{r},t)$ are dimensionless spatial-temporal coordinates such that the thickness of the layer is equal to $4\pi\delta$. The time-dependence is determined by the factors $\exp(-in\omega t)$, where $\omega := \kappa c$ is the dimensionless circular frequency and κ is a dimensionless frequency parameter such that $\kappa = \omega/c := 2\pi/\lambda$. This parameter characterises the ratio of the true thickness h of the layer to the free-space wavelength λ, i.e. $h/\lambda = 2\kappa\delta$. $c = (\varepsilon_0\mu_0)^{-1/2}$ denotes a dimensionless parameter, equal to the absolute value of the speed of light in the medium containing the layer ($\mathfrak{Im}\, c = 0$). ε_0 and μ_0 are the material parameters of the medium. The absolute values of the true variables $\mathbf{r}', t', \omega'$ are given by the formulas $\mathbf{r}' = h\mathbf{r}/4\pi\delta$, $t' = th/4\pi\delta$, $\omega' = \omega 4\pi\delta/h$.

3. Quasi-homogeneous electromagnetic fields in a transversely inhomogeneous non-linear dielectric layered structure and the excitation by wave packets

The scattered and generated field in a transversely inhomogeneous, non-linear dielectric layer excited by a plane wave is *quasi-homogeneous* along the coordinate y, hence it can be represented as

(C1) $E_1(\mathbf{r}, n\kappa) =: E_1(n\kappa; y, z) := U(n\kappa; z)\exp(i\phi_{n\kappa} y), \quad n = 1, 2, 3.$

Here $U(n\kappa; z)$ and $\phi_{n\kappa} := n\kappa \sin\varphi_{n\kappa}$ denote the complex-valued transverse component of the Fourier amplitude of the electric field and the value of the longitudinal propagation constant (longitudinal wave-number) at the frequency $n\kappa$, respectively, where $\varphi_{n\kappa}$ is the given angle of incidence of the exciting field of frequency $n\kappa$ (cf. Fig. 1).
Furthermore we require that the following *condition of the phase synchronism of waves* is satisfied:

(C2) $\phi_{n\kappa} = n\phi_\kappa, \quad n = 1, 2, 3.$

Then the permittivity of the non-linear layer can be expressed as

$$\begin{aligned}
&\varepsilon_{n\kappa}(z, \alpha(z), E_1(\mathbf{r}, \kappa), E_1(\mathbf{r}, 2\kappa), E_1(\mathbf{r}, 3\kappa)) \\
&= \varepsilon_{n\kappa}(z, \alpha(z), U(\kappa; z), U(2\kappa; z), U(3\kappa; z)) \\
&= \varepsilon^{(L)}(z) + \alpha(z)\left[|U(\kappa; z)|^2 + |U(2\kappa; z)|^2 + |U(3\kappa; z)|^2\right. \\
&\quad + \delta_{n1}|U(\kappa; z)||U(3\kappa; z)|\exp\{i[-3\arg(U(\kappa; z)) + \arg(U(3\kappa; z))]\} \\
&\quad \left. + \delta_{n2}|U(\kappa; z)||U(3\kappa; z)|\exp\{i[-2\arg(U(2\kappa; z)) + \arg(U(\kappa; z)) + \arg(U(3\kappa; z))]\}\right], \\
&\qquad\qquad n = 1, 2, 3.
\end{aligned} \tag{19}$$

For the the components of the non-linear polarisation $P_1^{(G)}(\mathbf{r}, n\kappa)$ (playing the role of the sources generating radiation in the right-hand sides of the system (16)) we have that

$$\begin{aligned}
-4\pi\kappa^2 P_1^{(G)}(\mathbf{r}, \kappa) &= -\alpha(z)\kappa^2 U^2(2\kappa; z)\overline{U}(3\kappa; z)\exp(i\phi_\kappa y), \\
-4\pi(3\kappa)^2 P_1^{(G)}(\mathbf{r}, 3\kappa) &= -\alpha(z)(3\kappa)^2\left\{\frac{1}{3}U^3(\kappa; z) + U^2(2\kappa; z)\overline{U}(\kappa; z)\right\}\exp(i\phi_{3\kappa} y).
\end{aligned}$$

A more detailed explanation of the condition (C2) can be found in (Angermann & Yatsyk, 2011, Sect. 3). In the considered case of spatially quasi-homogeneous (along the coordinate y) electromagnetic fields (C1), the condition of the phase synchronism of waves (C2) reads as

$$\sin\varphi_{n\kappa} = \sin\varphi_\kappa, \quad n = 1, 2, 3.$$

Consequently, the given angle of incidence of a plane wave at the frequency κ coincides with the possible directions of the angles of incidence of plane waves at the multiple frequencies $n\kappa$. The angles of the wave scattered by the layer are equal to $\varphi_{n\kappa}^{\text{scat}} = -\varphi_{n\kappa}$ in the zone of reflection $z > 2\pi\delta$ and $\varphi_{n\kappa}^{\text{scat}} = \pi + \varphi_{n\kappa}$ and in the zone of transmission of the non-linear layer $z < -2\pi\delta$, where all angles are measured counter-clockwise in the (y, z)-plane from the z-axis (cf. Fig. 1).
The conditions (C1), (C2) allow a further simplification of the system (16). Before we do so, we want to make a few comments on specific cases which have already been discussed in the literature. First we mention that the effect of a weak quasi-homogeneous electromagnetic field (C1) on the non-linear dielectric structure such that harmonics at multiple frequencies are not generated, i.e. $E_1(\mathbf{r}, 2\kappa) = 0$ and $E_1(\mathbf{r}, 3\kappa) = 0$, reduces to find the electric field component $E_1(\mathbf{r}, \kappa)$ determined by the first equation of the system (16). In this case, a diffraction problem

for a plane wave on a non-linear dielectric layer with a Kerr-type non-linearity $\varepsilon_{n\kappa} = \varepsilon^{(L)}(z) + \alpha(z)|E_1(\mathbf{r},\kappa)|^2$ and a vanishing right-hand side is to be solved, see Angermann & Yatsyk (2008); Kravchenko & Yatsyk (2007); Serov et al. (2004); Shestopalov & Yatsyk (2007); Smirnov et al. (2005); Yatsyk (2006; 2007). The generation process of a field at the triple frequency 3κ by the non-linear dielectric structure is caused by a strong incident electromagnetic field at the frequency κ and can be described by the first and third equations of the system (16) only. Since the right-hand side of the second equation in (16) is equal to zero, we may set $E_1(\mathbf{r},2\kappa) = 0$ corresponding to the homogeneous boundary condition w.r.t. $E_1(\mathbf{r},2\kappa)$. Therefore the second equation in (16) can be completely omitted, see Angermann & Yatsyk (2010).

A further interesting problem consists in the investigation of the influence of a packet of waves on the generation of the third harmonic, if a strong incident field at the basic frequency κ and, in addition, weak incident quasi-homogeneous electromagnetic fields at the double and triple frequencies 2κ, 3κ (which alone do not generate harmonics at multiple frequencies) excite the non-linear structure. The system (16) allows to describe the corresponding process of the third harmonics generation. Namely, if such a wave packet consists of a strong field at the basic frequency κ and of a weak field at the triple frequency 3κ, then we arrive, as in the situation described above, at the system (16) with $E_1(\mathbf{r},2\kappa) = 0$, i.e. it is sufficient to consider the first and third equations of (16) only. For wave packets consisting of a strong field at the basic frequency κ and of a weak field at the frequency 2κ, (or of two weak fields at the frequencies 2κ and 3κ) we have to take into account all three equations of system (16). This is caused by the inhomogeneity of the corresponding problem, where a weak incident field at the double frequency 2κ (or two weak fields at the frequencies 2κ and 3κ) excites (resp. excite) the dielectric medium.

So we consider the problem of scattering and generation of waves on a non-linear, layered, cubically polarisable structure, which is excited by a packet of plane waves consisting of a strong field at the frequency κ (which generates a field at the triple frequency 3κ) and of weak fields at the frequencies 2κ and 3κ (having an impact on the process of third harmonic generation due to the contribution of weak electromagnetic fields)

$$\left\{E_1^{\text{inc}}(\mathbf{r},n\kappa) := E_1^{\text{inc}}(n\kappa;y,z) := a_{n\kappa}^{\text{inc}} \exp\left(i(\phi_{n\kappa}y - \Gamma_{n\kappa}(z-2\pi\delta))\right)\right\}_{n=1}^{3}, \quad z > 2\pi\delta, \tag{20}$$

with amplitudes $a_{n\kappa}^{\text{inc}}$ and angles of incidence $\varphi_{n\kappa}$, $|\varphi| < \pi/2$ (cf. Fig. 1), where $\phi_{n\kappa} := n\kappa \sin\varphi_{n\kappa}$ are the longitudinal propagation constants (longitudinal wave-numbers) and $\Gamma_{n\kappa} := \sqrt{(n\kappa)^2 - \phi_{n\kappa}^2}$ are the transverse propagation constants (transverse wave-numbers).

In this setting, if a packet of plane waves excites a non-magnetic, isotropic, linearly polarised (i.e.

$$\mathbf{E}(\mathbf{r},n\kappa) = (E_1(n\kappa;y,z),0,0)^\top, \ \mathbf{H}(\mathbf{r},n\kappa) = \left(0, \frac{1}{in\omega\mu_0}\frac{\partial E_1(n\kappa;y,z)}{\partial z}, -\frac{1}{in\omega\mu_0}\frac{\partial E_1(n\kappa;y,z)}{\partial y}\right)^\top$$

(E-polarisation)), transversely inhomogeneous $\varepsilon^{(L)} = \varepsilon^{(L)}(z) = 1 + 4\pi\chi_{11}^{(1)}(z)$ dielectric layer (see Fig. 1) with a cubic polarisability $\mathbf{P}^{(NL)}(\mathbf{r},n\kappa) = (P_1^{(NL)}(n\kappa;y,z),0,0)^\top$ of the medium, the complex amplitudes of the total fields

$$E_1(\mathbf{r},n\kappa) =: E_1(n\kappa;y,z) := U(n\kappa;z)\exp(i\phi_{n\kappa}y) := E_1^{\text{inc}}(n\kappa;y,z) + E_1^{\text{scat}}(n\kappa;y,z)$$

satisfy the system of equations (cf. (16) – (18))

$$\begin{cases} \nabla^2 E_1(\mathbf{r},\kappa) + \kappa^2 \varepsilon_\kappa(z,\alpha(z),E_1(\mathbf{r},\kappa),E_1(\mathbf{r},2\kappa),E_1(\mathbf{r},3\kappa))E_1(\mathbf{r},\kappa) \\ \qquad = -\alpha(z)\kappa^2 E_1^2(\mathbf{r},2\kappa)\overline{E}_1(\mathbf{r},3\kappa), \\ \nabla^2 E_1(\mathbf{r},2\kappa) + (2\kappa)^2 \varepsilon_{2\kappa}(z,\alpha(z),E_1(\mathbf{r},\kappa),E_1(\mathbf{r},2\kappa),E_1(\mathbf{r},3\kappa))E_1(\mathbf{r},2\kappa) = 0, \\ \nabla^2 E_1(\mathbf{r},3\kappa) + (3\kappa)^2 \varepsilon_{3\kappa}(z,\alpha(z),E_1(\mathbf{r},\kappa),E_1(\mathbf{r},2\kappa),E_1(\mathbf{r},3\kappa))E_1(\mathbf{r},3\kappa) \\ \qquad = -\alpha(z)(3\kappa)^2\left\{\frac{1}{3}E_1^3(\mathbf{r},\kappa) + E_1^2(\mathbf{r},2\kappa)\overline{E}_1(\mathbf{r},\kappa)\right\} \end{cases} \tag{21}$$

together with the following conditions, where $\mathbf{E}_{\text{tg}}(n\kappa;y,z)$ and $\mathbf{H}_{\text{tg}}(n\kappa;y,z)$ denote the tangential components of the intensity vectors of the full electromagnetic field $\{\mathbf{E}(n\kappa;y,z)\}_{n=1,2,3}$, $\{\mathbf{H}(n\kappa;y,z)\}_{n=1,2,3}$:

(C1) $E_1(n\kappa;y,z) = U(n\kappa;z)\exp(i\phi_{n\kappa}y), \quad n = 1,2,3$
(the quasi-homogeneity condition w.r.t. the spatial variable y introduced above),

(C2) $\phi_{n\kappa} = n\phi_\kappa, \quad n = 1,2,3,$
(the condition of phase synchronism of waves introduced above),

(C3) $\mathbf{E}_{\text{tg}}(n\kappa;y,z)$ and $\mathbf{H}_{\text{tg}}(n\kappa;y,z)$ (i.e. $E_1(n\kappa;y,z)$ and $H_2(n\kappa;y,z)$) are continuous at the boundary layers of the non-linear structure,

(C4) $E_1^{\text{scat}}(n\kappa;y,z) = \begin{Bmatrix} a_{n\kappa}^{\text{scat}} \\ b_{n\kappa}^{\text{scat}} \end{Bmatrix} \exp\left(i\left(\phi_{n\kappa}y \pm \Gamma_{n\kappa}(z \mp 2\pi\delta)\right)\right), \quad z \gtrless \pm 2\pi\delta, \quad n = 1,2,3$
(the radiation condition w.r.t. the scattered field).

The condition (C4) provides a physically consistent behaviour of the energy characteristics of scattering and guarantees the absence of waves coming from infinity (i.e. $z = \pm\infty$), see Shestopalov & Sirenko (1989). We study the scattering properties of the non-linear layer, where in (C4) we always have

$$\mathfrak{Im}\,\Gamma_{n\kappa} = 0, \quad \mathfrak{Re}\,\Gamma_{n\kappa} > 0. \tag{22}$$

Note that (C4) is also applicable for the analysis of the wave-guide properties of the layer, where $\mathfrak{Im}\,\Gamma_{n\kappa} > 0$, $\mathfrak{Re}\,\Gamma_{n\kappa} = 0$. The desired solution of the scattering and generation problem (21) under the conditions (C1) – (C4) can be represented as follows:

$$E_1(n\kappa;y,z) = U(n\kappa;z)\exp(i\phi_{n\kappa}y)$$
$$= \begin{cases} a_{n\kappa}^{\text{inc}}\exp(i(\phi_{n\kappa}y - \Gamma_{n\kappa}(z-2\pi\delta))) + a_{n\kappa}^{\text{scat}}\exp(i(\phi_{n\kappa}y + \Gamma_{n\kappa}(z-2\pi\delta))), & z > 2\pi\delta, \\ U(n\kappa;z)\exp(i\phi_{n\kappa}y), & |z| \le 2\pi\delta, \\ b_{n\kappa}^{\text{scat}}\exp(i(\phi_{n\kappa}y - \Gamma_{n\kappa}(z+2\pi\delta))), & z < -2\pi\delta, \end{cases} \tag{23}$$
$$n = 1,2,3.$$

Substituting this representation into the system (21), the following system of non-linear ordinary differential equations results, where "$'$" denotes the differentiation w.r.t. z:

$$U''(n\kappa;z) + \left\{\Gamma_{n\kappa}^2 - (n\kappa)^2\left[1 - \varepsilon_{n\kappa}(z,\alpha(z),U(\kappa;z),U(2\kappa;z),U(3\kappa;z))\right]\right\}U(n\kappa;z)$$
$$= -(n\kappa)^2\alpha(z)\left(\delta_{n1}U^2(2\kappa;z)\overline{U}(3\kappa;z) + \delta_{n3}\left\{\frac{1}{3}U^3(\kappa;z) + U^2(2\kappa;z)\overline{U}(\kappa;z)\right\}\right), \tag{24}$$
$$|z| \le 2\pi\delta, \quad n = 1,2,3.$$

The boundary conditions follow from the continuity of the tangential components of the full fields of diffraction $\{\mathbf{E}_{\text{tg}}(n\kappa; y, z)\}_{n=1,2,3}$ $\{\mathbf{H}_{\text{tg}}(n\kappa; y, z)\}_{n=1,2,3}$ at the boundary $z = 2\pi\delta$ and $z = -2\pi\delta$ of the non-linear layer (cf. (C3)). According to (C3) and the representation of the electrical components of the electromagnetic field (23), at the boundary of the non-linear layer we obtain:

$$\begin{aligned} U(n\kappa; 2\pi\delta) &= a_{n\kappa}^{\text{scat}} + a_{n\kappa}^{\text{inc}}, \quad U'(n\kappa; 2\pi\delta) = i\Gamma_{n\kappa}\left(a_{n\kappa}^{\text{scat}} - a_{n\kappa}^{\text{inc}}\right), \\ U(n\kappa; -2\pi\delta) &= b_{n\kappa}^{\text{scat}}, \qquad U'(n\kappa; -2\pi\delta) = -i\Gamma_{n\kappa} b_{n\kappa}^{\text{scat}}, \quad n = 1,2,3. \end{aligned} \tag{25}$$

Eliminating in (25) the unknown values of the complex amplitudes $\{a_{n\kappa}^{\text{scat}}\}_{n=1,2,3}$, $\{b_{n\kappa}^{\text{scat}}\}_{n=1,2,3}$ of the scattered field and taking into consideration that $a_{n\kappa}^{\text{inc}} = U^{\text{inc}}(n\kappa; 2\pi\delta)$, we arrive at the desired boundary conditions for the problem (21), (C1) – (C4):

$$\begin{aligned} i\Gamma_{n\kappa} U(n\kappa; -2\pi\delta) + U'(n\kappa; -2\pi\delta) &= 0, \\ i\Gamma_{n\kappa} U(n\kappa; 2\pi\delta) - U'(n\kappa; 2\pi\delta) &= 2i\Gamma_{n\kappa} a_{n\kappa}^{\text{inc}}, \quad n = 1,2,3. \end{aligned} \tag{26}$$

The system of ordinary differential equations (24) and the boundary conditions (26) form a semi-linear boundary-value problem of Sturm-Liouville type, see also Angermann & Yatsyk (2010); Shestopalov & Yatsyk (2007; 2010); Yatsyk (2007).

4. Existence and uniqueness of a weak solution of the non-linear boundary-value problem

Denote by $\mathbf{u} = \mathbf{u}(z) := (u_1(z), u_2(z), u_3(z))^\top := (U(\kappa; z), U(2\kappa; z), U(3\kappa; z))^\top$ the (formal) solution of (24)&(26) and let, for $\mathbf{w} = (w_1, w_2, w_3)^\top \in \mathbb{C}^3$,

$$\mathbf{F}(z, \mathbf{w}) := \begin{pmatrix} \{\Gamma_\kappa^2 - \kappa^2 [1 - \varepsilon_\kappa(z, \alpha(z), w_1, w_2, w_3)]\} w_1 + \alpha(z)\kappa^2 w_2^2 \overline{w}_3 \\ \{\Gamma_{2\kappa}^2 - (2\kappa)^2 [1 - \varepsilon_{2\kappa}(z, \alpha(z), w_1, w_2, w_3)]\} w_2 \\ \{\Gamma_{3\kappa}^2 - (3\kappa)^2 [1 - \varepsilon_{3\kappa}(z, \alpha(z), w_1, w_2, w_3)]\} w_3 + \alpha(z)(3\kappa)^2 \left\{\frac{1}{3} w_1^3 + \overline{w}_1 w_1^2\right\} \end{pmatrix}.$$

Then the system of differential equations (24) takes the form

$$-\mathbf{u}''(z) = \mathbf{F}(z, \mathbf{u}(z)), \quad z \in \mathcal{I} := (-2\pi\delta, 2\pi\delta). \tag{27}$$

The boundary conditions (26) can be written as

$$\mathbf{u}'(-2\pi\delta) + i\mathbf{G}\mathbf{u}(-2\pi\delta) = 0, \quad \mathbf{u}'(2\pi\delta) - i\mathbf{G}\mathbf{u}(2\pi\delta) = -2i\mathbf{G}\mathbf{a}^{\text{inc}}, \tag{28}$$

where $\mathbf{G} := \operatorname{diag}(\Gamma_\kappa, \Gamma_{2\kappa}, \Gamma_{3\kappa})$ and $\mathbf{a}^{\text{inc}} := (a_\kappa^{\text{inc}}, a_{2\kappa}^{\text{inc}}, a_{3\kappa}^{\text{inc}})^\top$. Taking an arbitrary complex-valued vector function $\mathbf{v} : \mathcal{I}^{\text{cl}} := [-2\pi\delta, 2\pi\delta] \to \mathbb{C}^3$, $\mathbf{v} = (v_1, v_2, v_3)^\top$, multiplying the vector differential equation (27) by the complex conjugate $\overline{\mathbf{v}}$ and integrating w.r.t. z over the interval $\mathcal{I}$, we arrive at the equation

$$-\int_{\mathcal{I}} \mathbf{u}'' \cdot \overline{\mathbf{v}}\, dz = \int_{\mathcal{I}} \mathbf{F}(z, \mathbf{u}) \cdot \overline{\mathbf{v}}\, dz.$$

Integrating formally by parts and using the boundary conditions (28), we obtain:

$$\begin{aligned} &-\int_{\mathcal{I}} \mathbf{u}'' \cdot \overline{\mathbf{v}}\, dz = \int_{\mathcal{I}} \mathbf{u}' \cdot \overline{\mathbf{v}}\, dz - (\mathbf{u}' \cdot \overline{\mathbf{v}})(2\pi\delta) + (\mathbf{u}' \cdot \overline{\mathbf{v}})(-2\pi\delta) \\ &= \int_{\mathcal{I}} \mathbf{u}' \cdot \overline{\mathbf{v}}' dz - i\left[((\mathbf{G}\mathbf{u}) \cdot \overline{\mathbf{v}})(2\pi\delta) + ((\mathbf{G}\mathbf{u}) \cdot \overline{\mathbf{v}})(-2\pi\delta)\right] + 2i(\mathbf{G}\mathbf{a}^{\mathrm{inc}}) \cdot \overline{\mathbf{v}}(2\pi\delta). \end{aligned} \tag{29}$$

Now we take into consideration the complex Sobolev space $H^1(\mathcal{I})$ consisting of functions with values in $\mathbb{C}$, which together with their weak derivatives belong to $L_2(\mathcal{I})$. According to (29) it is natural to introduce the follwing forms for $\mathbf{w}, \mathbf{v} \in \mathbf{V} := \left[H^1(\mathcal{I})\right]^3$:

$$\begin{aligned} a(\mathbf{w}, \mathbf{v}) &:= \int_{\mathcal{I}} \mathbf{w}' \cdot \overline{\mathbf{v}}' dz - i\left[((\mathbf{G}\mathbf{w}) \cdot \overline{\mathbf{v}})(2\pi\delta) + ((\mathbf{G}\mathbf{w}) \cdot \overline{\mathbf{v}})(-2\pi\delta)\right], \\ b(\mathbf{w}, \mathbf{v}) &:= \int_{\mathcal{I}} \mathbf{F}(z, \mathbf{w}) \cdot \overline{\mathbf{v}} dz - 2i(\mathbf{G}\mathbf{a}^{\mathrm{inc}}) \cdot \overline{\mathbf{v}}(2\pi\delta). \end{aligned}$$

So we arrive at the following weak formulation of boundary-value problem (24):

$$\text{Find} \quad \mathbf{u} \in \mathbf{V} \quad \text{such that} \quad a(\mathbf{u}, \mathbf{v}) = b(\mathbf{u}, \mathbf{v}) \qquad \forall \mathbf{v} \in \mathbf{V}. \tag{30}$$

The space $\mathbf{V}$ is equipped with the usual norm and seminorm, resp.:

$$\|\mathbf{v}\|_{1,2,\mathcal{I}}^2 := \sum_{n=1}^{3} \left[\|v_n\|_{0,2,\mathcal{I}}^2 + \|v_n'\|_{0,2,\mathcal{I}}^2\right], \quad |\mathbf{v}|_{1,2,\mathcal{I}}^2 := \sum_{n=1}^{3} \|v_n'\|_{0,2,\mathcal{I}}^2,$$

where $\|v\|_{0,2,\mathcal{I}}$, for $v \in L_2(\mathcal{I})$, denotes the usual $L_2(\mathcal{I})$-norm. If $\mathbf{v} \in [L_2(\mathcal{I})]^3$, we will use the same notation, i.e. $\|\mathbf{v}\|_{0,2,\mathcal{I}}^2 := \sum_{n=1}^{3} \|v_n\|_{0,2,\mathcal{I}}^2$. Then the above norm and seminorm in $\mathbf{V}$ can be written in short as

$$\|\mathbf{v}\|_{1,2,\mathcal{I}}^2 := \|\mathbf{v}\|_{0,2,\mathcal{I}}^2 + \|\mathbf{v}'\|_{0,2,\mathcal{I}}^2, \quad |\mathbf{v}|_{1,2,\mathcal{I}} := \|\mathbf{v}'\|_{0,2,\mathcal{I}}. \tag{31}$$

Analogously, we will not make any notational difference between the absolute value $|\cdot|$ of a (scalar) element of $\mathbb{C}$ and the norm $|\cdot|$ of a (vectorial) element of $\mathbb{C}^3$.
On $\mathbf{V}$, the following norm can be introduced:

$$\|\mathbf{v}\|_{\mathbf{V}}^2 := \sum_{n=1}^{3} \left[|v_n(-2\pi\delta)|^2 + |v_n(2\pi\delta)|^2 + \|v_n'\|_{0,2,\mathcal{I}}^2\right] = |\mathbf{v}(-2\pi\delta)|^2 + |\mathbf{v}(2\pi\delta)|^2 + |\mathbf{v}|_{1,2,\mathcal{I}}^2. \tag{32}$$

Corollary 1. *The norms defined in (31) and (32) are equivalent on* $\mathbf{V}$, *i.e.*

$$C_- \|\mathbf{v}\|_{1,2,\mathcal{I}} \le \|\mathbf{v}\|_{\mathbf{V}} \le C_+ \|\mathbf{v}\|_{1,2,\mathcal{I}} \qquad \forall \mathbf{v} \in \mathbf{V}$$

with $C_- := 1/\sqrt{16\pi^2\delta^2 + 1}$, $C_+ := \sqrt{\max\left\{\frac{1}{2\pi\delta} + 1; 2\right\}}$.

Proof. It is not difficult to verify the following inequality for any (scalar) element $v \in H^1(\mathcal{I})$ (see, e.g., (Angermann & Yatsyk, 2008, Cor. 4)):

$$\|v\|_{0,2,\mathcal{I}}^2 \le 4\pi\delta\left[|v(-2\pi\delta)|^2 + |v(2\pi\delta)|^2\right] + 16\pi^2\delta^2 \|v'\|_{0,2,\mathcal{I}}^2. \tag{33}$$

Consequently, by (31), $\|\mathbf{v}\|^2_{1,2,\mathcal{I}} \le 4\pi\delta[|\mathbf{v}(-2\pi\delta)|^2 + |\mathbf{v}(2\pi\delta)|^2] + (16\pi^2\delta^2+1)\|\mathbf{v}'\|^2_{0,2,\mathcal{I}}$. Since $4\pi\delta < 16\pi^2\delta^2 + 1$, we immediately obtain the left-hand side of the desired estimate:

$$\|\mathbf{v}\|^2_{1,2,\mathcal{I}} \le (16\pi^2\delta^2+1)\|\mathbf{v}\|^2_{\mathbf{V}}.$$

On the other hand, a trace inequality (see, e.g., (Angermann & Yatsyk, 2008, Cor. 5)) says that we have the following estimate for any element $v \in H^1(\mathcal{I})$:

$$|v(-2\pi\delta)|^2 + |v(2\pi\delta)|^2 \le \left(\frac{1}{2\pi\delta}+1\right)\|v\|^2_{0,2,\mathcal{I}} + \|v'\|^2_{0,2,\mathcal{I}}.$$

Thus $\|\mathbf{v}\|^2_{\mathbf{V}} \le \left(\frac{1}{2\pi\delta}+1\right)\|\mathbf{v}\|^2_{0,2,\mathcal{I}} + 2\|\mathbf{v}'\|^2_{0,2,\mathcal{I}}$, that is $C_+^2 := \max\left\{\frac{1}{2\pi\delta}+1;2\right\}$. ◄

Lemma 1. *If the matrix* $\mathbf{G}$ *is positively definite, then the form* a *is coercive and bounded on* $\mathbf{V}$*, i.e.*

$$C_K\|\mathbf{v}\|^2_{1,2,\mathcal{I}} \le |a(\mathbf{v},\mathbf{v})|, \qquad |a(\mathbf{w},\mathbf{v})| \le C_b\|\mathbf{w}\|_{1,2,\mathcal{I}}\|\mathbf{v}\|_{1,2,\mathcal{I}}$$

for all $\mathbf{w},\mathbf{v} \in \mathbf{V}$ *with* $C_K := \frac{\sqrt{2}}{2}\min\{1;\Gamma_\kappa;\Gamma_{2\kappa};\Gamma_{3\kappa}\}C_-^2$, $C_b := \max\{1;\Gamma_\kappa;\Gamma_{2\kappa};\Gamma_{3\kappa}\}C_+^2$.

Remark 1. *Due to (22), the assumption of the lemma is satisfied.*

Proof of the lemma: Obviously,

$$|a(\mathbf{v},\mathbf{v})| = \sqrt{|\Re\mathfrak{e}\, a(\mathbf{v},\mathbf{v})|^2 + |\Im\mathfrak{m}\, a(\mathbf{v},\mathbf{v})|^2} \ge \frac{\sqrt{2}}{2}\left[|\Re\mathfrak{e}\, a(\mathbf{v},\mathbf{v})| + |\Im\mathfrak{m}\, a(\mathbf{v},\mathbf{v})|\right]$$

$$= \frac{\sqrt{2}}{2}\sum_{n=1}^{3}\left[\|v_n'\|^2_{0,2,\mathcal{I}} + \Gamma_{n\kappa}|v_n(-2\pi\delta)|^2 + \Gamma_{n\kappa}|v_n(2\pi\delta)|^2\right] \ge \frac{\sqrt{2}}{2}\min\{1;\Gamma_\kappa;\Gamma_{2\kappa};\Gamma_{3\kappa}\}\|\mathbf{v}\|^2_{\mathbf{V}}, \tag{34}$$

where we have used the convention $\Gamma_{1\kappa} := \Gamma_\kappa$. By Corollary 1, this estimate implies the coercivity of a on $\mathbf{V}$. The proof of the continuity runs in a similar way:

$$\begin{aligned}|a(\mathbf{w},\mathbf{v})| &\le \max\{1;\Gamma_\kappa;\Gamma_{2\kappa};\Gamma_{3\kappa}\}\sum_{n=1}^{3}\big[\|w_n'\|_{0,2,\mathcal{I}}\|v_n'\|_{0,2,\mathcal{I}} \\ &\quad + |w_n(-2\pi\delta)||v_n(-2\pi\delta)| + |w_n(2\pi\delta)||v_n(2\pi\delta)|\big] \\ &\le \max\{1;\Gamma_\kappa;\Gamma_{2\kappa};\Gamma_{3\kappa}\}\|\mathbf{w}\|_{\mathbf{V}}\|\mathbf{v}\|_{\mathbf{V}},\end{aligned}$$

where the last estimate is a consequence of the Cauchy-Schwarz inequality for finite sums. From Corollary 1 we obtain the above expression for C_b. ◄

Corollary 2. *Under the assumption of Lemma 1, given an antilinear continuous functional* $\ell : \mathbf{V} \to \mathbb{C}$*, the problem to find an element* $\mathbf{u} \in \mathbf{V}$ *such that*

$$a(\mathbf{u},\mathbf{v}) = \ell(\mathbf{v}) \qquad \forall \mathbf{v} \in \mathbf{V} \tag{35}$$

is uniquely solvable and the following estimate holds:

$$\|\mathbf{u}\|_{1,2,\mathcal{I}} \le C_K^{-1}\|\ell\|_*, \qquad \textit{where} \quad \|\ell\|_* := \sup_{\mathbf{v}\in\mathbf{V}}\frac{|\ell(\mathbf{v})|}{\|\mathbf{v}\|_{1,2,\mathcal{I}}}.$$

Proof. This general result is well-known (see, e.g., (Showalter, 1994, Thm. 2.1)). ◀

Corollary 3. *If the antilinear continuous functional* $\ell : \mathbf{V} \to \mathbb{C}$ *has the particular structure*

$$\ell(\mathbf{v}) := \int_{\mathcal{I}} \mathbf{f} \cdot \overline{v}\, dz + \gamma_- \cdot \overline{\mathbf{v}}(-2\pi\delta) + \gamma_+ \cdot \overline{\mathbf{v}}(2\pi\delta),$$

where $\mathbf{f} \in [L_2(\mathcal{I})]^3$ *and* $\gamma_-, \gamma_+ \in \mathbb{C}^3$ *are given, then*

$$\|\ell\|_* \le C_+ \sqrt{\max\{4\pi\delta + 1; 16\pi^2\delta^2\}} \left\{ \|\mathbf{f}\|^2_{0,2,\mathcal{I}} + |\gamma_-|^2 + |\gamma_+|^2 \right\}^{1/2}.$$

Proof. By the Cauchy-Schwarz inequality for finite sums, we see that

$$\begin{aligned} |\ell(\mathbf{v})| &\le \|\mathbf{f}\|_{0,2,\mathcal{I}} \|\mathbf{v}\|_{0,2,\mathcal{I}} + |\gamma_-||\mathbf{v}(-2\pi\delta)| + |\gamma_+||\mathbf{v}(2\pi\delta)| \\ &\le \left\{ \|\mathbf{f}\|^2_{0,2,\mathcal{I}} + |\gamma_-|^2 + |\gamma_+|^2 \right\}^{1/2} \left\{ \|\mathbf{v}\|^2_{0,2,\mathcal{I}} + |\mathbf{v}(-2\pi\delta)|^2 + |\mathbf{v}(2\pi\delta)|^2 \right\}^{1/2}. \end{aligned}$$

Using the estimate (33), it follows

$$|\ell(\mathbf{v})| \le \sqrt{\max\{4\pi\delta + 1; 16\pi^2\delta^2\}} \left\{ \|\mathbf{f}\|^2_{0,2,\mathcal{I}} + |\gamma_-|^2 + |\gamma_+|^2 \right\}^{1/2} \|\mathbf{v}\|_{\mathbf{V}}. \tag{36}$$

It remains to apply Corollary 1. ◀

Remark 2. *Combining Corollary 2 and Corollary 3, we obtain the following estimate for the solution* $\mathbf{u}$ *of (35):*

$$\|\mathbf{u}\|_{1,2,\mathcal{I}} \le \frac{C_+ \sqrt{2\max\{4\pi\delta + 1; 16\pi^2\delta^2\}}}{C_-^2 \min\{1; \Gamma_\kappa; \Gamma_{2\kappa}; \Gamma_{3\kappa}\}} \left\{ \|\mathbf{f}\|^2_{0,2,\mathcal{I}} + |\gamma_-|^2 + |\gamma_+|^2 \right\}^{1/2}.$$

The obtained constant suffers from the twice use of the norm equivalence in the proofs of Lemma 1 and Corollary 3, respectively. It can be improved if we start from the estimate (34). Namely, setting $\mathbf{v} := \mathbf{u}$ *in (35), we obtain from (34) and (36):*

$$\begin{aligned} &\frac{\sqrt{2}}{2} \min\{1; \Gamma_\kappa; \Gamma_{2\kappa}; \Gamma_{3\kappa}\} \|\mathbf{v}\|^2_{\mathbf{V}} \le |a(\mathbf{u}, \mathbf{u})| = |\ell(\mathbf{u})| \\ &\le \sqrt{\max\{4\pi\delta + 1; 16\pi^2\delta^2\}} \left\{ \|\mathbf{f}\|^2_{0,2,\mathcal{I}} + |\gamma_-|^2 + |\gamma_+|^2 \right\}^{1/2} \|\mathbf{u}\|_{\mathbf{V}}. \end{aligned}$$

Therefore, by Corollary 1,

$$\|\mathbf{u}\|_{1,2,\mathcal{I}} \le C_N \left\{ \|\mathbf{f}\|^2_{0,2,\mathcal{I}} + |\gamma_-|^2 + |\gamma_+|^2 \right\}^{1/2} \quad \text{with } C_N := \frac{\sqrt{2\max\{4\pi\delta + 1; 16\pi^2\delta^2\}}}{C_- \min\{1; \Gamma_\kappa; \Gamma_{2\kappa}; \Gamma_{3\kappa}\}}. \tag{37}$$

The identity

$$\mathcal{A}\mathbf{w}(\mathbf{v}) := a(\mathbf{w}, \mathbf{v}) \qquad \forall \mathbf{w}, \mathbf{v} \in \mathbf{V}$$

defines a linear operator $\mathcal{A} : \mathbf{V} \to \mathbf{V}^*$, where $\mathbf{V}^*$ is the dual space of $\mathbf{V}$ consisting of all antilinear continuous functionals acting from $\mathbf{V}$ to $\mathbb{C}$. By Lemma 1 and Corollary 2, $\mathcal{A}$ is a bounded operator with a bounded inverse $\mathcal{A}^{-1} : \mathbf{V}^* \to \mathbf{V}$:

$$\|\mathbf{w}\|_{1,2,\mathcal{I}} \le C_K^{-1} \|\mathcal{A}\mathbf{w}\|_* \qquad \forall \mathbf{w} \in \mathbf{V}.$$

Lemma 2. *If $\varepsilon^{(L)}, \alpha \in L_\infty(\mathcal{I})$, then the formal substitution $\mathcal{N}(\mathbf{w})(z) := \mathbf{F}(z, \mathbf{w}(z))$ defines a Nemyckii operator $\mathcal{N} : \mathbf{V} \to [L_2(\mathcal{I})]^3$, and there is a constant $C_S > 0$ such that*

$$\|\mathcal{N}(\mathbf{w})\|_{0,2,\mathcal{I}} \leq \kappa^2 \left[9\|\varepsilon^{(L)} - \sin^2 \varphi_\kappa\|_{0,\infty,\mathcal{I}} + \sqrt{170}\, C_S^2 \|\alpha\|_{0,\infty,\mathcal{I}} \|\mathbf{w}\|_{1,2,\mathcal{I}}^2\right] \|\mathbf{w}\|_{0,2,\mathcal{I}}.$$

Proof. It is sufficient to verify the estimate. According to the decomposition $\mathbf{F}(z, \mathbf{w}) := \mathbf{F}^{(L)}(z, \mathbf{w}) + \mathbf{F}^{(NL)}(z, \mathbf{w})$ with

$$\mathbf{F}^{(L)}(z, \mathbf{w}) := \begin{pmatrix} \{\Gamma_\kappa^2 - \kappa^2 \left[1 - \varepsilon^{(L)}(z)\right]\} w_1 \\ \{\Gamma_{2\kappa}^2 - (2\kappa)^2 \left[1 - \varepsilon^{(L)}(z)\right]\} w_2 \\ \{\Gamma_{3\kappa}^2 - (3\kappa)^2 \left[1 - \varepsilon^{(L)}(z)\right]\} w_3 \end{pmatrix},$$

$$\mathbf{F}^{(NL)}(z, \mathbf{w}) := \begin{pmatrix} F_1^{(NL)}(z, \mathbf{w}) \\ F_2^{(NL)}(z, \mathbf{w}) \\ F_3^{(NL)}(z, \mathbf{w}) \end{pmatrix} := \alpha(z) \begin{pmatrix} \kappa^2 \left[|\mathbf{w}|^2 w_1 + \overline{w}_1^2 w_3 + w_2^2 \overline{w}_3\right] \\ (2\kappa)^2 \left[|\mathbf{w}|^2 w_2 + w_1 \overline{w}_2 w_3\right] \\ (3\kappa)^2 \left[|\mathbf{w}|^2 w_3 + \frac{1}{3} w_1^3 + \overline{w}_1 w_2^2\right] \end{pmatrix}$$

(cf. (19), (24)), it is convinient to split $\mathcal{N}$ into a linear and a non-linear part as $\mathcal{N}(\mathbf{w})(z) := \mathcal{N}^{(L)}(\mathbf{w})(z) + \mathcal{N}^{(NL)}(\mathbf{w})(z)$, where $\mathcal{N}^{(L)}(\mathbf{w})(z) := \mathbf{F}^{(L)}(z, \mathbf{w}(z))$ and $\mathcal{N}^{(NL)}(\mathbf{w})(z) := \mathbf{F}^{(NL)}(z, \mathbf{w}(z))$. Now, by the definition of the wave-numbers (see Section 3),

$$\Gamma_{n\kappa}^2 - (n\kappa)^2 \left[1 - \varepsilon^{(L)}(z)\right] = (n\kappa)^2 \left[\varepsilon^{(L)}(z) - \sin^2 \varphi_{n\kappa}\right] = (n\kappa)^2 \left[\varepsilon^{(L)}(z) - \sin^2 \varphi_\kappa\right], \ n = 1, 2, 3,$$

where the last relation is a consequence of the condition (C2). Therefore,

$$\|\mathcal{N}^{(L)}(\mathbf{w})\|_{0,2,\mathcal{I}} \leq (3\kappa)^2 \|\varepsilon^{(L)} - \sin^2 \varphi_\kappa\|_{0,\infty,\mathcal{I}} \|\mathbf{w}\|_{0,2,\mathcal{I}}. \tag{38}$$

Next, since $H^1(\mathcal{I})$ is continuously embedded into $C(\mathcal{I}^{\text{cl}})$ by Sobolev's embedding theorem (see, e.g., (Adams, 1975, Thm. 5.4)), there exists a constant $C_S > 0$ such that

$$\|\mathbf{w}\|_{0,\infty,\mathcal{I}} := \sup_{z \in \mathcal{I}} |\mathbf{w}(z)| = \sup_{z \in \mathcal{I}} \left\{ \sum_{n=1}^{3} |w_n(z)|^2 \right\}^{1/2} \leq C_S \|\mathbf{w}\|_{1,2,\mathcal{I}}. \tag{39}$$

Using this fact we easily obtain the following triple of estimates:

$$\begin{aligned}
\|F_1^{(NL)}(\cdot, \mathbf{w})\|_{0,2,\mathcal{I}} &\leq \kappa^2 \|\alpha\|_{0,\infty,\mathcal{I}} \left[\||\mathbf{w}|^2 w_1\|_{0,2,\mathcal{I}} + \|\overline{w}_1^2 w_3\|_{0,2,\mathcal{I}} + \|w_2^2 \overline{w}_3\|_{0,2,\mathcal{I}}\right] \\
&\leq \kappa^2 \|\alpha\|_{0,\infty,\mathcal{I}} \Big[\|\mathbf{w}\|_{0,\infty,\mathcal{I}}^2 \|w_1\|_{0,2,\mathcal{I}} \\
&\quad + \|w_1\|_{0,\infty,\mathcal{I}}^2 \|w_3\|_{0,2,\mathcal{I}} + \|w_2\|_{0,\infty,\mathcal{I}} \|w_3\|_{0,\infty,\mathcal{I}} \|w_2\|_{0,2,\mathcal{I}}\Big] \\
&\leq \kappa^2 \|\alpha\|_{0,\infty,\mathcal{I}} \|\mathbf{w}\|_{0,\infty,\mathcal{I}}^2 \left[\|w_1\|_{0,2,\mathcal{I}} + \|w_2\|_{0,2,\mathcal{I}} + \|w_3\|_{0,2,\mathcal{I}}\right] \\
&\leq \sqrt{3}\, \kappa^2 C_S^2 \|\alpha\|_{0,\infty,\mathcal{I}} \|\mathbf{w}\|_{1,2,\mathcal{I}}^2 \|\mathbf{w}\|_{0,2,\mathcal{I}},
\end{aligned}$$

$$\begin{aligned}
\|F_2^{(NL)}(\cdot,\mathbf{w})\|_{0,2,\mathcal{I}} &\le (2\kappa)^2\|\alpha\|_{0,\infty,\mathcal{I}}\left[\||\mathbf{w}|^2 w_2\|_{0,2,\mathcal{I}} + \|w_1\overline{w}_2 w_3\|_{0,2,\mathcal{I}}\right] \\
&\le (2\kappa)^2\|\alpha\|_{0,\infty,\mathcal{I}}\|\mathbf{w}\|^2_{0,\infty,\mathcal{I}}\left[\|w_2\|_{0,2,\mathcal{I}} + \|w_3\|_{0,2,\mathcal{I}}\right] \\
&\le \sqrt{2}\,(2\kappa)^2 C_S^2\|\alpha\|_{0,\infty,\mathcal{I}}\|\mathbf{w}\|^2_{1,2,\mathcal{I}}\|\mathbf{w}\|_{0,2,\mathcal{I}},
\end{aligned}$$

$$\begin{aligned}
\|F_3^{(NL)}(\cdot,\mathbf{w})\|_{0,2,\mathcal{I}} &\le (3\kappa)^2\|\alpha\|_{0,\infty,\mathcal{I}}\left[\||\mathbf{w}|^2 w_3\|_{0,2,\mathcal{I}} + \frac{1}{3}\|w_1^3\|_{0,2,\mathcal{I}} + \|\overline{w}_1 w_2^2\|_{0,2,\mathcal{I}}\right] \\
&\le (3\kappa)^2\|\alpha\|_{0,\infty,\mathcal{I}}\|\mathbf{w}\|^2_{0,\infty,\mathcal{I}}\left[\frac{1}{3}\|w_1\|_{0,2,\mathcal{I}} + \|w_2\|_{0,2,\mathcal{I}} + \|w_3\|_{0,2,\mathcal{I}}\right] \\
&\le \sqrt{\frac{5}{3}}\,(3\kappa)^2 C_S^2\|\alpha\|_{0,\infty,\mathcal{I}}\|\mathbf{w}\|^2_{1,2,\mathcal{I}}\|\mathbf{w}\|_{0,2,\mathcal{I}}.
\end{aligned}$$

These estimates immediately imply that

$$\|\mathcal{N}^{(NL)}(\mathbf{w})\|^2_{0,2,\mathcal{I}} = \sum_{n=1}^{3}\|F_n^{(NL)}(\cdot,\mathbf{w})\|^2_{0,2,\mathcal{I}} \le 170\kappa^4 C_S^4\|\alpha\|^2_{0,\infty,\mathcal{I}}\|\mathbf{w}\|^4_{1,2,\mathcal{I}}\|\mathbf{w}\|^2_{0,2,\mathcal{I}}. \tag{40}$$

Putting the estimates (38) and (40) together, we obtain the desired estimate. ◄
As a consequence of Lemma 2, the following non-linear operator $\mathcal{F}:\ \mathbf{V}\to\mathbf{V}^*$ can be introduced:

$$\mathcal{F}(\mathbf{w})(\mathbf{v}) := b\,(\mathbf{w},\mathbf{v}) = \int_{\mathcal{I}}\mathcal{N}(\mathbf{w})\cdot\overline{\mathbf{v}}dz - 2i(\mathbf{G}\mathbf{a}^{\mathrm{inc}})\cdot\overline{\mathbf{v}}(2\pi\delta) \qquad \forall\mathbf{w},\mathbf{v}\in\mathbf{V}.$$

Then the problem (30) is equivalent to the operator equation $\mathcal{A}\mathbf{u} = \mathcal{F}(\mathbf{u})$ in $\mathbf{V}^*$. Furthermore, by Lemma 1, this equation is equivalent to the fixed-point problem

$$\mathbf{u} = \mathcal{A}^{-1}\mathcal{F}(\mathbf{u}) \qquad \text{in} \quad \mathbf{V}. \tag{41}$$

Theorem 1. *Assume there is a number $\varrho > 0$ such that*

$$C_N\kappa^2\left[9\|\varepsilon^{(L)} - \sin^2\varphi_\kappa\|_{0,\infty,\mathcal{I}} + 3\sqrt{514}\,C_S^2\|\alpha\|_{0,\infty,\mathcal{I}}\varrho^2\right] \le \frac{\sqrt{2}}{2} \quad \text{and} \quad C_N|\mathbf{G}\mathbf{a}^{inc}| \le \frac{\sqrt{2}}{4}\varrho.$$

Then the problem (41) has a unique solution $\mathbf{u}\in K_\varrho^{cl} := \{\mathbf{v}\in\mathbf{V} : \|\mathbf{v}\|_{1,2,\mathcal{I}} \le \varrho\}$.

Proof. Obviously, K_ϱ^{cl} is a closed nonempty subset of $\mathbf{V}$. We show that $\mathcal{A}^{-1}\mathcal{F}(K_\varrho^{\mathrm{cl}}) \subset K_\varrho^{\mathrm{cl}}$. By (37) with the particular choice $\mathbf{f} := \mathcal{N}(\mathbf{w})$, $\gamma_- := 0$, $\gamma_+ := -2i\mathbf{G}\mathbf{a}^{inc}$, for $\mathbf{w}\in K_\varrho^{\mathrm{cl}}$ we have that

$$\begin{aligned}
&\|\mathcal{A}^{-1}\mathcal{F}(\mathbf{w})\|_{1,2,\mathcal{I}} \le C_N\left\{\|\mathcal{N}(\mathbf{w})\|^2_{0,2,\mathcal{I}} + 4|\mathbf{G}\mathbf{a}^{inc}|^2\right\}^{1/2} \\
&\le C_N\left\{\kappa^4\left[9\|\varepsilon^{(L)} - \sin^2\varphi_\kappa\|_{0,\infty,\mathcal{I}} + \sqrt{170}\,C_S^2\|\alpha\|_{0,\infty,\mathcal{I}}\|\mathbf{w}\|^2_{1,2,\mathcal{I}}\right]^2\|\mathbf{w}\|^2_{0,2,\mathcal{I}} + 4|\mathbf{G}\mathbf{a}^{inc}|^2\right\}^{1/2} \\
&\le C_N\left\{\kappa^4\left[9\|\varepsilon^{(L)} - \sin^2\varphi_\kappa\|_{0,\infty,\mathcal{I}} + \sqrt{170}\,C_S^2\|\alpha\|_{0,\infty,\mathcal{I}}\varrho^2\right]^2\varrho^2 + 4|\mathbf{G}\mathbf{a}^{inc}|^2\right\}^{1/2} \le \varrho.
\end{aligned}$$

Next, from (37) with the choice $\mathbf{f} := \mathcal{N}(\mathbf{w}) - \mathcal{N}(\mathbf{v}),\ \gamma_- := \gamma_+ := 0$ we conclude that

$$\begin{aligned}\|\mathcal{A}^{-1}\mathcal{F}(\mathbf{w}) - \mathcal{A}^{-1}\mathcal{F}(\mathbf{v})\|_{1,2,\mathcal{I}} &\le C_N \|\mathcal{N}(\mathbf{w}) - \mathcal{N}(\mathbf{v})\|_{0,2,\mathcal{I}} \\ &\le C_N \left\{ \|\mathcal{N}^{(L)}(\mathbf{w}) - \mathcal{N}^{(L)}(\mathbf{v})\|_{0,2,\mathcal{I}} + \|\mathcal{N}^{(NL)}(\mathbf{w}) - \mathcal{N}^{(NL)}(\mathbf{v})\|_{0,2,\mathcal{I}} \right\}.\end{aligned}$$

The linear term can be estimated as in the proof of Lemma 2 (cf. (38)):

$$\begin{aligned}\|\mathcal{N}^{(L)}(\mathbf{w}) - \mathcal{N}^{(L)}(\mathbf{v})\|_{0,2,\mathcal{I}} &= \|\mathcal{N}^{(L)}(\mathbf{w} - \mathbf{v})\|_{0,2,\mathcal{I}} \\ &\le (3\kappa)^2 \|\varepsilon^{(L)} - \sin^2\varphi_\kappa\|_{0,\infty,\mathcal{I}} \|\mathbf{w} - \mathbf{v}\|_{0,2,\mathcal{I}}.\end{aligned}$$

To treat the non-linear term, we start with the following estimates:

$$\begin{aligned}&\|F_1^{(NL)}(\cdot,\mathbf{w}) - F_1^{(NL)}(\cdot,\mathbf{v})\|_{0,2,\mathcal{I}} \\ &\le \kappa^2 \|\alpha\|_{0,\infty,\mathcal{I}} \left[\||\mathbf{w}|^2 w_1 - |\mathbf{v}|^2 v_1\|_{0,2,\mathcal{I}} + \|\overline{w}_1^2 w_3 - \overline{v}_1^2 v_3\|_{0,2,\mathcal{I}} + \|w_2^2 \overline{w}_3 - v_2^2 \overline{v}_3\|_{0,2,\mathcal{I}} \right],\end{aligned}$$

$$\begin{aligned}&\|F_2^{(NL)}(\cdot,\mathbf{w}) - F_2^{(NL)}(\cdot,\mathbf{v})\|_{0,2,\mathcal{I}} \\ &\le (2\kappa)^2 \|\alpha\|_{0,\infty,\mathcal{I}} \left[\||\mathbf{w}|^2 w_2 - |\mathbf{v}|^2 v_2\|_{0,2,\mathcal{I}} + \|w_1 \overline{w}_2 w_3 - v_1 \overline{v}_2 v_3\|_{0,2,\mathcal{I}} \right],\end{aligned}$$

$$\begin{aligned}&\|F_3^{(NL)}(\cdot,\mathbf{w}) - F_3^{(NL)}(\cdot,\mathbf{v})\|_{0,2,\mathcal{I}} \\ &\le (3\kappa)^2 \|\alpha\|_{0,\infty,\mathcal{I}} \left[\||\mathbf{w}|^2 w_3 - |\mathbf{v}|^2 v_3\|_{0,2,\mathcal{I}} + \frac{1}{3}\|w_1^3 - v_1^3\|_{0,2,\mathcal{I}} + \|\overline{w}_1 w_2^2 - \overline{v}_1 v_2^2\|_{0,2,\mathcal{I}} \right].\end{aligned}$$

The subsequent collection of simple estimates shows that the absolute value of all terms appearing in the $L_2(\mathcal{I})$-terms of the right-hand sides above can be bounded by one and the same upper bound. Namely, since

$$\begin{aligned}|\mathbf{w}|^2 w_n - |\mathbf{v}|^2 v_n &= |\mathbf{w}|^2 (w_n - v_n) + v_n (|\mathbf{w}|^2 - |\mathbf{v}|^2) \\ &= |\mathbf{w}|^2 (w_n - v_n) + v_n (|\mathbf{w}| + |\mathbf{v}|)(|\mathbf{w}| - |\mathbf{v}|)\end{aligned}$$

and

$$\big||\mathbf{w}| - |\mathbf{v}|\big| \le |\mathbf{w} - \mathbf{v}|,$$

we obtain

$$\big||\mathbf{w}|^2 w_n - |\mathbf{v}|^2 v_n\big| \le \left[|\mathbf{w}|^2 + |\mathbf{w}||\mathbf{v}| + |\mathbf{v}|^2\right] |\mathbf{w} - \mathbf{v}|, \quad n = 1, 2, 3.$$

Similarly,

$$\begin{aligned}|\overline{w}_1^2 w_3 - \overline{v}_1^2 v_3| &\le \left[|\mathbf{w}|^2 + |\mathbf{w}||\mathbf{v}| + |\mathbf{v}|^2\right] |\mathbf{w} - \mathbf{v}|, \\ |w_1 \overline{w}_2 w_3 - v_1 \overline{v}_2 v_3| &\le \left[|\mathbf{w}|^2 + |\mathbf{w}||\mathbf{v}| + |\mathbf{v}|^2\right] |\mathbf{w} - \mathbf{v}|, \\ |w_1^3 - v_1^3| &\le \left[|\mathbf{w}|^2 + |\mathbf{w}||\mathbf{v}| + |\mathbf{v}|^2\right] |\mathbf{w} - \mathbf{v}|.\end{aligned}$$

Therefore

$$\begin{aligned}
&\|F_1^{(NL)}(\cdot,\mathbf{w}) - F_1^{(NL)}(\cdot,\mathbf{v})\|_{0,2,\mathcal{I}} \\
&\le 3\kappa^2\|\alpha\|_{0,\infty,\mathcal{I}}\left[\|\mathbf{w}\|^2_{0,\infty,\mathcal{I}} + \|\mathbf{w}\|_{0,\infty,\mathcal{I}}\|\mathbf{v}\|_{0,\infty,\mathcal{I}} + \|\mathbf{v}\|^2_{0,\infty,\mathcal{I}}\right]\|\mathbf{w}-\mathbf{v}\|_{0,2,\mathcal{I}} \\
&\le 3\kappa^2\|\alpha\|_{0,\infty,\mathcal{I}}C_S^2\left[\|\mathbf{w}\|^2_{1,2,\mathcal{I}} + \|\mathbf{w}\|_{1,2,\mathcal{I}}\|\mathbf{v}\|_{1,2,\mathcal{I}} + \|\mathbf{v}\|^2_{1,2,\mathcal{I}}\right]\|\mathbf{w}-\mathbf{v}\|_{0,2,\mathcal{I}},
\end{aligned}$$

$$\begin{aligned}
&\|F_2^{(NL)}(\cdot,\mathbf{w}) - F_2^{(NL)}(\cdot,\mathbf{v})\|_{0,2,\mathcal{I}} \\
&\le 2(2\kappa)^2\|\alpha\|_{0,\infty,\mathcal{I}}C_S^2\left[\|\mathbf{w}\|^2_{1,2,\mathcal{I}} + \|\mathbf{w}\|_{1,2,\mathcal{I}}\|\mathbf{v}\|_{1,2,\mathcal{I}} + \|\mathbf{v}\|^2_{1,2,\mathcal{I}}\right]\|\mathbf{w}-\mathbf{v}\|_{0,2,\mathcal{I}}
\end{aligned}$$

$$\begin{aligned}
&\|F_3^{(NL)}(\cdot,\mathbf{w}) - F_3^{(NL)}(\cdot,\mathbf{v})\|_{0,2,\mathcal{I}} \\
&\le \frac{7}{3}(3\kappa)^2\|\alpha\|_{0,\infty,\mathcal{I}}C_S^2\left[\|\mathbf{w}\|^2_{1,2,\mathcal{I}} + \|\mathbf{w}\|_{1,2,\mathcal{I}}\|\mathbf{v}\|_{1,2,\mathcal{I}} + \|\mathbf{v}\|^2_{1,2,\mathcal{I}}\right]\|\mathbf{w}-\mathbf{v}\|_{0,2,\mathcal{I}}.
\end{aligned}$$

It follows that

$$\begin{aligned}
&\|\mathcal{N}^{(NL)}(\mathbf{w}) - \mathcal{N}^{(NL)}(\mathbf{v})\|^2_{0,2,\mathcal{I}} = \sum_{n=1}^{3}\|F_n^{(NL)}(\cdot,\mathbf{w}) - F_n^{(NL)}(\cdot,\mathbf{v})\|^2_{0,2,\mathcal{I}} \\
&\le 514\kappa^4\|\alpha\|^2_{0,\infty,\mathcal{I}}C_S^4\left[\|\mathbf{w}\|^2_{1,2,\mathcal{I}} + \|\mathbf{w}\|_{1,2,\mathcal{I}}\|\mathbf{v}\|_{1,2,\mathcal{I}} + \|\mathbf{v}\|^2_{1,2,\mathcal{I}}\right]^2\|\mathbf{w}-\mathbf{v}\|^2_{0,2,\mathcal{I}}.
\end{aligned}$$

Hence, for $\mathbf{w},\mathbf{v} \in K_\varrho^{\mathrm{cl}}$: $\|\mathcal{N}^{(NL)}(\mathbf{w}) - \mathcal{N}^{(NL)}(\mathbf{v})\|_{0,2,\mathcal{I}} \le 3\sqrt{514}\kappa^2C_S^2\|\alpha\|_{0,\infty,\mathcal{I}}\varrho^2\|\mathbf{w}-\mathbf{v}\|_{0,2,\mathcal{I}}$.
In summary, by assumption we arrive at the estimate

$$\begin{aligned}
&\|\mathcal{A}^{-1}\mathcal{F}(\mathbf{w}) - \mathcal{A}^{-1}\mathcal{F}(\mathbf{v})\|_{1,2,\mathcal{I}} \\
&\le C_N\kappa^2\left[9\|\varepsilon^{(L)} - \sin^2\varphi_\kappa\|_{0,\infty,\mathcal{I}} + 3\sqrt{514}C_S^2\|\alpha\|_{0,\infty,\mathcal{I}}\varrho^2\right]\|\mathbf{w}-\mathbf{v}\|_{0,2,\mathcal{I}} \le \frac{\sqrt{2}}{2}\|\mathbf{w}-\mathbf{v}\|_{1,2,\mathcal{I}}.
\end{aligned}$$

By Banach's fixed-point theorem, the problem (41) has a unique solution $\mathbf{u} \in K_\varrho^{\mathrm{cl}}$. ◄

5. The non-linear problem and the equivalent system of non-linear integral equations

The problem (21), (C1) – (C4) can be reduced to finding solutions of one-dimensional non-linear integral equations w.r.t. the components $U(n\kappa; z)$, $n = 1,2,3$, $z \in [-2\pi\delta, 2\pi\delta]$, of the fields scattered and generated in the non-linear layer. Similar to the results of the papers Angermann & Yatsyk (2011), Angermann & Yatsyk (2010), Shestopalov & Yatsyk (2010), Yatsyk (2007), Shestopalov & Yatsyk (2007), Kravchenko & Yatsyk (2007), Shestopalov & Sirenko (1989), we give the derivation of these equations for the case of excitation of the non-linear structure by a plane-wave packet (20).
Taking into account the representation (23), the solution of (21), (C1) – (C4) in the whole space $Q := \{q = (y,z) : |y| < \infty, |z| < \infty\}$ is obtained using the properties of the canonical Green's function of the problem (21), (C1) – (C4) (for the special case $\varepsilon_{n\kappa} \equiv 1$) which is defined, for

$Y > 0$, in the strip $Q_{\{Y,\infty\}} := \{q = (y, z) : \ |y| < Y, \ \ |z| < \infty\} \subset Q$ by

$$\begin{aligned} G_0(n\kappa; q, q_0) &:= \frac{i}{4Y} \exp\left\{ i\left[\phi_{n\kappa}\left(y - y_0\right) + \Gamma_{n\kappa}|z - z_0|\right]\right\} / \Gamma_{n\kappa} \\ &= \exp\left(\pm i\phi_{n\kappa} y\right) \frac{i\pi}{4Y} \int_{-\infty}^{\infty} H_0^{(1)}\left(n\kappa\sqrt{\left(\breve{y} - y_0\right)^2 + (z - z_0)^2}\right) \exp\left(\mp i\phi_{n\kappa}\breve{y}\right) d\breve{y}, \quad n = 1, 2, 3, \end{aligned} \tag{42}$$

where $H_0^{(1)}$ as usual denotes the Hankel function of the first kind of order zero (cf. Shestopalov & Sirenko (1989); Sirenko et al. (1985)).

The system of non-linear integral equations is obtained by means of an iterative approach Angermann & Yatsyk (2011), Yatsyk (2007), Shestopalov & Yatsyk (2007), Shestopalov & Sirenko (1989), Titchmarsh (1961). Denote both the scattered and the generated full fields of diffraction at each frequency $n\kappa$, $n = 1, 2, 3$, i.e. the solution of the problem (21), (C1) – (C4), by $E_1\left(n\kappa; q|_{q=(y,z)}\right) = U(n\kappa; z) \exp\left(i\phi_{n\kappa} y\right)$ (cf. (23)), and write the system (21) in the form

$$\begin{aligned} \left(\nabla^2 + (n\kappa)^2\right) E_1(n\kappa; q) &= \left[1 - \varepsilon_{n\kappa}\left(q, \alpha(q), E_1(\kappa; q), E_1(2\kappa; q), E_1(3\kappa; q)\right)\right] (n\kappa)^2 E_1(n\kappa; q) \\ &\quad - \delta_{n1}\alpha(q)(n\kappa)^2 E_1^2(2\kappa; q)\overline{E}_1(3\kappa; q) \\ &\quad - \delta_{n3}\alpha(q)(n\kappa)^2 \left\{\frac{1}{3}E_1^3(\kappa; q) + E_1^2(2\kappa; q)\overline{E}_1(\kappa; q)\right\}, \quad n = 1, 2, 3. \end{aligned} \tag{43}$$

At the right-hand side of the system (43), the first term outside the layer vanishes, since, by assumption, the permittivity of the medium in which the non-linear layer is situated is equal to one, i.e. $1 - \varepsilon_{n\kappa}\left(q, \alpha(q), E_1(\kappa; q), E_1(2\kappa; q), E_1(3\kappa; q)\right) \equiv 0$ for $|z| > 2\pi\delta$.

The excitation field of the non-linear structure can be represented in the form of a packet of incident plane waves $\left\{E_1^{\text{inc}}(n\kappa; q)\right\}_{n=1,2,3}$ satisfying the condition of phase synchronism (C2), where

$$E_1^{\text{inc}}(n\kappa; q) = a_{n\kappa}^{\text{inc}} \exp\left\{i\left[\phi_{n\kappa} y - \Gamma_{n\kappa}(z - 2\pi\delta)\right]\right\}, \quad n = 1, 2, 3. \tag{44}$$

Furthermore, in the present situation described by the system (43), we assume that the excitation field $E_1^{\text{inc}}(\kappa; q)$ of the non-linear structure at the frequency κ is sufficiently strong (i.e. the amplitude a_κ^{inc} is sufficiently large such that the third harmonic generation is possible), whereas the amplitudes $a_{2\kappa}^{\text{inc}}$, $a_{3\kappa}^{\text{inc}}$ corresponding to excitation fields $E_1^{\text{inc}}(2\kappa; q)$, $E_1^{\text{inc}}(3\kappa; q)$ at the frequencies 2κ, 3κ, respectively, are selected sufficiently weak such that no generation of multiple harmonics occurs.

In the whole space Q, for each frequency $n\kappa$, $n = 1, 2, 3$, the fields $\left\{E_1^{\text{inc}}(n\kappa; q)\right\}_{n=1,2,3}$ of incident plane waves satisfy a system of homogeneous Helmholtz equations:

$$\left(\nabla^2 + (n\kappa)^2\right) E_1^{\text{inc}}(n\kappa; q) = 0, \quad q \in Q, \quad n = 1, 2, 3. \tag{45}$$

For $z > 2\pi\delta$, the incident fields $\left\{E_1^{\text{inc}}(n\kappa; q)\right\}_{n=1,2,3}$ are fields of plane waves approaching the layer, while, for $z < 2\pi\delta$, they move away from the layer and satisfy the radiation condition (since, in the representation of the fields $E_1^{\text{inc}}(n\kappa; q)$, $n = 1, 2, 3$, the transverse propagation constants $\Gamma_{n\kappa} > 0$, $n = 1, 2, 3$ are positive).

Following Angermann & Yatsyk (2011), we construct a sequence $\left\{E_{1,s}(n\kappa; q)\right\}_{s=0}^{\infty}$, $n = 1, 2, 3$, of functions in the region Q (where each function, starting with the index $p = 1$, satisfies the conditions (C1) – (C4)) such that the limit functions $E_1(n\kappa; q) = \lim\limits_{s\to\infty} E_{1,s}(n\kappa; q)$ at the frequencies $n\kappa$, $n = 1, 2, 3$, satisfy (21), (C1) – (C4), i.e.

$$\begin{aligned}
&\left(\nabla^2 + (n\kappa)^2\right) E_{1,0}(n\kappa; q) = 0\,,\\
&\left(\nabla^2 + (n\kappa)^2\right) E_{1,1}(n\kappa; q) = \left[1 - \varepsilon_{n\kappa}\left(q, \alpha(q), E_{1,0}(\kappa; q), E_{1,0}(2\kappa; q), E_{1,0}(3\kappa; q)\right)\right]\\
&\qquad \times (n\kappa)^2 E_{1,0}(n\kappa; q) - \delta_{n1}\alpha(q)(n\kappa)^2 E_{1,0}^2(2\kappa; q)\overline{E}_{1,0}(3\kappa; q)\\
&\qquad - \delta_{n3}\alpha(q)(n\kappa)^2 \left\{\tfrac{1}{3}E_{1,0}^3(\kappa; q) + E_{1,0}^2(2\kappa; q)\overline{E}_{1,0}(\kappa; q)\right\}, \ldots,\\
&\left(\nabla^2 + (n\kappa)^2\right) E_{1,s+1}(n\kappa; q) = \left[1 - \varepsilon_{n\kappa}\left(q, \alpha(q), E_{1,s}(\kappa; q), E_{1,s}(2\kappa; q), E_{1,s}(3\kappa; q)\right)\right]\\
&\qquad \times (n\kappa)^2 E_{1,s}(n\kappa; q) - \delta_{n1}\alpha(q)(n\kappa)^2 E_{1,s}^2(2\kappa; q)\overline{E}_{1,s}(3\kappa; q)\\
&\qquad - \delta_{n3}\alpha(q)(n\kappa)^2 \left\{\tfrac{1}{3}E_{1,s}^3(\kappa; q) + E_{1,s}^2(2\kappa; q)\overline{E}_{1,s}(\kappa; q)\right\}, \ldots,\\
&\qquad n = 1, 2, 3.
\end{aligned} \tag{46}$$

The system of equations (46) is formally equivalent to the following one:

$$\begin{aligned}
E_{1,0}(n\kappa; q) &:= E_1^{\mathrm{inc}}(n\kappa; q),\\
E_{1,1}(n\kappa; q) &= -(n\kappa)^2 \iint_{Q_\delta} G_0(n\kappa; q, q_0)\times\\
&\times \left[1 - \varepsilon_{n\kappa}\left(q_0, \alpha(q_0), E_{1,0}(\kappa; q_0), E_{1,0}(2\kappa; q_0), E_{1,0}(3\kappa; q_0)\right)\right] E_{1,0}(n\kappa; q_0) dq_0\\
&+ \delta_{n1}(n\kappa)^2 \iint_{Q_\delta} G_0(n\kappa; q, q_0)\alpha(q_0) E_{1,0}^2(2\kappa; q_0)\overline{E}_{1,0}(3\kappa; q_0) dq_0\\
&+ \delta_{n3}(n\kappa)^2 \iint_{Q_\delta} G_0(n\kappa; q, q_0)\alpha(q_0) \left\{\frac{1}{3}E_{1,0}^3(\kappa; q_0) + E_{1,0}^2(2\kappa; q_0)\overline{E}_{1,0}(\kappa; q_0)\right\} dq_0\\
&+ E_{1,0}(n\kappa; q), \ldots,\\
E_{1,s+1}(n\kappa; q) &= -(n\kappa)^2 \iint_{Q_\delta} G_0(n\kappa; q, q_0)\times\\
&\times \left[1 - \varepsilon_{n\kappa}\left(q_0, \alpha(q_0), E_{1,s}(\kappa; q_0), E_{1,s}(2\kappa; q_0), E_{1,s}(3\kappa; q_0)\right)\right] E_{1,s}(n\kappa; q_0) dq_0\\
&+ \delta_{n1}(n\kappa)^2 \iint_{Q_\delta} G_0(n\kappa; q, q_0)\alpha(q_0) E_{1,s}^2(2\kappa; q_0)\overline{E}_{1,s}(3\kappa; q_0) dq_0\\
&+ \delta_{n3}(n\kappa)^2 \iint_{Q_\delta} G_0(n\kappa; q, q_0)\alpha(q_0) \left\{\frac{1}{3}E_{1,s}^3(\kappa; q_0) + E_{1,s}^2(2\kappa; q_0)\overline{E}_{1,s}(\kappa; q_0)\right\} dq_0\\
&+ E_{1,0}(n\kappa; q), \ldots, \qquad q \in Q, \quad n = 1, 2, 3.
\end{aligned} \tag{47}$$

Here $Q_\delta := \{q = (y, z) : |y| < \infty, |z| \le 2\pi\delta\}$ denotes the strip filled by the non-linear dielectric layer. The extension of the permitted values $q \in Q_{\{Y,\infty\}} \subset Q$ from the strip $Q_{\{Y,\infty\}}$ (where the Green's function (42) is defined) to the whole space Q is realised by passing to the limit $Y \to \infty$ (where this procedure is admissible because of the free choice of the parameter Y and the asymptotic behaviour of the integrands as $\mathcal{O}\left(Y^{-1}\right)$, see (42)). Letting s tend to infinity in (47), we obtain the integral representations of the unknown diffraction fields in the region Q:

$$\begin{aligned}
E_1(n\kappa; q) &= -(n\kappa)^2 \iint_{Q_\delta} G_0(n\kappa; q, q_0) \times\\
&\times \left[1 - \varepsilon_{n\kappa}\left(q_0, \alpha(q_0), E_1(\kappa; q_0), E_1(2\kappa; q_0), E_1(3\kappa; q_0)\right)\right] E_1(n\kappa; q_0) dq_0\\
&+ \delta_{n1}(n\kappa)^2 \iint_{Q_\delta} G_0(n\kappa; q, q_0)\alpha(q_0) E_1^2(2\kappa; q_0)\overline{E}_1(3\kappa; q_0) dq_0\\
&+ \delta_{n3}(n\kappa)^2 \iint_{Q_\delta} G_0(n\kappa; q, q_0)\alpha(q_0) \left\{\frac{1}{3}E_1^3(\kappa; q_0) + E_1^2(2\kappa; q_0)\overline{E}_1(\kappa; q_0)\right\} dq_0\\
&+ E_1^{\mathrm{inc}}(n\kappa; q), \quad q \in Q, \qquad n = 1, 2, 3.
\end{aligned} \tag{48}$$

Now, substituting the representation (42) for the canonical Green's function G_0 into the system (48) and taking into consideration the expressions for the permittivity

$$\varepsilon_{n\kappa}\left(q_0,\alpha(q_0),E_1(\kappa;q_0),E_1(2\kappa;q_0),E_1(3\kappa;q_0)\right)=\varepsilon_{n\kappa}\left(z_0,\alpha(z_0),U(\kappa;z_0),U(2\kappa;z_0),U(3\kappa;z_0)\right),$$

we get the following system w.r.t. the unknown quasi-homogeneous fields
$E_1\left(n\kappa;q|_{q\equiv(y,z)}\right)=U(n\kappa;z)\exp\left(i\phi_{n\kappa}y\right),\, n=1,2,3,\ |z|\le 2\pi\delta$:

$$\begin{aligned}
&U(n\kappa;z)\exp\left(i\phi_{n\kappa}y\right)\\
&=-\lim_{Y\to\infty}\left(\frac{i(n\kappa)^2}{4Y\Gamma_{n\kappa}}\exp(i\phi_{n\kappa}y)\int_{-2\pi\delta}^{2\pi\delta}\int_{-Y}^{Y}\exp(i\Gamma_{n\kappa}|z-z_0|)\times\right.\\
&\left.\times\left[1-\varepsilon_{n\kappa}\left(z_0,\alpha(z_0),U(\kappa;z_0),U(2\kappa;z_0),U(3\kappa;z_0)\right)\right]U(n\kappa;z_0)\,dy_0dz_0\right)\\
&+\lim_{Y\to\infty}\left(\delta_{n1}\frac{i(n\kappa)^2}{4Y\Gamma_{n\kappa}}\exp(i\phi_{n\kappa}y)\times\right.\\
&\qquad\left.\times\int_{-2\pi\delta}^{2\pi\delta}\int_{-Y}^{Y}\exp(i\Gamma_{n\kappa}|z-z_0|)\alpha(z_0)U^2(2\kappa;z_0)\overline{U}(3\kappa;z_0)dy_0dz_0\right)\\
&+\lim_{Y\to\infty}\left(\delta_{n3}\frac{i(n\kappa)^2}{4Y\Gamma_{n\kappa}}\exp(i\phi_{n\kappa}y)\times\right.\\
&\left.\times\int_{-2\pi\delta}^{2\pi\delta}\int_{-Y}^{Y}\exp(i\Gamma_{n\kappa}|z-z_0|)\alpha(z_0)\left\{\frac{1}{3}U^3(\kappa;z_0)+U^2(2\kappa;z_0)\overline{U}(\kappa;z_0)\right\}dy_0dz_0\right)\\
&+U^{\text{inc}}(n\kappa;z)\exp(i\phi_{n\kappa}y),\quad |z|\le 2\pi\delta,\quad n=1,2,3.
\end{aligned}$$

Integrating in the region Q_δ w.r.t. the variable y_0, we arrive at a system of non-linear Fredholm integral equations of the second kind w.r.t. the unknown functions $U(n\kappa;\cdot)\in L_2(-2\pi\delta,2\pi\delta)$:

$$\begin{aligned}
&U(n\kappa;z)+\frac{i(n\kappa)^2}{2\Gamma_{n\kappa}}\int_{-2\pi\delta}^{2\pi\delta}\exp(i\Gamma_{n\kappa}|z-z_0|)\times\\
&\qquad\times\left[1-\varepsilon_{n\kappa}\left(z_0,\ \alpha(z_0),U(\kappa;z_0),U(2\kappa;z_0),U(3\kappa;z_0)\right)\right]U(n\kappa;z_0)dz_0\\
&=\delta_{n1}\frac{i(n\kappa)^2}{2\Gamma_{n\kappa}}\int_{-2\pi\delta}^{2\pi\delta}\exp(i\Gamma_{n\kappa}|z-z_0|)\alpha(z_0)U^2(2\kappa;z_0)\overline{U}(3\kappa;z_0)dz_0\\
&+\delta_{n3}\frac{i(n\kappa)^2}{2\Gamma_{n\kappa}}\int_{-2\pi\delta}^{2\pi\delta}\exp(i\Gamma_{n\kappa}|z-z_0|)\alpha(z_0)\left\{\frac{1}{3}U^3(\kappa;z_0)+U^2(2\kappa;z_0)\overline{U}(\kappa;z_0)\right\}dz_0\\
&+U^{\text{inc}}(n\kappa;z),\quad |z|\le 2\pi\delta,\quad n=1,2,3.
\end{aligned}\tag{49}$$

Here $U^{\text{inc}}(n\kappa;z)=a^{\text{inc}}_{n\kappa}\exp\left[-i\Gamma_{n\kappa}(z-2\pi\delta)\right],\, n=1,2,3.$
The solution of the original problem (21), (C1) – (C4), represented as (23), can be obtained from (49) using the formulas

$$U(n\kappa;2\pi\delta)=a^{\text{inc}}_{n\kappa}+a^{\text{scat}}_{n\kappa},\quad U(n\kappa;-2\pi\delta)=b^{\text{scat}}_{n\kappa},\quad n=1,2,3,\tag{50}$$

(cf. (C3)). The derivation of the system of non-linear integral equations (49) shows that (49) can be regarded as an integral representation of the desired solution of (21), (C1) – (C4) (i.e. solutions of the form $E_1\left(n\kappa;y,z\right)=U(n\kappa;z)\ \exp\left(i\phi_{n\kappa}y\right),\ n=1,2,3$, see (23)) for points located outside the non-linear layer: $\left\{(y,z):\ |y|<\infty,\ |z|>2\pi\delta\right\}$. Indeed, given the solution of non-linear integral equations (49) in the region $|z|\le 2\pi\delta$, the substitution into the integrals of (49) leads to explicit expressions of the desired solutions $U(n\kappa;z)$ for points $|z|>2\pi\delta$ outside the non-linear layer at each frequency $n\kappa$, $n=1,2,3$.

6. A sufficient condition for the existence of solutions of the system of non-linear equations

In the case of a linear system (49), i.e. if $\alpha \equiv 0$, the problem of existence and uniqueness of solutions has been investigated in Sirenko et al. (1985), Shestopalov & Sirenko (1989). In the general situation, the system of non-linear integral equations can have a unique solution, no solution or several solutions, depending on the properties of the kernel and the right-hand side.

We start with the derivation of sufficient conditions for the existence of solutions of the system (49) (cf. Shestopalov & Yatsyk (2010), Shestopalov & Yatsyk (2007), Kravchenko & Yatsyk (2007)). To do so, in the region $|z| \le 2\pi\delta$ we consider two sequences of solutions $\{U_s(n\kappa; z),\ n = 1,2,3\}_{s=0}^{\infty}$ and $\{\Psi_s(n\kappa; z),\ n = 1,2,3\}_{s=0}^{\infty}$ of the following systems of integral equations:

$$\begin{aligned}
&U_{s+1}(n\kappa; z) + \frac{i(n\kappa)^2}{2\Gamma_{n\kappa}} \int_{-2\pi\delta}^{2\pi\delta} \exp(i\Gamma_{n\kappa}|z - z_0|) \\
&\qquad \times [1 - \varepsilon_{n\kappa}(z_0, \alpha(z_0), U_s(\kappa; z_0), U_s(2\kappa; z_0), U_s(3\kappa; z_0))] U_s(n\kappa; z_0) dz_0 \\
&= \delta_{n1} \frac{i(n\kappa)^2}{2\Gamma_{n\kappa}} \int_{-2\pi\delta}^{2\pi\delta} \exp(i\Gamma_{n\kappa}|z - z_0|) \alpha(z_0) U_s^2(2\kappa; z_0) \overline{U}_s(3\kappa; z_0) dz_0 \\
&+ \delta_{n3} \frac{i(n\kappa)^2}{2\Gamma_{n\kappa}} \int_{-2\pi\delta}^{2\pi\delta} \exp(i\Gamma_{n\kappa}|z - z_0|) \\
&\times \alpha(z_0) \left\{ \frac{1}{3} U_s^3(\kappa; z_0) + U_s^2(2\kappa; z_0) \overline{U}_s(\kappa; z_0) \right\} dz_0 + U^{\text{inc}}(n\kappa; z), \quad |z| \le 2\pi\delta,\ n = 1,2,3,
\end{aligned}$$

$$\begin{aligned}
&\Psi_s(n\kappa; z) + \frac{i(n\kappa)^2}{2\Gamma_{n\kappa}} \int_{-2\pi\delta}^{2\pi\delta} \exp(i\Gamma_{n\kappa}|z - z_0|) \\
&\qquad \times [1 - \varepsilon_{n\kappa}(z_0, \alpha(z_0), U_s(\kappa; z_0), U_s(2\kappa; z_0), U_s(3\kappa; z_0))] \Psi_s(n\kappa; z_0) dz_0 \\
&= \delta_{n1} \frac{i(n\kappa)^2}{2\Gamma_{n\kappa}} \int_{-2\pi\delta}^{2\pi\delta} \exp(i\Gamma_{n\kappa}|z - z_0|) \alpha(z_0) U_s^2(2\kappa; z_0) \overline{U}_s(3\kappa; z_0) dz_0 \\
&+ \delta_{n3} \frac{i(n\kappa)^2}{2\Gamma_{n\kappa}} \int_{-2\pi\delta}^{2\pi\delta} \exp(i\Gamma_{n\kappa}|z - z_0|) \\
&\times \alpha(z_0) \left\{ \frac{1}{3} U_s^3(\kappa; z_0) + U_s^2(2\kappa; z_0) \overline{U}_s(\kappa; z_0) \right\} dz_0 + U^{\text{inc}}(n\kappa; z), \quad |z| \le 2\pi\delta,\ n = 1,2,3.
\end{aligned} \tag{51}$$

The first system of equations (51) coincides with the iterative scheme (47) for the solution of the non-linear system (49). The second system w.r.t. $\Psi_s(n\kappa; z)$, $n = 1,2,3$, is nothing else than the linearisation of the non-linear system (49) around $U_s(n\kappa; z)$, $n = 1,2,3$.

In the case that the functions $\Psi_s(n\kappa; z)$, $n = 1,2,3$, are not eigen-functions of the linearised problem under consideration with the induced permittivity of the layer (cf. 19))

$$\begin{aligned}
&\varepsilon_{n\kappa}(z, \alpha(z), U_s(\kappa; z), U_s(2\kappa; z), U_s(3\kappa; z)) \\
&= \varepsilon^{(L)}(z) + \varepsilon_{n\kappa}^{(NL)}(\alpha(z), U_s(\kappa; z), U_s(2\kappa; z), U_s(3\kappa; z)) \\
&= \varepsilon^{(L)}(z) + \alpha(z)\{|U_s(\kappa; z)|^2 + |U_s(2\kappa; z)|^2 + |U_s(3\kappa; z)|^2 \\
&\quad + \delta_{n1}|U_s(\kappa; z)|\,|U_s(3\kappa; z)| \exp[i\{-3\arg U_s(\kappa; z) + \arg U_s(3\kappa; z)\}] \\
&\quad + \delta_{n2}|U_s(\kappa; z)|\,|U_s(3\kappa; z)| \exp[i\{-2\arg U_s(2\kappa; z) + \arg U_s(\kappa; z) + \arg U_s(3\kappa; z)\}]\}, \\
&\qquad |z| \le 2\pi\delta,\ n = 1,2,3,
\end{aligned} \tag{52}$$

a solution of the second system in (51) exists uniquely (Sirenko et al. (1985), Shestopalov & Sirenko (1989)) and can be represented as

$$\Psi_s(n\kappa;z) = \Psi(n\kappa;z,\alpha(z),U_s(\kappa;z),U_s(2\kappa;z),U_s(3\kappa;z)), \quad n = 1,2,3. \tag{53}$$

Moreover, at each iteration step (i.e. for any iteration parameter $s \in \{0,1,2,\ldots\}$) the solution (53) which is caused by the exciting wave packet $\{|U^{\rm inc}(n\kappa;z)| = a^{\rm inc}_{n\kappa}\}_{n=1}^{3}$, satisfies the estimate

$$|\Psi_s(n\kappa;z)|^2 \leq \sum_{m=1}^{3}(a^{\rm inc}_{m\kappa})^2, \quad \forall s \in \{0,1,2,\ldots\},\ n = 1,2,3 \tag{54}$$

due to energy relations. In particular,

$$\begin{aligned}&|\varepsilon_{n\kappa}(z,\alpha(z),U_s(\kappa;z),U_s(2\kappa;z),U_s(3\kappa;z))|\\ &\leq |\varepsilon^{(L)}(z)| + |\alpha(z)|(3+\delta_{n1}+\delta_{n2})\sum_{m=1}^{3}(a^{\rm inc}_{m\kappa})^2, \quad \forall s \in \{0,1,2,\ldots\},\ n = 1,2,3.\end{aligned} \tag{55}$$

The analysis of appropriate convergence criteria for the sequences $\{U_s(n\kappa;z),\ n = 1,2,3\}_{s=0}^{\infty}$ and $\{\Psi_s(n\kappa;z),\ n = 1,2,3\}_{s=0}^{\infty}$ given by (51) provides a sufficient condition for the existence and uniqueness of solutions of the non-linear integral equations (49). Since the kernels of the integral equations (51) are identical, it is easy to estimate the distance between the elements $U_{s+1}(n\kappa;z)$ and $\Psi_s(n\kappa;z)$:

$$\begin{aligned}
&\varrho(U_{s+1}(n\kappa;z),\Psi_s(n\kappa;z)) = \left[\int_{-2\pi\delta}^{2\pi\delta}|U_{s+1}(n\kappa;z)-\Psi_s(n\kappa;z)|^2dz\right]^{1/2}\\
&= \left|\frac{i(n\kappa)^2}{2\Gamma_{n\kappa}}\right|\left[\int_{-2\pi\delta}^{2\pi\delta}\left|\int_{-2\pi\delta}^{2\pi\delta}\exp(i\Gamma_{n\kappa}|z-z_0|)\right.\right.\\
&\quad\times\left.\left.[1-\varepsilon_{n\kappa}(z_0,\alpha(z_0),U_s(\kappa;z_0),U_s(2\kappa;z_0),U_s(3\kappa;z_0))](U_s(n\kappa;z_0)-\Psi_s(n\kappa;z_0))dz_0\right|^2dz\right]^{1/2}\\
&= \frac{(n\kappa)^2}{2\Gamma_{n\kappa}}\left[\int_{-2\pi\delta}^{2\pi\delta}\left|\int_{-2\pi\delta}^{2\pi\delta}[1-\varepsilon_{n\kappa}(z_0,\alpha(z_0),U_s(\kappa;z_0),U_s(2\kappa;z_0),U_s(3\kappa;z_0))]\right.\right.\\
&\quad\times\left.\left.(U_s(n\kappa;z_0)-\Psi_s(n\kappa;z_0))dz_0\right|^2dz\right]^{1/2}\\
&\leq \frac{(n\kappa)^2}{2\Gamma_{n\kappa}}\left[\int_{-2\pi\delta}^{2\pi\delta}\int_{-2\pi\delta}^{2\pi\delta}|1-\varepsilon_{n\kappa}(z_0,\alpha(z_0),U_s(\kappa;z_0),U_s(2\kappa;z_0),U_s(3\kappa;z_0))|^2dz_0dz\right]^{1/2}\\
&\quad\times\left[\int_{-2\pi\delta}^{2\pi\delta}|U_s(n\kappa;z_0)-\Psi_s(n\kappa;z_0)|^2dz_0\right]^{1/2}\\
&\leq \frac{(n\kappa)^2}{2\Gamma_{n\kappa}}4\pi\delta\max_{z\in[-2\pi\delta,2\pi\delta]}\left(|1-\varepsilon^{(L)}(z)|+|\varepsilon_{n\kappa}(z,\alpha(z),U_s(\kappa;z),U_s(2\kappa;z),U_s(3\kappa;z))|\right)\\
&\quad\times\varrho(U_s(n\kappa;z),\ \Psi_s(n\kappa;z))\\
&\leq \frac{(n\kappa)^2}{2\Gamma_{n\kappa}}4\pi\delta\max_{z\in[-2\pi\delta,2\pi\delta]}\left(|1-\varepsilon^{(L)}(z)|+4|\alpha(z)|\sum_{m=1}^{3}(a^{\rm inc}_{m\kappa})^2\right)\varrho(U_s(n\kappa;z),\Psi_s(n\kappa;z)),\\
&\qquad n = 1,2,3.
\end{aligned} \tag{56}$$

The last inequality in (56) is a consequence of (55). The estimate (56) shows that the iterative process defined by the first system of equations (51) converges to a unique solution determined by the second system of equations (51) if in (56) the factor in front of $\varrho(U_s(n\kappa;z), \Psi_s(n\kappa;z))$ satisfies the condition

$$\frac{(n\kappa)^2}{2\Gamma_{n\kappa}} 4\pi\delta \max_{z\in[-2\pi\delta,2\pi\delta]} \left(|1-\varepsilon^{(L)}(z)| + 4|\alpha(z)| \sum_{m=1}^{3} (a_{m\kappa}^{\text{inc}})^2 \right) < 1, \quad n=1,2,3.$$

Taking into account the expressions for the transverse propagation constants $\Gamma_{n\kappa} = ((n\kappa)^2 - \phi_{n\kappa}^2)^{1/2} = ((n\kappa)^2 - (n\kappa \sin\varphi_{n\kappa})^2)^{1/2} = n\kappa\cos\varphi_{n\kappa}$, $n=1,2,3$, and the condition of phase synchronism (C2) $\varphi_\kappa = \varphi_{n\kappa}$, $n=1,2,3$, these inequalities can be represented as

$$n\kappa\, 2\pi\delta \max_{z\in[-2\pi\delta,2\pi\delta]} \left(|1-\varepsilon^{(L)}(z)| + 4|\alpha(z)| \sum_{m=1}^{3} (a_{m\kappa}^{\text{inc}})^2 \right) < \cos\varphi_\kappa, \quad n=1,2,3. \tag{57}$$

In summary, we have proved the following result.

Theorem 2. *The condition (57) is a sufficient condition for the existence of solutions of the non-linear integral equations (49). Such a solution can be obtained by using the iterative process given the first system of equations in (51), or by using the equivalent iterative process that can be built on the basis of the second system of equations in (51). The solution* $\Psi_s(n\kappa;z)$, $n=1,2,3$, *should be regarded as an* $(s+1)$*st approximation* $U_{s+1}(n\kappa;z) := \Psi_s(n\kappa;z)$ *to the desired solution* $U(n\kappa;z)$, $n=1,2,3$:

$$\begin{aligned} &U_{s+1}(n\kappa;z) + \frac{i(n\kappa)^2}{2\Gamma_{n\kappa}} \int_{-2\pi\delta}^{2\pi\delta} \exp(i\Gamma_{n\kappa}|z-z_0|) \\ &\quad\times [1-\varepsilon_{n\kappa}(z_0,\alpha(z_0),U_s(\kappa;z_0),U_s(2\kappa;z_0),U_s(3\kappa;z_0))]U_{s+1}(n\kappa;z_0)dz_0 \\ &= \delta_{n1}\frac{i(n\kappa)^2}{2\Gamma_{n\kappa}} \int_{-2\pi\delta}^{2\pi\delta} \exp(i\Gamma_{n\kappa}|z-z_0|)\alpha(z_0)U_s^2(2\kappa;z_0)\overline{U}_s(3\kappa;z_0)dz_0 \\ &+ \delta_{n3}\frac{i(n\kappa)^2}{2\Gamma_{n\kappa}} \int_{-2\pi\delta}^{2\pi\delta} \exp(i\Gamma_{n\kappa}|z-z_0|) \\ &\times\ \alpha(z_0)\{\frac{1}{3}U_s^3(\kappa;z_0) + U_s^2(2\kappa;z_0)\overline{U}_s(\kappa;z_0)\}dz_0 + U^{inc}(n\kappa;z), \quad |z|\le 2\pi\delta,\ n=1,2,3. \end{aligned} \tag{58}$$

Moreover, if the elements $\Psi_s(n\kappa;z) \equiv U_{s+1}(n\kappa;z)$, $n=1,2,3$, *are not eigen-functions of the linearised problem (49) with the induced permittivity of the layer (52) (i.e. solutions of the homogeneous system (58)), then this solution is unique.*

7. A self-consistent approach to the numerical analysis of the non-linear integral equations

According to Angermann & Yatsyk (2011), Angermann & Yatsyk (2010), the application of suitable quadrature rules to the system of non-linear integral equations (49) leads to a system of complex-valued non-linear algebraic equations of the second kind:

$$\begin{cases} (\mathbf{I} - \mathbf{B}_\kappa(\mathbf{U}_\kappa, \mathbf{U}_{2\kappa}, \mathbf{U}_{3\kappa}))\mathbf{U}_\kappa = \mathbf{C}_\kappa(\mathbf{U}_{2\kappa}, \mathbf{U}_{3\kappa}) + \mathbf{U}_\kappa^{\text{inc}}, \\ (\mathbf{I} - \mathbf{B}_{2\kappa}(\mathbf{U}_\kappa, \mathbf{U}_{2\kappa}, \mathbf{U}_{3\kappa}))\mathbf{U}_{2\kappa} = \mathbf{U}_{2\kappa}^{\text{inc}}, \\ (\mathbf{I} - \mathbf{B}_{3\kappa}(\mathbf{U}_\kappa, \mathbf{U}_{2\kappa}, \mathbf{U}_{3\kappa}))\mathbf{U}_{3\kappa} = \mathbf{C}_{3\kappa}(\mathbf{U}_\kappa, \mathbf{U}_{2\kappa}) + \mathbf{U}_{3\kappa}^{\text{inc}}, \end{cases} \tag{59}$$

where $\{z_l\}_{l=1}^{N}$ is a discrete set of nodes such that $-2\pi\delta =: z_1 < z_2 < ... < z_l < ... < z_N =: 2\pi\delta$.
$\mathbf{U}_{n\kappa} := \{U_l(n\kappa)\}_{l=1}^{N} \approx \{U(n\kappa; z_l)\}_{l=1}^{N}$ denotes the vector of the unknown approximate solution values corresponding to the frequencies $n\kappa$, $n = 1, 2, 3$. The matrices are of the form

$$\mathbf{B}_{n\kappa}(\mathbf{U}_{\kappa}, \mathbf{U}_{2\kappa}, \mathbf{U}_{3\kappa}) = \{A_m K_{lm}(n\kappa, \mathbf{U}_{\kappa}, \mathbf{U}_{2\kappa}, \mathbf{U}_{3\kappa})\}_{l,m=1}^{N}$$

with entries

$$\begin{aligned}
&K_{lm}(n\kappa, \mathbf{U}_{\kappa}, \mathbf{U}_{2\kappa}, \mathbf{U}_{3\kappa}) := -\frac{i(n\kappa)^2}{2\Gamma_{n\kappa}} \exp\left(i\Gamma_{n\kappa}|z_l - z_m|\right)\Big[1 - \Big\{\varepsilon^{(L)}(z_m)\\
&+ \alpha(z_m)\Big(|U_m(\kappa)|^2 + |U_m(2\kappa)|^2 + |U_m(3\kappa)|^2\\
&+ \delta_{n1}\,|U_m(\kappa)|\,|U_m(3\kappa)|\,\exp\left\{i\left[-3\arg U_m(\kappa) + \arg U_m(3\kappa)\right]\right\}\\
&+ \delta_{n2}\,|U_m(\kappa)|\,|U_m(3\kappa)|\,\exp\left\{i\left[-2\arg U_m(2\kappa) + \arg U_m(\kappa) + \arg U_m(3\kappa)\right]\right\}\Big)\Big\}\Big].
\end{aligned} \tag{60}$$

The numbers A_m are the coefficients determined by the quadrature rule, $\mathbf{I} := \{\delta_{lm}\}_{l,m=1}^{N}$ is the identity matrix, and δ_{lm} is Kronecker's symbol.
The right-hand side of (59) is defined by

$$\mathbf{U}_{n\kappa}^{\text{inc}} := \{a_{n\kappa}^{\text{inc}} \exp[-i\Gamma_{n\kappa}(z_l - 2\pi\delta)]\}_{l=1}^{N},$$

$$\mathbf{C}_{\kappa}(\mathbf{U}_{2\kappa}, \mathbf{U}_{3\kappa}) := \left\{\frac{i\kappa^2}{2\Gamma_{\kappa}} \sum_{m=1}^{N} A_m \exp(i\Gamma_{\kappa}|z_l - z_m|)\alpha(z_m) U_m^2(2\kappa)\overline{U}_m(3\kappa)\right\}_{l=1}^{N},$$

$$\mathbf{C}_{3\kappa}(\mathbf{U}_{\kappa}, \mathbf{U}_{2\kappa}) := \left\{\frac{i(3\kappa)^2}{2\Gamma_{3\kappa}} \sum_{m=1}^{N} A_m \exp(i\Gamma_{3\kappa}|z_l - z_m|)\alpha(z_m)\left[\frac{1}{3}U_m^3(\kappa) + U_m^2(2\kappa)\overline{U}_m(\kappa)\right]\right\}_{l=1}^{N}.$$

The solution of (59) is approximated by means of the following iterative method:

$$\left\{\begin{aligned}
&\Big\{\Big[\mathbf{I} - \mathbf{B}_{\kappa}\Big(\mathbf{U}_{\kappa}^{(s-1)}, \mathbf{U}_{2\kappa}^{(S_2(q))}, \mathbf{U}_{3\kappa}^{(S_3(q))}\Big)\Big]\mathbf{U}_{\kappa}^{(s)}\\
&\quad = \mathbf{C}_{\kappa}\Big(\mathbf{U}_{2\kappa}^{(S_2(q))}, \mathbf{U}_{3\kappa}^{(S_3(q))}\Big) + \mathbf{U}_{\kappa}^{\text{inc}}\Big\}_{s=1}^{S_1(q):\ \|\mathbf{U}_{\kappa}^{(S_1(q))} - \mathbf{U}_{\kappa}^{(S_1(q)-1)}\|/\|\mathbf{U}_{\kappa}^{(S_1(q))}\| < \xi}\\
&\Big\{\Big[\mathbf{I} - \mathbf{B}_{2\kappa}\Big(\mathbf{U}_{\kappa}^{(S_1(q))}, \mathbf{U}_{2\kappa}^{(s-1)}, \mathbf{U}_{3\kappa}^{(S_3(q))}\Big)\Big]\mathbf{U}_{2\kappa}^{(s)}\\
&\quad = \mathbf{U}_{2\kappa}^{\text{inc}}\Big\}_{s=1}^{S_2(q):\ \|\mathbf{U}_{2\kappa}^{(S_2(q))} - \mathbf{U}_{2\kappa}^{(S_2(q)-1)}\|/\|\mathbf{U}_{2\kappa}^{(S_{2q})}\| < \xi}\\
&\Big\{\Big[\mathbf{I} - \mathbf{B}_{3\kappa}\Big(\mathbf{U}_{\kappa}^{(S_1(q))}, \mathbf{U}_{2\kappa}^{(S_2(q))}, \mathbf{U}_{3\kappa}^{(s-1)}\Big)\Big]\mathbf{U}_{3\kappa}^{(s)}\\
&\quad = \mathbf{C}_{3\kappa}\Big(\mathbf{U}_{\kappa}^{(S_1(q))}, \mathbf{U}_{2\kappa}^{(S_2(q))}\Big) + \mathbf{U}_{3\kappa}^{\text{inc}}\Big\}_{s=1}^{S_3(q):\ \|\mathbf{U}_{3\kappa}^{(S_3(q))} - \mathbf{U}_{3\kappa}^{(S_3(q)-1)}\|/\|\mathbf{U}_{3\kappa}^{(S_{3q})}\| < \xi}
\end{aligned}\right\}_{q=1}^{Q}, \tag{61}$$

where, for a given relative error tolerance $\xi > 0$, the terminating index $Q \in \mathbb{N}$ is defined by the requirement

$$\max\left\{\|\mathbf{U}_{\kappa}^{(Q)} - \mathbf{U}_{\kappa}^{(Q-1)}\|/\|\mathbf{U}_{\kappa}^{(Q)}\|, \|\mathbf{U}_{2\kappa}^{(Q)} - \mathbf{U}_{2\kappa}^{(Q-1)}\|/\|\mathbf{U}_{2\kappa}^{(Q)}\|, \|\mathbf{U}_{3\kappa}^{(Q)} - \mathbf{U}_{3\kappa}^{(Q-1)}\|/\|\mathbf{U}_{3\kappa}^{(Q)}\|\right\} < \xi.$$

8. Eigen-modes of the linearised problems of scattering and generation of waves on the cubically polarisable layer

The solution of the system of non-linear equations (49) is approximated by the solution of the linearised system of non-linear equations (58), for given values of the induced dielectric permittivity and of the source functions at the right-hand side of the system. The solution can be found by the help of algorithm (61), where at each step a system of linearised non-linear complex-valued algebraic equations of the second kind is solved iteratively. The analytic continuation of the linearised non-linear problems into the region of complex values of the frequency parameter allows us to switch to the analysis of spectral problems. That is, the eigen-frequencies and the corresponding eigen-fields of the homogeneous linear problems with an induced non-linear permittivity are to be determined. The results of the development of a spectral theory of linear problems for structures with non-compact boundaries can be found in Yatsyk (2000), Shestopalov & Yatsyk (1997), Sirenko et al. (1985), Shestopalov & Sirenko (1989), Sirenko et al. (2007), Sirenko & Ström (2010).

As mentioned above, the classical formulation of the problem of scattering and generation of waves, described by the *system* of boundary value problems (21), (C1) – (C4), can be reformulated as a *set* of independent spectral problems in the following way:

Find the eigen-frequencies κ_n and the corresponding eigen-functions $E_1(\mathbf{r},\kappa_n)$ (i.e. $\left\{\kappa_n \in \Omega_{n\kappa} \subset \mathsf{H}_{n\kappa}, E_1(\mathbf{r},\kappa_n)\right\}_{n=1}^{3}$, where $\Omega_{n\kappa}$ are the sets of eigen-frequencies lying on the two-sheeted Riemann surfaces $\mathsf{H}_{n\kappa}$, see Fig. 2 and the more detailed explanations below) satisfying the equations

$$\nabla^2 E_1(\mathbf{r},\kappa_n) + \kappa_n^2 \varepsilon_{n\kappa}\left(z,\alpha(z),E_1(r,\kappa),E_1(\mathbf{r},2\kappa),E_1(\mathbf{r},3\kappa)\right) E_1(\mathbf{r},\kappa_n) = 0, \quad n = 1,2,3, \qquad (62)$$

together with the following conditions:

(CS1) $E_1(\kappa_n;y,z) = U(\kappa_n;z)\exp(i\phi_{n\kappa}y), \quad n = 1,2,3$
(the quasi-homogeneity condition w.r.t. the spatial variable y),

(CS2) $\phi_{n\kappa} = n\phi_\kappa, \quad n = 1,2,3$ (the condition of phase synchronism of waves),

(CS3) $\mathbf{E}_{\mathrm{tg}}(\kappa_n;y,z)$ and $\mathbf{H}_{\mathrm{tg}}(\kappa_n;y,z)$ (i.e. $E_1(\kappa_n;y,z)$ and $H_2(\kappa_n;y,z)$) are continuous at the boundary layers of the structure with the *induced* permittivity $\varepsilon_{n\kappa}$ for $\kappa := \kappa^{\mathrm{inc}}, \quad n = 1,2,3,$

(CS4) $E_1(\kappa_n;y,z) = \begin{Bmatrix} a_{\kappa_n} \\ b_{\kappa_n} \end{Bmatrix} \exp\left(i\left(\phi_{n\kappa}y \pm \Gamma_{\kappa_n}(\kappa_n,\phi_{n\kappa})(z \mp 2\pi\delta)\right)\right), \; z \gtrless \pm 2\pi\delta, \quad n = 1,2,3$
(the radiation condition w.r.t. the eigen-field).

For real values of the parameters κ_n and $\phi_{n\kappa}$, the condition (CS4) meets the physically reasonable requirement of the absence of radiation fields of waves coming from infinity $z = \pm\infty$:

$$\mathfrak{Im}\,\Gamma_{\kappa_n}(\kappa_n,\phi_{n\kappa}) \geq 0,\ \mathfrak{Re}\,\Gamma_{\kappa_n}(\kappa_n,\phi_{n\kappa})\,\mathfrak{Re}\,\kappa_n \geq 0 \quad \text{for} \quad \mathfrak{Im}\,\phi_{n\kappa} = 0,\ \mathfrak{Im}\,\kappa_n = 0, \quad n = 1,2,3. \qquad (63)$$

The non-trivial solutions (eigen-fields) of problem (62), (CS1) – (CS4) can be represented as

$$E_1(\kappa_n;y,z) = U(\kappa_n;z)\exp(i\phi_{n\kappa}y) = \begin{cases} a_{\kappa_n}\exp(i(\phi_{n\kappa}y + \Gamma_{\kappa_n}(\kappa_n,\phi_{n\kappa})(z - 2\pi\delta))), & z > 2\pi\delta, \\ U(\kappa_n;z)\exp(i\phi_{n\kappa}y), & |z| \leq 2\pi\delta, \\ b_{\kappa_n}\exp(i(\phi_{n\kappa}y - \Gamma_{\kappa_n}(\kappa_n,\phi_{n\kappa})(z + 2\pi\delta))), & z < -2\pi\delta, \end{cases}$$
$$\kappa_n \in \Omega_{n\kappa} \subset \mathsf{H}_{n\kappa}, \quad n = 1,2,3, \qquad (64)$$

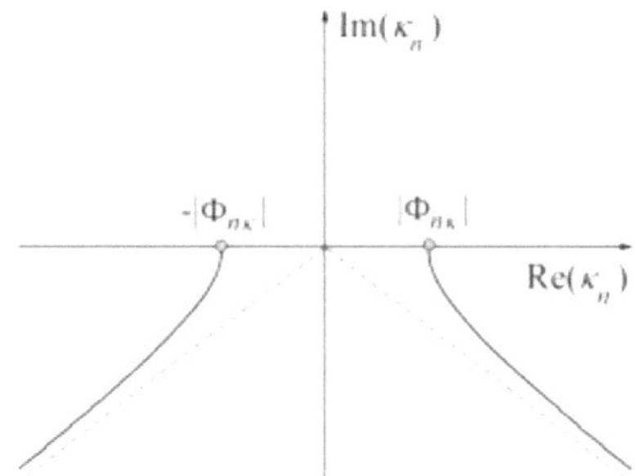

Fig. 2. The geometry of the two-sheeted Riemann surfaces $H_{n\kappa}$

where $\kappa := \kappa^{\mathrm{inc}}$ is a given constant equal to the value of the excitation frequency of the non-linear structure, $\Gamma_{\kappa_n}(\kappa_n, \phi_{n\kappa}) := (\kappa_n^2 - \phi_{n\kappa}^2)^{1/2}$ are the transverse propagation *functions* depending on the complex frequency spectral variables κ_n, $\phi_{n\kappa} := n\kappa \sin(\varphi_{n\kappa})$ denote the given real values of the longitudinal propagation constants,
$\varepsilon_{n\kappa} = \varepsilon_{n\kappa}(z, \alpha(z), E_1(\mathbf{r}, \kappa), E_1(\mathbf{r}, 2\kappa), E_1(\mathbf{r}, 3\kappa))$ are the induced dielectric permittivities at the frequencies $n\kappa$ corresponding to the parameter $\kappa := \kappa^{\mathrm{inc}}$, $\Omega_{n\kappa}$ are the sets of eigen-frequencies and $\mathsf{H}_{n\kappa}$ are two-sheeted Riemann surfaces (cf. Fig. 2), $n = 1, 2, 3$. The range of the spectral parameters $\kappa_n \in \Omega_{n\kappa}$ is completely determined by the boundaries of those regions in which the analytic continuation (consistent with the condition (63)) of the canonical Green's functions

$$G_0(\kappa_n; q, q_0) = \frac{i}{4Y} \exp\left\{i\left[\phi_{n\kappa}(y - y_0) + \Gamma_{\kappa_n}(\kappa_n, \phi_{n\kappa})\,|z - z_0|\right]\right\} / \Gamma_{\kappa_n}(\kappa_n, \phi_{n\kappa}), \quad n = 1, 2, 3,$$

(cf. (42)) into the complex space of the spectral parameters κ_n of the unperturbed problems (62), (CS1) – (CS4) (i.e. for the special case $\varepsilon_{n\kappa} \equiv 1$, $n = 1, 2, 3$) is possible. These complex spaces are two-sheeted Riemann surfaces $\mathsf{H}_{n\kappa}$ (see Fig. 2) with real algebraic branch points of second order $\kappa_n^{\pm} : \Gamma_{\kappa_n}(\kappa_n^{\pm}, \phi_{n\kappa}) = 0$ (i.e. $\kappa_n^{\pm} = \pm|\phi_{n\kappa}|$, $n = 1, 2, 3$) and with cuts starting at these points and extending along the lines

$$(\mathfrak{Re}\,\kappa_n)^2 - (\mathfrak{Im}\,\kappa_n)^2 - \phi_{n\kappa}^2 = 0, \quad \mathfrak{Im}\,\kappa_n \leq 0,\ n = 1, 2, 3. \tag{65}$$

The first, "physical" sheets (i.e. the pair of values $\{\kappa_n, \Gamma_{\kappa_n}(\kappa_n, \phi_{n\kappa})\}$) on each of the surfaces $\mathsf{H}_{n\kappa}$, $n = 1, 2, 3$, are completely determined by the condition (63) and the cuts (65). At the first sheets of $\mathsf{H}_{n\kappa}$ the signs of the pairs $\{\kappa_n, \mathfrak{Re}\,\Gamma_{\kappa_n}\}$ and $\{\kappa_n, \mathfrak{Im}\,\Gamma_{\kappa_n}\}$ are distributed as follows: $\mathfrak{Im}\,\Gamma_{\kappa_n} > 0$ for $0 < \arg \kappa_n < \pi$, $\mathfrak{Re}\,\Gamma_{\kappa_n} \geq 0$ for $0 < \arg \kappa_n < \pi/2$ and $\mathfrak{Re}\,\Gamma_{\kappa_n} \leq 0$ for $\pi/2 \leq \arg \kappa_n < \pi$. For points κ_n with $3\pi/2 \leq \arg \kappa_n \leq 2\pi$, the function values (where $(\mathfrak{Re}\,\kappa_n)^2 - (\mathfrak{Im}\,\kappa_n)^2 - \phi_{n\kappa^2} > 0$) are determined by the condition $\mathfrak{Im}\,\Gamma_{\kappa_n} < 0$, $\mathfrak{Re}\,\Gamma_{\kappa_n} > 0$, for the remaining points κ_n the function $\Gamma_{\kappa_n}(\kappa_n, \phi_{n\kappa})$ is determined by the condition $\mathfrak{Im}\,\Gamma_{\kappa_n} > 0$, $\mathfrak{Re}\,\Gamma_{\kappa_n} \leq 0$. In the region $\pi < \arg \kappa_n < 3\pi/2$ the situation is similar to the previous one up to the change of the sign of $\mathfrak{Re}\,\Gamma_{\kappa_n}$. The second, "unphysical" sheets of the surfaces $\mathsf{H}_{n\kappa}$, $n = 1, 2, 3$ are different from the "physical" ones in that, for each κ_n, the signs of both $\mathfrak{Re}\,\Gamma_{\kappa_n}$ and $\mathfrak{Im}\,\Gamma_{\kappa_n}$ are reversed.
The qualitative analysis of the eigen-modes of the linearised problems (62), (CS1) – (CS4) is carried out using the equivalent formulation of spectral problems for the linearised non-linear integral equations (49). It is based on the analytic continuation of (49) (see also (58)) into the space of spectral values $\kappa_n \in \Omega_{n\kappa} \subset \mathsf{H}_{n\kappa}$, $n = 1, 2, 3$.

The spectral problem reduces to finding non-trivial solutions $U(\kappa_n;z)$ of a set of homogeneous (i.e. with vanishing right-hand sides), linear (i.e. linearised equations (49)) integral equations with the induced dielectric permittivity at the frequencies $n\kappa$ of excitation and generation:

$$\begin{gathered} U(\kappa_n;z)+\frac{i\kappa_n^2}{2\Gamma_{\kappa_n}(\kappa_n,\phi_{n\kappa})}\int_{-2\pi\delta}^{2\pi\delta}\exp(i\Gamma_{\kappa_n}(\kappa_n,\phi_{n\kappa})|z-z_0|) \\ \times\ [1-\varepsilon_{n\kappa}(z_0,\alpha(z_0),U(\kappa;z_0),U(2\kappa;z_0),U(3\kappa;z_0))]\,U(\kappa_n;z_0)dz_0=0; \\ |z|\leq 2\pi\delta,\quad \kappa:=\kappa^{\text{inc}},\quad \kappa_n\in\Omega_{n\kappa}\subset \mathrm{H}_{n\kappa},\quad n=1,2,3. \end{gathered} \tag{66}$$

The solution of the problem (62), (CS1) – (CS4) can be obtained from the solution of the equivalent problem (66), where – according to (CS3) – in the representation of the eigen-fields (64) the following formulas are used:

$$U(\kappa_n;2\pi\delta)=a_{\kappa_n},\quad U(\kappa_n;-2\pi\delta)=b_{\kappa_n},\quad n=1,2,3. \tag{67}$$

The analyticity w.r.t. the argument $\kappa_n\in\mathrm{H}_{n\kappa}$, $n=1,2,3$, and the compactness of the operator functions (cf. (66)) $B_{n\kappa}(\kappa_n)\,[U(\kappa_n;\cdot)]:\ L_2(-2\pi\delta,2\pi\delta)\to L_2(-2\pi\delta,2\pi\delta)$, $n=1,2,3$, where

$$\begin{gathered} B_{n\kappa}(\kappa_n)\,[U(\kappa_n;z)]=-\frac{i\kappa_n^2}{2\Gamma_{\kappa_n}(\kappa_n,\phi_{n\kappa})}\int_{-2\pi\delta}^{2\pi\delta}\exp(i\Gamma_{\kappa_n}(\kappa_n,\phi_{n\kappa})|z-z_0|) \\ \times\ [1-\varepsilon_{n\kappa}(z_0,\alpha(z_0),U(\kappa;z_0),U(2\kappa;z_0),U(3\kappa;z_0))]\,U(\kappa_n;z_0)dz_0,\quad \kappa:=\kappa^{\text{inc}},\ n=1,2,3, \end{gathered} \tag{68}$$

are necessary conditions in the analytic Fredholm theorem (see (Reed & Simon, 1980, Thm. VI.14)). Taking into account that the resolvent set of (66) is non-empty in $\mathrm{H}_{n\kappa}$, the theorem implies that the resolvent operator $(I-B_{n\kappa}(\kappa_n))^{-1}$ (where I is the identity operator) exists and is a holomorphic operator function of the parameters $\kappa_n\in\mathrm{H}_{n\kappa}$, with the exception of not more than countable sets of isolated points $\Omega_{n\kappa}$, $n=1,2,3$ (i.e. sets that have no acumulation points in the finite part of each of the surfaces $\mathrm{H}_{n\kappa}$, $n=1,2,3$). In this case $(I-B_{n\kappa}(\kappa_n))^{-1}$ is meromorphic in $\mathrm{H}_{n\kappa}$, the residues at the poles are operators of finite rank and, if $\kappa_n\in\Omega_{n\kappa}$, then the equation (66) $B_{n\kappa}(\kappa_n)U=U$ has a non-trivial solution in $\mathrm{H}_{n\kappa}$, $n=1,2,3$. Summarizing the above discussion, we obtain the following result.

Theorem 3. *The spectra $\Omega_{n\kappa}$ of the problem (62), (CS1) – (CS4), and also of the equivalent problem (66) for the dielectric layer with the induced piecewise continuous permittivity at the frequencies $n\kappa$ of excitation and generation, consist of not more than countable sets of isolated points, without accumulation points in the finite part of each of the surfaces $\mathrm{H}_{n\kappa}$, $n=1,2,3$. The resolvents of the spectral problems at these points are poles of finite order.*

9. Algorithm for the numerical analysis of the eigen-modes of the linearised problems

The qualitative analysis of the spectral characteristics allows to develop algorithms for solving the spectral problems (62), (CS1) – (CS4) by reducing them to the equivalent spectral problem of finding non-trivial solutions of the integral equations (66), see Shestopalov & Yatsyk (1997), Yatsyk (2000). The solvability of (66) follows from an analysis of the basic qualitative characteristics of the spectra. Applying to the integral equations (66) appropriate quadrature formulas, we obtain a set of independent systems of linear algebraic equations of second kind depending non-linearly on the spectral parameter: $(\mathbf{I}-\mathbf{B}_{n\kappa}(\kappa_n))\mathbf{U}_{\kappa_n}=\mathbf{0}$, where $\kappa_n\in\mathrm{H}_{n\kappa}$, $\kappa:=\kappa^{\text{inc}}$, $n=1,2,3$. Consequently, the spectral problem of finding the eigen-frequencies

$\kappa_n \in \Omega_{n\kappa} \subset \mathrm{H}_{n\kappa}$ and the corresponding eigen-fields (i.e. the non-trivial solutions of the integral equations (66)) reduces to the following algorithm:

$$\begin{cases} f_{n\kappa}(\kappa_n) := \det(\mathbf{I} - \mathbf{B}_{n\kappa}(\kappa_n)) = 0, \\ (\mathbf{I} - \mathbf{B}_{n\kappa}(\kappa_n))\mathbf{U}_{\kappa_n} \qquad = \mathbf{0}, \\ \qquad \kappa := \kappa^{\text{inc}}, \quad \kappa_n \in \Omega_{n\kappa} \subset \mathrm{H}_{n\kappa}, \quad n = 1,2,3. \end{cases} \tag{69}$$

Here we use a similar notation to that in Section 7. κ_n are the desired eigen-frequencies, and $\mathbf{U}_{\kappa_n} = \{U(\kappa_n; z_l)\}_{l=1}^N := \{U_l(\kappa_n)\}_{l=1}^N$ are the vectors of the unknown approximate solution values corresponding to the frequencies κ_n. The matrices are of the form

$$\mathbf{B}_{n\kappa}(\kappa_n) := \mathbf{B}_{n\kappa}(\kappa_n; \mathbf{U}_\kappa, \mathbf{U}_{2\kappa}, \mathbf{U}_{3\kappa}) = \{A_m K_{lm}(\kappa_n, \mathbf{U}_\kappa, \mathbf{U}_{2\kappa}, \mathbf{U}_{3\kappa})\}_{l,m=1}^N \tag{70}$$

with *given* values of the vectors of the scattered and generated fields $\mathbf{U}_{n\kappa} = \{U(n\kappa; z_l)\}_{l=1}^N := \{U_l(n\kappa)\}_{l=1}^N$, $n = 1,2,3$. The numbers A_m are the coefficients determined by the quadrature rule, and the entries $K_{lm}(\kappa_n, \mathbf{U}_\kappa, \mathbf{U}_{2\kappa}, \mathbf{U}_{3\kappa})$ are calculated by means of (60), where the first argument $n\kappa$ is replaced by κ_n. The eigen-frequencies $\kappa_n \in \Omega_{n\kappa} \subset \mathrm{H}_{n\kappa}$, $n = 1,2,3$, i.e. the characteristic numbers of the dispersion equations of the problem (69), are obtained by solving the corresponding dispersion equations $f_{n\kappa}(\kappa_n) := \det(\mathbf{I} - \mathbf{B}_{n\kappa}(\kappa_n)) = 0$ by the help of Newton's method or its modifications. The non-trivial solutions $\mathbf{U}_{\kappa_n}$ of the homogeneous systems of linear algebraic equations (69) corresponding to these characteristic numbers are the eigen-fields (64) of the linearised non-linear layered structures with an induced dielectric constant (52). Since the solutions $\mathbf{U}_{\kappa_n}$ are unique up to multiplication by an arbitrary constant, we require $U(\kappa_n; 2\pi\delta) = a_{\kappa_n} := 1$ (cf. (64)). According to (70), the matrix entries in (69) depend on the dielectric permittivities. The latter are defined by the scattered and generated fields $\mathbf{U}_\kappa, \mathbf{U}_{2\kappa}, \mathbf{U}_{3\kappa}$ of the problem (49) by means of the algorithm (61). This defines the basic design of the implemented numerical algorithm. The investigation of the eigen-modes of the linearised non-linear structures (69) should always precede the solution of the non-linear scattering and generation problem in the self-consistent formulation (61). Note that, in the analysis of the linear structures, the problem of excitation (scattering) and the spectral problem can be solved independently.

In physical applications, a very useful theorem (see (Sánchez-Palencia, 1980, Thm. 7.2)) asserts the continuous dependence of the operator (68) (or (70)) of the spectral problem on some non-spectral parameter τ of the problem under consideration, i.e. $B_{n\kappa}(\kappa_n, \tau)$ (or $\mathbf{B}_{n\kappa}(\kappa_n, \tau)$). In particular, this theorem implies that the characteristic numbers $\kappa_n(\tau)$, i.e. the poles of $(I - B_{n\kappa}(\kappa_n, \tau))^{-1}$ (or $(\mathbf{I} - \mathbf{B}_{n\kappa}(\kappa_n, \tau))^{-1}$) continuously depend on τ and, therefore, they may appear or disappear only at the boundary of any given open, connected region $D \subset \mathrm{H}_{n\kappa}$, $n = 1,2,3$. Further, we interpret $\kappa_n(\tau) \in \Omega_{n\kappa}(\tau) \subset \mathrm{H}_{n\kappa}$ as a branch of the dispersion curves in the eigen-fields $U(\kappa_n(\tau), z)$ of the problem under investigation. Finally we mention that the classification of scattered, generated or eigen-fields of the dielectric layer by the $H_{m,l,p}$-type adopted in our paper is identical to that given in Shestopalov & Sirenko (1989), Shestopalov & Yatsyk (1997), Yatsyk (2000). In the case of E-polarisation, see (12), $H_{m,l,p}$ (or $TE_{m,l,p}$) denotes the type of polarisation of the wave field under investigation. The subscripts indicate the number of local maxima of $|E_1|$ (or $|U|$, as $|E_1| = |U|$, see (23), (64)) along the coordinate axes x, y and z (see Fig. 1). Since the considered waves are homogeneous along the x-axis and quasi-homogeneous along the y-axis, we study actually fields of the type $H_{0,0,p}$ (or $TE_{0,0,p}$), where the subscript p is equal to the number of local maxima of the function $|U|$ of the argument $z \in [-2\pi\delta, 2\pi\delta]$.

10. Numerical analysis. Third-harmonic generation by resonant scattering of a wave on a layer with negative and positive values of the cubic susceptibility

Consider the excitation of the non-linear structure by a strong electromagnetic field at the basic frequency only (see (20)), i.e.

$$\{E_1^{\mathrm{inc}}(\kappa;q) \neq 0,\ E_1^{\mathrm{inc}}(2\kappa;q) = 0,\ E_1^{\mathrm{inc}}(3\kappa;q) = 0\}, \quad \text{where} \quad \{a_\kappa^{\mathrm{inc}} \neq 0,\ a_{2\kappa}^{\mathrm{inc}} = a_{3\kappa}^{\mathrm{inc}} = 0\}. \quad (71)$$

In this case, the number of equations in the systems can be reduced. The second equations in all the systems (21), (24) and (49), corresponding to a problem at the double frequency 2κ with a trivial right-hand side, can be eliminated by setting $E_1(\mathbf{r}, 2\kappa) := 0$ (cf. Angermann & Yatsyk (2010), Angermann & Yatsyk (2011)). The dielectric permittivity of the non-linear layer (cf. (19)) in the case (71) simplifies to

$$\begin{aligned}
&\varepsilon_{n\kappa}(z, \alpha(z), E_1(\mathbf{r},\kappa), 0, E_1(\mathbf{r}, 3\kappa)) = \varepsilon_{n\kappa}(z, \alpha(z), U(\kappa; z), U(3\kappa; z)) \\
&=: \varepsilon^{(L)}(z) + \varepsilon_{n\kappa}^{(NL}(\alpha(z), U(\kappa; z), U(3\kappa; z)) \\
&= \varepsilon^{(L)}(z) + \alpha(z)\left[|U(\kappa; z)|^2 + |U(3\kappa; z)|^2\right] \\
&\quad + \delta_{n,1}\alpha(z)|U(\kappa; z)||U(3\kappa; z)| \exp\left[i\left\{-3\arg U(\kappa; z) + \arg U(3\kappa; z)\right\}\right], \quad n = 1, 3.
\end{aligned} \quad (72)$$

The desired solution of the scattering and generation problem (21), (C1) – (C4) (or of the equivalent problems (24) and (49)) can be represented as follows (cf. (23)):

$$\begin{aligned}
&E_1(n\kappa; y, z) = U(n\kappa; z)\exp(i\phi_{n\kappa} y) \\
&= \begin{cases} \delta_{n1} a_{n\kappa}^{\mathrm{inc}} \exp(i(\phi_{n\kappa} y - \Gamma_{n\kappa}(z - 2\pi\delta))) + a_{n\kappa}^{\mathrm{scat}} \exp(i(\phi_{n\kappa} y + \Gamma_{n\kappa}(z - 2\pi\delta))), & z > 2\pi\delta, \\ U(n\kappa; z)\exp(i\phi_{n\kappa} y), & |z| \leq 2\pi\delta, \\ b_{n\kappa}^{\mathrm{scat}} \exp(i(\phi_{n\kappa} y - \Gamma_{n\kappa}(z + 2\pi\delta))), & z < -2\pi\delta, \end{cases} \\
&\qquad n = 1, 3,
\end{aligned} \quad (73)$$

where $U(\kappa; z)$, $U(3\kappa; z)$, $|z| \leq 2\pi\delta$, are the solutions of the reduced systems (24) or (49). According to (25) we determine the values of complex amplitudes $\{a_{n\kappa}^{\mathrm{scat}}, b_{n\kappa}^{\mathrm{scat}} : n = 1, 3\}$ in (73) for the scattered and generated fields by means of the formulas

$$U(n\kappa; 2\pi\delta) = \delta_{n1} a_{n\kappa}^{\mathrm{inc}} + a_{n\kappa}^{\mathrm{scat}}, \quad U(n\kappa; -2\pi\delta) = b_{n\kappa}^{\mathrm{scat}}, \quad n = 1, 3. \quad (74)$$

According to the results of Section 7,the solution of (21), (C1) – (C4) reduces in the case of (71) to the following system (cf. (59)):

$$\begin{cases} (\mathbf{I} - \mathbf{B}_\kappa(\mathbf{U}_\kappa, \mathbf{U}_{3\kappa}))\mathbf{U}_\kappa = \mathbf{U}_\kappa^{\mathrm{inc}}, \\ (\mathbf{I} - \mathbf{B}_{3\kappa}(\mathbf{U}_\kappa, \mathbf{U}_{3\kappa}))\mathbf{U}_{3\kappa} = \mathbf{C}_{3\kappa}(\mathbf{U}_\kappa). \end{cases} \quad (75)$$

The system (75) is written taking into account (71), i.e. $\mathbf{U}_{2\kappa}^{\mathrm{inc}} = \mathbf{0}$, $\mathbf{U}_{3\kappa}^{\mathrm{inc}} = \mathbf{0}$, $\mathbf{U}_{2\kappa} = \mathbf{0}$. Here (cf. (59)) $\mathbf{B}_{n\kappa}(\mathbf{U}_\kappa, \mathbf{U}_{3\kappa}) = \mathbf{B}_{n\kappa}(\mathbf{U}_\kappa, \mathbf{0}, \mathbf{U}_{3\kappa})$, $n = 1, 3$, denote the matrices of the complex-valued non-linear algebraic equations, and $\mathbf{U}_\kappa^{\mathrm{inc}}$, $\mathbf{C}_{3\kappa}(\mathbf{U}_\kappa) = \mathbf{C}_{3\kappa}(\mathbf{U}_\kappa, \mathbf{0})$, $\mathbf{C}_\kappa(\mathbf{0}, \mathbf{U}_\kappa) = \mathbf{0}$ are the right-hand side vectors. The solution of (75) is obtained by means of successive approximations using the self-consistent approach based on the iterative algorithm (61).

In order to describe the scattering and generation properties of the non-linear structure in the zones of reflection $z > 2\pi\delta$ and transmission $z < -2\pi\delta$, we introduce the following notation:

$$R_{n\kappa} := |a_{n\kappa}^{\text{scat}}|^2 / |a_{\kappa}^{\text{inc}}|^2 \quad \text{and} \quad T_{n\kappa} := |b_{n\kappa}^{\text{scat}}|^2 / |a_{\kappa}^{\text{inc}}|^2.$$

The quantities $R_{n\kappa}$, $T_{n\kappa}$ are called *reflection, transmission* or *generation coefficients* of the waves w.r.t. the intensity of the excitation field.
We note that in the considered case of the excitation $\{a_{\kappa}^{\text{inc}} \neq 0, a_{2\kappa}^{\text{inc}} = 0, a_{3\kappa}^{\text{inc}} = 0\}$ and for non-absorbing media with $\mathfrak{Im}\left[\varepsilon^{(L)}(z)\right] = 0$, the energy balance equation $R_{\kappa} + T_{\kappa} + R_{3\kappa} + T_{3\kappa} = 1$ is satisfied. This equation represents the law of conservation of energy (Shestopalov & Sirenko (1989), Vainstein (1988)). The quantity $W_{3\kappa}/W_{\kappa}$, which characterises the portion of energy generated in the third harmonic in comparison to the energy scattered in the non-linear layer, is of particular interest. Here by $W_{n\kappa} = |a_{n\kappa}^{\text{scat}}|^2 + |b_{n\kappa}^{\text{scat}}|^2$ we denote the total energy of the scattered and generated fields at the frequencies $n\kappa$, $n = 1, 3$.
The spectral characteristics of the linearised non-linear problems (62), (CS1) – (CS4) with the induced dielectric permittivity (72) at the frequency κ of excitation and the frequency 3κ of generation were calculated by means of the algorithm (69). In the graphical illustration of the eigen-fields $\mathbf{U}_{\kappa_n}$ in the representation (64) we have set $a_{\kappa_n} := 1$ for $\kappa_n \in \Omega_{n\kappa} \subset \mathsf{H}_{n\kappa}$, $n = 1, 3$.
In what follows we want to discuss some results of the numerical analysis of scattering and generation properties of cubic non-linear polarisable layers with both negative and positive values of the cubic susceptibility of the medium. We consider non-linear dielectric layers (see Fig. 1) with a dielectric permittivity $\varepsilon_{n\kappa}\left(z, \alpha(z), U(\kappa; z), U(3\kappa; z)\right) = \varepsilon^{(L)}(z) + \varepsilon_{n\kappa}^{(NL}$ of the form (72), where

$$\left\{\varepsilon^{(L)}(z), \alpha(z)\right\} = \left\{\varepsilon^{(L)} = 16,\ \alpha = \mp 0.01,\ z \in [-2\pi\delta, 2\pi\delta]\right\}$$

with the parameter $\delta := 0.5$, the excitation frequency $\kappa^{\text{inc}} := \kappa := 0.375$, the generation frequency of the third harmonic field $\kappa^{\text{gen}} := 3\kappa := 1.125$, and the angle of incidence of the plane wave $\varphi_{\kappa} \in [0°, 90°)$.

10.1 A non-linear layer with a negative value of the cubic susceptibility of the medium

The results of the numerical analysis of scattering and generation properties as well as the eigen-modes of the dielectric layer with a negative value of the cubic susceptibility of the medium ($\alpha = -0.01$) are presented in Fig. 3 – Fig. 9.

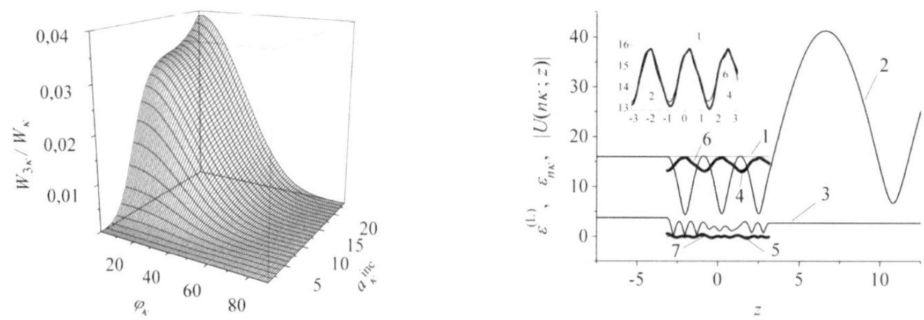

Fig. 3. The portion of energy generated in the third harmonic (left) and some graphs describing the properties of the non-linear layer at $a_{\kappa}^{\text{inc}} = 24$ and $\varphi_{\kappa} = 0°$ (right): #1 ... $\varepsilon^{(L)}$, #2 ... $|U(\kappa; z)|$, #3 ... $|U(3\kappa; z)|$, #4 ... $\mathfrak{Re}(\varepsilon_{\kappa})$, #5 ... $\mathfrak{Im}(\varepsilon_{\kappa})$, #6 ... $\mathfrak{Re}(\varepsilon_{3\kappa})$, #7 ... $\mathfrak{Im}(\varepsilon_{3\kappa}) \equiv 0$

Fig. 3 (left) shows the dependence of $W_{3\kappa}/W_\kappa$ on the angle of incidence φ_κ and on the amplitude a_κ^{inc} of the incident field. It describes the portion of energy generated in the third harmonic by the non-linear layer when a plane wave at the excitation frequency κ and with the amplitude a_κ^{inc} is passing the layer under the angle of incidence φ_κ. In particular, $W_{3\kappa}/W_\kappa = 0.039$ at $a_\kappa^{\mathrm{inc}} = 24$ and $\varphi_\kappa = 0°$, i.e. $W_{3\kappa}$ amounts to 3.9% of the total energy W_κ scattered at the frequency of excitation κ. Fig. 3 (right) displays some graphs characterising the scattering and generation properties of the non-linear structure. Graphs #4 and #5 show the real and imaginary parts of the permittivity at the frequency of excitation, while graphs #6 and #7 display the corresponding values at the generation frequency. The figure also shows the absolute values $|U(\kappa;z)|$ of the amplitudes of the full scattered field $H_{0,0,4}$ at the frequency of excitation κ (graph #2) and $|U(3\kappa;z)|$ of the generated field of the $H_{0,0,9}$-type at the frequency 3κ (graph #3). The values $|U(\kappa;z)|$ and $|U(3\kappa;z)|$ are given in the non-linear layered structure ($|z| \le 2\pi\delta$) and outside it (i.e. in the zones of reflection $z > 2\pi\delta$ and transmission $z < -2\pi\delta$).

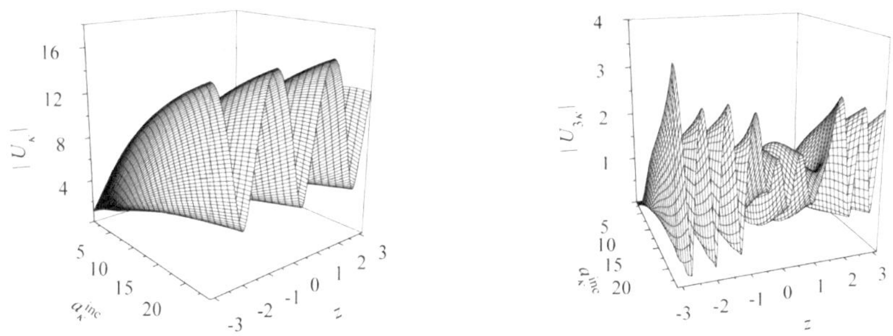

Fig. 4. Graphs of the scattered and generated fields in the non-linear layered structure for $\varphi_\kappa = 0°$: $\left|U_\kappa\left[a_\kappa^{\mathrm{inc}}, z\right]\right|$ (left), $\left|U_{3\kappa}\left[a_\kappa^{\mathrm{inc}}, z\right]\right|$ (right)

Figs. 4 and 5 show the numerical results obtained for the scattered and the generated fields in the non-linear structure and for the non-linear dielectric permittivity of the layered structure in dependence on the amplitude a_κ^{inc} at normal incidence $\varphi_\kappa = 0°$ of the plane wave.
Fig. 4 shows the graphs of $\left|U_\kappa\left[a_\kappa^{\mathrm{inc}}, z\right]\right|$ and $\left|U_{3\kappa}\left[a_\kappa^{\mathrm{inc}}, z\right]\right|$ demonstrating the dynamic behaviour of the scattered and the generated fields $|U(\kappa;z)|$ and $|U(3\kappa;z)|$ in the non-linear layered structure in dependence on an increasing amplitude a_κ^{inc} at normal incidence $\varphi_\kappa = 0°$ of the plane wave of the frequency κ. We mention that, in the range of amplitudes $a_\kappa^{\mathrm{inc}} \in (0, 24]$ under consideration, the scattered field is of the type $H_{0,0,4}$, see Fig. 4 (left). The generation of the third harmonic field can be observed within the range $a_\kappa^{\mathrm{inc}} \in [4, 24]$, see Fig. 4 (right). The generated field has the type $H_{0,0,10}$ for $a_\kappa^{\mathrm{inc}} \in [4, 23)$, and $H_{0,0,9}$ for $a_\kappa^{\mathrm{inc}} \in [23, 24]$. *The change of type* of the generated field from $H_{0,0,10}$ to $H_{0,0,9}$ for an increasing amplitude a_κ^{inc} is due to the loss of one local maximum of the function $|U(3\kappa;z)|$, $z \in [-2\pi\delta, 2\pi\delta]$, at $a_\kappa^{\mathrm{inc}} = 23$ (see the point with coordinates ($a_\kappa^{\mathrm{inc}} = 23$, $z = 1.15$, $|U_{3\kappa}| = 1.61$) in Fig. 4 (right)).
The non-linear parts $\varepsilon_{n\kappa}^{(NL)}$ of the dielectric permittivity at each frequency κ and 3κ depend on the values $U_\kappa := U(\kappa;z)$ and $U_{3\kappa} := U(3\kappa;z)$ of the fields, see (72). The variation of the non-linear parts $\varepsilon_{n\kappa}^{(NL)}$ of the dielectric permittivity for an increasing amplitude a_κ^{inc} of the incident field are illustrated by the behaviour of $\mathfrak{Re}\left(\varepsilon_\kappa\left[a_\kappa^{\mathrm{inc}}, z\right]\right)$ (Fig. 5 (top left)) and $\mathfrak{Im}\left(\varepsilon_\kappa\left[a_\kappa^{\mathrm{inc}}, z\right]\right)$ (Fig. 5 (top right)) at the frequency κ, and by $\varepsilon_{3\kappa}\left[a_\kappa^{\mathrm{inc}}, z\right]$ at the triple frequency 3κ (Fig. 5 (bottom left)). In Fig. 5 (top right) the graph of $\mathfrak{Im}\left(\varepsilon_\kappa\left[a_\kappa^{\mathrm{inc}}, z\right]\right)$ for a given amplitude a_κ^{inc} characterises the *loss of energy* in the non-linear medium (at the frequency of excitation κ)

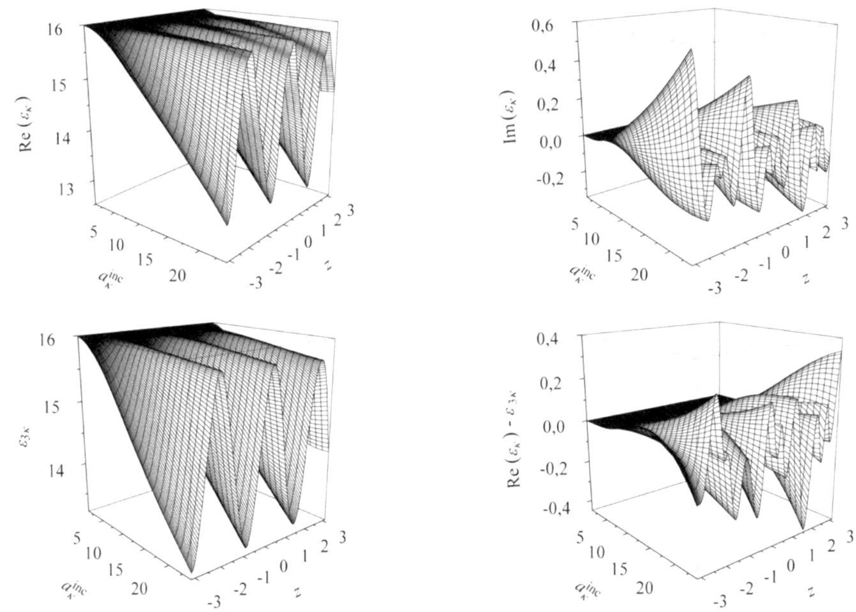

Fig. 5. Graphs characterising the non-linear dielectric permittivity at normal incidence of the plane wave $\varphi_\kappa = 0^\circ$: $\mathfrak{Re}\left(\varepsilon_\kappa\left[a_\kappa^{\text{inc}}, z\right]\right)$ (top left), $\mathfrak{Im}\left(\varepsilon_\kappa\left[a_\kappa^{\text{inc}}, z\right]\right)$ (top right), $\varepsilon_{3\kappa}\left[a_\kappa^{\text{inc}}, z\right]$ (bottom left), $\mathfrak{Re}\left(\varepsilon_\kappa\left[a_\kappa^{\text{inc}}, z\right]\right) - \varepsilon_{3\kappa}\left[a_\kappa^{\text{inc}}, z\right]$ (bottom right)

caused by the *generation* of the electromagnetic field of the third harmonic (at the frequency 3κ). In our case $\mathfrak{Im}\left[\varepsilon^{(L)}(z)\right] = 0$ and $\mathfrak{Im}\left[\alpha(z)\right] = 0$, therefore, according to (72),

$$\mathfrak{Im}(\varepsilon_\kappa) = \alpha(z)|U(\kappa;z)||U(3\kappa;z)|\mathfrak{Im}\left(\exp\left[i\left\{-3\arg U(\kappa;z) + \arg U(3\kappa;z)\right\}\right]\right). \qquad (76)$$

From Fig. 5 (top right) we see that a small value of a_κ^{inc} induces a small amplitude of the function $\mathfrak{Im}(\varepsilon_\kappa)$, i.e. $|\mathfrak{Im}(\varepsilon_\kappa)| \approx 0$. The increase of a_κ^{inc} corresponds to a strong field excitation and leads to the generation of a third harmonic field $U(3\kappa;z)$. Fig. 5 (top right) shows the dynamic behaviour of $\mathfrak{Im}(\varepsilon_\kappa)$. It can be seen that the values of $\mathfrak{Im}(\varepsilon_\kappa)$ may be *positive or negative* along the height of the non-linear layer, i.e. in the interval $z \in [-2\pi\delta, 2\pi\delta]$. The zero values of $\mathfrak{Im}(\varepsilon_\kappa)$ are determined by the phase relation between the scattered and the generated fields $U(\kappa;z)$, $U(3\kappa;z)$ in the non-linear layer, see (76), $-3\arg U(\kappa;z) + \arg U(3\kappa;z) = p\pi, \quad p = 0, \pm 1, \ldots$ We mention that the behaviour of both the quantities $\mathfrak{Im}(\varepsilon_\kappa)$ and

$$\mathfrak{Re}(\varepsilon_\kappa) - \varepsilon_{3\kappa} = \alpha(z)|U(\kappa;z)||U(3\kappa;z)|\mathfrak{Re}\left(\exp\left[i\left\{-3\arg U(\kappa;z) + \arg U(3\kappa;z)\right\}\right]\right)$$

plays an essential role in the process of third harmonic generation because of the presence of the last term in (72). Fig. 5 (bottom right) shows the graph describing the behaviour of $\mathfrak{Re}\left(\varepsilon_\kappa\left[a_\kappa^{\text{inc}}, z\right]\right) - \varepsilon_{3\kappa}\left[a_\kappa^{\text{inc}}, z\right]$.

The scattering and generation properties of the non-linear structure in the range $\varphi_\kappa \in [0^\circ, 90^\circ)$, $a_\kappa^{\text{inc}} \in [1, 24]$ of the parameters of the excitation field are presented in Figs. 6 – 7. The graphs show the dynamics of the scattering ($R_\kappa\left[\varphi_\kappa, a_\kappa^{\text{inc}}\right]$, $T_\kappa\left[\varphi_\kappa, a_\kappa^{\text{inc}}\right]$, see Fig. 6 (top)) and generation ($R_{3\kappa}\left[\varphi_\kappa, a_\kappa^{\text{inc}}\right]$, $T_{3\kappa}\left[\varphi_\kappa, a_\kappa^{\text{inc}}\right]$, see Fig. 6 (bottom)) properties of the structure. Fig. 7

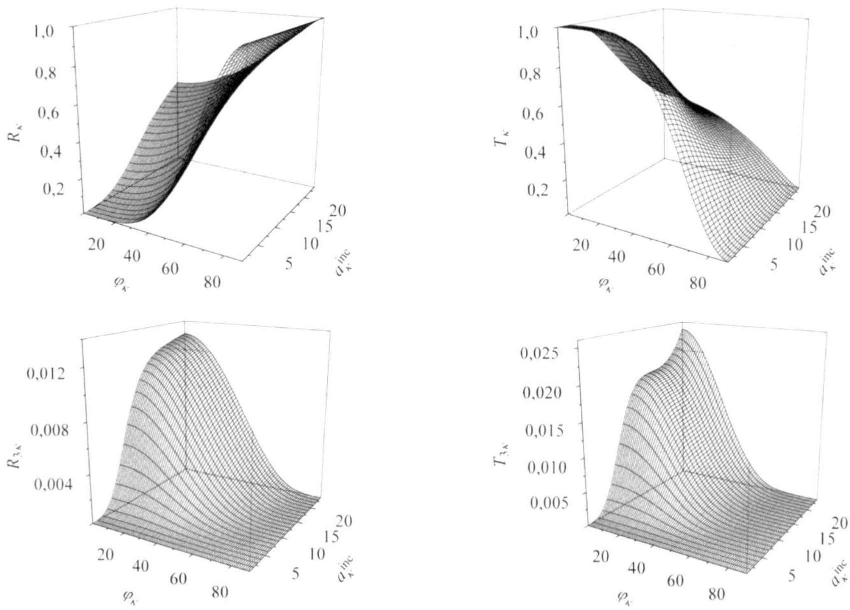

Fig. 6. The scattering and generation properties of the non-linear structure: $R_\kappa\left[\varphi_\kappa, a_\kappa^{\text{inc}}\right]$ (top left), $T_\kappa\left[\varphi_\kappa, a_\kappa^{\text{inc}}\right]$ (top right), $R_{3\kappa}\left[\varphi_\kappa, a_\kappa^{\text{inc}}\right]$ (bottom left), $T_{3\kappa}\left[\varphi_\kappa, a_\kappa^{\text{inc}}\right]$ (bottom right)

shows cross sections of the graphs depicted in Figs. 6 and 3 by the planes $\varphi_\kappa = 0^\circ$ and $a_\kappa^{\text{inc}} = 20$.

In the resonant range of wave scattering and generation frequencies, i.e. $\kappa^{\text{scat}} := \kappa^{\text{inc}} = \kappa$ and $\kappa^{\text{gen}} = 3\kappa$, resp., the dynamic behaviour of the characteristic quantities depicted in Figs. 6 – 7 has the following causes. The scattering and generation frequencies are close to the corresponding eigen-frequencies of the linear ($\alpha = 0$) and linearised non-linear ($\alpha \neq 0$) spectral problems (62), (CS1) – (CS4). Furthermore, the distance between the corresponding eigen-frequencies of the spectral problems with $\alpha = 0$ and $\alpha \neq 0$ is small. Thus, the graphs in Fig. 7 can be compared with the dynamic behaviour of the branches of the eigen-frequencies of the spectral problems presented in Fig. 8. The graphs of the eigen-fields corresponding to the branches of the considered eigen-frequencies are shown in Fig. 9.

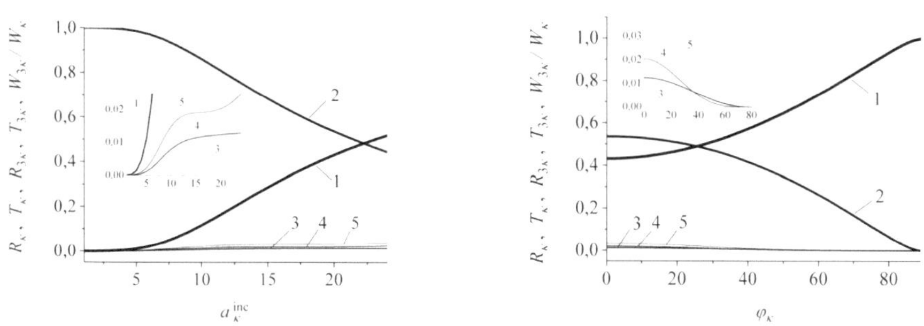

Fig. 7. The curves R_κ (#1), T_κ (#2), $R_{3\kappa}$ (#3), $T_{3\kappa}$ (#4), $W_{3\kappa}/W_\kappa$ (#5) for $\varphi_\kappa = 0^\circ$ (left) and $a_\kappa^{\text{inc}} = 20$ (right)

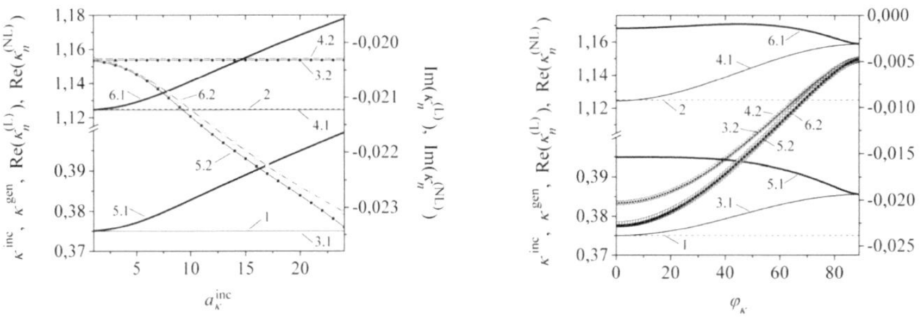

Fig. 8. The curves $\kappa := \kappa^{\text{inc}} := 0.375$ (#1), $3\kappa = \kappa^{\text{gen}} = 3\kappa^{\text{inc}} = 1.125$ (#2), the complex eigen-frequencies $\mathfrak{Re}(\kappa_1^{(L)})$ (#3.1), $\mathfrak{Im}(\kappa_1^{(L)})$ (#3.2), $\mathfrak{Re}(\kappa_3^{(L)})$ (#4.1), $\mathfrak{Im}(\kappa_3^{(L)})$ (#4.2) of the linear problem ($\alpha = 0$) and $\mathfrak{Re}(\kappa_1^{(NL)})$ (#5.1), $\mathfrak{Im}(\kappa_1^{(NL)})$ (#5.2), $\mathfrak{Re}(\kappa_3^{(NL)})$ (#6.1), $\mathfrak{Im}(\kappa_3^{(NL)})$ (#6.2) of the linearised non-linear problem ($\alpha = -0.01$) for $\varphi_\kappa = 0°$ (left) and $a_\kappa^{\text{inc}} = 20$ (right)

Fig. 8 illustrates the dispersion characteristics of the linear ($\alpha = 0$) and the linearised non-linear ($\alpha = -0.01$) layer $\varepsilon_{n\kappa} = \varepsilon^{(L)} + \varepsilon_{n\kappa}^{(NL}$, $n = 1, 3$, see (72). The non-linear components of the permittivity at the scattering (excitation) frequencies $\kappa^{\text{scat}} := \kappa^{\text{inc}} = \kappa$ and the generation frequencies $\kappa^{\text{gen}} := \kappa$ depend on the amplitude a_κ^{inc} and the angle of incidence φ_κ of the incident field. This is reflected in the dynamics of the behaviour of the complex-valued eigen-frequencies of the linear and the linearised non-linear layer. Comparing the results shown in Fig. 8 and Fig. 7, we note the following. The dynamics of the change of the scattering properties R_κ, T_κ of the non-linear layer (compare the behaviour of curves #1 and #2 in Fig. 7) depends on the magnitude of the distance between the curves #3.1 and #5.1 in Fig. 8. Decanelising properties of the layer occur when $\alpha < 0$. A previously transparent (Fig. 7 (left)) or reflective (Fig. 7 (right)) structure loses its properties. It becomes transparent and the reflection and transmission coefficients become comparable. The greater the distance between the curves #4.1 and #6.1 (see Fig. 8), the greater the values of $R_{3\kappa}$, $T_{3\kappa}$, $W_{3\kappa}/W_\kappa$, characterising the generating properties of the non-linear layer, see Fig. 7. The magnitudes of the absolute

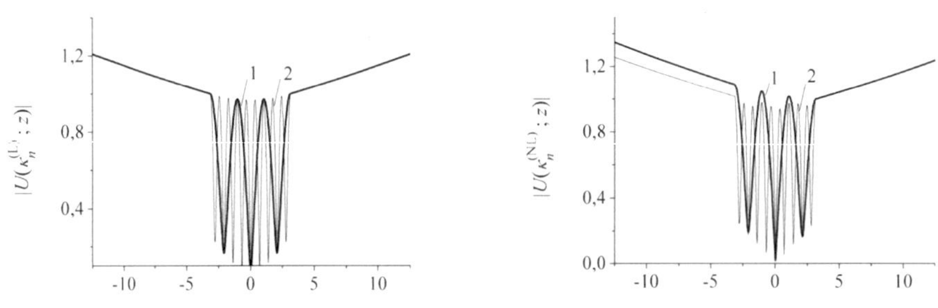

Fig. 9. The graphs of the eigen-fields of the layer for $\varphi_\kappa = 0°$, $a_\kappa^{\text{inc}} = 20$. The linear problem ($\alpha = 0$, left figure): $|U(\kappa_1^{(L)}; z)|$ with $\kappa_1^{(L)} = 0.3749822 - i\,0.02032115$ (#1), $|U(\kappa_3^{(L)}; z)|$ with $\kappa_3^{(L)} = 1.124512 - i\,0.02028934$ (#2), the linearised non-linear problem ($\alpha = -0.01$, right figure): $|U(\kappa_1^{(NL)}; z)|$ with $\kappa_1^{(NL)} = 0.3949147 - i\,0.02278218$ (#1), $|U(\kappa_3^{(NL)}; z)|$ with $\kappa_3^{(NL)} = 1.168264 - i\,0.02262382$ (#2)

values of the eigen-fields shown in Fig. 9 correspond to the branches of the eigen-frequencies of the linear and the linearised non-linear spectral problems, see Fig. 8. The curves in Fig. 9 are labeled by #1 for an eigen-field of type $H_{0,0,4}$ and by #2 for an eigen-field of type $H_{0,0,10}$. The loss of symmetry in the eigen-fields with respect to the z-axis in Fig. 9 (right) is due to the violation of the symmetry (w.r.t. the axis $z = 0$) in the induced dielectric permittivity at both the scattering (excitation) and the oscillation frequencies, see Fig. 5.

10.2 A non-linear layer with a positive value of the cubic susceptibility of the medium

The results of the numerical analysis of scattering and generation properties as well as the eigen-modes of the dielectric layer with a positive value of the cubic susceptibility of the medium ($\alpha = +0.01$) are presented in Fig. 10 – Fig. 16.

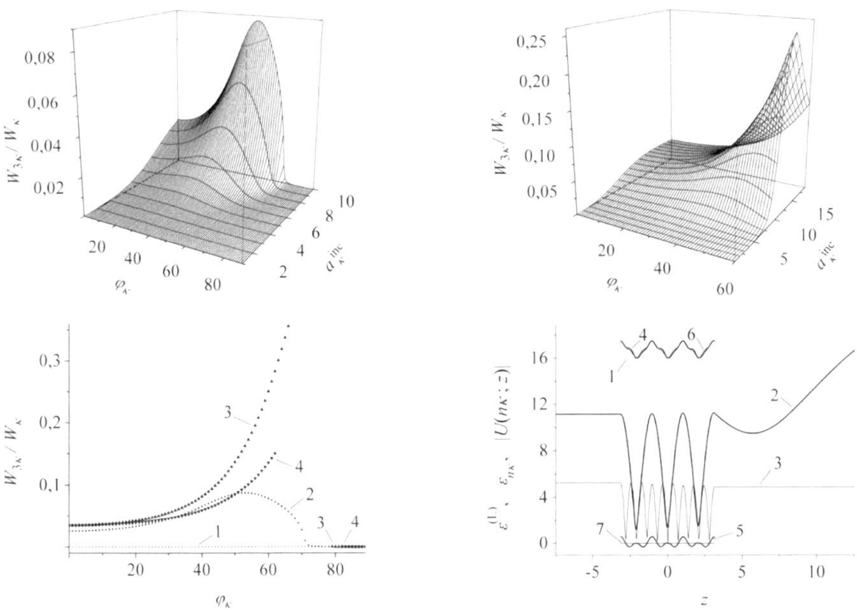

Fig. 10. The portion of energy generated in the third harmonic (top left/right and bottom left): #1 ... $a_\kappa^{\text{inc}} = 1$, #2 ... $a_\kappa^{\text{inc}} = 9.93$, #3 ... $a_\kappa^{\text{inc}} = 14$, #4 ... $a_\kappa^{\text{inc}} = 19$, and some graphs describing the properties of the non-linear layer for $a_\kappa^{\text{inc}} = 14$ and $\varphi_\kappa = 66^\circ$ (bottom right): #1 ... $\varepsilon^{(L)}$, #2 ... $|U(\kappa;z)|$, #3 ... $|U(3\kappa;z)|$, #4 ... $\mathfrak{Re}(\varepsilon_\kappa)$, #5 ... $\mathfrak{Im}(\varepsilon_\kappa)$, #6 ... $\mathfrak{Re}(\varepsilon_{3\kappa})$, #7 ... $\mathfrak{Im}(\varepsilon_{3\kappa}) \equiv 0$

The results shown in Fig. 10 (top left/right and bottom left) allow us to track the dynamic behaviour of the quantity $W_{3\kappa}/W_\kappa$ characterising the ratio of the generated and scattered energies. In particular, the value $W_{3\kappa}/W_\kappa = 0.3558$ for $a_\kappa^{\text{inc}} = 14$ and $\varphi_\kappa = 66^\circ$ (see the graph #3 in Fig. 10 (bottom left)) indicates that $W_{3\kappa}$ is 35.58% of W_κ. This is the maximal value of $W_{3\kappa}/W_\kappa$ that has been achieved. The numerical analysis of the processes displayed by the curves #3 in the range of angles $\varphi_\kappa \in (66^\circ, 79^\circ)$ and #4 in the range of angles $\varphi_\kappa \in (62^\circ, 82^\circ)$ did not lead to the convergence of the computational algorithm. Among the results shown in Fig. 10 (bottom right) we mention that the curve #2 describes the scattered field of type $H_{0,0,4}$, and the curve #3 the generated field of type $H_{0,0,10}$.

The results of Fig. 11 (left) show that, in the range $a_\kappa^{\rm inc} \in (0, 22]$ of the amplitude of the incident field and for an incident angle $\varphi_\kappa = 60^\circ$ of the plane wave, the scattered field has the type $H_{0,0,4}$. The generated field, observed in the range $a_\kappa^{\rm inc} \in [5, 22]$, is of the type $H_{0,0,10}$, see Fig. 11 (right). The surfaces presented in Fig. 12 characterise the non-linear dielectric permittivity of the layer (72) induced by the scattered and generated fields shown in Fig. 11. Here, as in Subsection 10.1, the quantity $\mathfrak{Im}\,(\varepsilon_\kappa)$ takes both positive and negative values along the height of the non-linear layer (i.e. in the interval $z \in [-2\pi\delta, 2\pi\delta]$), see Fig. 12 (top right). For a given amplitude $a_\kappa^{\rm inc}$, the graph of $\mathfrak{Im}\left(\varepsilon_\kappa\left[a_\kappa^{\rm inc}, z\right]\right)$ characterises the *loss of energy* in the non-linear layer at the excitation frequency caused by the *generation* of the electromagnetic field of the third harmonic.

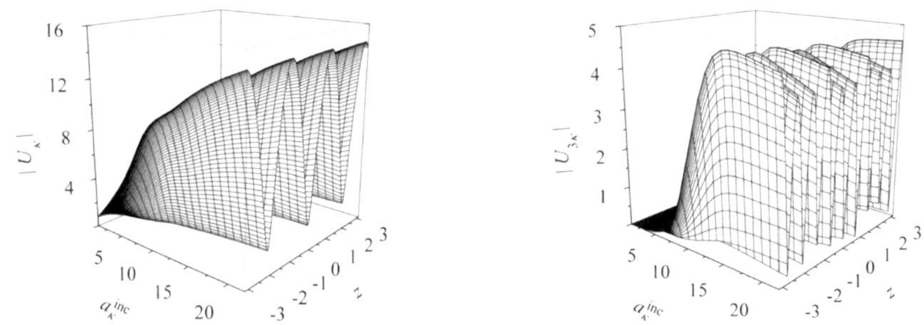

Fig. 11. Graphs of the scattered and generated fields in the non-linear layered structure for $\varphi_\kappa = 60^\circ$: $\left|U_\kappa\left[a_\kappa^{\rm inc}, z\right]\right|$ (left), $\left|U_{3\kappa}\left[a_\kappa^{\rm inc}, z\right]\right|$ (right)

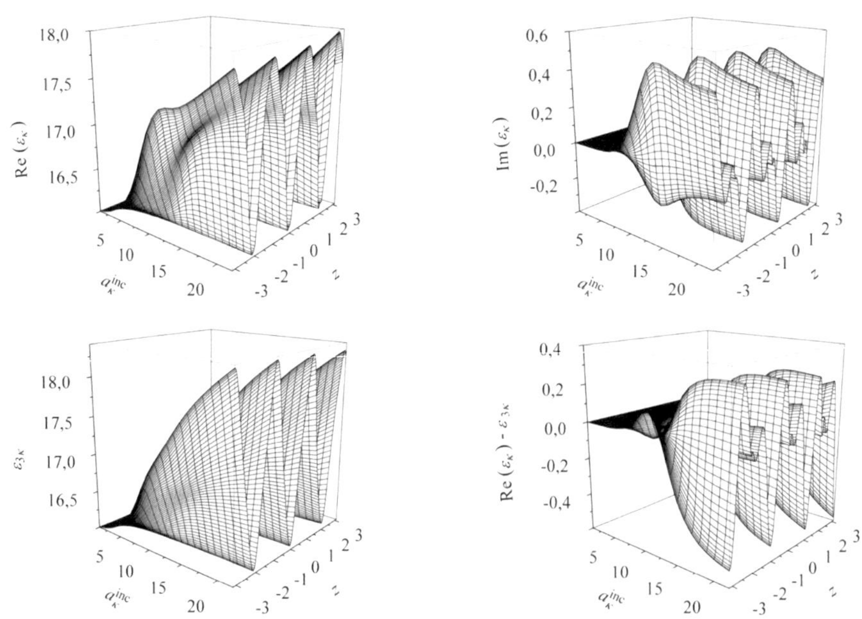

Fig. 12. Graphs characterising the non-linear dielectric permittivity for $\varphi_\kappa = 60^\circ$: $\mathfrak{Re}\left(\varepsilon_\kappa\left[a_\kappa^{\rm inc}, z\right]\right)$ (top left), $\mathfrak{Im}\left(\varepsilon_\kappa\left[a_\kappa^{\rm inc}, z\right]\right)$ (top right), $\varepsilon_{3\kappa}\left[a_\kappa^{\rm inc}, z\right]$ (bottom left), $\mathfrak{Re}\left(\varepsilon_\kappa\left[a_\kappa^{\rm inc}, z\right]\right) - \varepsilon_{3\kappa}\left[a_\kappa^{\rm inc}, z\right]$ (bottom right)

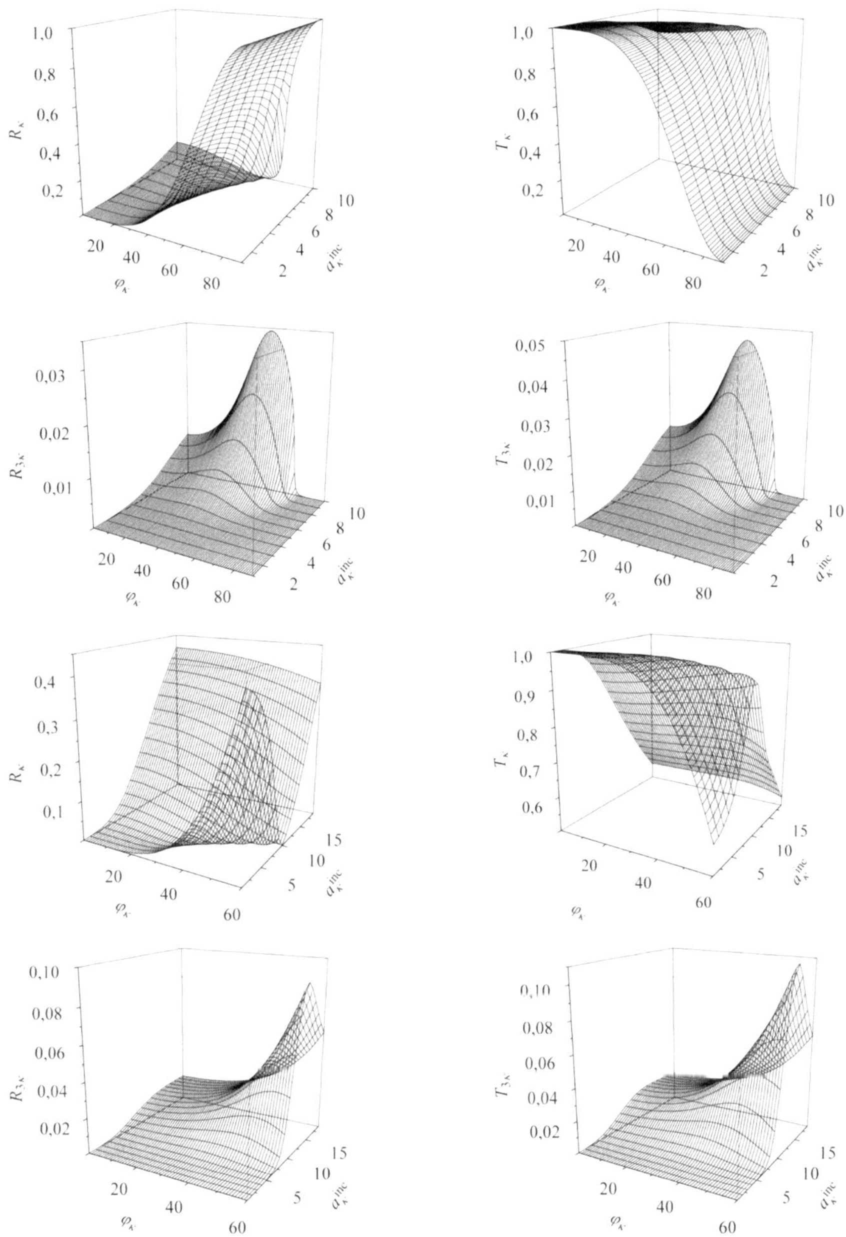

Fig. 13. The scattering and generation properties of the non-linear structure: $R_\kappa\left[\varphi_\kappa, a_\kappa^{\text{inc}}\right]$ (top left, second to the last left), $T_\kappa\left[\varphi_\kappa, a_\kappa^{\text{inc}}\right]$ (top right, second to the last right), $R_{3\kappa}\left[\varphi_\kappa, a_\kappa^{\text{inc}}\right]$ (second from top left, bottom left), $T_{3\kappa}\left[\varphi_\kappa, a_\kappa^{\text{inc}}\right]$ (second from top right, bottom right)

The scattering and generation properties of the non-linear structure in the ranges $\varphi_\kappa \in [0^\circ, 90^\circ)$, $a_\kappa^{\text{inc}} \in [1, 9.93]$ and $\varphi_\kappa \in [0^\circ, 60^\circ]$, $a_\kappa^{\text{inc}} \in [1, 19]$ of the parameters of the excitation field are presented in Fig. 13 (top 4) and (last 4), respectively. Fig. 14 shows cross sections of the surfaces depicted in Fig. 13 and of the graph of $W_{3\kappa}/W_\kappa\left[\varphi_\kappa, a_\kappa^{\text{inc}}\right]$ (see Fig. 10 (top)) by

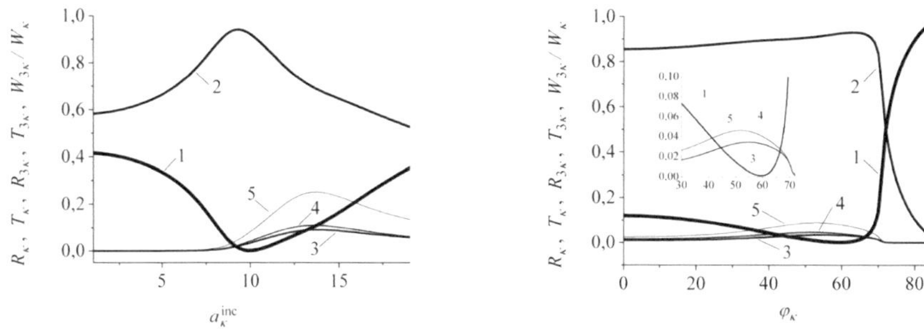

Fig. 14. The curves R_κ (#1), T_κ (#2), $R_{3\kappa}$ (#3), $T_{3\kappa}$ (#4), $W_{3\kappa}/W_\kappa$ (#5) for $\varphi_\kappa = 60°$ (left) and $a_\kappa^{inc} = 9.93$ (right)

the planes $\varphi_\kappa = 60°$ and $a_\kappa^{inc} = 9.93$. The dynamic behaviour of the characteristic quantities depicted in Figs. 13 and 14 is caused by the fact that the corresponding eigen-frequencies of the problems (62), (CS1) – (CS4) with $\alpha = 0$ and with $\alpha \neq 0$ are close together. They also depend on the proximity of the corresponding eigen-frequencies to the scattering (excitation) and generation frequencies $\kappa^{scat} := \kappa^{inc} = \kappa$ and $\kappa^{gen} := 3\kappa$ of the waves.

We start the analysis of the results of our calculations with the comparison of the dispersion relations given by the branches of the eigen-frequencies (curves #3.1, #3.2 and #5.1, #5.2) near the scattering frequency (curve #1, corresponding to the excitation frequency) and (curves #4.1, #4.2, #6.1, #6.2) near the oscillation frequency (line #2) in the situations presented in Fig. 8 (where $\alpha < 0$) and Fig. 15 (where $\alpha > 0$). We point out that the situations shown in Fig. 8 and Fig. 15 are fundamentally different. In the case of Fig. 8 ($\alpha < 0$), the graph #5.1 lies above the graph #3.1 and the graph #6.1 above the graph #4.1 in the vicinity of the lines #1 and #2, respectively. This is the typical for the case of decanalisation, see Subsection 10.1.

In the situation of Fig. 15 ($\alpha > 0$) we observe a different behaviour. Here, near the lines #1 and #2, respectively, the graph #5.1 lies below the graph #3.1 and the graph #6.1 below the graph #4.1. That is, canalising properties (properties of transparency) of the non-linear layer occur if $\alpha > 0$. This case is characterised by the increase of the angle of transparency of the non-linear structure at the excitation frequency with an increasing amplitude of the incident field (see

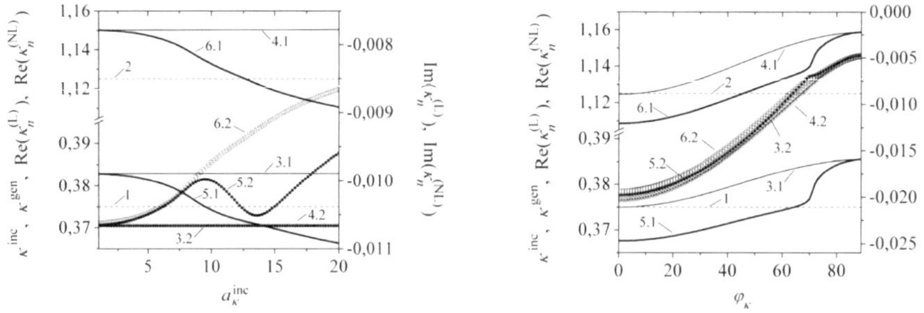

Fig. 15. The curves $\kappa := \kappa^{inc} := 0.375$ (#1), $3\kappa = \kappa^{gen} = 3\kappa^{inc} = 1.125$ (#2), the complex eigen-frequencies $\mathfrak{Re}(\kappa_1^{(L)})$ (#3.1), $\mathfrak{Im}(\kappa_1^{(L)})$ (#3.2), $\mathfrak{Re}(\kappa_3^{(L)})$ (#4.1), $\mathfrak{Im}(\kappa_3^{(L)})$ (#4.2) of the linear problem ($\alpha = 0$) and $\mathfrak{Re}(\kappa_1^{(NL)})$ (#5.1), $\mathfrak{Im}(\kappa_1^{(NL)})$ (#5.2), $\mathfrak{Re}(\kappa_3^{(NL)})$ (#6.1), $\mathfrak{Im}(\kappa_3^{(NL)})$ (#6.2) of the linearised non-linear problem ($\alpha = +0.01$) for $\varphi_\kappa = 60°$ (left) and $a_\kappa^{inc} = 9.93$ (right)

Fig. 13 (top left), (second to the last left), there where the reflection coefficient is close to zero). The analysis of the eigen-modes of Fig. 15 ($\alpha > 0$) allows us to explain the mechanisms of the canalisation phenomena (transparency) (see Fig. 13 (top left), (second to the last left), Fig. 14) and wave generation (see Fig. 13 (second from top), (bottom), Fig. 14).
Comparing the results shown in Fig. 14 and Fig. 15 we note the following. The intersection of the curves #1 and #5.1 in Fig. 15 defines certain parameters, in the neighborhood of which the canalisation effect (transparency) of the non-linear structure can be observed in Fig. 14. For example, in Fig. 15 (left) the curves #1 and #5.1 intersect at $a_{\kappa}^{\text{inc}} = 9.5$, also here the curve #5.2 achieves a local maximum. Near this value, we see the phenomenon of canalisation (transparency) of the layer in Fig. 14 (left). If we compare the Figs. 14 (right) and 15 (right), we detect a similar situation. The intersection of the curves #1 and #5.1 defines the parameter $\varphi_{\kappa} = 64^{\circ}$, near which we observe the canalisation effect in Fig. 15 (right). The same is true – to some extent – for the description of the wave generation processes. For example, for similar values of the imaginary parts of the branches of the eigen-frequencies #5.2 and #6.2 in Fig. 15 (right), the intersection of the curves #2 and #6.1 defines the parameter $\varphi_{\kappa} = 45^{\circ}$. Near this value, stronger generation properties of the layer can be observed, see Fig. 14 and Fig. 13 (second from top), at $\varphi_{\kappa} = 45^{\circ}$. Let us also consider the situation in Fig. 15 (left). Here, at the point of intersection of the curves #2 and #6.1, the graph #5.2 starts to decrease monotonically in some interval. The intersection of the curves #2 and #6.1 defines the parameter $a_{\kappa}^{\text{inc}} = 12.6$, which falls into the range $[9.5, 13.6]$ of values of the amplitudes at which the curve #5.2 is monotonically decreasing. This leads to a shift in the imaginary part of the eigen-frequency of the scattering structure (graph #5.2) with respect to the eigen-frequency of the generating structure (graph #6.2). The magnitude of the shift depends on the distance between the curves of #6.2 and #5.2 at the given value a_{κ}^{inc}. The maximal distance between the graphs #6.2 and #5.2 is achieved at the local minimum of the graph #5.2 at $a_{\kappa}^{\text{inc}} = 13.6$. Right from this point, i.e. with an increasing amplitude a_{κ}^{inc}, the distance between the graphs #6.2 and #5.2 shows no significant change. The maximum value of the generation is achieved at an amplitude close to the intersection of curves #2 and #6.1, but shifted to the point of the local minimum of the curve #5.2, see $R_{3\kappa}$, $T_{3\kappa}$, $W_{3\kappa}/W_{\kappa}$ in Fig. 14 (left), Fig. 13 (bottom) and Fig. 10 (top right).

Fig. 16. The graphs of the eigen-fields of the layer for $\varphi_{\kappa} = 60^{\circ}$, $a_{\kappa}^{\text{inc}} = 14$. The linear problem ($\alpha = 0$, left figure): $|U(\kappa_1^{(L)}; z)|$ with $\kappa_1^{(L)} = 0.3829155 - i\,0.01066148$ (#1), $|U(\kappa_3^{(L)}; z)|$ with $\kappa_3^{(L)} = 1.150293 - i\,0.01062912$ (#2), the linearised non-linear problem ($\alpha = +0.01$, right figure): $|U(\kappa_1^{(NL)}; z)|$ with $\kappa_1^{(NL)} = 0.3705110 - i\,0.01049613$ (#1), $|U(\kappa_3^{(NL)}; z)|$ with $\kappa_3^{(NL)} = 1.121473 - i\,0.009194824$ (#2)

Fig. 16 presents the characteristic distribution of the eigen-fields corresponding to the branches of the eigen-frequencies under consideration. The graphs of the eigen-fields of type $H_{0,0,4}$ are labeled by #1, the graphs of the eigen-fields of type $H_{0,0,10}$ by #2.
The numerical results presented in this paper were obtained using an approach based on the description of the wave scattering and generation processes in a non-linear, cubically polarisable layer by a system of non-linear integral equations (49), and of the corresponding spectral problems by the non-trivial solutions of the integral equations (66). We have considered an excitation of the non-linear layer defined by the condition (71). For this case we passed from (49) to (75) and from (66) to (69) by the help of Simpson's quadrature rule. The numerical solution of (75) was obtained using the self-consistent iterative algorithm (61). The problem (69) was solved by means of Newton's method. In the investigated range of parameters, the dimension of the resulting systems of algebraic equations was $N = 301$, and the relative error of calculations did not exceed $\xi = 10^{-7}$.

11. Conclusion

We have investigated the problem of scattering and generation of waves on a non-linear, layered, cubically polarisable structure, which is excited by a packet of waves, in the range of resonant frequencies. The theoretical and numerical results complement the previously presented investigations from Angermann & Yatsyk (2011), Angermann & Yatsyk (2010), Shestopalov & Yatsyk (2010). The mathematical description of the wave scattering and generation processes on a non-linear, layered, cubically polarisable structure reduces to a system of non-linear boundary-value problems. This classical formulation of the problem is equivalent to a system of boundary-value problems of Sturm-Liouville type and to a system of one-dimensional non-linear Fredholm integral equations of the second kind. In this paper, for each of these problems we have obtained sufficient conditions for existence and uniqueness of the solution and we have developed self-consistent algorithms for the numerical analysis. Within the framework of the self-consistent approach we could show that the variation of the imaginary part of the permittivity of the layer at the excitation frequency can take both positive and negative values along the height of the non-linear layer. This effect is caused by the energy consumption in the non-linear medium at the frequency of the incident field which is spent for the generation of the electromagnetic field of the third harmonic. It was shown that layers with negative and positive values of the coefficient of cubic susceptibility of the non-linear medium have fundamentally different scattering and generation properties in the range of resonance. So, for the considered here layer with a negative value of the susceptibility, the maximal portion of the total energy generated in the third harmonic was observed in the direction normal to the structure and amounted to 3.9% of the total dissipated energy. For a layer with a positive value of the susceptibility it was possible to reach such intensities of the excitation field under which the maximum of the relative portion of the total energy was 36% and was observed near the angle of transparency which increasingly deviates from the direction normal to the layer with increasing intensity of the incident field.
The approximate solution of the non-linear problems was obtained by means of solutions of linear problems with an induced non-linear dielectric permeability. The analytical continuation of these linear problems into the region of complex values of the frequency parameter allowed us to switch to the analysis of spectral problems. In the frequency domain, the resonant scattering and generation properties of non-linear structures are determined by the proximity of the excitation frequencies of the non-linear structures to the complex

eigen-frequencies of the corresponding homogeneous linear spectral problems with the induced non-linear dielectric permeability of the medium.

12. References

Adams, R. (1975). *Sobolev spaces*, Vol. 65 of *Pure and Applied Mathematics (Amsterdam)*, Academic Press, New York–London.

Agranovich, V. & Ginzburg, V. (1966). *Spatial Dispersion in Crystal Optics and the Theory of Excitons*, Interscience, London.

Akhmediev, N. & Ankevich, A. (2003). *Solitons*, Fizmatlit, Moscow.

Angermann, L. & Yatsyk, V. (2008). Numerical simulation of the diffraction of weak electromagnetic waves by a Kerr-type nonlinear dielectric layer, *Int. J. Electromagnetic Waves and Electronic Systems* 13(12): 15–30.

Angermann, L. & Yatsyk, V. (2010). Mathematical models of the analysis of processes of resonance scattering and generation of the third harmonic by the diffraction of a plane wave through a layered, cubically polarisable structure, *Int. J. Electromagnetic Waves and Electronic Systems* 15(1): 36–49. In Russian.

Angermann, L. & Yatsyk, V. (2011). Generation and resonance scattering of waves on cubically polarisable layered structures, *in* L. Angermann (ed.), *Numerical Simulations – Applications, Examples and Theory*, InTech, Rijeka/Vienna, Croatia/Austria, pp. 175–212.

Butcher, P. (1965). Nonlinear optical phenomena, *Bulletin 200*, Ohio State Univ., Columbus.

Kivshar, Y. & Agrawal, G. (2005). *Optical Solitons : From Fibers To Photonic Crystals*, FizMatLit, Moscow. Russian translation of the 2003 English original edition.

Kleinman, D. (1962). Nonlinear dielectric polarization in optical media, *Phys. Rev.* 126(6): 1977–1979.

Kravchenko, V. & Yatsyk, V. (2007). Effects of resonant scattering of waves by layered dielectric structure with Kerr-type nonlinearity, *Int. J. Electromagnetic Waves and Electronic Systems* 12(12): 17–40.

Miloslavsky, V. (2008). *Nonlinear Optics*, V.N. Karazin Kharkov National Univ., Kharkov.

Reed, M. & Simon, B. (1980). *Methods of Modern Mathematical Physics, I: Functional Analysis*, 2nd, revised and enlarged edn, Academic Press, San Diego–London.

Sánchez-Palencia, E. (1980). *Non-Homogeneous Media and Vibration Theory*, Vol. 127 of *Lecture Notes in Physics*, Springer-Verlag, Berlin–Heidelberg–New York.

Schürmann, H. W., Serov, V. & Shestopalov, Y. (2001). Reflection and transmission of a TE-plane wave at a lossless nonlinear dielectric film, *Physica D* 158: 197–215.

Serov, V., Schürmann, H. & Svetogorova, E. (2004). Integral equation approach to reflection and transmission of a plane te-wave at a (linear/nonlinear) dielectric film with spatially varying permittivities, *J. Phys. A: Math. Gen.* 37: 3489–3500.

Shestopalov, V. & Sirenko, Y. (1989). *Dynamical Theory of Gratings*, Naukova, Dumka, Kiev.

Shestopalov, V. & Yatsyk, V. (1997). Spectral theory of a dielectric layer and the Morse critical points of dispersion equations, *Ukrainian J. of Physics* 42(7): 861–869.

Shestopalov, Y. & Yatsyk, V. (2007). Resonance scattering of electromagnetic waves by a Kerr nonlinear dielectric layer, *Radiotekhnika i Elektronika (J. of Communications Technology and Electronics)* 52(11): 1285–1300.

Shestopalov, Y. & Yatsyk, V. (2010). Diffraction of electromagnetic waves by a layer filled with a Kerr-type nonlinear medium, *J. of Nonlinear Math. Physics* 17(3): 311–335.

Showalter, R. (1994). *Hilbert space methods for partial differential equations*, Electronic Monographs in Differential Eqs., San Marcos, TX. Electronic reprint of the 1977 original.

Sirenko, Y., Shestopalov, V. & Yatsyk, V. (1985). Elements of the spectral theory of gratings, *Preprint 266*, IRE NAS Ukraine, Kharkov.

Sirenko, Y. & Ström, S. (eds) (2010). *Modern Theory of Gratings. Resonant Scattering: Analysis Techniques and Phenomena*, Springer-Verlag, New York. Springer Series in Optical Sciences, Vol. 153.

Sirenko, Y., Ström, S. & Yashina, N. (2007). *Modeling and Analysis of Transient Processes in Open Resonant Structures. New Methods and Techniques.*, Springer-Verlag, New York.

Smirnov, Y., Schürmann, H. & Shestopalov, Y. (2005). Propagation of TE-waves in cylindrical nonlinear dielectric waveguides, *Physical Review E* 71: 0166141–10.

Sukhorukov, A. P. (1988). *Nonlinear wave interactions in optics and radio physics*, Nauka, Moskva. (In Russian).

Titchmarsh, E. (1961). *Razlozheniya po sobstvennym funktsiyam, svyazannye s differentsialnymi uravneniyami vtorogo poryadka. Chast II*, Izdat. Inostr. Lit., Moscow. (Russian translation of: E.C. Titchmarsh. *Eigenfunction expansions associated with second-order differential equations.* Vol. 2, Clarendon Press, Oxford 1958).

Vainstein, L. (1988). *Electromagnetic Waves*, Radio i Svyas, Moscow. In Russian.

Vinogradova, M., Rudenko, O. & Sukhorukov, A. (1990). *Wave Theory*, Nauka, Moscow.

Yatsyk, V. (2000). A constructive approach to construction of local equations of irregular dispersion and evolution of fields in a quasi-homogeneous electrodynamic structure, *Usp. Sovr. Radioelektroniki* 10: 27–44. Translated in: *Telecommunications and Radio Engineering*, 56(8&9): 89-113, 2001.

Yatsyk, V. (2006). Diffraction by a layer and layered structure with positive and negative susceptibilities of Kerr-nonlinear media, *Usp. Sovr. Radioelektroniki* 8: 68–80.

Yatsyk, V. (2007). About a problem of diffraction on transverse non-homogeneous dielectric layer of Kerr-like nonlinearity, *Int. J. Electromagnetic Waves and Electronic Systems* 12(1): 59–69.

16

Cholesteric Elastomers with Mechanical Control of Optical Spectra

J. Adrián Reyes[1], Laura O. Palomares[2] and Carlos G. Avendaño[3]
[1]*Universidad Autónoma Metropolitana (on leave from Instituto de Fisica, UNAM)*
Ixtapalapa, Apartado,
[2]*Research Institute for Solid State Physics and Optics of the*
Hungarian Academy of Sciences, H-1525 Budapest,
[3]*Universidad Autónoma de la Ciudad de México,*
[1,3]*México*
[2]*Hungary*

1. Introduction

Elastomers are elastic media which mixes as no one, also, they have three important properties: orientational order of large range in amorphous soft materials, macroscopic susceptibility to the molecular shape, and quenching to the topological constraints. Classical liquid crystals are fluids typically composed by rigid molecules, which with a continuous model, are represented by bars and exhibit an orientational order of large range. The simplest order displayed by these systems is the nematic, for which, all the molecules are aligned in average. Complementary, the polymeric long chains embodying anisotropic rigid units can be nematically aligned and may form polymeric liquid crystals.

However, the long chains are elongated when their rigid monomeric components are oriented, giving rise to an anisotropic material. If additionally, the polymeric chains are joint to a backbone in such way that their topology is restrained, hence the melt condenses in a very elastic solid or rubber. It is convenient to mention that in general, within the rubbers, the nematic monomeric molecules retain the same mobility as in a liquid phase.

These soft constrictions make the resulting material, which is then a solid very extensible. Rubbers resist mechanic deformations since the polymeric chains reach their maximum entropy when they stay in their natural state without deformation. The polymerization of these compounds creates links between the chains which joint to the backbone formed collectively among themselves.

It is to be expected that in this process, the anisotropic rigid units of nematic character, for instance (nematogens) which lie in the inner of the medium, form spontaneously domains distributed in all the rubber, whose preferred orientation is to be in different directions. This variety of domains causes light scattering giving rise to a macroscopic turbid appearance to the material. One very important advance in the design of these materials was managed by Finkelman, by developing a procedure for obtaining samples which form a single domain. The basic idea consists in applied electric field to the melt substance in order to align the anisotropic monomeric units while the polimerization is taking place and/or the temperature

is decreased to maintain the orientational order of the oriented nematic elements and in turn attain a monodomain.
A material prepared in this way is called as a liquid crystal elastomer and has the amazing property of being deformable, within certain interval of elongations, investing in this a practically negligible amount of energy. This is caused essentially by the reorientation and accommodation of the anisotropic structure formed by the rigid nematic monomers or other mesophases in the inner of long polymeric chains while the material is distorted in such way that the energy utilized is minimized when the mentioned structure turns.
Nowadays the liquid crystal elastomers are synthesized to generate phases of the same variety obtained with classical liquid crystals. It can be created nematics, smectic and chiral liquid crystals. Similarly to any polymeric rubber, the materials are very deformable since their dimensions can be changed under the influence of external stresses as much as 300% of their original sizes. They can be easily oriented by electrical fields as the ordinary liquid crystals. Both features make of these new materials excellent candidates to design artificial muscles. On the other hand, their transparency in the monodomain phase and the fact that as solids do not require a container make them excellent candidates to be used as electro optic devices.

2. Polymers

2.1 Polymer configuration

A polymer is a very long chain formed by many repeated molecular units, as much as thousands, called monomers. All polymeric chains possess a characteristic length l, at which the chains can be bended. This length can contain various monomers when the total number of monomers in the chain N is much larger than the number of monomers per length l and then we can affirm that there exist various possible spatial configurations for the polymer. Thus, it is possible to employ a Gaussian distribution to describe adequately the system. Here we shall assume that this is the case.

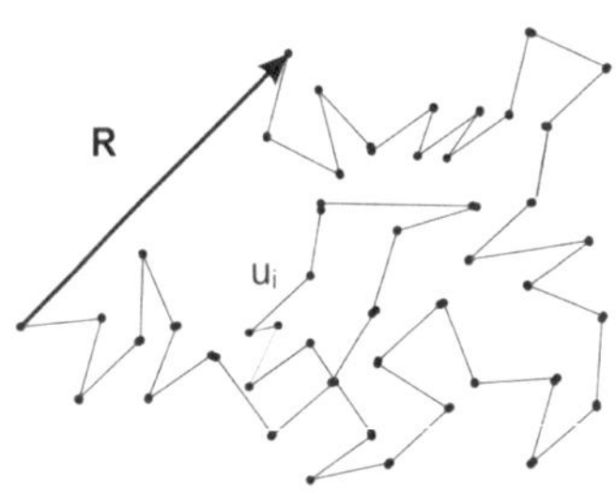

Fig. 1. A chain composed by N bar of length a freely joint is statistically equivalent to the path of a random walk with fixed step a.

A chain composed by N bar of length a freely joint like the one shown in Fig. 1 is statistically equivalent to the path of a random walk with fixed step a. The average mean square of the end to end vector $\mathbf{u}_l$ formed after this walk of N steps (Warner & Terentjev, 2003)

$$\langle \mathbf{R}_x^2 \rangle = \langle \mathbf{R}_y^2 \rangle = \langle \mathbf{R}_z^2 \rangle = \frac{1}{3}\langle \mathbf{R}^2 \rangle = \frac{1}{3}a^2 N \equiv \frac{1}{3}aL, \tag{1.1}$$

where $L = Na$ is the total length of the chain. In terms of the end to end vector which joint the monomers $\mathbf{u}_i$ of length a, the distance between the two edges of the chain is given by the magnitude of the vector: $\mathbf{R} = \Sigma_i \mathbf{u}_i$. This quantity is very important since provides a notion of the spatial configuration of the polymer chains and from the polymer chain distribution we can derive the free energy characterizing the system. The vectors $\mathbf{u}_i$ are not correlated each other so that $\langle \mathbf{u}_i \mathbf{u}_j \rangle = \delta_{ij} a^2$ which is consistent with Eq. (1.1).

A polymer is a thermodynamical system in equilibrium interacting with its surrounding at volume and temperature constants. Hence, the total number of possible conformations of one given chain Z_N (or the number of possible random walk without restrictions) must be equal to the partition function of the chain: $Z_N = \Sigma_{config} e^{-H/kT}$, where H is the energy of the configuration, k is the Boltzmann constant and T is the temperature. If we take a random walk for which the two ends of the walks are fixed, we expect that the number of possible configurations is to be smaller than that of the system just mentioned. Thus the number of configurations for fixed boundary conditions is given by

$$Z_N(\mathbf{R}) = p_N(\mathbf{R}) Z_N , \tag{1.2}$$

where $p_N(\mathbf{R})$ is the probability of having a configuration with an end to end vector $\mathbf{R}$. Since Z_N is the partition function of the system, the probability of distribution must be Gaussian and as a consequence its explicit normalized expression is given by (Kac et al., 1976):

$$p_N(\mathbf{R}) = \left(\frac{3}{2\pi R_o^2} \right)^{3/2} e^{-3\mathbf{R}^2/2R_o^2} , \tag{1.3}$$

Where R_o is the variance of the system and from Eq. (1.1) we get $R_o^2 = aL$.

The fact that the polymer chain is kept at volume and temperature constants allow us to use the Helmholtz free energy to describe the system which in agreement with the statistical physics is given by $F = -k_B T ln Z_N(\mathbf{R})$; substitution of Eqs. (1.2) and (1.3) in this expression, yields

$$F(\mathbf{R}) = F_o + k_B T \left(\frac{3\mathbf{R}^2}{2R_o^2} \right) + C , \tag{1.4}$$

where $F_o = -k_B T ln Z_N$ is the free energy of the polymeric chain without restrictions in their ends and C is the constant coming from the normalization of the distribution P_N, F_o and C are constants independent of $\mathbf{R}$, which only determine the reference point, thus their values are irrelevant to find the minimum of the free energy of the system. We remark that the energy given in Eq. (1.4) is purely entropic since only depends on the number of possible configuration of the systems. To obtain this expression we assume that all the possible configurations contribute with the same energy and we neglect the chemical energy caused by the electromagnetic repulsion between the molecules. If we take into account the internal energy associated with chemical processes, $U(\mathbf{R})$, the free energy of the system is given by

$$F(\mathbf{R}) = U(\mathbf{R}) - TS(\mathbf{R}) , \tag{1.5}$$

where the entropy per molecule is represented $S(\mathbf{R})$. Strictly speaking, this term should be considered however it has been shown that the entropy (Warner & Terentjev, 2003) dominates the free energy and then we can neglect it, $S(R) = -3k_B \mathbf{R}^2 / 2R_o^2$.

2.2 Polymeric liquid crystals

A polymeric liquid crystal combines the spontaneous orientation of the liquid crystals with the elasticity governed by entropy discussed above. It is necessary a delicate balance in these properties to create a polymeric liquid crystal.

2.2.1 Polymeric liquid crystals shape

The average shape of the polymeric main chain is crucial since this is responsible of the equilibrium elastic response of the network it belongs. Some ordinary polymers are isotropic or spheric so that only one dimension is enough to characterize these materials. In contrast, the nematic polymers may adopt diverse shapes due to the fact that the average backbone is distorted by the reorientation suffered by the molecular bars guided by the director $\mathbf{n}$, that is, the nematic order modifies the backbone form of the polymer. Hence, nematic polymers require more than one direction to describe their anisotropic form.
To characterize the chain form and its probability distribution the quadratic mean square of shape is given by

$$\langle R_i R_j \rangle = \frac{1}{3} l_{ij} L, \tag{1.6}$$

where we have defined l_{ij} as the effective length steps in distinct directions. For uniaxial polymers $\langle R_i R_j \rangle$, is the same for every perpendicular direction to $\mathbf{n}$. Thus, if $\mathbf{n}$ is along the z-axis, where $R_x = R_y = R_\perp$ and the tensor $\boldsymbol{l}$ for this case is

$$\mathbf{l}_o = \begin{pmatrix} l_\perp & 0 & 0 \\ 0 & l_\perp & 0 \\ 0 & 0 & l_{\|} \end{pmatrix}, \tag{1.7}$$

where $l_\perp$ and $l_{\|}$ are the length steps in the directions parallel and perpendicular to $\mathbf{n}$, respectively. In general, when $\mathbf{n}$ is not necessarily aligned with one of the axes of our reference system, the matrix $\mathbf{l}$ is not necessarily diagonal but uniaxial and has the form:

$$\mathbf{l}_o = l_\perp(\boldsymbol{\delta} + [r-1]\mathbf{nn}) \quad \text{and} \quad \mathbf{l}_o^{-1} = l_\perp(\boldsymbol{\delta} + [\frac{1}{r} - 1]\mathbf{nn}), \tag{1.8}$$

where we have defined the radius r, as the ratio between parallel and perpendicular effective length steps as $r = l_{\|} / l_\perp$.

If $r > 1$, means that we have a prolate backbone, which is larger along the direction $\mathbf{n}$; Instead for $r < 1$, we have an oblate, whose backbone is larger in perpendicular plane to $\mathbf{n}$.

The probability of find a configuration with end to end vector: $\mathbf{R}$, Eq. (1.3). On the other hand the probability of find certain configuration for the anisotropic case, is obtained by using Eq. (1.1) from which it can be derived $R_o^2 = l_{ij} L$ and $\mathbf{R}^2 = R_i R_j$. By substituting these expressions in Eq. (1.3), we get the probability for the anisotropic case:

$$p(\mathbf{R}) = \left[\left(\frac{3}{2\pi L} \right)^3 \frac{1}{Det[\mathbf{l}]} \right]^{1/2} e^{\left(\frac{-3}{2L} R_i l_{ij}^{-1} R_j \right)}. \tag{1.9}$$

3. Nematic rubber elasticity

Nematic elastomer systems can be quite elastic, that is they can be extended or compressed for large proportions. The difference between nematic and isotropic rubbers is the molecular shape anisotropy induced by liquid crystal order. The simplest description of nematic rubbers is arising from extension of the molecular theory of rubbers just discussed in the foregoing section and is known as the neoclassic theory.

The number of configurations in one thread connecting two crosslinkings separated by a distance $\mathbf{R}$ in one nematic rubber is proportional to the anisotropic Gaussian distribution given by Eq. (1.9).

The step length vector $\mathbf{l}$ reflexes the actual nematic order in the rubber. In contrast, in the formation state the end to end vector is $\mathbf{R}_f$ and the shape of the chains in that state is similarly described by Gaussian distribution whose form is the same as Eq. (1.9) but with step length vector $\mathbf{l}_o$. One reason for having a different distribution is that the temperature changes and thus the nematic order as well. If the starting state is nematic, then an orientation change can also modified the distribution. As before, let us consider that a total deformation $\boldsymbol{\eta}_t$ affine leads from the formation state $\mathbf{R}_f$ to the actual situation $\mathbf{R} = \boldsymbol{\eta}_t \cdot \mathbf{R}$. The free energy, is obtained by averaging over a ensemble of initial condition whose energy of formation are the same

$$\mathcal{F} = -k_B T \langle \ln p(\mathbf{R}) \rangle_{p_0(\mathbf{R}_f)} = \frac{3k_B T}{2L} \langle \mathbf{R} \cdot \mathbf{l}^{-1} \cdot \mathbf{R} \rangle + \frac{k_B T}{2} \ln\left(\frac{Det[\mathbf{l}]}{a^3} \right), \tag{1.10}$$

the term in $Det[\mathbf{l}]$ is due to the normalization factor containing the information of the nematic order by means of the step length tensor $\mathbf{l}$. The free energy can be rewritten as

$$\mathcal{F} = \frac{3k_B T}{2L} \langle \mathbf{R}_f \cdot \boldsymbol{\eta}^T \cdot \mathbf{l}^{-1} \cdot \boldsymbol{\eta} \cdot \mathbf{R}_f \rangle_{p_0} \tag{1.11}$$

The average over the set of formation states can be performed directly using the relation $\langle \mathbf{R}_f \mathbf{R}_f \rangle_{p_0(\mathbf{R}_f)} - \mathbf{l}_o L / 3$, so that the average energy per thread is finally

$$\mathcal{F} = \frac{k_B T}{2} Tr\left(\mathbf{l}_o \cdot \boldsymbol{\eta}^T \cdot \mathbf{l}^{-1} \cdot \boldsymbol{\eta} \right) + \frac{k_B T}{2} \ln\left(\frac{Det[\mathbf{l}]}{a^3} \right). \tag{1.12}$$

This expression is a generalization of the classical free energy for the elasticity for a thread for which is known as the neoclassic free energy. To obtain the whole free energy of the rubber we need to count the number of threads per volume n_s, that is $F = n_s \mathcal{F}$ and since the linear shear modulus of a rubber is $\mu = n_s k_B T$. The free energy density is given by

$$F = \frac{\mu}{2} Tr\left(\mathbf{l}_o \cdot \boldsymbol{\eta}^T \cdot \mathbf{l}^{-1} \cdot \boldsymbol{\eta} \right) \tag{1.13}$$

This expression is valid for all the deformation, including the larger ones, but it cannot describe those deformations which can stretch totally the polymeric chains. This expression involves the orientational information about the initial state $\mathbf{n}_0$ and actual state $\mathbf{n}$ of the elastomer by means of $\mathbf{l}_o$ and $\mathbf{l}$. By contrary the free energy of ordinary nematic liquid crystal only depends on the actual state of distortion. Eq. (1.13) exhibits a complex structure since the distortions appear expressed in terms of the combination $\boldsymbol{\eta}^T \cdot \mathbf{l}^{-1} \cdot \boldsymbol{\eta}$.

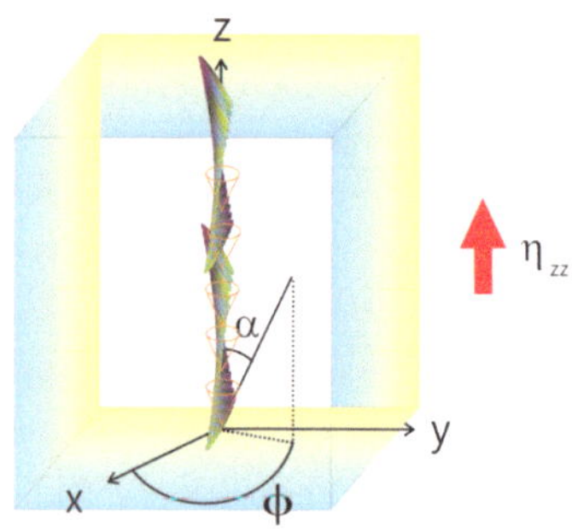

Fig. 2. A cholesteric elastomer submitted to distortion along its chiral axis, which in this case we chose to be the z – axis.

4. Chiral elastomer under an axial deformation

Let us consider a cholesteric elastomer submitted to distortion along its chiral axis, which in this case we chose to be the z-axis, as shown in Fig. 2.
The deformation tensor, in its simplest, can be expressed as (Warner & Terentjev, 2003):

$$\boldsymbol{\eta} = \begin{pmatrix} 1/\sqrt{\eta} & 0 & \eta_{xz} \\ 0 & 1/\sqrt{\eta} & \eta_{yz} \\ 0 & 0 & \eta \end{pmatrix}, \tag{1.14}$$

where we have simplified the notation by the convention $\eta_{zz} = \eta$, as we discussed above, to keep the volume fixed , it is necessary to ask for $Det\boldsymbol{\eta} = 1$, which is straightforwardly fulfil by this expression since $Det\boldsymbol{\eta} = \frac{1}{\sqrt{\eta}}\frac{1}{\sqrt{\eta}}\eta = 1$. The terms η_{xz} and η_{yz}, are coming from the fact that we allow the director to rotate in the plane x-y which will be modified by expanding the elastomer along the z-axis.

The initial elastomer vector without deformation is given by: $\mathbf{n}_o = \{\cos q_o z, \sin q_o z, 0\}$, where $q_o = 2\pi / p$ is the helix wave number and p is the spatial periodicity or pitch. After the deformation, the director will be rotated by the angle $\omega = \frac{\pi}{2} - \alpha$, in such way that it lies over the surface of a cone Fig. 2; the new vector after deformation can be expressed as $\mathbf{n} = \{\sin\alpha\cos qz, \sin\alpha\sin qz, \cos\alpha\}$, where $q = q_0 / \eta$ is the new helix wave number scaled by the strain η; the initial step length vector without deformation, $\mathbf{l}_o$, is defined in agreement with Eq. (1.8) in terms of $\mathbf{n}_o$ and the inverse tensor is similar to Eq. (1.8) by replacing the vectors $\mathbf{n}$ by $\mathbf{n}_o$.

The Helmholtz free energy of the system is given by Eq. (1.13) in terms of $\mathbf{l}_o$ and $\mathbf{l}^{-1}$ of Eq. (1.8). Upon expansion of the Helmholtz free energy we can express this as

$$F = \mathbf{x}^T \cdot \mathbf{A} \cdot \mathbf{x} + \mathbf{b} \cdot \mathbf{x} + c \tag{1.15}$$

where

$$\mathbf{A} = \begin{pmatrix} \dfrac{1 + 3r + (r-1)\cos 2\alpha - 2(r-1)\cos 2qz \sin^2 \alpha}{4r} & -\dfrac{(r-1)\sin 2qz \sin^2 \alpha}{2r} \\ -\dfrac{(r-1)\sin 2qz \sin^2 \alpha}{2r} & \dfrac{1 + 3r + (r-1)\cos 2\alpha + 2(r-1)\cos 2qz \sin^2 \alpha}{4r} \end{pmatrix},$$

$$\mathbf{b} = \{-\frac{(r-1)\eta \cos qz \sin 2\alpha}{r}, \frac{(r-1)\eta \sin qz \sin 2\alpha}{r}\},$$

$$c = \frac{\eta^3 + r(3 + r + \eta^3) + (r-1)(r - \eta^3)\cos 2\alpha}{2r\eta}$$

and $\mathbf{x} = \{\eta_{xz}, \eta_{yz}\}$. Eq. (1.15) represents the free energy of our system; it contains the information of the deformation. To find the state of deformation after a relaxation it is necessary to find the minimum of the free energy. When we extend an elastomer, first the strains are relaxed and after that the molecules will reorient. The minimum of energy is found by searching the values η_{xz} and η_{yz} which minimize the energy and then the optimum value of α.

To this purpose, we find the minimum with respect to η_{xz} and η_{yz} by diagonalizing the matrix A and translating to the principal axes system of the deformation (where we have denoted the strains by: η'_{xz} and η'_{yz}). Using the fact the minimum of system's energy is the same in any frame. Once doing this we find

$$\frac{2}{\mu}F = \eta^2 + \frac{1}{\eta} + \eta'^2_{yz} - \frac{1}{2r}([r + 1 + (r-1)\cos 2\alpha](\eta^2 - \frac{r}{\eta} - \eta'^2_{xz}) + 2(r-1)\eta\eta'_{xz}\sin 2\alpha - 2\eta^2, \tag{1.16}$$

This free energy depends on the strains and the angle. These variables used to be coupled to keep constant the free energy and cause a soft elasticity for which the elastomer is to be deformed at no energy cost. For this reason we minimize the energy by using two steps. We first find the minimum of Eq. (1.16) with respect to the strains and then minimize the resulting expression with respect to the angle. Performing the first step, we get from (1.16):

$$\eta'_{yz} = 0 \quad \text{and} \quad \eta'_{xz} = \eta \frac{(r-1)\sin 2\alpha}{(r+1) + (r-1)\cos 2\alpha}. \tag{1.17}$$

Notice that this expression allows to write the strains in terms of the orientation angle. Thus, after substitution of this expression in Eq. (1.16) we obtain the following expression which only depends on the angle α:

$$F = \frac{1}{2}\mu\left(\frac{2\eta^2}{r + 1 + (r-1)\cos 2\alpha} + \frac{3 + r + 1 + (r-1)\cos 2\alpha}{2\eta}\right). \tag{1.18}$$

It remains to know the value of α that minimizes the latter expression. The director vector **n** will incline towards the z-direction with an angle ω (see Fig. 2) after the deformation, to find the value of ω, we minimize the free energy Eq. (1.18) we respect to ω, we get:

$$\cos 2\omega = \frac{r+1-2\eta^{3/2}}{r-1}; \qquad \omega(\eta) = \arcsin\sqrt{\frac{\eta^{3/2}-1}{r-1}}. \tag{1.19}$$

The latter equation indicates which is the degree of reorientation of the cholesteric, the maximum deformation is reached when the director vector aligns totally parallel to the z-axis, that is, when $\omega = \pi/2$, we find that $\eta = r^{2/3}$; when there is no deformation $\eta = 1$ and $\omega = 0$.

5. Propagation of waves in a layered medium

Optical propagation in layered media can be studied by conveniently writing Maxwell's equations in a 4×4 matrix. First, we show that in this matrix representation the boundary conditions of waves impinging on material can be imposed in a simpler way in such a way the transfer and scattering matrix formalism can be used in a natural way to obtain the transmittances and reflectances (Chuang, 2009; Hecht & Zajac, 1986). Next, we analytically solve the problem of axial propagation of an electromagnetic wave through a cholesteric elastomer by solidly rotating the laboratory reference system along the axial direction in the same way as the director $\mathbf{n}$. Finally, we compute the optical spectra of a cholesteric elastomer under the influence of an externally induced mechanical strain.

5.1 4 x 4 Matrix representation

The transversality of electromagnetic waves suggest to rewrite the well known Maxwell equations in a representation which permits to analyze, at the same time, the behaviour of the four transversal components (2 components for electric field $\bar{E}$ and 2 components for magnetic field $\bar{H}$). This formalism is frequently referred to as Marcuvitz-Schwinger representation (Marcuvitz & Schwinger, 1951). If we define the four-vector

$$\Psi(x,y,z) = \boldsymbol{\beta}(z)e^{i(k_x x+k_y y)-i\omega t} = \begin{pmatrix} e_x \\ e_y \\ h_x \\ h_y \end{pmatrix} e^{i(k_x x+k_y y)}e^{-i\omega t}, \tag{1.20}$$

with ω the angular frequency of the propagating wave and k_x, k_y the transversal components of wavevector. Maxwell's equations, inside a non-magnetic medium, can be written as:

$$\frac{\partial \boldsymbol{\beta}}{\partial z} = i\boldsymbol{A}(z).\boldsymbol{\beta}, \tag{1.21}$$

where the 4×4 matrix $\boldsymbol{A}(z)$ has the particular elements

$$\boldsymbol{A}(z)=\begin{pmatrix} -\dfrac{k_x\varepsilon_{zx}}{\varepsilon_{zz}} & -\dfrac{k_x\varepsilon_{zy}}{\varepsilon_{zz}} & \dfrac{k_xk_y}{k_0\varepsilon_{zz}} & k_0\left(1-\dfrac{k_x^2}{k_0^2\varepsilon_{zz}}\right) \\ -\dfrac{k_y\varepsilon_{zx}}{\varepsilon_{zz}} & -\dfrac{k_y\varepsilon_{zy}}{\varepsilon_{zz}} & k_0\left(-1+\dfrac{k_y^2}{k_0^2\varepsilon_{zz}}\right) & -\dfrac{k_xk_y}{k_0\varepsilon_{zz}} \\ k_0\left(\dfrac{k_xk_y}{k_0^2}-\varepsilon_{yx}+\dfrac{\varepsilon_{yz}\varepsilon_{zx}}{\varepsilon_{zz}}\right) & k_0\left(-\dfrac{k_x^2}{k_0^2}-\varepsilon_{yy}+\dfrac{\varepsilon_{yz}\varepsilon_{zy}}{\varepsilon_{zz}}\right) & -\dfrac{k_y\varepsilon_{yz}}{\varepsilon_{zz}} & \dfrac{k_x\varepsilon_{yz}}{\varepsilon_{zz}} \\ k_0\left(-\dfrac{k_y^2}{k_0^2}+\varepsilon_{xx}-\dfrac{\varepsilon_{xz}\varepsilon_{zx}}{\varepsilon_{zz}}\right) & k_0\left(-\dfrac{k_xk_y}{k_0^2}+\varepsilon_{xy}-\dfrac{\varepsilon_{xz}\varepsilon_{zy}}{\varepsilon_{zz}}\right) & \dfrac{k_y\varepsilon_{xz}}{\varepsilon_{zz}} & -\dfrac{k_x\varepsilon_{xz}}{\varepsilon_{zz}} \end{pmatrix} \tag{1.22}$$

ε_{ij}, with $i,j=x,y,z$ the elements of dielectric tensor and $k_0=2\pi/\lambda_0$ the wavenumber in free space. It is worth to mention that, in writing expressions (1.20), we have defined the dimensionless electric $\boldsymbol{e}$ and magnetic $\boldsymbol{h}$ fields related to $\boldsymbol{E}$ and $\boldsymbol{H}$ fields as follows:

$$\boldsymbol{e}=Z_0^{-1/2}\boldsymbol{E},\qquad \epsilon_0\boldsymbol{d}=Z_0^{-1/2}\boldsymbol{D},\qquad \boldsymbol{h}=Z_0^{1/2}\boldsymbol{H},\qquad \mu_0\boldsymbol{b}=Z_0^{1/2}\boldsymbol{B}, \tag{1.23}$$

with $Z_0=\sqrt{\mu_0/\epsilon_0}$ the free space impedance and ε_0 and μ_0 the permittivity and permeability of free space, respectively.

5.2 Boundary condition

Let us consider a cholesteric elastomer confined between two planes at $z = 0$ and $z = d$ where the optical properties continuously depend on z and the surrounding medium is air. This implies that the elements of dielectric tensor ε_{ij} depend only on the $z-$ coordinate. An incident electromagnetic wave, having wavevector $\boldsymbol{k}_t = (k_x, k_y)$, impinges from the left side of the cholesteric elastomer. This electromagnetic wave propagates through the sample and it is transmitted and reflected outside the medium having the structure

$$\boldsymbol{\beta}(x,y,z)=\boldsymbol{\beta}(z)\exp[k_xx+k_yy], \tag{1.24}$$

because the phase matching condition implies the continuity of the tangential components of $\boldsymbol{k}$.

The general solution of the differential equation (1.21) for electromagnetic waves propagating in homogeneous media is the superposition of four plane waves, two left-going and two right-going waves. With this in mind, we state the procedure to find the amplitudes of the transmitted (at $z = d$) and reflected waves in terms of incident waves at z=0. This implies the definition of the following quantities (Altman & Sucky, 1991):

i. The propagation matrix $U(0, z)$, that is implicitly defined by the equations

$$\boldsymbol{\beta}(z)=\boldsymbol{U}(0,z).\boldsymbol{\beta}(0),\qquad \boldsymbol{U}(0,0)=\mathbf{1}, \tag{1.25}$$

where $\mathbf{1}$ is the identity matrix and $\boldsymbol{U}(0,z)$ satisfies the same propagation equation (1.21) found for $\boldsymbol{\beta}$:

$$\partial_z \boldsymbol{U}(0,z) = i\boldsymbol{A}(z).\boldsymbol{U}(0,z) \; ; \tag{1.26}$$

the propagation matrix gives the right-side field amplitudes of the cholesteric elastomer as function of the left-side ones.

ii. For a specific value d , the transfer matrix is defined as $\boldsymbol{U}(0,z)$.

iii. The scattering matrix $\boldsymbol{S}$, that gives the output field as function of the incident one. The matrix $\boldsymbol{S}$ is defined as:

$$\boldsymbol{\alpha}_{out} = \boldsymbol{S}.\boldsymbol{\alpha}_{in} \tag{1.27}$$

where $\boldsymbol{\alpha}_{in}$ and $\boldsymbol{\alpha}_{out}$ are the amplitudes of the in-going and out-going waves.

To find out $\boldsymbol{S}$, we must express the field, in any one of the external media, as a superposition of planes waves, by setting:

$$\boldsymbol{\beta} = \mathbf{T}.\boldsymbol{\alpha}; \qquad \boldsymbol{U}_{\alpha}(0,d) = \boldsymbol{T}^{-1}.\boldsymbol{U}(0,d).\boldsymbol{T}, \tag{1.28}$$

where

$$\boldsymbol{\alpha} = (a_1^+, a_2^+, a_1^-, a_2^-)^T \, . \tag{1.29}$$

The relation $\boldsymbol{\beta} = \boldsymbol{T}.\boldsymbol{\alpha}$ can be interpreted as a basis change in the four dimensional space of the *state vectors* $\boldsymbol{\beta}$. The columns of T are the $\boldsymbol{\beta}$ vectors representing the four plane waves generated by the incident waves in the two external medium (here we assume as identical). The elements of vector $\boldsymbol{\alpha}$ are the amplitudes of the four plane wave. The choice of the new basis could be different depending on the particular problem. By setting

$$\boldsymbol{U}(\boldsymbol{\alpha}) = \begin{pmatrix} \boldsymbol{U}_{ff} & \boldsymbol{U}_{bf} \\ \boldsymbol{U}_{fb} & \boldsymbol{U}_{bb} \end{pmatrix}, \tag{1.30}$$

the scattering matrix writes:

$$S = \begin{pmatrix} \boldsymbol{U}_{ff} - \boldsymbol{U}_{bf}\boldsymbol{U}_{bb}^{-1}\boldsymbol{U}_{fb} & \boldsymbol{U}_{bf}\boldsymbol{U}_{bb}^{-1} \\ -\boldsymbol{U}_{bb}^{-1}\boldsymbol{U}_{fb} & \boldsymbol{U}_{bb}^{-1} \end{pmatrix}. \tag{1.31}$$

In equations (1.29) and (1.30) the symbols + and f (- and b) mean forward (backward) propagating waves.

We point out that the methods of transfer and scattering matrices are very useful in studying the plane wave transmission and reflection from surfaces or a multilayered medium.

Eq. (1.21) can be formally integrated over a certain distance d of the cholesteric

$$\boldsymbol{\beta}(d) = e^{i\int_0^d \boldsymbol{A}(z')dz'} \cdot \boldsymbol{\beta}(0) \, , \tag{1.32}$$

and by straight comparison of Eqs. (1.25) and (1.32), the transfer matrix $\boldsymbol{U}(0,z)$ is defined as:

$$\boldsymbol{U}(0,d) = e^{i\int_0^d \boldsymbol{A}(z')dz'}, \tag{1.33}$$

where plane waves are incident and reflected in the half-space $z < 0$ and a plane waves are transmitted on the half-space $z > d$.

It can be seen immediately that the problem of finding $\boldsymbol{U}(0,z)$ is reduced to find a method to integrate expression (1.33) on the whole cholesteric elastomer. Because of the non-homogeneity of the medium proposed here, we consider it as broken up into many thin parallel layers and treating each as if it had homogeneous anisotropic optical parameters (Berreman & Scheffer, 1970). In this way, $\boldsymbol{U}(0,z)$ is obtained by multiplying iteratively the matrix for each sublayer from $z = 0$ to $z = d$.

In next section, we will show that for axial propagation in a cholesteric elastomer, and by choosing appropriately a reference system, the system (1.21) and the transfer matrix U have completely analytical solutions.

Now, we proceed to give the explicit form of the four-vector $\boldsymbol{\beta}$ for the surrounding medium (free space). As said above, the general solution of the differential equation (1.21) for electromagnetic waves propagating in homogeneous media is the superposition of forward and backward propagating waves. We consider an incident wave from left-half space with wavevector, $\boldsymbol{k} = (k_x, k_y, k_z) = k_0 n_d(\sin\theta\cos\varphi, \sin\theta\sin\varphi, \cos\theta)$, where n_d is the refractive index of surrounding medium, θ is the angle made between $\boldsymbol{k}$ and $z-axis$ and φ is the angle made between $\boldsymbol{k}$ and $x-axis$ in the $xy-plane$. For an arbitrary polarization state the solutions of (1.21) can be expressed as (Lakhtakia & Reyes, 2006; Espinosa-Ortega & Reyes, 2008):

$$\boldsymbol{\beta}(z) = \begin{pmatrix} \left(a_L \dfrac{i\hat{u} - \hat{v}_+}{\sqrt{2}} - a_R \dfrac{i\hat{u} + \hat{v}_+}{\sqrt{2}}\right) e^{ik_z z} + \left(-r_L \dfrac{i\hat{u} - \hat{v}_-}{\sqrt{2}} + r_R \dfrac{i\hat{u} + \hat{v}_-}{\sqrt{2}}\right) e^{-ik_z z} \\ -in_d \left(a_L \dfrac{i\hat{u} - \hat{v}_+}{\sqrt{2}} + a_R \dfrac{i\hat{u} + \hat{v}_+}{\sqrt{2}}\right) e^{ik_z z} + in_d \left(r_L \dfrac{i\hat{u} - \hat{v}_-}{\sqrt{2}} + r_R \dfrac{i\hat{u} + \hat{v}_-}{\sqrt{2}}\right) e^{-ik_z z} \end{pmatrix}, \quad \text{for } 0 \le z, \tag{1.34}$$

where a_L, a_R represent the amplitude of incident propagating waves and r_L, r_R denote the reflection amplitude of propagating waves. The subscript indexes R and L correspond to right- and left-circularly polarized wave, respectively. The unit vectors $\hat{u}$ and $\hat{v}$ are defined as

$$\hat{u} = -\hat{x}\sin\varphi + \hat{y}\cos\varphi \quad \text{and} \quad \hat{v}_\pm = \mp(\hat{x}\cos\varphi + \hat{y}\sin\varphi)\cos\theta + \hat{z}\sin\theta, \tag{1.35}$$

with $\hat{x}$, $\hat{y}$, $\hat{z}$ the unit vectors parallel to the $x-, y-, z-axis$, respectively. In the region $z \ge d$, we write the transmitted field as

$$\boldsymbol{\beta}(z) = \begin{pmatrix} \left(t_L \dfrac{i\hat{u} - \hat{v}_+}{\sqrt{2}} - t_R \dfrac{i\hat{u} + \hat{v}_+}{\sqrt{2}}\right) e^{ik_z(z-d)}, \\ -in_d \left(t_L \dfrac{i\hat{u} - \hat{v}_+}{\sqrt{2}} + t_R \dfrac{i\hat{u} + \hat{v}_+}{\sqrt{2}}\right) e^{ik_z(z-d)} \end{pmatrix}, \quad \text{for } z \ge d. \tag{1.36}$$

At the tangential components of $\boldsymbol{e}$ and $\boldsymbol{h}$ must be continuous across the planes $z=0$ and $z=d$, the boundary values $\boldsymbol{\beta}(0)$ and $\boldsymbol{\beta}(d)$ can be fixed as:

$$\boldsymbol{\beta}(0)=\frac{1}{\sqrt{2}}\boldsymbol{Q}\cdot\begin{pmatrix} a_R \\ a_L \\ r_R \\ r_L \end{pmatrix} \quad \text{and} \quad \boldsymbol{\beta}(d)=\frac{1}{\sqrt{2}}\boldsymbol{Q}\cdot\begin{pmatrix} t_R \\ t_L \\ 0 \\ 0 \end{pmatrix}, \tag{1.37}$$

where

$$\boldsymbol{Q}=\begin{pmatrix} \cos\theta\cos\varphi & \cos\theta\cos\varphi & \cos\theta\cos\varphi & \cos\theta\cos\varphi \\ \cos\theta\sin\varphi & \cos\theta\sin\varphi & \cos\theta\sin\varphi & \cos\theta\sin\varphi \\ in_d\cos\theta\cos\varphi & -in_d\cos\theta\cos\varphi & in_d\cos\theta\cos\varphi & -in_d\cos\theta\cos\varphi \\ in_d\cos\theta\sin\varphi & -in_d\cos\theta\sin\varphi & in_d\cos\theta\sin\varphi & -in_d\cos\theta\sin\varphi \end{pmatrix} + \begin{pmatrix} i\sin\varphi & -i\sin\varphi & -i\sin\varphi & i\sin\varphi \\ -i\cos\varphi & i\cos\varphi & i\cos\varphi & -i\cos\varphi \\ -n_d\sin\varphi & -n_d\sin\varphi & n_d\sin\varphi & n_d\sin\varphi \\ n_d\cos\varphi & n_d\cos\varphi & -n_d\cos\varphi & -n_d\cos\varphi \end{pmatrix}. \tag{1.38}$$

If we restrict our analysis to the case when the electromagnetic wave is incident parallel to $z-axis$, the angles φ and θ equal zero and expression for $\boldsymbol{Q}$ is reduced to

$$\boldsymbol{Q}_0=\begin{pmatrix} 1 & 1 & 1 & 1 \\ -i & i & i & -i \\ in_d & -in_d & in_d & -in_d \\ n_d & n_d & -n_d & -n_d \end{pmatrix}. \tag{1.39}$$

Using Eqs. (1.32), (1.33) and (1.37) the problem of reflection-transmission can be established as follows

$$\begin{pmatrix} t_R \\ t_L \\ 0 \\ 0 \end{pmatrix}=\boldsymbol{M}\cdot\begin{pmatrix} a_R \\ a_L \\ r_R \\ r_L \end{pmatrix} \tag{1.40}$$

where $\boldsymbol{M}=\boldsymbol{Q}^{-1}\cdot\boldsymbol{U}(0,d)\cdot\boldsymbol{Q}$ and $U(0,d)$ are defined in (1.33). Notice that the matrix equation (1.40) gives a set of coupled equations relating amplitudes a_L, a_R, r_L and r_R (from $z\leq 0$) to transmitted amplitudes t_L, t_R (for $z\geq d$).

The scattering matrix $\boldsymbol{S}$ relates amplitudes t_L, t_R, r_L and r_R with the incident amplitudes a_L, a_R. This relation can be expressed in terms of matrix M as follows (Avendaño et al., 2005)

$$\begin{pmatrix} t_R \\ t_L \\ r_R \\ r_L \end{pmatrix} = S\cdot\begin{pmatrix} a_R \\ a_L \end{pmatrix} \quad \text{with} \quad S = \begin{pmatrix} t_{RR} & t_{RL} \\ t_{LR} & t_{LL} \\ r_{RR} & r_{RL} \\ r_{LR} & r_{LL} \end{pmatrix}, \tag{1.41}$$

Co-polarized coefficients have both subscripts identical, and cross-polarized have different subscripts. The square of the amplitudes of t and r is the corresponding transmittance and reflectance; thus, $T_{RR} = |t_{RR}|^2$ is the transmittance corresponding to the transmission coefficient t_{RR}, and so on. In the absence of dissipation of energy inside the sample, the principle of conservation of energy must be satisfied, this means that

$$T_{RR} + T_{LR} + R_{RR} + R_{LR} = 1 \quad \text{and} \quad T_{RL} + T_{LL} + R_{RL} + R_{LL} = 1. \tag{1.42}$$

Before ending this section, we mention that an alternative to find the transmission and reflection coefficients is using the expressions given by (1.30) and (1.31). Also, the system of equations (1.40) can be solved numerically to find the scattering matrix.

5.3 The Oseen transformation

Using a numerical procedure the set of coupled differential equations (1.21) can be solved in a straight way. Nevertheless, the intrinsic helical symmetry of a cholesteric elastomer allows to suggest the possibility of finding a reference system, for normally incident wave, for which the solution can be obtained analytically. For this aim, it is convenient to write the coupled equations in a frame of reference in which the matrix A is diagonal and not dependent on z propagation coordinate. This can be realized by rotating solidly and uniformly the four-vector $\boldsymbol{\beta}$ around $z-$ axis, with the principal axes of ε_{ij} making constant angles with z. The required transformation can be realized by setting:

$$\tilde{\boldsymbol{\beta}} = \begin{pmatrix} \tilde{e}_x \\ \tilde{e}_y \\ \tilde{h}_x \\ \tilde{h}_y \end{pmatrix} = \mathcal{R}(-qz).\boldsymbol{\beta} \quad \text{and} \quad \boldsymbol{H} = \mathcal{R}(-qz).\boldsymbol{A}(z).\mathcal{R}(qz) \tag{1.43}$$

with $\mathcal{R}(qz)$ the rotation matrix defined as

$$\mathcal{R}(qz) = \exp(\mathbf{R}qz) \equiv \cos qz\mathbf{1} + \sin qz\mathbf{R}, \qquad \mathbf{R} = \begin{pmatrix} 0 & -1 & 0 & 0 \\ 1 & 0 & 0 & 0 \\ 0 & 0 & 0 & -1 \\ 0 & 0 & 1 & 0 \end{pmatrix}, \tag{1.44}$$

$q = q_o / \eta$ and 1I the 4×4 identity matrix. This transformation is known as Oseen's transformation (Oseen, 1933). For axial propagation and by considering the explicit form of dielectric tensor ε_{ij} for the cholesteric elastomer, the Eqs. (1.21) are reduced to

$$\frac{d\tilde{\boldsymbol{\beta}}}{dz'} = i\boldsymbol{H}.\tilde{\boldsymbol{\beta}}, \tag{1.45}$$

the matrix $\boldsymbol{H}$ is given by

$$\boldsymbol{H} = \begin{pmatrix} 0 & -iq & 0 & 2\pi\lambda^{-1} \\ iq & 0 & -2\pi\lambda^{-1} & 0 \\ 0 & -2\pi\epsilon_{\perp}\lambda^{-1} & 0 & -iq \\ 2\pi\epsilon_{\parallel}\lambda^{-1}\epsilon(\eta) & 0 & iq & 0 \end{pmatrix} \tag{1.46}$$

and

$$\epsilon(\eta) = \frac{\epsilon_{\perp}}{\epsilon_{\perp}\sin^2\alpha(\eta) + \epsilon_{\parallel}\cos^2\alpha(\eta)} \tag{1.47}$$

λ is the wavelength in free space. Here $\epsilon_{\parallel}$ and $\epsilon_{\perp}$ represent the principal values of ϵ and μ in the rotating frame with axes x_1, x_2, z. In what follows, and to simplify notation, we omit the symbol ($\sim$) of four-vector $\tilde{\boldsymbol{\beta}}$, $\tilde{\mathbf{e}}$ and $\tilde{\mathbf{h}}$. Unless we say the contrary, by writing $\boldsymbol{\beta}$, $\boldsymbol{e}$ and $\boldsymbol{h}$, we will always mean the fields in the rotating frame and z the dimensionless variable.
Since the system matrix $\boldsymbol{H}$ is $z-$ independent, the propagation equation (1.45) admits four solutions having the form of plane waves

$$\boldsymbol{\beta}^j(z) = \boldsymbol{t}^j \exp(in_j z), \tag{1.48}$$

where $n_j, \boldsymbol{t}^j$ are the eigenvalues and eigenvectors of $\boldsymbol{H}$, respectively. They are given by:

$$\begin{aligned} n_{1,2}^2 &= \frac{4\pi^2\epsilon_m \mp 2\pi u}{\lambda^2} + q^2, \\ \boldsymbol{t}_1^{\pm} &= c_1\left(\pm n_1 u_1\lambda, i\lambda q(4\pi\epsilon_m + u_1), \mp 2iq\epsilon_m n_1\lambda^2, 2\epsilon_{\parallel}\epsilon(\eta)\pi u_1 + 2\epsilon_m q^2\lambda^2\right), \\ \boldsymbol{t}_2^{\pm} &= c_2\left(\pm n_2 u_2\lambda, i\lambda q(4\pi\epsilon_m + u_2), \mp 2iq\epsilon_m n_2\lambda^2, 2\epsilon_{\parallel}\epsilon(\eta)\pi u_2 + 2\epsilon_m q^2\lambda^2\right), \end{aligned} \tag{1.49}$$

where

$$\begin{aligned} u &= 2\sqrt{a_c^2\pi^2 + \epsilon_m q^2\lambda^2}, \qquad u_{1,2} = 2a_c\pi \mp u, \\ c_k &= \left|4n_k\lambda\left(\pi\epsilon_{\parallel}\epsilon(\eta)u_k^2 + 2\epsilon_m q^2\lambda^2(2\pi\epsilon_m + u_k)\right)\right|^{-1/2}, \quad (k = 1,2) \\ a_c &= \frac{\epsilon_{\parallel}\epsilon(\eta) - \epsilon_{\perp}}{2}, \qquad \epsilon_m = \frac{\epsilon_{\parallel}\epsilon(\eta) + \epsilon_{\perp}}{2}. \end{aligned} \tag{1.50}$$

As shows (1.48) the internal field can be represented as a superposition of the four eigenwaves (*amplitude representation*), by setting

$$\boldsymbol{\beta}(z) = a_j \boldsymbol{t}^j \exp(in_j z) \equiv \mathbf{T}.\boldsymbol{\alpha}(z), \tag{1.51}$$

where $\boldsymbol{T}$ is the matrix whose $j-th$ column coincides with t^j and $\boldsymbol{\alpha}(z)$ is the 4-vector with components $a_j \exp(in_j z)$. Obviously, $\boldsymbol{\alpha}$ and $\boldsymbol{\beta} = \boldsymbol{T}.\boldsymbol{\alpha}$ represent the same state in two different sets of basis vectors. The metrization of the state space is obtained by defining a metric tensor G and a scalar product $\beta_1^{\dagger}.G_{\beta}.\beta_2 \equiv \alpha_1^{\dagger}.G_{\alpha}.\alpha_2$, where G_{β} and $G_{\alpha} = T^{\dagger}.G_{\beta}.T$ are the matrices representing G in the two sets of basis vectors. Setting

$$G_{\beta} = \begin{pmatrix} 0 & 0 & 0 & 1 \\ 0 & 0 & -1 & 0 \\ 0 & -1 & 0 & 0 \\ 1 & 0 & 0 & 0 \end{pmatrix}, \tag{1.52}$$

the norm of the state vector represents the time average of the z – component of the Poynting vector, and the tensor G satisfies the relation $G \equiv G^{\dagger} \equiv G^{-1}$.

In lossless media the z – derivative of the norm is identically zero and the matrix $G_{\beta}.H$ is self-adjoint: $G_{\beta}.H = (G_{\beta}.H)^{\dagger} \equiv H^{\dagger}.G_{\beta}$. This property and the fact that the eigenvalue equation for H is biquadratic imply that the eigenvalues are $n_1, n_2, n_3 = -n_1, n_4 = -n_2$, with n_j real or purely imaginary. It is worth mentioning that the normalization constants c_1, c_2 were obtained using the metric tensor G; this means that $c_k = \left| t_k^* G t_k \right|^{-1/2}$ with $k = 1, 2$, and t_k^* is the conjugated complex of eigenvectors t_k.

Only the modes n_1 show a band gap for λ within the interval defined for the positive roots of equation $n_1 = 0$. Band edges are given by $\lambda_1 = 2\pi\eta\sqrt{\epsilon_{\perp}} / q_o$ and $\lambda_2 = 2\pi\eta\sqrt{\epsilon_{\parallel}\epsilon(\eta)} / q_o$. Here, the modes n_1 are pure imaginary and their corresponding eigenvectors $1^{\pm}$ define evanescent and linearly polarized standing waves. The central wavelength of the bandgap is

$$\lambda_c = \frac{\lambda_1 + \lambda_2}{2} = \frac{p\eta\sqrt{\epsilon_{\perp}}}{2} \left(1 + \sqrt{\frac{\epsilon_{\parallel}}{\epsilon_{\perp} + \epsilon_a \left(\eta^{3/2} - 1 \right) / (r - 1)}} \right) \tag{1.53}$$

where Eqs. (1.19) and (1.47) were substituted in the last equation. This equation demonstrates clearly that for a positively anisotropic elastomer, the reflected wavelength λ_c increases by stretching the sample along the helix axis. This behaviour is in qualitative agreement with the biaxial extension experiments performed by Finkelmann et al. (Finkelmann et al., 2001) in which λ_c decreases due to an effective compression along the helix axis. We finally observe that within the gap the polarization of the propagating eigenwaves $2^{\pm}$ is nearly circular and, in general, they are elliptically polarized.

6. Reflection bands of distorted elastomers with and without defects

It is a known fact that structural chiral materials presents the circular Bragg phenomenon in a wavelength regime (Hirota et al., 2008), where normal incident electromagnetic plane waves of left- and right-circular polarization states are reflected and transmitted differently, i. e., light of right handedness is highly reflected in a right helical structure whereas a similar plane wave but of the reverse handedness is not. Thus, structural chiral materials are circular-polarization rejection filters in optics (Avendaño et al., 2005; de Gennes & Prost, 1993; Macleod, 2001). In addition, cholesteric elastomers are very sensitive to external stimuli as electric fields, temperature and mechanical stress. Therefore, it is possible to control the Bragg regime with these sorts of stimuli. Cholesteric elastomers are formed by monomers of liquid crystals cross-linked to polymeric chains that produces a flexible material whose molecular order is similar to cholesteric liquid crystals with the advantage that in this new material it is feasible to change the optical properties by means of macroscopic deformations. In this section we focus in the control of circular Bragg phenomenon under the influence of an externally induced mechanical strain applied parallel to the helical axis of a slab of cholesteric elastomer.

One or more defects in a periodic structure may give rise to resonant modes inside the photonic band gaps, namely standing waves with huge energy density localized in the proximity of the defects (defect modes). A conventional 1D structure with only one defect can be considered as a Fabry-Perot interferometer in which the reflecting layers at the two sides of the cavity are constituted by 1D crystals whose thickness is comparable with the attenuation length of the standing waves within the gap, which are exponentially attenuated. The cavity acts here as a defect in the periodic structure. Very interesting optical properties are obtained by considering: 1) anisotropic periodic structures, which display two different sets of band gaps for light with different polarization states, and 2) samples with more than one defect.

In this section we also consider theoretically light propagation along the helix axis of samples in which the periodic structures are cholesteric elastomers and the thickness of the cavities goes to zero. Any defect reduces therefore to a simple discontinuity plane within the periodic structure. Such samples can be obtained as follows: 1) we consider first an cholesteric elastomer without defects between planes orthogonal to the helix axis; 2) then we cut the sample in such a way to obtain two or more layers between parallel planes; and 3) we finally rotate any layer with respect to the preceding one around their common helix axis by a given angle ξ (twist angle) (Lakhtakia, 2000; Schmidtke et al., 2003; Ozaki et al., 2003; Song et al., 2004).

Some interesting numerical results have already been found recently for helical photonic crystals with only one twist defect (Becchi et al., 2004; Hodgkinson et al., 2000; Kopp & Genack, 2002; Schimdtke & Stille, 2003; Wang & Lakhtakia, 2003; Oldano, 2003; Kopp & Genack, 2003). Here we present a theoretical and analytical approach for samples with any number of twist defects under the action of axial strain.

6.1 Mechanical control of optical spectra in a cholesteric elastomer without defects

The reflectances and transmittances obtained from the scattering matrix or the transfer matrix are in terms of the elongation η, the wavelength λ and the incidence angles of light, θ (angle between the light direction and the helical axis) and φ (angle between the light direction and the $x-$axis). By numerically solving the set of equations (1.40) for oblique incidence we obtained these optical spectra for a sample of siloxane backbone chain reacting with 90% mol and 10% of the flexible difunctional cross-linking groups (di-11UB). The rod like mesogenic groups are present in the proportion 4:1 between the nematic 4-pentylphenyl-4'-(4-buteneoxy) benzoate (PBB) and the derivative of chiral cholesterol penteonate (ChP) (Cicuta et al., 2002). The behaviour of the optical spectra for another material is expected to be qualitatively similar to the presented here; the material parameters are: $r = 1.16$, $L = 10.7\mu m$, $p/2 = 214nm$, $\epsilon_{\perp} = 1.91$, $\epsilon_{\parallel} = 2.22$, $\mu = 1$.

The optical spectra show a circular Bragg regime in the R_{RR} co-polarized transmittances and reflectances, which depends on the axial elongation of the cholesteric and the incidence angles of light. These spectra are consistent with the circular Bragg phenomenon for which the right

circularly polarized wave impinging a right-handed elastomer, is highly reflected, while the left circularly polarized wave is transmitted, as we can see in Fig. 3, for reflectances
We see in this figure that the center of the bandgap blue-shifts as the incidence angle increases, as it occurs in the absence of stress. We also observe that by increasing the elongation, the band width decreases as can be seen by comparing the right and the left hand columns of this figure, that correspond to $\eta = 1$ (elastomer under no deformation) and $\eta_m = 1 + (\eta_M - 1) / 2 = 1.052$ (elastomer submitted to half of its critical elongation). Moreover, when the strain is the critical $\eta_M = 1.162^{2/3} = 1.1040$, the bandgap disappears due to the fact the cholesteric director is completely aligned with the helical axis as can be observed in Fig. 4. This effect opens up the door for proposing novel devices to mechanically control the light flow, since it allows to switch off a bandgap by applying a mechanical stress to the elastomer. This is clearly illustrated in Fig. 4 where the bandwidth diminishes as a function of the deformation for normal incidence. Further results confirm the displacement of the band reflection for R_{RR} for larger incidence angles as η get larger (Espinosa-Ortega & Reyes, 2008). Therefore, these results show the possibility of mechanically control the circular Bragg phenomenon for tuning and switching applications. On the other hand, analytical results show that the reflected wavelength at normal incidence red-shifts by stretching the elastomer along the helical axis.

6.2 Optical spectra of elastomers with defects under axial strain

In the last section the mechanical control of optical spectra in a slab of cholesteric elastomer was studied. The light impinges on the slab at normal and oblique incidence. The solutions are found performing a numerical integration to find the transmission and reflection coefficients as a function of the mechanical elongation and the incidence angles. However the strongest results are in normal incidence where an analytical solution of the problem is found. Particularly, if a twist defect with $\xi = \pi / 2$ is introduced in the center of the slab, it causes a break of the symmetry that gives rise to a peak in transmission, in the middle of the Bragg regime. Thus, in the case of normal incidence, the wavelength of this peak, as a function of the elongation η, is easy to find from the results for a slab without defect. As said above the band gap and transmittance of a cholesteric elastomer can be controlled by elongating the material along its helix axis, then, the elastically control of the lasing in a cholesteric elastomer is deduced.
A result that was shown in the last section is that the axial elongation in a cholesteric elastomer close the band gap when the cholesteric director is fully oriented along the helix axes. This effect is a consequence of the presence of optical axis in the locally anisotropic material forming the sample, which in turn originates the existence of a pseudoisotropic curve. In this section we explore the optical properties around this curve and its effect over the dwell time by means of the analytical results for normal incidence. The enhancement of the dwell time is desirable for optical application, hence, the conditions in which this occurs shall widely explored (Mota et al., 2010). Finally, the mechanical tuning of two or more defect modes in the cholesteric elastomer is explored by taking into account the coupling and interference among the various defect modes.

6.2.1 Singlets

We consider now a sample between the planes $z = -\ell$ and $z = +\ell$ with only one twist defect at $z = 0$, which divides the sample in two regions referred to as a for $z < 0$ and b for $z > 0$. Next, we approach the problem quantitatively by means of exact equations for normal incidence.

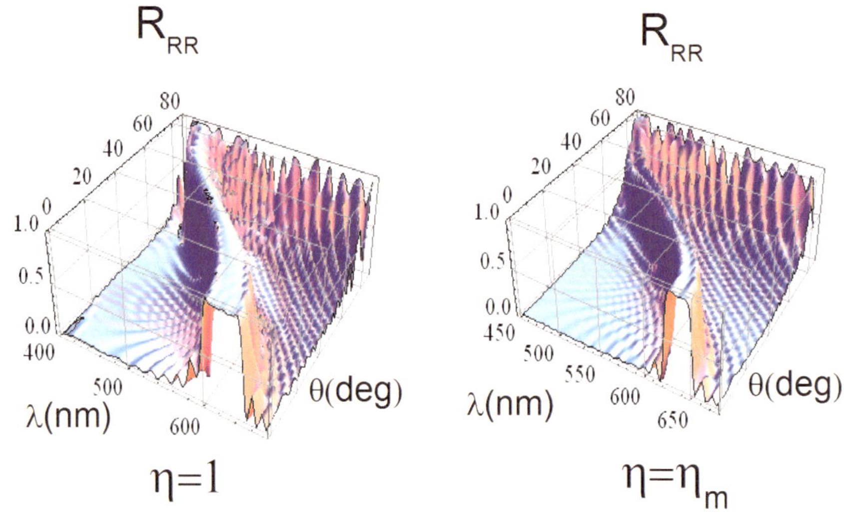

Fig. 3. Co-polarized reflectance R_{RR} versus the wavelength λ and the incidence angle θ for the elongations $\eta = 1$ and $\eta = \eta_m$.

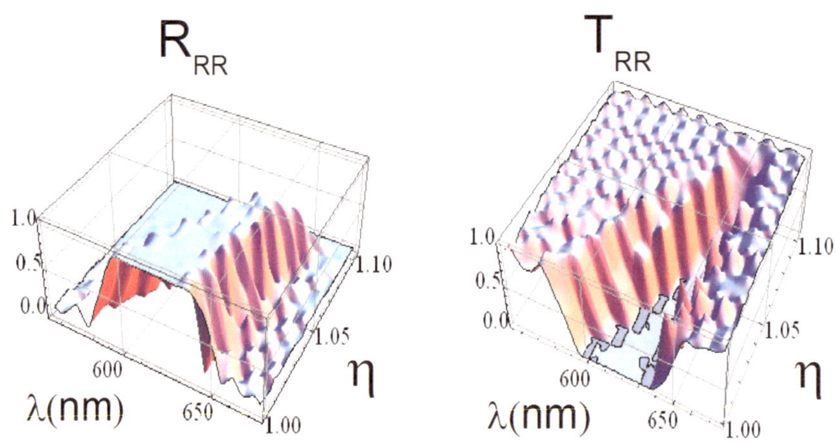

Fig. 4. Co-polarized transmittances T_{RR} and reflectances R_{RR} as a function of the wavelength λ and the axial elongation η for normal incidence.

6.2.1.1 Thick samples

Let us first consider an unbounded structure, i.e., the limit $l \to \infty$. Any solution of Maxwell equations can be written as $\boldsymbol{\beta}(z) = \boldsymbol{\beta}_a(z)\Theta(z) + \boldsymbol{\beta}_b(z)\Theta(-z)$, where $\Theta(z)$ is the Heaviside step function and $\boldsymbol{\beta}_k(z) = \boldsymbol{T}_k \boldsymbol{\alpha}_k(z)$ $(k = a,b)$ is a linear combination of the eigenwaves $\boldsymbol{t}_k^j \exp(in_j z)$ defined above in Eq. (1.48). Again, we mention that the choice of vector α_k is the same as expressed in (1.29). The eigenvectors t_a^j and t_b^j are given by Eq. (1.49) in two different frames a and b whose axes x_{2a}, x_{2b} make an angle ξ . In a frame having as axis x_2 the bisector of x_{2a}, x_{2b}, such vectors are obtained by applying the rotation matrices $\mathcal{R}(-\xi/2)$ and $\mathcal{R}(\xi/2)$ to the vectors t^j defined by Eq. (1.49). The corresponding matrices $\boldsymbol{T}_a$ and $\boldsymbol{T}_b$ are therefore given by

$$\boldsymbol{T}_a = \mathcal{R}(-\xi/2).\boldsymbol{T}, \qquad \boldsymbol{T}_b = \mathcal{R}(\xi/2).\boldsymbol{T}\,. \tag{1.54}$$

In the limit $\ell \to \infty$ the exponential factors of the first component of $\boldsymbol{\alpha}_a = \boldsymbol{T}_a^{-1}.\boldsymbol{\beta}_a$ and of the third component of $\boldsymbol{\alpha}_b = \boldsymbol{T}_b^{-1}.\boldsymbol{\beta}_b$ (corresponding to the eigenwaves 1^+ in the region a and 1^- in the region b), diverge for z going to $-\infty$ and $+\infty$, respectively. Their amplitude must therefore be zero. The tangential continuity of the vectors $\boldsymbol{e}$ and $\boldsymbol{h}$ at $z = 0$, give four homogeneous equations for the other six components. It is found that the amplitude of the evanescent modes is much larger than that of the propagating ones at the defect frequency ω_d and the defect mode has approximately the profile $\exp(-|z|/l_d)$ whose line width is $l_d \equiv 1/n_1(\omega_d)$ (Becchi et al., 2004).
Particularly, if the twist angle is $\pi/2$, the frequency of the peak created by this defect in the middle of the circular Bragg regime is: $\omega_d \equiv \dfrac{\omega_1 + \omega_2}{2}$ whose band width is $\Delta\omega_d \equiv \omega_1 - \omega_2$.
Also, the angle $\gamma_{em}(\omega_d)$ between the electric and magnetic field of the evanescent modes at the defect frequency, may be obtained from the components of their corresponding eigenvectors given by expressions (1.49).
Thus, the defect frequency depends on the axial elongation η and the fractional shape anisotropy r as

$$\omega_d(\eta, r) = \frac{\omega_1 + \omega_2}{2} = \frac{cq_0}{\eta}\left(\frac{1}{\sqrt{\epsilon_\perp}} + \frac{1}{\sqrt{\epsilon_\parallel \epsilon(\eta, r)}}\right), \tag{1.55}$$

where c is the speed of light in vacuum and $\epsilon(\eta, r)$ is given by Eq. (1.47), and the defect wavelength $\lambda_c = 2\pi c/\omega_d$ at the center of the bandgap is given by Eq. (1.53). Fig. 5 shows λ_c versus the axial elongation η and fractional shape anisotropy r. It shows that λ_c redshifts by enlarging both η and r for oblate (a) and prolate materials (b) which allows the possibility of mechanically tune the defect mode.
In Fig. 6 the inverse line width l_d^{-1}, band width $\Delta\omega_d$, angle between the electric and magnetic field of the evanescent modes at the defect frequency $\gamma_{em}(\omega_d)$ are plotted in different scales versus η and r. In this figure we observe a locus where $\eta = \eta_M$, l_d^{-1} (l_d diverges), $\Delta\omega_d$ and $\gamma_{em}(\omega_d)$ vanish. In this locus the attenuation for the defect mode is null since $l_d^{-1} = 0$, the circular Bragg regime closes $\Delta\omega_d = 0$, and the Poynting vector is null since $\gamma_{em}(\omega_d) = 0$. We shall call this curve in the $\eta - r$ space the pseudoisotropic curve.

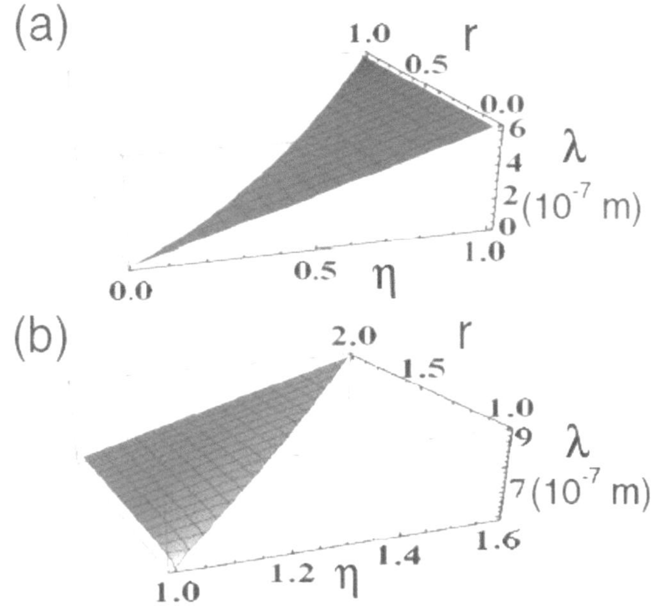

Fig. 5. Defect wavelength λ_d versus η and r for a) oblate ($0 < r < 1$) and b) ($r > 1$) prolate cholesteric elastomers axially elongated. Other parameters are, $p/2 = 214\ nm$, $\epsilon_{\perp} = 1.91$, $\epsilon_{\parallel} = 2.22$. These correspond to a typical cholesteric elastomer.

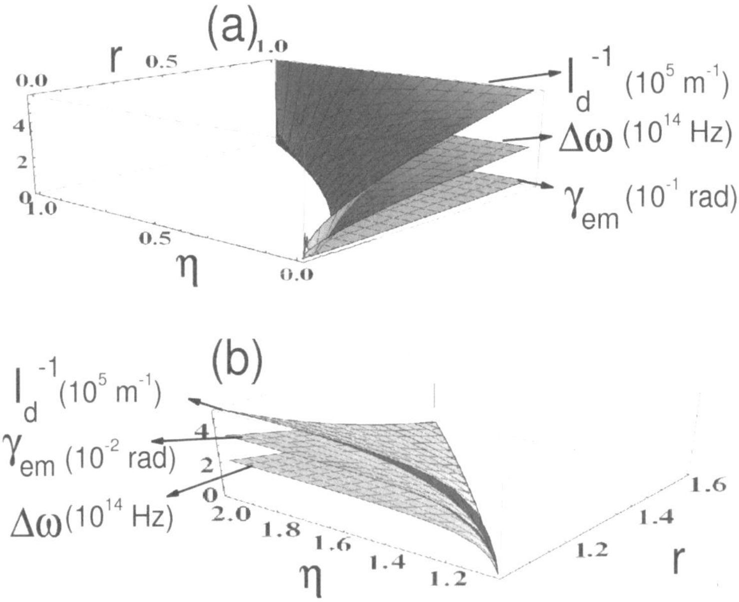

Fig. 6. Line width $l_d^{-1}(\omega_d)$, band width $\Delta\omega$ and angle between magnetic and electric field $\gamma_{em}(\omega_d)$ for the evanescent mode versus η and r. a) Oblate and b) prolate cholesteric elastomers. For the same system as Fig. 3.

This slab of cholesteric elastomer with a twist defect has potential optical applications as, for instance, Fabry-Perot resonators in lasers, hence, the enhancement of the photonic dwell time is very attractive. On the other hand, Kopp and Genack have shown that the photonic dwell time saturates versus thickness of the slab after a certain crossover thickness. It has been shown (Espinosa-Ortega & Reyes, 2008) that the deformation of the elastomer may close the band gap when the cholesteric director is fully oriented along the helix axis. This effect is a consequence of the presence of optical axis in the locally anisotropic material forming the elastomer, which in turn originates the existence of a pseudoisotropic curve (Abdulhalim, 1999). We carefully review the implications of the existence of a pseudoisotropic curve in the physical system and its effect over the photonic dwell time.

Above, it was shown that the cholesteric elastomer with a twist defect presents a pseudoisotropic curve where $\eta = \eta_M$, the line width (l_d) diverges and the band width ($\Delta\omega_d$) and angle between the electric and magnetic field ($\gamma_{em}(\omega_d)$), at the frequency defect, are null.

Therefore, for values of the axial elongation η and the fractional shape anisotropy r on the pseudoisotropic curve the following facts are found (Mota et al., 2010):

i. the transport of waves within the sample having the defect frequency is much more efficient since the attenuation for the defect mode is null, $l_d^{-1} = 0$,

ii. the circular Bragg regime closes since $\Delta\omega_d = 0$,

iii. there are no energy leaks in the sample since $\gamma_{em}(\omega_d) = 0$ amounts to a null Poynting vector.

It is worth mentioning that a pseudoisotropic curve in the $\eta - r$ space can be found when $\eta = \eta_M$ (Mota et al., 2010) with the same implications.

On the other hand, the photonic dwell time τ of this defect mode is given by the ratio between the electromagnetic energy stored by the sample and the total power of the outgoing waves. By neglecting the contribution of the propagating waves to the stored energy the dwell time dependent on the thickness of the slab, l, is found $\tau(l)$ as

$$\tau(l) \equiv \frac{l_d}{4c}\sqrt{(2\epsilon_\perp + \epsilon_\parallel)/3}\,\frac{1-\exp(-2l/l_d)}{\sin^2(\gamma_{em}/2)+\exp(-2l/l_d)} \tag{1.56}$$

whose asymptotic value for a very large sample is

$$\tau(\infty) = (l_d/(4c\sin^2(\gamma_{em}/2)))\sqrt{(2\epsilon_\perp + \epsilon_\parallel)/3}\,. \tag{1.57}$$

If, however, the CE is tuned on the pseudo-isotropic curve, $\tau(l)$ reduces to

$$\tau(l) = (l/2c)\sqrt{(2\epsilon_\perp + \epsilon_\parallel)/3}\,, \tag{1.58}$$

which now depends proportionally on the sample thickness and hence never gets bounded as it occurs for a point outside of the mentioned curve. Hence, the photon dwell time can be prolonged without limit by enlarging the sample thickness l, that opposes the behaviour reported by Kopp and Genak for cholesteric liquid crystals (Kopp & Genack, 2002).

In Fig. 7 it is depicted the inverse relative line width $\tau\omega_d$ at the defect frequency (dimensionless dwell time) which is proportional to the dwell time τ, versus the sample thickness l and η for a) an oblate, $r = 0.5$, and b) a prolate, $r = 1.16$ cholesteric elastomer, with wave number $q_0 = \pi / (214 \times 10^{-9})\ m^{-1}$ and shows how a sharp protrusion increases its amplitude around the pseudoisotropic curve, whereas the other points in the $l - \eta$ plane remains at the smaller value of $\tau\omega_d \sim 10^7$. It is worth mentioning that the density of photonic states D at the defect frequency (Ashcroft & Mermin, 1976): $D(\omega_d) \equiv l\partial n_1(\omega_d) / \partial\omega$ can be analytically obtained from Eq. (1.49). It is easy to show that $D(\omega_d)$ diverges at the pseudo-isotropic curve. Moreover, a similar behaviour in the dwell time, around the pseudoisotropic curve, is found when it is plotted versus the axial elongation and the wave number for an oblate and a prolate cholesteric elastomer (Mota et al., 2010).

Therefore, a slab of cholesteric elastomer presents a pseudoisotropic curve where the propagation of waves in the sample with the defect frequency is extremely efficient since the attenuation for the defect mode is quite small, and the energy loses in the sample are negligible due to an almost vanishing Poynting vector. These facts support the following main results: around a pseudoisotropic curve the behaviour of the photonic dwell time has a tremendous variation and the density of photonic states diverges there.

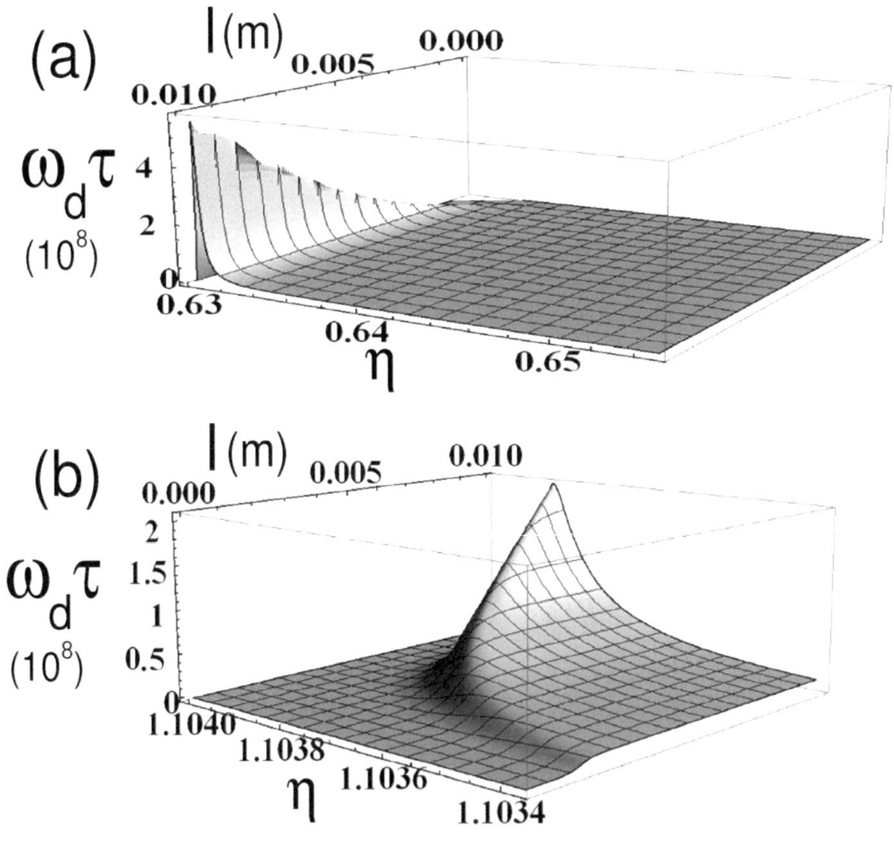

Fig. 7. $\tau\omega_d$ versus l (m) and η. For a) an oblate ($r = 0.5$) and b) a prolate ($r = 1.16$) cholesteric elastomer. For the same system as Fig. 3.

6.2.1.2 Thin samples

When the thickness $2l$ of the sample is comparable with the attenuation length l_d of the modes $1^{\pm}$, all the eigenwaves can reach the defect plane. According to the transfer matrix method seen above, we consider first the transfer matrix $\boldsymbol{U}$, which is, implicitly defined by the equation $\boldsymbol{\alpha}_b(0^+) = \boldsymbol{U}.\boldsymbol{\alpha}_a(0^-)$. Taking into account Eq. (1.54), the relation $\boldsymbol{\beta} = \boldsymbol{T}.\boldsymbol{\alpha}$, and the continuity of $\boldsymbol{\beta}(z)$ at the defect site $z = 0$, one obtains:

$$\boldsymbol{U} = \boldsymbol{U}\left(0^+,0^-\right) = \boldsymbol{T}^{-1}.\exp(-\boldsymbol{R}\xi).\boldsymbol{T} \equiv \cos\xi\mathbf{1} - \sin\xi\boldsymbol{R}_S\,, \tag{1.59}$$

where $\boldsymbol{R}_S = \boldsymbol{T}^{-1}.\boldsymbol{R}.\boldsymbol{T}$. Straightforward calculations give

$$\boldsymbol{R}_S = \begin{pmatrix} r_1 & -r_5 & -r_3 & -r_5^* \\ -ir_5 & -ir_2 & -ir_5^* & ir_4 \\ r_3 & r_5^* & -r_1 & r_5 \\ ir_5^* & -ir_4 & -ir_5 & ir_2 \end{pmatrix}, \tag{1.60}$$

where the quantities r_j ($j = 1,\dots,5$) are

$$\begin{aligned} r_{1,3} &= 4q\lambda c_1^2[\mp\pi\epsilon_{\|}\epsilon(\eta)u_1(4\epsilon_m\pi + u_1) + \epsilon_m\lambda^2(n_1^2u_1 \mp q^2(4\epsilon_m\pi + u_1))], \\ r_{2,4} &= 4q\lambda c_2^2[\pm\pi\epsilon_{\|}\epsilon(\eta)u_2(4\epsilon_m\pi + u_2) + \epsilon_m\lambda^2(n_2^2u_2 \pm q^2(4\epsilon_m\pi + u_2))], \\ r_5 &= q\lambda c_1c_2\left[4\epsilon_{\|}\epsilon(\eta)\pi(u_1u_2 + 2\epsilon_m(u_1 + u_2)) + 2\epsilon_m\lambda^2(in_1n_2(u_1 + u_2) + q^2(8\epsilon_m\pi + u_1 + u_2))\right]. \end{aligned} \tag{1.61}$$

It is convenient to write the 4×4 matrix $\boldsymbol{U}$ in the form (1.30) and to consider also the matrix $\boldsymbol{U}^{-1} \equiv \cos\xi\mathbf{1} + \sin\xi\boldsymbol{R}_S$. The 2×2 submatrix S_{bb} is the inverse of $(\boldsymbol{U}^{-1})_{bb}$, that is given by

$$(\boldsymbol{U}^{-1})_{bb} = \begin{pmatrix} \cos\xi - ir_2\sin\xi & -r_5\sin\xi \\ ir_5\sin\xi & \cos\xi + r_1\sin\xi \end{pmatrix}. \tag{1.62}$$

As mentioned above, four eigenmodes from region a (or b) reach the defect plane. Thus, the scattering matrix S_{bb} it relates the amplitudes of backward waves in region a (the transmitted ones) to the backward waves from region b (the incident ones).

It is found numerically that, for $\lambda = \lambda_d$ the eigenwave 2^- impinging from region b is totally reflected at the defect plane. This means that the element (2,2) of S_{bb} is equal to zero. Therefore, the quantity λ_d is implicitly defined by the equation

$$\xi = \cot^{-1}\left[-r_1(\lambda_d)\right], \tag{1.63}$$

which is completely equivalent to expression (1.55) for the particular case $\xi = \pi/2$.

On the other hand, it is found that, for $\lambda = \lambda_d$ the eigenwave 1^- in region b generates at the other side of the layer an eigenwave 1^- with a huge amplitude.

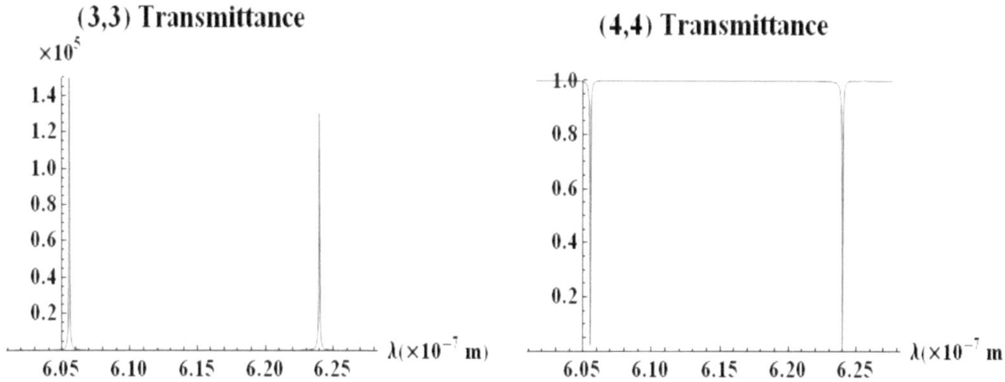

Fig. 8. Square amplitudes of the elements (3,3) and (4,4) of the scattering matrix $\boldsymbol{S}$. Here, $\theta = \pi/2$, $\epsilon_{\parallel} = 2.22$, $\epsilon_{\perp} = 1.91$, $p = 428nm$, $\eta = 1$ and $r = 1.16$.

6.2.2 Doublets

Now, we consider an unbounded sample with two identical twist defects at $z = 0$ and $z = z_1$. The transfer matrix relating the $\boldsymbol{\alpha}$ – vectors at $z = 0^-$ and $z = z_1^+$ is given by

$$\boldsymbol{U} = \boldsymbol{U}\left(z_1^+, 0^-\right) = \boldsymbol{U}_1\left(\exp(i\boldsymbol{N}z_1)\right)\boldsymbol{U}_1, \tag{1.64}$$

where $\boldsymbol{U}_1$ is the transfer matrix for each one of the defect planes, given by (1.59), and $\boldsymbol{N}$ is the diagonal matrix with diagonal elements equal to n_j. The matrix $\exp(i\boldsymbol{N}z_1)$ is the transfer matrix $\boldsymbol{U}\left(z_1^-, 0^+\right)$ of the layer between the two twist defects.

The analysis presented in previous section is also valid for obtaining the transmittances and reflectances. The curves giving these quantities are quite similar to the corresponding for a single twist defect, except for the fact that they have the structure of a doublet with two defect frequencies $\lambda_{d1}, \lambda_{d2}$. The curves (3,3) and (4,4) giving the transmittance for the eigenmodes 1 and 2 are plotted in Fig. 8.

As show the plot 8 there exist two defect frequencies $\lambda_{d1}, \lambda_{d2}$ for which the eigenwave 2^- impinging from right side is totally reflected at the defect planes. This means that the element (2,2) of S_{bb} is equal to zero. Therefore, the quantity λ_d is implicitly defined by

$$\xi = \cot^{-1}\left[\sqrt{r_3^2 e^{-2z_1/l_d} - i r_5^2 e^{-z_1(in_2+1/l_d)} + i r_5^{*2} e^{-z_1(-in_2+1/l_d)}} - r_1\right]. \tag{1.65}$$

Notice that the square root terms act here as perturbing terms whose asymptotic values vanish for $z_1 \gg l_d$ and expression (1.65) becomes identical to (1.63). When $z_1 \gg l_d$ the components peaked at $z = 0$ and $z = z_1$ do not have "interaction" between them and they act independently. Thus, the defect modes become degenerated, since $\lambda_d = \lambda_{d2} = \lambda_{d1}$. By solving the Eq. (1.65) we can obtain a 3D plot of the difference $\Delta\lambda_d = \lambda_{d2} - \lambda_{d1}$ as function of shear strain η and fractional shape anisotropy r for three values of defect separation z_1. In Fig. 9 it can be seen that the difference $\Delta\lambda_d$ of the defect frequencies vanish along the pseudoisotropic curve for any value of l_d and goes to zero as z_1 gets much bigger than l_d; under this conditions $\lambda_d = \lambda_{d2} = \lambda_{d1}$. On the other side, notice how, at a fixed value r,

$\Delta\lambda_d$ reaches its maximum value at $\eta = 1$ and its minimum one at the pseudoisotropic locus; this fact suggests the possibility of mechanically tune the two defect modes and the interaction between them by controlling η.

Before ending this section we mention that for both, oblate and prolate cholesteric elastomers, the defect frequencies redshift as η and r augment (curves are not shown).

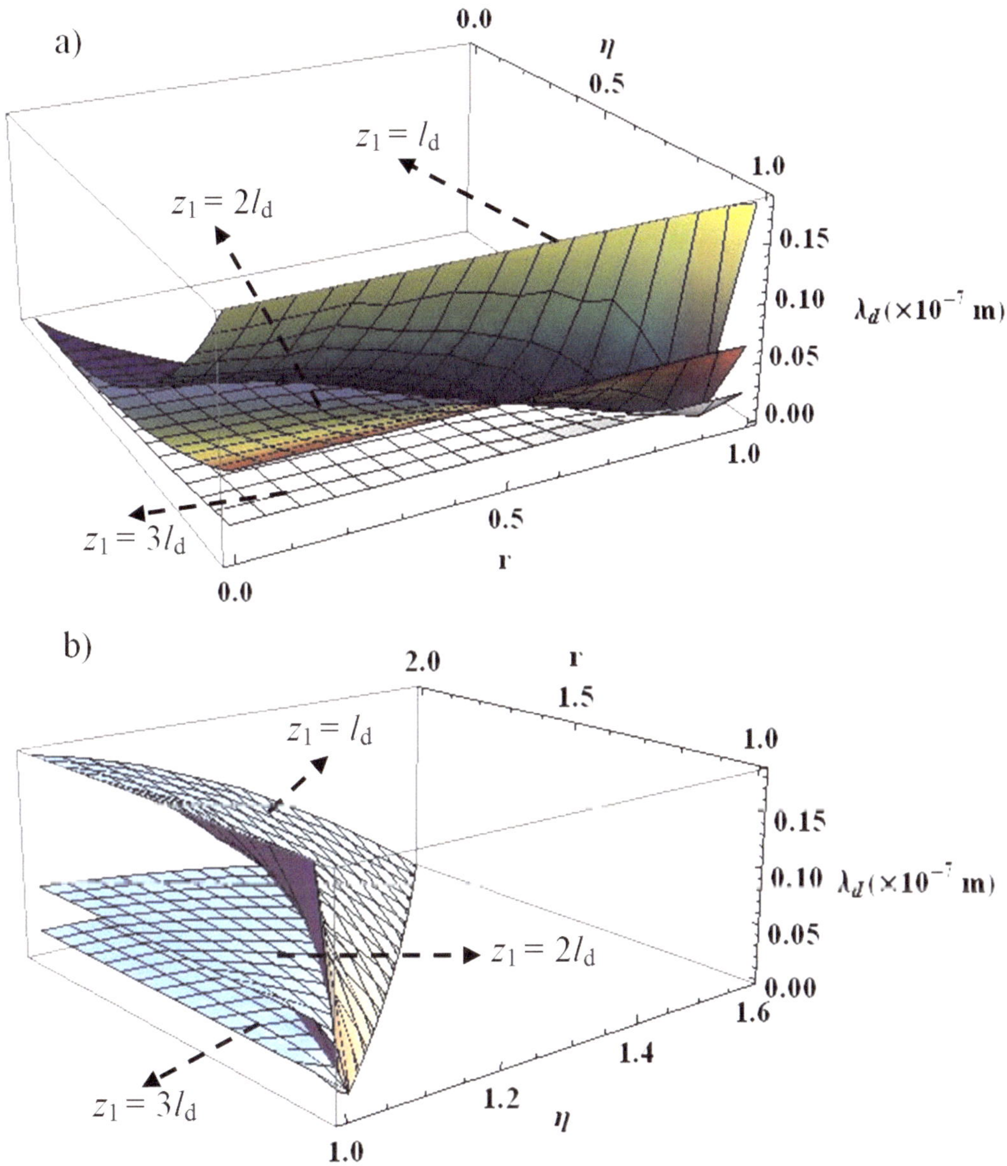

Fig. 9. Plot of the difference $\Delta\lambda_d = \lambda_{d2} - \lambda_{d1}$ as function of shear strain η and fractional shape anisotropy r for a) oblate and b) prolate cholesteric elastomers. The parameter values are the same as Fig. 8.

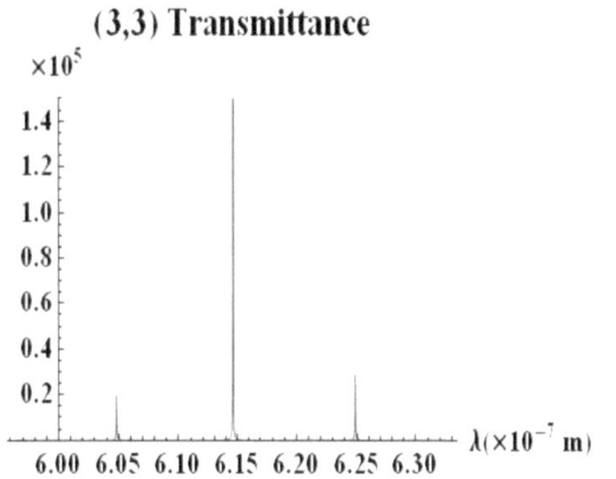

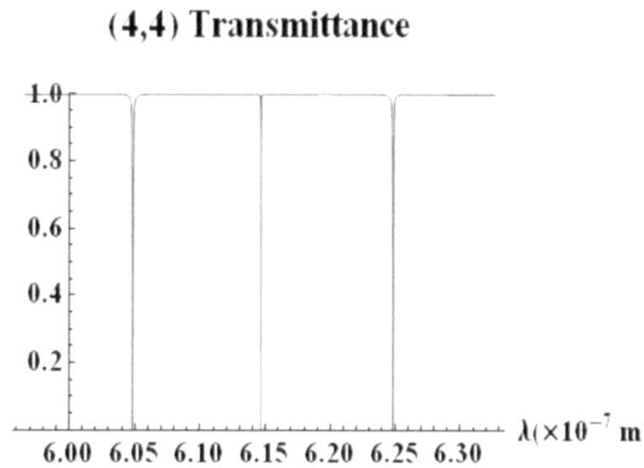

Fig. 10. Square amplitudes of the elements (3,3) and (4,4) of the scattering matrix S . Here, $\theta = \pi / 2$, $\epsilon_{\|} = 2.22$, $\epsilon_{\perp} = 1.91$, $p = 428nm$, $\eta = 1$ and $r = 1.16$.

6.2.3 Multiplets

The equation giving the transfer matrix for n_0 identical and equidistant twist defects at $z = nz_1$, where $n = 0,1,2,...,n_0 - 1$, can be written as

$$\boldsymbol{U} = \boldsymbol{U}\left((n_0 - 1)z_1^+, 0^-\right) = \boldsymbol{U}_1\left[\left(\exp(i\boldsymbol{N}z_1)\right)\boldsymbol{U}_1\right]^{n_0 - 1}. \tag{1.66}$$

For $n_0 = 2$, Eq. (1.66) becomes identical to equation (1.64). The square amplitudes of the elements (3,3) and (4,4) of the scattering matrix are plotted in Fig. 10 for $N = 3$, $\xi = \pi / 2$ and $z_1 = l_d$.

The figure shows three different twist defects at the wavelengths $\lambda_{d1}, \lambda_{d2}, \lambda_{d3}$, where λ_{d2} is nearly equal to the average value of $\lambda_{d1}, \lambda_{d3}$. At the defect frequencies, the structure reflects totally the eigenwave 2. Also, it is shown that, at the defect frequencies, the eigenwave 1^- in region b generates at the other side of the layer an eigenwave 1^- with a huge amplitude.

7. References

Warner, M. & Terentjev, E. M. (2003). *Liquid Crystals Elastomers*, Oxford University Press, ISBN 0198527675, Oxford

Kac, M.; Uhlenbeck, G. E.; Hibbs, A. R. & van der Pol, B. (1976). *Probability and Related Topics in Physical Sciences*, American Mathematical Society, ISBN 0-8218-0047-7, USA

Chuang, S. L. (2009). *Physics of Photonic Devices*, (2nd. Ed.), John Wiley & Sons, ISBN 978-0-470-29319-5, New Jersey USA

Hecht, E. & Zajac, A. (1986). *Optica*, Addison-Wesley Iberoamericana, ISBN 0-201-02839-5 Delaware USA

Marcuvitz, N. & Schwinger, J. (1951). On the representation of electric and magnetic field produced by currents and discontinuities in waveguides. *Journal of Applied Physics*, 22, 806-819

Altman, C. & Sucky, K. (1991). *Reciprocity, Spatial Mapping and Time Reversal in Electromagnetics*, Kluwer Academic Publishers, ISBN 0792313399, Derdrecht The Netherlands

Berreman, D. W. & Scheffer, T. J. (1970). Bragg reflection of light from single-domain cholesteric liquid-crystal films. *Physical Review Letters*, 25, 577-581

Lakhtakia, A. & Reyes, J. A. (2008). Theory of electrically controlled exhibition of circular Bragg phenomenon by and obliquely excited structurally chiral material.*OPTIK-International Journal for Light and electron Optics*, 119, 6, 269-275

Espinosa-Ortega, T. & Reyes, J. A. (2008). Mechanically controlled bandgap in cholesteric elastomers. *Optics Communications*, 281, 5830-5837

Avendaño, C. G.; Ponti, S.; Reyes, J. A. & Oldano, C. (2005). Multiplet structure of the defect modes in 1D helical photonic crystals with twist defects. *Journal of Physics A: Mathematical and General*, 38, 8821-8840

Oseen, C. W. (1933). The theory of liquid crystals. *Transactions of the Faraday Society*, 29, 883-899

Finkelmann, H.; Kim, S. T.; Muñoz, A.; Palffy Muhoray, P. & Taheri, B. (2001). *Advanced Materials*, 13, 14, 1069-1072

Hirota, Y.; Ji, Y.; Serra, F.; Tajbakhsh, A. & Terentjev, E. (2008). Effect of crosslinking on the photonic bandgap in deformable cholesteric elastomers. *Optics Express*, 16, 5320-5331

de Gennes, P. G. & Prost, J. (1993). *The Physics of Liquid Crystals*, (2nd. Ed.), Clarendon Press, ISBN 0198520247, Oxford, UK

Macleod, H. A. (2001). *Thin-Film Optical Filters*, (3rd Ed.), Institute of Physics Bristol, ISBN 9780750306881, U. K.

Lakhtakia, A. (2000). Shear axial modes in a PCTSCM. Part V: transmission spectral holes. *Sensors and Actuators A*, 80, 216-223

Schmidtke, J.; Stille W. & Finkelmann, H. (2003). Defect mode emission of a dye doped cholesteric polymer network. *Physical Review Letters*, 90, 83902-83905

Ozaki, M.; Ozaki, R.; Matsui, T. & Yoshino, K. (2003). Twist-defect-mode lasing in photopolymerized cholesteric liquid crystal. *Japanese Journal of Applied Physics*, 42, L472-L475

Song, M. H.; Shin, K. C.; Park, B.; Takanishi, Y.; Ishikawa, K.; Watanabe, J.; Nishimura, S.; Toyooka, T.; Zhu, Z. G.; Swager, T. M. & Takezoe, H. (2004). Polarization characteristics of phase retardation defect mode in polymeric cholesteric liquid crystals. *Science and technology of Advanced Materials*, 5, 437-441

Becchi, M.; Ponti, S.; Reyes, J. A. & Oldano, C. (2004). Defect Modes in helical photonic crystals: an analytic approach. *Physical Review B*, 70, 033103-033106

Hodgkinson, J. J.; Wu, Q. H.; Thorn, K. E.; Lakhtakia, A. & McCall, M. W. (2000). Spacerless circular-polarization spectral-hole filters using chiral sculptured thin films: theory and experiment. *Optics Communications*, 184, 57-66

Kopp, V. I. & Genack, A. Z. (2002). Twist defect in chiral photonic structures. *Physical Review Letters*, 89, 033901-033904

Schmidtke, J. & Stille, W. (2003). Photonic defect modes in cholesteric liquid crystal films. *The European Physical Journal E*, 12, 553-564

Wang, F. & Lakhtakia, A. (2003). Specular and nonspecular, thicknessdependent spectral holes in a slanted chiral sculptured thin film with a central twist defect. *Optics Communications*, 215, 79-92

Oldano, C. (2003). Comment on "Twist defect in chiral photonic structures". *Physical Review Letters*, 91, 259401

Kopp, V. I. & Genack, A. Z. (2003). Kopp and Genack Reply:. *Physical Review Letters*, 91, 259402

Cicuta, P.; Tajbakhsh, A. R. & Terentjev, E. M.. (2002). Evolution of photonic structure on deformation of cholesteric elastomers. *Physical Review E*, 65, 051704-051709

Mota, A. E.; Palomares, L. O. & Reyes, J. A. (2010). Mechanically controlled defect mode in cholesteric elastomers. *Applied Physics Letters*, 96, 081906-081908

Abdulhalim, I. (1999), Point of ultra-sensitivity to perturbations for axial propagation in helicoidal bianisotropic structures, *Europhysics Letters*. 48, 2, 177

Ashcroft, N. W. & Mermin, N. D. (1976). *Solid State Physics*, Harcourt-Saunders, ISBN 0-03-083993-9, Philadelphia

17

Time Domain Reflectometry: Temperature-dependent Measurements of Soil Dielectric Permittivity

Wojciech Skierucha
Institute of Agrophysics, Polish Academy of Sciences
Poland

1. Introduction

The purpose of this study is to determine the temperature influence on the soil bulk dielectric permittivity, ε_b, calculated from the measurement of the electromagnetic wave velocity of propagation along the parallel waveguide in a TDR probe, i.e. a probe working in Time Domain Reflectometry technique. The experimental evidence shows that the existing models do not completely describe the temperature effect. However, it has been confirmed that the observed temperature effect is the result of two competing phenomena; ε_b increases with temperature following the release of bound water from soil solid particles and ε_b decreases with temperature increase following the temperature effect of free water molecules. It has been found that there is a soil type characteristic moisture value, θ_{eq}, named the equilibrium water content, having the specific temperature property. The temperature effect for this moisture is not present, which means that for soils with the moisture value equal to θ_{eq} the both competing phenomena mentioned earlier compensate each other. The equilibrium water content, θ_{eq}, decrease is correlated with the soil specific surface area. The temperature correction formula adjusting the soil moisture determined by TDR, θ_{TDR}, at various temperatures to the corresponding value at 25°C, based on knowledge of θ_{eq}, decreases the standard deviation of the absolute measurement error of soil moisture θ_{TDR} by the factor of two as compared to the uncorrected values.

The majority of measurements of physical, chemical and biological properties of porous materials including soil should be accompanied with the measurement of soil water content and temperature. These parameters determine almost all processes in natural environment. It seems obvious to accompany the sensors for the measurement of soil salinity, oxygenation, content of nutrients, soil water potential, and others with the temperature and moisture sensors having the same measurement volume and performing measurements at the same time. Temperature sensors of various accuracy and size are easily available and together with the necessary electronics they can fit into the desired sensor enclosure. More problems are encountered with moisture sensors of porous materials because they do not measure water content directly, but use other parameters of the measured matter that indirectly and selectively determine its moisture. Such a property of porous materials accompanied with moisture is dielectric permittivity.

Measurement of soil moisture using time domain reflectometry (TDR) has become increasingly popular because of simplicity of operation, satisfactory accuracy and fast result available, the process of measurement is non-destructive, portable systems are available and the method gives ability for measurement automation and probes multiplexing (Dirksen & Dasberg, 1993; Malicki et al., 1996; Topp et al., 1980). Beside above advantages there are some drawbacks of this measurement technique including the requirement of excellent contact of the probe to soil and the dependence of dielectric permittivity of soil on temperature affecting the TDR soil moisture readout.

In first applications of the TDR technique for soil moisture determination, the influence of temperature on the TDR determined soil bulk dielectric permittivity, ε_b, was neglected. The significant fluctuation of measured data, which was obviously correlated with soil temperature, was noticed with the introduction of soil moisture field monitoring systems based on reflectometric meters. Also, it was found (Pepin et al., 1995; Halbertsma et al., 1995) that ε_b decreased with temperature increase for sandy, silt and peat soils and it did not change for the tested clay soil.

The text below uses two terms that need explanation: (i) soil free water - composed from water particles that rotation in the electric filed is not hampered, and (ii) soil bound water - composed from water particles so close to solid phase that their rotation in the electric field is hampered by surface charge on the solids. The real part of the complex dielectric permittivity of free water is about 80 at room temperature and is much smaller for bound water (Boyarskii et al., 2002) reaching the minimal value of ice (about 3.2) for the first layer of water particles adsorbed on the clay soil surface.

2. Basics of TDR technique

Time Domain Reflectometry (TDR) technique for the measurement of water content of isotropic and homogenous media becomes popular for the simplicity of operation, accuracy and the non-destructive, as compared to other methods, way of measurement (Malicki, 1999; O'Connor & Dowding, 1999). This measurement technique takes advantage of four physical phenomena characteristic to the porous materials including soil:

- for soils with negligible magnetic properties and electrical conductivity, which is true for the majority of arable soils, and in the frequency range of about 1 GHz, the complex dielectric permittivity of the soil can be approximated by its real value and the electromagnetic wave propagation velocity, v, in the soil can be calculated from:

$$v \approx \frac{c}{\left[\varepsilon(\theta)\right]^{1/2}} = \frac{c}{n} = \frac{2L}{\Delta t} \tag{1}$$

  where: c is a velocity of light in free space, $\varepsilon(\theta)$ is a real part of the complex dielectric permittivity dependent on its water content, θ; $n = [\varepsilon(\theta)]^{1/2}$ is the medium refractive index; L is the length of TDR probe rods inserted into the soil; Δt is the time distance between the reflections of TDR pulse from the beginning and the end of the probe rods, inserted into the medium,
- the dielectric permittivity of the medium liquid phase has much higher value than the other medium phases, *i.e.* about 80 against 2÷4 for the solid and 1 for the gas phase,

- the relation between the water content of the medium and its dielectric permittivity is highly correlated for the majority of the porous media (Davis et al, 1977; Malicki & Skierucha, 1989; Topp et al., 1980),
- the attenuation of the amplitude of electromagnetic wave traveling along the parallel transmission line inserted into the medium, measured from the amplitudes of the pulse before (U_{in}) and after (U_{out}) attenuation caused by the pulse travel twice a distance of the probe length, L (Fig. 1) depends on its bulk electrical conductivity, EC_b (Dalton et al., 1984):

$$EC_b\left(\mathrm{Sm}^{-1}\right)=\frac{(\varepsilon)^{1/2}}{120\pi L}\ln\left(\frac{U_{in}}{U_{out}}\right) \tag{2}$$

Therefore, water content of the soil, which is assumed to be an isotropic and homogenous medium, is the main reason determining its bulk dielectric permittivity.
The simplified formula (1) for the determination of the electric pulse propagation velocity in porous medium is derived from more general one:

$$v=\frac{c}{\left(\frac{\varepsilon'}{2}\left(1+(1+\mathrm{tg}^2\delta)^{1/2}\right)\right)^{1/2}} \tag{3}$$

where: c is the light velocity of propagation in free space, ε' is the real value of the complex dielectric permittivity of the medium, $\mathrm{tg}\delta$ is the dielectric loss defined as:

$$\mathrm{tg}\delta=\frac{\varepsilon''+\frac{EC_b}{2\pi f\varepsilon_0}}{\varepsilon'} \tag{4}$$

The complex dielectric permittivity of the medium, ε, is:

$$\varepsilon=\varepsilon'-j\left(\varepsilon''+\frac{EC_b}{2\pi f\varepsilon_0}\right) \tag{5}$$

where: ε'' represents dielectric loss connected with the dielectric polarization of soil particles, EC is the medium electrical conductivity, f is the frequency of the electromagnetic field, ε_0 is dielectric permittivity of free space, j jest equal to $(-1)^{1/2}$.
The idea of simultaneous measurement soil water content and electrical conductivity as well as the respective hardware setup are presented in Fig. 1.
The TDR probe consists of two waveguides connected together: a coaxial one, called the feeder, and a parallel one, called the sensor, made of two or more parallel metal rods, one of which is connected to the cable hot wire the others to the shield. The sensor should be fully inserted into the measured medium, that by definition should be homogeneous. The initial needle pulse travels from the generator by the feeder towards the sensor. A fast sampling oscilloscope registers this pulse in time. In the junction between the feeder and the sensor, there is a rapid change in geometry of the electromagnetic wave travel path. At this point some energy of the pulse is reflected back to the generator, like in radar, and the remaining travels along the parallel waveguide to be reflected completely from the rods ending. The

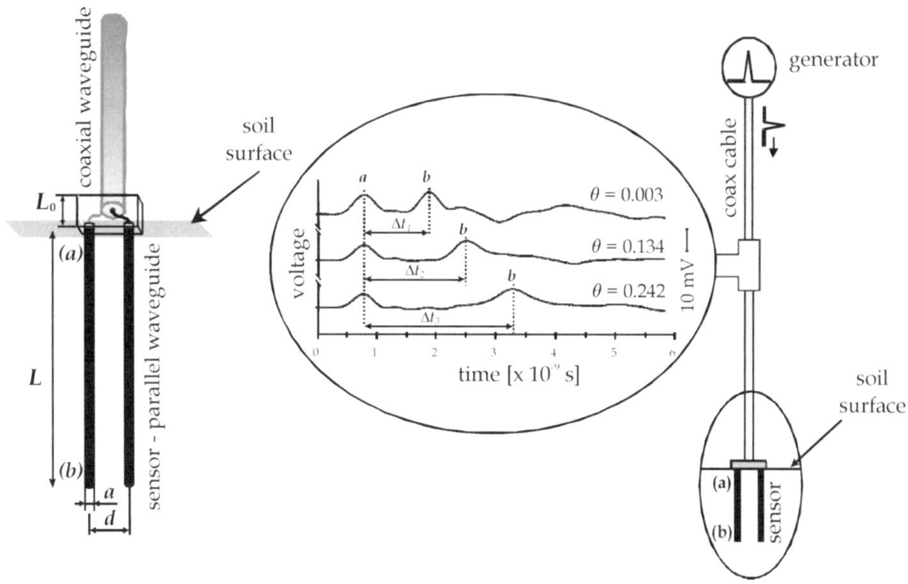

Fig. 1. Hardware setup for simultaneous measurement of soil water content and electrical conductivity using Time Domain Reflectometry technique

successive reflections are recorded for the calculation of the time distance between the two reflections (*a*) and (*b*). Three reflectograms representing pictures from the oscilloscope screen (voltage as a function of time at the chosen point in the feeder) are drawn in Fig. 1. They represent cases when the sensor was placed in dry, wet and water saturated soil. The time Δt_i necessary for the pulse to cover the distance equal to the double length of the metal rods in the soil increases with the soil dielectric permittivity, thus water content. The reason for that is the change of electromagnetic propagation velocity in media of different dielectric permittivity, according to (1). Also, the amplitude of the pulse at the point (*b*) decreases with the increase of soil electrical conductivity, according to (2).

2.1 Temperature effect of soil free water and electrical conductivity

Corrections of the TDR determined moisture data related to the temperature effect of dielectric permittivity of free water was examined by Pepin et al. (1995) and Halbertsma et al. (1995) and the influence of soil texture on the observed temperature effect was reported. Temperature dependence of dielectric permittivity of free water can be described as (Wheast, 1979):

$$n_{fw}(T) = n_{fw} \cdot d(T) = 8.851 \cdot d(T) \tag{6}$$

$$\left[d(T)\right]^2 = 1 - 0.4536 \cdot 10^{-2}(T-25) + 0.9319 \cdot 10^{-7}(T-25)^2 \tag{7}$$

where $n_{fw} = (\varepsilon_{fw})^{1/2} = 8.851$ is free water refractive index at 25°C for the frequency in the range of 10^8 Hz, ε_{fw} is dielectric permittivity of free water at 25°C, and T is the temperature in °C.

If bulk soil dielectric permittivity had a temperature dependence related solely to free water, measurements of soil ε_b would show a negative correlation with temperature that would increase with water content. Results from Pepin et al. (1995) show this negative correlation, but the overall changes are smaller than predicted from a dielectric mixing model that integrated the temperature change in dielectric permittivity of water. The same was observed by Halbertsma et al. (1995) who made measurements of sand and silt soil samples. Also, he reported the lack of the discussed temperature effect for clay soil. It was obvious that the physical processes involving other soil phases beside the liquid phase should be taken into account when interpreting the temperature effect on the soil ε_b.

Literature reports (Nadler et al., 1999; Or & Wraith, 1999) show that soil electrical conductivity does not influence the TDR determined soil bulk dielectric permittivity, ε_b, in the soil salinity range where the majority of plants are growing, i.e. up to the corresponding values of soil solution from 1600 to 2600 mSm^{-1}. Following the discussion in Or & Wraith (1999) it is possible to present the frequency dependence of the real and imaginary parts of a salt-water mixture's dielectric permittivity, ε' and ε'', as well as the velocity of propagation, v, of electromagnetic wave along the TDR probe rods inserted in this mixture for different temperatures and electrical conductivities, σ_{dc}. The increase of temperature of the salt-water mixture increases the frequency range around the frequency 1 GHz, where the velocity of propagation of an electromagnetic wave along the waveguide, used in TDR technique, does not change. Therefore the temperature effect on the electrical conductivity of soil, resulting mainly from ionic conductivity of soil electrolyte should not influence the bulk temperature effect of soil dielectric permittivity.

2.2 Temperature effect of soil bound water

The discussion of the release of bound water from the solids with the temperature is presented by Or & Wraith (1999), where the authors applied the Debye model for polar liquids and liquid viscosity dependence on temperature, T, for bound water molecules at the distance x from the solid surface to describe their relaxation frequency. They reported that ε_b increased with temperature for a silt loam soil for all soil moisture, however another silt loam soil showed the increase of ε_b for relatively low water contents, whereas it decreased with temperature at higher water contents. The change of ε_b with temperature is not fully explained yet but they discuss its two reasons: ε_b decreases with temperature because the real part of soil complex dielectric permittivity decreases following the temperature effect of soil free water and ε_b increases with temperature following the release of bound water molecules.

Water molecules that are adsorbed to the solid surface are less mobile in the imposed EM (electromagnetic) field as the not adsorbed ones. For a given temperature, the more distant the water molecules are from the solids, the higher is their relaxation frequency. The increase of temperature increases their kinetic energy that raises the relaxation frequency and they become more mobile in the imposed EM field. Bound water molecules released from the solid surface become now free with higher value of the real part of the complex dielectric permittivity.

The release of water molecules from the solids results in the increase of the real part of the soil dielectric permittivity and decreases its imaginary part describing the loss tangent of dielectric material. This phenomenon leads to a new equilibrium with more free water molecules and less bound water ones.

According to recent studies (Or and Wraith, 1999; Boyarskii et al., 2002), only a few layers of water covering soil particles are subjected to the change of relaxation time in relation to relaxation time of free water. The analysis of nuclear resonance spectra of bound water films in clay shows the approximated relation between relaxation time of bound water, and thickness of the film covering soil particles (Boyarskii et al., 2002). The resonant frequency of hydrogen in water particles, measured by NMR, is different for various layers of water films on the solids and consequently different frequency response of their dielectric permittivity. The relaxation time of bound water drops rapidly with the number of layers covering the particles and it seems that only one or two layers have significantly longer relaxation times than that of free water.

On the base of Debye model (Debye, 1929; Hasted, 1973) for polar liquids and using relations showing dependence of viscosity of liquid molecules on temperature T and distance x from the solid phase, the relaxation frequency f_{rel} of water molecules can be expressed as (Or & Wraith, 1999):

$$f_{rel}(x,T) = \frac{kT\exp\left(-\frac{1}{T}\left(\frac{a}{x}+d\right)\right)}{8\pi^2 r^3 c} \tag{8}$$

where: k is Boltzman constant ($1.38 \cdot 10^{-23}$ JK^{-1}), T is temperature in Kelvin degrees, $a = 1621 \cdot 10^{-10}$ mK, $d = 2.047 \cdot 10^{3}$ K and $c = 9.5 \cdot 10^{-7}$ Pas are constants as the consequences of applied simplifications, $r = 1.8 \div 2.5 \times 10^{-10}$ m is the radius of water molecule.

The equation (8) shows that moving closer to solid phase, water particles are increasingly hampered to rotate in the electromagnetic field of high frequency, which is manifested by decreasing relaxation frequency for these molecules. Also, the temperature of water particles will have some effect on the f_{rel}, which is simulated in Fig. 2A.

The dependence of dielectric permittivity (real part) for free and bound water particles on the frequency can be described by Cole-Cole formula (9), where $\omega=2\pi f$ is the angular frequency of the external electrical field, $\varepsilon_{low}= 81$ and $\varepsilon_{hi}= 4.23$ are relative dielectric permittivity values for free and bound water particles in the frequency values lower and higher, respectively, from 18 GHz (relaxation frequency of free water particles), $\tau=1/(2\pi f_{rel})$ is the relaxation time of water particles, h=0.013 is a parameter describing interaction of water dipoles (Hasted, 1973).

$$\varepsilon = \varepsilon_{hi} + \frac{(\varepsilon_{low}-\varepsilon_{hi})\left[1+(\omega\tau)^{1-h}\sin\frac{h\pi}{2}\right]}{1+(\omega\tau)^{2(1-h)}+2(\omega\tau)^{1-h}\sin\frac{h\pi}{2}} \tag{9}$$

Three curves in Fig. 2B represent water particles located at various distances from solid phase and the highest one relates to free water with relaxation frequency of 18 GHz, while the lowest to bound water particles close to solid phase.

The frequency range of an electric field, for which the values of ε determined from equation (9) are fixed at about 81, is the largest for free water. Adsorbed water molecules, by virtue of proximity to the solid phase are characterized by lower relaxation frequency and are less mobile in an external electric field. Consequently, according to the formula of Cole-Cole (9), the value of ε for frequencies below the relaxation frequencies are lower than for free water. The decreasing number of adsorbed water molecules, due to increasing temperature, increases the amount of free water molecules. Fig. 2A shows that moist soil temperature increase from 278 K to 338 K increases the relaxation frequency of water molecules located $2 \cdot 10^{-10}$ m away from the solid phase from 0.1 GHz to 0.7 GHz. As a result, this temperature increase also increases the effective value of the dielectric permittivity of adsorbed water. Or and Wraith (1999) recognized the frequency limit of 1 GHz of the TDR method as the cut-off frequency, f^*, that distinguishes water molecules between free and bound. Water layers having dielectric relaxation frequencies lower than the cut-off frequency were considered as bound water having smaller dielectric permittivity than for free water. It was shown (Dirksen & Dasberg, 1993) that the quantity of bound water increases with the volume of clay fraction in the soil due to large specific surface area of clay.

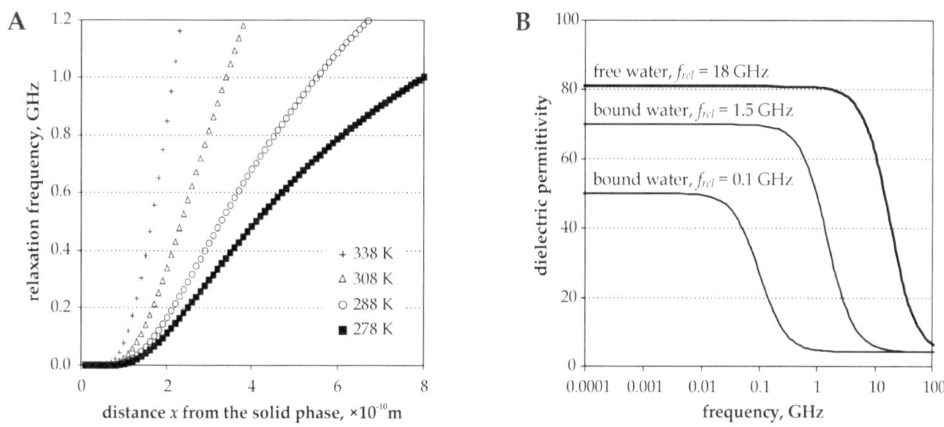

Fig. 2. A - relaxation frequency for water particles bound to solids according to (8) and related to its distance x from the solid phase; B - example relation between dielectric permittivity of water molecules and relaxation frequency for bound and free water particles

3. Materials and methods

TDR measurements were performed in the laboratory on mineral soils with negligible organic content, collected from Lublin region, Poland (Table 1) mainly from the topsoil layer. The soils were spread in layers of about 1 cm thickness on a flat surface in laboratory for drying at room temperature and then they were grinded to destroy big aggregates and passed through 2 mm sieve.

Dry soils were mixed with appropriate amount of distilled water to obtain five soil samples for each analyzed soil, with moisture values from air dry to saturation with regular differences in moisture content, taking care to get homogeneous distribution of water in the soil samples. Eight containers with soil samples covered with plastic foil to minimize evaporation were placed in a specially constructed rack. The volume of the soil containers was 0.5 dm^2 and their shape assured that the TDR probes sphere of influence was in the measured soil samples. The gravimetric moisture content, θ_g, and bulk density, ρ, were determined for each soil sample directly after completion of the TDR measurements. The values of ρ in Table 1 are the mean for all applied moistures for each tested soil. Soil texture was determined by standard Bouyoucos method (Pansu & Gautheyrou, 2006). The values of soil specific surface area, S, were measured by water vapor adsorption method (Oscik, 1983).

Soil No.	Soil type	Specific surface S $\times 10^3$ m^2kg^{-1}	Bulk density ρ $\times 10^3$ kgm^{-3}	ISSS soil texture		
				sand 2-0.02 mm	silt 0.02-0.002 mm	clay < 0.002 mm
611	brown soil	9	1.59	94	5	1
566	rendsina	10	1.45	92	7	1
589	soil lessive	10	1.70	88	11	1
605	brown soil	10	1.69	95	4	1
597	soil lessive	11	1.53	94	5	1
604	brown soil	12	1.65	97	2	1
593	brown soil	19	1.35	77	20	3
560	rendsina	20	1.46	87	11	2
569	brown soil	21	1.33	73	23	4
606	muck soil	23	1.42	97	2	1
591	brown soil	25	1.40	70	27	3
601	chernozem	31	1.38	64	31	5
568	brown soil	34	1.35	60	30	10
570	brown soil	35	1.40	63	26	11
623	chernozem	36	1.39	62	25	13
622	chernozem	37	1.40	61	33	6
621	chernozem	42	1.33	60	34	6
562	rendsina	65	1.44	52	35	13
619	chernozem	69	1.16	87	12	1
565	rendsina	83	1.04	77	18	5

Table 1. Selected physical parameters of the tested soils

The applied TDR probes had two parallel metal rods of 10 cm length and they were enhanced with the electronics (microcontroller, digital output temperature sensor, analog-to-digital converter and serial interface) for independent measurement of the probe temperature and soil electrical conductivity. The construction of such a "smart sensor" is presented in Fig. 3.

The electrical conductivity of the soil sample, EC_b, is measured from the voltage drop on the reference resistor R_1 connected in series with the soil equivalent resistor R_2. Low frequency conductivity, EC_b, of the soil samples were determined from the formula: $EC_b = C/R_2$, where C is a calibration constant determined individually for each TDR probe by the measurement in NaCl solution of known conductivities.

The source voltage for electrical conductivity measurement was a square wave, generated by the microcontroller, of 100 kHz frequency that does not polarize the electrode-soil system, the inductance L is for separation of high frequency TDR signal from much lower frequencies.

Reflectometric measurements for determination of the change of soil samples bulk dielectric permittivity with temperature, moisture and electrical conductivity were performed using the setup that is shortly described below as well as in Fig. 4 and Fig. 5.

It consists of three functional modules controlled by a program running on a PC compatible computer: (i) oscilloscope frame HP54120B with the TDR test set HP54121A, (ii) self-designed and manufactured interface connecting TDR probes (two wire waveguide, 10 cm long and spaced 1.5 cm) to the TDR unit by eight position 0.01 - 2.4 GHz multiplexer and reading selected temperature sensors and controlling the temperature chamber, (iii) temperature chamber consisting of a freezer, a fan-heater inside it and an additional fan to minimize temperature gradients inside the chamber.

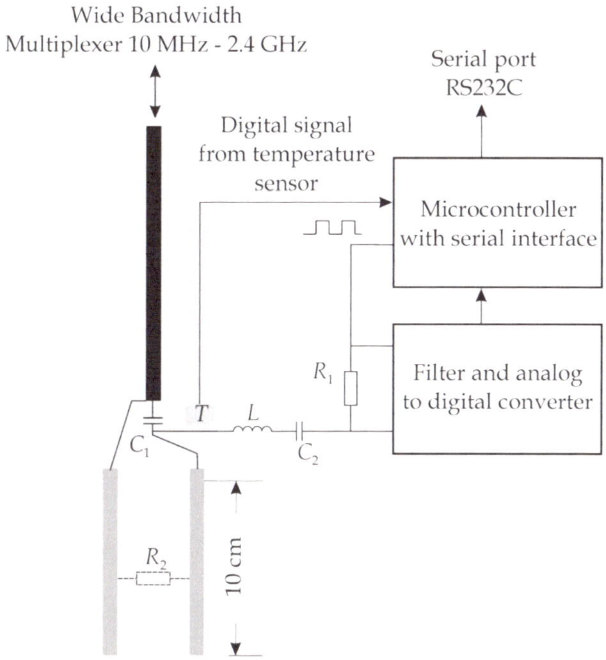

Fig. 3. Block diagram of the TDR probe with the electronics for soil electrical conductivity and temperature measurements

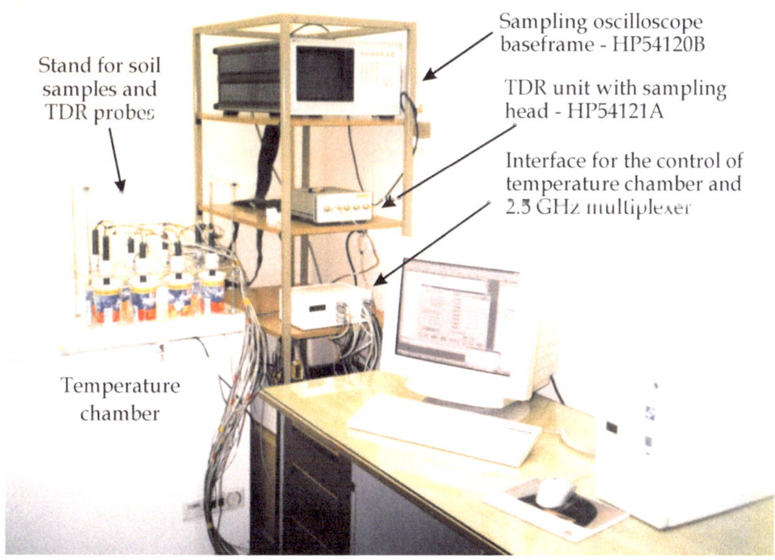

Fig. 4. Laboratory setup for the determination of the temperature variability of the soil dielectric permittivity

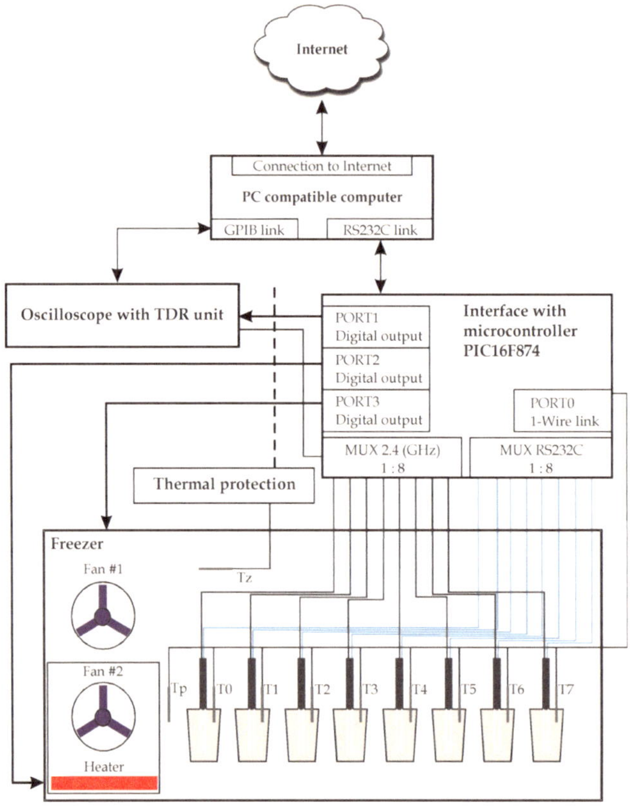

Fig. 5. Block diagram of the laboratory setup for the determination of temperature effect on soil dielectric permittivity

PC compatible computer controlling the dedicated interface by means of the serial RS232C interface and the oscilloscope by universal GPIB link, was connected to Internet to monitor remotely the performance of the experiment. The application software provides a user friendly interface for the operator. Temperature of the soil samples was controlled by switching on and off the fan-heater located at the bottom of the temperature chamber and the freezer, only one working at a time. For security reasons, an independent temperature sensor connected to the main security switch was applied to disconnect all power devices in the system in case of reaching upper limit temperature, T_z=65°C, inside the chamber.

The complete measurement cycle at six temperature values, from 5 °C to 55 °C in 10 °C (±1°C) increments, for the set of eight soil samples took about 12 hours. Starting from the 4th cycle, the measurable decrease of TDR moisture values of soil samples was noticed, which was attributed to the loss of water evaporating from the not perfectly sealed holes of plastic foil where the TDR probe rods entered the soil samples.

The collected data in ASCII format representing reflectograms were processed by proprietary software to calculate travel times along the TDR probe rods, and soil bulk dielectric permittivity, ε_b, according to the "flat tangent" approach (Fig. 6) described by Wraith & Or (1999).

The bound water volume fraction of the soil θ_{bw} is the product of solid phase specific surface S, temperature dependent thickness of bound water layer $x(T)$ and bulk density ρ:

$$\theta_{bw}(T) = x(T) S \rho \tag{10}$$

and $x(T)$ can be presented after Or & Wraith (1999) from reorganizing (8) as:

$$x(T) = \frac{a}{-d + T \ln\left(\frac{kT}{8\pi^2 r^3 c f^*}\right)} \tag{11}$$

where $f^* = 1$ GHz is the cut-off frequency, that distinguishes water molecules between free and bound. Having determined the dependence of bound water volume fraction of soil, $\theta_{bw}(T)$, on temperature using (10), it was possible to find the overall temperature dependence of bulk soil dielectric permittivity, ε_b, by application of dielectric mixing models. Among many dielectric mixing models describing soil as the mixture of solids, liquid, gas and also bound water phase, there are two applied in the presented study: alpha model (Birchak et al., 1974) and the model of de Loor (1990), given by the (12) and (13), respectively:

$$\varepsilon_b^\alpha = (1-\phi)\varepsilon_s^\alpha + (\phi - \theta)\varepsilon_a^\alpha + (\theta - \theta_{bw})\varepsilon_{fw}^\alpha + \theta_{bw}\varepsilon_{bw}^\alpha \tag{12}$$

$$\varepsilon_b = \frac{3\varepsilon_s + 2\theta_{fw}(\varepsilon_{fw} - \varepsilon_s) + 2\theta_{bw}(\varepsilon_{bw} - \varepsilon_s) + 2\theta_a(\varepsilon_a - \varepsilon_s)}{3 + \theta_{fw}\left(\frac{\varepsilon_s}{\varepsilon_{fw}} - 1\right) + \theta_{bw}\left(\frac{\varepsilon_s}{\varepsilon_{bw}} - 1\right) + \theta_a\left(\frac{\varepsilon_s}{\varepsilon_a} - 1\right)} \tag{13}$$

where: ε_b – bulk dielectric permittivity determined by TDR method; ϕ - soil porosity (m^{3}m^{-3}); ε_s, ε_a, ε_{fw} and ε_{bw} are dielectric permittivity of dry soil, air, free water and bound water, respectively; θ, θ_{bw} and θ_a (m^{3}m^{-3}) are volume fractions of water in the soil (free and bound water), only bound water, and air, respectively.
To perform analysis of the models (12) and (13) the following values were assumed: $\varepsilon_{bw} = 3.2$ (the same as for ice), $\varepsilon_s = 5$ as used by Or & Wraith (1999), $\varepsilon_a = 1$.

4. Results and discussion

The experimental setup described earlier produced reflectograms like in Fig. 6 - upper curves, which after processing by dedicated software were converted to data pairs (ε_b,T) representing bulk dielectric permittivity and temperature of respective soil samples. Lower curve in Fig. 6 is produced by analysing software application software and it represents one differentiated curve, which local maxima localize inclination points in reflectograms for application the "flat tangent" algorithm.
All tested soils were divided into three groups with the soil specific surface areas below 12·10^3 m^{2}kg^{-1}, between 20 and 35·10^3 m^{2}kg^{-1} and above 37·10^3 m^{2}kg^{-1}. The calibration curves taken for the tested soils are trend lines of 2nd degree polynomials fitted to the data pairs (θ, $\varepsilon_b(T)$) collected experimentally at six applied temperature values. The first group of soils had TDR calibration curves close to the equation given by Topp et al. (1980).

With the increase of the soil specific surface area, S, as well as the decrease of bulk density, ρ, the bulk dielectric permittivity values, ε_b, were below the Topp's calibration. This is in agreement with other reports (Dirksen & Dasberg, 1983) showing the influence of S and ρ on the calibration of the TDR method for soil moisture determination. The biggest temperature effect on ε_b was observed for the soil 569 at its highest moisture content; with the temperature increase of 50°C its value decreased by 3.49. This temperature change of the TDR determined bulk dielectric permittivity corresponds to a decrease of the calculated soil moisture of 0.037 m^3m^{-3}, using the calibration from Topp et al. (1980).

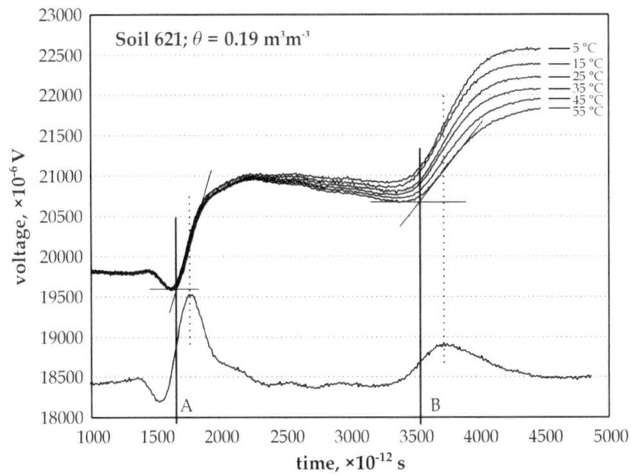

Fig. 6. Set of reflectograms registered by the measurement setup for the same soil sample in different temperatures

The calibration curves for each soil taken for different soil temperatures meet at a characteristic moisture content value θ_{eq}, where the physical phenomena responsible for the temperature effect of soil dielectric permittivity equalize (Fig. 7). This moisture content is named the "equilibrium water content" in this study. For water contents below θ_{eq}, soil bulk dielectric permittivity ε_b measured at 5°C is smaller than the one measured at 55°C and for water contents above θ_{eq} the change of ε_b with temperature is opposite. The observed temperature effect of soil dielectric permittivity confirms the theory of Or & Wraith (1999), which explains it by the temperature caused exchange of water particles between free and bound phases.

The value of ε_b for air-dry soils does not depend on temperature. With the increase of soil moisture from air dry, there is an increase of ε_b with temperature. All soils except one (soil 562) have higher values of ε_b for 5°C than 55°C at high water contents, and the biggest difference is observed for the soils having medium values of S (Table 1). For the soil 562, there was no equilibrium water content observed in the analysed temperature range, although this soil does not have the highest value of specific surface area from all the tested soils. The bulk dielectric permittivity for this soil is higher at 55°C than for 5°C in the whole range of moisture from air dry state to almost saturation. The soil bulk electrical conductivity in low frequency range, σ_{dc}, for the soil 562 and for example the soil 565 at different temperatures and high moisture was compared.

Bulk electrical conductivity for the soil 562 ranged from 47 mSm^{-1} at 5°C to 192 mSm^{-1} at 55°C for $\theta = 0.3$ m^3m^{-3}, while for the soil 565 the respective values were even bigger, i.e. they ranged from 50 mSm^{-1} at 5°C to 221 mSm^{-1} at 55°C for θ = 0.4 m^{3}m^{-3}, and still the temperature behaviour of the soil 565 dielectric permittivity was typical. This confirms the literature reports (Nadler et al., 1999; Or & Wraith, 1999) that the increase of soil bulk electric conductivity does not increase the TDR readout of ε_b in the presented variability range of σ_{dc}. Further studies are needed to diagnose the temperature effect of the soil 562.

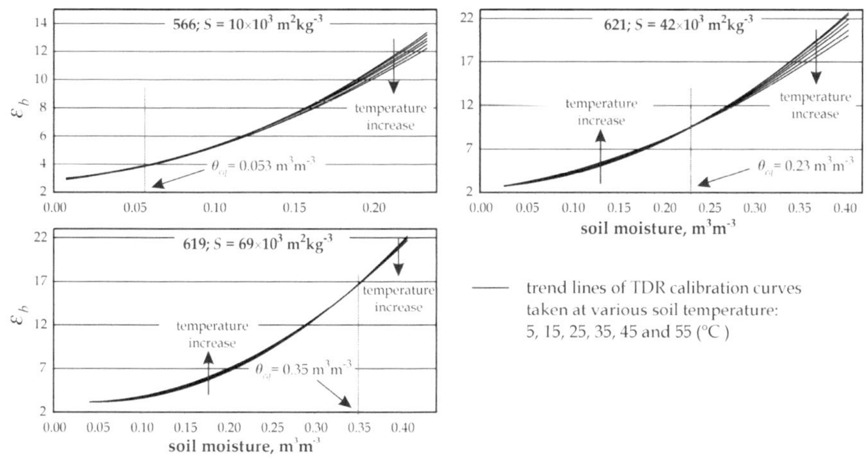

Fig. 7. TDR calibration curves for three tested soils at different temperatures, S is the soil specific surface area, θ_{eq} is the equilibrium moisture where the temperature effect caused by the described two competing physical phenomena is compensated

The value of soil moisture at the equilibrium point, θ_{eq}, depends on the amount of bound water attracted by the soil, which is described by equation (10). The relation between the equilibrium water content, θ_{eq}, and the specific surface for the tested 19 soils, except the soil 562, is presented in Fig. 8.

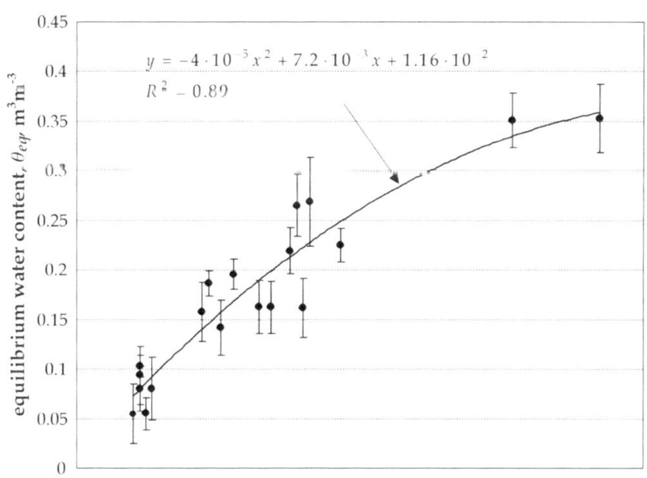

Fig. 8. Empirical relation between the equilibrium water content, θ_{eq}, and soil specific surface, S, for the tested soils

The good correlation between the soil specific surface S and θ_{eq} confirms the assumed physical description of the processes involving the temperature effect of soil dielectric permittivity.

The temperature dependence of the soil bulk dielectric permittivity, ε_b, of the selected soils is presented in Fig. 9. The majority of tested soils show similar trends that confirm other experimental data (Pepin et al., 1995; Or & Wraith, 1999). For small and medium moisture values there is a negligible temperature effect and the linear trend lines in Fig. 9 are almost in parallel to the horizontal axis representing no or a small positive temperature change.

This is especially evident for the two trend lines representing the lowest moisture values in $\varepsilon_b(T)$ relations for all soils in Fig. 9. For higher moisture values there is a tendency to decrease ε_b with temperature proving that the dominant reason for this behaviour is the decrease of free water dielectric permittivity with the temperature increase. However, as expected for the soils with large specific surface area this tendency is much smaller or has the opposite direction, as for the soil no. 562.

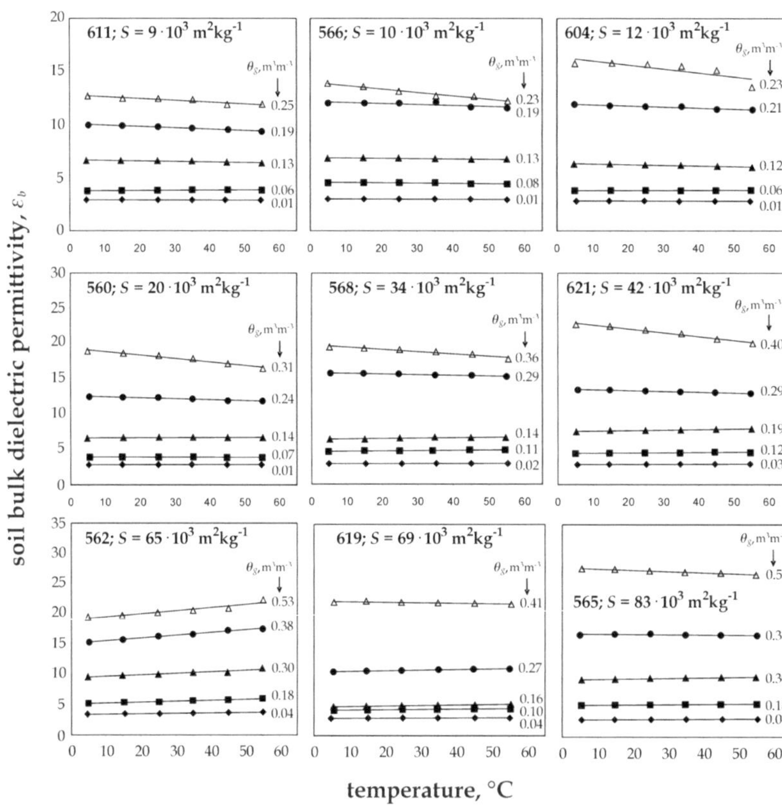

Fig. 9. Bulk dielectric permittivity temperature dependence of selected soils bulk dielectric permittivity for different soil volumetric water contents (θ_g stands for thermo-gravimetrically determined soil water content)

Other soils from the same group as the soil no. 562, i.e. no. 619 and no. 565 having larger values of soil specific surface area, $69 \cdot 10^3$ $m^2 \cdot kg^{-1}$ and $83 \cdot 10^3$ $m^2 \cdot kg^{-1}$ respectively, show typical temperature effect of soil dielectric permittivity, i.e. for moisture below θ_{eq}, ε_b

increases with increasing temperature, and for moisture above θ_{eq}, ε_b decreases with increasing temperature. For all tested soils, except the soil no. 562, the slope was small and positive for low and negative for high soil moisture.
The applied models: 4-phase alpha and de Loor models do not follow the measured data $\varepsilon_b(T)$ for all soils and for all moisture values. The examples of the performance of these models are presented in Fig. 10, for the soil no. 562 (S = 65·10^3 m^{2}kg^{-1}).

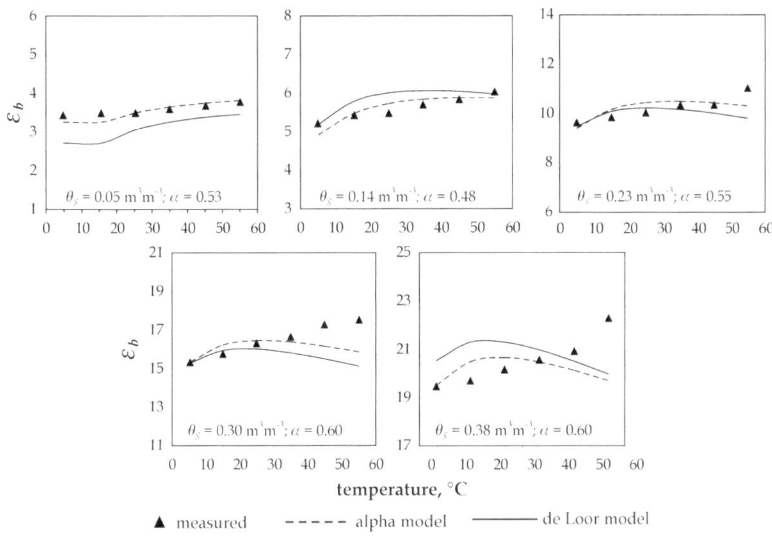

Fig. 10. Performance of applied 4-phase soil dielectric mixing models for the soil no. 562. The assumed values for dielectric permittivity for soil solid phase ε_s=5 and bound water ε_{bw}=3.2

The alpha model has a fitting parameter representing the geometry of modelled medium (Roth et al., 1990) and enabling to adjust model data to the measured ones, as it was done in Fig. 10. The values of ε_b from the model data are generally higher than the TDR determined from the measurement. Therefore the applied correction of the temperature effect on the soil dielectric permittivity is based on empirical data and it accounts for the observed property of θ_{eq}. The applied correction turns the slope of the linear trend lines of the measured values of $\varepsilon_b(T)$ to zero. The turning centre is the soil temperature of 25°C (Fig. 9).
For each soil sample the temperature dependent value of soil moisture $\theta_{TDR}(T)$ was determined basing on the individual TDR soil calibration function giving the bulk dielectric permittivity values at different temperatures. Temperature corrected values of volumetric water content θ_{TDR}^{corr}, were determined from the relationship between $\theta_{TDR}(T)$ and θ_{eq}, as follows:

$$\text{For } \theta_{TDR}(T) \le \theta_{eq}: \ \theta_{TDR}^{corr} = \theta_{TDR}(T) \cdot \left[1 - A \cdot \left(\theta_{eq} - \theta_{TDR}(T)\right) \cdot (T - 25)\right] \tag{14}$$

$$\text{For } \theta_{TDR}(T) > \theta_{eq}: \theta_{TDR}^{corr} = \theta_{eq} + \frac{\theta_{TDR}(T) - \theta_{eq}}{1 + n_{25} \cdot B \cdot \left[d(T) - 1\right] \cdot \frac{\partial \theta_{TDR}}{\partial n}} \tag{15}$$

where: $n = (\varepsilon_b)^{1/2}$ is the soil refractive index, n_{25} is its value at 25°C, $\partial\theta_{TDR}/\partial n$ is the slope of TDR calibration, for example the one taken from Topp et al. (1980) is 0.127.
Formula (14) refers to TDR determined soil moisture values not exceeding equilibrium water content, θ_{eq}, for which the temperature effect of the soil bulk dielectric permittivity increases its value, which is due to release of water molecules adsorbed by the surface of the soil solid phase. Formula (15) applies TDR determined soil moisture values exceeding equilibrium water content, θ_{eq}, when the temperature effect of the soil bulk dielectric permittivity is mainly from the temperature effect on free water, resulting in declining soil bulk dielectric permittivity with increasing temperature.
Parameters $A = 0.008$ and $B = 0.55$ are empirically adjusted to minimize the slope $\Delta\varepsilon_b/\Delta T$ for all examined soils. Because of good correlation between soil specific surface, S, and θ_{eq} (Fig. 8), the former can be applied from the fitted line.

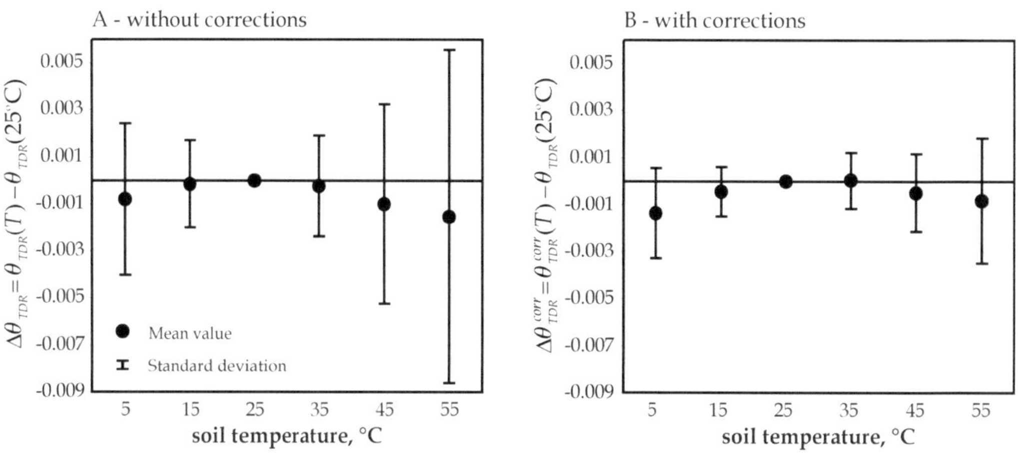

Fig. 11. Errors of TDR determined soil moisture at various temperature values referenced to the respective values at 25°C: A - without temperature corrections, B - with temperature corrections based on (14) and (15)

The comparison of mean values and standard deviations of absolute errors for measured data, $\Delta\theta_{TDR}$, and temperature corrected data, $\Delta\theta_{TDR}^{corr}$, according to the temperature corrections (14) and (15) is presented in Fig. 11. The mean values for the both errors are calculated for 20 tested soils at six values of temperature from 5°C to 55°C with 10°C steps between.
The mean values of TDR determined soil moistures at tested soil temperatures are below the respective values at 25°C. The absolute measurement error of soil moisture measured by TDR, defined as double the standard deviation from the mean value does not exceed 1.5% of measured value. After applying the empirical correction given by (14) and (15) the absolute measurement error decreased almost three times to the value not exceeding 0.54%.

5. Summary

The temperature effect of the examined mineral soils' bulk dielectric permittivity, ε_b, determined by TDR method, confirms the theory presented by Or and Wraith (1999), describing it as the result of two competing phenomena; ε_b increases with temperature

increase following the release of bound water from soil solid particles and ε_b decreases with temperature increase following the temperature effect of free water molecules.
It has been found that there is a soil type characteristic moisture value, θ_{eq}, named the equilibrium water content, having the specific temperature property. The temperature effect for this soil water content value is not present, which means that at θ_{eq} the both competing phenomena mentioned earlier compensate each other. The equilibrium water content, θ_{eq}, is correlated with the soil specific surface area. For soils with water content below θ_{eq}, the temperature effect of soil dielectric permittivity is positive, i.e. it increases with temperature, and for soils with water content above θ_{eq}, the temperature effect of soil dielectric permittivity is negative, i.e. it decreases with temperature.
The temperature correction formula adjusting the soil moisture determined by TDR at various temperature values to the corresponding value at 25°C, based on knowledge of θ_{eq}, decreases the standard deviation almost three times as compared to uncorrected values of θ_{TDR}. The electrical conductivity of the examined soils does not show any influence on the observed temperature effect of soil dielectric permittivity.

6. References

Birchak, J.R., Gardner, C.G., Hipp, J.E. & Victor J.M. (1974). High dielectric constant microwave probes for sensing soil moisture, *Proceedings of the IEEE*, 62(1), 93-98.

Boyarskii D.A., Tikhonov, V.V., Komarova, N.Yu. (2002). Model of dielectric constant of bound water in soil for applications of microwave remote sensing. *Progress In Electromagnetic Research*, 35, 251-269.

Dalton, F.N., Herkelrath, W.N., Rawligs, D.S. & Rhoades, J.D. (1984). Time-domain reflectometry simultaneous measurement of soil water content and electrical conductivity with a single probe, *Science*, 224, 989-990.

Davis, J.L. & Annan A.P. (1977). Electromagnetic detection of soil moisture: progress report I, *Canadian Journal of Remote Sensing*, 3, 76-86.

Debye, P. (1929). *Polar Molecules*. Dover, Mineola, New York.

de Loor, G. P. (1990). Dielectric properties of heterogeneous mixtures. BCRS Rep. No. 90-13. TNO Physics and Electronics Lab., The Hague.

Dirksen, C. & Dasberg, S. (1993). Improved calibration of Time Domain Reflectometry soil water content measurements, *Soil Science Society of America Journal*, 57, 660-667.

Halbertsma, J.; van den Elsen, E.; Bohl, H. & Skierucha W. (1995). Temperature effects on TDR determined soil water content. Proceedings of the Symposium: Time Domain Reflectometry Applications in Soil Science. Research Center Foulum, Sept. 16, 1995. SP Report 11, 35-37, Danish Institute of Plant and Soil Sci., Lyngby, Denmark, 1995.

Hasted, J.B. (1973). *Aqueus dielectrics*. Chapman and Hall, London.

Malicki, M.A. (1999). Methodical questions of monitoring of water status in selected biological materials (in Polish), *Acta Agrophysica*, 19, 1-108.

Malicki, M.A., Plagge, R. & Roth, C.H. (1996). Reduction of soil matrix effect on TDR dielectric moisture determination by accounting for bulk density or porosity, *European Journal of Soil Science*, 47(3), 357-366.

Malicki, M.A. & Skierucha, W. (1989) A manually controlled TDR soil moisture meter operating with 300 ps rise-time needle pulse. *Irrigation Science*, 10, 153-163.

Nadler, A.; Gamliel, A. & Peretz, I. (1999). Practical aspects of salinity effect on tdr-measured water content: a field study. *Soil Science Society of America Journal*, 63, 1070-1076.

O'Connor, K.M. & Dowding C.H. (1999). *Geomeasurements by pulsing TDR cables and probes*, CRC Press, ISBN 0-8493-0586-1, Boca Raton.

Or, D. & Wraith, J.M. (1999). Temperature effects on soil bulk dielectric permittivity measured by time domain reflectometry: A physical model, *Water Resources Research*, 35(2), 371-383.

Oscik, J. (Ed.) (1983). *Adsorption*. PWN. Warszawa.

Pansu, M. & Gautheyrou, J. (2006). *Handbook of soil analysis; mineralogical, organic and inorganic methods*, Springer-Verlag Berlin Heidelberg.

Pepin, S.; Livingston, N.J. & Hook, W.R. (1995). Temperature-dependent measurement errors in time domain reflectometry determinations of soil water, *Soil Science Society of America Journal*, 59, 38-43.

Roth, K.; Schulin, R.; Flühler, H. & Attinger W. (1990). Calibration of Time Domain Reflectometry for water content measurement using a composite dielectric approach, *Water Resources Research*, 26, 2267-2273.

Skierucha, W. (2002). Temperature effect on soil dielectric permittivity: description of laboratory setup and applied software (in Polish), *Acta Agrophysica*, 72, 125-133.

Topp, G.C.; Davis, J.L. & Annan A.P. (1980). Electromagnetic determination of soil water content: measurements in coaxial transmission lines, *Water Resources Research*, 16, 574-582.

Wheast, R.C. (Ed.) (1979). *CRC handbook of chemistry and physics*, CRC Press Inc. Boca Raton. Florida, USA.

Wraith, J.M. & Or, D. (1999). Temperature effects on soil bulk dielectric permittivity measured by time domain refiectometry: Experimental evidence and hypothesis development, *Water Resources Research*, 35 (2), 361-369.

18

The Temperature Behavior of Resonant and Non-resonant Microwave Absorption in Ni-Zn Ferrites

Raúl Valenzuela
Departamento de Materiales Metálicos y Cerámicos,
Instituto de Investigaciones en Materiales,
México

1. Introduction

The magnetic response of Ni-Zn ferrites at microwave frequencies has been recently investigated by means of resonance techniques, by several authors. In this chapter, we present a review of recent results obtained on the resonant microwave absorption (electron paramagnetic resonance, EPR, and ferromagnetic resonance, FMR) in the X-band (9.5 GHz), of polycrystalline Ni-Zn ferrites ($Zn_xNi_{1-x}Fe_2O_4$) for several temperature ranges. We begin at high temperatures in the paramagnetic state ($T > T_C$, where T_C is the Curie point); as temperature decreases, the onset of magnetic ordering is investigated, with its effects on the main FMR parameters. When experiments are carefully carried out, magnetic transitions can be detected as critical points in plots of the thermal behavior of the resonance line width.
We investigate also the behavior of nonresonant properties by means of the low-field microwave absorption (LFMA). This absorption, which occurs at applied fields of the same order of magnitude than the anisotropy field, H_K, of the sample, is providing valuable information concerning the magnetization processes. LFMA is typically measured in the -1 kOe $< H_{DC} <$ +1 kOe field range. LFMA is associated with the nonresonant microwave absorption occurring during the magnetization processes from the unmagnetized state up to the approach to saturation. We provide here a short review of this particular measuring technique. Then, we propose to begin the study of LFMA in Ni-Zn ferrites also by decreasing the measuring temperature from the Curie transition. Clearly, LFMA is absent at $T > T_C$ since it depends on the magnetization processes in the ordered phase. For the 200 K $< T < T_C$ temperature range, a direct comparison of the anisotropy field calculated from LFMA and a calculation by using results of a direct measurement of H_K on a ferrite single crystal. A very good agreement is obtained, thus confirming that LFMA is strongly dependent of the total anisotropy (magnetocrystalline, magnetoelastic and shape anisotropies) of the sample.
We use as well a novel nonresonant microwave absorption technique known as magnetically modulated microwave absorption spectroscopy, MAMMAS. This technique is particularly well adapted to detect phase transitions of many types, as it is based on the change of microwave absorption regime during a change of crystalline, magnetic or electronic structure. MAMMAS is briefly described and applied to Ni-Zn ferrites.

2. Ferrites

Ferrites, also known as magnetic ceramics, are a very well established group of magnetic materials (Valenzuela, 2005a). Ferrites possess three different crystal structures: spinels, garnets, which belong to cubic systems, and hexagonal, which can be considered as derived from magnetoplumbite. In this review, the focus will be on spinel ferrites, and in particular on the Ni-Zn "family" which will be taken as an example. A brief review of crystal structure, magnetic structure and magnetic properties of these ferrites is given.

2.1 Spinel structure

The spinel structure is a cubic structure extremely stable, with a dominant ionic character. In addition to charge compensation, the cation/anion ratio is ¾. More than 140 oxides and 80 sulphides have been systematically studied (Hill & al 1979). Most of the commercially important spinels are synthetic, but the most important and probably the oldest one with practical applications, magnetite, Fe_3O_4, is a natural oxide. Magnetite has also the remarkable feature of the simultaneous presence of ferrous (Fe^{2+}) and ferric (Fe^{3+}) iron on equivalent crystal sites, which provides unusual electrical and magnetic properties. In addition to the 2,3 spinels (2,3 refers to divalent and trivalent cations, respectively), formed by a combination of one divalent and two trivalent cations to balance the 8 negative charges provided by the oxygen in the formula $D^{+2}T^{+3}{}_2O^{-2}{}_4$, there are other combinations with spinel structure, which provide 3 cations with a total valency of 8, such as 2,4 (Co_2GeO_4), 1,3,4 ($LiFeTiO_4$), 1,3 ($Li_{0.5}Fe_{2.5}O_4$), 1,2,5 ($LiNiVO_4$), and 1,6 (Na_2WO_4) spinels.
The crystal structure, belonging to the Fd3m space group, can be described as a close-packed (fcc) arrangement of oxygens, which includes tetrahedral and octahedral interstitial sites. One-half of the interstitial octahedral sites and one-eighth of the tetrahedral sites are occupied by cations. They are known also as "A" sites (tetrahedral) and "B" sites (octahedral).
The unit cell is formed by eight formula units AB_2O_4, with eight A sites, 16 B sites and 32 oxygen. This unit cell can be divided into octants of edge $a/2$ (a = unit cell parameter) for a better view of the two sites, Fig. 2.1. In this representation, a tetrahedral cation is taken as the origin of the cell. The nearest neighbors of both sites are illustrated in Fig. 2.2.
When divalent cations occupy the A sites and trivalent cations enter the B sites, the spinel is known as having a "normal" cation distribution. This arrangement can be represented as (D^{+2}) $[T^{3+}{}_2]$. A variant of this structure is the "inverse" spinel, where A sites contain a trivalent cation, while B sites contain the divalent and the remaining trivalent cation, (T^{3+}) $[D^{2+}\ T^{3+}]$. In some cases, an intermediate distribution can be achieved by playing with thermal treatments, leading to $(D_{1-\delta}T_\delta)[D_\delta T_{2-\delta}]$, where δ is the "degree of inversion". The distribution of cations on the two spinel sites depend on a complex interplay of cation radius, electrostatic energy, crystal field energy, and polarization effects (covalency contribution, for instance).
A remarkable feature of stability of spinel structure is that it can form an extremely large variety of total solid solutions. Some conditions apply; first, electrical neutrality, i.e., the addition of the charge of all cations should balance oxygen total charge (-8 for a formula); second, the ratio of cations/oxygen should remain ¾, and finally, there should be relatively small differences between cation radii. In solid solutions, composition can be changed on a continuous basis, leading also to continuous variations in the physical properties. This allows a very precise tailoring of magnetic properties, which is a major advantage for any application. Divalent cation in the 2,3 spinel formula can be formed by any combination of

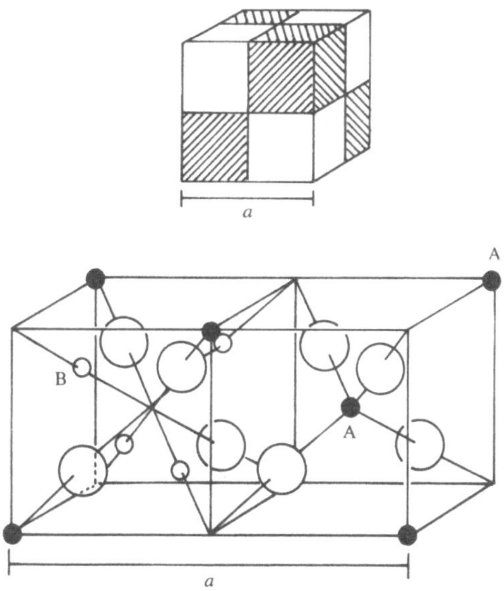

Fig. 2.1. Unit cell of the spinel structure. Cations on A sites are represented by small black circles, cations on octahedral B sites by small open circles, and large circles are oxygens. The unit cell parameter is *a*.

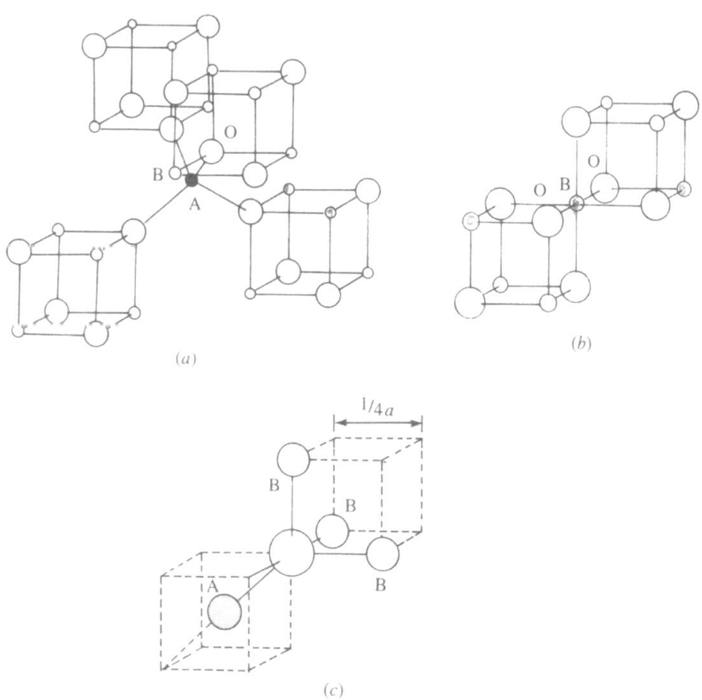

Fig. 2.2. Nearest neighbors of a) A site, b) B site and c) oxygen site.

divalent Ni^{2+}, Co^{2+}, Mn^{2+}, Fe^{2+}, Cu^{2+}, Zn^{2+}, Cd^{2+}, Mg^{2+}, Ca^{2+}. Ferric ions can also be substituted, or combined with Al^{3+}, V^{3+}, Cr^{3+}, Mn^{3+}, Ga^{3+}, In^{3+}, etc. One of the most interesting and representative solid solution is Ni-Zn ferrites, with formula $Ni_{1-x}Zn_xFe_2O_4$, with $0 \le x \le 1.0$ (Ravindranathan & Patil, 1987).

2.2 Nickel-zinc ferrites

In spite of having a large cation radius, Zn^{2+} has a strong preference for A sites, which are smaller than B sites. Ferric ions manifest no preference for A or B sites. Therefore, zinc ferrite $ZnFe_2O_4$ is a normal spinel. In contrast, divalent nickel shows a strong tendency to occupy B sites. This means that nickel ferrite, $NiFe_2O_4$ tends to be an inverse spinel. Ni-Zn solid solutions (when prepared by solid state reaction with a slow cooling from the sintering temperature) exhibit therefore a cation distribution which is normal with respect to Zn, and inverse for Ni. This means that Zn will occupy A sites (with ferric ions completing the "filling" of A sites), while nickel and the remaining ferric cations share B sites:

$$(Zn_xFe_{1-x})\ [Ni_{1-x}\ Fe_{1+x}]$$

The cell parameter, Fig. 2.3a, shows a linear dependence with composition x. Since Zn^{2+} is a relatively large cation occupying the small A sites, the cell parameter increases with Zn content. The Curie temperature exhibits a strong decrease with zinc concentration, Fig. 2.3b. For $x = 1$, zinc ferrite ($ZnFe_2O_4$) manifests an antiferromagnetic behavior with a Néel temperature of 9 K. While the increase in cell parameter is quite linear, the decrease in T_C is more rapid. This fact can be understood by recalling that as Zn content increases, in addition to the expansion of the unit cell (and therefore, cations become far apart), there is an effect of dilution, since Zn cations are diamagnetic. However, there is also a change in magnetic structure, since for the very high content of Zn, the ferrite changes from ferrimagnetic (with a high T_C = 858 K for x = 0), to an antiferromagnetic arrangement with a very low Néel temperature. This result will be briefly discussed below.

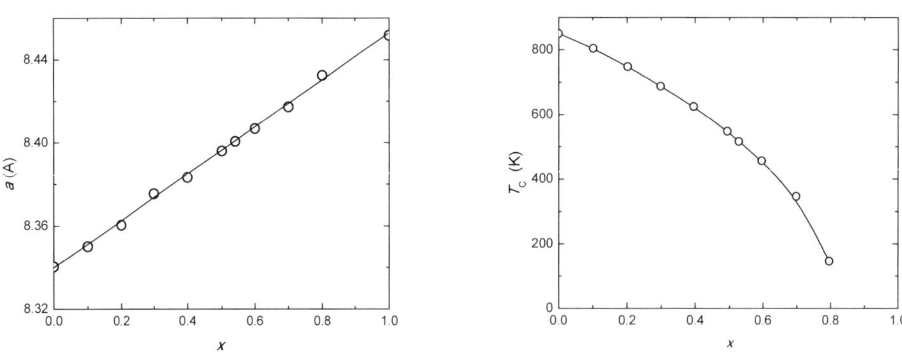

(a) (b)

Fig. 2.3. Variation in the cell parameter, (a), and the Curie transition, (b), both as a function of Zn content x (Adapted from Valenzuela, 2005a).

The ferrimagnetic order in ferrites is the result of superexchange interactions. The 3d unpaired spins of transition metals exhibit an antiparallel arrangement which occurs through anions, as schematically shown in Fig. 2.4. This interaction takes place by means of p orbitals of oxygen. Since p orbitals are linear, this interaction sensitively depends not only on the distance between cations and anion, but also on the angle between them. It is expected to be a maximum for a 180° angle. The first discussion on superexchange interactions was proposed by Anderson (1959).

The main superexchange interactions in spinels are the A-O-B and the B-O-B interactions. The former takes place between a cation in an A site, which becomes antiparallel to cations on the nearest B site. The latter consists on the antiparallel arrangement between two cations on neighboring B sites. The A-O-B interaction is expected to be significantly stronger than the B-O-B one, since the angle between these sites is close to 180° (see Fig. 2.2 (c)); the B-O-B geometry involves a 90° angle, quite different from the linear geometry of p orbitals.

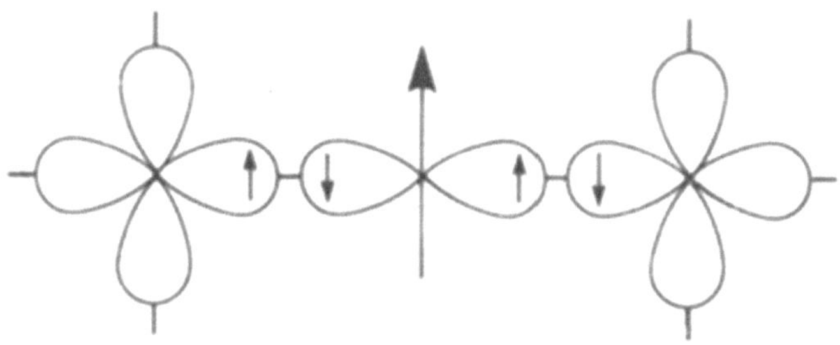

Fig. 2.4. Schematical representation of the superechange interactions in oxides. The spins in the unfilled 3d orbitals of transitions metals, on the sides, can interact with cation nearest neighbors through the 2p oxygen orbitals, in the center. This interaction can be extremely strong, leading to high Curie temperatures.

For $x = 0$, the cation distribution is as follows: (Fe)[NiFe]. By assuming that A-O-B interaction is dominant, the iron in the A site will be aligned in an antiparallel direction with respect to spins of cations on B sites. If we simplify the magnetic structure of Fig. 2.2 (c) and represent one A site and two B sites around an oxygen anion (in the basic formula, the ratio of A to B sites is ½), and if all of them are assumed to be on the same plane, we can draw a cartoon like the one on Fig. 2.3. Nickel ferrite, with one Fe^{3+} on the A site, the other one on a B site and the Ni on the other B site should have a magnetic structure like the one in Fig. 2.5 (a). The interaction Fe(A)-O-Fe(B) is among the strongest in spinels, as Fe^{3+} has a 3d orbital half-filled and the angle between sites is close to 180°. Accordingly, the Curie temperature is maximum for this family (858 K), and it has the same value for most inverse spinels (such as $CoFe_2O_4$, for instance).

For zinc ferrite ($x = 1$), the site occupancy is: (Zn) [Fe_2]. The A site contains only Zn ions (with no magnetic moment) and therefore the only interaction in the system is B-O-B. Irons on both B sites become antiparallel and the ferrite is antiferromagnetic, with a Néel temperature of 9 K. This low value of superexchange interaction is explained mostly by the angle between interacting cations (90°), and also by the expansion of the unit cell, as a consequence of the larger size of Zn cations [Fig. 2.3.(a)]. For compositions in the $0.5 < x < 0.8$ range, with a distribution: $(Zn_xFe_{1-x})[Ni_{1-x}Fe_{1+x}]$, where both interactions become comparable, the magnetic structure can be represented by a triangular arrangement known as the Yafet-Kittel structure, first proposed by these authors (Yafet and Kittel 1952).

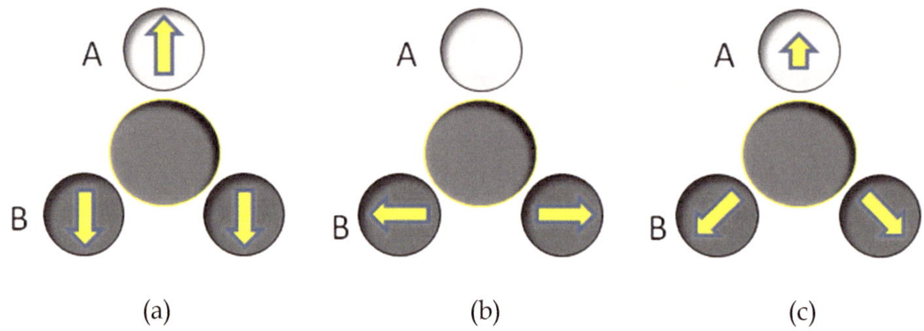

Fig. 2.5. Simplified representation of an A and two B sites around an oxygen. Arrows represent the spins as they can be expected for (a) nickel ferrite ($x = 0$), (b) zinc ferrite ($x = 1$), and (c) a composition rich in Zn ($0.5 < x < 0.8$).

A plot of saturation magnetization (at low temperatures) as a function of the composition starts at $\sigma_s \sim 2.33$ Bohr magneton/formula unit, since the ferric cations are in opposition (Fig. 2.5 (a)) leaving only the nickel magnetic moment as a result, as shown in Fig. 2.6. If the A-O-B interaction were dominant on all the composition range, the total magnetic moment would exhibit an increase with x up to a value of 10 Bohr magnetons for $x = 1$ (broken line in Fig. 2.6), a condition with all A sites occupied by Zn (with no magnetic moment) and both B sites with Fe, and spins in a parallel orientation. But the weakening of this interaction results in the competition of B-O-B interaction, leading to the antiparallel arrangement on sites B, with the variations in saturation magnetization illustrated in Fig. 2.6.

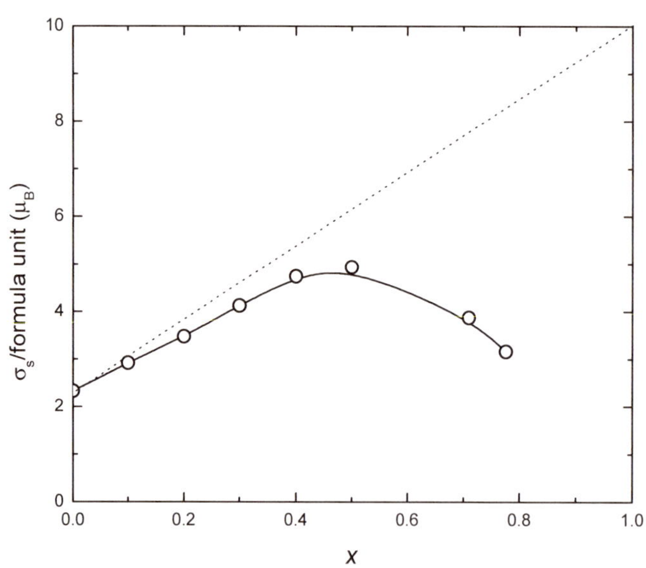

Fig. 2.6. Behavior of saturation magnetization of $Ni_{1-x}Zn_xFe_2O_4$ ferrites at very low temperature, as a function of Zn content.

After many years, NiZn ferrites remain as an excellent system to study magnetic properties of solids.

3. Microwave absorption

Microwave absorption has become a very powerful investigation and characterization tool in the study of magnetic materials, both in the paramagnetic, disordered state (electron paramagnetic resonance, EPR) and the ferri or ferromagnetic, ordered phase (electron ferromagnetic resonance, FMR) (see, for instance, Kittel 2005, Pilbrow 1990). The radiation emerging from interaction with a solid possesses changes (with respect to the incident radiation) that in principle, allow deducing the structural and magnetic properties of the material. To simplify, we can consider the interaction of a spin with a constant magnetic field. If the magnetic moment is originated only by the spin,

$$\mu = g\mu_B\mathbf{S} = \gamma\mathbf{S} \tag{3.1}$$

where g is the gyromagnetic factor (in general depending on **L** and **S**, the quantum mechanical numbers of orbital and spin momenta), γ is the total gyromagnetic ratio. In an external field, $\mathbf{H}_0 = H_0\mu_z$, and energy is expressed as:

$$E = -\,\mu\bullet\mathbf{H} = -\,\mu_z H = -\,\gamma m_s\hbar H \tag{3.2}$$

with the spin $m_s = \pm\,½$, corresponding to the two orientation of the magnetic moment, i.e., parallel ($m_s = -\,½$), or antiparallel ($m_s = +\,½$) to the magnetic field; the population of both levels is given by the Boltzmann statistics,

$$f = N^+/N^- = \exp\{-\Delta E/k_BT\} \tag{3.3}$$

where $N = N^+ + N^-$ is the total population of atoms with spin parallel (N^-) and antiparallel (N^+) to the magnetic field, k_B is the Boltzmann constant, and $-\Delta E$ the energy difference between the two levels. The net magnetic moment per atom is then:

$$\mu_z = (g\mu_B/2)[(N^+ + N^-)/N] = (g\mu_B/2)[(1-f)/(1+f)] \tag{3.4}$$

A series expansion of (3.4) for not so low temperatures ($k_BT >> \Delta E$) leads to the Curie law,

$$\langle\mu_z\rangle = (g\mu_B/3)(\Delta E/k_BT) = CH/T \tag{3.5}$$

with $C = g\mu_B^2/3k_B$. It is possible to induce transitions between the two spin states by application of electromagnetic radiation of the relevant frequency, which satisfies the Bohr condition,

$$\begin{aligned}\Delta E &= \gamma\hbar H = \hbar\omega \\ \omega &= \gamma H\end{aligned} \tag{3.6}$$

This shows the resonance conditions. Equation (3.6) is also known as the Larmor resonance condition.

In the case of magnetic materials with a spontaneous magnetization (ferri and ferromagnetic materials), H includes the internal field, in most cases leading to a lower external field needed to attain the resonance conditions. Both EPR and FMR have been used to investigate a wide variety of materials such as ferrites (Montiel et al 2004, Wu et al 2006) and amorphous alloys (Valenzuela et al 2005b, Montiel et al 2006).

In addition to these methods, nonresonant microwave absorption, or low field microwave absorption (LFMA) has been observed in many materials, such as amorphous metallic thin

films (Rivoire & Suran 1995), amorphous ribbons (Medina et al 1999), glass coated amorphous microwires (Chiriac et al 2000), ferrites (Montiel et al 2004), multilayer thin films (de Cos et al 2007). LFMA is strongly associated with magnetic order since in all cases it is present only below the transition temperature between the paramagnetic-ferrimagnetic (or para-ferromagnetic) phases. LFMA has also shown to be sensitive to mechanical stresses (Montiel et al 2006). In this chapter, we show that LFMA can also be used to detect changes in the magnetic structure. From the experimental point of view, LFMA needs an accurate measurement of the magnetic field for low fields, and the possibility to reverse the field, i.e., typically in the $-1000 < H < +1000$ Oe. This can be challenging in the case of large electromagnets, which tend to keep a non negligible remanent field.
Another nonresonant method recently proposed for the investigation of magnetic transition is the method known as magnetically modulated microwave absorption spectroscopy (MAMMAS) (Alvarez & Zamorano 2004, Alvarez et al 2007), which is based on a simple idea: the nonresonant microwave absorption regime in a given material changes when a phase transition occurs. Since the microwave absorption depends on the wide definition of structure (crystalline, electronic, magnetic, etc.), virtually any phase change can be detected, with the significant advantage that microwave absorption is extremely sensitive. Experimentally, the sample is subjected to a low magnetic field (clearly lower than the resonance field in the temperature range), and the microwave absorption is measured as the sample temperature is slowly varied. Phase transitions appear typically as a minimum in a dP/dH vs T plot.

4. Microwave absorption in ferrites

In this Section, we discuss the microwave absorption of polycrystalline Ni-Zn ferrites as a function of measuring temperature. The resonant mode is first considered. The description and analysis of these properties is quite useful, as NiZn ferrites offer a wide variety of magnetic structures and phenomena. The study of non resonant absorption also sheds light on magnetic structure phenomena of ferrites.

4.1 High temperatures ($T > T_C$)

By "high temperature" we mean a temperature higher than the transition from the ordered (ferrimagnetic) phase to the disordered (paramagnetic) phase. This transition is the Curie temperature, T_C, and it is an intrinsic property, depending entirely on the ferrite composition (except for ferrite nanoparticles, where the Curie temperature might depend on the nanoparticle size). The effects of temperature are shown in Fig. 4.1 for a polycrystalline NiZn ferrite with $x = 0.65$. Most of spectra exhibit an additional absorption in the low field range ($H < 0.75$ kOe). This non-resonant absorption is the low field microwave absorption (LFMA, Section 3) and will be discussed in Section 5. We focus now on the resonance phenomena.
The reported Curie temperature for this composition is ~ 435 K (Globus et al 1977); the spectrum at 460 K corresponds therefore to the paramagnetic state. In these conditions, the thermal energy is high enough to overwhelm the internal field that results in the long range order of spins, and they are free to interact with the DC field, H_{DC}, and the microwave field, h_{AC}. In the Larmor relation (Eq. 3.6),

$$\omega = \gamma H \tag{4.1}$$

(where ω is the resonance frequency, γ is the gyromagnetic factor), H is the total field on the spins. In the absence of any internal field, H is simply the external applied field H_{res}. The microwave absorption at 460 K, Fig. 4.1, appears as a narrow, symmetric line. This is an EPR (electron spin resonance) spectrum.

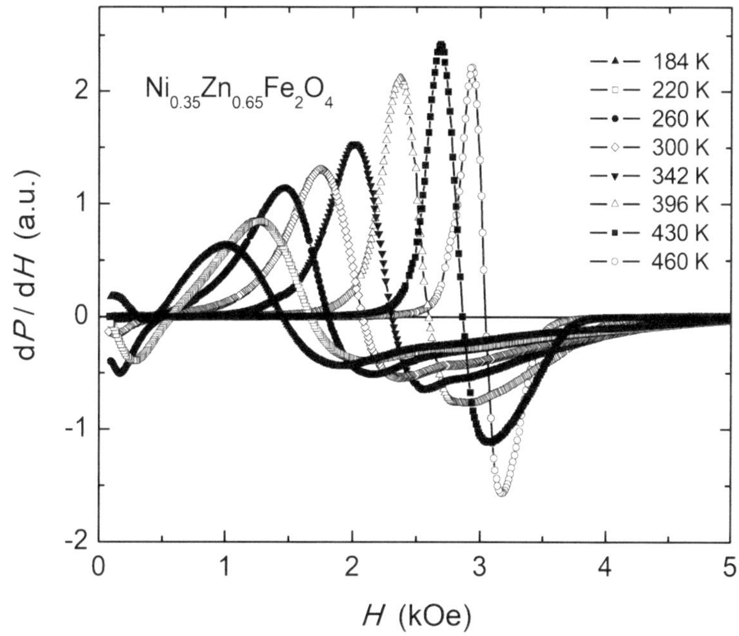

Fig. 4.1. Ferromagnetic resonance as a function of measuring temperature ($184 \leq T \leq 460$ K), on a polycrystalline ferrite with composition $Ni_{0.35}Zn_{0.65}Fe_2O_4$ (adapted from Montiel *et al.*, 2004).

4.2 Intermediate temperatures ($T < T_C$)

As temperature decreases below the Curie transition, several changes are apparent in the resonance spectra. First, the resonance field (usually taken as the intercept of the line with the field axis) decreases. Second, the lineshape becomes broader as T decreases, especially in the $H > H_{res}$ field region. These changes are due to the rise of the internal field, H_i, leading to the long range order of magnetic moments. For $T < T_C$, H_i possesses a larger energy than the thermal energy of the ferrite. Accordingly, the total field in the Larmor relation should include now the contribution from the internal field,

$$H = H_{res} + H_i \tag{4.2}$$

The internal field is the combination of all the factors associated with the long range order in the ferrite: the exchange field, H_{ex}, the anisotropy field, H_K, the demagnetization field, H_d, the porosity field, H_p (which is the field due to the appearance of magnetic dipoles on pores), etc. An additional source of inhomogeneity in ferrites is associated with differences, as well as with disorder, in site occupancy by the cations. As discussed in Section 2, transition metal cations have different stabilization energies on sites with diverse symmetry, as the tetrahedral and octahedral sites of spinels. While some of them exhibit a clear "preference" for one of the sites (i.e., Zn^{2+} for tetrahedral or A sites, Co^{2+}, Ni^{2+}, and Fe^{2+} for octahedral or B sites), other cations can be found equally on both sites (Fe^{3+}). To make things more complicated, it is possible to change the cation distribution by means of thermal treatments. The resonance phenomenon can be therefore slightly different when this occupancy of sites is not strictly homogeneous, since some terms of the internal field are not exactly the same for all the microwave absorbers.

The other source of inhomogeneity in the internal field is the disorder in the site occupancy. Even if the occupancy of sites is well determined (i.e., in Ni-Zn ferrite, *all* Zn cations on A sites, *all* Ni cations on B sites), there can be an inhomogeneous distribution of each of them on the sites. A simple example could be nickel ferrite, $NiFe_2O_4$, with all Ni^{2+} on B sites (and of course, Fe^{3+} on both sites). An extreme arrangement would be a long range order of Ni^{2+} and Fe^{3+} on octahedral sites; the cation nearest neighbor of any Ni^{2+} is then one Ni^{2+} and two Fe^{3+}, and viceversa (see Fig. 2.2). On the other extreme, the "disordered" spinel would be the one with Ni^{2+} and Fe^{3+} randomly distributed on B sites. Obviously, the cation nearest neighbor of a given Ni^{2+} could be, on equal probability another Ni^{2+} or a Fe^{3+}. Internal fields would not be strictly the same for each situation. These two sources of line broadening in FMR in ferrites depending on cation distribution could be written as H_{dist}. To our knowledge, this contribution has not been discussed in literature.

The internal field can therefore be expressed as:

$$H_i = H_{ex} + H_K + H_d + H_p + H_{dis} \tag{4.3}$$

Figure 4.2 shows the behavior of the resonance field, H_{res}, as a function of temperature. H_{res} increases as temperature increases because the internal field decreases until it is overwhelmed by thermal vibrations at T_C. For higher temperatures, the magnetic field needed to satisfy the Larmor relation has to be supplied entirely by the external field.

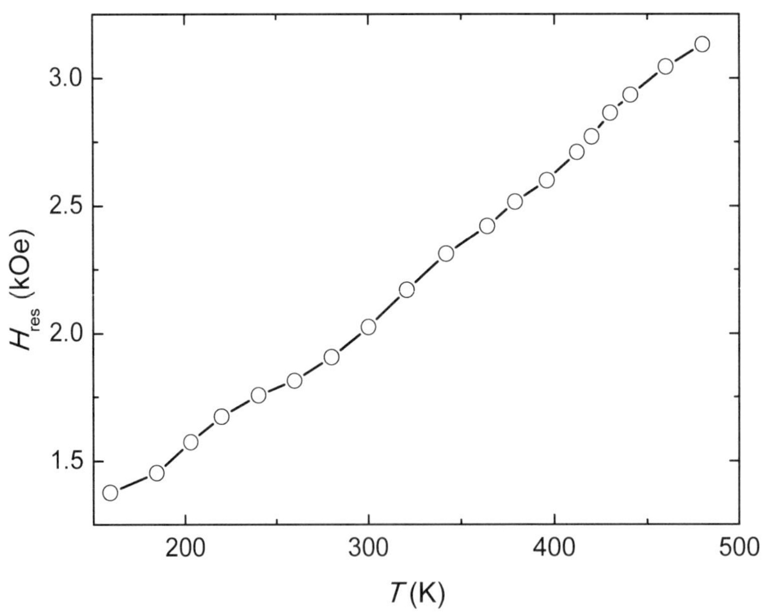

Fig. 4.2. Variation of the resonance field, H_{res}, with temperature for $Ni_{0.35}Zn_{0.65}Fe_2O_4$ ferrites (adapted from Alvarez *et al*, 2010).

The total linewidth, ΔH (taken as the field between the maximum and the minimum in the resonance signal), has also an additive character in polycrystalline materials, and can be written:

$$\Delta H = \Delta H_p + \Delta H_K + \Delta H_{eddy} + \Delta H_d + \Delta H_{dist} \tag{4.4}$$

Where ΔH_p is the linewidth broadening associated with porosity, ΔH_K is due to magnetic anisotropy, ΔH_{eddy} is related with eddy currents, ΔH_d is the linewidth broadening produced by demagnetizing fields, and ΔH_{dist} is the linewidth broadening originated by variations in cation distribution on the A and B sites of ferrite. It appears that anisotropy, and in particular magnetocrystalline anisotropy has a strong contribution to total linewidth. By measuring nickel ferrite with Co^{2+} substitutions, Sirvetz & Saunders (1956) observed a minimum in linewitdth for the composition corresponding to the compensation of anisotropies (x = 0.025 in $Co_xNi_{1-x}Fe_2O_4$), since nickel ferrite has a small negative contribution (single-ion contribution to anisotropy), while cobalt cations provide a strong positive contribution to the total magnetocrystalline anisotropy. More recently, Byun *et al* (2000) showed that in the case of Co-substituted NiZnCu ferrites, ΔH increases for a Co composition higher than the magnetocrystalline anisotropy compensation point. Another source of linewidth broadening is certainly related with the polycrystalline nature of most samples. By modeling one ensemble of single domain nanoparticles, Sukhov *et al* (2008) have shown that the random distribution of anisotropy axis is directly associated with the broadening of the FMR signal.
Figure 4.3 shows the behavior of linewidth with temperature for the same sample than Figs. 4.2 and 4.1. A clear change in slope can be observed at about 430 K, and a smooth variation is also apparent at about 250 K. The former is associated with the Curie transition, which for this Ni/Zn ratio is ~ 430 K (Valenzuela, 2005a), and the latter with a change in magnetic structure which will be discussed later. By comparison with Fig. 4.2 it appears that linewidth, ΔH, is more sensitive to structural changes than the resonance field, H_{res}.

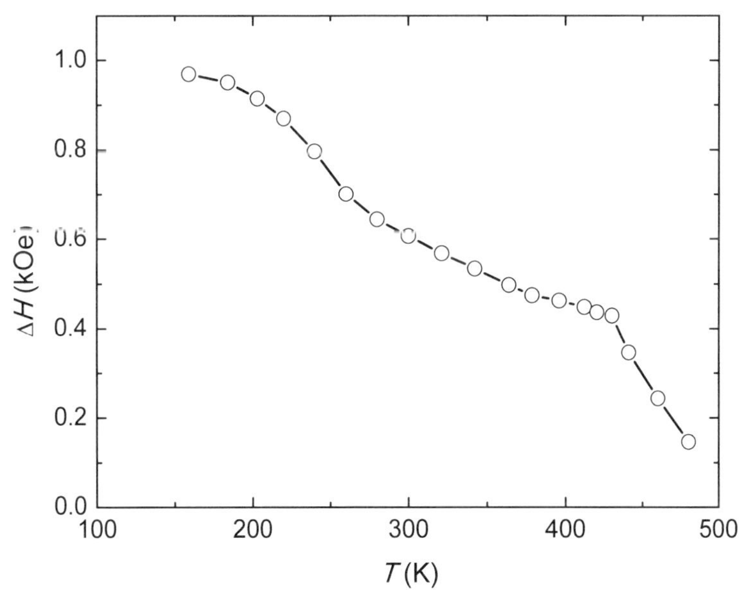

Fig. 4.3. Variations in linewidth with temperature, for $Ni_{0.35}Zn_{0.65}Fe_2O_4$ ferrites (adapted from Alvarez *et al*, 2010).

The increase in resonance field as temperature rises is due to the fact that internal field decreases (exchange interaction, anisotropy field, and the fields associated with magnetization, i.e., demagnetization fields on surfaces including the ones created by porosity). In contrast, linewidth decreases with temperature, essentially because one of the major contributions to ΔH is originated by magnetocrystalline anisotropy, and this contribution is proportional to this parameter (Byun et al 2000). At $T > T_C$, as discussed in Section 4.1, the resonance line becomes narrow and symmetrical, as the spectrum for $T = 460$ K in Fig. 4.1.

4.3 Low temperatures (T << T_C)

Ni-Zn ferrites present other interesting phenomena at T below room temperature. These appear as small "bumps" in the thermal behavior of both the resonance field and the linewidth, at about 240 K. These phenomena are more evident in LFMA (Low-Field Microwave Absorption). In this absorption mode (see Section 3.), the system is far from the resonance conditions as stated in the Larmor relation, and it can be thought as the interaction between the microwave field and the ordered spins in the material as the magnetization state progresses from the demagnetized state toward magnetic saturation. In the simple case, LFMA appears as an antisymmetric signal at both sides of the $H = 0$, Fig. 4.4. LFMA also exhibits hysteresis by cycling the application of the magnetic field.

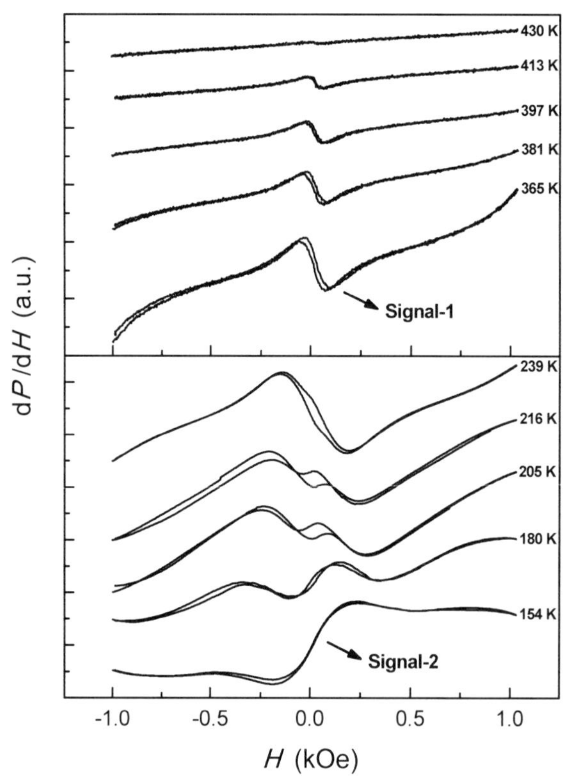

Fig. 4.4. LFMA measurements of $Ni_{0.35}Zn_{0.65}Fe_2O_4$ ferrites, for selected temperatures. The temperature range in the upper graph is 300-430 K, while in the lower one it is 154-239 K (adapted from Alvarez et al 2010).

Several features are significant in these plots. Beginning with the high temperatures (upper part of Fig. 4.4), it is evident that the amplitude of the signal at both sides of $H = 0$ decreases as T increases, leading to a flat response for $T \geq 430$ K, which is the Curie point. It is therefore confirmed that LFMA is associated with magnetization processes in the ordered phase.

Also, it can be observed that the field corresponding to the peak to peak magnetic field values, maxima (for negative fields) and the minima (for positive fields) increases as T decreases. By comparing with a direct calculation of the anisotropy field, H_K, Valenzuela et al (2011) were able to show that the amplitude between maxima-minima in LFMA is directly associated with H_K, upper part of Fig. 4.5.

By a comparison between the two sets of curves separated by $T \sim 250$ K, it appears that there is a continuous evolution of the antisymmetrical signal, from high T to low T, from a signal with the same phase as the FMR signal (Fig. 4.1) for $T > 250$ K, to the opposite, also antisymmetric, but minimum-maximum (Min-Max), or out-of-phase signal, which is clearly reached at $T \leq 150$ K. Both signals are centered on $H = 0$. The presence of such out-of-phase signal has been correlated with the occurrence of a ferromagnetic ordering (Owens 2001, 2005). In fact, an-out-of phase LFMA signal has been observed in many ferromagnetic systems (Montiel et al 2005, 2008, de Cos et al 2008). It can be assumed that a parallel, ferromagnetic arrangement of spins is related with this signal, while an antiparallel, ferrimagnetic structure leads to the opposite result (in phase signal). In the present case, the evolution of the signal when decreasing temperature should be associated with the appearance of a parallel arrangement of spins for $T \leq 150$ K. Due to the Yafet-Kittel triangular structure, Fig. 2.5 (c), there is effectively a ferromagnetic arrangement simply by considering the components of the canted spins of cations on B sites. Figure 4.5 shows the evolution of Curie temperature and Yafet-Kittel transition for NiZn ferrites; the latter was determined by neutron diffraction (Satya Murthy et al, 1969). The results on Fig. 4.4 were obtained for x = 0.65, leading to a Yafet-Kittel transition, T_{YK}, about 250 K, which is in very good agreement with these results.

The transition from the collinear arrangement to the Yafet-Kittel triangular structure can be detected (as temperature decreases) by means of MAMMAS experiments, as shown in Fig.

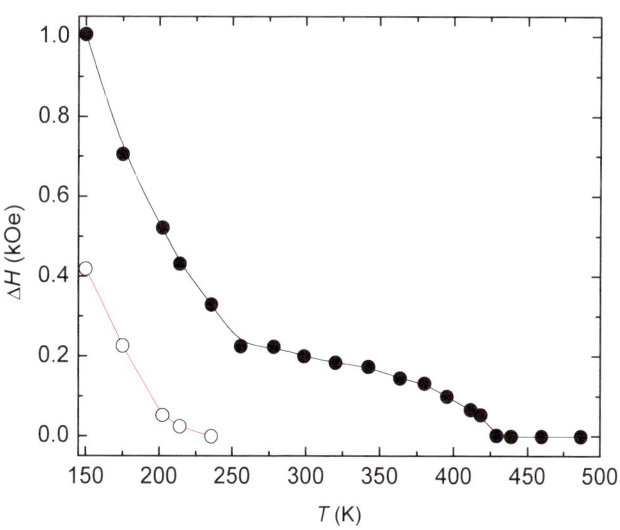

Fig. 4.5. Thermal variations of the peak-to-peak magnetic field of LFMA spectra for both signals.

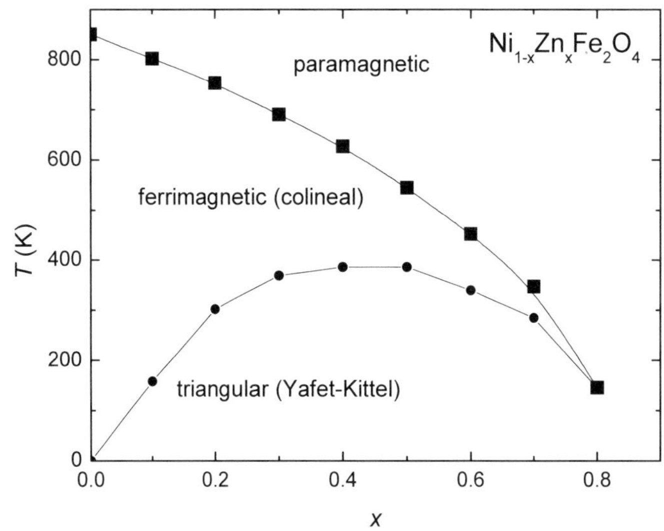

Fig. 4.6. Curie temperature and Yafet-Kittel temperature for NiZn ferrites (Adapted from Valenzuela 2005a and Satya Murthy et al 1969).

4.7. As explained in Section 3, the sample is subjected to a small magnetic field, and its microwave absorption is monitored as temperature is slowly changed. The MAMMAS response exhibits, from room temperature, a continuous decrease to a minimum value at about 240 K. Then, the absorption increases again as the temperature keeps decreasing. These features point to a change in the microwave absorption regime due to a change in the material structure. In this case, all evidence is associated with the transition from the

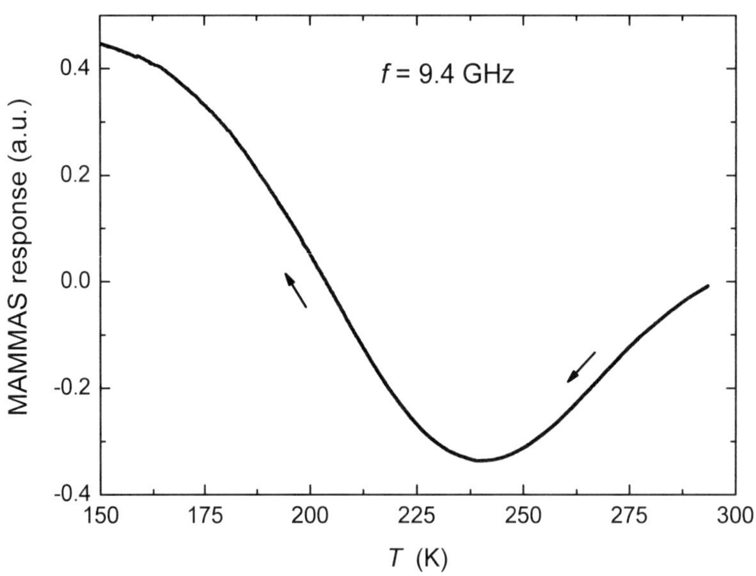

Fig. 4.7. Microwave absorption response of $Ni_{0.35}Zn_{0.65}Fe_2O_4$ ferrites in the MAMMAS experiment (Alvarez et al 2010).

collinear ferromagnetic structure with iron in A sites of the spinel coupled by a superexchange interaction with iron cations (and nickel cations) on B sites, for $T > 240$ K, to the triangular structure, where spins are no more collinear. Due to the weakening of the A-O-B interaction (as A sites become increasingly populated by zinc non magnetic ions), it becomes comparable to the B-O-B interaction which tends to establish an antiparallel geometry on the spins of ions on B sites.

5. Conclusions

As a conclusion, we can state that ferrites are complex materials: they offer a crystal complexity, with complex magnetic structures and complex magnetic properties. However, complexity can always be a rich source of knowledge. In addition to the well known ferromagnetic resonance methods, some significant steps can be done by investigating the microwave response of ferrites, particularly by using two novel research techniques, based on nonresonant absorption: low field and magnetically modulated microwave abosorption, which provide an original insight into these materials.

6. Acknowledgments

This work was partially supported by an ANR (France)-CONACyT (Mexico) grant (#139292).

7. References

Alvarez G. & Zamorano R. (2004). Characteristics of the magnetosensitive non-resonant power absorption of microwave by magnetic materials. *Journal of Alloys and Compounds*, Vol. 369, No. 1-2, (April 2004), pp. 231-234, ISSN: 0925-8388

Alvarez G., Montiel H., de Cos D., Zamorano R., García-Arribas A., Barandiaran J.M. & Valenzuela R. (2007). Experimental and theoretical correlation between low-field power absorption and magnetoimpedance in amorphous materials. *Journal of Non Crystalline Solids*, Vol. 353, No. 8-10, (April 2007), pp. 902-904, ISSN: 0022-3093

Alvarez G., Montiel H., Barrón J.F., Gutierrez M.P., Zamorano R. (2010). Yafet-Kittel-type magnetic ordering in $Ni_{0.35}Zn_{0.65}Fe_2O_4$ ferrite detected by magnetosensitive microwave absorption measurements. *Journal of Magnetism and Magnetic Materials* Vol. 322 No. 3 (September 2009), pp. 348-352, ISSN: 0304-8853

Anderson P.W. (1959). New approach to the theory of superechange interactions. *Physical Review*, Vol. 115 No. 1 (February 1959) pp. 2-13, ISSN: 0031-899X

Byun T.Y., Byeon S.C., Hong K.S. and Kim C.K. Origin of line broadening in Co-substituted NiZnCu ferrites. (2000). *Journal of Applied Physics* Vol. 87 No. 9, (May 2000), pp. 6220-6222, ISSN: 0021-8979

Chiriac H., Colesniuc C.N. & Ovari T.-A. (2000). FMR Investigation of the nanocrystalline FeCuNbSiB glass-covered wires. *Journal of Magnetism and Magnetic Materials*, Vol. 215-216, No. 1 (June 2000), pp. 407-409, ISSN: 0304-8853

De Cos D., García-Arribas A., Alvarez G., Montiel H., Zamorano R., Barandiaran J.M. & Valenzuela R. (2007). Low field sensitivity for gigahertz magneto-impedance sensors. *Sensor Letters* Vol. 5, No. 1 (September 2007), pp. 73-76, ISSN: 1546-198X

Globus A., Pascard H., and Cagan V. (1977). Distance between ions and fundamental properties in ferrites. *Journal de Physique*, Vol. C1-38, (month 1977), pp. C1-163-C1-168, ISSN: 0449-1947

Hill R.J., Craig J.R. and Gibbs G.V. (1979). Systematics of the spinel structure type. *Physics and Chemistry of Minerals*, Vol. 4 (April 1979) pp. 317-339, ISSN: 0342-1791

Kittel C. (2005). *Introduction to Solid State Physics*, 8th Edition, John Wiley and Sons, N.Y. ISBN: 0-471-41526-X

Medina A.N., Knobel M., Salem-Sugui S., & Gandra F.G. (1999). Resonant microwave cavity response of amorphous ribbons. *Journal of Applied Physics*, Vol. 79, No. 8 (September 1999), pp. 5462-5464, ISSN: 1054-1500

Montiel H., Alvarez, G., Gutiérrez M.P., Zamorano, R., and Valenzuela R. (2004). Microwave absorption in Ni-Zn ferrites through the Curie Transition. *Journal of Alloys and Compounds*, Vol. 369, No. 1-2, (April 2004), pp. 141-143, ISSN: 0925-8388

Montiel H., Alvarez, G., Betancourt I., Zamorano, R., and Valenzuela R. (2005). Correlation between low-field microwave absorption and magnetoimpedance in Co-based amorphous ribbons. *Applied Physics Letters*, Vol. 86 No 7 (February 2005) paper 072503, pp. 1-3, ISSN: 0003-6951

Montiel H., Alvarez G., Gutiérrez M.P., Zamorano R. & Valenzuela R. (2006). The effect of metal-to-glass ratio on the low-field microwave absorption at 9.4 GHz of glass-coated CoFeBSi microwires. *IEEE Transactions on Magnetics* Vol. 42, No. 10, (September 2006), pp. 3380-3382 ISSN: 0018-9464

Owens F.J. (2001). Resonant and nonresonant microwave absorption study of ferromagnetic transition in $RuSr_2Gd_{0.5}Eu_{0.5}Cu_2O_8$ superconductor. *Physica C* vol. 353 No. 3, 4 (May 2001), pp. 265-269, ISSN: 0921-4534

Owens F.J. (2005). Ferromagnetism above room temperature in bulk sintered gallium phosphide doped with manganese. *Journal of Physics and Chemistry of Solids* Vol. 66 No. 5 (May 2005), pp. 793-790, ISSN: 0022-3697

Pilbrow J.R. (1990). *Transition ion electron paramagnetic resonance*, Clarendon Press, Oxford. ISBN: 0-19-855214-9

Ravindranathan P. & Patil K.C. (1987). *Journal of Materials Science* Vol. 22, No. 9, (September 1987), pp. 3261- 3264, ISSN: 0022-2461

Rivoire M. & Suran G. (1995). Magnetization of thin films with in-plane uniaxial anisotropy studied by microwave absorption. *Journal of Applied Physics*, Vol. 78, No. 3 (***1995), pp. 1899-1905 , ISSN: 0021-8979

Satya Murthy N.S., Natera M.G., Youssef S.J., Begum R.J., Srivastava C.M. (1969). Yafet-Kittel Angles in Zinc-Nickel Ferrites. *Physical Review* Vol. 181 No. 2 (May 1969), pp. 969-977, ISSN: 1098-0121

Sirvetz M.H. and Saunders J.H. (1956). Resonance widths in polycrystalline nickel-cobalt ferrites. *Physical Review* Vol. 102 No. 2 (April 1956), pp. 366-367, ISSN: 1098-0121

Valenzuela R. (2005a). *Magnetic Ceramics,* Cambridge University Press, Cambridge, UK (September 2005) ISBN: 0-521-01843-9

Valenzuela R., Montiel H., Gutierrez M.P., Betancourt I. (2005b) Characterization of soft ferromagnetic materials by inductance spectroscopy and magnetoimpedance, *Journal of Magnetism and Magnetic Materials*, Vol. 294 No. 2 (July 2005), pp. 239-244, ISSN: 0304-8853

Valenzuela R., Gutiérrez M.P., Vázquez G. & Acevedo U. (2011). (To be published).

Wu K.H., Shin Y.M., Yang C.C., Wang G.P., & Horng D.N. (2006). *Materials Letters* Vol. 60, No. (** 2006) pp. 2707- , ISSN:

Yafet Y. & Kittel C. (1952) Antiferromagnetic Arrangements in Ferrites. *Physical Review*, Vol. 87, No. 2, (March 1952), pp. 290-294, ISSN: 1098-0121

19

Complex Permittivity Measurement of High Loss Liquids and its Application to Wine Analysis

Z.E. Eremenko[1], V.N. Skresanov[1], A.I. Shubnyi[1],
N.S. Anikina[2], V.G. Gerzhikova[2] and T.A. Zhilyakova[2]
[1]*Usikov Institute of Radiophysics and Electronics of National Academy of Science of Ukraine,*
[2]*National Institute of Vine and Wine "Magarach" of Ukrainian Academy of Agricultural Sciences, Ukraine*

1. Introduction

The microwave dielectrometry of high loss liquids is discussed in the present work. For such liquids the ratio of the imaginary to the real complex permittivity parts is of the order of one or more. High loss liquids are water and water solutions of organic and non-organic substances. Water is the most investigated liquid during long time. The reason for the impressive interest to water is due to its ubiquity in our environment. Usually water is used as the reference liquid for differential measurement of water-containing liquids under test, for instance, grape wines and musts. The present paper is directed to the control of naturalness (authenticity) of grape wines and musts by their complex permittivity determination. This complex permittivity determination will be performed in the millimeter wave band where the frequency dispersion of complex permittivity of water has a maximum.

It is necessary to underline that the wine trade is a multi-million-income industry in which frauds can have enormous financial repercussions and the quality of products directly impacts on human health. A universal wine and must authenticity identification method does not exist. Each of known methods (the methods of analytical chemistry, wine sampling, physical-chemical methods) have own advantages and disadvantages and assume a definite area of usage [Organisation internationale de la Vigne et du Vin, 2009]. Thus, the development of alternative methods of wine product quality determination remains actual.

One feature of the wine trade in Ukraine and other former Soviet Union countries is the increase of fraud. The main types of fraud are unregulated sugar use, substitution of the sort of vine, not keeping the ripening terms for the wines, water addition, blending of cheap wines instead the best wines, addition of artificial sweeteners, dyes, aromatizers, and fake wine production without vine processing. In Europe and USA there is fraud growth by bottling inferior wine under a quality label - with the wine originating from a different region or even country than that it is stated.

Despite the rigor of wine quality demands fake wine production did not decrease due to lack of reliable methods for identification and suitable reference data [Ezhov et al., 1999].

The main part of fraud (up to 90-95%) is associated with wines that result of the fermentation of sugar-water or diluted grape must solution. Organoleptic determination of such wines is possible but is not reliable. For reliable examination there are both classical analytical methods (the determination of the density, extract, mineral substance and so on) and methods based on isotope patterns and nuclear magnetic resonance. The major inconvenience of these techniques is that they are generally time-consuming and/or expensive and not suitable for continuous monitoring. Therefore, the development of the reliable methods of wines and fraud identification is still actual.
The authenticity control of wine and fruit musts is a quite complicate problem. At present the great number of parameters is used for that purpose, such as the alcohol content, sugar, acidity, and extractable substances with and without sugar etc. Additional parameters used are the number of anions (Cl^-, SO_4^{2-}) and cations (K^+, Na^+, Ca^{++}, Mg^{++}), organic acids (tartaric, malic, lactic, citric), pH etc. In grape, fruit must, and wine production the measurement of density, pH, buffer capacity, conductivity, viscosity, and optical density are used to manage the quality as well.
The great number of parameters required determining the authenticity and the time for their determination make the testing of wine production during the technological process is really complicated. European scientists have known for almost a decade that analysis of various chemical elements in wine can help to identify roughly the geographic origin of wines, but the tests have been too complex and not conclusive enough to be used on a large scale.
The necessity to measure small changes of complex permittivity of high loss liquids at millimeter wave band arises at biophysical research of native albuminous molecule conformation. At the preparation of calibration data during distance sounding of water surfaces and interpretation of sounding results, at physical-chemical study of properties of water solutions and so on. One of the important practical usages of dielectrometry of high loss liquids is the identification of wines and musts and their authenticity. Differential complex permittivity measurement has an advantage towards to absolute complex permittivity measurement when it is necessary to measure small complex permittivity changes in liquids under study.

2. Techniques for high loss liquid complex permittivity measurement

Numerous techniques for complex permittivity determination of high loss liquids are applied at microwave frequencies [Sato & Buchner, 2004; Masaki at al., 2007; Agilent 85070E, 2008; Buckmaster et al., 1985; Hu, et al., 1994; Cherpak et al., 2004]. Among of them it is well known that both resonator [Afsar & Ding, 2001] and waveguide [Afsar & Suwanvisan, 2005] methods are used for the complex permittivity measurement of high loss liquids.
The resonator methods use the complex resonant frequency of a resonator to obtain complex permittivity of liquid. And the waveguide methods use the complex wave propagation factor in a waveguide segment under study. The significant limitation of the above-mentioned methods is a high attenuation observed in the high loss liquid measurements. Moreover, the high loss liquid has large real and imaginary complex permittivity parts (much greater than 1, specifically, for water) and these values are of the same order. Therefore, measured parameters (the resonant frequency and Q-factor in a resonator or the wave phase and wave attenuation in a waveguide) depend simultaneously both on real and imaginary complex permittivity parts of high loss liquids. Owing to this fact the complex permittivity determination becomes considerably complicated while using measured

parameters. For the resonator method, the Q-factor is usually rather small (it is approximately 10-100 for enough complex permittivity measurement sensitivity) and as a result it is difficult to determine the resonant frequency with needed accuracy. Concerning to the waveguide method, the wave attenuation is high at the propagation of the wave even via a small length (compare to skin layer in liquid) waveguide section at presence of high loss liquid like water or water solutions.

There are a number of papers where the complex permittivity determination of high loss liquids was studied using different resonator methods. For example, in Cherpak's papers whispering-gallery modes in cylindrically shaped dielectric resonators were studied to obtain complex permittivity of high loss liquids. A resonator as a radially two-layered dielectric disc placed between conducting endplates with an internal layer filled with air or loss liquid such as water, ethyl alcohol, benzene and aqueous solutions of ethyl alcohol was studied [Barannik et al., 2007]. In the paper [Shaforost, 2009] it was described a novel approach of high sensitivity liquid analysis for volumes in the nanolitre range with challenging perspectives for practical sensor applications in chemistry, biology and medicine. Whispering-gallery modes in cylindrically shaped dielectric disks machined from low-loss single crystalline materials such as sapphire or quartz allow having very high quality factors. The interaction of extremely small volumes of the liquid under test with the evanescent field located in the vicinity of the dielectric disk surface at micro-to-millimeter wave frequencies was employed for the investigation of aqueous solutions with relevance to biological applications. Based on this resonator type, three different liquid sensing approaches were developed and analyzed at 10, 35 and 170 GHz with emphasis on the determination of the complex permittivity of liquids of nanolitre volumes. In [Cherpak, 2006] electrodynamics' properties of quasi-optical dielectric resonators of two types with liquid-filled small cavities have been studied. One of the quasi-optical dielectric resonators types is a two semi-disc resonator with a diametrical slot in which a cavity with a thin ($0.01 \div 0.1$ mm) flat liquid layer is placed and the other one is a disc resonator with a small diameter ($0.3 \div 2$ mm) cylindrical capillary. Measurements have been carried out at room temperature in K_a-waveband by using the resonators made of Teflon material. The obtained results allow to conclude that the given approaches to the development of measuring technique for characterization of liquids (for example, water and aqueous solutions) are quite perspective. In [Shaforost, 2007] they showed that whispering-gallery mode dielectric resonators from 10 GHz to 3 THz are attractive for highly sensitive liquid detection and identification of small droplets down to volumes of picolitres. Since droplets are usually generated by computer controlled microinjection pipettes being moved on a 2D scanning table, free access to the sensitive resonator surface from above is essential.

To obtain small complex permittivity differences of various substances, in particular, high loss liquids differential measurement methods are widely used. The necessity of small complex permittivity difference registration for high loss liquids appears, for instance, during biological study of conformation changes of native protein molecules [Pethig, 1992] or reference data preparation for the device calibration of water surface remotely sensing [Ellison, 2007]. The dielectric properties of free water are well described by the Debye formula of the second order. Dissolved molecules in water can change the relaxation time both due to coupling of part of free water molecules (hydration) and due to the rebuilding of hydrogen bonds in water clusters. The latter mechanism is common for water - ethanol mixtures [Sato & Buchner, 2004]. Owing to mentioned reasons of complex permittivity measurement of water solutions is sensitive to the variation of their chemical composition at Ka band.

Known waveguide-differential complex permittivity measurement methods distinguished by type of the measurement cell usage due to dielectric properties of liquids and measurement conditions. So, the cells based on waveguides completely filled with high loss liquid [Buckmaster et al., 1985], and also cells with waveguides with capillary with liquid put inside of the waveguide [Masaki et al., 2007] are used. It is known the usage of dielectric waveguides for complex permittivity measurement of high loss liquid [Meriakri & Parkhomenko, 2000]. In a set of Buckmaster's papers for the development and improvement of complex permittivity measurement technique for light and heavy water at 9.335 GHz it is shown the possibility to reach high accuracy of absolute measurements [Buckmaster et al.,1985; Hu et al., 1994]. The measurement setup in [Hu et al., 1994] was a microwave bridge contained a waveguide cavity with a precise length change of a part of rectangular waveguide filled with water. They used thermo stabilization for the cavity with the error of the order of 0.005°C and a high sensitivity superheterodyne receiver with double frequency transformation. After this transformation the precise measurements carried out at frequency 1 kHz. As a result they reached the relative measurement error for the real and imaginary complex permittivity parts the order of 0.1% and 0.2%, respectively. The 1σ standard deviations both for real and for imaginary complex permittivity parts do not exceed 0.02%. According to own metrological characteristics the setup in [Hu et al., 1994] can be used for study the dielectric properties of wines and musts with needed accuracy. It is necessary to have laboratory conditions and the cavity of this device does not correspond to carry out mass measurements.

The upper limit frequency for waveguide methods usage with reasonable errors is 140 GHz [Jain & Voss, 1994]. The majority of such methods are based on the impedance measurement of waveguide system where a sample under study is placed. Possible study at wide frequency band both solid and liquid matter. The most suitable objects are media with average dielectric losses ($0.1<\text{tg}\delta<1$, where δ is angle of dielectric losses). The measurement error real complex permittivity part ε' is often 0.5%, and $\text{tg}\delta$ - 3-5%. It is known several types of waveguide methods. So, completely filling with liquid here the waves propagation is studied in waveguide [Brandt, 1963]. If a sample is absent in the line there is pure standing wave. In [Van Loon & Finsy, 1974] they used the similar method to measure liquid samples at the frequency range from 5 to 40 GHz and from 60 to 150 GHz. At that they underlined the practical difficulties at waveguide usage at the frequencies up 40 GHz due to not enough mechanical power and attenuation increase in waveguide. The authors saw two main obstacles to reach high accuracy: there is undesirable reflection and attenuation in waveguide walls and there are inhomogeneities where higher modes are excited. In the papers [Afsar et al., 2005] and [Fuchs & Kaatze, 2002] the authors are also paid great attention on the analysis of possible errors and recommendations to increase the measurement accuracy. For long wave part of millimeter wave band they decreased the error of $\Delta\varepsilon' / \varepsilon'$ and $\Delta\varepsilon'' / \varepsilon''$ up to 0.02. In the paper [Alekseev & Ziskin, 2001] measured the reflection, propagation and attenuation of electromagnetic waves at the interaction with film samples (water and water-ethanol solutions). At 42.25 GHz and 53.35 GHz it is studied the influence of film thickness and the character of attenuation in the cell depending on the thickness and design of isolated interlayer as well. The authors found out that optimal thickness of a sample when the straight-line attenuation of electromagnetic waves takes place is interval of 0.28 - 0.33 mm.

Alison and Sheppard [Alison & Sheppard, 2001] designed a dielectrometer where the liquid layer was changed using short-cut plunger that moved into the liquid. The layer thickness was measured with the step of 0.06 mm. Complex permittivity was obtained at 29 - 44 GHz using the measurement the reflection coefficient for two different thicknesses. The liquid separated from the empty part of a waveguide by means of windows made of organic glass. The appearance of higher modes in transmission section the authors did not observed. With the same technique the authors measured complex permittivity of human blood at 29 - 90 GHz [Alison & Sheppard, 1993]. They found out the existence of the field of additional (except Debay's ones) high frequency dispersion.
In the paper [Zanfolin, 1983] they described the measurement device that allows to obtain complex permittivity of high loss liquid at millimeter wave band. The basis of a dielectrometer is waveguide interferometer (bridge) where in measurement shoulder there is a cell with liquid that is irradiated from open waveguide end. The thickness of a sample is changed using a plunger moving by micrometer screw. As a result, an input signal in the bridge is a function of the thickness of liquid. The authors give the results obtained for ethanol, methanol and pure water at 20°C and 70 GHz. Later the dielectrometer was improved. New variant [Buckmaster et al, 1985] contained two channel superheterodyne system with twice frequency transformation. The authors analyzed the origin of errors and elaborated the ways of their decrease. For that purpose they carried out maximum exact measurement of length of a sample, there was increased the frequency and stability of the signal.
In spite of the various considered methods of high loss liquids complex permittivity measurements, the problem now still actual due to the difficulties of measurement accuracy and it must be at least the order of 0.01% as our measurement showed to identification of water solutions of chemical and biological media and their differences between each others.
One of the possible practical usages of the dielectrometry of high loss liquids is wine and must identification in order to determine their authenticity [Watanabe, 2009]. As it was mentioned above, wines and musts are high loss liquids, because of mass fraction of water in their content is the order of 75-90%. It was proposed to use complex permittivity measurement data to control the stages of sake fermentation - the main biochemical process of wine-alcohol industry [Masaki et al., 2007]. As known the change of quantitative chemical wine composition, for example, for dry (table) natural wines in specification standard product limit influences on complex permittivity values within the scope the order of some percents. Therefore, for the solving of wine identification problem it is necessary to have a dielectrometer that has much higher sensitivity in comparison with complex permittivity deviation for high loss liquids. It is desirable to have a device with differential sensitivity at least the order of 0.1% or higher. Authors in [Watanabe, 2009] could not identify the change of different samples of dry wines components except ethanol at Ka waveband just because of small differential sensitivity. The experimental tool in [Watanabe, 2009] was standard probes for the measurement of complex reflection coefficient in liquids by means of vector analyzer [Agilent 85070E, 2008].
Our goal is to develop a dielectrometer for carrying out the express wine product analysis at industrial chemical laboratories by staff with average technical secondary education level. The device must possess enough high differential sensibility (at least the order of 0.1%). High accuracy of absolute wines and musts complex permittivity measurement does not obligate, because the main requirement is the measurement of wave propagation coefficients difference in two measurement liquid-filled cells with similar dielectric properties. We

reached the assigned task by the use of a measurement cavity [Ganapolskii et al., 2009] and made complete automated measurements.

There are no restrictions for liquid under test for a volume size during wine and must complex permittivity measurement i.e., the cavities of rather big volume can be used and the cavity clearing can be made by means of the liquid under test as well. As a result, measurement errors caused by a cavity with small size or inhomogeneity of cavity filling are minimized. The decrease of accuracy at the measurement in small volume of the liquid is common for a set of known cavities, for instance, a capillary with a liquid under test in a rectangular waveguide [Masaki et al., 2007] or a thin ring belt cavity with a liquid under test in a shielded cylinder dielectric resonator [Eremenko et al., 2009].

A measurement waveguide type cavity proposed in [Ganapolskii et al., 2009] consists of two open measurement cells with a dielectric (quartz) cylinder rod inside of them. These two cells are filled with a reference liquid and a liquid under test, respectively. The operating principle of these cells is based on the dependence of electromagnetic wave propagation coefficients on dielectric properties of a surrounded dielectric rod liquid. The difference between the attenuation and phase coefficients of the wave propagation in two measurement cells is the measured data. The attenuation coefficient of the wave propagated along the quartz rod immersed into water is rather small (the order of 10 dB/cm). The influence of the reflection off a cell body can be neglected during the wines and musts complex permittivity measurements, if the diameter of the cell is two times more than a rod diameter [Ganapolskii et al., 2009]. These reasons create comfortable conditions for the measurement open cavity design to carry out precise express complex permittivity measurements of high loss liquids.

The main goal of our measurements is to detect small complex permittivity differences of two liquids with similar dielectric properties. For such a goal the knowledge of absolute complex permittivity liquid values has reference character only and we can limit ourselves by small difference of two wave propagation coefficients values for two cells. It is preferentially to minimize measurement errors. The errors of indirect complex permittivity measurements can be calculated if we know the errors of direct measurements of wave attenuation and phase coefficients in the measurement cell. For example, relative errors of the real and imaginary complex permittivity parts are associated with absolute standard deviations of independent measurements of phase and attenuation coefficients $\delta h'$ and, $\delta h''$ respectively. So, we obtain for 10% ethanol in water: $\delta\varepsilon' / \varepsilon' = 0.31\%$ and $\delta\varepsilon'' / \varepsilon'' = 0.046\%$; for 30% ethanol solution - $\delta\varepsilon' / \varepsilon' = 0.27\%$ and $\delta\varepsilon'' / \varepsilon'' = 0.097\%$. Here the indirect complex permittivity measurement errors are in several times higher than direct measurement errors of wave propagation coefficients. The errors of indirect complex permittivity measurement in [Hu et al., 1994] are also higher than the errors of wave propagation during the direct measurement.

We proposed an original measurement cell with an innovation part as a dielectric rod immersed into high loss liquid [Ganapolskii et al., 2009]. The dielectric rod is an open waveguide system where the complex electromagnetic wave propagation constant depends on complex permittivity of liquid. The electrodynamic waveguide structure similar to the proposed one was studied during the electromagnetic wave propagation in tunnels [Holloway et al., 2000]. We elaborated and tested the laboratory sample of a waveguide differential dielectrometer where the designed cell was used. The device, analysis of its performance, and measurement results are present in this report. We analyzed the influence of different reasons that impact on the measurement errors and we elaborated the conditions to minimize these errors. As a result the error was decreased in ten times in comparison with the data presented in [Jain & Voss, 1994].

3. High precision computer-aided dielectrometer for high loss liquid

Fig.1 presents the appearance of the computer-aided dielectrometer. An operator only fills in and out liquids in the cavity cells. We designed this instrument in order to make up routine high precision measurement for qualitative control of wines and must in biochemical laboratories [Eremenko & Skresanov, 2010]. The determination of high loss liquid complex permittivity is done by means of a computer program. An initial data for computation are the measurement difference of wave attenuation and phase coefficients in two dielectrometer cavity cells. These cells filled with a liquid under test and a reference liquid, e.g. the distilled water.

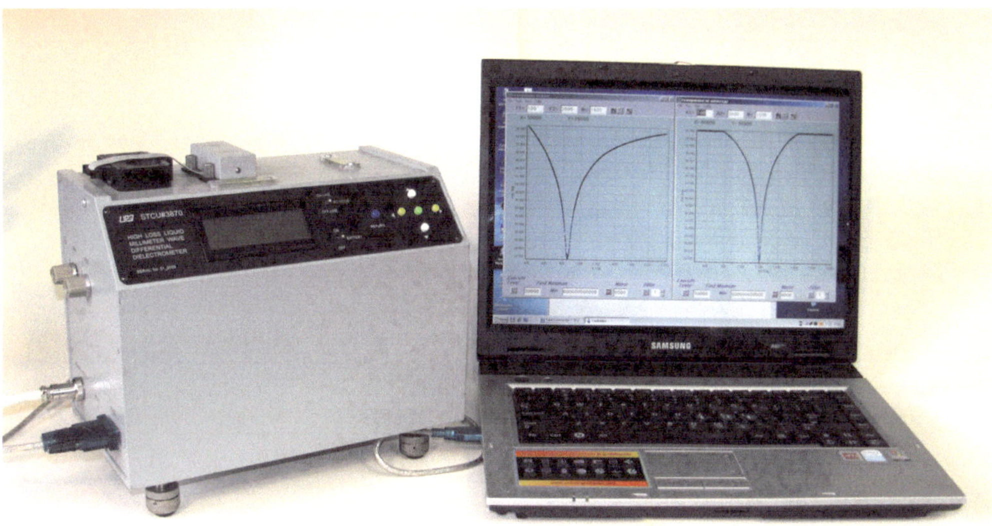

Fig. 1. The appearance of the differential dielectrometer connected with a notebook.

For such measurements we proposed the novel differential cavity [Ganapolskii et al., 2009]. This cavity consists of two identical cells that are cylinder glasses of diameter $D = 20$ mm that are made of a common copper body 1 (Fig.2). Quartz rods 2 were inserted perpendicularly to the side walls. The glasses were filled with liquids 3 (the water solution under test and its solvent - distilled water) at open thermo isolated covers 4 up to the level of overflow holes 5. After measurements liquids were poured out using draining holes 6. The temperature sensors 7 placed into covers 4 are for the temperature control in the cells. Microwave power comes to measurement cells via standard millimeter wave rectangular waveguide 8. The cells are matched with rectangular waveguides by round waveguides sections 9 filled with Teflon. The standing-wave ratio does not exceed 1.05 for the cells input and output at operating frequency 31.82 GHz. The cavity principle of operation is based on the dependence of the wave propagation coefficients in quartz rods on dielectric properties of an outer medium [Ganapolski et al., 2009]. The attenuation coefficient of the wave propagated along the quartz rods inserted into high loss liquid is rather small (the order of

10 dB/cm). The influence of the glass side faces can be neglected at the distance at least of the order of wavelength from the rod surface [Eremenko & Skresanov, 2010].

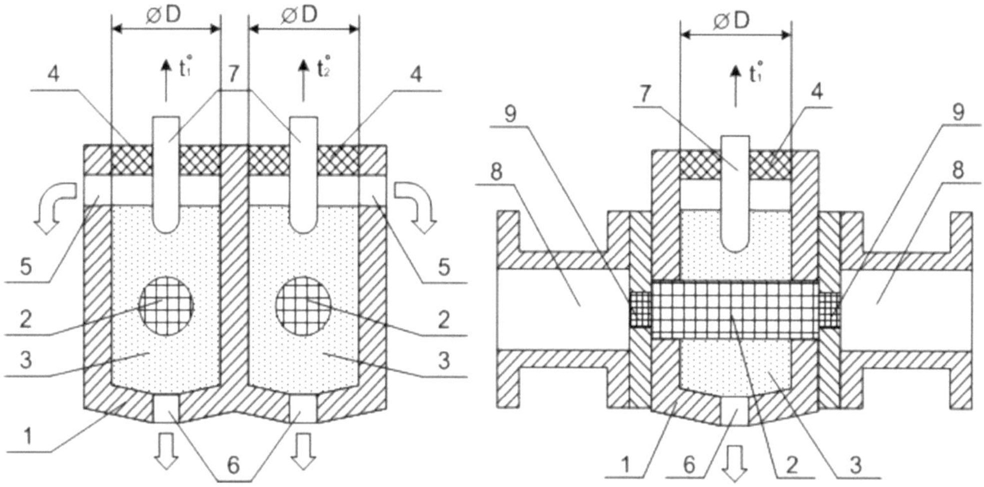

Fig. 2. The schematic picture of dielectrometer differential measurement cavity: the cross section of the cavity (left) and longitudinal section of one of the cavity cells (right). 1- the differential cavity body; 2 - the quartz cylinders; 3 - the liquids; 4 - the covers; 5 - the overflow holes; 6 - the drain holes; 7- the temperature sensors; 8 - the rectangular waveguide sections; 9 - the round waveguides section filled with Teflon.

We can obtain the complex permittivity values from the characteristic equation for the infinite rod in high loss medium [Ganapolskii et al., 2009]

$$\left(\frac{J'_m(g_1a)}{g_1aJ_m(g_1a)}-\frac{H'_m(g_2a)}{g_2aH_m(g_2a)}\right)\left(\frac{\varepsilon_1J'_m(g_1a)}{g_1aJ_m(g_1a)}-\frac{\varepsilon_2H'_m(g_2a)}{g_2aH_m(g_2a)}\right)=\frac{m^2h^2(g_2^2-g_1^2)^2}{k^2g_1^4g_2^4a^4}, \quad (1)$$

where $J_m(x), H_m(x)$ - Bessel and Hankel of the order of m and of the first kind; $g_1=\sqrt{k^2\varepsilon_1\mu_1-h^2}$, $g_2=\sqrt{k^2\varepsilon_2\mu_2-h^2}$; k - the wave number in vacuum; h - the longitudinal wave number; $\varepsilon_1=\varepsilon'_1+j\varepsilon''_1$, $\varepsilon_2=\varepsilon'_2+j\varepsilon''_2$ - the complex permittivity of the rod and liquid, respectively; a - the radius of the rod; $\mu_1,\mu_2=1$ - the permeability of the rod and liquid. The typical relations between real and imaginary complex permittivity parts at our measurements are the following

$$\varepsilon''_1<<\varepsilon'_1, \quad \varepsilon'_1<\varepsilon'_2, \quad \varepsilon''_1<<\varepsilon''_2, \quad \varepsilon''_2/\varepsilon'_2\geq1. \quad (2)$$

The ratio between the impedance of dielectric material of the rod and the liquid (water solutions) $\zeta=\sqrt{\varepsilon'_1}/\sqrt{\varepsilon'_2}$ is approximately equal to 0.5. We can obtain a set of complex roots h_{mn} at definite operating frequency, complex permittivity of liquid, radius of the rod, and its complex permittivity. The azimuth index m is equal to the number of half-wavelengths placed along azimuth coordinate φ from 0 to 2π; the radial index n is equal to the number of half-wavelengths placed along radial coordinate inside the rod from 0 to $r=a$. The analysis of electromagnetic field distribution for the corresponding wave number h_{mn} shows that a set of four wave types can be excited in the dielectric rod immersed into high loss liquid. Two of them are as follows. The transverse-electric waves (TE_{on}) and transverse-magnetic waves (TM_{on}) have no z- field component of electric or magnetic field, respectively. Two other types have non-zero z- field components and they are quasi-TE_{mn} or quasi-TM_{mn}. In general, any of the mentioned above waves can be used for our measurement. We used quasi-TE_{11} wave type, because it can be easily excited in the rod by a rectangular or a round waveguide with the basic wave types H_{10} or H_{11}.

The technique of complex permittivity determination is as follows. The wave attenuation $h_r''[\text{dB/cm}]$ and phase $h_r'[\text{rad/cm}]$ coefficients are calculated from the characteristic equation (1) using known complex permittivity of solvent for the cell with solvent (the reference liquid). We measure the difference of attenuation coefficients $\Delta h''[\text{dB/cm}]$ and phase coefficients $\Delta h'[\text{rad/cm}]$ for the cells with solvent and the liquid under test. The attenuation coefficient $h_t''[\text{dB/cm}]$ and the phase coefficient $h_t'[\text{rad/cm}]$ of the wave in the cell with solvent are calculated using formulas: $h_t'' = h_r'' + \Delta h''$ and $h_t' = h_r' + \Delta h'$. And, finally, using equation (1) the complex permittivity of the liquid under test ($\varepsilon_t = \varepsilon_t' + i\varepsilon_t''$) is calculated with the help of obtained h_t'' and h_t'.

It is suitable to use the distilled water as the reference liquid at the complex permittivity determination of water solutions of wines and wine model liquids. In [Ellison, 2007] there is a formula to calculate complex permittivity of water at 0-25 THz and at the temperature band 0-100°C. We use this formula for the complex permittivity calculation of the distilled water at known room temperature as a liquid in the cavity.

We use the principle of differential measurement of the difference in wave attenuation coefficient $\Delta L = \Delta h'' l_{eff}^{(1)}$ and phase shift coefficient $\Delta\varphi = \Delta h' l_{eff}^{(2)}$ in the cavity cells as it was done in [Ganapolskii et al, 2009]. So-called "cell effective lengths" $l_{eff}^{(1)}$ and $l_{eff}^{(2)}$ approximately equal to glass diameter D of the cells. The measurement scheme (Fig. 3.) is a microwave bridge. The signal splitting between bridge arms is done at an oscillator output by means of E-joint, and the signal summation is done at a detector input by means of H-joint. The local oscillator at 31.82 GHz is a phase-locked loop transistor VCO at the frequency 7955 MHz with a reference quartz frequency standard and further multiplication by four. A power amplifier on the basis of the chip CHA3093c was implemented. To increase a signal to noise ratio, the amplitude modulation of a microwave carrier with frequency 100 kHz and a synchronous demodulation were used.

The amplitude of the signal at one of the bridge arms is controlled by the measurement P-I-N attenuator. The high precision short-circuiting plunger was designed for the phase shifter. This plunger is controlled by a step motor. The discrete step of the plunger motion is 2.5μm that corresponds to phase change 0.144°. The tuning of the attenuator and the phase shifter in the bridge arms is done in accordance with a microcontroller program. The microcontroller block was worked out on the basis of AT90USB1287 chip. Its main function is amplitude and phase level control in the microwave bridge arms. Besides, we measure the signal level at the receiver output, the temperature of the liquids in the cells and the temperature of the P-I-N attenuator body. We also control the level of output oscillator signal by the controller. The microcontroller block provides a user interface in manual mode and the data exchange with PC.

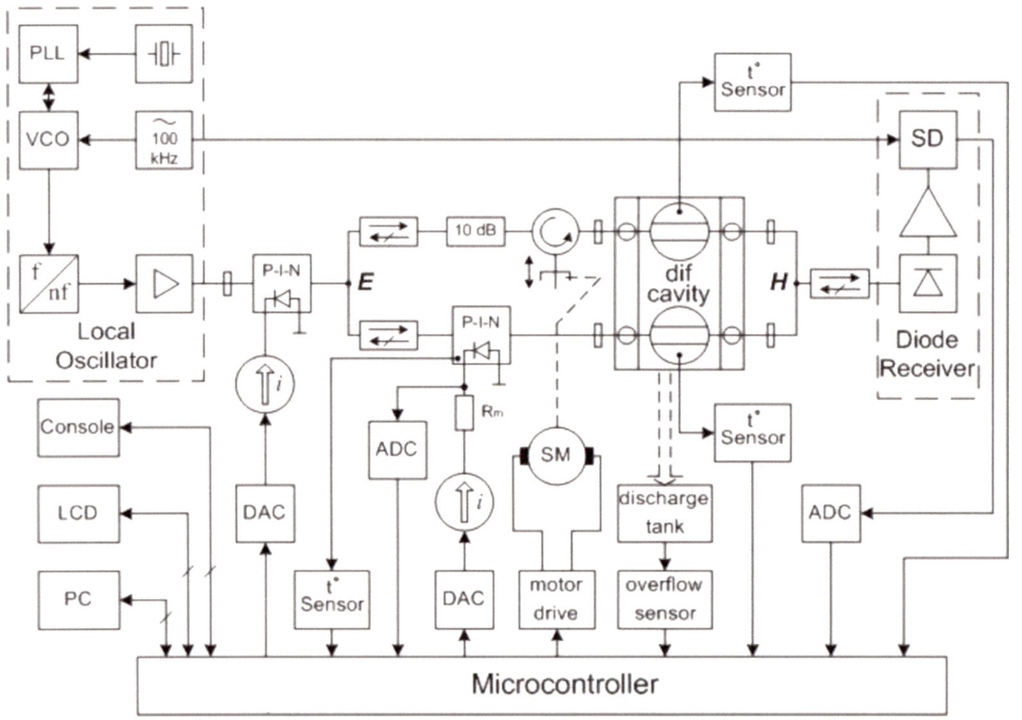

Fig. 3. The structural scheme of the differential dielectrometer.

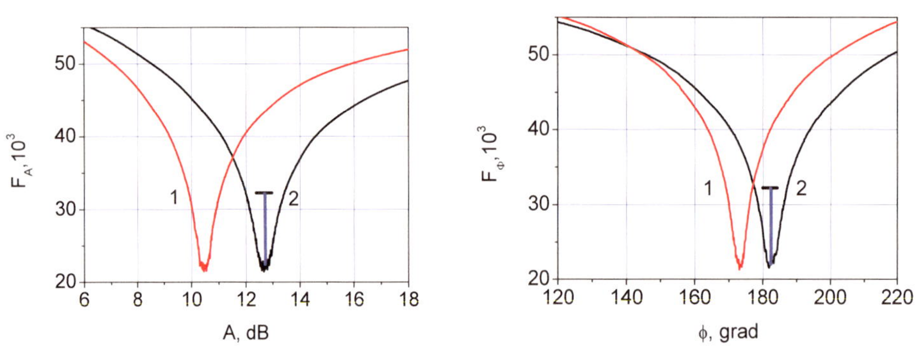

Fig.4. The amplitude (left) and phase (right) dielectrometer functions for the distilled water in two cavity cells (1) for the distilled water and (2) for the table wine in the different cells. The vertical line is the position of a minimum using "bracket" technique.

The readings $F(\delta L, \delta\varphi)$ of an analog-to-digital converter of the receiver in logarithmic units as functions of the differences in amplitude δL[dB] and phase $\delta\varphi[\text{grad}]$ at the bridge arms can be written

$$F(\delta L, \delta\varphi) = \frac{\partial F_A(L_0, \varphi_0)}{\partial L}\delta L + \frac{\partial F_\Phi(L_0, \varphi_0)}{\partial \varphi}\delta\varphi \,, \tag{3}$$

where L_0 and φ_0 are the attenuation and phase in the arms of the balanced bridge.

The PC program algorithm for the recording F_A and F_Φ functions is as follows. During the first iteration the phase scanning is carried out at arbitrary fixed amplitude; the phase function minimum F_Φ is calculated; the phase shifter is returned to a minimum position; the amplitude scanning is carried out; the amplitude function minimum F_A is calculated; the attenuator is returned to a minimum position. That is the end of the first iteration. Our testing showed that in order to reach maximum accuracy of the bridge balancing it is necessary to do three iterations. In Fig.4 we present the amplitude and phase functions of the dielectrometer. These plots are displayed on PC screen in a real time scale. After curves registration the digital low frequency data filtration is made and the minimum position is calculated according to the "bracket" technique. The minimum position is the average attenuation (phase shift) at the instrumental function slopes where the signal-to-noise ratio is of the order of 10 dB. In Fig.4 the calculated minimum position for a dry table wine with respect to the distilled water is shown by vertical lines.

We carried out the detail analysis of origins and values of random and systematic measurement errors of attenuation and phase coefficients $\delta h'$ and $\delta h''$ for the designed dielectrometer. The random errors determine so-called differential sensitivity of our device i.e., the ability to recognize minimal possible differences of phase $\Delta h' \approx 2\delta(\Delta\varphi)$ or attenuation $\Delta h'' \approx 2\delta(\Delta L)$ coefficients of two liquids with close complex permittivity values. The systematic errors determine the absolute complex permittivity measurement errors.

In the designed dielectrometer we have made a number of schematic and design improvements in order to minimize random measurement errors. They are as follows: 1) the usage of the high power signal oscillator (of the order of 100 mW); 2) the usage of a high modulation frequency (100 kHz); 3) the usage of synchronous detection at the modulation frequency; 4) the usage of a low noise current controller of a P-I-N attenuator; 5) the realization of play-free mechanism of the short-circuiting plunger moving by the small discrete step. The dynamic technique of the minimum position determination of the instrumental functions of the dielectrometer leads to the minimization of random measurement errors as well. The mentioned above steps provide root-mean-square random measurement errors of attenuation ΔL and phase shift $\Delta\varphi$ that are of the order of $\delta(\Delta L) \approx \pm 0.001 dB$ and $\delta(\Delta\varphi) \approx \pm 0.05°$, respectively. This error values were estimated by recoverable measurements with the same liquid at stable ambient conditions. As a result these random errors determine the limit of differential sensibility R_h of our dielectrometer. For the liquid with dielectric properties close to the distilled water ($h' = 11.1$ rad/cm and $h'' = 8.8$ dB/cm) the differential sensibility $R_{h'} = 2\delta(\Delta\varphi)/h' \cdot 100\% \approx 0.02\%$ for the phase shift values and $R_{h''} = 2\delta(\Delta L)/h'' \cdot 100\% \approx 0.02\%$ for the attenuation ones.

Another origin of random errors is random temperature deviation for liquids in the cells. The measured mean-square temperature difference in the cells during entire measurement cycle does not exceed 0.1°C after thermal balance achievement. The entire measurement cycle consists of the microwave bridge balancing with the solvent in two cells, the replacement of the solvent in one of the cell by the liquid under test, thermal equality reaching, and one more microwave bridge balancing. The approximate time of entire cycle is about 3 minutes. The direct calculation of temperature coefficients of real and imaginary parts of the complex wave propagation coefficient h was made. It gives $\partial h' / \partial T = -0.00566(\mathrm{rad/cm})/{}^0\mathrm{C}$ and $\partial h'' / \partial T = -0.0462(\mathrm{dB/cm}) / {}^0\mathrm{C}$ in the cell with distilled water at the operating frequency. Thus, the differential sensibility caused by the temperature fluctuation in the cells will be $R_{h'} \approx 0.01\%$ for the phase coefficient and $R_{h''} \approx 0.09\%$ for the attenuation coefficient. Several measurement sets of the wave propagation coefficients in the cells with water and with 10% ethanol solutions in water were made. Each measurement was made according to the entire measurement cycle. We found out that $1\,\sigma$ standard deviations both for $\Delta h'$ and $\Delta h''$ does not exceed 0.06 grad/cm and 0.02 dB/cm, respectively, in absolute units or $R_{h'} \approx 0.05\%$ and $R_{h''} \approx 0.2\%$ in relative units. Obtained measurement data approximately correspond to the given theoretical estimation. If we use a cavity thermostat for the temperature of liquid stabilization, for example, with the accuracy of the order of ±0.01°, then the differential measurement sensitivity will be of the order of 0.01% both for real and imaginary parts of complex wave propagation coefficient.

The absolute complex permittivity measurement error consists of mean-square random errors mentioned above and a number of systematic errors. We analyzed the following systematic errors: 1) a method error $(\delta h)_1$ due to uncertainty of effective length of the cavity. This error exists owing to diffraction effects at excitation of the quartz cylinder in the liquid by the waveguide; 2) an error of absolute calibration $(\delta h)_2$ of the attenuator and the phase shifter; 3) an error $(\delta h)_3$ due to ambient space temperature deviation; 4) an error $(\delta h)_4$ due to parasite phase (attenuation) deviations at attenuation (phase) turning in the microwave bridge arms. One more origin of a method error $(\delta h)_5$ does not have direct connection to quality of measurements. This is the statistical complex permittivity uncertainty of the reference liquid (the distilled water).

The key contribution in absolute measurement accuracy is the error of the uncertainty of the effective length of the cell, which was estimated numerically by 'CST Microwave Studio'. We obtained $(\delta h')_1 / h' \leq 1\%$ for the phase coefficient and $(\delta h'')_1 / h'' \leq 0.5\%$ for the attenuation coefficient at whole measurement range of any table wines and musts. But this error does not impact on the differential sensibility of our device for the liquids under test with complex permittivity values difference is less than 5 units. The measured value of the temperature attenuation coefficient of the P-I-N attenuator does not exceed 0.03 dB/°C. In order to minimize $(\delta h)_3$ we inserted a temperature numerical correction by the PC program based on a measured temperature deviation of the attenuator body. The final calibration P-I-N attenuator error does not exceed 0.1% at the total attenuation deviation range and the ambient temperature. The most essential origins of the systematic error of phase shift measurement are parasite deviation of the wave phase passed via the P-I-N attenuator at the attenuation control. It is minimized by our PC program as well. According to our estimations the maximal phase shift measurement error due to all reasons does not exceed 0.4° or 0.06%. Summing up all systematic errors $(\delta h)_i$, $i = 1,2,3,4$ we obtain the total absolute phase coefficient measurement error $(\delta h')_{\Sigma} \approx 6.2^{0}$ or 1.1% and the total absolute attenuation coefficient measurement error $(\delta h'')_{\Sigma} \approx 0.05\,\text{dB}$ or 0.6%.

4. Results of complex permittivity measurement of wine and wine model liquids

All results presented in this section were obtained by means of our designed dielectrometer. We carried out a set of complex permittivity measurement wines and musts (some results were published in [Eremenko, 2009, Anikina, 2010]. More than 100 dry table wines samples were under test. As an example, the measurement results are presented in Fig.5. We obtained histograms for the increment of real $\Delta h'$ and imaginary $\Delta h''$ parts of complex wave propagation in the cell with dry table wines and musts relative to the wave propagation in the cell with the distilled water. In Fig.5 the calculation results of absolute complex permittivity values for the same wine and must samples are presented as well. All wines satisfying to the nowadays quality standard for the dry natural wines were made of musts-self-flowing using the following types of grapes: Chardonnay, Aligote, Riesling Rhine and Rkatsiteli of 2007 harvest that were obtained using microvinification technique.
We observed small but valid distinctions in the complex permittivity and the complex wave propagation coefficients for various sample wines (musts). We also obtained 100 % correlation of the complex permittivity and the wave propagation coefficients of wine samples and corresponding samples of musts. It is interesting to note, that we can recognize distinctions in the complex permittivity and the wave propagation coefficients for wines and musts of the same sort of grapes (Riesling Rhine) with different vintage dates. The additional study has been shown that it can be explained by different sugar content in these musts.
We carried out quantitative analysis of wines and musts chemical content. The essential correlation between the complex permittivity and wines (musts) chemical content was obtained. The possibility to identify wines according to grapes growing regions or a wine sample with wrong production technology was shown. For complex permittivity measurement method it is necessary to have the data of complex permittivity of model liquids: water solution of chemical wine composition elements that are combined in different proportions. The complex permittivity measurement of model liquids allows establishing cause-and-effect relations between concentrations of the solution components and complex permittivity of solutions.
As an example of the complex permittivity of model liquids in Fig.6 we present the measurement results of the differences between the complex permittivity of water and water solutions of glucose, glycerol, and ethanol at 31.82 GHz at temperature 25°C. We apply the complex permittivity of the distilled water (25.24+i31.69) at the same conditions. The concentration of solution components is presented in the mole ratio, i.e. the number of diluted substance molecules on one molecule of a solvent (water). The confidence measurement interval is ±0.007 dB/cm for the attenuation coefficient and is ±0.05 grad/cm for the phase coefficient. Errors of substances concentrations in solutions are higher, but they do not exceed some tenth of percents. It is necessary to note that we compared with other authors complex permittivity data of water-ethanol solutions presented in [Ganapolskii et al, 2009].

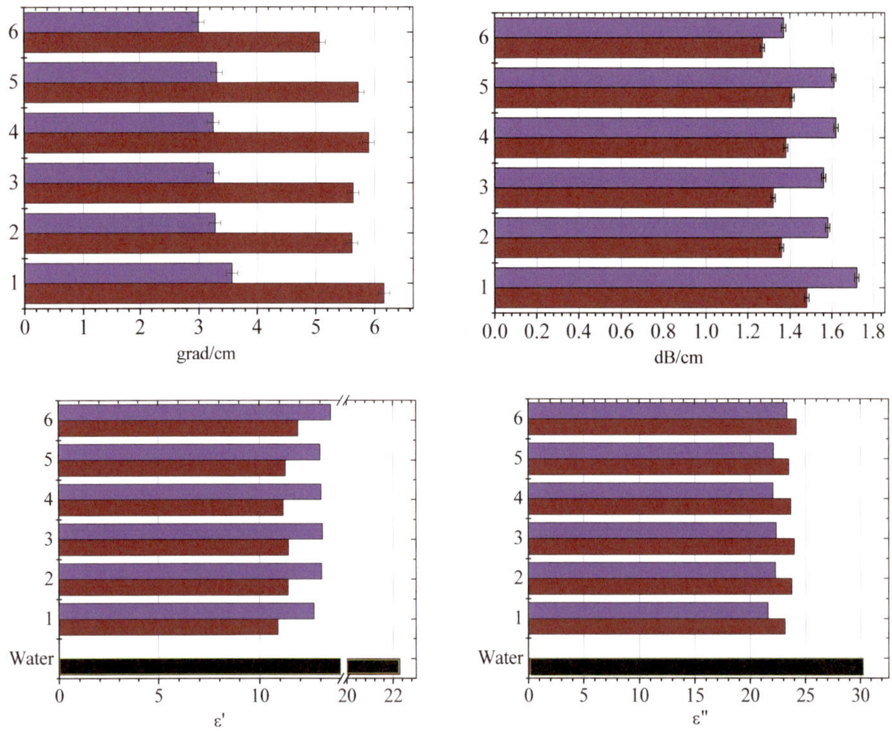

Fig. 5. The increment of the wave phase (upper left) and the attenuation (upper rigth) coefficients in the cell with water and in the cell with table wines (musts) with respect to the distilled water are presented. There are the real (bottom left) and imaginary (bottom rigth) complex permittivity parts of wines and musts samples, respectively. In blue there are data for musts, in brown there are data for wines. The data of grapes vintage are shown on the vertical axis such as 1 - Chardonnay 8 Sept. 07, 2 - Aligote 14 Sept. 07, 3 - Riesling Rhine12 Sept. 07, 4 - Riesling Rhine 19 Sept. 07, 5 - Riesling Rhine 20 Sept. 07, 6 - Rkatsiteli 27 Sept. 07.

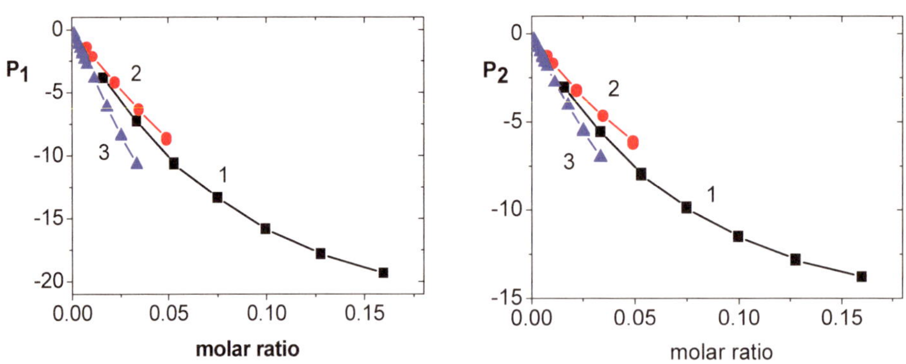

Fig. 6. The differences of the real (left) and imaginary (right) complex permittivity parts of water and water solutions of ethanol, glycerol, and glucose on their concentration in mole ratio. $P_1 = \varepsilon'_x - \varepsilon'_{water}$, $P_2 = \varepsilon''_x - \varepsilon''_{water}$, x is one of components of solutions. The numbers denote 1- ethanol, 2- glycerol, 3 - glucose.

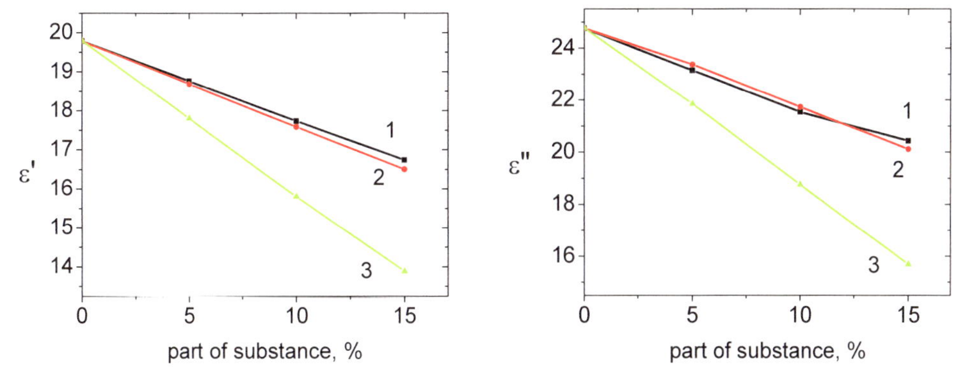

Fig. 7. The influence of wine components on the real (left) and imaginary (right) complex permittivity parts of 10% water - ethanol solutions with additive components: 1 - saccharose, 2 - glycerol, 3 -saccharose and glycerol mixed.

The values of the real and imaginary complex permittivity parts of their water solutions are reduced at the concentration increase of any of three substances. This reduction is approximately linear at small concentrations. Therefore, at mole ratios $r < 0.05$ there is a summation of the contributions of different complex permittivity components of wines and musts (hypothesis of additivity). In Fig.7 there are dependences of complex permittivity of water solutions of saccharose, glycerol, and also their mixture. It validates the hypothesis of additivity. The concentration of quantity of substances is in mass percents.

The water, ethanol, sugars (glucose, saccharose, fructose), and glycerol are chemical components that have the strongest impact on complex permittivity of wines and musts at 8- millimeter wave band in comparison with the other wine components. For instance, in Fig.8 there are dependences of complex permittivity of malic, tartaric, and citric acids diluted with 10% water - ethanol solutions on mass concentration of acids. It presents that the influence of organic acids concentration change on the complex permittivity of wines in several times less than the influence of the mentioned above wine components and these dependences have non-monotonic behavior.

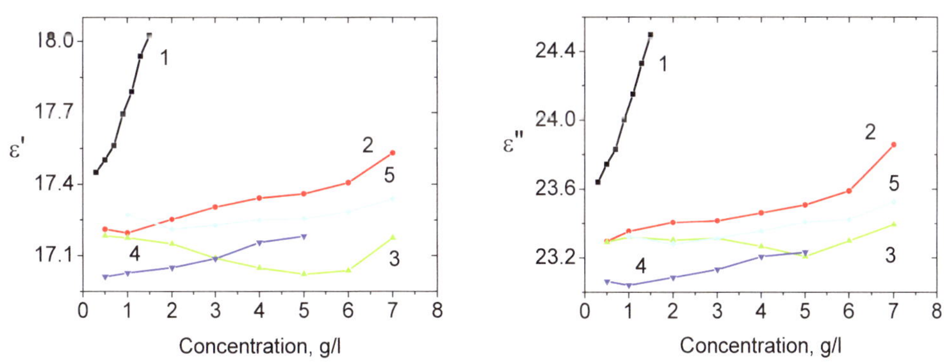

Fig. 8. The dependences of the real (left) and the imaginary (right) complex permittivity parts of organic acids diluted with 10% water - ethanol solutions on mass concentration of organic acids. The numbers denote: 2 - malic acid, 3 - tartaric acid, 4 - citric acid, 5 - tartaric, and malic acids; they are in equal amount. The curve 1 is the dependence of the complex permittivity of potassium diluted with 10% water - ethanol solutions on mass potassium concentration (g/l).

The deviation of cations concentration has enough strong influence on the complex permittivity of wines (the dependence for potassium cations is in Fig.8). However, their absolute quantity in wines and musts is small. Apparently, the influence of cations on the complex permittivity of wines is the reason to have the application possibility of the correlations between complex permittivity and a region of wine-growing.
The results of experimental complex permittivity determination of wines and musts with a different quantity of added water are presented as well. Our objects of research were samples of the natural and diluted with water musts and wines made of the grapes of the following grades: Aligote, Riesling Rhine, Rkatsitely, Cabernet-Sauvignon. It was a crop of 2007-2008. The modeling samples of wines were received by entering water and sugars in the must and squash before the fermentation. Diluted must samples were made by adding the water in the must from 10 % up to 50 %. Diluted wine samples were made by adding the water in natural wine from 5 % up to 30 %.
We defined the following parameters of musts and wines samples: the volume fraction of ethanol, mass concentration of sugars, the total extract, total acidity, viscosity, conductivity, pH, buffer capacity, mass concentration of chlorides, sulfates, potassium, sodium, magnesium, calcium, glycerol, glucose, and saccharose. It was done by the methods accepted in winemaking. Glycerol and separate sugars were defined by high-performance liquid chromatography (HPLC) method on liquid chromatograph Shimadzu LC-20AD. Cation of metals were defined by the method of nuclear absorption on spectrophotometer C115-M1. The viscosimetry, densitometry, titrometry, conductimetry, and pH methods were used for other parameters.
The chemical composition of water added in trial samples, and the monitoring samples of musts and table wines made of grapes growing in a foothill zone of Crimea are presented in Table 1 as well.

Parameters	Riesling Rhine (must)	Aligote (wine)	Cabernet-Sauvignon (wine)	Water
Ethanol, vol. %	-	10.9	13	-
Mass concentration, g/l: Sugars	215	0.81	2.5	-
Total acidity	10.1	8.3	9.8	-
Total extract	239.0	19.0	28.0	-
Phenolic substances	0.524	0.174	1.940	-
Glycerol	0.43	7.4	8.7	-
Chlorides	0.026	0.011	0.03	0.004
Sulfates (K_2SO_4)	0.179	0.289	0.283	0.102
Magnesium	0.136	0.92	0.104	0.007
Calcium	0.140	0.12	0.063	0.104
Potassium	1.200	0.400	0.512	0.001
Sodium	0.066	0.015	0.047	0.008
Buffer capacity, mg-eq/l	64	38	44	9
pH	2.9	2.9	3.1	8.1
Specific conductivity, mS/cm	2.32	1.67	1.95	0.38
Kinematic viscosity, mm^2/s	1.94	1.54	1.73	1.06

Table 1. Physical and chemical parameters of water and natural grape musts and wines.

We performed the analysis of interrelationship between the degree of dilution, chemical composition, and physic-chemical parameters of the diluted by water musts and wines. These results show that the part of bivalent cations grows, and the content of other components decreases with the increasing of the amount of added in wine water.
The change of componential structure in must and wine diluted with water influence on their dielectric properties. Fig. 9 shows the dependence of the complex permittivity on the water added in a must and table wine at temperature 25 °C (31.82 GHz).

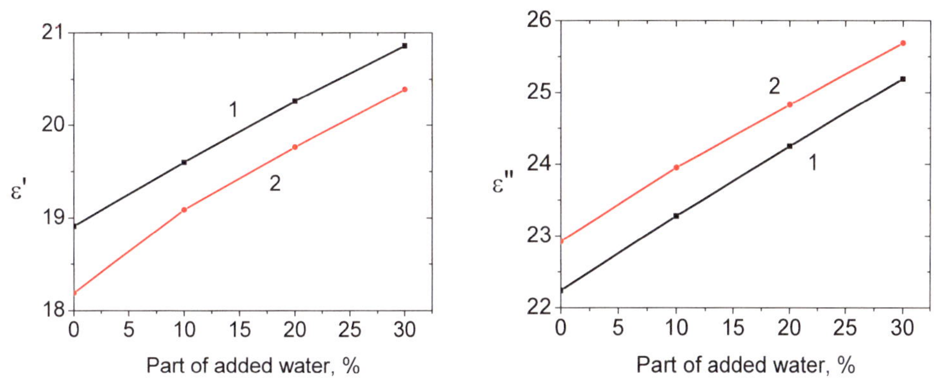

Fig. 9. The complex permittivity of the grape table wine (Rkatsitely) and must (from white Rkatsitely grape) with different degree of water dilution. The real (left) and the imaginary (right) complex permittivity parts. The numbers denote: 1 - must, 2 - wine.

The increasing of the added water part results in the growth of the real and imaginary parts of complex permittivity of wine and must samples under test. The similar effect is observed when the addition of water in must or squash before a fermentation takes place. Close correlation of the complex permittivity with the part of water, added in must and wine indicates the high sensitivity of this parameter to the amount of added water in comparison with traditional parameters of wine composition (Table 2).

Sample	The additive of water in	ε′	ε''	Ethanol	Sugar	Total extract	Total acidity	Phenolic substan-ces
Must	Must	0.999	1.000	-	0.991	0.999	0.997	0.996
Wine	Wine	0.990	0.995	0.996	0.993	0.968	0.983	0.993
Wine	Must (squash)	0.999	0.998	0.990	0.930	0.977	0.918	0.983

Table 2. Correlation coefficient between the complex permittivity and the basic components of must and wine contents with the amount of the added water.

The differential sensibility of dielectometry method is higher than the sensibility of traditional used chemical methods at the determination of the added water (more than 5%.) The small dielectometry analysis duration is also attractive. These facts allows using the difference in complex permittivity of the wine sample under test relative to the complex permittivity of the control sample of natural wine or must for identification of its authenticity. To solve this problem we started the formation of database of complex permittivity for musts and table wines of Crimea.

5. Conclusion

We presented the results of the design of the simple in operation differential 8 - millimeter wave range dielectrometer for the express analysis of high loss liquids. Our dielectrometer has the measurement cavity with two identical cells filled with the reference liquid and the liquid under test. The measurements are based on the dependence of wave propagation along quarts rod immersed into high loss liquid on its complex permittivity. The complex permittivity measurement is computer-aided and the entire measurement cycle does not exceed 3 minutes. The differential sensibility is 0.05% for the real complex permittivity part of liquid under test and 0.2% – for its imaginary complex permittivity part. In particular, it allows solving the natural table wine and must identification problem i.e., fraud detection by means of added water of the order of 0.1%. We presented complex permittivity measurement data of water and water - ethanol solutions of a number of substances that are wine components and selected those of them which have strong impact on the complex permittivity of liquid under test value. These data can be in use during the development of techniques of dielectrometry usage in wine industry. This device can be used in biochemical laboratories.

Our future research is as follows: The reduction of the measurement errors using thermostat, design the cavity with the cells of different lengths to remove uncertainty due to cell effective length error determination. It is necessary to note that other type of wine and must frauds can be detected using our dielectrometer, not only the wine dilution by water. Our dielectrometer can be used in wine manufacturing process as well.

6. References

Afsar M. and Ding H. (2001). A novel open-resonator system for precise measurement of permittivity and loss-tangent. *IEEE Trans. Inst. & Meas.*,Vol. 50, pp. 402–405.

Afsar M., Suwanvisan N., Yong Wang. (2005). Permittivity measurement of low and high loss liquids in the frequency range of 8 to 40 GHz using waveguide transmission line technique. *Microwave and Optical Technology Letters*, Vol. 48, №2, pp. 275–281.

Afsar M., Suwanvisan N., Yong Wang. (2005). Permittivity measurement of low and high loss liquids in the frequency range of 8 to 40 GHz using waveguide transmission line technique. *Microwave and Optical Technology Letters*, Vol. 48, №2, pp. 275–281.

Agilent 85070E (2008). Dielectric Probe Kit 200 MHz – 50 GHz. Technical Overview Agilent Technologies, Inc. 2003-2008. Printed in USA, March 28, 5989-0222EN.

Alekseev S.I., Ziskin M.C. (2001). Reflection and absorption of millimetre waves by thin absorbing films. *Bioelectromagnetics,* Vol. 21, pp. 264–271.

Alison J.M. Sheppard R.J. (1993). Dielectric properties of human blood at microwave frequencies. *J. Phys. Med. Biol.,* Vol. 38, pp. 971–978.

Alison J.M. Sheppard R.J. (2001). A precision waveguide system for the measurement of complex permittivity of lossy liquids and solid tissues in the frequency range 29 GHz to 90 GHz-Part III. Meas. Sci. Technol., Vol. 2, pp. 975–979.

Anikina N.S., Gerzhikova V.G., Zhiljakova T.A., Eremenko Z.E., SkresanovV.N. (2010) Usage of dielectrometry of millimeter wave band for the determination authenticity of grapes musts and table wines. *WinoGrad*, No.1-2, pp.45-47 (in Russian).

Barannik A. A., Cherpak N. T., Prokopenko Yu. V., Filipov Yu. F., Shaforost E. N. and Shipilova I. A. (2007). Two-layered disc quasi-optical dielectric resonators: electrodynamics and application perspectives for complex permittivity measurements of lossy liquids. *Meas. Sci. Technol.*, Vol. 18, pp. 2231–2238.

Brandt. A.A. (1963). The study of dielectrics at microwaves. Moscow, Nauka, 404 p. (in Russian)

Buckmaster H.A., Hansen C.H., Zaghloul H. (1985). Complex permittivity instrumentation for high - loss liquids at microwave frequencies. *IEEE Trans. Microwave Theory Tech*, Vol. 33, No. 9, pp. 822 - 824.

Cherpak N.T. et al. (2004). A new technique of dielectric characterization of liquids. In the book: *Nonlinear Dielectric Phenomena in Complex Liquids.* Kluwer Academic Publisher. Netherlands.

Cherpak N.T., Lavrinovich A.A., Shaforost E.N. (2006). Quasi-optical dielectric resonators with small cuvette and capillary filled with ethanol-water mixtures. *Int. Journ. of Infr.& mm waves,* Vol. .27, No. 1, pp. 115-133.

Compendium of International Methods of Wine and Must Analysis. (2009). *Organisation internationale de la Vigne et du Vin*, Paris, Vol. 1, 2.

Ellison W.J. (2007) Permittivity of pure water, at standard atmospheric pressure over the frequency range 0-25 THz and the temperature range 0-100°C. *J. Phys. Chem. Ref. Data*, Vol. 36, No. 1, pp.1-18.

Eremenko Z.E., Ganapolskii E.M., Skresanov V.N. Vasilchenko V.V., Gerzhikova V.G., Zhiljakova T.A., Anikina N.S. (2009). The cylinder resonator with radial cavity for the complex permittivity determination of table wines. *Radiophysics and Electronics*. Vol. 14, No. 2, pp.239-249. (in Russian)

Eremenko Z.E., Skresanov V.N. (2010) High loss liquids permittivity measurement using millimeter wave differential dielectrometer. *Proc. 40th European Microwave Conference*, 28-30 September, Paris, France, pp. 1532-1535.

Eremenko Z.E., Skresanov V.N. Gerjikova V.G., Zhilyakova T.A., Anikina N.S. (2009) Correlation of complex permittivity and physical-chemistry parameters of table wines. *Proceedings of 19 International Crimea conf. «SHF technique and telecommunications technologies»*, 14-18 September, Sevastopol, Crimea, pp. 843-844.

Eremenko Z.E., Skresanov B.N., Gerjikova V.G., Anikina N.S., Zhilyakova T.A. (2009). Research on dielectric properties of table wines by the method of millimeter dielectrometry. *Viticulture and wine-making*, Vol. XXXIX, pp.79-84. (in Russian)

Ezhov V.N., Gerzhikova V.G., Anikina N.S., Zhilyakova T.A. (1999) The Ways of the detection to falsifications of the wines. Proceedings. of the 1-th conf. "Identification quality and safety to alcoholic product". - Moscow, 1-4 March. (in Russian)

Fuchs K., Kaatze U. (2002). Dielectric spectra of mono- and disaccharide aqueous solutions. *J. Chem. Phys.*, Vol. 116, No. 16, pp. 7137–7144.

Ganapolskii E.M., Eremenko Z.E., Skresanov V.N. (2009). A millimeter wave dielectrometer for high loss liquids based on the Zenneck wave. *Meas. Sci. Technol.*, Vol. 20, 055701 (8pp).

Holloway C.L., Hill D.I., Dalke R.A., Hufford G.A. (2000). Radio Wave Propagation Characteristics in Lossy Circular Waveguides Such as Tunnels, Mine Shafts, and Boreholes. *IEEE Trans. On Antennas and Propagations*, Vol. 48, No.9, pp. 1354-1366.

Hu X., Buckmaster H.A., Baralas O. (1994). The 9.355 GHz complex permittivity of light and heavy water from 1 to 90°C using an improved high-precision instrumentation system. *J. of Chemical and Engineering Data*, Vol. 39, No.4, pp. 625-638.

Jain R. C., Voss W. A. G. (1994) Dielectric measurement methods for industrial scientific and medical applications in the microwave frequency range. *IETE Technical Rev.*, Sept & Dec. Vol. 11, No.5 & 6, pp. 297–311.

Kazenelenbaum V.Z., (1966). Highfrequency electrodynamics, Moscow, Nauka, 240 p. (in Russian)

Masaki K., Atsuhiro N., Kaori F., Shunsuke M. (2007). Complex Permittivity Measurement at Millimetre-Wave Frequencies During Fermentation Process of Japanese Sake. J. Phys. D: Appl. Phys., Vol. 40, pp.54-60.

Meriakri V.V., Parchomenko M. P. (2000). The usage of dielectric waveguide for the control of water content in ethanol. *Electromagnetic waves and systems,* Vol. 5, No.1, pp. 32-40.

Pethig R. (1992). Protein-water interactions determined by dielectric method. *Anny. Rev. Phys. Chem.*, Vol. 43, pp. 177-205.

Sato T., Buchner R. (2004). Dielectric Relaxation Processes in Ethanol/Water Mixtures. *J. Phys. Chem. A*, Vol. 108, No. 23, pp. 5007-5015.

Shaforost O.N., Klein N., Vitusevich S.A., Barannik A.A., Cherpak N.T. (2009). High-sensitive microwave characterization of organic molecule solutions of nanolitre volume. *Applied Physics Letters*, Vol. 94, pp. 112901-4.

Shaforost, E. Barannik, A. Klein, N. (2007). Whispering-Gallery Mode Resonators for Liquid Droplet Detection, *Conference Proceedings of the Sixth International Kharkov Symposium, Physics and Engineering of Microwaves, Millimeter and Sub-millimeter Waves and Workshop on Terahertz Technologies*, MSMW '07, Vol. 2, pp. 919 - 921.

Van Loon R. and Finsy R. (1974). Measurement of complex permittivity of liquids at frequencies from 60 to 150 GHz. *Rev. Sci. Instrum.*, Vol. 45, No. 4, pp. 523–525.

Watanabe K., Taka Y., Fujiawara O. (2009). Measurement of Cole-Cole Plot for Quality Evaluation of Red Wine. *Measurement Science Review*, Vol. 9, No. 5, pp. 113-116.

Zanforlin L. (1983). Permittivity Measurements of Lossy Liquids at Millimeter-Wave Frequencies. *IEEE Trans. MTT.*, Vol. 83, No. 5, pp. 417–419

Part 6

Applications of Plasma

20

EMI Shielding using Composite Materials with Plasma Layers

Ziaja Jan and Jaroszewski Maciej
Wrocław University of Technology,
Institute of Electrical Engineering Fundamentals
Poland

1. Introduction

Electromagnetic compatibility (EMC), including the engineering of materials used for shielding,, is currently one of the most extensively developing field of applications of composite materials (Bula et al., 2006; Jaroszewski & Ziaja, 2010; Koprowska et al., 2004, 2008; Sarto et al. 2003, 2005; Wei et al., 2006; Ziaja et al., 2008, 2009, 2010). The development of lightweight, mechanically resistant, shielding materials is possible by using plasma technology. Due to rapid increase in the number of sources generating the electromagnetic (EM) fields, e.g. radio broadcasting, television, radio communication, cellular networks, continuously extending range of applied frequencies, and increasing power generated by PEM sources, the shielding design is getting more and more challenging. These challenges stem from the fact that complex EM power engineering systems are built of miniaturized electronic circuits. The progressing miniaturization reduces the resistance of the electronic circuits to electromagnetic exposure. Therefore, the choice of suitable materials for the shields and their appropriate arrangement has an essential meaning.

2. Criteria of selection of materials and fabrication technologies for shielding materials

Materials used in the technique of electromagnetic field shielding must meet following conditions:

- have a suitably high coefficient of the shielding effectiveness SE,
- be resistant to mechanical impact and easy to handle (rigidity, elasticity, gravity, the way of installation, sealing),
- be resistant to harmful influence of external environment (oxidation, corrosion),
- durable,
- homogenous,
- easy to form the shield,
- low costs of production.

Shields made as metal sheets or foil, and metal mesh are characterised by a good EM field shielding effectiveness coefficient. However they are characterised by low resistance to environmental impact. Their fundamental disadvantage is weight. They are primarily used in low frequency electromagnetic field shielding.

The alternative to metal shields is the use of composite shields, that have lately found a wide application in the EM field shielding technology (Wojkiewicz et al., 2005; Holloway et al., 2005; Ziaja et al., 2008a, 2008b), and they are interchangeable equivalents for metals. Main advantages these materials are: good mechanical and chemical properties, lower weight, higher rigidity and strength, resistance to corrosion, lower costs of treatment and easier processing. Based on their structure composite materials are divided into two groups:

- laminar systems,
- inclusions in substrate (admixture materials in a form of fibres, flakes or particles).

The first group is made in the special fabric form (fabrics with entwined conductive fibres e.g. Zn, C, Cu, and Ag): nonwoven fabrics with conductive layers deposited or laminates with pressed in conductive layers. Conductive layers may be put on by using the following methods: silk-screen printing (Wang et al., 2009); vacuum evaporation or magnetron sputtering (Ziaja, 2007; Ziaja et al., 2008). They are characterised by easy deposition on different kind of surfaces and the shielding efficiency coefficient above 50 dB. However not every material is suitable for conductive coating. It is determined by surface properties of the substrate material.

The second group covers conductive glues or dyes.

While designing composite shielding material, it is necessary to take into account, that efficiency of shielding depends on the following factors:

- volume fraction of inclusions,
- electrical and magnetic properties,
- shape and size of inclusions, and the way of their orientation,
- EM field frequency,
- number and sequence of layers.

Choice and optimisation of appropriate composite structure is made on the basis of assumed mathematical models (Sarto et al., 2004; Szhulz et al., 1988; Pospieszna, 2006). These models must assume effective application of different mechanisms of shielding:

- effectiveness of the reflection mechanism on interfacial surfaces, which among other things depends on their size. In case of composite materials this effect is obtained by implementing extenders with expanded specific surfaces,
- losses caused by multiple reflections are negligibly small, when a distance between successive reflection surfaces (interfaces) are big in comparison to the depth of penetration,
- because of the skin effect, for effective use of whole cross section area of the elementary unit of extender, its size should be comparable or less than a depth of penetration,
- effective absorption of radiation ensures material with a high permittivity and / or with a high magnetic permeability.

The distribution of inner field is mainly determined by the orientation of fibres and the polarisation of the incident field. In order to achieve a relatively high and constant value of the shielding effectiveness SE, it is necessary to use anisotropic materials with certain number of layers displaying different orientation of fibres.

The surface of composite materials, which reflects the EM wave must be electrically conductive, although its high conductivity is not required. It means, that in a case of composite material, containing conductive extender, a conductive connection between extender units is not required, even though such connection improves shielding. Good shielding effectiveness values are obtained in polymer composites containing carbon powder (Kim et al., 2004; Wu et al., 2003; Huang et al., 2007) or silver powder (Hong et al., 2001).

Coating of materials by thin layers (metallic, ceramic and polymer) meets wide use in many branches of technique, both in laboratory scale and in industry. The winning of materials with unique properties, a possibility of miniaturization, improvement of the lifespan and reliability, and a high energetic efficiency of electronic and optoelectronic devices are main driving force of thin-layer technique development to produce EM composite shileding. The most often used methods of layers winning are: the chemical deposition from the gas phase (CVD) and the physical deposition from the gas phase (PVD).

PVD technologies are based on deposition of a conductive layer on the substrate from the gas phase using physical phenomena, such as: metal evaporation (vacuum sputtering), sputtering in vacuum, gases and metallic vapour ionisation. For most of them the common feature is a process of deposition of a layer from stream of gas containing ionised molecules. Any differences between PVD methods are due to the way of winning of ionised metal vapours.

Nowadays, the most often used method of the obtaining different kinds of layers is magnetron sputtering (glow discharge, plasmatic). This method is based on bombarding of surface of the sputtering electrode (the target) with high energetic ions of working gases. Ions knock out atoms or molecules of material from the target, which next are deposited on substrates. This way metallic layers, dielectric semiconductor layers and high-melting layers can be deposited. Sputtering devices using an electrical and a magnetic field are called magnetron or plasmatic guns. In electrical and magnetic fields, ions from working gases obtain very high kinetic energy, and to the layers can be deposited at a very high speed,, which for metals is a few μm/min and for dielectric layers is in the order of 10-100 nm/min. Such high speeds of putting on layers are not available in other methods. The magnetron structure differentiates from other atomiser structures by a dominating presence of secondary electron ionisation (surface ionisation effect). Thanks to that, current density flowing through the target is at least one order of magnitude greater than in other structures.

Additionally during a process of sputtering, plasma cleans and activates surface of substrates. Therefore with magnetron methods one can do metallization of materials such as: PTFE (polytetrafluoroethylene) or PP (polypropylene), which can not be transformed by metallization with other methods, because of their surface properties.

Universality of magnetron sputtering devices gives opportunity of using them in both production lines on an industrial scale and also in small scientific laboratories. The placement of the target with regard to substrates being coated is almost optional and the deposition area is virtually unlimited. Magnetrons of rectangle shape with targets up to 4000 mm long are used to cover wide glass sheets with layers with different optical properties. It even ensures coating of a layer with the retention of stoichiometry of winning compounds and the thickness repeatability of single multilayer structures. The measure of usefulness of magnetron systems, is among other things the ability to realise precise requirements for optical coatings (e.g. $Ti0_2$, $Si0_2$, Si_3N_4, $Sn0_2$, ZnO) for needs of automotive and spacecraft industry.

In the project for winning electrical field shielding layers, on textile carriers, magnetron gun WMK-100 was used. The deposition experiment was prepared by the Laboratory of Vacuum and Plasma Technology of Wrocław University of Technology. This magnetron gun is characterised by a power densities of (~ 300 Wcm^{-2}) striking targets, which are not obtainable in standard devices. Additional advantages of the magnetron gun are:

- the easiness of target exchange,
- the possibility of target sputtering of thickness from 1 to 6 mm,
- the possibility of spraying of dielectric target.

Magnetron guns are powered by direct (DC) and high frequency alternating (AC) current sources. DC currents are primarily used for metallic targets. Increasing the supply current frequency allows to sputter dielectric materials. Sources with radio frequency (RF) are also used; however, they didn't meet broader application in industry.

In recent years impulse current sources have been developed for magnetron gun power supply. In standard systems, the power emitted on a target is controlled by the magnetron supplying current value, but in impulse systems the power is controlled by the length of impulse time. On Fig. 1 types of power supply modulation of magnetron sputtering devices are shown. During the impulse duration t1 sputtered material is accumulated in the plasma area. During the intervals between impulses (t2-t1), the deposition of atoms on the substrates surface takes place. Applying uni-polar pulses to metallic targets (Fig.1 b) and bipolar ones to dielectric ones (Fig.1 a) is a common practice. The advantage of this type of solution is a possibility of making stable reaction processes. This stability stems from elimination of non controlled arc discharges. Since in a single impulse, there is always the maximum sputtering speed, the impulse modulation lets to increase the ratio of sputtered material to impurities.

Describing the power supplied we use two electrical quantities: the real power P and the circulate power Pc. The first defines directly the effectiveness of the layers coating, the second is a certain property of the power supply system. It reflects changes of plasma parameters (change of impedance) connected with a change of concentration and type of charged particles. It is especially seen during sputtering in oxygen.

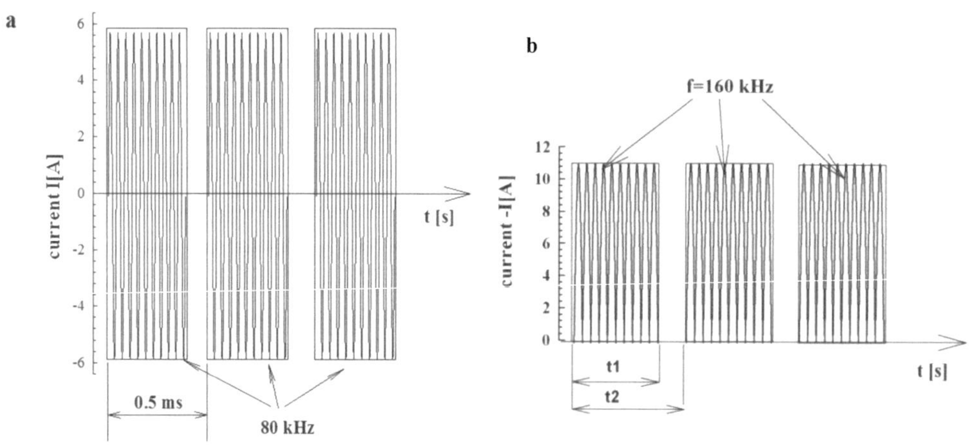

Fig. 1. Types of impulse modulation: a - alternating current (bipolar), b - variable unipolar current

Recent current source solutions let not only to change the length of impulse time t1, but also the current bursts repeating frequency 1/t2.

This solution is useful during executing reactive processes, because the ratio of both times t1/t2 is a characteristic for coming into existence of each chemical compound. By changing this ratio, one can change chemical composition of emerging compounds. In reactive processes (winning of oxides or nitrides) the dwelling time of atomised material in the plasma area plays the key role in a winning of layers with the pre-set chemical composition. If times t1 are too short, monoxide layers obtained, may not have the required stoichiometric composition. Even when sputtering metallic titanium in pure oxygen, one can obtain metallic layers. Therefore the proportion t1/t2 is a very important technological parameter. Processes of sputtering titanium (99.9 %) and zinc (99.9%) targets were realised by use of magnetron launcher WMK-100 with the magnetic system of diameter f= 75 mm. The argon pressure (99.999 %) used as a working gas was changed from $7*10^{-3}$ to $3*10^{-2}$ Tr. It was established, that pressure of $1*10^{-2}$ Tr was optimal, because the etching area of the target surface is then the widest. It ensures 85 % utilisation of the target area. This technological parameter is very important from the industry point of view, because it ensures the most advantageous use of target material. In this case, it may be increased using the magnetic system of a wider diameter. For this magnetron the optimal diameter of magnetic system is 100 mm. The distance between the target and nonwoven fabrics has been set to 100 mm. The magnetron gun was energized by impulse current source type DPS (DORA POWER SYSTEM) which works in AC-M and DC-M modes controlling power up to P= 12 kW. This source can control group frequency in the range of fg=50 Hz - 5 kHz.

Fig. 2. The zinc target installed on the WMK-100 magnetron

3. Barrier composites on textile substrates

When designing of shielding materials in a composite structure polymer-conductive layer, it is necessary to take into account not only electrical parameters (resistivity ρ, permittivity ε), but also the thickness of layers and the morphology of their surface. The layer thickness may be controlled by both: the power emitted on atomised electrode and by the time of atomisation.

The appropriate choice of these parameters guarantees obtaining of uniform layers with the expanded specific surface and numerous conductive bridges. The volume resistivity ρ_V plays the significant role in the shielding mechanism, however it is not possible to measure this quantity in these type of composites.. Good solution seems to be a measurement of the surface resistivity ρ_s and its influence on a value of the shielding coefficient SE.

The ρ_s value of metallic layers is determined by the surface morphology of used nonwoven fabric and by the number of conductive bridges.

Titanium layers were deposited on polypropylene nonwoven fabric of the base weight of 60 g/m^2 and 150 g/m^2 , also on nonwoven fabric with consistence of viscose 30%-polyester 70 % and the base weight of 70g/m^2 (VISC30+PES70). To establish the structure of obtained metallic layers, the surface of cross section areas of samples were examined at magnification 300x. On Fig. 3 nonwoven fabric surfaces before and after the metallization process are shown.

Nonwoven fabric VIS30%+PES70% is characterised by the non-uniform expanded surface with clear seen fibres. Also for propylene, the influence of the basis weight on the morphology of nonwoven fabric clearly appears. PP samples of the base weight of 60 g/m^2 still have apparent single fibres. PP of the base weight of 180 g/cm^2 has the most uniform surface.

The power P emitted by the gun was varied from 0,5 do 3 kW. Increasing of power above 3 kW causes heating of nonwoven fabric, which may even lead to melting . Heating of nonwoven fabric results in desorption of gases from its volume or chemical dissolution causing arising of reactive gases in the magnetron chamber. These gases react chemically with atomised titanium and form its compounds, which are characterised by other electrical properties than metallic titanium. Time of depositing was varied from 0,5 to 20 min.

The investigation of the surface morphology was conducted with the use of scanning microscope Quanta 200 in the low vacuum mode (without covering of preparation surface with a gold layer), The sample surface was in its natural state. The acceleration voltage 15,0 kV of an electron bunch and the SSD detector were used.

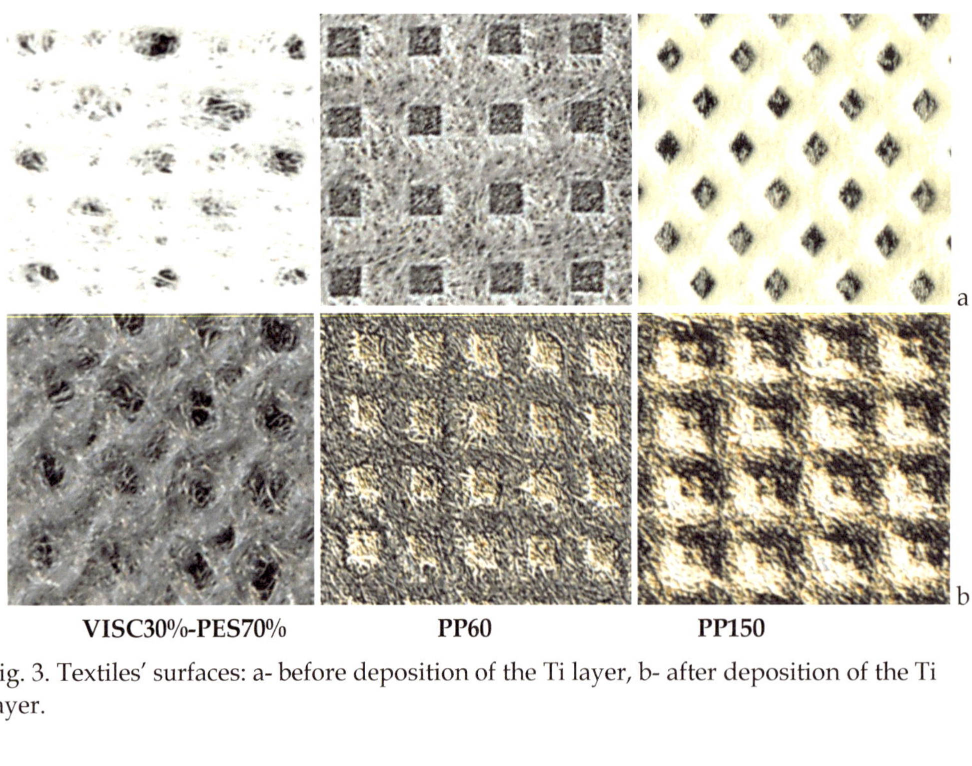

Fig. 3. Textiles' surfaces: a- before deposition of the Ti layer, b- after deposition of the Ti layer.

The microscope examination (Fig. 4) has shown, that deposited layers are not uniform, what is probably caused by penetration of metal into areas between fibres.
In covering metallic layers, there are few cracks and splits, and their adhesion to substrate is very good. For PP nonwoven fabric, titanium layers also form themselves into areas between interleaves. Increasing the base weight of the fabrics causes forming of more solid layers and uniform metallic layers. The covering process of layers Ti begins with forming of nuclei of crystallization centres (for depositing time less than 10s) on passing holes of nonwoven fabric (Fig. 3). X-ray radiography examination shown, that Ti layer is formed by crystallites of dimensions from 8 to 13 nm. Increase in the depositing time causes the spreading of layers so that a continuous structure can be obtained.
Interesting part is also building over areas between passing holes of nonwoven fabric and deposition of Ti layer on fibres, taking place inside nonwoven fabric (Fig. 4). Such extension is characterised by a big specific surface, which increases the reflection and the dispersion of the electromagnetic wave effect.

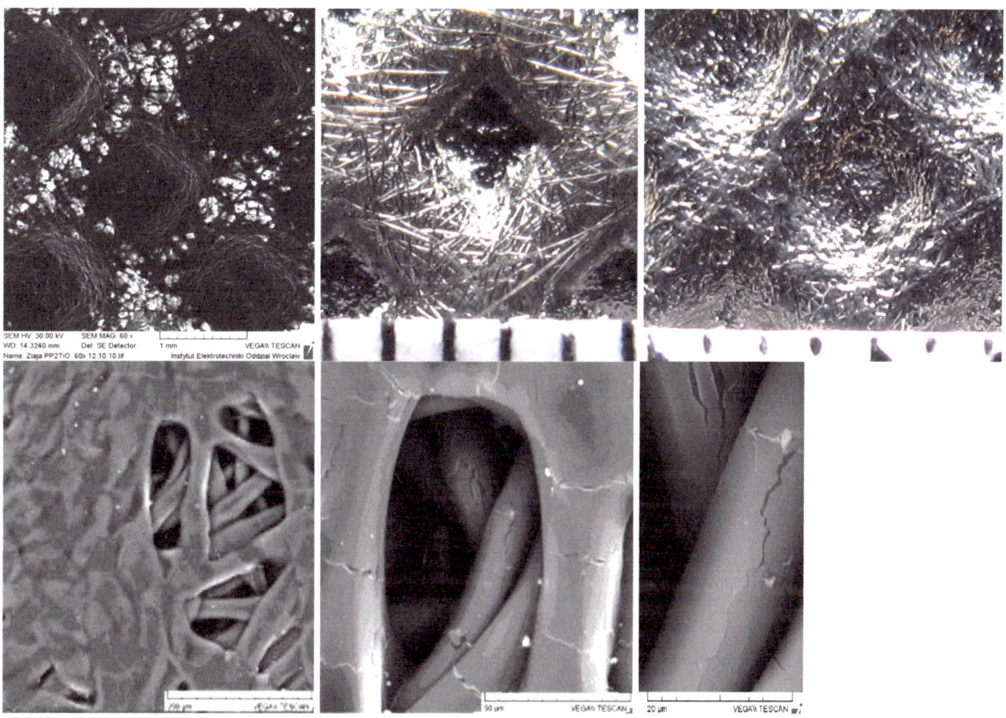

Fig. 4. Microscopic image of the Ti surface at different magnifications.

Values of surface resistivity ρ_s of titanium layers depend on morphology of the nonwoven fabric surface. The precise microscopic analysis has shown obvious differences in the surface morphology of either side of the nonwoven fabric PP150 (Fig. 5). The one side of the surface is almost smooth, consistent and without any expanded porous structure; whereas the second surface is scabrous and porous, with visible micro-fibres forming a spongy microstructure. Such a surface makes it impossible to form a nucleus of crystallization centres, which form conductive metallic bridges.

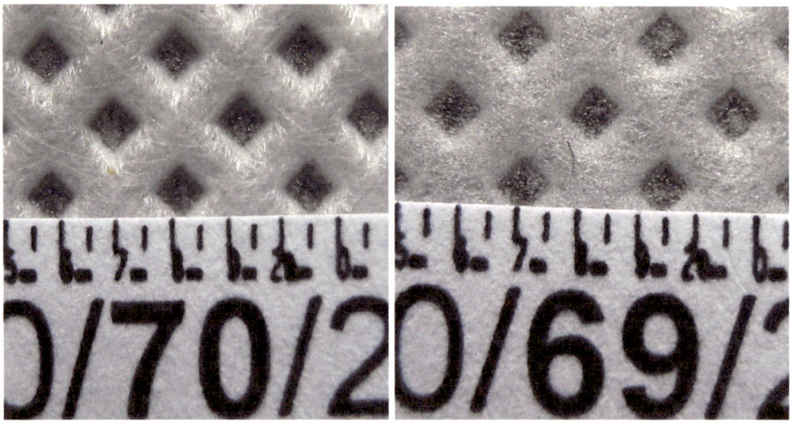

a b

Fig. 5. The surface morphology of each side of polypropylene textile of the base weight of g=180gcm^{-2} , a- smooth surface, b- scabrous surface.

The process of winning of zinc layers and zinc-bismuth layers on PP onwoven fabric was made by sputtering metallic targets with the weight consistence of 0.9Zn-0.1Bi and 0.96Zn-0.04Bi, and metallic zinc with the purity of 99.9 %. Metallic layers were deposited on polypropylene nonwoven fabric of the different base weight, alumina ceramics, CORNIG glass covered with conductive and transparent layer of ITO (the solid solution SnO_2 and In_2O_3).

To identify the structure of the obtained metallic layer the surface of the samples cross section was examined at magnifications: 300x, 1000x (Fig. 6). In covered layers Zn there are few cracks and splits, and their adhesion to substrate is very good and similar to titanium layers. In magnified figures below (Fig.6), one can observe well formed, conducting zinc bridges.

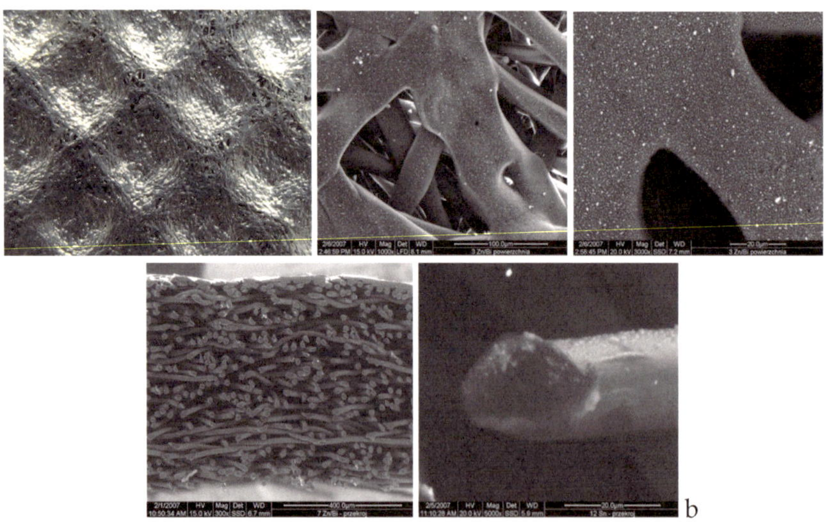

a

b

Fig. 6. The microscopic pictures of (a) the surface and (b) the cross-section of Zn layer.

During the deposition of Zn and Zn-Bi target, in contrast to titanium ones, one is not allowed to use high values of power. Exceeding the value of power over 2 kW (for magnetic system use), causes uncontrolled arcing discharges on defects and impurities of the target surface, which leads to overheating. Further increasing of power heats the target in such a way, that process of melting follows it, and then it causes evaporation of the target material. This effect leads to damage of the target and eventually to water leak from cooling system into vacuum chamber.

Zinc and zinc-bismuth layers, just like titanium, are half-amorphous. The X-ray analysis (Fig. 7) has shown, that the average dimension of crystallites is of the order of 10 nm. Lines characteristic for Zn are characterised by a small intensity and a big half-width.

In Fig. 8 the dependence of group frequency of a magnetron gun power supply and types of nonwoven fabric, also their base weight on a surface resistivity ρ_s of titanium layers is presented. It was determined, that the value of the surface resistivity depends on the substrate base weight. PP150 nonwoven fabric is characterised by the lowest value ρ_s of the order of 10^3 Ω. It is a big value and according to the expanded surface of nonwoven fabric, also numerous of resistance bridges, in spite of that the magnification picture of cross section suggests continuity of the layer. For nonwoven fabric PP60 the surface resistivity is one level higher. The highest value ρ_s was obtained for VISC30+PES70, which is above 10^7Ω. This is the result of lack of the metallic continuity of layers.

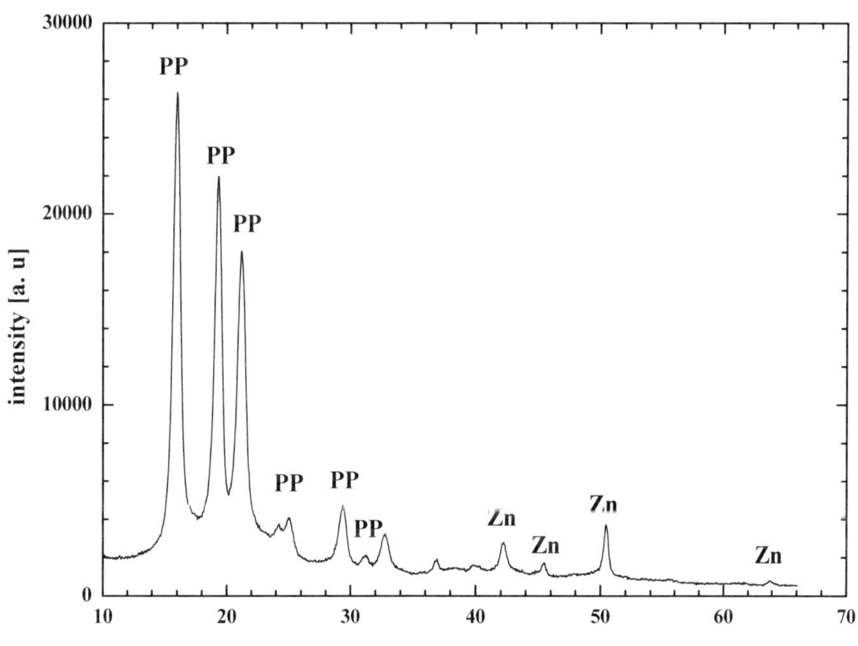

Fig. 7. X-ray picture of PP textile with covering Zn layer for radiation wavelength Co.

It was determined, that with increase in group frequency, value of ρ_s was increased slightly. It is the result of the longer time of sputtering t1 and the longer dwelling time of the substrate in the area of plasma.

Probably a chemical dissolution of substrates occurs as a result of the effect of the higher temperature. Substances are released, which react with titanium layer changing its electrical properties.

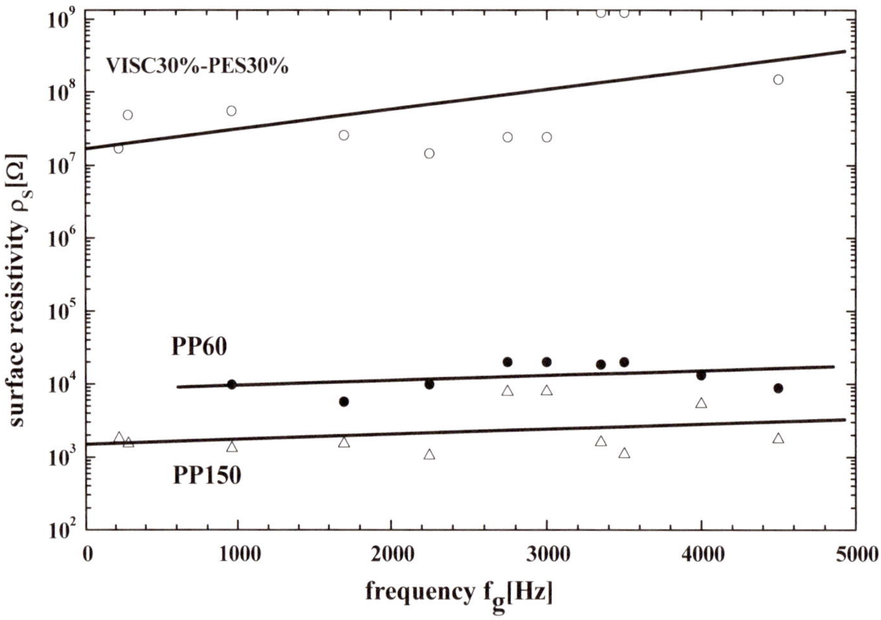

Fig. 8. Influence of the group frequency fg on the surface resistivity ρ_s of titanium layers, P_{Ar} =2*10-2, power emitted on the target P=1 kW, the depositing time t=10 min.

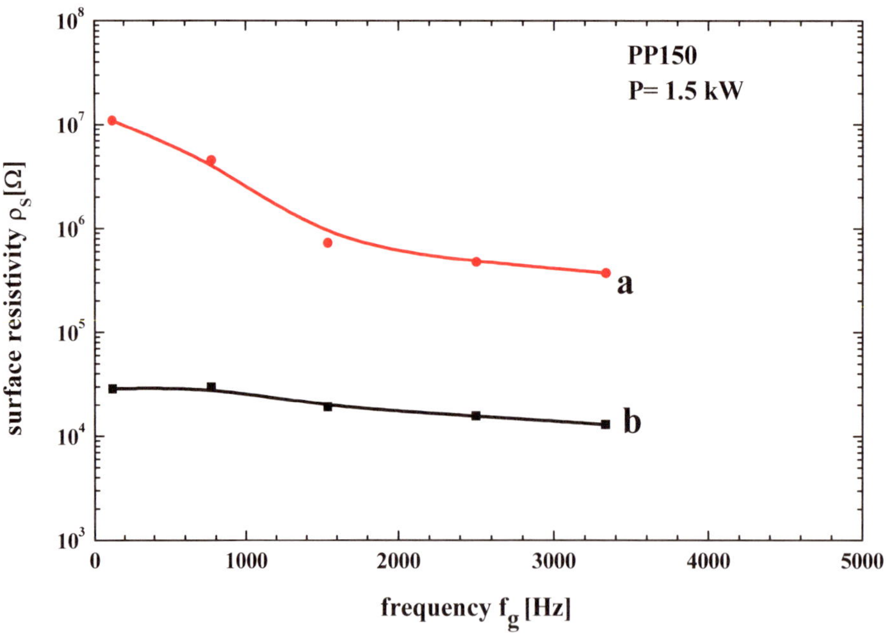

Fig. 9. Influence of the group frequency fg on a surface resitivity ρ_s of titanium layers, P_{Ar}=2*10-2, the depositing time t=3 min. in case of unipolar DC-M supply

The surface microstructure has the decisive impact on the value of the surface resistivity ρ_s. The porous side of PP nonwoven fabric with an extended structure is characterised by a very high surface resistivity of the order of 10^7 Ω (Fig. 9) and it is two levels higher than on the smooth side ($\rho_s = 3*10^4$ Ω). With increasing of group frequency of the power supply, ρ_s gets lower into value $3*10^5$ Ω, but even then it is one level higher than on a smooth side.
The value of ρ_s depends also on the time of depositing metallic layers. Increasing the time, the covering layer not only increases its thickness, but it gets more solid and uniform. After obtaining the specified thickness, value ρ_s does not change.

4. Shielding properties of PP/Ti and PP/Zn composites

Measurements of shielding attenuation were realised on the test setup prepared in the Institute of Telecommunications, Teleinformatics and Acoustics of Wrocław University of Technology in accordance with the method of ASTM D4935-99. The test diagram is presented on Fig. 10. The test settup consists of network analyser - model HP 8711A of Hewlett-Packard firm and measuring adapter, which is a section of an air concentric line of characteristic impedance 50 Ohm. Measuring uncertainty is approximately 2 dB. Measurements of attenuation were made using standard sample. Shielding attenuation is calculated as a difference between transmittances or between attenuation of a standard and an examined sample.

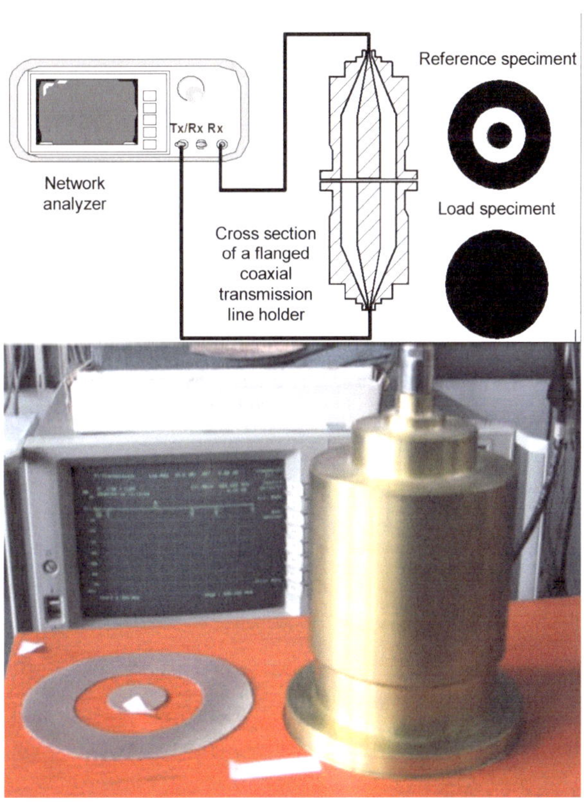

Fig. 10. The diagram of the test setup of shielding attenuation

Examinations performed have shown an influence of types of used substrates on a value of shielding attenuation of metallic layers. On Fig. 11, increasing of shielding attenuation Ti layers of 10 dB for substrates of smooth surface (CORNIG glass) in comparison to the substrate of extended specific surface (nonwoven fabric PP) is observed. This observation points to a possibility of obtaining higher values of SE for nonwoven fabrics.
The highest values of shielding attenuation coefficient (SE>50 dB) are obtained for composites PP/Zn (Fig. 12). The value of SE may increase by doping zinc layers with bismuth. Increasing bismuth concentration from 4 to 10 % of atomic, causes increase in SE by approximately 20 dB (Fig. 13 and 14). Further increase in concentration of Bi does not cause any increment of SE. Observed changes of SE value as a function of power released on the target, are caused by increase in Zn layer thickness.

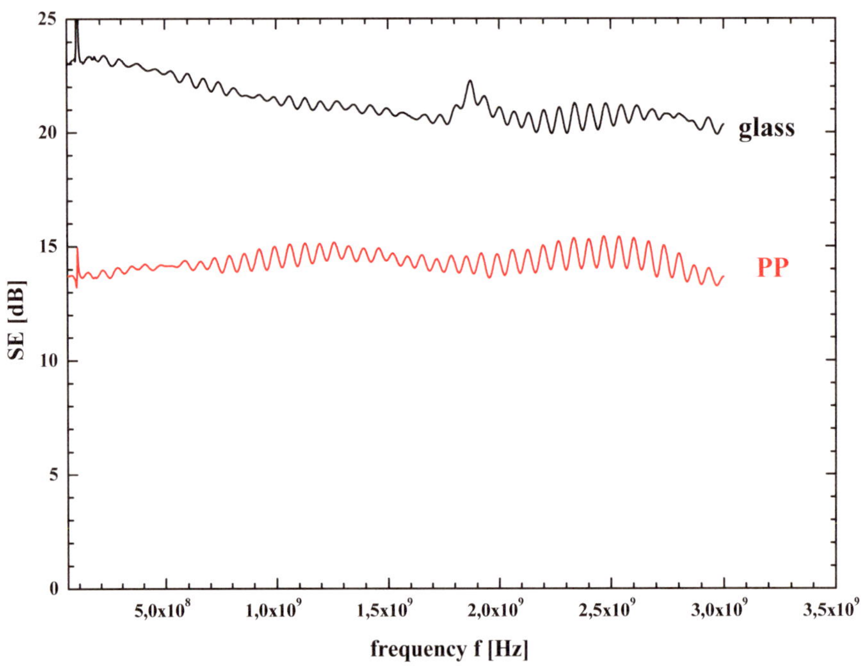

Fig. 11. The coefficient of shielding efficiency SE versus frequency f for PP/Ti and glass/ Ti composites.

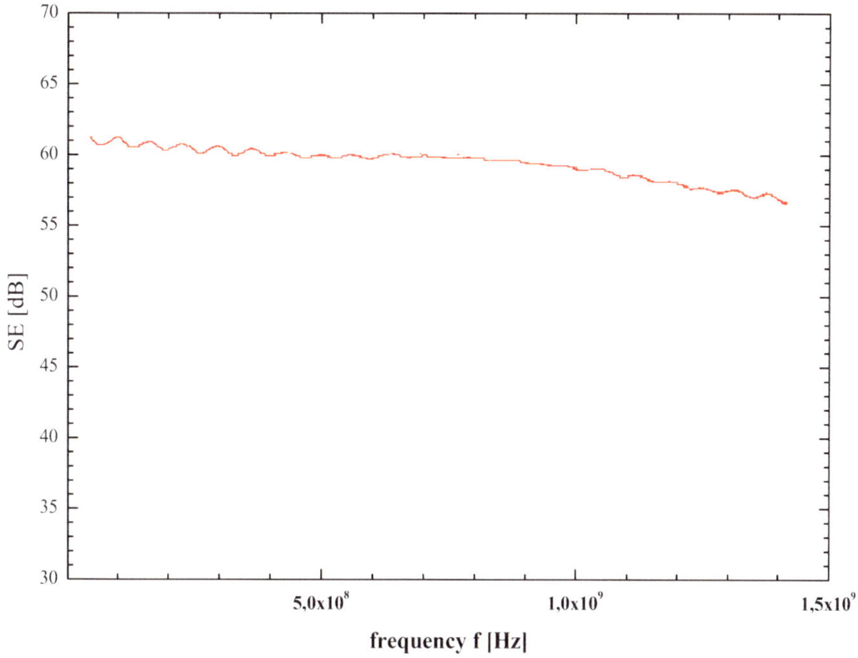

Fig. 12. The coefficient of shielding efficiency SE versus frequency f for PP/Zn composites.

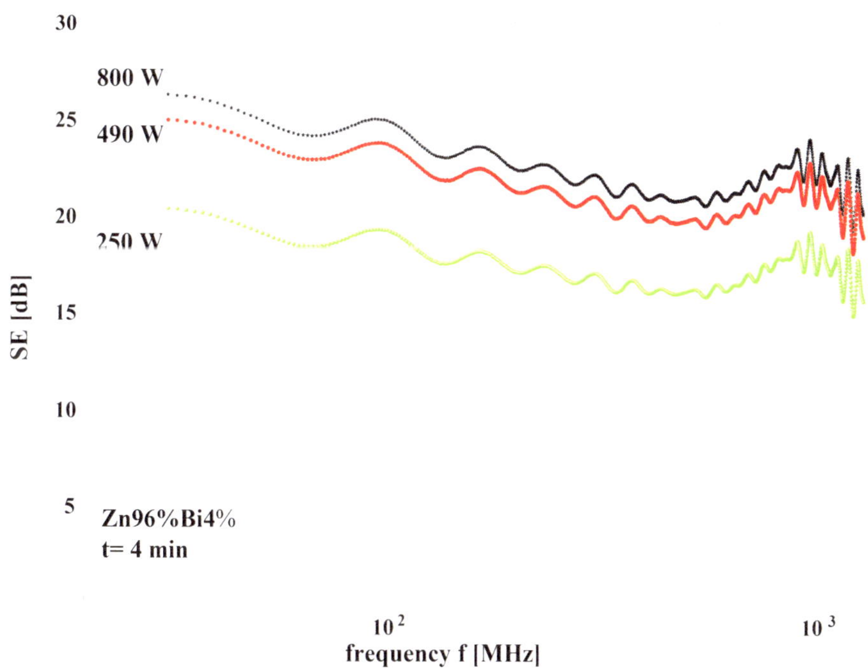

Fig. 13. The coefficient of shielding efficiency SE versus frequency f, for different powers released on the Zn96%Bi4% target in case of unipolar DC-M supply.

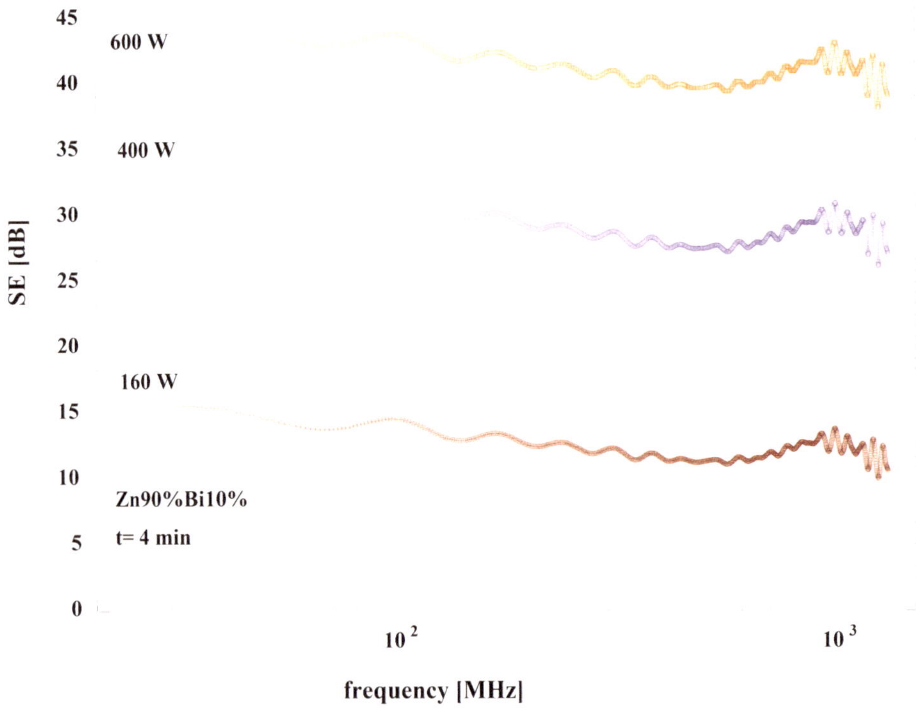

Fig. 14. The coefficient of shielding efficiency SE versus frequency f, for different powers released on the Zn96%Bi4%target in case of bipolar DC-M supply.

Despite of the initial assumption, that the surface resistivity may be a good tool for testing composites PP/Me (Me- metal) for evaluation of their shielding properties, it has been found, that deciding role is played by the thickness of layers. In case of composites PP/Zn and PP/Ti, for the same ρ_s, different values of shielding coefficient were obtained. On Fig. 15 , over twofold increase in SE value, for the same values of surface resitivity, was observed. In this case, Zn layer is characterised by higher thickness and uniformity than Ti layer. Higher uniformity and continuity of Zn layers causes increase in the reflecting effect of electromagnetic wave from this surface.

5. Assessment of barrier shielding materials with plasma layers deposited on the fabrics and parameters of technological process with the impedance spectroscopy method

Shielding effectiveness of composite material polymer-metal is not only determined by its surface resistivity. There are other physical phenomena occurring both in each composite layer and on interphases between them that play very important roles.

Composite materials working in alternating electrical fields are characterised by a presence, beside conductivity, also polarisation phenomena, that is they get a certain electrical moment m. The total dielectric polarisation is a superposition of three polarisation mechanisms:

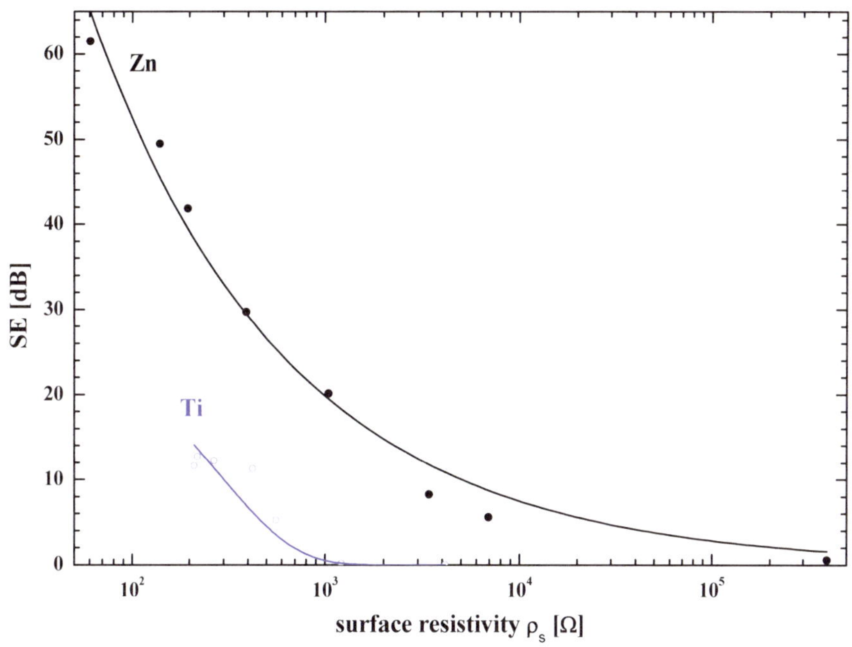

Fig. 15 . The coefficient of shielding efficiency SE versus frequency f, for different powers released on the Zn96%Bi4%target in case of unipolar DC-M supply.

- distortion polarisation, which occurs both in dipole dielectrics and in non dipole dielectrics, and being a sum of:
  - electron polarisation resulting from a deformation of electron shell, and
  - atomic polarisation resulting from a displacement of atoms from their original positions,
- orientation (dipole) polarisation, which occurs only in dielectrics with permanent dipole moments resulting from an alignment of dipole axes along direction of outer electric field lines,
- ion polarisation, which occurs in dielectrics with ion bonds, resulting from displacement of ions in relation to each other from the place of balance.

Generally the polarisation phenomenon is that an arising on a dielectric surface, inserted into electric field, induced electric charges, which are bonded charges ρ_p. We can treat them as charges of dipoles chain ends, directed by the outer electric field. If we take into consideration the surface charge density for a capacitor not filled with dielectric ρ_0, then the total charge density for a capacitor with dielectric is a sum of the bounded charge ρ_p and the free charge ρ_0. Inserting of dielectric into electric field (flat capacitor configuration) causes increasing of the capacitance from value C_0 to value C. The ratio of the capacitor capacitance increasing to its capacitance C_0 is defined as the electric susceptibility of the dielectric χ, but the proportion of the capacitor capacitance with a dielectric to its capacitance without a dielectric, as the electrical permeability ε. Thus we can define the density of the total capacitor charge as:

$$\rho = \rho_0 + \rho_p = \rho_0 + \chi\rho_0 = \rho_0(\chi + 1) \quad (1)$$

A measure of this charge is the dielectric displacement vector (the vector of an electrical flux density, the induction vector) D:

$$D \cdot n = \rho = \rho_0(\chi + 1) = \varepsilon_0 \cdot E \cdot n + P \cdot n \tag{2}$$

where:
n - is the normal unit vector directed in the dielectric direction,
ε_0 - is permittivity of vacuum.
Polarisation P, directly bonded with ρp surface density of polarisation charge, can be therefore presented in the form:

$$P = D - \varepsilon_0 E = (\chi + 1)\varepsilon_0 E - \varepsilon_0 E = \varepsilon\varepsilon_0 E - \varepsilon_0 E = (\varepsilon - 1)\varepsilon_0 E = \chi\varepsilon_0 E \tag{3}$$

A basement of the impedance spectroscopy method (IS) is a measurement of a linear electrical answer of an examined material to an excitation with a relatively small voltage signal of the sinusoidal form u(t) = Um sin (ω t + ψ_u), in a wide frequency range f (ω = 2 π f). An answer to a sinusoidal voltage signal is a sinusoidal form of current i(t) = Im sin (ω t + ψ_i) with the same angle speed ω. This current is a sum of conductive current and polarisation (displacement) current (Walter S. Zaengl, 2003):

$$i(t) = \rho_0 E(t) + \partial D(t)/\partial t = \rho_0 E(t) + \varepsilon_0 \partial E(t)/\partial t + \partial P(t)/\partial t \tag{4}$$

Using the Fourier transform to both sides of equation (6) we get an answer of dielectric in the frequency domain (A.K. Jonscher, 1983; A. Bouaïcha et.al, 2009):

$$\underline{I}(\omega) = \rho_0 \underline{E}(\omega) + j\omega \underline{D}(\omega) = \{\rho_0 + \varepsilon_0 \omega \chi''(\omega) + j\omega\varepsilon_0[1 + \chi'(\omega)]\}\underline{E}(\omega) \tag{5}$$

where: χ′ i χ″ are respectively the real and the imaginary component of the complex electric susceptibility χ.
The nature of examination of material electrical answer is a measure of the effective current's value I and its φ phase displacement in relation to forcing voltage (of the effective value U), whereby $\varphi = \psi_u - \psi_i$.
Based on these measures the T(ω) spectral transmittance is calculated, which characterises dependence between the forcing and the answer phase displaced for the same ω pulsation:

$$\underline{T}(\omega) = \left|\underline{T}(\omega)\right| \cdot e^{j\psi(\omega)} \tag{6}$$

where: $\left|\underline{T}(\omega)\right|$ is a modulus and ψ (ω) is an argument of the spectral transmittance $\underline{T}(\omega)$.
In the impedance spectroscopy, the spectral transmittance usually has a form of the Z(ω) complex impedance or Y(ω)admittance, which are defined as:

$$\left|\underline{T}(\omega)\right| = \underline{Z}(\omega) = \frac{\underline{U}(\omega)}{\underline{I}(\omega)} = \left|\underline{Z}(\omega)\right| e^{j\phi(\omega)} = \operatorname{Re}\underline{Z} + j\operatorname{Im}\underline{Z} = Z'(\omega) + jZ''(\omega) \tag{7}$$

$$\left|\underline{T}(\omega)\right| = \underline{Y}(\omega) = \underline{Z}^{-1}(\omega) = \frac{1}{Z}e^{-j\varphi(\omega)} = Ye^{-j\varphi} = \operatorname{Re}\underline{Y} + j\operatorname{Im}\underline{Y} = Y'(\omega) + jY''(\omega) \tag{8}$$

where

$$|\underline{Z}| = Z = \sqrt{\left(Z^{'}\right)^2 + \left(Z^{''}\right)^2}\ ,\ |\underline{Y}| = Y = \sqrt{\left(Y^{'}\right)^2 + \left(Y^{''}\right)^2}\ ,\ \phi(\omega) = arctg\frac{Z^{''}(\omega)}{Z^{'}(\omega)}$$

In the IS method we are not limited only to frequency analysis of impedance or admittance of examined material (their amplitude and phase characteristics or dependence of real and imaginary parts of the complex transmittance upon frequency), but we can also use other quantities, such as the complex capacitance $\underline{C}$, or the complex permittivity $\underline{\varepsilon}$. These quantities are bonded with the complex admittance by frequency and geometrical parameters of sample.
The complex admittance, which can be associated to a simple equivalent scheme in a parallel connection configuration of two ideal elements - resistance R and capacitance C, is explained by a formula:

$$\underline{Y} = \frac{1}{R} + j\omega C = G + j\omega C \tag{9}$$

Transformation of expression (9) allows to introduce a term of the $\underline{C}$ complex capacitance:

$$\underline{C}(\omega) = \frac{\underline{Y}(\omega)}{j\omega} = C' - jC'' \tag{10}$$

If we assume, that examined material of d thickness is placed between two flat parallel electrodes of area S, that is the geometrical capacitance of capacitor electrodes configuration $Co = \varepsilon_o S/d$ is given, then a knowledge of this parameter allows to define the relative permittivity in a complex form:

$$\varepsilon(\omega) = C(\omega) \ /\ C_o = C(\omega)\ d\ /\ \varepsilon_o S = \varepsilon' + j\varepsilon'' \tag{11}$$

The capacitance and the relative permittivity in a complex form are the most often used form of the spectral transmittance in the IS method. Their frequency analysis shows multiple phenomena, which take place simultaneously in an examined material.
Below the examinations of an answer of dielectric composite materials made of nonwoven fabrics covered by thin plasmatic layers of titanium and titanium monoxide as a representative for composites described in this chapter are shown. For these examinations the measurement system from the High Voltage Group of Wrocław University of Technology was used. Its main elements are the precise impedance analyser Agilent 4294A and the measuring cell Agilent 16451B consisting of the three-electrode measuring capacitor. Table 1 includes the description of samples made of alternately covered titanium and titanium monoxide layers on nonwoven fabric.
Bode diagrams of examined materials are shown on figure 16. The frequency spectrum of the modulus of impedance does not indicate significant differences between examined samples. It varies for all examined samples in the same way, and additionally for samples 3TiO and Ti overlaps in almost entire spectrum of frequencies.

	Ti	TiO	Ti	TiO	Ti	TiO
3TiO	•	•	•	•	•	
2TiO		•	•	•		
1TiO		•	•	•	•	•
Ti	•					

Table 1. Description of samples made of titanium and titanium monoxide layers alternately deposited on a textile

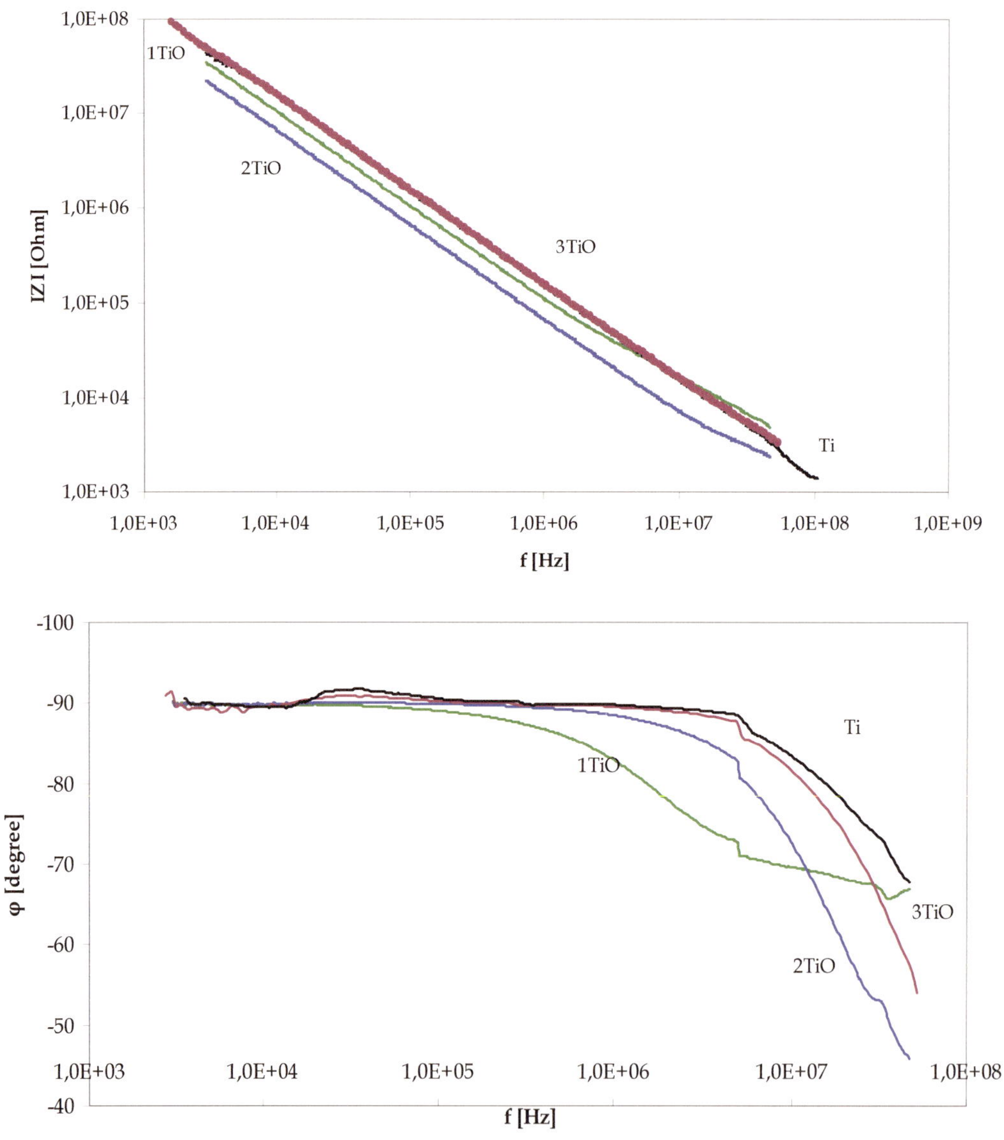

Fig. 16. The Bode diagram of examined composite structures.

The Bode diagram analysis suggests a possibility to use the equivalent scheme consisted of a parallel connection RC and also a presence of the Maxwell-Wagner polarisation to describing the polarisation phenomena. It is possible to find analogies to examinations outcomes of some laminar structures (Nitsch, 1999).

More pieces of information about dielectric properties may be given by a form analysis of the complex capacitance presented as a function of frequency and on the complex plane (the Cole-Cole diagram). Figures 17 and 18 present frequency spectrums of examined complex capacitance. An incline of real component curves of the complex capacitance for 1TiO and 2TiO samples can be seen there, what suggests a presence of at least one relaxation mechanism in the examined frequency range.

Diagram forms for Ti and 3TiO samples suggest a possibility of a presence of relaxation processes in frequencies above the measuring range. It is confirmed on diagrams of the imaginary component of the complex capacitance presented on figure 18. We can observe the presence of a relaxation pick for 1TiO sample for frequency under 10^7 Hz also for 2TiO sample for frequency approximately $5x10^7$ Hz.

The Cole-Cole diagram of the complex capacitance of examined composite structures is presented on figure 19. The diagram has a form of deformed semicircles with clearly displaced centres. Each of the examined samples distinguishes then with a specific relaxation process of different time-constants. Furthermore we can observe a presence of the second semicircle for each of examined composites, what clearly suggests the presence of another relaxation process at higher frequencies.

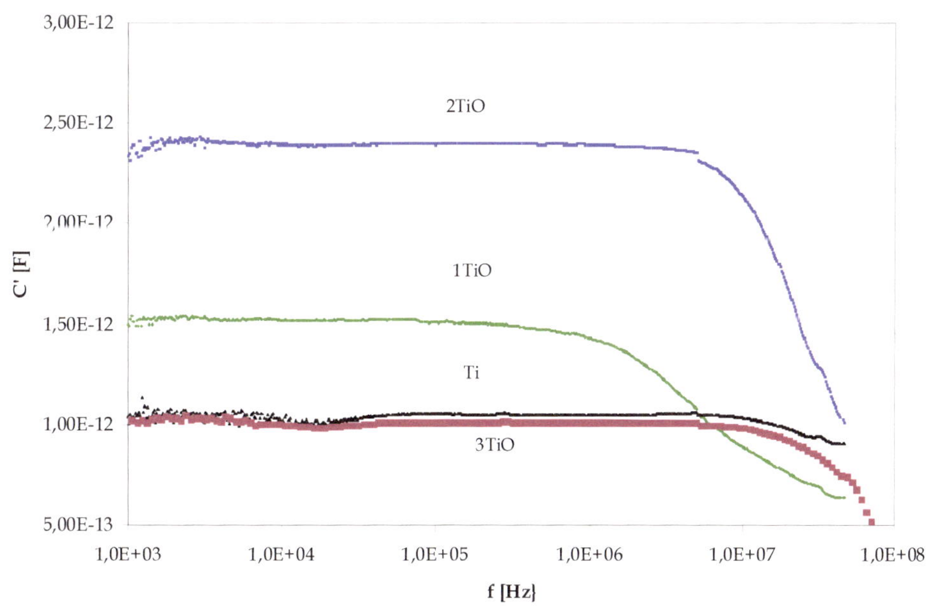

Fig. 17. Frequency spectrums of real component of complex capacitance.

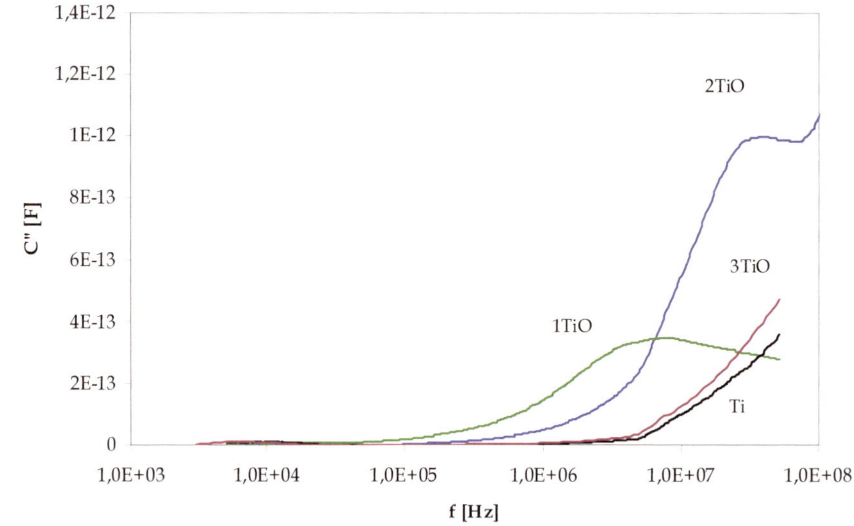

Fig. 18. Frequency spectrums of imaginary component of complex capacitance.

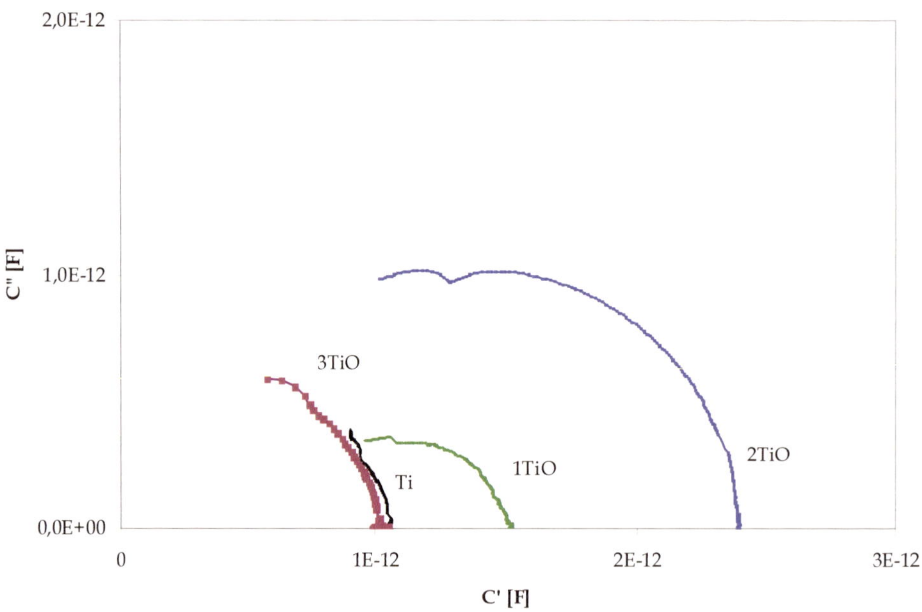

Fig. 19. The Cole-Cole diagram of the complex capacitance

In the outcomes presented above, one can notice some general rules. We can observe a clear separation between dielectric properties of composites, in which the outer layer is formed by

titanium (Ti), and of composites with the outer layer formed by a titanium monoxide (TiO). For composites with outer titanium layer (Ti and 3TiO samples) in the measuring frequency range, we observe very little differences in frequency spectrums of measured parameters and in the Cole-Cole diagrams independently from a number of layers Ti-TiO.
In the diagram form of the imaginary component of the complex capacitance as a function of frequency, we do not observe a relaxation pick. At the same time, the diagram form of this characteristic suggests the presence of a relaxation pick at frequencies higher than the measuring range. Quite a different situation for composites with the outer layer formed by a titanium monoxide (1TiO and 2TiO samples) appears. In this case we can observe the strong dependence of dielectric composite properties upon a number of formed Ti-TiO layers.
Frequency diagram forms of the imaginary component of the complex capacitance shows, in the examined measuring frequency range, the presence of a relaxation pick and a possibility of a presence of the second relaxation phenomenon at higher frequencies. The value and frequency of a relaxation pick presence are strictly depended on a number of Ti-TiO layers forming a composite. Increasing a number of layers results in reducing of a relaxation pick value and in displacement in the lower frequencies direction. It is confirmed by Cole-Cole diagrams of the complex capacitance, in which there is a clear presence of a displacement of the semicircle centre to the right for 1TiO sample.
The capability of composite materials to shield electromagnetic fields is coherently associated with their dielectric properties in a wide frequency band. The method of impedance spectroscopy allows one to connect the measured frequency characteristics with the physical structure of tested material and the alternations in the structure.
The method has been used by the authors to determine the connection between surface structure of a fabric being a substrate and dielectric properties of obtained composite fabric -carbon (Jaroszewski et al., 2010) and to evaluate the correlation between dielectric response of the system and surface resistance of the carbon layer (Pospieszna et al., 2010, Pospieszna & Jaroszewski, 2010). The possibilities to design desired electric properties of composite materials are also used to improve the shielding properties of the materials. Thus, the connection of the impedance spectroscopy method with those properties.

6. Summary

It should be noted that the performed studies and collected experience in the field of modern technologies of shielding have already solved a lot of actual problems but there is still a challenge for further work to improve the efficiency of shielding and to develop new designs of electromagnetic shields. They can also be used in the shielding of power engineering systems, where a compatibility with environment in a wide sense of this meaning is the main problem (i.e. not only in the aspect of emission and electromagnetic disturbances). In the light of the latest experiences it seems that the future in the area of EM field shielding is connected with the application of modern technologies to fabricate thin-film composite coatings, including nano-composites. The materials are capable to fulfil all conditions of effective shielding from EM fields and to eliminate all undesired occurrences associated with operation of the shielded systems. The results of our investigations, presented above, point out the possibility of industrial fabrication of the composite shielding materials with the coefficient of shielding efficiency exceeding 50 dB. Good mechanical properties and high resistance to environmental effects are additional advantages of such materials.

7. Acknowledgment

This publication was prepared with the key project - POIG no. 01.03.01-00-006/08 co-financed from the founds of European Regional Development Found within the framework of the Operational Programme Innovative Economy.

8. References

Bula K., Koprowska J., Janukiewicz J. (2006). *Application of Cathode Sputtering for Obtaining Ultra-thin Metallic Coatings on Textile Products*, Fibres & Textiles in EE, Vol. 14, No. 5 (59) (2006) pp.75 - 79

Holloway C. L., Sarto M. S., Johansson M., (2005). *Analyzing Carbon-Fiber Composite Materials with Equivalent-Layer Models*, IEEE Trans. on Electromagnetic Compatibility, 47, n.4, 833-844,

Huang Yi, Ning Li, Yanfeng Ma, Feng Du, Feifei Li , Xiaobo He ,Xiao Lin, Hongjun Gao, Yongsheng Chen, (2007). *The influence of single-walled carbon nanotube structure on the electromagnetic interference shielding efficiency of its epoxy composites*, Carbon 45 1614–1621

Hong Y.K., Lee C.Y., Jeong C.K., Sim J.H., Kim K., J oo J ., Kim M.S., Lee J.Y., Jeong S.H., Byun S.W., (2001). *Electromagnetic interference shielding characteristics of fabric complexes coated with conductive polypyrrole and thermally evaporated Ag*, Current Applied Physics 1 439–442

Jaroszewski M., Ziaja J. (2010). *Zinck-unvowen fabric composite obtained by magnetron sputtering*, Proceedings of Twelfth International Conference on Plasma Surface Engineering; September 13 - 17, 2010, PSE 2010, Garmisch-Partenkirchen, Germany, PSE 2010

Jaroszewski M., Pospieszna J., Ziaja J. (2010). *Dielectric properties of polypropylene fabrics with carbon plasma coatings for applications in the technique of electromagnetic field shielding*, J. Non-Cryst. Solids, Volume 356, Issues 11-17, 2010, 625-628

Kim H.M, Kim K., Lee S.J. ; Joo J.,*, Yoon H.S., Cho S.J., Lyu S.C., Lee C.J., (2004). *Charge transport properties of composites of multiwalled carbon nanotube with metal catalyst and polymer: application to electromagnetic interference shielding*, Current Applied Physics 4 577–580

Koprowska J., Ziaja J., Janukiewicz J. (2008). *Plasma Metallization Textiles as Shields for Electromagnetic Fields*, EMC Europe 2008, Hamburg, Germany, September 8-12, 2008, pp. 493-496

Koprowska J., Pietranik M., Stawski W. (2004). *New Type of Textiles with Shielding Properties*, Fibres &Textiles in Eastern Europe, vol. 12, (2004), n.3 (47), 39-42

Ning Li, Yi Huang, Feng Du, Xiaobo He, Xiao Lin, Hongjun Gao, Yanfeng Ma, Feifei Li, Yongsheng Chen, Peter C. Eklund, (2006) *Electromagnetic Interference (EMI) Shielding of Single-Walled Carbon Nanotube Epoxy Composites*, Nano Letters, 2006,, Vol. 6, No 5, 1141-1145

Nitsch K. (1999). *Application of impedance spectroscopy in the study of electronic materials*, Wroclaw University of Technology Press, in Polish, ISBN 83-7085-417-6

Pospieszna J., Jaroszewski M., Bretuj W. (2010). Tchórzewski M. *Influence of surface and volume electrical resistivity on dielectric properties of carbon-polypropylene fabric composite obtained by plasma deposition*, Electrotech. Rev. 2010, R. 86, nr 5, pp. 275-278

Pospieszna J., Jaroszewski M., Szafran G. (2010). *Influence of substratum on dielectric properties of plasma carbon films*, presented at X Symposium on High-Voltage Engineering IW2010, 7-9 Jun 2010, Poznań-Będlewo

Pospieszna J., Material advances in electromagnetic field shielding technology. (2006) Electrotechnical Review, n. 1, 2006, 205-207,

Sarto F., Sarto M.S., Larciprete M.C., Sibilia C. (2003). *Transparent films for electromagnetic shielding of plastics*, Rev. Adv. Mater. Sci., (2003), n.5, 329-336

Sarto F., Sarto M. S., Larciprete M.C., Sibilia C., (2004). *Electromagnetics of nanolayered transparent metals*, Conference materials URSI EMTS 2004, 683-684

Sarto M. S., Li Voti R., Sarto F., Larciprete M. C. (2005). *Nanolayered Lightweight Flexible Shields with Multidirectional Optical Transparency*, IEEE Trans. on EMC, vol. 47, No 3, (2005) pp.602- 611

Schulz R. B., Plantz V. C., Brusch D. R., *Shielding Theory and Practice* ;(1998). IEEE Transactions On Electromagnetic Compatibility, VOL. 30, NO. 3, AUGUST 1988, 187-201

Tzong-Lin Wu, Wern-Shiarng Jou, S. G. Dai, Wood-Hi Cheng,(2006). *Effective Electromagnetic Shielding of Plastic Packaging in Low-Cost Optical Transceiver Modules*, Journal of Lightwave Technology, VOL. 21, NO. 6, JUNE 2003, 1536-1542

Wang Li-Li, Tay Beng-Kang, See Kye-Yak, Sun Zhuo, Tan Lin-Kin, Lua Darren (2009). *Electromagnetic interference shielding effectiveness of carbon-based materials prepared by screen printing*. Carbon 47, s. 1905-1910

Wei Q. F., Xu W. Z., Ye H., Huang F. L. (2006). *Surface Functionalization of Polymer Fibres by Sputtering Coating*, J. Industrial Textiles , Vol. 35 No. 4 (2006) pp.287-294

Wojkiewicz J. L., Hoang N. N., Redon N., Miane J. L., (2005). *Intrinsically Conducting Nanocomposites: High Performance Electomagnetic Shielding Materials*, VIth Int. Symp. on Electromagnetic Compatibility and Electromagnetic Ecology, St Petersburg, Russia ; pp. 58-61

Ziaja J., Ozimek M., Janukiewicz J. (2010). *Application of thin films prepared by impulse magnetron sputtering for shielding of electromagnetic fields*, Electrotech. Rev. 2010, R. 86, nr 5, pp. 222-224

Ziaja J., Ozimek M., Koprowska J. (2009). *Metallic and oxide Zn and Ti layers on textile as shields for electromagnetic fields*, EMC Europe 2009 Workshop, Athens, Greece, 11-12 June 2009, pp. 30-33

Ziaja J., Koprowska J., Janukiewicz J. (2008). *The use of plasma metallization in the manufacture of textile screens for protection against electromagnetic fields*, Fibres & Textiles in Eastern Europe. 2008, vol. 16, nr 5, pp. 64-66

Ziaja J., Koprowska J., Janukiewicz J., (2008a). *Using of plasma metallization for fabrication of fabric screens against electromagnetic field*, FIBRES & TEXTILES in Eastern Europe 5, s. 70-72

Ziaja J., Koprowska J., Żyłka P. (2008b). *Influence of nonwoven structures on surface resistivity of plasma titanium films*. Proceedings of 6th International Conference ELMECO-6 :

electromagnetic devices and processes in environment protection joint with 9th Seminar "Applications of Superconductors" AoS-9, Nałęczów, Poland, June 24-27, 2008. s. 95-96

Ziaja J. *ZnO thin film deposition with pulsed magnetron sputtering*. (2007). Electrotechnical Review. 2007, R. 83, nr 11, s. 235-239

21

Reduction of Reflection from Conducting Surfaces using Plasma Shielding

Çiğdem Seçkin Gürel and Emrah Öncü
[1]*Department of Electrical and Electronics Engineering, Hacettepe University,*
[2]*Communication Systems Group, TUBITAK Space Technologies Research Institute,*
Turkey

1. Introduction

Plasma mediums have taken considerable interest in recent studies due to their tunable characteristics offering some advantages in radio communications, radio astronomy and military stealth applications. Special plasma mediums have been used as electromagnetic wave reflectors, absorbers and scatterers. Reflection, absorbtion and transmission of electromagnetic waves by a magnetized nonuniform plasma slab are analysed by different authors using different methods in literature. It is known that plasma parameters such as length, collision frequency and electron density distribution function considerably affect plasma response. Among those, especially the electron density distribution considerably affects the frequency selectivity of the plasma (Gurel & Oncu, 2009a, 2009b, 2009c). Conducting plane covered with plasma layer has been considered and analysed in literature for some specific density distribution functions such as exponential and hyperbolic distributions (Shi et al., 2001; Su et al., 2003 J. Zhang & Z. Liu, 2007). The effects of external magnetic field applied in different directions to the plasma are also important and considered in those studies.

In order to analyze the characteristics of electromagnetic wave propagation in plasma, many theoretical methods have been developed. Gregoire et al. have used W.K.B approximate method to analyze the electromagnetic wave propagation in unmagnetized plasmas (Gregoire et al., 1992) and Cao et al. used the same method to find out the absorbtion characteristics of conductive targets coated with plasma (Cao et al., 2002). Hu et al. analyzed reflection, absorbtion, and transmission characteristics from nonuniform magnetized plasma slab by using scattering matrix method (SMM) (Hu et al., 1999). Zhang et al. and Yang et al. used the recursion formula for generalized reflection coefficient to find out electromagnetic wave reflection characteristics from nonuniform plasma (Yang et al., 2001; J. Zhang & Z. Liu, 2007). Liu et al. used the finite difference time domain method (FDTD) to analyze the electromagnetic reflection by conductive plane covered with magnetized inhomogeneous plasma (Liu et al., 2002).

The aim of this study is to determine the effect of plasma covering on the reflection characteristics of conducting plane as the function of special electron density distributions and plasma parameters. Plasma covered conducting plane is taken to model general stealth application and normally incident electromagnetic wave propagation through the plasma medium is assumed. Special distribution functions are chosen as linearly varying electron density distribution having positive or negative slopes and purely sinusoidal distribution which have shown to provide wideband frequency selectivity characteristics in plasma shielding applications in recent studies (Gurel & Oncu, 2009a, 2009b, 2009c). It is shown that linearly varying profile with positive and negative slopes can provide adjustable reflection or absorbtion performances in different frequency bands due to proper selection of operational parameters. Sinusoidally-varying electron distribution with adjustable phase shift is also important to provide tunable plasma response. The positions of maximums and minimums of the electron number density along the slab can be changed by adjusting the phase of the sinusoid as well as the other plasma parameters. Thus plasma layer can be tuned to behave as a good reflector or as a good absorber. In this study, plasma is taken as cold, weakly ionized, steady state, collisional, nonuniform while background magnetic field is assumed to be uniform and parallel to the magnetized slab.

2. Physical model and basic theory

There are several theoretical methods as mentioned in the previous section for the analysis of electromagnetic wave propagation through the plasma which will be summarized in this part.

2.1 Generalized reflection coefficient formula

Firstly two successive subslabs of plasma layer are considered as shown in Fig. 1.

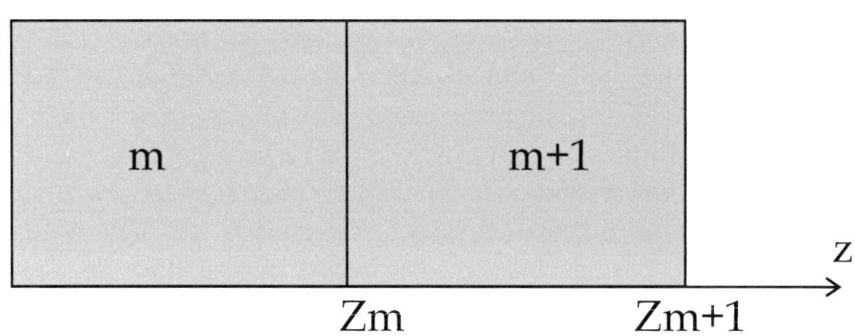

Fig. 1. Two successive plasma subslabs.

The incident and reflected field equations for the mth subslab can be written as

$$E_i(m,z) = e_i(m)\exp\{-jk(m)(z - Z_m)\} \tag{1}$$

$$E_r(m,z) = e_r(m)\exp\{jk(m)(z - Z_m)\} \tag{2}$$

where E_i is the incident field and E_r is the reflected field. Then, incident and reflected field equations for the $(m+1)^{th}$ subslab can be similarly given as

$$E_i(m+1,z)=e_i(m+1)\exp\{-jk(m+1)(z-Z_{m+1})\} \quad (3)$$

$$E_r(m+1,z)=e_r(m+1)\exp\{-jk(m+1)(z-Z_{m+1})\} \quad (4)$$

Then total electric field in the mth subslab is

$$E_y(m,z)=e_i(m)\exp\{-jk(m)(z-Z_m)\}+e_r(m)\exp\{jk(m)(z-Z_m)\} \quad (5)$$

and for $(m+1)^{th}$ subslab

$$E_y(m+1,z)=e_i(m+1)\exp\{-jk(m+1)(z-Z_{m+1})\}+e_r(m+1)\exp\{jk(m+1)(z-Z_{m+1})\} \quad (6)$$

After getting the electric field equations, magnetic field equations are obtained as follows

$$H_x(m,z)=\frac{1}{\eta_m}\left[-e_i(m)\exp\{-jk(m)(z-Z_m)\}+e_r(m)\exp\{jk(m)(z-Z_m)\}\right] \quad (7)$$

$$H_x(m+1,z)=\frac{1}{\eta_{m+1}}\left[-e_i(m+1)\exp\{-jk(m+1)(z-Z_{m+1})\}\right]+ \\ +e_r(m+1)\exp\{jk(m+1)(z-Z_{m+1})\} \quad (8)$$

where η_m and η_{m+1} are the intrinsic impedances of mth and $(m+1)^{th}$ subslabs respectively. The intrinsic impedance for the mth subslab is

$$\eta_m=\sqrt{\frac{\mu_0\mu_{r_m}}{\varepsilon_o\varepsilon_{r_m}}} \quad (9)$$

To match the boundary conditions at $z=Z_m$, following two equations can be written

$$E_y(m,z)=E_y(m+1,z) \quad (10)$$

$$H_x(m,z)=H_x(m+1,z) \quad (11)$$

Then (5) and (6) are inserted into equation (10), and it is obtained that

$$e_i(m)\exp\{-jk(m)(z-Z_m)\}+e_r(m)\exp\{jk(m)(z-Z_m)\}=e_i(m+1). \\ .\exp\{-jk(m+1)(z-Z_{m+1})\}+e_r(m+1)\exp\{jk(m+1)(z-Z_{m+1})\} \quad (12)$$

Since $z=Z_m$ at the boundary, equation (12) can be arranged as below

$$e_i(m)+e_r(m)=\exp\{jk(m+1)d_{m+1}\}\left[e_i(m+1)\exp\{-2jk(m+1)d_{m+1}+e_r(m+1)\}\right] \quad (13)$$

where d_{m+1} is the thickness of the $(m+1)^{th}$ subslab.
By using the following equalities,

$$\exp\left\{jk(m+1)d_{m+1}\right\} = A \tag{14}$$

$$\exp\left\{-2jk(m+1)d_{m+1}\right\} = B \tag{15}$$

Equation (13) becomes

$$e_i(m) + e_r(m) = A\left[e_i(m+1)B + e_r(m+1)\right] \tag{16}$$

By inserting equations (7) and (8) into (11), it is obtained that

$$\begin{aligned}&\frac{1}{\eta_m}\left[-e_i(m)\exp\left\{-jk(m)(z-Z_m)\right\} + e_r(m)\exp\left\{jk(m)(z-Z_m)\right\}\right] = -\frac{1}{\eta_{m+1}}e_i(m+1).\\&\exp\left\{-jk(m+1)(z-Z_{m+1})\right\} + \frac{1}{\eta_{m+1}}e_r(m+1)\exp\left\{jk(m+1)(z-Z_{m+1})\right\}\end{aligned} \tag{17}$$

Then by replacing $z = Z_m$ in (17) it is obtained that

$$\frac{1}{\eta_m}\left[-e_i(m) + e_r(m)\right] = \frac{1}{\eta(m+1)}\left[-e_i(m+1)\exp\begin{pmatrix}-jk(m+1)d_{m+1} + e_r(m+1).\\ \exp\left(jk(m+1)d_{m+1}\right)\end{pmatrix}\right] \tag{18}$$

By relating the intrinsic impedance to permittivity and arranging the equation (18),

$$-e_i(m) + e_r(m) = \sqrt{\frac{\varepsilon_r(m+1)}{\varepsilon_r(m)}}\exp\left(jk(m+1)d_{m+1}\begin{bmatrix}e_i(m+1)\exp\left(-2jk(m+1)d_{m+1}\right)+\\ +e_r(m+1)\end{bmatrix}\right) \tag{19}$$

Now, the final equation is obtained as

$$-e_i(m) + e_r(m) = A\left[-e_i(m+1)B + e_r(m+1)\right]C \tag{20}$$

where

$$C = \sqrt{\frac{\varepsilon_r(m+1)}{\varepsilon_r(m)}} \tag{21}$$

Then equations (16) and (20) are combined and expressed in matrix form as

$$\begin{bmatrix}e_r(m)\\ e_i(m)\end{bmatrix} = \frac{1}{2}A\begin{bmatrix}B(1+C) & (1-C)\\ B(1-C) & (1+C)\end{bmatrix}\begin{bmatrix}e_r(m+1)\\ e_i(m+1)\end{bmatrix} \tag{22}$$

$$S_m = \frac{1}{2}A\begin{bmatrix}B(1+C) & (1-C)\\ B(1-C) & (1+C)\end{bmatrix} \tag{23}$$

For $m = n-1$

$$\begin{bmatrix}e_r(n-1)\\ e_i(n-1)\end{bmatrix} = S_{n-1}\begin{bmatrix}e_r(n)\\ e_i(n)\end{bmatrix} \tag{25}$$

where n is the last boundary of the plasma slab which is located before conductive target. For $m = n - 2$

$$\begin{bmatrix} e_r(n-2) \\ e_i(n-2) \end{bmatrix} = S_{n-2} \begin{bmatrix} e_r(n-1) \\ e_i(n-1) \end{bmatrix} \tag{25}$$

When we continue to write the field equations iteratively until m=0 which means the boundary between free space and the first subslab of plasma, we have

$$\begin{bmatrix} e_r(0) \\ e_i(0) \end{bmatrix} = -S_0 S_1 S_2 S_3 ....S_{n-1} \begin{bmatrix} e_r(n) \\ e_i(n) \end{bmatrix} \tag{26}$$

This can be written in the following compact form,

$$\begin{bmatrix} e_r(0) \\ e_i(0) \end{bmatrix} = \left(\prod_{m=0}^{n-1} S_m \right) \begin{bmatrix} e_r(n) \\ e_i(n) \end{bmatrix} \tag{27}$$

Letting

$$\left(\prod_{m=0}^{n-1} S_m \right) = \begin{bmatrix} M_1 & M_2 \\ M_3 & M_4 \end{bmatrix} = M \tag{28}$$

Then by inserting equation (28) into equation (27)

$$\begin{bmatrix} e_r(0) \\ e_i(0) \end{bmatrix} = \begin{bmatrix} M_1 & M_2 \\ M_3 & M_4 \end{bmatrix} \begin{bmatrix} e_r(n) \\ e_i(n) \end{bmatrix} \tag{29}$$

Hence

$$e_r(0) = M_1 e_r(n) + M_2 e_i(n) \tag{30}$$

$$e_i(0) = M_3 e_r(n) + M_4 e_i(n) \tag{31}$$

By dividing the both sides of the equations by $e_i(n)$, the following two equations are obtained

$$\frac{e_r(0)}{e_i(n)} = M_1 \frac{e_r(n)}{e_i(n)} + M_2 \tag{32}$$

$$\frac{e_i(0)}{e_i(n)} = M_3 \frac{e_r(n)}{e_i(n)} + M_4 \tag{33}$$

Then by dividing these two equations side by side, we get (J. Zhang & Z. Liu, 2007)

$$\frac{e_r(0)}{e_i(0)} = \frac{M_1 \Gamma(n) + M_2}{M_3 \Gamma(n) + M_4} \tag{34}$$

where $\Gamma(n)$ is the reflection coefficient of the conductive target. In order to calculate the total reflection coefficient, the matrix M is needed to be computed.

2.2 Wentzel-Kramers-Brillouin (WKB) approximate method

It is known that WKB method is used for finding the approximate solutions to linear partial differential equations that have spatially varying coefficients. This mathematical approximate method can be used to solve the wave equation that defines the electromagnetic wave propagation in a dielectric plasma medium.
Let us write the wave equation as

$$\frac{d^2E}{dz^2} + k_z^{\ 2}E_z = 0 \tag{35}$$

Then WKB method can be applied to derive an approximate solution (Gregoire et al., 1992),

$$E_Z = E_0 \exp\left(j\int_0^z k(z')dz' \right) \tag{36}$$

This solution is valid in any region where

$$\frac{1}{k^2}\frac{dk}{dz} \prec\prec 1 \tag{37}$$

The pyhsical meaning of (37) is that the wavenumber of the propagating electromagnetic wave changes very little over a distance of one wavelength.
It is assumed that the electromagnetic wave enters the plasma at $z = z_0$ and reflects back at $z = z_1$. Then total reflected power can be found by WKB approximation (Gregoire et al., 1992) as

$$\overline{P}_r = \exp\left(-4\int_{z_0}^{z_1} z_1 \operatorname{Im}\left(k(z')\right)dz' \right) \tag{38}$$

where $\overline{P}_r$ is the normalized total reflected power.

2.3 Finite-difference time-domain analysis

Finite-difference time-domain analysis have been extensively used in literature to solve the electromagnetic wave propagation in various media (Hunsberger et al., 1992; Young, 1994, 1996; Cummer, 1997; Lee et al., 2000; M. Liu et al., 2007). When the electromagnetic wave propagates in a thin plasma layer, the W.K.B method may not accurately investigate the wave propagation (X.W. Hu, 2004; S. Zhang et al., 2006). The reason is the plasma thickness is near or less than the wavelength of the plasma exceeds the wavelength of the incident wave, the variation of the wave vector with distance cannot be considered as weak (M. Liu et al., 2007).
In the analysis electric field is considered in the x direction and propagation vector is in z direction and the electromagnetic wave enters normally into the plasma layer.

Lorentz equation (electron momentum equation) and the Maxwell's equations can be written as

$$n_e m_e \frac{\partial \vec{v}_e}{\partial t} = -en_e E - n_e m_e \nu_{cl} \vec{v}_e \tag{39}$$

$$\frac{\partial \vec{H}}{\partial t} = -\frac{1}{\mu_0} \nabla \times \vec{E} \tag{40}$$

$$\frac{\partial \vec{E}}{\partial t} = \frac{1}{\varepsilon_0}\left(\nabla \times \vec{H} - \vec{J}\right) \tag{41}$$

$$\vec{J} = -en_e \vec{v}_e \tag{42}$$

where $\vec{E}$ is the electric field vector, $\vec{H}$ is the magnetic field vector, ε_0 is the permittivity, μ_0 is the magnetic permeability of free space, $\vec{J}$ is the current density, n_e is the denstiy of electron, m_e and $\vec{v}_e$ are the mass and velocity vector of the electron, respectively and ν_{cl} is collision frequency. Then FDTD algorithm of equations (39), (40), (41) and (42) can be written as (Chen et al., 1999; Jiang et al., 2006; Kousaka & Ono, 2002; M.H. Liu et al., 2006)

$$H^{n+\frac{1}{2}}_{y_i+\frac{1}{2},j,k+\frac{1}{2}} = H^{n-\frac{1}{2}}_{y_i+\frac{1}{2},j,k+\frac{1}{2}} - \frac{\Delta t}{\mu} \frac{E^n_{x_i+\frac{1}{2},j,k+1} - E^n_{x_i+\frac{1}{2},j,k}}{\Delta z} \tag{43}$$

$$E^{n+1}_{x_i+\frac{1}{2},j,k} = E^n_{x_i+\frac{1}{2},j,k} - \frac{\Delta t}{\varepsilon}\left(\frac{H^{n+\frac{1}{2}}_{y_i+\frac{1}{2},j,k+\frac{1}{2}} - H^{n+\frac{1}{2}}_{y_i+\frac{1}{2},j,k-\frac{1}{2}}}{\Delta z} + J^{n+\frac{1}{2}}_{x_i+\frac{1}{2},j,k} \right) \tag{44}$$

$$J^{n+\frac{1}{2}}_{x_i+\frac{1}{2},j,k} = -en^{n+\frac{1}{2}}_{e_i+\frac{1}{2},j,k} v_e{}^{n+\frac{1}{2}}_{x_i+\frac{1}{2},j,k} \tag{45}$$

$$v_e{}^{n+\frac{1}{2}}_{x_i+\frac{1}{2},j,k} = \frac{1}{\left(2+\nu_{cl}.\Delta t\right)}\left[\left(2-\nu_{cl}.\Delta t\right) v_e{}^{n-\frac{1}{2}}_{x_i+\frac{1}{2},j,k} - \frac{2e\Delta t}{m_e} E^n_{x_i+\frac{1}{2},j,k}\right] \tag{46}$$

where Δz is the spatial discretization and Δt is the time step. By using equations (43) to (46), the electromagnetic wave propagation in a plasma slab can be simulated in time domain (M. Liu et al., 2007).

2.4 Scattering matrix method (SMM) analysis

This analytical technique is the manipulation of the 2x2 matrix approach which was presented by Kong (Kong, 1986). SMM analysis gives the partial reflection and transmission

coefficients in the subslabs. This makes it easy to analyze the partial absorbed power in each subslab of the plasma (Hu et al., 1999).
Let us write the incident and reflected fields as follows

$$E_y{}^i = E_0 \exp\left(-jk_z^0 z\right) \tag{47}$$

$$E_y{}^r = AE_0 \exp\left(jk_z{}^0 z\right) \tag{48}$$

where k_z^0 is the z component of the free space wave number and A is the reflection coefficient for the first subslab.
The total electric field in incidence region is

$$E_y^0 = E_0\left(\exp\left(-jk_z^0 z\right) + A\exp\left(jk_z^0 z\right)\right) \tag{49}$$

In the same manner, we can write the total electric field in m^{th} layer as

$$E_y^m = E_0\left(B_m \exp\left(-jk_z^m z\right) + C_m \exp\left(jk_z^m z\right)\right) \tag{50}$$

where B_m and C_m are the unknown coefficients.
After the last subslab there is only transmitted wave that travels in free space. The electric field for this region is

$$E_y^p = DE_0 \exp\left(-jk_z^p z\right) \tag{51}$$

where D is the unknown coefficient. After writing the total electric fields in each subslab, boundary conditions can be applied.
For the first boundary

$$\begin{pmatrix} B_1 \\ C_1 \end{pmatrix} = S_1 \begin{pmatrix} A \\ 1 \end{pmatrix} \tag{52}$$

where

$$S_1 = \frac{1}{2k_z^1}\begin{pmatrix} k_z^1 - k_z^0 & k_z^1 + k_z^0 \\ k_z^1 + k_z^0 & k_z^1 - k_z^0 \end{pmatrix} \tag{53}$$

For the mth boundary,

$$\begin{pmatrix} B_m \\ C_m \end{pmatrix} = S_m \begin{pmatrix} B_{m-1} \\ C_{m-1} \end{pmatrix} \tag{54}$$

where

$$S_m = \begin{pmatrix} \exp\left(-jk_z^m d_m\right) & \exp\left(jk_z^m d_m\right) \\ k_z^m \exp\left(-jk_z^m d_m\right) & -k_z^m \exp\left(jk_z^m d_m\right) \end{pmatrix}^{-1} \times \begin{pmatrix} \exp\left(-jk_z^{m-1} d_m\right) & \exp\left(jk_z^{m-1} d_m\right) \\ k_z^{m-1} \exp\left(-jk_z^{m-1} d_m\right) & -k_z^{m-1} \exp\left(jk_z^{m-1} d_m\right) \end{pmatrix} \tag{55}$$

Lastly, for the last boundary

$$\begin{pmatrix} B_n \\ C_n \end{pmatrix} = V_p.D \tag{56}$$

where

$$V_p = \frac{1}{2k_z^n} \begin{pmatrix} \left(k_z^n + k_z^p\right)\exp\left(j\left(k_z^n - k_z^p\right)d_p\right) \\ \left(k_z^n - k_z^p\right)\exp\left(-j\left(k_z^n + k_z^p\right)d_p\right) \end{pmatrix} \tag{57}$$

By using equations (52) and (54), equation (56) can be written as,

$$S_g \begin{pmatrix} A \\ 1 \end{pmatrix} = V_p D \tag{58}$$

where S_g is the global scattering matrix and can be written as,

$$S_g = \left(S_{g1}, S_{g2}\right) \tag{59}$$

where S_{g1} represents the first column vector and S_{g2} represents the last column vector of the global scattering matrix. Then equation (58) can be written (Hu et al., 1999) as

$$\begin{pmatrix} A \\ D \end{pmatrix} = -\left(S_{g1} - V_p\right)^{-1} S_{g2} \tag{60}$$

By using equation (60), A and D coefficients can be computed. The coefficient A represents total reflection coefficient and the coefficient B represents total transmission coefficient. Absorbed power values for each subslab and the total absorbed power inside the plasma can be obtained by the help of equations (52), (54) and (56).

2.5 Formulation of reflection from plasma covered conducting plane

In this chapter another method is presented to analyze the characteristics of electromagnetic wave propagation in a plasma slab. This method is simple, accurate and provides less computational time as compared to other methods mentioned in previous sections.

Normally incident electromagnetic wave propagation through a plasma slab is assumed as shown in Fig. 2. In the analysis, inhomogenous plasma is divided into sufficiently thin, adjacent subslabs, in each of which plasma parameters are constant. Then starting with Maxwell's equations, reflected, absorbed and transmitted power expressions are derived. Here, plasma layer is taken as cold, weakly ionized, steady state and collisional. Background magnetic field is assumed to be uniform and parallel to the magnetized slab.

For a magnetized and source free plasma medium, plasma permittivity is in tensor form. This tensor form permittivity can be approximated by a scaler permittivity. Let us give the details of this approximation.

The equation of motion for an electron of mass m is

$$-mw^2\vec{r} + mv_{cl}jw\vec{r} = e\vec{E} + ejw\vec{r} \times \vec{B} \tag{61}$$

where w is the angular frequency, $\vec{r}$ is the distance vector, v_{cl} is the collison frequency and $\vec{B}$ is the magnetic field vector.

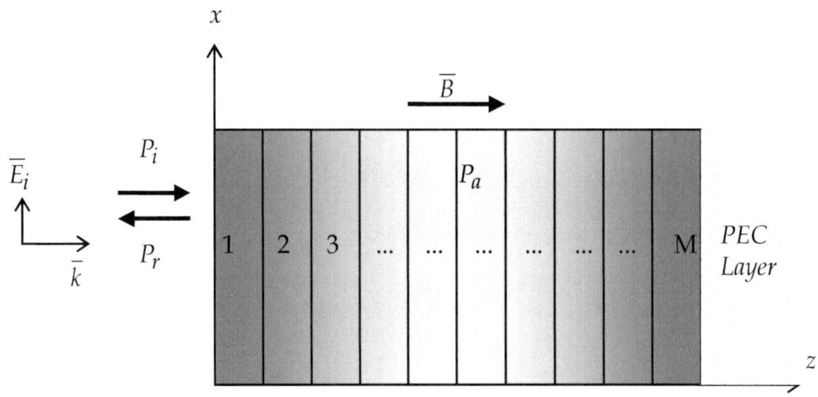

Fig. 2. Electromagnetic wave propagation through a plasma (with subslabs) covered conducting plane.

Then we insert polarization vector $\vec{P}$ into equation (61) and we get,

$$-w^2 m\vec{P} + jmv_{cl}w\vec{P} = Ne^2\vec{E} + jwe\vec{P}\times\vec{B} \tag{62}$$

Now inserting the following terms into equation (62)

$$X = w_N \ / \ w^2 = Ne^2 \ / \ (\varepsilon_0 m w^2) \tag{63}$$

$$\vec{Y} = \frac{e\vec{B}}{mw} \tag{64}$$

$$Z = \frac{v}{w} \tag{65}$$

Then, it is obtained that

$$\varepsilon_o X\vec{E} = \vec{P}(1 - jZ) - j\vec{Y}\times\vec{P} \tag{66}$$

In cartesian coordinates, equation (66) can be written as

$$-\varepsilon_0 X \begin{pmatrix} E_X \\ E_Y \\ E_Z \end{pmatrix} = \begin{pmatrix} U & jl_Z Y & -jl_Y Y \\ -jl_Z Y & U & jl_X Y \\ jl_Y Y & -jl_X Y & U \end{pmatrix} \begin{pmatrix} P_X \\ P_Y \\ P_Z \end{pmatrix} \tag{67}$$

where

$$U = 1 - jZ \tag{68}$$

In equation (69), l_x , l_y and l_z are the direction cosines of $\vec{Y}$. We can take the polarization matrix by using equation (67)

$$\vec{P} = \varepsilon_o [M] \vec{E} \tag{69}$$

where $[M]$ is the susceptibility tensor. Then dielectric tensor is obtained as

$$[\varepsilon] = \varepsilon_0 \{1 + [M]\} \tag{70}$$

If coordinate system is oriented such that $\vec{Y}$ is parallel to the xz plane, equation (67) becomes

$$-\varepsilon_0 X \begin{pmatrix} E_X \\ E_Y \\ E_Z \end{pmatrix} = \begin{pmatrix} U & jY_l & 0 \\ -jY_l & U & jY_t \\ 0 & -jY_t & U \end{pmatrix} \begin{pmatrix} P_X \\ P_Y \\ P_Z \end{pmatrix} \tag{71}$$

where Y_l is the longitudinal component and Y_t is the transverse component of $\vec{Y}$. In the direction of the electromagnetic wave we have,

$$D_Z = \varepsilon_0 E_Z + P_Z = 0 \tag{72}$$

where D_z is the z component of electric flux density vector.
By using equation (71)

$$-\varepsilon_0 X E_Z = -jY_t P_Y + UP_Z \tag{73}$$

If E_Z is eliminated by using equations (72) and (73), it is obtained that

$$(U - X)P_Z = jY_t P_Y \tag{74}$$

By using equation (69) and the following equations

$$\vec{D} = \varepsilon_O \vec{E} + \vec{P} \tag{75}$$

$$\rho = \frac{E_Y}{E_X} = -\frac{H_X}{H_Y} = \frac{D_Y}{D_X} \tag{76}$$

$$P_Y = \rho P_X \tag{77}$$

where ρ is the polarization ratio, we have

$$-\varepsilon_0 X E_X = (U + j\rho Y_l) P_X \tag{78}$$

$$-\varepsilon_0 X E_Y = \left\{ \rho U - jY_l - \rho Y^2{}_t (U - X)^{-1} \right\} P_X \tag{79}$$

When equations (78) and (79) are divided side by side, it is obtained that

$$\rho^2 - j\rho Y^2{}_t \left[(U - X) Y_l^2 \right]^{-1} + 1 = 0 \tag{80}$$

The solution of the equation (80) is given by

$$\rho = \frac{jY_t^2}{2(U-X)Y_l} \pm j\left[1 + \frac{Y_t^4}{4(U-X)^2 Y_l^2}\right]^{1/2} \tag{81}$$

By using equations (72) and (74)

$$E_Z = -\frac{1}{\varepsilon_0}\frac{jY_t}{U-X}\rho P_X \tag{82}$$

By the help of equation (72)

$$P_X = D_X - \varepsilon_0 E_X = \varepsilon_0(\mu^2 - 1)E_X \tag{83}$$

Then if it is inserted into equation (78)

$$\mu^2 - 1 = \frac{-X}{U + jY_l\rho} \tag{84}$$

By using equation (81)

$$\mu^2 = 1 - \frac{X(U-X)}{U(U-X) - 1/2Y_t^2 \pm \left\{ 1/4Y^4{}_t + Y_l^2(U-X)^2 \right\}^{1/2}} \tag{85}$$

By inserting the followings into equation (83)

$$X = w_p{}^2 / w^2 \tag{86}$$

$$Z = \frac{\nu}{w} \tag{87}$$

$$U = 1 - jZ \tag{88}$$

$$Y_t = \frac{w_{ce}}{w}\sin\theta \tag{89}$$

$$Y_l = \frac{w_{ce}}{w}\cos\theta \tag{90}$$

Finally Appleton's formula (Heald & Wharton, 1978) results as

$$\tilde{\varepsilon}_r = 1 - \frac{(\omega_p/\omega)^2}{\left[1 - j\frac{v_{en}}{\omega} - \frac{(\omega_{ce}/\omega)^2\sin^2\theta}{2\left(1 - \frac{\omega_p^2}{\omega^2} - j\frac{v_{en}}{\omega}\right)}\right] \pm \left[\frac{(\omega_{ce}/\omega)^4\sin^4\theta}{4\left(1 - \frac{\omega_p^2}{\omega^2} - j\frac{v_{en}}{\omega}\right)^2} + \frac{\omega_{ce}^2}{\omega^2}\cos^2\theta\right]^{1/2}} \tag{91}$$

where
$\tilde{\varepsilon}$ is the complex permittivity of the plasma,
ω_p is the plasma frequency,
ω is the angular frequency of the electromagnetic wave,
v_{en} is the collision frequency,
ω_{ce} is the electron gyrofrequency,
θ is the incident angle of the electromagnetic wave.
Plasma frequency ω_p and electron-cyclotron frequency ω_{ce} are given (Ginzburg, 1970) as

$$w_p{}^2 = e^2 \frac{N}{m\varepsilon_0} \tag{92}$$

$$w_{ce} = \frac{eB}{m} \tag{93}$$

where e is the charge of an electron , N is the electron number density, m is the mass of an electron, ε_0 is the permittivity of free space and B is the external magnetic field strength.
The presence of the $\pm$ sign in the Appleton's formula is due to two separate solutions for the refractive index. In the case of propagation parallel to the magnetic field, the '+' sign represents a left-hand circularly polarized mode, and the '-' sign represents a right-hand circularly polarized mode.
For an incident electromagnetic wave (that is $\theta = 0^o$), complex dielectric constant of the plasma can be simply determined from equation (91) as

$$\tilde{\varepsilon}_r = 1 - \frac{(\omega_p / \omega)^2}{1 - j\frac{v_{en}}{\omega} - \frac{\omega_{ce}}{\omega}} \tag{94}$$

Now, in order to analyze electromagnetic wave propagation in a plasma slab in Fig. 2, multiple reflections are taken into consideration as shown in Fig. 3.
Reflection coefficient $\Gamma(j,z)$ at the jth interface and total reflection coefficient at $z = -d$ interface for normal incidence case can be obtained as (Balanis, 1989)

$$\Gamma_{in}(j = 1, z = -d) = \Gamma_{12} + \frac{T_{12}T_{21}\Gamma_{23}e^{-2\gamma d}}{1 - \Gamma_{21}\Gamma_{23}e^{-2\gamma d}} \tag{95}$$

The relations between the reflection and transmission coefficients are given as

$$\Gamma_{21} = -\Gamma_{12} \tag{96}$$

$$T_{12} = 1 + \Gamma_{21} = 1 - \Gamma_{12} \tag{97}$$

$$T_{21} = 1 + \Gamma_{12} \tag{98}$$

When multiple reflections are ignored due to highly lossy plasma, by taking $|\Gamma_{12}| << 1$ and $|\Gamma_{23}| << 1$, reflection coefficient on the $(j+1)^{th}$ interface is obtained as (Balanis, 1989)

$$\Gamma_{in}(j+1) = \frac{\sqrt{\tilde{\varepsilon}_r(j)} - \sqrt{\tilde{\varepsilon}_r(j+1)}}{\sqrt{\tilde{\varepsilon}_r(j)} + \sqrt{\tilde{\varepsilon}_r(j+1)}} \tag{99}$$

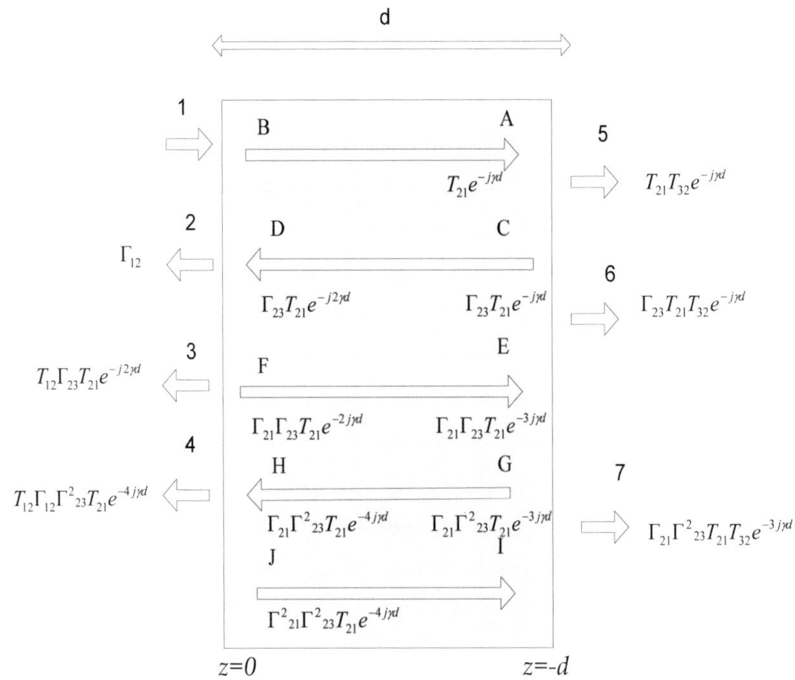

Fig. 3. Representation of multiple reflection in a subslab of the plasma layer (Balanis, 1989).

While the electromagnetic wave propagates through the plasma slab as seen in Fig. 2, reflection occurs at each interface of subslabs. The reflected electromagnetic wave from the first interface propagates in free space. Hence, there is no attenuation, reflected wave is given as

$$P_{r1} = P_i \times \Gamma(1)^2 \tag{100}$$

The power of the transmitted wave can be computed as

$$P_i = \Gamma(1)^2 \times P_i + P_{t1} \Rightarrow P_{t1} = P_i\left(1 - \Gamma(1)^2\right) \tag{101}$$

While reaching the second interface, the wave attenuates inside the slab

$$P_{t2} = e^{-2\alpha(1)d} P_i\left(1 - \Gamma(1)^2\right) \tag{102}$$

Some portion of the electromagnetic wave reflects back and some portion continues to propagate inside the plasma.

$$P_{r2} = P_i(1 - \Gamma(1)^2)\Gamma(2)^2 e^{-2\alpha(1)d} \tag{103}$$

The reflected portion of the wave attenuates until it reaches the free space. Hence, the power of the reflected wave is,

$$P_{r''} = P_i e^{-4\alpha(1)d}(1-\Gamma(1)^2)\Gamma(2)^2 \tag{104}$$

The transmitted portion also attenuates inside the plasma until it reaches the third interface

$$P_{t''} = P_i e^{-2\alpha(1)d}(1-\Gamma(1)^2)e^{-2\alpha(2)d}(1-\Gamma(2)^2) \tag{105}$$

Some portion reflects from the third interface as

$$P_{r'''} = P_i e^{-2\alpha(1)d}(1-\Gamma(1)^2)e^{-2\alpha(2)d}(1-\Gamma(2)^2)\Gamma(3)^2 \tag{106}$$

The reflected wave attenuates until it leaves the slab and reaches to the free space.

$$P_{r''''} = P_i e^{-4\alpha(1)d}(1-\Gamma(1)^2)e^{-4\alpha(2)d}(1-\Gamma(2)^2)\Gamma(3)^2 \tag{107}$$

The reflected waves from other interfaces can be written in the same manner. Hence, total reflected power is written as

$$P_r = P_i\Gamma(1)^2 + P_i e^{-4\alpha(1)d}(1-\Gamma(1)^2)\Gamma(2)^2 + P_i e^{-4\alpha(1)d}(1-\Gamma(1)^2)(1-\Gamma(2)^2)e^{-4\alpha(2)d}\Gamma(3)^2 + \ldots \tag{108}$$

Equation (108) can be arranged as follows (Tang et al., 2003)

$$P_r = P_i\left\{|\Gamma(1)|^2 + \sum_{j=2}^{M}\left(|\Gamma(j)|^2\prod_{i=1}^{j-1}(\exp[-4\alpha(i)d](1-|\Gamma(i)|^2))\right)\right\} \tag{109}$$

where d is the thickness of each subslabs.
In order to obtain total transmitted wave power, attenuation inside the plasma slab must be considered. The total transmitted power can be computed from

$$P_t = e^{-2\alpha(1)d}(1-\Gamma(1)^2)e^{-2\alpha(2)d}(1-\Gamma(2)^2)e^{-2\alpha(3)d}\ldots\ldots e^{-2\alpha(M)d}(1-\Gamma(M)^2) \tag{110}$$

From equation (110), we get (Tang et al., 2003),

$$P_t = P_i\prod_{i=1}^{M}\left\{\exp[-2\alpha(i)d](1-|\Gamma(i)|^2)\right\} \tag{111}$$

The absorbed power by the plasma slab can be computed from

$$P_a = P_i - P_r - P_t \tag{112}$$

Due to perfect electric conductor after the plasma slab, $|\Gamma(M)| = 1$ and thus equation (112) becomes

$$P_a = P_i - P_r \tag{113}$$

This mentioned model is acceptable as the first approximation under the assumption that the plasma properties vary slowly along the wave propagation path (Heald & Wharton, 1978).

3. Results and discussion

In this part, reflection of electromagnetic wave power from plasma coated conducting plane is analysed by considering three different electron density distribution functions. These functions are selected as linear distribution function with positive slope, linear distribution function with negative slope and sinusoidal distribution function.

Linearly varying electron density distribution function with positive and negative slopes is defined as

$$N = \begin{cases} N_m z / L \\ N_m (L - z) / L \end{cases} \tag{114}$$

respectively. Sinusoidal electron density distribution function is written as

$$N = N_m \left(0.505 + 0.5\cos(pd\pi / 12 + \phi) \right) \tag{115}$$

where N_m is the maximum electron number density value, L is the thickness of the plasma, p is the sinusoid frequency parameter taken as 2 and ϕ is the phase shift introduced to the sinusoidal distribution. In this part, the plasma length L is taken as 12 cm. In Fig. 4, it is shown that linearly increasing distribution reduces the reflected power much more than the other two distributions in 1-18 GHz range.

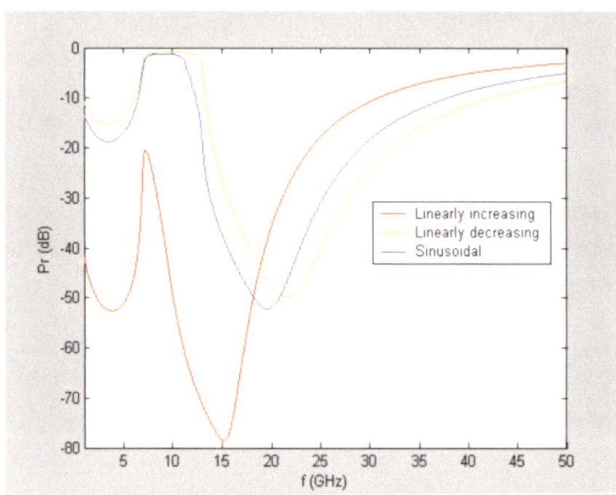

Fig. 4. Reflected power for $N_m = 1 \times 10^{18}\ m^{-3}$, $\nu_{en} = 1\ GHz$, $B = 0.25\ T$.

After 20 GHz, the reflected power increases as the frequency increases due to mismatch between the free space and the plasma slab. After 35 GHz the reflected power for linearly increasing case is below -10 dB which means that plasma slab behaves as a transparent medium for incident electromagnetic wave propagation. For other distribution functions, reflection coefficient is nearly equal to 1 for 7.5 - 13 GHz range thus all wave power is reflected back. But as the frequency increases, plasma absorbes more wave power.

Fig. 5. shows the results when maximum electron number density is increased to $5 \times 10^{18}\ m^{-3}$ while the other parameters are remained same.

In this case, linearly increasing distribution reduces the reflected power much more than the other distribution functions up to 33 GHz. For 35-50 GHz range, sinusoidal and linearly decreasing functions are more useful in terms of small reflection.

Fig. 6. shows the results when maximum electron number density is decreased to 5×10^{17} m^{-3}. It is observed that linearly increasing distribution considerably minimizes the reflected power in 1-20 GHz range. The other two distributions show nearly no attenuation for 7.5-11.5 GHz range. It can be seen that the zero attenuation region for sinusoidal and linearly decreasing distribution functions become narrower as the maximum electron number density decreases.

As shown in Fig. 7, when maximum electron number density is decreased to 1×10^{17} m^{-3} small reflection band of the plasma slab becomes narrower for all distribution functions. Up to 15 GHz, reflected power is below -10 dB for all three cases.

In Fig. 8. the reflected power characteristics when effective collision frequency is increased to 60 GHz while the other parameters are the same with Fig. 4. Considerably reduced reflection is observed for all cases with respect to the previous cases. Thus it can be concluded that by increasing the effective collision frequency, reflection from plasma covered conducting plane can be considerably reduced in critical applications.

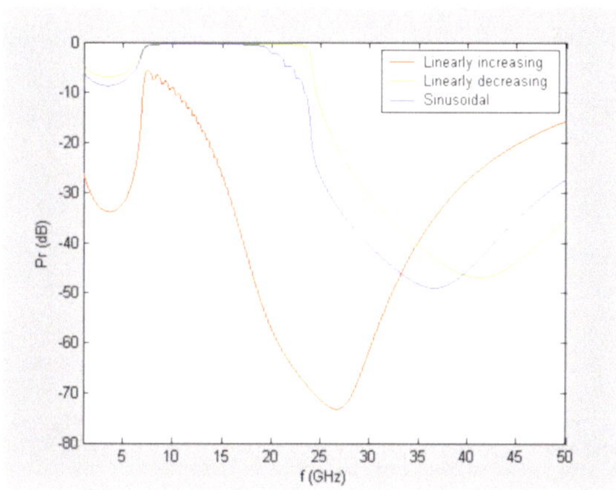

Fig. 5. Reflected power for $N_m = 5 \times 10^{18}$ m^{-3}, $\nu_{en} = 1\ GHz$, $B = 0.25\ T$.

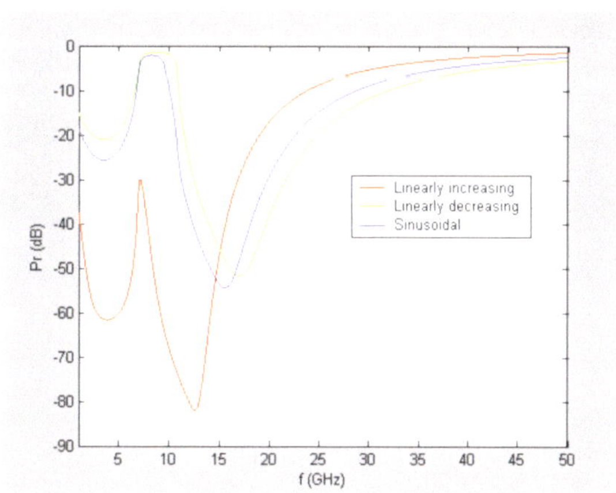

Fig. 6. Reflected power for $N_m = 5 \times 10^{17} m^{-3}$, $\nu_{en} = 1 GHz$, $B = 0.25T$.

The following two graphs are given to determine the effect of magnetic field on the wave propagation in the plasma slab. The magnetic field strength is taken as 0.5 T and 0.75 T in Fig.9 and in Fig.10, respectively. Other parameters are the same with Fig. 4. It is seen from Fig. 9 and Fig. 10 that the reflected power characteristics shift to right as the magnetic field increases. Hence, magnetic field can be used for tuning of the reflection and the passbands of the power reflectivity characteristics of the plasma slab. It must also be noted that adjustment of magnetic field slightly affects the bandwidth and amplitudes of the power reflection characteristics while providing frequency shift.

In the last figure, Fig.11, the effect of plasma slab thickness on the reflected wave power is shown. In this figure, thickness of the plasma slab is doubled and taken as 24 cm. The other parameters are the same with Fig.4. The reflected power decreases due to increased plasma thickness as seen from the figure. This is an expected result since the absorbtion of the electromagnetic wave power increases due to increased propagation path along the plasma.

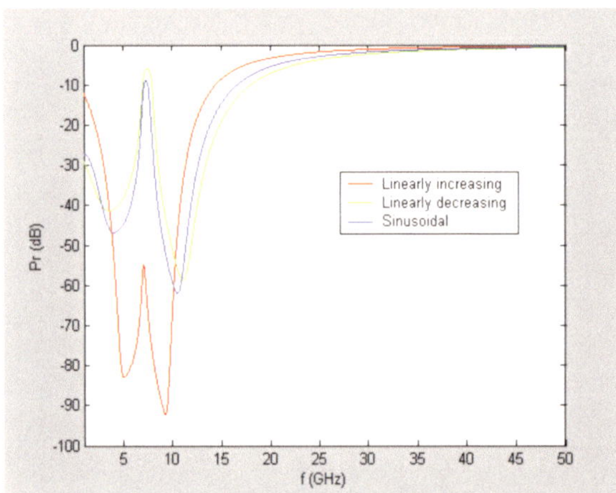

Fig. 7. Reflected power for $N_m = 1 \times 10^{17}\ m^{-3}$, $\nu_{en} = 1\ GHz$, $B = 0.25T$.

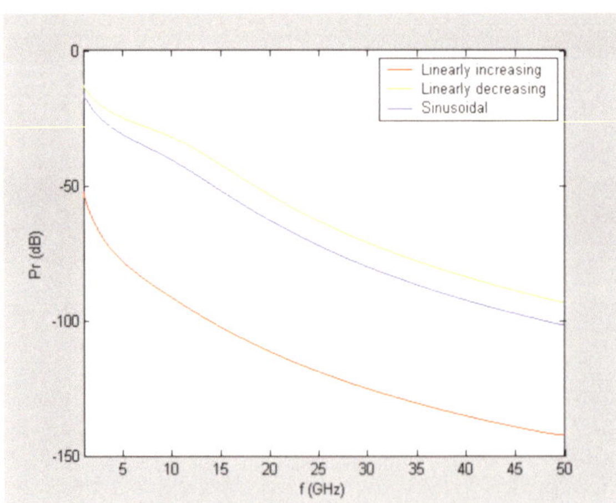

Fig. 8. Reflected power for $N_m = 1 \times 10^{18}\ m^{-3}$, $\nu_{en} = 60 GHz$, $B = 0.25T$.

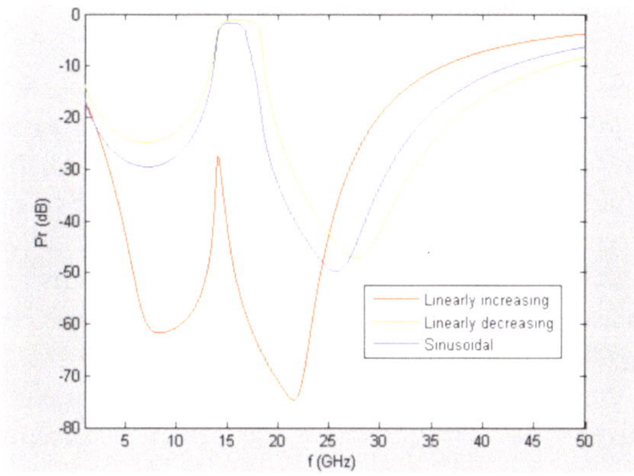

Fig. 9. Reflected power for $N_m = 1\times10^{18}\ m^{-3}$, $\nu_{en} = 1\ GHz$, $B = 0.5\ T$.

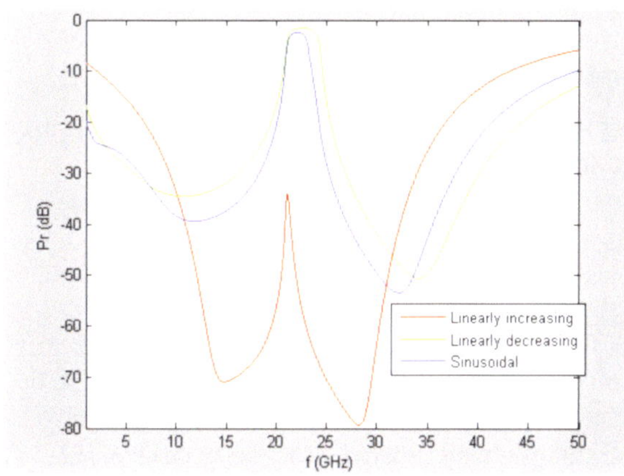

Fig. 10. Reflected power for $N_m = 1\times10^{18}\ m^{-3}$, $\nu_{en} = 1GHz$, $B = 0.75T$.

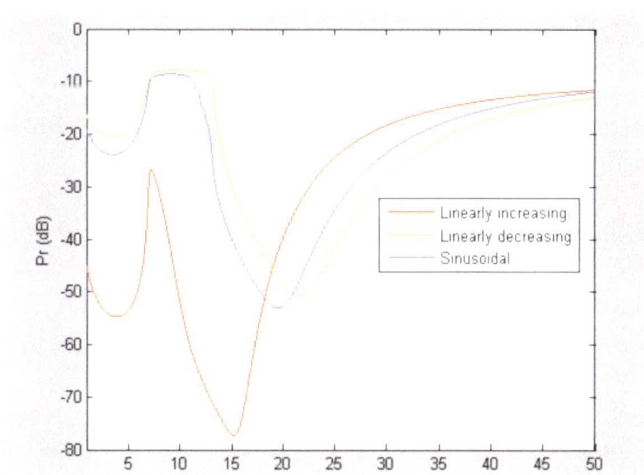

Fig. 11. Reflected power for L=24 cm, $N_m = 1\times10^{18}\ m^{-3}$, $\nu_{en} = 1GHz$, $B = 0.25T$.

4. Conclusion

In this chapter, some of the methods presented in literature for the analysis of electromagnetic wave propagation through the plasma slab is explained briefly and the stealth characteristics of the plasma covered conducting plane is analysed for three different electron density distribution functions. The selected distributions show tunable stealth characteristics in different frequency ranges depending on the adjustment of the plasma parameters. It is seen that especially the linearly increasing density distribution function shows better stealth characteristics considerably reducing the reflected power for 1-20 GHz range. Above 20 GHz, other two functions show better characteristics up to 50 GHz due to adjustment of plasma parameters. It must be noted that the maximum value of electron density function, effective collision frequency, the length of the plasma slab and the external magnetic field strength considerably effect the stealth characteristics of the plasma covered conducting plane and must be carefully adjusted in special applications. In the following studies other distribution functions for electron density will be analysed to obtain an improved performance.

5. Acknowledgement

This study is supported by TUBİTAK-Turkish Scientific and Technological Research Council under contract EEEAG-104E046.

6. References

Balanis, C., A.; (1989). *Advanced Engineering Electromagnetics, John Wiley*, 0-47162194-3, USA.

Cao, J.; Li, J.; Chen, G. (2002). Absorbtion characteristics of conductive targets coated with plasma. *Chinese Journal of Radio Science,* Vol. 17, No. 2, 2002, pp. 125-128.

Chen, Q.; Aoyagi, P.; Katsurai, M. (1999). Numerical Analysis of surface wave excitation in a planar type nonmagnetized plasma processing device. *IEEE Trans. Plasma Sci.,* Vol. 27, pp. 164-170.

Cummer, S., A. (1997). An analysis of new and existing FDTD methods for isotropic cold plasma and a method for improving their accuracy. *IEEE Trans. Antennas Propag.,* Vol.45, pp. 392-400.

Ginzburg, V.; L. (1970). *The Propagation of Electromagnetic Waves in Plasmas.* Pergamon Press, New York.

Gregoire, D., J.; Santoru, J.; Schumacher, R., W. (1992). Electromagnetic-wave propagation in Unmagnetized Plasmas. AD250710, 1992.

Gurel, C., S; Oncu, E., (2009). Frequency Selective Characteristics of a Plasma Layer with Sinusoidally Varying Electron Density Profile, *Int J Infrared Millimeter Waves,* Vol. 30, pp.589-597, 2009.

Gurel, C., S; Oncu, E., (2009). Interaction of electromagnetic wave and plasma slab with partially linear and sinusoidal electron density profile, *Electromagn. Research Lett. PIERL,* Vol. 12, pp.171-181, 2009.

Gurel, C., S; Oncu, E., (2009). Characteristics of electromagnetic wave propagation through a magnetised plasma slab with linearly varying electron density, *Progress In Electromagnetics Research B,* Vol. 21, 385-398, 2010.

Heald, M.; A., Wharton, C.; B. (1978). *Plasma Diagnostics with Microwaves.* Krieger, 1978, pp. 71-94.

Hu, B.; J.; Wei, G.; Lai, Sh.; L. (1999). SMM analysis of reflection, absorbtion, and transmission from nonuniform magnetized plasma slab. *IEEE Trans. Plasma Sci.,* Vol.17, No.4, 1999, pp. 1131-1136.

Hu, X., W.; Liu, M.; H. (2004). Propagation of an electromagnetic wave in a mixing of Plasma- dense Neutral gas . *Plasma Sci. Technol.,* Vol. 6, pp. 2564-2567.

Hunsberger, F.; Luebbers, R.; J; Kunz, K.; S. (1992). Finite difference time domain analysis of gyrotropic media I. Magnetized Plasma. *IEEE Trans. Antennas Propag.,* Vol. 40, pp.1489-1495.

Jiang, Z.; H.; Hu, X.; W; Liu, M.; H.;Lan, C.; H.; Zhang, S.; Pan, Y.; (2006). Propagation of Electromagnetic TM (S-Polarization) Mode in two-dimensional Atmospheric Plasma. *Plasma Sci. Technol.,* Vol. 8, pp. 297-300.

Kong, J., A. (1986). *Electromagnetic Wave Theory.* Wiley, New York.

Kousaka, H; Ono, K. (2002). Numerical Analysis of the Electromagmetic Fields in a Microwave Plasma Source Excited by Azimuthally Symmetric Surface Waves. *Japan J. Appl. Phys.,* Vol.41, pp. 2199-2206.

Lee, J., L.; Kaulluri, D., K.; Nigg, G., C. (2000). FDTD Simulation of Electromagnetic Wave Transformation in a dynamic magnetized Plasma. *Int. J. Infrared Millim. Waves,* Vol. 21, No. 8, pp. 1223-1253.

Liu, M.; Hu, W.; Jiang, Z., Zhang, S.; Lan, C.; Pan, Y. (2007). Reflection of a wave from a thin plasma layer attached to a metal plate by finite-difference time-domain analysis, *Plasma Sources Science and Technology,* Vol. 16, pp. 614-618.

Liu, M., H., Jiang, Z., H.; Zhang, S.; Pan, Y. (2006). Finite-difference time-domain analysis of wave propagation in a Thin Plasma Layer. *Chin. Phys. Lett.,* Vol. 23, No. 2, pp. 410-413.

Liu, S.; Mo., J.; Yuan, N. (2002). FDTD analysis of electromagnetic reflection by conductive plane covered with magnetized inhomogeneous plasma, *Int. J. Infrared Milli. Waves,* Vol. 23, No. 12, pp. 1803-1815.

Shi, J.; Ling, Y.; Qiu, L.; (1995). Scattering cross-section of a an inhomogeneous plasma cyclinder, *Int. J. Infrared Milli. Waves,* Vol. 16, No. 11, pp. 1927-1934.

Su, W.; Yang, J.; Wei, K.; Mao, G. ; He, H.; (2003). Calculation and analysis on the wave reflected characteristics of plasma before conductor plate, *Acta Phys. Sin.,* Vol. 52, No. 12, pp. 3102-3107.

Tang, D., L.; Sun, A., P., Qiu, X., M.; Chu, P., K., (2003). Interaction of electromagnetic waves with a magnetized nonuniform plasma slab, *IEEE Transactions on plasma Science,* Vol. 31, No. 3, pp.405-410.

Yang, H.; Shi, J.;Ling, Y.;Fan, X. (2001). Evaluation of reflection coefficients for microwave oblique incidence on a nonuniform plasma, *Journal of microwaves,* Vol. 17, No. 2, pp.67-71.

Young, J. L. (1994). A full Finite-Difference time domain implementation for radio wave propagation in a plasma., *Radio Sci.,* Vol. 29, No. 6, pp. 1513-1522.

Young, J. L. (1996). A higher order FDTD method for EM propagation in a collisionless cold plasma., *IEEE Transaction Antennas Propagation,* Vol. 44, No. 9, pp. 1283-1289.

Zhang, J.; Liu, Z., Q. (2007). Electromagnetic reflection from conductive plate coated with nonuniform plasma, *International Journal of Infrared and Millimeter Waves*, Vol. 28, 71-78, 2007.

Zhang, S.; Hu, X., W.; Jiang, H.; Liu, M., H.; He, Y. (2006), Propagation of an Electromagnetic Wave in an Atmospheric Pressure Plasma: Numerical Solutions. *Phys. Plasmas,* Vol. 13., No. 1, pp. 13502-13509.

Part 7

Biological Effects and Medical Imaging

22

Electromagnetic Waves and Human Health

Feyyaz Özdemir[1] and Aysegül Kargi[2]
[1]The Black Sea University Medical Faculty, Medical Oncology Department Trabzon
[2]Denizli Goverment Hospital, Medical Oncology Department Denizli,
Turkey

1. Introduction

Electromagnetic waves are produced by the motion of electrically charged particles. These waves are also called electromagnetic radiation because they radiate from the electrically charged particles. They travel through empty space as well as through air and other substances. Electromagnetic waves at low frequencies are referred to as electromagnetic fields and those at very high frequencies are called electromagnetic radiations (1,2).

2. Classification of electromagnetic waves

According to their frequency and energy, electromagnetic waves can be classified as either ionizing radiations or non-ionizing radiations (NIR).

Ionizing radiations are extremely high frequency electromagnetic waves (X-rays and gamma rays), which have enough photon energy to produce ionization by breaking the atomic bonds that hold molecules in cells together.

Non-ionizing (NIR) is a term for that part of the electromagnetic spectrum which has photon energies too weak to break atomic bonds. They include ultraviolet radiation, infrared radiation, radiofrequency and microwave fields.

NIR can not cause ionization however have been shown to produce other biological effects, for instance by heating, altering chemical reactions or inducing electrical currents in tissues and cells.

There are four subgroups of electromagnetic radiation fields with frequency and intensity. This electromagnetic spectrum begins at a frequency of 1 Hertz (Hz), which is 1 wave per second (1,2,3).

2.1 Static electric

Stationary electric charge that is built up on the surfaces and materials. Electric fields are associated with the presence of electric charge, magnetic fields result from the physical movement of electric charge. Human body can not feel less than 2000 volts of static discharge. Magnetic fields can exert physical forces on electric charges when charges are in motion. The magnetic flux density measured in teslas (T), is accepted as the most relevant quantity for relating to magnetic field effects (4). A summary of sources of exposure to static fields in Table 2.

Type	Frequency range	Source
Static	0 Hz	Natural Video MRI Industrial electrolysis
Extremely low frequency (ELF)	(0< f ≤300 Hz),	Powerlines Domestic distribution Electric engines in cars, train and tramway
Intermediate frequency (IF)	300 Hz < f≤100 kHz	Monitors, Anti theft devices in shops, Hands free access control systems, Card readers Metal detectors
Radio frequency (RF)	100 kHz< f≤ 300 GHz	Broadcasting and TV; Mobile telephony Microwave oven Radar Portable and stationary radio transceivers, Personal mobile radio.

*Adopted from: Possible effects of Electromagnetic Fields (EMF) on Human Health. Scientific Committee On Emerging And Newly Identified Health Risks (SCENIHR) 19 July 2006 MRI: Magnetic Resonance Imaging

Table 1. Classification and sources of electromagnetic radiation fields*.

Sources	flux density
Typical electric fields	
Video Display Unit, Tv	20 kV/m
Under 500 Kv Transmission Line	30 kV/m
Atmospher	12-150 V/m
Typical magnetic fields	
Geomagnetic Field	0,03-0,07 mT
Magnetic Resonance Imaging (MRI)	2,5 T
Industrial DC Equipment	50 mT
Small Bar Magnets	1-10 mT
Magnetic Levitation Train	50 mT

Table 2. Sources of exposure to static fields and their flux densities.

2.2 Extremely Low Frequency (ELF)
Extremely low frequency is a term used to describe radiation frequencies below 300 Hertz (Hz). ELF fields are oscillating fields and very important for public health because of the widespread use of electrical power at 50-60 Hz in most countries (1,5).

2.3 Intermediate Frequency (IF)
Intermediate Frequency is a term to describe radiation frequency between 300 Hz and 100 kHz. There are experimental and epidemiological data from the IF range. Therefore, assessment of acute health risks in the IF range is currently based on known hazards at lower frequencies and higher frequencies. Proper evaluation and assessment of possible health effects from long term exposure to IF fields are important because human exposure to such fields is increasing due to new and emerging technologies. Typical examples are: computer and tv screens with use cathode ray tubes, compact fluorescent lamps, as well as radio transmitters, anti theft devices in shops, hands free access control systems, card readers and metal detectors. It is also used in electrosurgery (1,2).

2.4 Radio Frequency (RF)
RF is includes the frequencies between 100 kHz and 300 GHz of the electromagnetic spectrum. RF sources is widespread used in whole world. Majority examples are mobile phones, broadcasting, medical and industrial applications. The RF sources are used in different frequency bands and subdivided in different categories:

2.4.1 Sources operated close to the human body
Main examples of this type are mobile RF transmitters. One of the examples is mobile phones; more than 1.5 billion people are using mobile phones worldwide. In addition to mobile phones, other wireless applications like cordless phones, e.g. DECT, or WLAN systems are very common. The maximum peak power level of a DECT system is 250 mW, of a WLAN system 200 mW.

2.4.2 Sources operated far away from the human body
Such sources are fixed installed RF transmitters. An example is base stations that are an essential part of mobile communication networks.

2.4.3 Medical applications
Some medical applications use electromagnetic fields in the RF range. Therapeutic applications such as soft tissue healing appliances, hyperthermia for cancer treatment, or diathermy expose the patient well above the recommended limit values to achieve the intended biological effects (1,5).

3. Effects on biological systems of electromagetic fields

In 1935 Burr and Northrop examined and published the effects of stable voltage gradients on various biological systems. They were followed by a lot of scientists who found that stable voltage gradients led to many drastic changes in the organism, including growth and local injury. Studies have shown that these effects were associated with changes in distribution of ions (6).

According to some authors, there is connection with electromagnetic fields and disappearance of bees known as colony collapse disorder in Europe and the US, and that it could also interfere with bird migration (7,8).

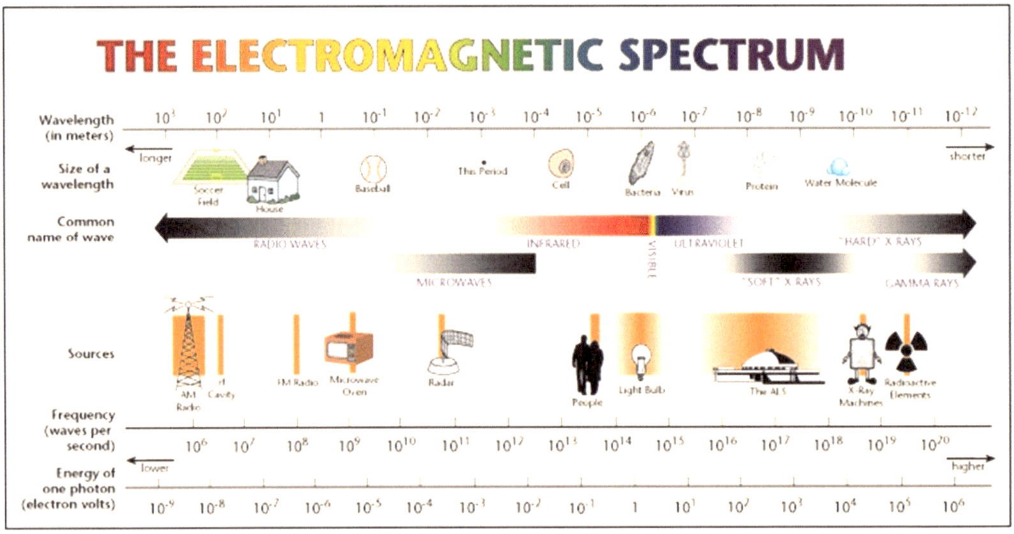

Fig. 1. The Electromagnetic waves spectrum. Adopted from Electromagnetic cellular interactions Cifra M, Fields JZ, Farhadi A.

4. Effects of human health

While the positive aspect of technologic innovation makes the life easier, it may also involve components that impair the quality of life via its certain negative effects. A discussion about the adverse effects of electromagnetic waves on the biological life has been ongoing since the discovery of electricity in the 19th century (6).

Electromagnetic waves generated by many natural and human-made sources can travel for long distances and play a very important role in daily life. In particular, the electromagnetic fields in the Radiofrequency (RF) zone are used in communications, radio and television broadcasting, cellular networks and indoor wireless systems. Resulting from the technological innovations, the use of electromagnetic fields gradually increases and thus people are exposed to electromagnetic waves at levels much higher than those present in the nature (1,2,5). Along with the widespread use of technological products in daily life, the biological effects of electromagnetic waves started to be discussed.

Particularly, the dramatically increasing number of mobile phones users rise significant concerns due to its potential damage on people exposed by radiofrequency waves.

Since mobile phones are used in positions very close to the human body and require a large number of base station antennas, the public and the scientists have question marks in their mind about the impact of mobile phone networks on health (9).

4.1 Evidence for cellular effects of electromagnetic fields

The general opinion is that there is no direct evidence of hazardous effects on human health incurred by low-frequency radiofrequency waves. Studies at the cellular level, which uses relatively higher frequencies, demonstrate undesirable effects (10-11). Some studies revealed that different dimensions of electromagnetic waves have not shown any DNA damage on different cell lines. For example, in a comprehensive review published, Brusick et al have reported no evidence regarding the direct mutagenic effect of radiofrequency signals on cells (12).

On the other hand, there are a lot of contrary study published in recent years. Most of them concerned about evidence of biochemical or cellular effects of electromagnetic fields. Marino and Becker have shown that static or very low-frequency electromagnetic fields may lead to biological effects associated with redistribution of ions. Furthermore, many studies demonstrated that biological effects of low-frequency magnetic fields may penetrate into deeper tissues (13).

Foletti et al. showed that ELF-EMF may have an effect on several cellular functions such as cell proliferation and differentiation, which was followed by many other researchers such as Tian et al. who showed its effect on apoptosis, Takahashi et al. on DNA synthesis, Goodman et al. on RNA transcription, Goodman and Henderson on protein expression, Zrimec et al. on ATP synthesis, Paksy et al. on hormone production, Kula et al. on antioxidant enyzme systems, Milani et al. on metabolic activity, and Wolf et al. on NFkB and cell destruction (14,15,16,17,18,19,2021,22,23).

Giladi et al. demonstrated that EMF of intermediate frequency was effective in arresting the growth of cells. Kirson et al. indicated that this direct inhibitory effect on cell growth can be used for therapeutic purposes in the treatment of cancer (24,25).

EMF of very high frequency has thermal and non-thermal effects on the biological systems. This thermogenic effect is mainly associated with the intensity of EMF, which is expressed as specific absorption rate (SAR). Thermal effect or increased temperature lead to various changes in the cellular functions, which may result in cell destruction (26,27,28). Morrissey et al. showed that biological effects may occur even at very small temperature changes in in-vitro experimental models (29).

There are many papers showing that a weak EMF has no significant effect on biological systems. However, it appears that these studies have a poor design in general, and they lack appropriate control groups, and they are also accompanied by confounding factors (27,30).

The fact that no significant evidences were detected in the above epidemiological trials supporting the suspicions that exposure to electromagnetic waves could result in cancer is in line with the in vitro studies. The effects of electromagnetic fields on different cell lines were studied in the last 30 years and no evidence on their direct or indirect DNA damage were detected. Maes (31) and Vijayalaxmi (32) exposed peripheral blood cells to 935 and 2450 MHz electromagnetic field and reported no DNA damage in cells after 2-hours periods. Malyapa studied the effects of 2450 MHz electromagnetic signals on human gliablastoma cells and mouse fibroblast cell lines and detected no DNA damage in cells, including the 24-hour period (33,34). In a similar study, Tice et al demonstrated that 837 and 1909.8 MHz radiofrequency waves did not result in a significant DNA damage in leukocytes as a result of 3 and 24 hour exposures (35).

Atasoy et al. examined the effects of electromagnetic fields on peripheral mononuclear cells in-vitro. The primary objective of this study was to analyze the changes in the cell viability, rates of apoptosis, proliferation indices and cell surface antigenic structures resulting from 2-, 6- and 24-hour exposure of mononuclear cells isolated from the peripheral blood to 450, 900 and 1784 MHz electromagnetic waves. Data obtained showed that electromagnetic waves didn't have any effect on cell viability, rates of apoptosis and proliferation index.

Author	Year	Studied subject	Frequencies	Results
Goodman et al.	1983	RNA transciption	Pulsed EMF	increased activity of mRNA
Takashi et al.	1986	DNA synthesis	10-100 Hz	DNA synthesis is not repressed
Goodman and Henderson	1988	salivary gland cells	1,5-72 Hz ELF	alters polypeptide synthesis
Maes et al.	1997	Peripheral blood cells and Mitomycin C	935.2 MHz	combined exposure revealed weak effect
Malyapa RS et al.	1997	Human blastoma cells	835,62 and 847,74	No DNA damaged
Malyapa RS et al.	1997	cultured mammalian cells	continous 2450 MHz	No DNA damaged
Brusick et al.	1998	Nucleic acids	800-3000 mHz	Not directly mutagenic, predominantly hyperthermia
Vijayalaxmi et al	2000	Peripheral blood cells	pulsed 2450 mHz	No DNA damaged
Milani et al.	2001	human lymphocytes	EMF	deviation of metabolic activity
Tian et al.	2002	Apoptosis	ELF and X-Ray	suppress apoptosis
Zyrmec, Jerman, Lahajnar	2002	E. Coli ATP syntesis	100 Hz, alternate	stimulate ATP synthesis
Tice et al.	2002	Leukocytes	837 and 1909,8 MHz	No DNA damaged
Wolf et al.	2005	NfkB and cell destruction	50 Hz,0,5-1 mT ELF-EMF	influences cell proliferation and DNA damage
Giladi et al.	2008	Cell growth	10 Mhz	IF arrests cell growth
Kirson ED et al.	2009	human carcinoma cell series	TT Fields + chemo	increase of chemo efficacy
Atasoy A et al.	2009	peripheral mononuclear cells	450, 900, 1784 MHz	No effect cell viability, effect of functional capacity
Coskun S et al.	2009	plasma liver brain specimens of pigs	50 Hz, 1,5 mT EMF	Intermittant EMF effective on plasma lipid peroxydation
Akan A et al.	2010	Monocyt derived macrophage	50 Hz, 1 mT ELF-EMF	supressed caspase 9 activation
Martinez-Samano J et al.	2010	antioxidant system liver kidney and plasma	60 Hz, 2,4 mT	Decreased SOD and GSH

Table 3. Some investigations and their results about cellular effects on electromagnetic fields.

While electromagnetic waves didn't change HLADR and CD11b expression in the peripheral blood mononuclear cells, they decreased the CD11a expression and increased the CD49d expression. These data suggest that electromagnetic signals could affect the functional capacity of the peripheral blood mononuclear cells by changing their adhesion ability. Maybe these alterations are a sign of the immune system modulation (36).

Akan Z et al. evaluated the immune response of monocyte-derived macrophages to pathogenes in extremely low frequency electromagnetic fields. In this study, human monocytic leukemia cell line were cultured and 1 mT EMF was applied for 4–6 h to cells induced with *Staphylococcus aureus* or interferon gamma/lipopolysaccharide (IFγ/LPS). Alterations in nitric oxide (NO), inducible nitric oxide synthase (iNOS) levels, heat shock protein 70 levels (hsp70), cGMP levels, caspase-9 activation, and the growth rate of *S. aureus* were determined. The growth curve of exposed bacteria was found to be lower than the control. Field application increased NO levels, and this increase was more prominent for *S. aureus*-induced cells and appeared earlier than the increase in cells without field application. A slight decrease was observed in iNOS levels whereas there was an increase in cGMP levels. A time-dependent increase was observed in hsp70 levels. When cells were induced with *S. aureus* or IFγ/LPS, field application produced higher levels of hsp70, and suppressed the caspase-9 activation. These data showed that ELF-EMF affect the response of immune system, which suggests that it can be considered for beneficial uses (37).

Another hypothesis of effects related with ELF-EMF is that it changes the free radical levels in the organism. Free radicals in the body are eliminated through two pathways. The first pathway is the non-enzymatic pathway including glutathione, vitamins, carotenoids and flavonoids, while the second pathway relies on the activity of the enzyme, which is the most effective pathway. The key enzymes include catalase and superoxide dismutase. ELF-EMF convert free radicals into less active molecules and eliminate them (38,39). There is a balance between production and elimination of free radicals. An imbalance can promote oxidative stress, eventually resulting in cell destruction. One of the markers indicating destruction is malondialdehyde, the end product of lipid peroxidation (40). Coskun et al. exposed guinea pigs to 50 Hz, 1.5 mT ELF-EMF for 4 days. And, they found that it increases malondialdehyde, nitric oxide and myeloperoxidase activity, and decreased Glutatlon S transferase levels (41).

Martinez et al. evaluated the effects of exposure to ELF-EMF on the antioxidant systems in liver, kidney and plasma in Wistar rats. They found that two hours of 60 Hz EMF exposure led to early changes in free radical levels, and superoxide dismutase (SOD) activity in plasma and glutathione (GSH) content in heart and kidney were decreased, but there was no change in the lipid peroxidation (42).

4.2 Heavy metals exposure and electromagnetic hypersensitivity

Some people are more susceptible to exposure with electromagnetic fields from others. It is referred as Electrohypersensitivy (EHS). The pathophysiology of EHS is unknown. Some authors claimed it is concerned with heavy metal exposure. Heavy metals bound the proteins within tissues and organs are thought to have low toxicity. Mortazavi and co-workers have found that static magnetic field as well as microwave radiation emitted from mobile phones may induced the mercury vapor release from dental amalgam, increasing concentration of dissolved mercury in saliva among amalgam bearers (43,44,45).

4.3 Electromagnetic fields and blood-brain barrier

The blood-brain barrier (BBB) in mammalians is composed of endothelial cells with tight junctions including pericytes and extracellular matrix. Transmembrane proteins form a physical barrier (43). BBB tightness is provided by the connective tissue cells called pericytes and the extracellular matrix of the basement membrane (44). These cells, extracellular components and surrounding neurons are all called 'neurovascular unit' (45). BBB is not available in certain regions of the brain, which include the median eminence, the area postrema and nucleus tractus solitarius in the brain stem, the posterior pituitary, subfornical organ in the hypothalamus, organum vasculosum, subcommissural organ and pineal gland (45).

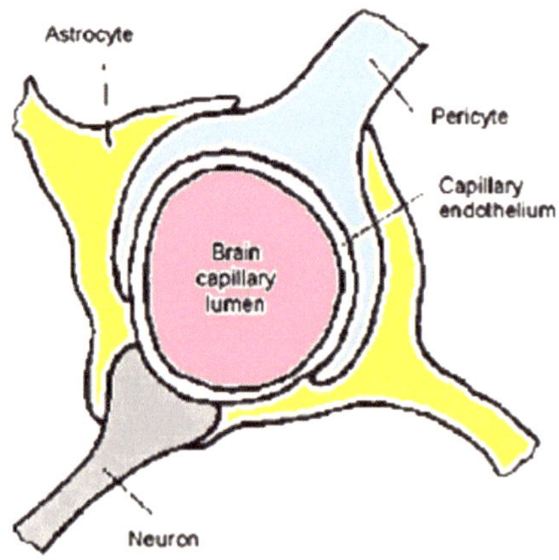

Fig. 2. Scheme of the blood brain Barrier.

4.3.2 Physiology of the blood-brain barrier

BBB allows for a more restricted Exchange of cells and molecules between the blood and the brain parenchyma. Transcellular and paracellular transport can ocur nat only via the blood vessel wall, but also via cranial and spinal nevre roots (46). Lipophilic compounds have unrestricted Access to the brain by passive diffusion through the endohtelial cell membranes. Charged and hydrophilic molecules which are essential for brain metabolism, such as ions, amino acids, glucose and nucleic acid constituents pass the BBB through specialised channels or carriers. Water molecules can pass the BBB through protein channels called aquaporins or carriers (47). The transport of hydrophilic molecules such as proteins and peptides that do nat have a specific transport system (48,49).

4.3.3 Thermal effects of EMF exposure on permeability

Environmental heat in excess of the mammalian thermoregulatory capacity can increase the permeability of the BBB to macromolecules (50). Neuronal albumin uptake in various brain regions was shown to be dose dependently related to brain temperature, with effects becoming apparent with temperature increases of 1 ° C or more (51). Thus, albumin bounded drugs uptake increases (52,53). In the study by Moriyama et al exposure of the sraque-dawley rats head at microwave frequencies (at 2,5-3,2 GHz) that leads to a brain temperature above 40 ° C can increase BBB permeability (54). The degree of increased permeability depend on the degree of temperature rise and hence on the SAR of RF energy, on exposure duration and on the rate of heat distribituon. Quock and co-workers assessed

permeability of capillary endothel cells after 2.45 GHz microwave irradiation cerebral cortex in albino rats (55). Quock and co-workers also demonstrated some hydrofilic drugs such as acetycholine antagonist methylatropine, dopamin antagonist domperidone, and the chemotherapeutic drug methotrexate uptake can be increased with microwave induced hyperthermia (55,56).
Exposure to microwaves at thermal levels may make the brain more vulnerable for infections. Following microwave exposure at 2.5 GHz with SAR between 24-98 W/kg, increased BBB permeability to Horse radish proteins (HRP) was accompanied by increased lethality of japanase encephalitis virus (57).

4.4 Effects on nervous system and psycologic disorders

Due to mobile phones used close the brain tissue, electromagnetic waves affects it the most. Numerous studies have investigated the effect of exposure to radiofrequency electromagnetic waves from the mobile phone base stations on nervous system and behaviours (58). Röösli and co-workers conducted a systematic review of these studies, analysing 17 reports. Five of them were randomized human laboratory trials, and 12 were epidemiological studies. Most of these reports evaluated non-specific disease symptoms. Most of these studies investigated if there was an association between mobile phone base station (MPBS) radiation and development of acute symptoms during or shortly after exposure, and none of them found such an association. Consequently, based on these randomized, blinded, human laboratory trials, it can be concluded that there is good evidence for non-association between MPBS exposure up to 10 volt and development of symptoms. However, no sufficient data is available to draw conclusions about health effects of long-term low level exposure, which occurs in daily environment (9).
Ntzouni MP et al. investigated the effect of mobile phone radiation on short-term memory in mice. They evaluated the effects of mobile phone electromagnetic fields on non-spatial memory task (Object Recognition Task- ORT) that requires entorhinal cortex function. They applied the task to three groups of mice Mus musculus C57BL/6 (exposed, sham-exposed and control) combined with 3 different radiation exposure protocols. In the first protocol of acute exposure, mice 45 days old (postnatal day 45) were exposed to mobile phone radiation (SAR value 0.22W/kg) during the habituation, the training and the ORT test sessions (except the 10 minute inter-trial interval (ITI) with consolidation of stored object information). In the second protocol of chronic exposure-I, the same mice were exposed for 17 days for 90 minutes per day starting at post-natal day 55 to the same MP radiation. ORT recognition memory was only present during the ITI phase, and it was performed at post natal day 72 with radiation. In the third protocol of chronic exposure-II, mice received daily radiation under the same conditions for 31 days up to post natal day 86. Ona day later, the ORT test was performed without any irradiation. A major effect was observed on the chronic exposure-I by the ORT-derived discrimination indices in three exposure protocols. It suggests a possible serious interaction between EMF and consolidation phase of the recognition memory processes. This may imply that the primary EMF target may be the information transfer pathway connecting the entorhinal-parahippocampal regions which participate in the ORT memory task (59).
A study by Heinrich S et al. has led to increasing concerns on the fact that increased number of mobile phone users, exposure to radiofrequency electromagnetic fields (RF EMF) may have potential adverse effects on acute health, particularly in children and adolescents. The authors assessed this potential relationship using personal dosimeters (60).

This population-based cross-sectional study conducted in Germany between 2006 and 2008, a 24-hour exposure profile was generated in 1484 children and 1508 adolescents. Personal interview data on socio-demographic characteristics, self-reported exposure and potential confounders were collected. Acute symptoms were evaluated twice during the study day using a symptom diary. Only a limited part of many associations assessed were found to be statistically significant. During noon time, adolescents with a measured exposure in the highest quartile during morning hours reported a statistically significant higher intensity of headache. During bedtime, adolescents with a measured exposure in the highest quartile during afternoon hours reported a statistically significant higher intensity of irritation in the evening while children reported a statistically significant higher intensity of concentration problems.

A limited number of statistically significant results, which were not consistent along the two time points, were observed. Furthermore, they couldn't confirm the significant results of the main analysis when 10% of the participants with the highest exposure. Based on the pattern of these results, they assumed that the few observed significant associations were not causal, but rather occurred by chance (60).

Sauter C et al. studied the potential effects of long-term exposure to Global System for Mobile Communications (GSM) 900 and Wideband Code Division Multiple Access (WCDMA) signals on attention and working memory. The results of studies showed the potential effects of electromagnetic waves emitted by mobile phones on cognitive functions are controversial. The sample consisted of 30 healthy male subjects, who were exposed to three exposure conditions in a randomly assigned and balanced order for nine days. All test were performed twice a day within a fixed timeframe on each test day. Univariate comparisons showed changes only in one parameter in vigilance test, and one parameter in divided attention test when subjects were exposed to GSM 900 compared to sham. In the WCDMA exposure condition, one parameter in the vigilance and one in the test on divided attention were altered compared to sham. Performance in the selective attention test and the n-back task was not affected by GSM 900 or WCDMA exposure. Time-of-day effects were evident for the tests on divided, selective attention, and working memory. Following the correction for multiple tests, only time of day effects remained significant for two tests. The authors concluded that results of their study did not provide any evidence of an EMF effect on human cognition, but they emphasize the necessity of control for time of day (61).

Lowden et al. examined the quality of sleep following an exposure to mobile phone in people who have symptoms associated with mobile phone use. Various studies showed increased activity for certain frequency bands (10-12 Hz) and for visually scored parameters during sleep after exposure to radiofrequency electromagnetic waves. Furthermore, shortening of REM duration has been reported. They evaluated the effects of a double-blind radiofrequency exposure (884 MHz, GSM signaling standard including non-DTX and DTX mode, time-averaged 10 g psSAR of 1.4 W/kg) on self-evaluated sleepiness and objective EEG measures during sleep. Forty-eight subjects with a mean age 28 years first underwent a 3 hours of controlled exposure prior to sleep (7:30–10:30 PM; active or sham), followed by a full-night polysomnographic recording in a sleep laboratory. The results following exposure showed that time in stages 3 and 4 decreased by 9.5 minutes (12%) while time in stage 2 increased by 8.3 minutes (4%). The latency to Stage 3 sleep was also prolonged by 4.8 min after exposure. Power density analysis indicated an enhanced activation in the frequency ranges 0.5–1.5 and 5.75–10.5 Hz during the first 30 min of Stage 2 sleep and 7.5–11.75 Hz elevation within the first hour of Stage 2 sleep, and bands 4.75–8.25 Hz elevated during the second hour of Stage 2 sleep. No pronounced power changes were observed in SWS or for the third hour of scored Stage 2 sleep. No differences were found between controls and subjects with prior complaints of mobile phone-related symptoms. The results confirm previous findings that RF exposure increased the EEG alpha range in the sleep EEG, and indicated moderate impairment of SWS. Furthermore, reported differences in sensitivity to mobile phone use were not reflected in sleep parameters (62).

Valentini et al. published a metanalysis which systematically reviewed the psychomotor effects of mobile phone electromagnetic fields. The authors indicate that during the last decade there has been increasing concern about the possible behavioral effects. This systematic review and meta-analysis focused on studies published since 1999 on the human cognitive and performance effects of mobile phone-related electromagnetic fields (EMF) with a search in the professional database of Pubmed, Biomed, Medline, Biological Sciences, Psychinfo, Psycarticles, Environmental Sciences and Pollution Management, Neurosciences Abstracts and Web of Sciences, and selection of 24 studies for metaanalysis. Each study had at least one psychomotor measurement result. Data were analysed using standardised mean difference (SMD) for measuring the effect size. Only three tasks (2-back, 3-back and simple reaction time (SRT)) displayed significant heterogeneity, but it didn't reach to a statistical significance. They concluded that mobile phone-like EMF did not seem to induce cognitive and psychomotor effects, and effects following chronic exposures should also be assessed (63).

Mohler et al. investigated the effect of every day radio frequency electromagnetic field exposure on sleep quality in a cross-sectional study. They assessed sleep disturbances and daytime sleepiness in a randomly selected population of 1375 subjects in Basel, Switzerland. They didn't observe any relationship between RF EMF exposure and sleep disturbances or excessive daytime sleepiness (64).

4.5 Effects on osteogenesis and chondrogenesis

Although extremely low electromagnetic fields have been shown to exert beneficial effecets on cartilage tissue (65,66), Lin and Lin investigated the effect of pulsed EMF exposure on osteoblast cells, associated with decreased proliferation and mineralization (67). Okudan, Suslu and co-workers reported the influences of 50 Hz and 0 Hz (static) electric fields (EF), on intact rat bones, as evaluated by dual energy X-ray absorbtion (DEXA) measurements on bone content and density when the animals were continuously exposed in utero and neonatally to EFs. Differences between 50 Hz and control groups were found to be significant for total bone mineral density (BMD). Differences between static EF and control groups were also found to be significant for BMD. These results have shown that both static and 50 Hz EFs influence the early development of rat bones. However, the influence of static EFs is more pronounced than that of the 50 Hz field (68).

4.6 Effects on tetsicle and spermatogenesis

Due to carrying mobile phones in the pockets, exposure of EMF on reproduction system has been growing interested. Tenorio showed in wistar rats, there were no change plasma testosterone levels but histopathological analyses showed testiculer degeneration after the 30 minutes a day 60 Hz and 1 mT EMF exposure (69). In contrast, Ozguner and co-workers showed 900 MHz EMF exposure for rats, lends no support to suggestions of adverse effect on spermatogenesis, and on germinal epithelium But there was a significant decrease in serum total testosterone level, and plasma LH and FSH levels in EMF group ($p<0.05$) (70).

4.7 Carcinogenesis and electromagnetic waves

Since the first observation by Wertheimer and Leeper in 1979, a lot of epidemiologic investigations done between magnetic fields exposure and cancer. Speculations that electromagnetic waves can be carcinogenic increased the number of relevant epidermiological and in vitro studies (71,72).

4.7.1 Lymphatic and hematopoetic cancers

Some epidemiological trials have published data stating that the exposure to high-frequency electromagnetic fields may be associated with lymphatic and hematopoetic cancer. A survey conducted in people living around the Vatican radio station reported more childhood leukemia cases than expected (73). Similar data were also obtained from another study performed by Hocking et al in Australia (74). Hocking et al reported a higher leukemia incidence among adults and children living 2 km around Television transmitter stations. However, in these studies, it s stated that a definite correlation can not be established due to the scarcity of leukemia cases and due to the fact that no measurements were performed in leukemia patients on exposure to radiofrequency waves. A study by Morgan et al conducted on 195 775 subjects working in units related to wireless device manufacturing, design and tests detected that mortality associated with brain cancer, leukemia and lymphoma is not higher in this population compared to the normal population (75). In a study performed in Denmark, the analysis of 450 085 mobile phone users revealed no increase in the brain cancer incidence (76).

Previous pooled analyses reported an association between magnetic fields and childhood leukemia. A pooled analysis was presented based on the primary data from studies on residential magnetic fields and childhood leukemia published after 2000. The analysis included 7 studies with a total of 10,865 cases and 12,853 controls. The main analysis focused on 24-hour magnetic field measurements or calculated fields in residences. In the combined results, risk increased with increase in exposure, but the estimates were imprecise. The odds ratios for exposure categories of 0.1-0.2 μT, 0.2-0.3 μT and ≥0.3 μT, compared with <0.1 μT, were 1.07 (95% CI 0.81-1.41), 1.16 (0.69-1.93) and 1.44 (0.88-2.36), respectively (77). With the exception of the most influential Brasil study, the odds ratio somewhat increased. Furthermore, a non-parametric analysis using a generalised additive model suggested an increasing trend (78).

According to Elliott et al., epidemiological evidences suggested that extremely low frequency magnetic field exposure with a chronic low intensity is associated with increased childhood leukemia. The causality of this association is uncertain. They conducted a national case control study regarding the relationship between average magnetic fields from high voltage overhead power lines in the address at birth and childhoood cancer using the National Grid records (79).

Draper et al observed 28,968 children born in England and Wales between 1962 and 1995, and received a diagnosis under 15 years of age. They found that the estimated relative risk for each 0.2 μT increase in magnetic field was 1.14 (95% confidence interval 0.57 to 2.32) for leukaemia, 0.80 (0.43-1.51) for CNS/brain tumours, and 1.34 (0.84-2.15) for other cancers. Although not statistically significant, their estimate for childhood leukaemia was similar to the results of comparable studies. The estimated attributable risk was below one case per year. They concluded that magnetic-field exposure during the year of birth was unlikely to be the whole cause of the association with distance from overhead power lines as previously reported (80).

	Brain tumours		Brain tumors short latency		Brain tumors longer latency	
Authors	**No. exp cases**	**RR estimate (95% CI)**	**No. exp cases**	**RR estimate (95% CI)**	**No. exp cases**	**RR estimate (95% CI)**
Hardell et al. 1999	78	1.0 (0.7-1.4)	78	1.0 (0.7-1.4) >1 yr	34 16	0.8 (0.5-1.4) >5 yr 1.2 (0.6-2.6) >10 yr
Muscat et al. 2000	66	0.8 (0.6-1.2)	28	1.1 (0.6-2.0) 2-3 yr	17	0.7 (0.4-1.4) >4 yr
Inskip et al. 2001	139	0.8 (0.6-1.1)	51	1.0 (0.6-1.6) 0.5-3 yr	54 22	1.0 (0.6-1.6) > 3 yr 0.7 (0.4-1.4) >5 yr
Johansen et al. 2001	154	1.0 (0.8-1.1)	87	1.1 (0.9-1.3) 1-4 yr	24	1.0 (0.7-1.6) >5 yr
Auvinen et al. 2002	40 analogue 16 digital	1.3 (0.9-1.8)	15 analogue 11 digital	1.2 (0.7-2.0) 1-2 yr	17 analogue 1 digital	1.5 (0.9-2.5) >2 yr
Hardell et al. 2002	188 analogue 224 digital	1.3 (1.0-1.6) 1.0(0.8-1.2)	188 analogue 224 digital	1.3 (1.0-1.6) >1 yr 1.0(0.8-1.2) >1 yr	46 analogue 33 digital	1.3 (0.8-2.3) >10 yr 0.9 (0.6-1.5) >5 yr
Lönn et al. 2005	214 glioma 118 meningioma	0.8 (0.6-1.0) 0.7 (0.5-0.9)	112 64	0.8 (0.6-1.1) 1-4 yr 0.6 (0.4-0.9) 1-4 yr	25 12	0.9 (0.5-1.5) >10 yr 0.9 (0.4-1.9) >10 yr
Christensen et al. 2005	47 low-grade glioma 59 high-grade glioma 67 meningioma	1.1 (0.6-2.0) 0.6 (0.4-0.9) 0.8 (0.5-1.3)	19 24 35	0.9 (0.4-1.8) 1-4 yr 0.6 (0.3-1.0) 1-4 yr 0.8 (0.5-1.3) 1-4 yr	6 8 6	1.6 (0.4-6.1) >10 yr 0.5 (0.2-1.3) >10 yr 1.0 (0.3-3.2) >10 yr
Hardell et al. 2005a, Hardell et al. 2005b	68 malignant, analogue 198 malignant, digital 35meningioma,analogue 151 meningioma, digital	2.6 (1.5-4.3) 1.9 (1.3-2.7) 1.7 (1.0-3.0) 1.3 (0.9-1.9)	20 analogue 100 digital 1 analogue 96 digital	1.8 (0.9-3.5) 6-10 yr† 1.6 (1.1-2.4) 1-5 yr 1.2 (0.1-12) 1-5 yr 1.2 (0.8-1.8) 1-5 yr	48 analogue 19 digital 20 analogue 8 digital	3.5 (2.0-6.4) >10 yr 3.6 (1.7-7.5) >10 yr 2.1 (1.1-4.3) >10 yr 1.5 (0.6-3.9) >10 yr
Hepworth et al. 2006	508 glioma	0.9 (0.8-1.1)	271 glioma	0.9 (0.7-1.1) 1.5-4yr	170 glioma 66 glioma	1.0 (0.8-1.3) 5-9 yr 0.9 (0.6-1.3) >10yr
Schüz et al. 2006	138 glioma 104 meningioma	1.0 (0.7 - 1.3) 0.8 (0.6 - 1.1)	82glioma 73meningioma	0.9 (0.6 – 1.2) 1–4 yr 0.9 (0.6 – 1.2) 1–4 yr	51 glioma 12 glioma 23meningioma 5meningioma	1.1 (0.8–1.7) >5yr 2.2 (0.9-5.1) >10yr 0.9 (0.5-1.5) >5yr 1.1 (0.4-3.4) >10yr

Table 4. Results of some epidemiological studies on mobile phone use and brain tumours. The table is modified from the report to the Swedish Radiation Protection board: Recent Research on EMF and Health Risks. Third annual report from SSI's Independent Expert Group on Electromagnetic Fields (SSI's Independent Group on Electromagnetic Fields 2005).

In a recent study by Cooke et al., they investigated if there was an increased risk of leukemia with mobile phone use. They evaluated a total of 806 leukemia cases with an age range of 18 to 59 years, who lived in southeastern England between 2003 and 2009 compared with 585 non-blood relatives as a control group. They found that mobile phone use for more than 15 years didn't statistically increase the risk for leukemia (81).
In conclusion, their results were consistent with the previous pooled analyses showing an association between magnetic fields and childhood leukemia. Generally, the association was weaker in the most recently conducted studies, but they were small and lack methodological improvements needed to resolve the apparent association. The authors concluded that recent studies on magnetic fields and childhood leukaemia did not alter the previous assessment that magnetic fields are possibly carcinogenic (79).

4.7.2 Brain tumors

Baldi I et al. indicate that the etiology of brain tumors mainly remains unknown, and among potential risk factors, electromagnetic field exposure is suspected. They analyzed the relationship between brain tumors and occupational or residential exposure in adults. They carried out a case control study in southwestern France between May 1999 and April 2001. The study included a total of 221 central nervous system tumors and 442 individually age- and sex-matched controls selected from the general population. Electromagnetic field exposure was assessed in occupational settings through expert judgement based on complete job calendar, and at home by assessing the distance to power lines with the help of a geographical information system. Confounders such as education, use of home pesticide, residency in a rural area and occupational exposure to chemicals were taken into account. Separate analyses were performed for gliomas, meningiomas and acoustic neurinomas. A nonsignificant increase in risk was found for occupational exposure to electromagnetic fields. It was found that the risk for meningioma was higher in subjects living in the vicinity of power lines when the increase was considered separately for ELF. These data suggested that occupational or residential exposure to ELF may play a role in the occurrence of meningioma (82).
The most recent review by Khurana et al. investigated the relationship of wireless phone use for more than 10 years with a risk of brain tumor. This review covering a total of 11 metaanalyses showed that the brain tumors, namely glioma and acoustic neuroma increased 2-fold in people using wireless phones for more than 10 years, achieving a statistical significance (83).

5. Conclusions

Although electronic devices and the development in communications makes the life easier, it may also involve negative effects. These negative effects are particularly important in the electromagnetic fields in the Radiofrequency (RF) zone which are used in communications, radio and television broadcasting, cellular networks and indoor wireless systems. Along with the widespread use of technological products in daily life, the biological effects of electromagnetic waves has began to be more widely discussed.
The general opinion is that there is no direct evidence of hazardous effects on human health incurred by low-frequency radiofrequency waves. Studies at the cellular level, which uses relatively higher frequencies, demonstrate undesirable effects. In recent years there are a lot

of studies about effects of EMF on cellular leve l; DNA, RNA molecules, some proteins, and hormones, intracellular free radicals, and ions are shown.
Particularly, the dramatically increasing number of mobile phones users rise significant concerns due to its potential damage on people exposed by radiofrequency waves. There are increasing number of in vivo, in vitro, and epidemiologic studies on the effects of mobile phones, base stations and other EMF sources in last decade.
Epidemiologic evidence compiled in the past ten years starts to indicate an increased risk, in particular for brain tumor, from mobile phone use. Because of mobile phones used close the brain tissue, electromagnetic waves affects it the most.The magnitude of the brain tumor risk is moderate.
A literature search on 'mobile phone use and cancer 'in Pubmed lists 350 studies. More than half of all of these studies is related to brain tumors. At present, evidence for a causal relationship between mobile phone use and brain tumors relies predominantly on epidemiology, in particular on the large studies on this subject. However, the etiopathogenesis of this causal relationship is not clear. The absence of this clear etiology even raise doubts about the cause itself. Weak evidence in favor of a causal relationship is provided by some animal and *in vitro* studies, but overall, genotoxicity assays, both ***in vivo*** and *in vitro*, are inconclusive to date.

6. References

[1] Possible effects of Electromagnetic Fields (EMF) on Human Health. (19 July 2010) Scientific Committee On Emerging And Newly Identified Health Risks (SCENIHR)

[2] http://pages.prodigy.net/unohu/electro.htm

[3] Cifra M, Fields JZ, Farhadi A. (2010) Electromagnetic cellular interactions. Progress In Biophysics and Molecular Biology. 1-24

[4] Guidelines On Limits Of Exposure To Static Magnetic Fields. In: International Commission On Non-Ionizing Radiation Protection ICNIRP Guidelines Health Physics April 2009, Volume 96, Number 4

[5] Exposure to high frequency electromagnetic fields, biological effects and health consequences (100 kHz-300 GHz).Review of the scientific evidence on dosimetry, biological effects, epidemiological observations, and health consequences concerning exposure to high frequency electromagnetic fields. Editors: Vecchia P, Matthes R, Ziegelberger G, Lin J, Saunders R, Swerdlow A. International Commission on Non-Ionizing Radiation Protection. ICNIRP 16/2009

[6] Burr HS, Northrop F.S.C. 1935. The electro-dynamic theory of life. The quarterly Review of Biology 10(3), 322-333.

[7] Ved Parkash Sharma, Neelima R. Kumar. (25 May 2010). Changes in honeybee behaviour and Biology under the influence of cellphone radiations. Current Science, Vol. 98, No. 10

[8] Balmori A. Electromagnetic pollution from phone masts. (2009 Mar 4). Effects on wildlife. Pathophysiology. 2009 Aug;16(2-3):191-9. Epub.

[9] Röösli M, Frei Patrizia, Mohler E, Hug K, (2010). Systematic review on the health effects of exposure to radiofrequency electromagnetic fields from mobile phone base stations. Bull World health Organ:88:887-896.

[10] Panagopoulos DJ, Margaritis LH. (2010). The effect of exposure duration on the biological activity of mobile telephone radiation. Mutation Research/genetic Toxicology and Environmental Mutagenesis. 17-22

[11] Schüz J, Elliott P, Auvinen A, et al. (2010.08.01). An International prospective cohort study of mobile phone users and health (Cosmos): Design considerations and enrolement. Cancer Epidemiology. 10.1016

[12] Brusick D, Albertini R, Mc Ree D, Peterson D et al. (1998) Genotoxicity of radiofrequency radiation. DNA/Genetox Espert Panel. Environ Mol Mutagen. 32(1):1-16

[13] Marino A, Becker R. (1977/9) Biological effects of extremely low frequency electric and magnetic fields: a review. Physiological Chemistry and Physics (2), 131-147.

[14] Foletti A, Lisi A, Ledda M et al. (2009). Cellular ELF signals as a possible tool in informative medicine. Electromagnetic Biology and Medicine. 28 (1), 71-79

[15] Tian F, Nakahara T, Yoshida M, Honda N, Hirose H, Miyakoshi J. (2002). Exposure to power frequency magnetic fields suppresses X-ray-induced apoptosis transiently in Ku80-deficient xrs5cells. Biocemical and Biophysical Research Communications 292 (2), 355-361.

[16] Takahashi K, Kaneko I, Date M, Fukada E. (1986). Effect of pulsing electromagnetic fields on DNA syntesis in mammalian cells in culture. Cellular and Molecular Life Sciences 42 (2), 185-186

[17] Goodman R, Basett C, Henderson A. (1983). Pulsing electromagnetic fields induce cellular transcription. Science 220(4603), 1283.

[18] Goodman R, Henderson A. (1988) Exposure of salivary gland cells to low-frequency electromagnetic fields alters polypeptide synhtesis. Proceedings of the national Academy of Sciences of the United States of America 85 (11), 3928.

[19] Zrimec A., Jerman I., Lahajnar G., (2002). Alternating electric fields stimulate ATP synthesis in Esherichia Coli. Cellular and Molecular Biology Letters 7(1), 172-175.

[20] Paksy K, Thuroczy G, Forgacs Z, Lazar P, Gaati I. (2000). Influence of sinusoidal 50-Hz magnetic field on cultured human ovarian granulosa cells. Electromagnetic Biology and Medicine 19(1), 95-99.

[21] Kula B, Sobczak A, Kuska R. (2000). Effects of static and ELF magnetic fields on free – radical processes in rat liver and kidney. Electromagnetic Biology and Medicine 19 (1), 99-105.

[22] Milani M, Balerini, M, Ferraro L, Zabeo M, Barberis M, Cannona M, Faleri M. (2001). Magnetic field effects on human lymphocytes. Electromagnetic field effects on human lymphocytes. Electromagnetic Biology and Medicine 20(1), 81-106

[23] Wolf FI, Torsello A, Tedesco B, Fasanella S, Boninsegna A, D'Ascenzo M, Grassi C, Azzena GB, Cittadini A. (2005): 50-Hz extremely low frequency electromagnetic fields enhance cell proliferation and DNA damage: possible involvement of a redox mechanism. Biochim Biophys Acta. 2005 Mar 22;1743(1-2):120-9.

[24] Giladi M, Porat Y, Blatt A, Wasserman Y, Kirson E, Dekel E, Palti Y. (2008). Microbial growth inhibition by alternating electrical fields. Antimicrobial Agents and Chemotherapy 52(10), 3517-3522.

[25] Kirson E, Schneiderman R, Dbaly V, Tovarys F. Et al. (2009). Chemotherapeutic treatment efficacy and sensitivity are increased by adjuvant alternating electric fields. BMC Medical Physics 9 (1),1.

[26] Banik S, Bandyopadhyay S, Ganguly S. (2003). Bioeffects of microwave-a brief review. Bioresource Technology. 87(2), 155-159.
[27] Belyaev I. (2005). Nonthermal biological effects of microwaves: current knowledge, further perspective, and urgent needs. Electromagnetic Biology and Medicine 24,375-403.
[28] Philips J, Singh N, lai H. (2009). Electromagnetic fields and DNA damage. Pathophysiology. 16 (2-3), 79-88
[29] Morrisey JJ. (2008). Possible mechanisms to explain biological effects from low level RF exposure.URSI GA in Chicago proceeding.
[30] Kumar A. (May 2003). Nonthermal effects of electromagnetic fields at microwave frequencies. CCECE 2003. In: Canadian conference on electrical and computer engineering. Vol.1 IEEE, pp. 285-288
[31] Maes A, Collier M, Van Gorp U, Vandoninck S, Verschaeve L. (1997). Cytogenetic effects of 935.2-MHz (GSM) microwaves alone and in combination with mitomycin C. Mutat Res; 393 (1 2): 151 156
[32] Vijayalaxmi, Leal BZ, Szilagyi M, Prihoda TJ, Meltz ML. (2009). Primary DNA damage in human blood lymphocytes exposed in vitro to 2450 MHz radiofrequency radiation. Radiat Res; 153 (4): 479 486.
[33] Malyapa RS, Ahern EW, Straube WL, Moros EG, Pickard WF, Roti Roti JL. (1997). Measurement of DNA damage after exposure to electromagnetic radiation in the cellular phone communication frequency band (835.62 and 847.74 MHz). Radiat Res; 148 (6): 618 627.
[34] Malyapa RS, Ahern EW, Straube WL, Moros EG, Pickard WF, Roti Roti JL. (1997). Measurement of DNA damage after exposure to 2450 MHz electromagnetic radiation. Radiat Res; 148 (6): 608 617.
[35] Tice RR, Hook GG, Donner M, McRee DI, Guy AW. (2002). Genotoxicity of radiofrequency signals. I. Investigation of DNA damage and micronuclei induction in cultured human blood cells. Bioelectromagnetics; 23 (2): 113 126.
[36] Atasoy A, Sevim Y, Kaya I, Yilmaz M, Durmuş A, Sönmez M, Ozdemir F, Ovali E. (2009). The effects of electromagnetic fields on peripheral blood mononuclear cell in vitro. Bratisl Lek Listy; 110(9):526-9.
[37] Akan Z, Aksu B, Tulunay A, ,Bilsel S, Inhan-Garip A. (2010). Extremely Low - Frequency Electromagnetic Fields Affect the Immune Response of Monocyte_derived Macrophages to Pathogens. Bioelectromagnetics 31:603-612.
[38] Genestra M., 2007. Oxyl radicals , redox- sensitive signaling cascades and antioxidants. Cellular Signalling 19: 1807-1819.
[39] Valko M., Leibfrietz D., Moncol J., Cronin MTD., Mazur M., Tesler J. 2007. Free radicals and antioxidants in normal physiological functions and human disease. The international Journal of Biochemistry & Cell Biology39:44-84.
[40] Moore K., Roberts JL., 1998. Measurement of lipid peroxidation . Free Radical Research 28:659-671.
[41] Coskun S., Balabanli B., Canseven A., Seyhan N., 2009. Effects of continous and intermittent magnetic fields on oxidative parameters. İn vivo. Neurochemical Research 34:238-243
[42] Martinez-Samano J., Torres-Duran V. P., Juarez -Oropeza M.A. , Elias-Vinas D., Verdugo-Diaz L. Effects of acute electromagnetic field exposure and movement

restraint on antioxidant system in liver , hearth , kidney and plasma of Wistar rats: A preliminary report. Int. J. Radiat. Biol. , Vol.86, No.12, December 2010 , pp.1088-1094

[43] Ghezel-Ahmadi D, Engel A, Weidermann J, Budnik LT, Baur X, Frick U, Hauser S, Dahmen N. Heavy metal exposure in patients suffering from electromagnetic hypersensitivity. 2010 jan15;408(4):774-8.

[44] Mortazavi SM, Daiee E, Yazdi A, et al. Mercury release from dental amalgam restorations after magnetic resonance imaging and following mobile phone use. Pak. J Biol Sci. 2008; 11 (8): 1142-6.

[45] Antonella C, Branca V. Heavy metals exposure and electromagnetic hypersensitivity. Science of the Total Enviroment 408 (2010) 4919-4920

[46] Begley D.J., Brightman M.W., 2003. Structural and functional aspects of the blood - brain barrier. In: Prokai L., Prokai-Tatrai K (Eds.)., *Peptide transport an delivery into the central nervous system. Birkhauser,* Basel.

[47] Hawkins B.T., Davis T.P., 2005. The blood-brain barrier / neurovasculer unit in health and disease. Pharmacol. Rev. 57,173-185.

[48] Hawkins B.T., Egleton R.D., 2008. Pathophysiology of the blood-brain barrier: animal models and methods. Curr Top. Dev. Biol. 80,277-309.

[49] Kastin AJ, Pan W. 2003 Peptide transport across the blood-brain barrier. In: Prokai L, Prokai-Tatrai K (Eds.) *Peptide transport and delivery into the central nervous system. Birkhauser,* Basel

[50] Lin J.C., Lin M.F.,1982. Microwave hyperthermia-induced blood-brain barrier alterations. Radiat. Res. 89, 77-87.

[51] Shivers R.R., Wijsman J.A., 1998. Blood-brain barrier permeability during hyperthermia. Prog. Brain Res.115,413-424.

[52] Lin J.C., Yuan P.M.K., Jung D.T., 1998. Enhancement of anticancer drug delivery to the brain by microwave induced hyperthermia. Bioelectroch. Bioener. 47, 259-264.

[53] Kiyatkin E.A., Sharma H.S., 2009. Permeability of the blood- brain barrier depends on brain temperature. Neuroscience 161,926-939.

[54] Moriyama E., Salcman M., Broadwell R.D., 1991. Blood-brain barrier alteration after microwave -induced hyperthermia is purely a thermal effect: I. Tempurature and power measurements. Surg. Neurol. ,35,177-182.

[55] Quock R.M., Fujimoto J.M., Ishii T.K., Lange D.G., 1986. Microwave facilitation of methylatropine antagonism of central cholinomimetic drug effects. Radiat. Res.105, 328-340.

[56] Quock RM, Kouchich FJ, Ishii TK, Lange DG, 1987. Microwave facilitation of domperidone antagonism of apomorhine -induced stereotypic climbing in mice. Bioelectromagnetics 8, 45-55.

[57] Lange D.G., Sedmak j, 1991. Japanese encephalitis virus (JEV) potentiation of lethality in mice by microwave radiation. Bioelectromagnetics 12, 335-348.

[58] Abdel-Rassoul G, El-Fateh OA, Salem MA, Michael A, Farahat F, El-Batanouny M, Salem E. (2007) Neurobehavioral effects among inhabitants around mobile phone base stations. Neurotoxicology. 2007 Mar;28(2):434-40. Epub 2006 Aug 1.

[59] Ntzouni M.P, Stamatakis A, Sytlianopoulo F, Margaritis LH, (2010). Short-term memory in mice is affected by mobile phone radiation Pathophysiology. -689.

[60] Heinrich S., Thomas S., Heumann C.,Von Kries R., Radon K., (2010,9). Association between exposure to radiofrequency electromagnetic fields assesed by dosimetry and acute symptoms in children and adolescents:a population based cross-sectional study. Environmental health:75

[61] Sauter C, Dorn H, Bahr A, et al. 2010. Effects of exposure to electromagnetic fields emitted by GSM 900 and WCDMA mobile phones on cognitive function in young male subjects. Bioelectromagnetics

[62] Lowden A, Akerstedt T, Ingre M et al: 2011. Sleep after mobile phone exposure in s ubjects with mobile phone related symptoms. Bioelectromagnetics 32:4-14

[63] Valentini E, Ferrara M, Presaghi F et al: Systematic rewiev and meta-analysis of Psycomotor effects of mobile phone electromagnetic fields. (2010). Occup. Environ. Med. .67: 708-716

[64] Mohler E, Frei P, Braun-Fahrländer C, Fröhlich J, Neubauer G, Röösli M; (2010 Sep). Effects of everyday radiofrequency electromagnetic-field exposure on sleep quality: a cross-sectional study. Radiat Res. 174(3):347-56.

[65] Mayer-Wagner S, Passberger A, Sievers B, Aigner J, Summer B, Schiergens TS, Jansson V, Müller PE Effects of low frequency electromagnetic fields on the chondrogenic differentiation of human mesenchymal stem cells. Bioelectromagnetics. 2011 May;32(4):283-90. doi: 10.1002/bem.20633. Epub 2010 Dec 22.

[66] Zhang D, Pan X, Ohno S, Osuga T, Sawada S, Sato K (2011) No effects of pulsed electromagnetic fields on expression of cell adhesion molecules (integrin, CD44) and matrix metalloproteinase-2/9 in osteosarcoma cell lines. Bioelectromagnetics. 2011 Apr 7. doi: 10.1002/bem.20647. [Epub ahead of print]

[67] Lin HY, Lin YJ In vitro effects of low frequency electromagnetic fields on osteoblast proliferation and maturation in an inflammatory environment. Bioelectromagnetics. 2011 Mar 29. doi: 10.1002/bem.20668. [Epub ahead of print]

[68] Okudan B, Keskin AU, Aydin MA, Cesur G, Cömlekçi S, Süslü H. (2006) DEXA analysis on the bones of rats exposed in utero and neonatally to static and 50 Hz electric fields. Bioelectromagnetics. 2006 Oct;27(7):589-92.

[69] Tenorio BM, Jimenez GC, Morais RN, Peixoto CA, Albuquerque Nogueira R, Silva Junior VA (2011) Evaluation of testicular degeneration induced by low-frequency electromagnetic fields. J Appl Toxicol. 2011 Mar 30. doi: 10.1002/jat.1680. [Epub ahead of print]

[70] Ozguner M, Koyu A, Cesur G, Ural M, Ozguner F, Gokcimen A, Delibas N (2005) Biological and morphological effects on the reproductive organ of rats after exposure to electromagnetic field. Saudi Med J. 2005 Mar;26(3):405-10.

[71] Wertheimer N, Leeper E. (1979 Mar). Electrical wiring configurations and childhood cancer. Am J Epidemiol. 109(3):273-84.

[72] Savitz DA. (1993 Apr). Overview of epidemiologic research on electric and magnetic fields and cancer. Am Ind Hyg Assoc J. 54(4):197-204.

[73] Michelozzi P, Capon A, Kirchmayer U et al. (2002). Adult and childhood leukemia near a high-power radio station in Rome, Italy. Am J Epidemiol; 155 (12): 1096 1103.

[74] Hocking B, Gordon IR, Grain HL, Hatfield GE. (1996). Cancer incidence and mortality and proximity to TV towers. Med J Aust; 165 (11 12): 601 605.

[75] Morgan RW, Kelsh MA, Zhao K, Exuzides KA, Heringer S, Negrete W. (2000). Radiofrequency exposure and mortality from cancer of the brain and lymphatic/hematopoietic systems. Epidemiology ; 11 (2): 118 127.

[76] Johansen C, Boice J Jr, McLaughlin J, Olsen J. (2001). Cellular telephones and cancer a nationwide cohort study in Denmark. J Natl Cancer Inst; 93 (3): 203 207.

[77] Kheifets L, Ahlbom A, Crespi CM et al. (2010) Pooled analysis of recent studies on magnetic fields and childhood leukemia. British Journal of Cancer. 03, 1128-1135

[78] Schüz J, Ahlbom A. (2008 Oct 16). Exposure to electromagnetic fields and the risk of childhood leukaemia: a review. Radiat Prot Dosimetry. 2008;132(2):202-11. Epub

[79] Elliott P, Toledano MB, Bennett J, Beale L, de Hoogh K, Best N, Brigggs DJ. (2010 June). Mobile phone base stations and early childhood cancers: case-control study. BMJ. 22;340.

[80] Draper G, Vincent T, Kroll ME, Swanson J. (2005 Jun 4). Childhood cancer in relation to distance from high voltage power lines in England and Wales: a case-control study. BMJ. ;330(7503):1290.

[81] Cooke R, Laing S, Swerdlow AJ. (2010 Nov. 23). A case- control study of risk of leukaemia in relation to mobile phone use. Br. J. Cancer . ;103 (11) :1729-35.

[82] Baldi I, Coureau G, Jaffre A, Gruber A, Ducamp S, Provost D, Leabilly P, Vital A, Loiseau H, Salamon R. (2010 Nov 12). Occupational and Residential Exposure to Electromagnetic Fields and Risk of Brain Tumours in adults: a case-control study in Gronde, France. International journal of cancer. Int J Cancer.

[83] Khurana VG, Teo C, Kundi M, Hardell L, Carlberg M. (2009) Sep Cell phones and brain tumors. Surg Neurol., 72 (3): 205-14.

Image Resolution and Sensitivity Improvements of a Molecular Imaging Technique Based on Magnetic Nanoparticles

Yasutoshi Ishihara[1], Tsuyoshi Kuwabara[2] and Naoki Wadamori[2]
[1]*Meiji University,*
[2]*Nagaoka University of Technology*
Japan

1. Introduction

The enhanced permeation and retention (EPR) effect (Matsumura & Maeda, 1986) caused by the leakage of internally administered nanoparticles from blood vessels and their accumulation in cancerous tissues can be used to diagnose cancer. Gleich and Weizenecker proposed the magnetic particle imaging (MPI) approach (Gleich & Weizenecker, 2005), whereby the positions of these magnetic nanoparticles (MNPs) accumulated in cancerous tissue can be detected by applying a local alternating magnetic field from a source positioned outside the body. In basic MPI, the local magnetic field distribution is scanned to encode the spatial information, and the magnetization signal with odd-order harmonics is detected from MNPs inside a selected region when an alternating magnetic field is applied to the MNPs. Furthermore, a fast data acquisition method by scanning spatial data along with a Lissajous trajectory was proposed (Gleich et al., 2008; Knopp et al., 2009), and real time image-data acquisition was achieved (Weizenecker et al., 2009). However, interference from the magnetization signal generated from the MNPs outside the selected region degraded the image resolution and signal sensitivity (signal-to-noise ratio).
We proposed an image reconstruction method for reducing these interference signals mainly generated by even harmonics, and a correction method to suppress the interference signals (Kusayama & Ishihara, 2007; 2009; Ishihara & Kusayama, 2009). This was achieved by taking into account the difference between the saturated waveform of the magnetization signal detected from the MNPs outside the selected region and that detected from the MNPs inside the region. We performed numerical analyses to prove that the image resolution in the molecular imaging technique can be improved by using our proposed image reconstruction method, which is based on the abovementioned ideas. Furthermore, a fundamental system was constructed and the numerical analyses were experimentally validated using MNPs with diameters of 10–50 nm. The detection sensitivity and the resolution were improved by the use of methods in the case of locally distributed MNPs. However, a reconstructed image with the correct distribution of MNPs may not be obtained when the MNPs are distributed continuously. This is because the abovementioned proposed method acts as an intense high-pass filter against the reconstructed image (Ishihara & Kusayama, 2011).

These problems in MPI originate from the characteristics of the MNPs and the imperfect distribution of the magnetic field applied to the MNPs, as discussed later. Gleich and Weizenecker concentrated on the fact that the observational data overlapped the system function reflecting the characteristics of the MNP and the applied magnetic field distribution, and proposed an image reconstruction method for improving image quality (Gleich & Weizenecker, 2005). The abovementioned method involved performing an inverse matrix operation (such as singular value decomposition; SVD) (Weizenecker et al., 2007) or an iterative operation (algebraic reconstruction technique; ART) (Weizenecker et al., 2009) on the obtained data. However, when an image matrix becomes large, the use of this method to reconstruct images, which is based on a matrix operation, may result in the reconstructed images being underspecified.
Here, we propose a new image reconstruction method with higher image resolution and signal intensity. Our method is based on information regarding the correlation between the observed signal and a system function, and it does not use the inverse-matrix method.

2. Principle

2.1 Magnetization response generated by a MNP

The static magnetization (*M*) of a MNP exposed to a magnetic field is described well by the Langevin theory of paramagnetism, which is defined in equation (1).

$$M = M_s \left[\coth\left(\frac{m\mu_0 H}{k_B T} \right) - \frac{1}{\frac{m\mu_0 H}{k_B T}} \right] \quad (1)$$

where M_s is the saturation magnetization of a MNP, μ_0 is the magnetic permeability of vacuum, m is the magnetic moment of a particle $m = \pi D^3 M_s / 6$, H is the applied field, k_B is Boltzmann's constant, and T is the absolute temperature (Vekas et al., 2000).
A magnetization response with higher-order harmonics corresponding to the nonlinear magnetization properties of the MNP is generated when an alternating magnetic field is applied to a MNP (Fig. 1, [A]). However, such harmonics are not generated when a local static magnetic field that is strong enough to saturate the magnetization of the MNPs is applied (Fig. 1, [B]). The harmonics can be extracted by Fourier transformation of the detected signals; therefore, the positions of the MNPs can be identified and imaged by scanning the local distribution of a magnetic field which has approximately zero strength in the desired region (the field-free point, FFP) and is strong enough to saturate the magnetization in regions other than the FFP (Gleich & Weizenecker, 2005).

2.2 Fundamental MPI system

A fundamental MPI model is shown in Fig. 2 (a). A magnetic field distribution with very high field strength that surrounds the selected region, in contrast to having a first-order gradient at the center, is achieved by applying a DC current I_{DC} to two sets of Maxwell coil pairs. Thus, an FFP is formed at the center of these coils (Fig. 2 (b)). The position of this FFP is scanned by applying an offset DC current to each coil (Fig. 2 (c)). The MNPs generate a magnetization response because of the alternating magnetic field created by the AC current I_{AC} in the Maxwell coil pair consisting of the top and bottom coils. The response is detected as an electromotive force induced by the receiver coil according to Faraday's law.

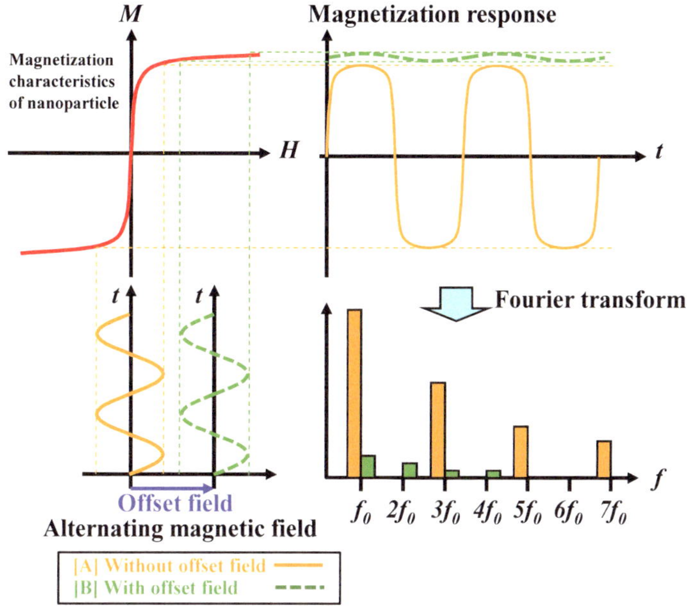

Fig. 1. Principle of MPI. MPI is based on the nonlinear magnetization properties of MNPs. When an alternating magnetic field with a frequency f_0 is applied to MNPs placed in a region where the static magnetic field strength is almost zero (FFP) [A], the MNPs generate a magnetization response with odd-order harmonics. In contrast, very few magnetization responses are detected from MNPs placed in a region where the magnetic field is strong enough to saturate their magnetization [B].

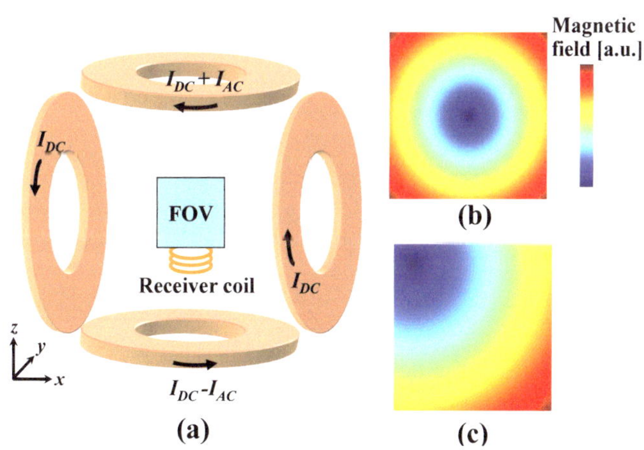

Fig. 2. Fundamental MPI model and generated magnetic field. Each Maxwell coil pair produces a magnetic field gradient along each axis. An FFP is formed at the center of the field of view (FOV), as shown in (b), when the DC currents applied to each coil are equal. The position of this FFP can be scanned by adjusting the ratio of currents applied to each coil (c). An alternating magnetization response is generated when an AC current is applied to the Maxwell coil pair consisting of the top and bottom coils. This response from the MNPs is recorded using a receiver coil.

2.3 Concept of image reconstruction by conventional method

As mentioned above, in MPI, image reconstruction is performed using the magnetization response waveform detected while scanning the FFP. In this approach, the shape of the FFP (the magnetic field distribution) applied to the MNPs has a significant influence on the resolution of the reconstructed image. For example, if the formed FFP is spatially localized by a steep magnetic field distribution as shown in Fig. 3 (a), image resolution comparable to the size of the FFP could be obtained. However, the local characteristics of the magnetic field distribution formed with the usual magnet are limited and imperfect (Fig. 3(b)). Moreover, the magnetizing properties (saturation characteristics) of the MNP also affect the spatial resolution. That is, the finite gradient of the magnetization curve limits the spatial resolution (Fig. 4). Under the influence of these two factors, an additional signal appears from the MNPs that are outside the boundary of the FFP, and interferes with the signal generated from the MNPs inside the FFP. This is shown in Fig. 5. When MNPs exist only in the center of the FOV, a signal is detected ideally only in the position of the desired FFP (*FFP-a,* shown in Fig. 5). However, when the gradient of the magnetic field distribution which forms the FFP is gently-sloping, or when the gradient of the magnetization curve of a MNP is finite, a signal is also detected from MNPs that are in a location other than the desired FFP (e. g., *FFP-b,* shown in Fig. 5).

Therefore, the conventional MPI image reconstruction method uses that the frequency spectrum of a magnetization response waveform would ideally consist only of odd harmonic components when the FFP is scanned at the point where the MNPs are located. Hence, when the FFP is in the two-dimensional plane (x-z plane) of $y = 0$ on Fig. 2, the signal strength in the reconstructed image is expressed by the following equation.

$$U(x,z) = \sum_{n=1}^{N_h} V\left[x,z,(2n+1)\right] \tag{2}$$

$U(x, z)$: reconstructed distribution in x-z plane

$V[x, z, n]$: n-th harmonic contained in the waveform at each FFP (x, z)

N_h: maximum harmonic order for reconstruction

However, because the amplitude is minute for the higher-order components, the information included in the higher-order components is not sufficiently utilized. Moreover, it is difficult to remove the interference caused by the imperfection of the two abovementioned factors using this method only, resulting in the appearance of a blur and an artifact in the reconstructed image.

In addition, from the MPI viewpoint, a larger MNP diameter is preferable according to the magnetization properties expressed in equation (1) (Yavuz et al., 2006), because a comparatively small alternating magnetic field is sufficient. On the other hand, because tens of nm is an effective distance for acquiring the EPR effect, a strong alternating magnetic field is required. This means that the influence of the abovementioned interference signal becomes large. Therefore, a method that successfully suppresses artifacts and improves detection sensitivity is indispensable.

3. Proposed methods

For typical MPI image-reconstruction, Weizenecker et al. (2009) proposed a method that performs an inverse-matrix operation on the obtained data by scanning the FFP along with a

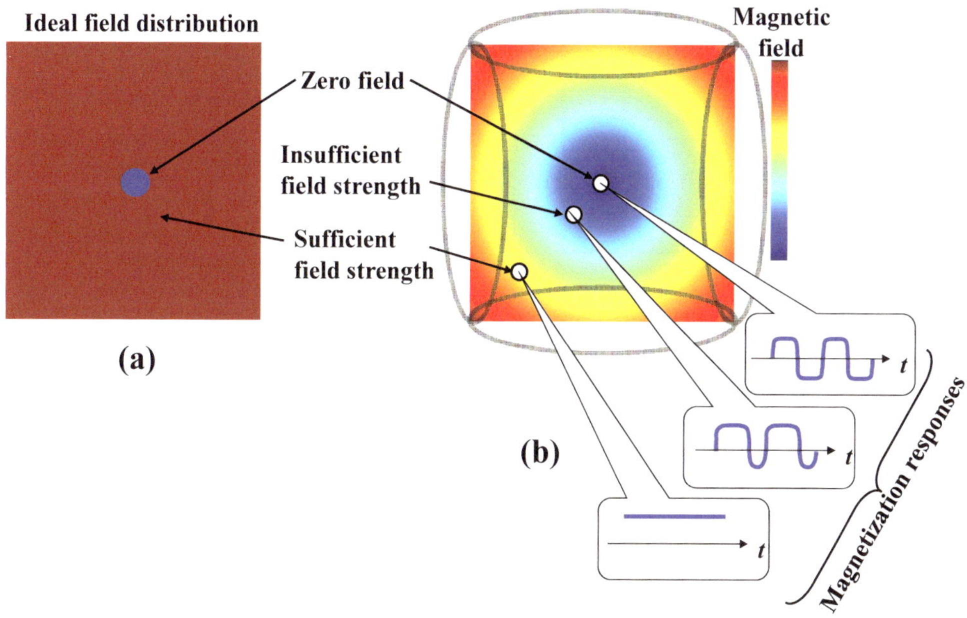

Fig. 3. Influence of FFP imperfection (magnetic field distribution). (a) Ideal FFP: The magnetic field strength is almost zero only in the area selected for acquiring a signal, in contrast to sufficient magnetization of MNP to reach saturation in other regions. (b) Actually formed FFP: the magnetic field distribution generated with Maxell coil pairs has a murky FPP boundary.

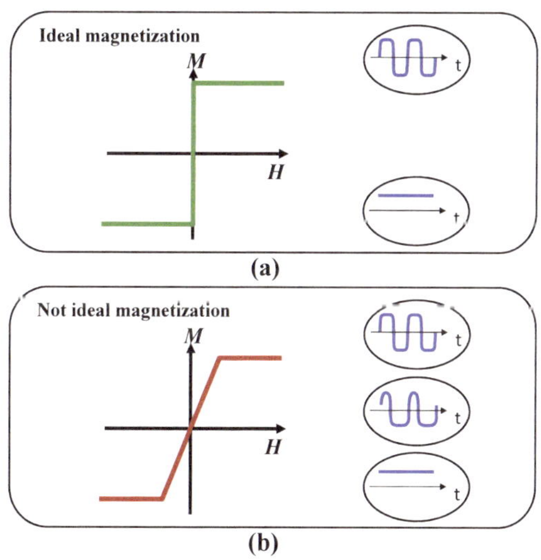

Fig. 4. Influence of imperfection in the MNP's magnetizing properties. (a) Ideal magnetization (saturation) property: when an alternating magnetic field is superimposed on an offset magnetic field, the AC component is not contained in the magnetization response of the MNP. (b) Not ideal magnetization (saturation) property: the AC component contained in the magnetization response of the MNP is affected by the offset magnetic field strength superimposed on an alternating magnetic field.

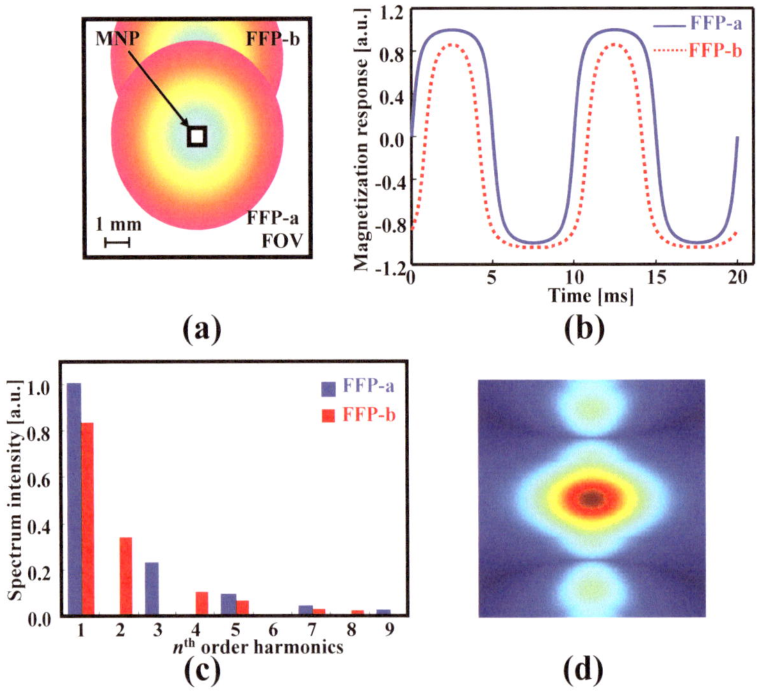

Fig. 5. Harmonic components detected at each FFP. A magnetization response composed of only odd harmonics is detected from the MNPs at the center of the FFP (FFP-a). Because the local magnetic field distribution that forms the FFP does not have a steep gradient, a magnetization response with even harmonics is also detected at a point where the FFP is at a certain distance from the MNPs (FFP-b).

Lissajous trajectory within the FOV, and delivered strong results. However, with this method, when the image size becomes large, a huge operation is needed, which raises concerns regarding the stability of the reconstructed image.

In this chapter, in order to reduce the image artifact and blurring, we first introduce the proposed method, and explain its efficiency and the problems it raises. Then, we propose an image reconstruction method that has excellent stability, image resolution, and detection sensitivity without performing inverse-matrix operations.

3.1 Adjusting the harmonic components

As explained in Fig. 5, in addition to the odd harmonics, considerable even harmonics are also detected in the signal when the FFP is set at a certain distance from a MNP owing to imperfections in the magnetic field distribution that forms the FFP. Although conventional MPI is reconstructed using only the odd harmonics based on equation (2), in the image reconstruction method proposed in this study, the odd harmonics are used, whereas the even harmonics are reduced, as defined in equation (3) (Kusayama & Ishihara, 2009). If the magnetic field distribution, which is formed as an FFP, and the magnetic characteristics of the MNP are known, the necessary components of the odd harmonics and the unnecessary component of the even harmonics can be determined. As a result, as shown in Fig. 6, the odd harmonics can be emphasized, and the even harmonics can be reduced.

$$U(x,z)=\sum_{n=1}^{N_h}\left\{V\left[x,z,(2n+1)\right]e^{\alpha(2n+1)}-V\left[x,z,(2n)\right]ke^{\beta(2n)}\right\} \tag{3}$$

$U(x, z)$: corrected distribution in x-z plane
$V[x, z, n]$: n-th harmonics contained in the waveform at each FFP (x, z)
N_h: maximum harmonic order for reconstruction
α, β: weighting factors for harmonics
k: arbitrary constants

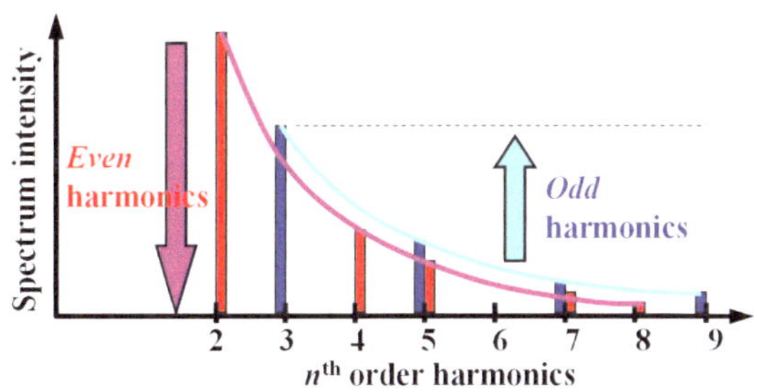

Fig. 6. Adjustment of harmonic components detected at each FFP. The odd harmonics were left unchanged and the even harmonics were reduced, in contrast to conventional image reconstruction methods.

3.2 Differentiating the waveforms obtained from inside and outside the FFP

Even if we use the method based on equation (3), the image resolution degrades because of the interference of the magnetization signals generated from several MNPs located outside the FFP boundary. For example, the signals generated from the MNPs located outside the FFP, as shown in Fig. 7 (a), may be composed of odd harmonics similar to those obtained from the MNPs placed inside the FFP. Because the effect of such interferences cannot be suppressed by the abovementioned procedure alone, the image artifact appears in the reconstructed image (Fig. 7 (d)), degrading the image resolution.

Here, a big difference is noticed when the waveforms generated from the MNPs inside the FFP region are compared with the interfering waveforms from outside the FFP region (Fig. 7 (b)). To distinguish between the interference signals, first, the offset component of the detected signal is corrected and normalization is performed. Second, the correlation between the corrected waveform and a waveform (system function) that is generated from a MNP at each FFP is evaluated. These processes are expressed in equation (4), and Fig. 8. Finally, this difference in waveforms is defined in terms of the correction factor, and suppression of unnecessary interference signals is attempted by multiplying their factors with the data determined from equation (3).

$$C_c=\left(1-N_s^{-1}\sum_{n=0}^{N_s-1}\left|\frac{w^i{}_n}{w^i{}_{\max}}-\frac{w_n-w_{mean}}{\left|w_{\max}-w_{mean}\right|}\right|\right)^{C_a} \tag{4}$$

N_s: number of sampling points
w_n, $w^i{}_n$: sampled interference and ideal waveform data
w_{max}, $w^i{}_{max}$: maximum value of each waveform
w_{mean}: mean value of interference waveform
C_a: arbitrary constant

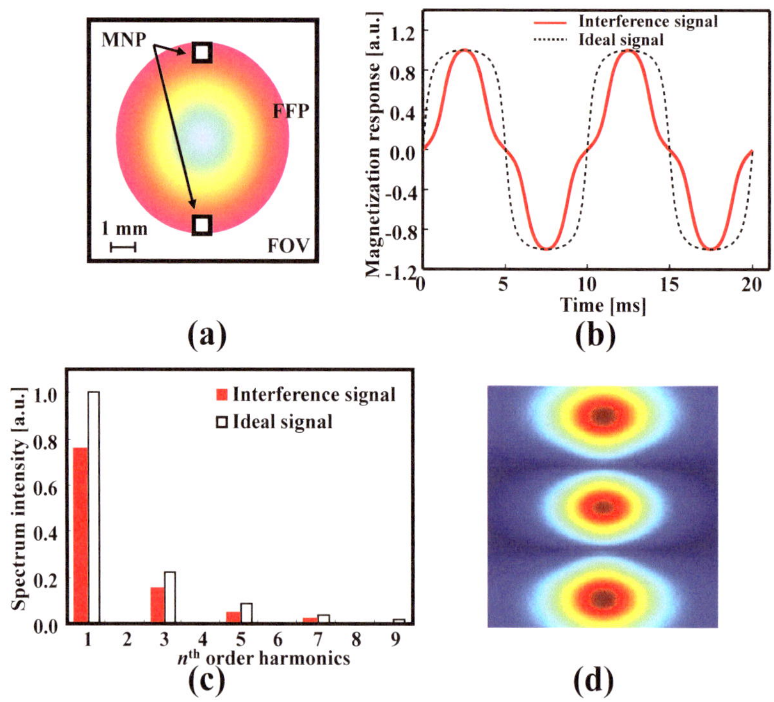

Fig. 7. Interference signal detected from regions outside the FFP. Because the magnetization response generated from the MNPs placed outside the FFP consists of odd harmonics at the FFP in the halfway point between MNPs, this interference cannot be eliminated by correction on the basis of the type of harmonics.

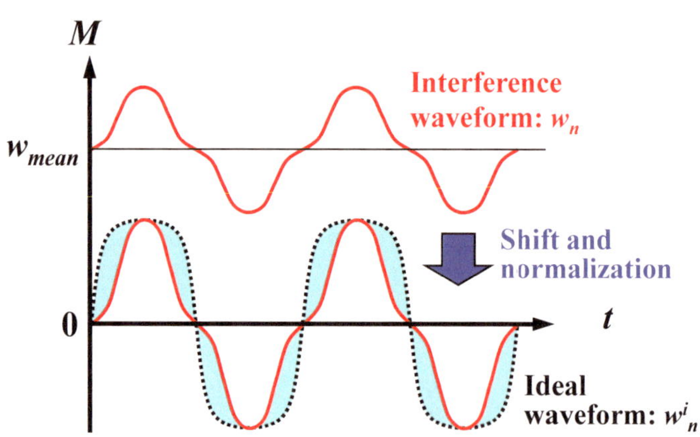

Fig. 8. Distinguishing the interference signal by using the correlation coefficient. In order to remove the interference of the magnetization signals, the following is performed: (1) correction of the offset component, (2) normalization of the waveform, (3) calculation of correlation coefficient, and (4) multiplication of correlation coefficient.

3.3 Correlating with the system function

The method proposed above is extremely effective in reducing blurring and artifacts in images, as will be shown later. However, this method emphasizes the signal from the isolated signal source that exists at the center of each FFP, and assumes that the signal generated from that circumference is an unnecessary interference. This is a problem when adjacent signal sources (MNPs) exist, as the intrinsic signal generated from these sources is recognized as an unnecessary interference to each other.

In order to overcome this unexpected effect, a method of reconstructing the exact spatial distribution of MNPs is proposed. Using the same method proposed by Weizenecker et al. (2009), the waveform generated from a MNP at each FFP is measured as a system function, but the correlation between this system function and a waveform generated by the unknown MNPs' distribution at each FFP is calculated without any inverse matrix operation. More specifically, the estimation of the MNPs' distribution is based on the correlation between the observed signal $S(x, z)$ from the unknown MNPs' distribution and the system function (*i.e.*, point-spread function) $G(i, j; x, z)$, which is a space-variant system determined by the interaction of the magnetic field and the MNPs' distribution. As shown in Fig. 9, this system function can be determined by measuring the waveforms at FFP points of (x, z) when a MNP is set at each point (i, j) within the FOV, and by connecting all these measured waveforms as one-dimensional data, which is sequentially arranged in an array of rows and columns (i, j). Consequently, the MNP distribution in the x-z plane is reconstructed using equation (5).

$$U(x,z)=\sum_{j=1}^{J}\sum_{i=1}^{I}S(x,z)G(i,j;x,z) \tag{5}$$

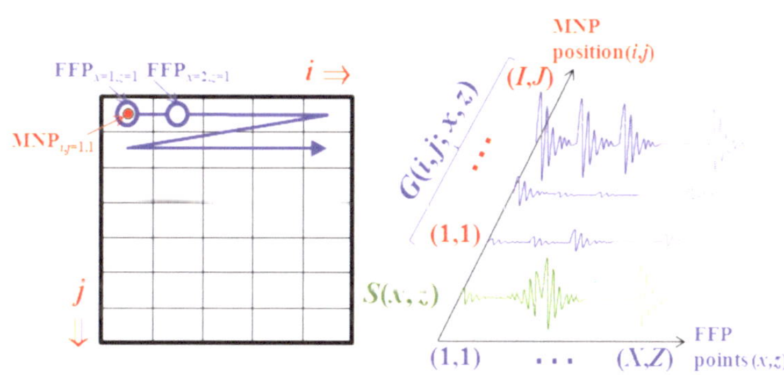

Fig. 9. Calculation of correlation coefficients: the system function $G(i, j; x, z)$ is defined as a space-variant system. It can be determined by measuring the waveforms at FFP points (x, z) when a MNP is set at each point (i, j) within the FOV. The MNP distribution is calculated by correlation with $S(x, z)$, which is the waveform measured by each FFP of (x, z) and arranged in a one-dimensional array.

4. Numerical simulation

4.1 Simulation methods

In order to examine the validity of the proposed method using the higher harmonic components appropriately, a numerical analysis using the system model shown in Fig. 2

was conducted. In this examination, two Maxell coil pairs (diameter: 50 mm, opposite distance: 50 mm) and a receiver coil (diameter: 16 mm, number of turns: 200) were used, and the FOV was set as 9 × 9 mm^2 with a matrix size of 64 × 32. A magnetic field distribution with a gradient magnetic field of about 5 T/m formed in the z direction at the MNPs with a particle diameter of 20 nm was applied as an FFP using this Maxell coil pair. In addition, an alternating magnetic field of 20 mT was applied in the same direction.

4.2 Simulation results

Figure 10 shows the reconstructed images of the MNP placed at the center of the FOV using the conventional method with equation (2) and the proposed method based on equation (3). The magnetic field was distributed over the region where the MNPs were positioned owing to the influence of the FFP formed as a result of the imperfection of the local magnetic field distribution. Therefore, because the alternating magnetic field was applied only in the *z*-direction, image blurring was observed particularly along the *z*-direction in the image reconstructed using the conventional method (Fig. 10(a)). On the other hand, such image blurring was reduced by suppressing the even harmonics on the basis of equation (3) and by using optimized parameters ($N_h = 7$, $\alpha = 0.19$, $\beta = 0.12$, $k = 1.39$) (Fig. 10(b)).
Furthermore, it was confirmed that a drastic reduction in the image artifact compared to the conventional method could be achieved by using the proposed method shown in Fig. 11 by differentiating between the interference of the signal from the MNPs placed outside and inside the region on the basis of equation (4) using the optimized parameters ($c_a = 10$).
Although this proposed method enables us to improve image resolution by suppressing the interference signal due to the even harmonic components generated from the MNPs and also to improve the sensitivity by emphasizing the odd harmonic components, the outer part of the reconstructed image was weighted excessively, as in a high-pass-filter effect. For example, the perimeter region is emphasized when the MNPs indicate continuous distribution, as shown in Fig. 12 (a) while there are fewer blurring and artifacts compared with the conventional method shown in Figs. 10 and 11 when the MNPs are separated.

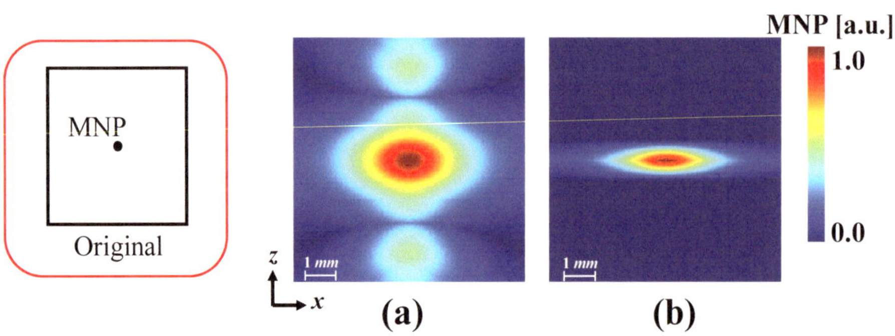

Fig. 10. Improvement of image resolution by adjusting the harmonic components based on equation (3). Overall image blurring due to the imperfection of the local magnetic field distribution formed as FFP is observed on the image reconstructed by the conventional method (a). The spread of the distribution along the *z*-direction was suppressed by using the proposed method with equation (3) (b).

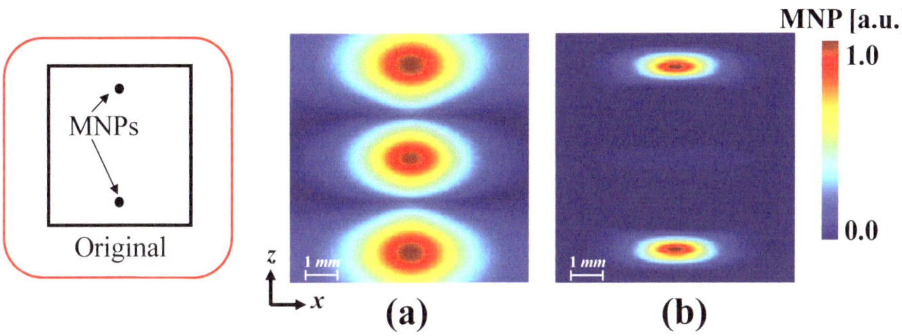

Fig. 11. Improvement of image resolution by suppressing the interference signals on the basis of equation (4). When using the conventional method, an image artifact was formed at the FFP between the nanoparticles placed symmetrically on either side of the FFP (a). With the proposed method, this image artifact was eliminated, and the spread of the field distribution in the z-direction was suppressed on the basis of equation (4) (b).

Consequently, the efficiency of the reconstruction method based on the newly proposed equation (5) was evaluated. Figure 13 illustrated the reconstructed results for the original image (Fig. 13(a)) using the conventional method (Fig. 13(b)), the method based on equations (3) and (4) (Fig. 13(c)), and the method based on equation (5) (Fig. 13(d)). With the conventional method, the reconstructed distribution was spread around the region where the MNPs were actually positioned, and an image artifact was observed. On the other hand, with the reconstruction method based on equation (5), although some image blurring is observed in the peripheral part of the image, a more exact reconstruction image with fewer artifacts is obtained, compared to other methods. Furthermore, it is confirmed from Fig. 14, which shows the profile of the central section of these images, that the sensitivity of this method based on equation (5) was high – about 20% compared with other methods.

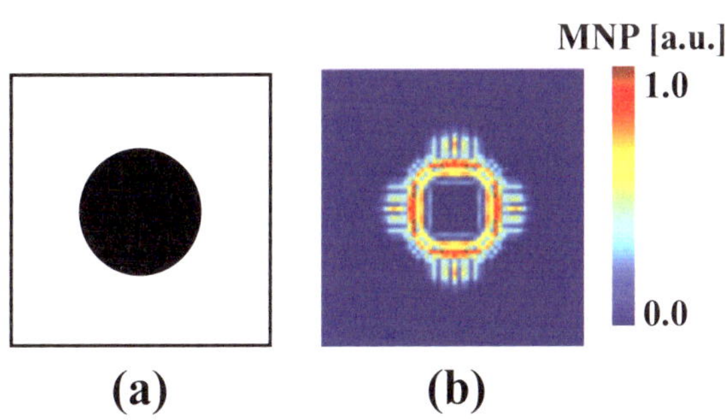

Fig. 12. The reconstruction result for a continuous distribution of MNP. The outer part of the reconstructed image is excessively emphasized, as in the effect of a high pass image filter.

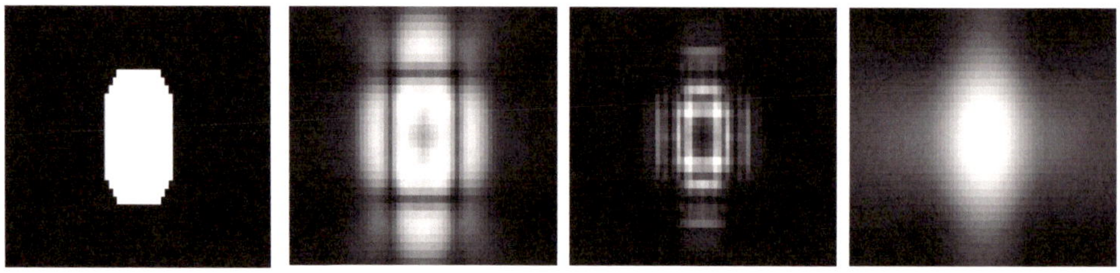

(a) (b) (c) (d)

Fig. 13. Reconstruction results of each method. (a) Original image. (b) With the conventional method, significant image blurring and an artifact appear. (c) With the proposed method based on equations (3) and (4); although the image is less blurred, the edges are emphasized excessively. (d) With the proposed method based on equation (5); although the image is somewhat blurred, the original image is reconstructed more accurately than with the other methods.

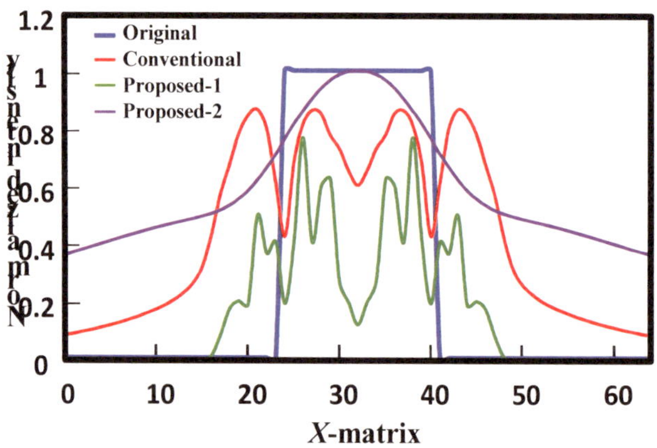

Fig. 14. Profile of a reconstructed image. The proposed method based on equation (5) reconstructs the MNP distribution more accurately than the other methods. In addition, this method has excellent sensitivity.

5. Experiment

5.1 Materials and methods

In order to confirm the validity of the proposed methods and numerical computation, the influence of the interference originating from the magnetization response waveform generated outside the target region was estimated by a fundamental experiment. The prototype Maxell coil pair (diameter: 180 mm, number of turns: 285 each, and opposite distance: 30–50 mm) (Toyojiki industry Co. Ltd., Niiza, Japan) used for the experiment is shown in Fig. 15.

5.1.1 Detecting magnetization response

When an alternating magnetic field was applied, the higher harmonic component contained in the magnetization response detected from the MNPs was evaluated, and the validity of the numerical analysis was confirmed. In this experiment, an alternating magnetic field with an amplitude of about 90 mT was generated in the center of the coil by applying an alternating current (frequency of 35 Hz, and amplitude 10.6 A) in the same direction to each coil. As a measuring object, a 2.0-g dry particle of iron oxide (nominal diameter of 10 nm), which has polar surface properties (EMG1500, Ferrotec Corp., Chiba, Japan)) was enclosed in the container, as shown in Fig. 15 (b).

The magnetization response waveform generated from the MNPs at the center of the Maxwell pair coil was detected using a receiver coil (diameter: 35 mm, number of turns: 40) which surrounds the phantoms. Here, in order to reduce a nonlinear error, intrinsic to the system, which originates from imperfections in the power supplies and the Maxell coil pair, difference processing between the signals observed with and without the existence of the measuring object was carried out.

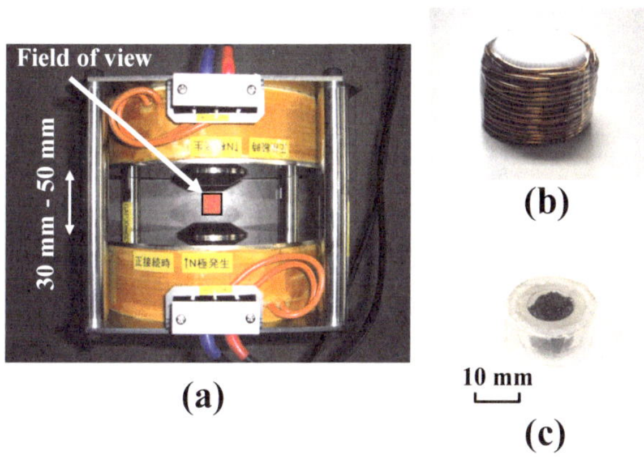

Fig. 15. Experimental modules for signal detection. A Maxwell coil pair (diameter: 180 mm, turns: 285 each, distance between coils: 30-50 mm) was used in the *z*-direction for generation of an FFP and an alternating magnetic field. The magnetization response from the dry particle of iron oxide enclosed within the container (c) was detected by a receiver coil (diameter: 35 mm, turns: 40) (b).

5.1.2 Differentiating the waveforms obtained from inside and outside of the FFP

Next, it was evaluated by the experiment that the interference of the magnetization response signal generated from MNPs outside an FFP can be decreased with the proposed method based on equations (3) and (4). In this experiment, an alternating magnetic field with an amplitude of about 20 mT was generated in the center of the coil by applying an alternating current (frequency of 80 Hz, and amplitude 4.7 A) in the same direction to each coil. At the same time, a direct current of 9.4 A was applied to each coil in the opposite direction and a gradient magnetic field of about 1.2 T/m was generated, and as a result, an FFP was generated around the center of the coil gap.

A 0.5-cc hydrophilic colloidal solution of superparamagnetic iron oxide (concentration about 500 mM/liter), which is coated with carboxydextran (Ferucarbotran, Fujifilm RI Pharma Co., Ltd., Tokyo, Japan), enclosed in the container shown in Fig. 16 (c) was used as a phantom. The magnetization response waveform generated from the MNPs at the center of the Maxwell coil pair was detected using a receiver coil (diameter: 35 mm, number of turns: 40) (Fig. 16 (a)) that surrounds the phantoms. In order to correspond to the numerical analysis (Fig. 11), the two phantoms were coaxially arranged, 20 mm apart, between the opposite coils (Fig. 16 (b)).

The FFP was set at the center between the coils, at an equal distance from the phantoms, and the magnetization response waveform generated from the phantoms was detected using the receiver coil (diameter: 23 mm, number of turns: 400) that surrounded the phantoms. The ideal waveforms (the system function) were determined by arranging a phantom at the center between the coils and measuring the signal, in order to calculate the correlation with the detected signal.

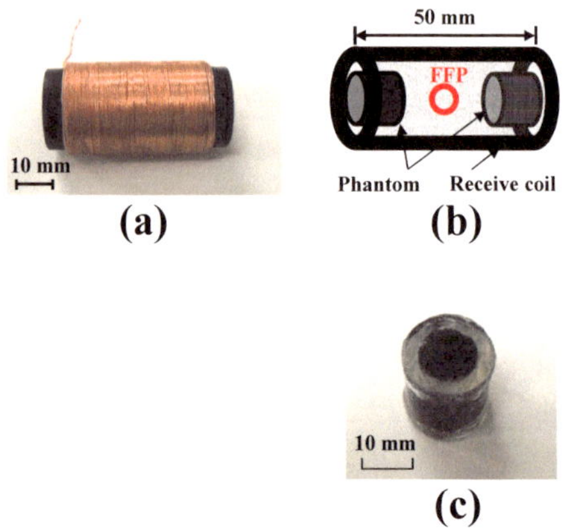

Fig. 16. Experimental modules for evaluation of the signal interference. The magnetization signal from the MNPs (MRI contrast agent: Ferucarbotran) (c), which are placed in two containers in order to reproduce the conditions of numerical analysis (b), was observed using a receiver coil (diameter: 23 mm, turns: 400) (a).

5.2 Experimental results

5.2.1 Detecting magnetization response

Figure 17 (a) shows the electromotive force induced by the receiver coil when applying the alternating magnetic field to the MNPs. By integrating over this electromotive force wave, the magnetization response waveform of the particles was obtained (Fig. 17 (b)) and is shown in Fig. 17 (c) with the result of the Fourier transform. These experimental results confirmed that the method of our numerical analysis is accurate.

5.2.2 Detecting magnetization response

The magnetization response waveform detected by the receiver coil and the ideal waveforms are shown in Fig. 18 (a). The influence of interference is reflected in the detected magnetization response waveform, and a similar wave shape to that shown in Fig. 7 was observed.

In order to confirm that the interference of the magnetization response waveform generated outside the FFP region can be suppressed by using the proposed method, the conventional method based on equation (2) and the proposed method based on equations (3) and (4) were applied to the magnetization response waveform obtained in the experiment. Here, the correction coefficients used in equations (3) and (4) were determined based on the characteristics of the ideal waveforms ($N_h = 7$, $\alpha = 0.10$, $\beta = 0.05$, $k = 1.22$, $c_a = 5.0$).

The signal strength reconstructed by each method is shown in Fig. 18 (c). For the conventional method, the signal strength, which reflects the interference of the magnetization response waveform generated from outside the FFP, (which corresponds to the interference signal detected at the center of the Maxwell coil pair) was about 79.0% of the ideal waveform signal. On the other hand, it was confirmed that this interference signal can be suppressed to about 9.0% by our proposed method based on equations (3) and (4).

The results of this fundamental experiment were well in agreement with the results of the numerical analysis, confirming the efficiency of the proposed methods and the computational process.

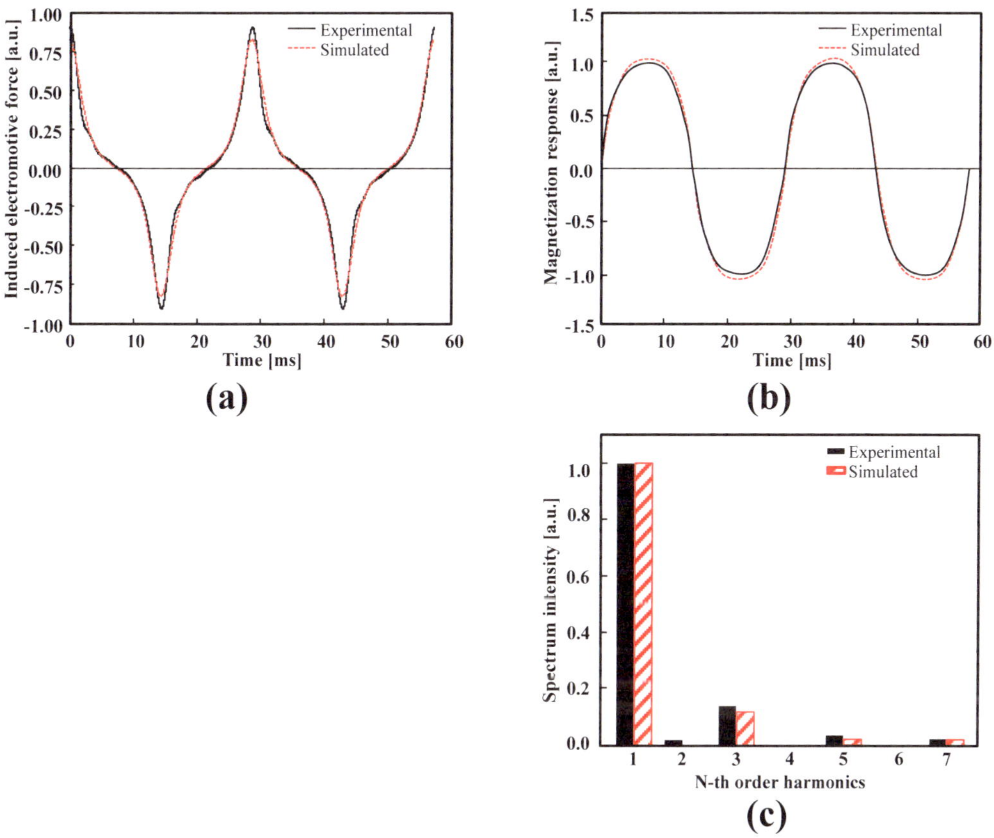

Fig. 17. Detected signal from MNPs. The signal from the MNPs is detected as electromotive force induced by the receiver coil (a). A magnetization response is obtained by integrating this signal (b), and the harmonics are computed by the Fourier transform of this response (c).

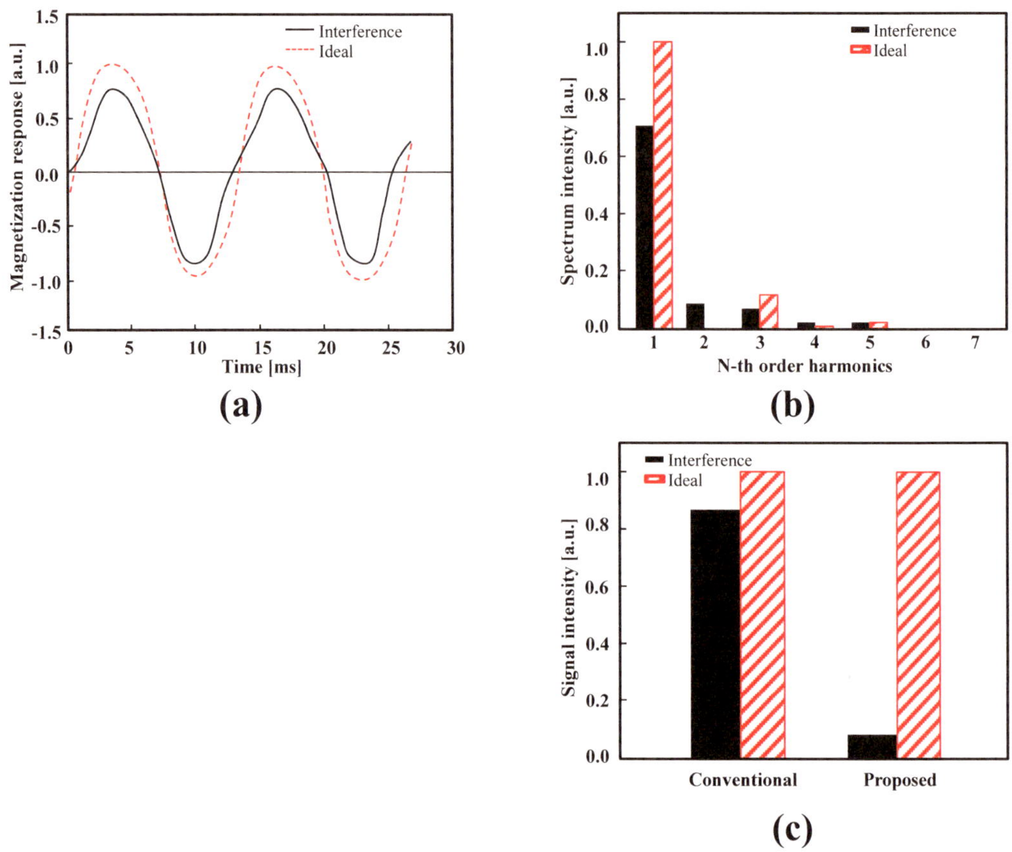

Fig. 18. Suppression effect of interference signal. It is confirmed that by using the proposed method, the interference signal detected from the center of the Maxwell coil (a) can be suppressed to less than 9.0% (c).

6. Conclusion

In MPI, interference of the magnetization signal generated from the MNPs outside the boundary of an FFP due to the nonlinear responses, results in degradation of the signal sensitivity. Although we proposed an image reconstruction method that suppresses the interference component while emphasizing the signal component using the property of the higher harmonic components generated from the MNPs, the perimeter of the reconstructed image was over-emphasized due to the high-pass-filter effect when using this method. We therefore proposed a new method based on the correlation information between the observed signal and a system function, and performed a numerical analysis. As a result, the image blurring was still visible, but we clearly showed that the detection sensitivity can be improved without the inverse-matrix operation used by the conventional image reconstruction method. In addition, although such proposed methods and numerical analyses could be demonstrated by a basic experiment, the reconstruction of an image by means of a phantom experiment should be evaluated in the future.

7. Acknowledgements

This study was supported by a Grant-in-Aid for Scientific Research (B), 20300155, 2008 from the Japan Society for the Promotion of Science (JSPS).

8. References

Gleich, B. & Weizenecker, J. (2005). Tomographic imaging using the nonlinear response of magnetic particles. *Nature*, Vol. 435, No. 7046, Jun. 2005, 1214-1217, 0028-0836

Gleich, B.; Weizenecker, J., & Borgert, J. (2008). Experimental results on fast 2D-encoded magnetic particle imaging. *Phys. Med. Biol.*, Vol. 53, No. 6, Mar. 2008, N81-N84, 0031-9155

Ishihara, Y. & Kusayama, Y. (2009). Resolution improvement of the molecular imaging technique based on magnetic nanoparticles. *Proceedings of SPIE Medical Imaging 2009*, pp. 72584I.1-72584I.8, 978-0-8194-7509-1, Orlando, USA, Feb. 2009

Ishihara, Y. & Kusayama, Y. (2011). Sensitivity improvement of a molecular imaging technique based on magnetic nanoparticles. *Proceedings of SPIE Medical Imaging 2011*, Orlando, USA, Feb. 2011 (in press).

Knopp, T.; Biederer, S., Sattel, T., Weizenecker, J., Gleich. B, Borgert, J. & Buzug, T. M. (2009). Trajectory analysis for magnetic particle imaging. *Phys. Med. Biol.*, Vol. 54, No. 2, Jan. 2009, 385-397, 0031-9155

Kusayama, Y. & Ishihara, Y. (2007). A preliminary study on the molecular imaging device using magnetic nanoparticles. *Technical Report of IEICE*, MBE, Vol. 107, No. 218, Sep. 2007, 15-18, 0913-5685

Kusayama Y. & Ishihara, Y. (2009). High-resolution image reconstruction method on the molecular imaging using magnetic nanoparticles. *IEICE Trans. D*, Vol. J92-D, No. 9, Sep. 2009, 1653-1662, 1880-4535

Matsumura, Y. & Maeda, H. (1986). A new concept for macromolecular therapeutics in cancer chemotherapy: mechanism of tumoritropic accumulation of proteins and the antitumor agent smancs. *Cancer Res.*, Vol. 46, No. 12, Dec. 1986, 6387-6392, 0008-5472

Vekas, L.; Rasa, M. & Bica, D. (2000). Physical properties of magnetic fluids and nanoparticles from magnetic and magneto-rheological measurements. *J. Colloid Interface Sci.*, Vol. 231, No. 2, Nov. 2000, 247-254, 0021-9797

Weizenecker, J.; Borgert, J. & Gleich, B. (2007). A simulation study on the resolution and sensitivity of magnetic particle imaging. *Phys. Med. Biol.*, Vol. 52, No. 21, Nov. 2007, 6363-6374, 0031-9155

Weizenecker, J.; Gleich, B., Rahmer, J., Dahnke, H. & Borgert, J. (2009). Three-dimensional real-time *in vivo* magnetic particle imaging. *Phys. Med. Biol.*, Vol. 54, No. 5, Mar. 2009, L1-L10, 0031-9155

Yavuz, C. T.; Mayo, J. T., Yu, W. W., Prakash, A., Falkner, J. C., Yean, S., Cong, L., Shipley, H. J., Kan, A., Tomson, M., Natelson, D. & Colvin, V. L. (2006). Low-field magnetic separation of monodisperse Fe_3O_4 nanocrystals. *Science*, Vol. 314, No. 5801, Nov. 2006, 964-967, 0036-8075